Grand Challenges in Biology and Biotechnology

Series editor
Pabulo H. Rampelotto

More information about this series at http://www.springer.com/series/13485

Armin Hallmann • Pabulo H. Rampelotto
Editors

Grand Challenges in Algae Biotechnology

Springer

Editors
Armin Hallmann
Department of Cellular and Developmental Biology of Plants
University of Bielefeld
Bielefeld, Germany

Pabulo H. Rampelotto
Federal University of Rio Grande do Sul
Porto Alegre, Rio Grande do Sul, Brazil

ISSN 2367-1017 ISSN 2367-1025 (electronic)
Grand Challenges in Biology and Biotechnology
ISBN 978-3-030-25232-8 ISBN 978-3-030-25233-5 (eBook)
https://doi.org/10.1007/978-3-030-25233-5

This Springer imprint is published by the registered company Springer Nature Switzerland AG.
The registered company address is: Gewerbestrasse 11, 6330 Cham, Switzerland

Preface

Algae reveal a phylogenetically diverse group of photosynthetic organisms that predominantly grow in all kinds of aquatic habitats whenever light is available. Appropriately evolved micro- or macroalgae species live in a wide range of salinities, climatic conditions, pH values, and light intensities. As a result, algae provide a vast array of biomolecules, cellular functions, and physiological characteristics. During the past decade, algae biotechnology has clearly demonstrated its significance for future-proof, sustainable business strategies. The future market share of algae biotechnology depends much on the development of production costs and the competitiveness of producers in relation to the developments in other technologies.

Currently, algae are already being used for sustainable production of healthy foods, nutraceuticals, and pharmaceuticals. Besides, there are a host of approaches for the development of further industrial products and processes. Due to the capacity to produce and store energy-rich oils, (micro)algae also receive considerable interest as a potential feedstock for producing sustainable transport biofuels. Moreover, some algal-based strategies also meet the requirements for use in bioremediation, biodegradation, or other environmental applications. Consequently, algae offer great potential to support the building of a bio-based economy, and they can contribute new solutions to some of the grand challenges of the twenty-first century.

Despite significant progress, algae biotechnology is yet far from fulfilling its potential. How to unleash this enormous potential is the challenge that this field is facing. New cultivation technologies and bioprocess engineering allow for the optimization of the operation strategy of state-of-the-art industrial-scale production systems, and they reduce the production costs. Parallel to this, new molecular technologies for genetic and metabolic engineering of (micro)algae develop quickly. The optimization of existing biochemical pathways or the introduction of pathway components makes high-yield production of specific metabolites possible. Novel screening technologies including high-throughput technologies enable testing of extremely large numbers of samples and, thus, allow for large-scale modelling of biomolecular processes, which would have not been possible in the past. Moreover, profitable production can demand for integrated biorefining, which combines consecutive processes and various feedstocks to produce transportation fuel, electric energy, and valuable chemicals.

In this book, researchers and practitioners working in the field present the major promises of algae biotechnology and they critically discuss the challenges arising from applications. Based on this assessment, the authors explore the great scientific, industrial, and economic potential opened up by algae biotechnology.

Part I of the book deals with topical aspects in (micro)algae cultivation systems. The issues covered are commercial microalgal cultivation systems (Chap. 1), various technologies for algal crop protection (Chap. 2), growth of microalgae in the heterotrophic mode (Chap. 3), and agronomic practices for photoautotrophic production of biomass (Chap. 4).

Part II of the book presents recent developments in key enabling technologies for genetic and metabolic engineering of (micro)algae, which are the driving force to unleash the enormous potential of algae biotechnology. The chapters of this part provide an overview of the advances in genetic engineering of microalgae (Chap. 5), address the optimization of the photosynthetic metabolism for increased biomass productivity (Chap. 6), and deal with metabolic engineering and synthetic biology approaches to enhance the production of long-chain polyunsaturated fatty acids (Chap. 7).

Part III of the book treats integrated approaches for process optimization in algae biotechnology. The chapters of this part cover integration concepts for process intensification in the production of algal biomolecules (Chap. 8), approaches to combine microalgae-based wastewater treatment with biofuel and bio-based production (Chap. 9), the use of microalgal consortia for wastewater treatment, biomass generation and bioenergy production (Chap. 10), and the utilization of downstream green processes for recovery of valuable compounds (Chap. 11).

Part IV of the book focuses on how practical applications of algae biotechnology may provide new solutions to some of the grand challenges of the twenty-first century. The chapters of this part explore the applications of algae as pharmaceuticals and nutraceuticals (Chap. 12), the use of algae in bioremediation of metal pollution (Chap. 13), the benefits of algal extracts in sustainable agriculture (Chap. 14), and the economic value from metabolites in cyanobacteria (Chap. 15). As lawmakers have recognized the importance of algae biotechnology for our future, national and international strategies and policies have evolved that enable entrepreneurs, businesses, scientists, investors, and government agencies to work together in order to explore and develop innovative algae biotech products. For this reason, legal aspects of algae biotechnology are discussed in the last chapter of this book (Chap. 16).

We hope our readers enjoy reading this fine collection of chapters written by the leading scientists in the field. In general, we expect that basic researchers, developers, industrial operators, and any other people interested in algae biotechnology will use the knowledge made available through this reliable reference. Particularly for leading authorities, the book also provides support in their efforts towards the development of an algae-based bioeconomy.

Bielefeld, Germany
Porto Alegre, Brazil

Armin Hallmann
Pabulo H. Rampelotto

Contents

Editors and Contributors

About the Editors

Pabulo Henrique Rampelotto is the founder and Editor-in-Chief of the Springer Book Series *Grand Challenges in Biology and Biotechnology*. He is also Editor-in-Chief, Associate Editor, Senior Editor, Guest Editor, and member of the editorial board of several scientific journals in the field of life sciences and biotechnology. Furthermore, Pabulo is member of four Scientific Advisory Boards of Lifeboat Foundation, along with several Nobel Laureates and other distinguished scientists, philosophers, educators, engineers, and economists. Most of his recent work has been dedicated to the editorial process of several scientific journals in life science and biotechnology, as well as on the organization of special issues and books in his fields of expertise. In his special issues and books, some of the most distinguished team leaders in the field have published their work, ideas, and findings, including Nobel laureates and several of the highly cited scientists according to the ISI Institute.

Armin Hallmann is professor and head of the Department of Cellular and Developmental Biology of Plants at Bielefeld University, Germany. For more than 25 years, he is utilizing eukaryotic microalgae as model systems for his research in the fields of developmental biology, molecular cell biology, biochemistry, and biotechnology. Before he came to Bielefeld University in 2003, he taught at the University of Regensburg, Germany, and at the Washington University in St. Louis, MO, USA. He is author, reviewer, and editor of numerous scientific publications (articles, reviews, book contributions) that appeared in international leading scientific journals.

Contributors

Gabriel Acién Department of Chemical Engineering, University of Almería, Almería, Spain

Gayane Avetisova SPC "Armbiotechnology" NAS RA, Yerevan, Armenia

Maria Barbosa Bioprocess Engineering, AlgaePARC, Wageningen University and Research, Wageningen, The Netherlands

John Beardall School of Biological Sciences, Monash University, Clayton, VIC, Australia

Mónica Bueno Laboratory of Foodomics, Institute of Food Science Research (CIAL, CSIC), Madrid, Spain

Michele Carone Genetics and Physiology of Microalgae, UR InBios/Phytosystems, University of Liège, Liège, Belgium

Wan-Loy Chu School of Postgraduate Studies, International Medical University, Kuala Lumpur, Malaysia

Pierre Collet IFP Energies nouvelles, Rueil-Malmaison, France

Roberta Congestri Laboratory of Biology of Algae, Department of Biology, University of Rome 'Tor Vergata', Rome, Italy

Amélie Corato Laboratory of Bioenergetics, UR InBios/Phytosystems, University of Liège, Liège, Belgium

Senka Curcin Faculty of Technology, Department of Biotechnology and Pharmaceutical Engineering, University of Novi Sad, Novi Sad, Serbia

Michael K. Danquah Department of Chemical Engineering, University of Tennessee, Chattanooga, TN, USA

Ly Thi Hai Dao School of Biological Sciences, Monash University, Clayton, VIC, Australia

Thomas Dauvrin Genetics and Physiology of Bacteria, UR InBios/CIP, University of Liège, Liège, Belgium

Lorenzo Durante Genetics and Physiology of Microalgae, UR InBios/Phytosystems, University of Liège, Liège, Belgium

Michel H. M. Eppink Bioprocess Engineering, AlgaePARC, Wageningen University and Research, Wageningen, The Netherlands

Alice Ferreira LNEG, National Laboratory of Energy and Geology I.P., Bioenergy Unit, Lisbon, Portugal

Carolyn L. Fisher Systems Biology Department, Sandia National Laboratories, Livermore, CA, USA

Fabrice Franck Laboratory of Bioenergetics, UR InBios/Phytosystems, University of Liège, Liège, Belgium

Claudio Fuentes-Grünewald Department of Biosciences, College of Science, Swansea University, Swansea, UK

Rocío Gallego Laboratory of Foodomics, Institute of Food Science Research (CIAL, CSIC), Madrid, Spain

Edgar Suarez Garcia Bioprocess Engineering, AlgaePARC, Wageningen University and Research, Wageningen, The Netherlands

Spyros Gkelis Department of Botany, School of Biology, Aristotle University of Thessaloniki, Thessaloniki, Greece

Ana L. Gonçalves Faculty of Engineering, LEPABE—Laboratory for Process Engineering, Environment, Biotechnology and Energy, University of Porto, Porto, Portugal

Luisa Gouveia LNEG, National Laboratory of Energy and Geology I.P., Bioenergy Unit, Lisbon, Portugal

Carolyn Greig Department of Biosciences, College of Science, Swansea University, Swansea, UK

Claudia Grewe SALATA AG, Ritschenhausen, Germany

Armin Hallmann Department of Cellular and Developmental Biology of Plants, University of Bielefeld, Bielefeld, Germany

Elena Ibáñez Laboratory of Foodomics, Institute of Food Science Research (CIAL, CSIC), Madrid, Spain

Jaison Jeevanandam Department of Chemical and Petroleum Engineering, Curtin University of Technology, Sarawak, Malaysia

Bernard Joris Genetics and Physiology of Bacteria, UR InBios/CIP, University of Liège, Liège, Belgium

Rahul Vijay Kapoore Department of Biosciences, College of Science, Swansea University, Swansea, UK

Inna Khozin-Goldberg Microalgal Biotechnology Laboratory, The French Associates Institute for Agriculture and Biotechnology for Drylands, J. Blaustein Institutes for Desert Research, Ben-Gurion University of the Negev, Beersheba, Israel

Bethan Kultschar Department of Biosciences, College of Science, Swansea University, Swansea, UK

Todd W. Lane Systems Biology Department, Sandia National Laboratories, Livermore, CA, USA

Philip A. Lee Department of Biology, Midland College, Midland, TX, USA
Pebble Labs, Inc., Los Alamos, NM, USA

Felix Leinemann Economy Sectors, Aquaculture and Maritime Spatial Planning, DG Maritime Affairs and Fisheries (MARE), European Commission, Brussels, Belgium

Carole A. Llewellyn Department of Biosciences, College of Science, Swansea University, Swansea, UK

Robert W. Lovitt College of Engineering, Swansea University, Swansea, UK

Valentina Mabilia Economy Sectors, Aquaculture and Maritime Spatial Planning, DG Maritime Affairs and Fisheries (MARE), European Commission, Brussels, Belgium

Lusine Melkonyan SPC "Armbiotechnology" NAS RA, Yerevan, Armenia

Jose A. Mendiola Laboratory of Foodomics, Institute of Food Science Research (CIAL, CSIC), Madrid, Spain

Paulina Mikulic School of Biological Sciences, Monash University, Clayton, VIC, Australia

Tomas Morosinotto Department of Biology, University of Padova, Padova, Italy

Raul Muñoz Institute of Sustainable Processes, Valladolid University, Valladolid, Spain

Miguel Olaizola OLAS-ALL THINGS ALGAE, LLC, Snowflake, AZ, USA

Giuseppe Olivieri Bioprocess Engineering, AlgaePARC, Wageningen University and Research, Wageningen, The Netherlands

Sharadwata Pan School of Life Sciences Weihenstephan, Technical University of Munich, Freising, Germany

Giorgio Perin Department of Biology, University of Padova, Padova, Italy
Department of Life Sciences, Imperial College London, London, UK

Siew-Moi Phang Institute of Biological Sciences, University of Malaya, Kuala Lumpur, Malaysia
Institute of Ocean and Earth Sciences (IOES), University of Malaya, Kuala Lumpur, Malaysia

José C. M. Pires Faculty of Engineering, LEPABE—Laboratory for Process Engineering, Environment, Biotechnology and Energy, University of Porto, Porto, Portugal

Rossella Pistocchi Department of Biological, Geological, and Environmental Sciences, University of Bologna, Ravenna, Italy

Alberto Reis LNEG, National Laboratory of Energy and Geology I.P., Bioenergy Unit, Lisbon, Portugal

J. Hans Reith Bioprocess Engineering, AlgaePARC, Wageningen University and Research, Wageningen, The Netherlands

Claire Remacle Genetics and Physiology of Microalgae, UR InBios/Phytosystems, University of Liège, Liège, Belgium

Francisca M. Santos Faculty of Engineering, LEPABE—Laboratory for Process Engineering, Environment, Biotechnology and Energy, University of Porto, Porto, Portugal

Olga Sayanova Department of Biological Chemistry and Crop Protection, Rothamsted Research, Harpenden, UK

Lolke Sijtsma Wageningen Food and Biobased Research, Wageningen University and Research, Wageningen, The Netherlands

Tung Le Thanh Laboratory of Bioenergetics, UR InBios/Phytosystems, University of Liège, Liège, Belgium

Corjan van den Berg Bioprocess Engineering, AlgaePARC, Wageningen University and Research, Wageningen, The Netherlands

Marian H. Vermuë Bioprocess Engineering, AlgaePARC, Wageningen University and Research, Wageningen, The Netherlands

Jelena Vladic Faculty of Technology, Department of Biotechnology and Pharmaceutical Engineering, University of Novi Sad, Novi Sad, Serbia

Rebecca L. White Qualitas Health, Inc, Houston, TX, USA
Pebble Labs, Inc., Los Alamos, NM, USA

René H. Wijffels Bioprocess Engineering, AlgaePARC, Wageningen University and Research, Wageningen, The Netherlands
Faculty of Biosciences and Aquaculture, Nord University, Bodø, Norway

Part I
Cultivation Systems

Chapter 1
Commercial Microalgal Cultivation Systems

Miguel Olaizola and Claudia Grewe

Abstract Commercial cultivation systems designed to produce microalgal biomass phototrophically (photobioreactors, PBRs) exist in many forms (e.g., open vs. closed, tubular vs. panels, vertical vs. horizontal). Independent of what they look like, their function is the same: to expose the largest fraction of the microalgal cells to optimal production conditions and to do it as economically as possible! Those conditions are dependent both on the target products and on the producing organism. In this chapter, the authors explore different photobioreactor types and how they may cope with changing growth conditions, especially outdoors. It is proposed that a simple productivity model can help to evaluate different photobioreactors and different production strategies. Finally, the authors believe that both open and closed PBRs, including new designs, will enable to close the gap between the reality and the potential of microalgae as economical providers of many products and services.

1.1 Introduction

1.1.1 Microalgae and Their Potential

As has been noted in later chapters in this volume and elsewhere, microalgae are an extremely diverse group of organisms that can provide many different useful products and services. "Microalgae" is not a phylogenetic but a functional term which includes, for example, both prokaryotes and eukaryotes. Microalgae, as a polyphyletic functional group, represent organisms that function as photoautotrophs (although some may also obtain nutrition and energy heterotrophically), are colorful (since they contain photosynthetic and secondary pigments), are aquatic (but can be

M. Olaizola (✉)
OLAS-ALL THINGS ALGAE, LLC, Snowflake, AZ, USA
e-mail: miguel@olasalgae.com

C. Grewe
SALATA AG, Ritschenhausen, Germany

A. Hallmann, P. H. Rampelotto (eds.), *Grand Challenges in Algae Biotechnology*, Grand Challenges in Biology and Biotechnology,
https://doi.org/10.1007/978-3-030-25233-5_1

found in most environments that contain a minimum of humidity) and are usually unicellular (although may also form colonies of mostly undifferentiated cells).

Microalgae can produce and accumulate many biochemicals with applications in human and animal nutrition, health, cosmetics, pharmaceuticals, analytics and biofuels. Microalgae can also provide services such as CO_2 capture and wastewater remediation. These products and services by microalgae are expected to represent "green" alternatives when compared to present sources of those biochemicals and services and offer environmental and economic benefits (Olaizola et al. 2018). The goal of an efficient and affordable energy source from microalgae is not expected to become a reality in the short term; instead, microalgae biotechnology is developing in the direction of already existing higher value markets.

It has also been noted that microalgal biotechnology has not yet fulfilled its potential and may even be controversial (Stephens et al. 2013). In those cases where microalgae have been used to commercialize novel products (e.g., *Spirulina* over the last 40 years or astaxanthin from *Haematococcus* over the last 20 years), a few commercial producers have found some success (while others have not survived). In our experience, the level of success has been dependent on the resources that different companies had available to introduce and market a new product that the public was unaware of.

In those cases where microalgae producers have attempted to introduce "green" replacement products such as biofuels, animal feeds, or nutritional fatty acids (e.g., EPA and DHA), we, as an industry, have discovered that the cost of manufacturing those products via microalgal photosynthetic pathways is much higher than harvesting traditional sources or produce them heterotrophically. Many factors contribute to this disparity (related to labor, nutrients, CO_2, water, harvesting and processing). One of them is the cost of cultivation, e.g., the need to contain the growth medium along with the microalgae (the culture). This containment function is provided by so-called photobioreactors: open and closed.

1.1.2 Industrial Production of Microalgae

Notwithstanding the tremendous potential of microalgae and the relatively lack of fulfilled potential stated above, we present here a snapshot of the state of progress in the industry. The five most important crops in terms of annual biomass production are *Spirulina* (*Arthrospira*), *Chlorella*, *Dunaliella*, *Haematococcus* and *Nannochloropsis*. The cyanobacteria *Nostoc* and AFA (*Aphanizomenon flos-aquae*) are also considered here but are distinguished as they are mainly collected from nature. Products are mainly whole biomasses, either frozen, spray-, or freeze-dried. Oleoresins are mainly being produced from the carotenoid rich genera *Haematococcus* and *Dunaliella*.

Table 1.1 summarizes mainly autotrophically operated global production sites (as of May 2017) that meet our arbitrary criteria of at least a hectare of open PBR (ponds and raceways) or at least a volume of 10 m^3 of closed PBR volume and a biomass output and active trading in the ton per year range. Thus, pilot and proof of

Table 1.1 Industrial microalgae producers with at least one hectare of open PBRs or a minimum of 10 m^3 of closed PBRs

Company	Country	City
AlgaeCan Biotech	Canada	Surrey
Algaetech International Sdn Bhd	Malaysia	Kuala Lumpur
Algalif	Iceland	Reykjanesbaer
Algamo	Czech Republic	Mostek
AlgaSpring	Netherlands	Almere
Algatechnologies	Israel	Ketura
Algosource	France	Saint-Nazaire
Algosud	France	Lunel
AlgaFarm	Portugal	Pataias
Alvita Saga	Japan	Saga
Atacama Bio Natural Products	Chile	Pozo Almonte
Ballarpur Industries	India	Nanjangud
BASF	Australia	Whyalla/Hutt Lagoon
BGG	China	Kunming
Bioreal	Sweden	Gustavsberg
BlueBioTech	Germany	Büsum
Chlorella Trebon	Czech Republic	Trebon
Cyanotech	USA	Kailua-Kona
Daesang Chlorella	South Korea	Seoul
Domain Traverse	France	La Crau
Dongtai City Spirulina Bio-engineering Co (C.B.N.)	China	Dongtai City
Dongying Haifu Fine Chemical Co	China	Dongying
Earthrise Nutritional	USA	Irvine
EID Parry	India	Chennai
Eparella GmbH	Austria	Bruck an der Leitha
Euglena Co Ltd	Japan	Okinawa
Far East Biotech Co (Febico)	Taiwan	Taipei
Fermentalg	France	Libourne
Fitoplancton Marino	Spain	Cádiz
Frutarom	Israel	Haifa
Fuqing King Dnarmsa Spirulina Co	China	Fuqing City
Genix Labiofam	Cuba	Jaruco
Greensea	France	Mèze
Hainan DIC Microalgae	China	Haikou City
Hainan Simai Pharmacy	China	Chengmai
Heliae	USA	Gilbert
Inner Mongolia Biomedical Eng	China	Otog Banner
Inner Mongolia Wushenzhao Ecologica Development Co	China	Ordos
June Pharmaceutical	Myanmar	Yangon

(continued)

Table 1.1 (continued)

Company	Country	City
Kemikler Köyü Mevkii	Turkey	Milas-Bodrum Muğla
Korea Chlorella Co	Korea	Gyeongsangnam-do
Mial GmbH	Germany	Bad Zwischenahn
Microalghe Camporosso	Italy	Camporosso
Microphyt	France	Baillargues
Monzon Biotech	Spain	Barcelona
Nan Pao International Biotech Co	Taiwan	Tainan City
Nature Beta Technologies (NBT)	Israel	Eilat
Necton	Portugal	Olhão
Omega Algae	Iceland	Hveragerði
PhycoBiotech	France	Montpellier Cedex 4
Phycom BV	Netherlands	Nijkerk
Proviron	Belgium	Hemiksem
Qualitas Health	USA	Columbus, NM
Reed Mariculture	USA	Campbell
Roquette	Germany	Klötze
Sagaing June Pharmaceutical & Foodstuff Industry	Myanmar	Sagaing
Salata	Germany	Ritschenhausen
Siam Algae Co	Thailand	Samut Prakan
Simris Alg	Sweden	Simrishamn
Solix Algredients	USA	Fort Collins
Spirulina Viva	Mexico	San Miguel de Allende
Spiruline des Iles d'Or	France	Hyères
Sun Chlorella Corporation	Japan	Kyoto
TAAU Australia	Australia	Berry Springs
Taiwan Chlorella Manufacturing Company	Taiwan	Taipei
Vedan Biotechnology Corporation	Taiwan	Taipei
Yaeyama Shokusan Co	Japan	Shiraho
Yantai Hairong Biology	China	Yantai
Yunnan Spirin Biotechnology Co	China	Yunnan

concept plants as well as experimental facilities are excluded. The list includes 69 production sites, excluding on-site production of algae biomass in hatcheries for aquaculture. There are traditional sites in Australia, Israel and Hawaii for *Dunaliella*, *Haematococcus* and *Spirulina* operating with low technology but consistent output and quality. In Europe, several new producers have established large-scale facilities, e.g., in Portugal AlgaFarm for *Chlorella*, in Iceland for *Haematococcus* and in Sweden for *Phaeodactylum*. In Southern France, we find *Spirulina* farmers with raceways in greenhouses as well as companies that produce microalgae and their extracts for cosmetics. In Germany, the biggest producer is Roquette with 700 m^3 autotrophic capacity and 100 m^3 of heterotrophic production

Table 1.2 Annual industrial microalgae production (2017): mass produced and value (B2B)

Genus	Amount [t/a]	B2B value [US$/kg]	Turnover [US$/a]
Mainly/ exclusively autotrophically produced			
Spirulina	18,000	13	234,000,000
Chlorella	9500	25	237,500,000
Dunaliella	1700	80	136,000,000
AFA	500	10	5,000,000
Haematococcus	300	120	36,000,000
Nannochloropsis	150	90	13,500,000
Euglena	50		38,500,000
Subtotal	**30,206**		**701,220,000**
Exclusively heterotrophically produced			
Ulkenia	950	20	19,000,000
Schizochytrium	31,000	20	620,000,000
Crypthecodinium	12,700	20	254,000,000
Subtotal	**44,650**		**893,000,000**
Total	**74,856**		**1,594,220,000**

For comparison purposes, we have included data on production and value of other crops by organisms that produce some products in common with microalgae but are produced strictly heterotrophically

volume. In the Netherlands, companies have been successfully founded that are using greenhouses from failed horticulture enterprises, such as AlgaSpring. Around the Mediterranean Sea, there are a few producers for aquaculture, the biggest in Turkey. There are also the traditional open pond producers in Southeast Asia for *Chlorella* and *Spirulina*, e.g., in Myanmar, Taiwan, South Korea and Japan. China has become a big exporter for all commercial relevant species, although the quality of the biomass with respect to purity, heavy metals and valuable content is very variable. The biggest companies are Yantai Hairong Biology, Hainan DIC and Inner Mongolia Biomedical Eng producing *Spirulina* in greenhouses. Here, *Spirulina* production received government support at large scale on arid land to help fight desertification. *Chlorella* has traditionally been produced in Taiwan and Japan. In the Americas, we have new *Haematococcus* producers in Canada and Chile. In the USA, open ponds for *Spirulina* production are operated by Cyanotech in Hawaii and Earthrise in California.

About 30,000 tons of biomass are produced annually across the globe mainly phototrophically. Table 1.2 lists the tonnage of the specific crops as well as their value. We have added production and value data for other microorganisms that, although may not be related to what are considered microalgae (Leyland et al. 2017), accumulate some common products and are produced exclusively heterotrophically (Barclay et al. 2010). *Spirulina* and *Chlorella* are the two largest microalgae crops in tonnage and in value. The economy of scale can be seen here; the higher the biomass price, the lower the traded amount, although this is also a question of production method. The value of the biomass lies in average between \$US13 and \$US120 kg^{-1} of dry matter, but there are also biomasses traded below \$US5 kg^{-1} and above US

\$120 kg^{-1}. Nonetheless, the global turnover for predominantly autotrophically produced biomass adds up to about US\$700 million per year led by *Chlorella.*

1.1.3 Function of a Photobioreactor

The function of a photobioreactor is to expose the largest fraction of the microalgal cells in the culture to optimal production conditions at the lowest possible cost (i.e., kg of product per unit cost input). These conditions are dependent on the species and desired products and include appropriate temperature and pH range, sufficient nutrients and CO_2, elimination of excess O_2 and as much light as possible without waste or cellular damage. Most importantly, these growth conditions must be distributed throughout the photobioreactor using appropriate levels of turbulence. Furthermore, at large scale, these production conditions need to be managed outdoors so that cultures can take advantage of "free" sunlight as efficiently as possible.

Microalgae biotechnology at commercial scale is clearly dominated by open PBRs; about 99% of the biomass is produced in ponds and raceways (Table 1.1). That means that the microalgae business today is more similar to modern agriculture than to biotechnology (see Chisti 2016). Although open PBRs have a larger presence in commercial production, many closed PBR designs have been produced over the last two to three decades, some of which have found a role in commercial microalgae production either as the final production unit or as the source of high-quality inoculum for the larger open PBRs.

Open and closed photobioreactors of many designs have been described in the literature (see next section). The authors' aim is not to offer a catalog of all PBRs available nor to select the "best" ones for different applications. One of the chapter's aims is to point out how available designs tackle the different challenges associated with producing microalgae at large scale (Sect. 1.2). Another aim is to point out that, independent of the PBR design, some aspects of the culture's management strategy have a profound effect on the performance of the reactor (Sect. 1.3). Finally (Sect. 1.4), the authors explore alternative approaches to PBR design that, although may still need to be tested at commercial scale, show new paths that may lead to more efficient and economically feasible PBRs.

1.2 Photobioreactor Design

Recent reviews have listed different types of closed and open photobioreactor designs available, and the reader is directed to those reviews for a snapshot of different approaches to microalgal containment while providing appropriate production conditions (e.g., Acien Fernandez et al. 2012; Chang et al. 2017; Chisti 2016; Huang et al. 2017; Johnson et al. 2018; Olivieri et al. 2014; Singh and Sharma 2012).

In general, the following can be considered as desirable characteristics of photobioreactors:

(a) Capable of high areal productivity (g m^{-2} d^{-1}), i.e., high productivity per footprint area since some costs are associated with the land area occupied by the photobioreactor
(b) Capable of high volumetric productivity (g L^{-1} d^{-1}), i.e., high productivity per volume of culture since some costs are associated with the volume of culture that needs to be managed during growth and at harvest time
(c) Large volume per photobioreactor (L per PBR) since some expenses scale with the number of photobioreactors installed (such as plumbing, probes and sampling)
(d) Inexpensive both in capital and operating expenses
(e) Reliable, which results in lower operating costs
(f) Ease of control of environmental parameters to maximize productivity of the desired product
(g) Ease of recovery from contamination events, including cleanup and sanitizing

Of course, some desirable characteristics may not be compatible with others, depending on the PBR design under consideration.

1.2.1 Growth Conditions

As mentioned earlier, large-scale photobioreactor productivity is dependent on the sufficient supply of inputs such as nutrients, CO_2 and sunlight at the appropriate pH and temperature conditions outdoors while minimizing the accumulation of O_2. One can argue that the most variable of these inputs outdoors is sunlight. Sunlight depends on the geographical location, time of year, time of day and associated weather events. Sunlight is also a driver for the demand of nutrients and CO_2: as more light supports higher productivity, it will also increase the demand on nutrients and CO_2. Photobioreactor designers have developed various solutions to solve different issues.

1.2.1.1 Nutrients

Nutrient control in photobioreactor cultures, whether open or closed, is relatively straightforward. Nutrients, as salts or previously dissolved, are added into the growth medium at the time of photobioreactor inoculation and thus are well distributed throughout the culture. For batch cultures, where the totality of the crop will be harvested at some point in time, no further nutrient management is necessary. However, in commercial photoautotrophic production of microalgae, semicontinuous production is state of the art, and culture age varies between a few weeks and up to several months. These cultures undergo repeated harvest cycles which requires replacement of the nutrients taken up by the biomass harvested. This can be carried out during harvest operations or, independently, driven by analytical results of nutrient concentrations in the growth medium. The frequency of nutrient additions will be dependent on the productivity of the culture.

In cases where biomass is harvested continuously, one can choose to add fresh growth medium and manage the cultures as turbidostats. However, the authors are not aware of any large-scale or commercial microalgal enterprises using this method at the present time.

1.2.1.2 Gas Exchange

Gas exchange refers to both incorporation of the CO_2 into the culture and removal of O_2. Incorporation of CO_2 provides carbon nutrition and provides pH modulation via the inorganic carbon cycle (Goldman et al. 1971). Removal of O_2 is necessary since high concentrations can decrease the productivity of the culture (Raso et al. 2012).

Gas exchange control is relatively simple: one needs to add CO_2 to the culture as needed which is easily done automatically via solenoid valves which respond to changes in culture pH. If the CO_2 is added mixed with air, the volume of gas input can aid in stripping excess O_2 from the culture and preserve its productivity. A simple and effective way to produce microalgae and have excellent control on gas exchange is to use airlift column photobioreactors (Fig. 1.1a), airlift panel PBRs (Fig. 1.1b) and thin vertical bubble column PBRs (Fig. 1.1c). Horizontal tubular PBRs (Fig. 1.1d) require a different approach which includes the use of special gas exchange zones (Torzillo and Chini Zittelli 2015). Nonetheless, loop lengths of up to 400 m have been realized and showed no significant decrease in biomass productivity compared to 100 m standard loops (personal observation). Serpentine tubular photobioreactors also require a separate area for gas exchange. A similar concern with carbon nutrition exists in open raceway photobioreactors where the loss of CO_2 to the atmosphere can represent a substantial cost; increasing the number CO_2 of injection points along the length of the raceway solves any concern for carbon nutrition of the culture.

1.2.1.3 Temperature

Careful choosing of the location of the microalgae production facility is critical; ambient temperature extremes are to be avoided. The range of temperature tolerated by the chosen organism is also critical. If microalgae with varying temperature ranges are produced throughout the year, cultures can be planned according to the suitable season. Still, at some frequency, temperature control of the cultures may be necessary to maintain the highest possible productivity.

Temperature control takes different forms on open vs. closed photobioreactors. In open systems, even those located in very hot areas, cooling is not necessary; evaporative cooling along with judicious selection of crop organism is all that is

Fig. 1.1 (**a**) Photograph of several 164 L polycarbonate airlift photobioreactor producing different microalgae (in use at Arizona Algae Products, LLC, USA). Gas exchange is accomplished by bubbling air from the bottom of the reactor, while CO_2 is added separately from a weighted diffuser from the top. (**b**) Flat Panel Airlift PBR developed at Fraunhofer IBG, Stuttgart, Germany, and at Subitec GmbH. These PVC PBRs (180 L single units) use the airlift-loop-principle for mixing and gas exchange. It was initially designed as a single-use bioreactor. (**c**) Thin vertical tubes connected via horizontal manifolds are also expected to provide excellent gas exchange (Copyright@ecoduna AG). The gas travels the full height of the photobioreactor providing turbulence and exposing a large volume of the culture to atmospheric air. (**d**) Photograph of a vertical-manifold glass tubular photobioreactor (15 to 45 m^3 individual units in use at Salata AG, Ritschenhausen, Germany). In

needed to support commercial production. In the Imperial Valley in southern California, one of the hottest areas of the state, summer ambient temperatures can reach above 45 °C with some frequency, but open ponds rarely exceed 32–33 °C (personal observation). Because of evaporative cooling, open photobioreactors tend to cool significantly when ambient temperature drops. In some cases, the temperature drops sufficiently that production activities cease during the coolest months (Belay 1997). Direct heating of large open photobioreactors would appear to be quite inefficient due to the evaporative heat loss. However, one approach that has worked in other cool areas is to cover the open photobioreactors with a greenhouse structure (plus a small amount of heat). This approach has permitted continuous production of *Nannochloropsis* and *Haematococcus* during the winter in northern Arizona (Fig. 1.2a). Besides temperature regulation, greenhouses offer the benefit of physically protecting the open photobioreactors (Fig. 1.2b) from blowing sand and dust (which results in a very high ash content of the crop, typical of unprotected raceways in the Southwest US), exclude birds and minimize the presence of insects.

Closed PBRs have the advantage that heat and evaporative loss are minimized during cooler periods. Because of the smaller heat loss, heating of the cultures is much more effective than in the case of open photobioreactors. Furthermore, large commercial closed photobioreactors can also be installed inside greenhouses (e.g., Fig. 1.1c, d). Because evaporative heat loss is nearly non-existing in closed photobioreactors (except at gas exchange zones), overheating is possible during warm periods. Heat exchangers can be installed to cool the culture in closed PBRs (common in laboratory-scale PBRs). At large, commercial scale two approaches have been used (either alone or in combination): the direct cooling of the tubes or the cooling of the greenhouse (if present).

On the one hand, useful with horizontal tubular PBRs, the solar collector itself can be laid down in a pool of water (Fig. 1.3). This was the approach used by Aquasearch, Inc to maintain 25,000 L closed PBRs at 18 °C even during the summer in Hawaii for the production of *Haematococcus* (Olaizola 2000). Aquasearch had the added advantage that the source water for the cooling baths originated well below the thermocline of the Pacific Ocean off the coast of Hawaii at a temperature of <5 °C (http://nelha.hawaii.gov/). For other types of closed PBRs, the warmer temperatures can be moderated by spraying or dripping water over the solar collector of the PBR using dripping lines (as they are used in horticulture). As the water evaporates, heat is removed from the system. Care must be taken to use water that is very low in dissolved solids; as the water evaporates, mineral deposits are left behind that lower the light transmission into the PBR and negatively affect productivity. Therefore, deionized water is used for that purpose.

Fig. 1.1 (continued) this approach, gas exchange is accomplished via the system tank in the head space. Thus, specialized gas exchange zones are created outside of the solar collector zone to accomplish gas exchange which, depending on conditions, may limit the length of the tube used as solar collector

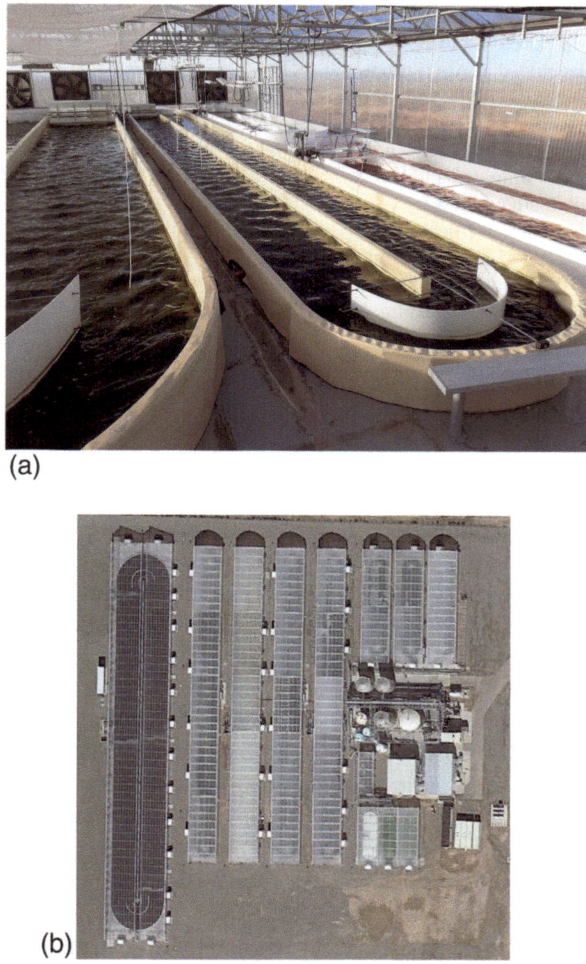

Fig. 1.2 (**a**) Photograph of open photobioreactors growing different crops under a greenhouse in northern Arizona (Arizona Algae products, LLC). (**b**) Satellite view of several greenhouse-covered open photobioreactors including a nearly 4000 m^2 unit near Phoenix, Arizona (Heliae, Inc)

The second approach is to cool the greenhouse. This is achieved by fully automated climate control via forced ventilation through adjustable ventilation flaps or by effective space cooling of the greenhouse using evaporation systems. The combination of these measures can drop the greenhouse temperature by several degrees (personal observation).

While avoiding suboptimal temperatures during the day by cooling, night temperature below 14 °C is desirable in order to reduce night loss of biomass caused by respiration (Edmundson and Huesemann 2015; Belay 1997). These biomass losses can account, depending upon the algae and the specific conditions, from 5 to 42% of day time productivity (Torzillo et al. 1993; Doucha and Lívanský 2009).

Fig. 1.3 Large (25,000 L) tubular closed PBR developed at Aquasearch, Inc in the 1990s and, more recently, used at Cellana, Inc. These polyethylene PBRs provided a low-cost alternative to more expensive, vertically stacked glass and plastic PBRs. Photograph originally published in Olaizola (2003) and is reproduced here with permission

1.2.1.4 Sunlight

In outdoor systems, sunlight availability ultimately limits the amount of production that can be accomplished photosynthetically. And light is quite variable from season to season, day to day and even minute to minute. Can one photobioreactor design be the best for all times?

The performance of the PBR with respect to light, and assuming all else is optimal, is reflected in the measured photosynthetic efficiency: the ratio of chemical energy in the form of biomass normalized to the amount of light energy intercepted by the land surface area occupied by the PBR. When considering a PBR for microalgae production directed toward a particular product or service, one should investigate whether this data is available and supported with long production runs (see, e.g., de Vree et al. 2015).

Mathematical models have been produced to understand the distribution of light in photobioreactors and how this is affected by the changes in light availability on the different time scales (e.g., Acien Fernandez et al. 2012; Endres et al. 2018; Gao et al. 2018; Molina et al. 2001), and technology is now available to test those models (e.g., Karam et al. 2018). Furthermore, expected cell trajectories within a PBR can be programed into illuminators to expose cultures to light conditions expected in different PBR designs under controlled conditions (Dodge et al. 2016; Sivakaminathan et al. 2018) so that results from laboratory studies may guide photobioreactor design.

The general recommendation regarding how to deal with the variable light field available within a PBR, open or closed, is to avoid very high light levels which might lead to photoinhibition (thus lowering the yield of the culture) while maintaining sufficient light penetration to promote efficient growth and biomass production. This is a problem.

Consider the following. Light availability decreases exponentially within a culture with distance from the illuminated surface. Some cells are exposed to high irradiance (which results in lower photosynthetic efficiency) some of the time, while others will be exposed to very low irradiance (even below the compensation irradiance, i.e., the irradiance at which enough photosynthesis is carried out to compensate for respiration losses). Fast mixing of the culture that transports the cells quickly (on the order of several milliseconds) between the high and low light zones helps increase the average photosynthetic efficiency of the culture (Perner-Nochta and Posten 2007). Energetic mixing can, however, produce cell damage and increases culture costs so one must balance the increased costs and risk of damage with realized increases in productivity.

However, even very well-mixed cultures will experience large changes in incoming irradiance over, for example, the length of the day. A population of cells in a PBR which has been established to utilize light efficiently at noon will be, on average, very dark in the morning and late afternoon. A population of cells in a PBR which has been established to utilize light efficiently in the morning will likely be photodamaged at noon and may not recover until the next day. Two possible solutions come to mind, both independent of the type of PBR: use removable shade cloths or modulate the population density in response to changes in available light. The metric that would trigger a change in light management (e.g., increase the amount of shading to protect a culture from photoinhibition) would be a decrease in the photosynthetic yield of photosystem II as measured by changes in the so-called F_v/F_m ratio (variable fluorescence normalized to the maximum fluorescence). This technique has been used for over 25 years in outdoor microalgae cultures (see, e.g., Vonshak et al. 1994), but the authors are not aware of widespread use among commercial producers.

1.2.1.5 Improving Commercial Microalgal Production

Effective management of the four factors listed in this section (nutrients, gas exchange, temperature and light) is critical to optimize microalgal production at large scale. As has been pointed out, different PBR designs, whether open or closed, use different approaches to increase the efficiency of input utilization. Summarizing the influencing factors on algal growth and productivity in commercial production PBRs, there are a number of factors that can be identified as being significant for the improvement of the production process.

Nutrient management may be considered a relatively easy task if the target is to provide sufficient nutrition. However, unless the growth medium is fully recycled after harvest, a significant amount of nutrients might be wasted. If medium recycling is not done (because of contamination concerns, e.g.), tight nutrient management is critical (to reduce costs): one needs to provide sufficient but not superfluous quantities. For tight nutrient management, a monitoring program needs to be implemented which also represents costs. The farm manager will need to choose between increased expenses from monitoring or increased expenses in wasted nutrients.

No matter what type of PBR is used, efficient gas exchange is critical for high productivity. Large surface area between the liquid and gas phases increases gas exchange to provide CO_2 and to remove O_2. Independently of the type of PBR used, a decrease in bubble diameter will enhance the bubble surface to volume ratio and therefore facilitate the gas exchange resulting in improved microalgal production. However, production of smaller bubbles will result in higher expenses since it requires higher energy expenditures. Again, the farm manager needs to determine whether the increase in production makes up for the increased costs.

As pointed out earlier, temperature management is critical to maximize microalgal production, but it can also represent a large cost. The examples given above illustrate how siting of the farm and PBR design affect the effectiveness of and requirements for temperature control. As with other factors, the decision must be made regarding the worthiness of increased expenditures versus the expected return in increased production, and this will be dependent on each individual situation.

The management of light is also a very important factor. Here the use of physiological parameters such as online measurements of chlorophyll fluorescence may guide the way to an improved production process. The tuning of times of light limitation versus light inhibition times requires the adoption of appropriate mixing rates and harvesting strategies, ideally fully process-automated. Changes in mixing rate may increase the demands for energy expenditures which, because of cost, should be as low as possible. Depending upon the light regime used, energy for circulation of the algal suspension can be reduced according to the natural light that is available at a given time point.

All the parameters discussed aim at the optimization of production costs and biomass quality. If they are managed to enhance biomass productivity at a given cost, the effect of a lowered production price per kg of dry matter will create a better margin and a faster return on investment. Falling below certain target market values will even allow new applications in food and feed sectors.

1.2.2 Materials

As indicated above, a multitude of PBR designs have been developed, and many different materials have been used for building PBRs (Table 1.3). The type of the PBR determines the selection of the possible materials. Both from an operator's and from a PBR developer's point of view, the following criteria for selection are of utmost importance:

(a) Type of PBR
(b) Alga to produce
(c) Target market of the final product

Open systems do not use the same construction materials as enclosed systems. The most important difference between open and closed PBRs is that no material is needed to contain the culture above its basin and the culture is directly exposed to the environment. Among closed PBRs, glass is used for thin tubular PBR arrangements. Flat plates and wider columns ($d > 5$ cm) at industrial scale are usually made out of

Table 1.3 Materials in use for industrial raceway and closed PBRs

Open PBR (raceways)		
Basin and baffle	Lining	Paddle wheel
Concrete	PVC	(Fiberglass-reinforced) plastics
Clay	(HD) PE	Fiberglass
Asphalt	PP	(Coated) metal, steel
Ground (compacted soil, sand, gravel)		Plywood
Corrugated sheet roof		
Fiberglass		
Closed PBR		
Solar collector	Connecting parts	Vessel/degasser
Borosilicate glass	PVC	PVC
Polyvinyl chloride (PVC)		PE
(Low-density (LD)) polyethylene (PE)		Steel
Acrylic glass (PMMA, e.g., plexiglass®)		
Polyethylene terephthalate glycol (PETG)		
Polycarbonate (PC)		
Silicone		

transparent synthetic polymers (plastics). Table 1.3 summarizes the materials in use for industrial raceway and closed PBRs.

Besides the PBR type itself, the target algae to be produced also influences the PBR material. If no ecological niche is available in terms of cultivation conditions (e.g., salinity, pH), closed PBRs are frequently favored, as well as if flexibility in the algal final product is needed. Algae that are known to secrete polysaccharides into the culture medium (e.g., *Porphyridium*, *Nostoc*) tend to stick to the solar collectors of closed PBRs, more so to plastics than to glass and especially if either mixing rates are insufficient or dead zones prevail. These conditions cause fouling and thereby decreased light penetration as well as product quality. In these cases, tubular glass shows higher culture stability as it has minimal surface roughness. Both the application (e.g., food, cosmetics) and the geographical location of the target market set quality parameters that need to be met for acceptance of the product, being, e.g., very restrictive for Japan. Thus, closed PBRs are frequently preferred for high value applications.

Raceway ponds, which in general consist of a circulation channel, either are dug in the ground or, more commonly, consist of an erected basin of various materials (see Table 1.3), with or without an additional lining of PVC (polyvinyl chloride), PE (polyethylene), or PP (polypropylene) in contact with the culture. The plastic lining material helps to reduce the Manning roughness coefficient of the pond. A relatively cheap setup makes use of a compacted ground construction lined with a 1–2-mm-thick plastic film (Chisti 2016). A textile material beneath the plastic lining can also be used. UV-resistant PVC lining has been proven durable for >20 years under dessert conditions (e.g., NBT in Israel). If basins are erected, concrete or clay blocks are often used, but also prefabricated concrete units are employed. The paddlewheel, which provides the energy supply for mixing, consists usually of eight straight or curved blades made out of metal, (reinforced) plastics, fiberglass, or even plywood.

In closed PBRs, the solar collector requires high transparency for visible light combined with low near-infrared and UV transmission and low cost. Among the materials used (see Table 1.3), the transparency of glass is highest (92%), followed by PMMA (polymethyl methacrylate) (91%) and UV-resistant PVC (67.6%) (Wintersteller 2018). Reflection at the transition of the interface between air and reactor and absorption by the PBR material result in the loss of photons. A few of the world's largest algae production facilities have been based upon borosilicate glass of a wall thickness of 1.6–3.2 mm, in order to withstand hydrostatic pressure, especially in vertical arrangements of the glass tubes. Both performance and durability have been demonstrated for well over 20 years (e.g., Algatechnologies, Israel, and Roquette, Germany). Transparent plastic materials have been investigated for their reduced density and cost. Here, either polymer "walls" as a replacement of glass are used, mainly PVC and PE, but also PC (polycarbonate), PETG (polyethylene terephthalate), PMMA and silicone. Plastic film can also be used, e.g., as "sleeves" or bags made out of low-density polyethylene (LDPE), with very low film thickness (180 μm). But even if UV-stabilized plastics are used, the maximum life of a LDPE film directly exposed to sunlight is 3 years (Burgess et al. 2007). Although the lifespan of the transparent materials is very different, due to UV-induced aging, it is always lower for plastics than for glass. The predicted lifetime ranges between 3 years for UV-stabilized LDPE (Hussain and Hamid 2004) and 20 years for PVC and PE (Chisti 2012), whereas the latter has still to be proven under outdoor conditions, in large-scale production. Additionally, the chemical and thermal resistance of plastics depends heavily upon the respective polymer used and any co-polymers added during manufacture. Bending strength and coefficient of elasticity are important for the installation of horizontally long tubes. While PMMA and PC are weather- and aging-resistant, the cheaper and more readily available PVC and PE dominate (Oeschger and Posten 2012). The dimensional stability under heat is sufficient for all the plastic polymers for algae cultivation, as heat sterilization is not an industry standard.

Regarding the safety of the materials used, three parameters must be taken into account: the possible inhibitory or even toxic effects of the material to the specific algae cells in culture, the permeation of media components into the material and its food safety in terms of possible migration of chemicals into the final product due to leaching. The latter can be excluded by migration tests that are standard for product touching polymers in food processing and packaging. Noxious or toxic effects on algal cultures have been documented for copper alloy, rubber, PVC, PP, nylon and silicon for specific species and conditions (Blankley 1973) but have not been extensively studied so far. Therefore, if in doubt, standardized methods for quantification of inhibitory or toxic effects should be employed. Heavy metals which can act as enzyme inhibitors and accumulate in algae biomass should be avoided

Fig. 1.4 Biofilm (fouling) formation in a PVC plate (left, Hamburg, Germany), a sleeve PBR (middle, Niederaußem, Germany) and a glass tube PBR (right, Potsdam, Germany) with a freshwater algae culture

wherever possible, as they can negatively influence both growth of culture and marketability of the final product.

Formation of biofilms, especially with species that produce high amounts of extracellular polysaccharides and proteins, is a major problem in closed PBRs as it diminishes an important advantage of tubular reactors: enhanced light introduction into phototrophic cultures by higher surface-to-volume ratios. This phenomenon occurs on both plastic and glass materials (see Fig. 1.4) depending upon culture age, turbulence, culture conditions (mainly light intensity) and algae species being cultivated. Although this issue has been and still is mostly ignored or neglected by PBR technology providers, our experience is that the immobilization of microalgae and their accompanying heterotroph microorganisms tends to be stronger on plastics compared to glass surfaces (see Fig. 1.4). Furthermore, in tubular PBRs, the zones where two sections of tube are joined can form small gaps where biomass may get caught and form the basis for biofilms. Clearly, closed PBRs that are easy to clean or that do not easily foul would have cost advantages (see next section).

1.2.3 Cleaning and Sanitizing

As mentioned in the previous section, algae are sticky and can develop biofilms on the surface areas of PBRs creating several problems:

(a) In closed PBRs, biofilms growing on the solar collectors reduce the amount of light available to the culture.
(b) In closed and open PBRs, biofilms can harbor contaminant organisms.
(c) In closed and open PBRs, biofilms can affect the flow of the culture thus changing the level of turbulence and, indirectly, affect gas exchange.

(d) For different reasons (usually contamination or decreased yield) at some point in time, the PBR needs to be emptied, cleaned, sanitized and re-inoculated.

It is not easy to find information on cleaning and sanitizing PBRs although this can be a substantial problem. For example, Arbib et al. (2013) report on an experiment comparing the yield of a closed PBR versus an open pond. Although they found that the closed PBR appeared to be more productive, the experiment had to be cut short because of biofouling.

As the inner surface of the closed PBRs cannot be cleaned directly (mechanically), either chemicals (where possible) are used or the photosynthetic units are disposed of (single-use PBRs). In horizontal tubular PBRs, biofilm formation can be reduced by enhancing the flow velocity to above 0.5 m s^{-1}, which enhances energy consumption but also the light availability per cell. Additionally, the use of plastic granulates of various densities has successfully been employed to reduce the formation of biofilms in tubular glass PBRs by mechanical means (author's personal experience). Other scouring materials (known as a pig unit) as large as the diameter of the PBR tube itself have been used (see, e.g., the unit described by Wilson et al. 2017) to attempt the elimination of biofilms. Some work has been done to try to understand biofilm formation in PBRs (reviewed in Zeriouh et al. 2017). The major factors affecting biofouling are the type of material used in the construction of the PBR, the ionic strength of the growth medium, properties of the cell surfaces and the water velocity and fluid dynamics within the PBR.

Stopping a production run to remove biofilm from a PBR can be a costly endeavor. Our personal experience indicates that, depending on the PBR and culture conditions, 2–4 days may be invested along with different chemicals and several water exchanges. Costs include:

(a) Lack of production while the PBR is being cleaned and sanitized
(b) Labor, chemicals and water
(c) The cost of producing inoculum for the next production run

The cleaning process itself may involve several steps. After the closed PBR is emptied of its culture, it needs to be refilled with water including cleaning solutions. The authors have had experience using citric acid dissolved in hot water (up to 80 °C) followed by a hydrogen peroxide treatment at final concentrations of 2–3%. We have also used dishwashing powders (non-foaming), NaOH solutions and ozone addition. Combining these chemical treatments with mechanical means such as the aforementioned pig units or plastic beads aid in breaking up the biofilms. Depending on the composition and thickness of the biofilm, this process may take several days and require being repeated.

After the biofilm has been completely removed, the PBR needs to be sanitized/ disinfected which can be accomplished with solutions of chlorine (at concentrations of up to 20 mg L^{-1}), hydrogen peroxide, or ozone injections.

One can argue that open PBRs such as raceways are simpler to clean. Biofilms in open reactors tend to form mostly at the bottom in so-called dead zones created by non-uniform flow (which can occur next to the middle berm following a turn) and along the interface of the water surface and the sidewalls of the PBR. Biomass also tends to accumulate and attach to the reactor bottom preferentially on the leg following the paddlewheel since the culture is deeper but slower than the return leg.

Because open PBRs are open, large mechanical means can be used to aid in the removal of attached or settled materials. Large manual brushes and even farm equipment with brushes attached can be used to scrub and sweep dislodged materials to the reactor's drain. Because of this approach, the open PBRs need relatively less water for cleaning than closed systems where several PBR volumes of water may be required.

Some closed PBR designs do not need to be cleaned of biofilms. Designs have been created using plastic films resulting in inexpensive solar collectors that can be thought of as single-use. One example is the Aquasearch Growth Module PBR created by Aquasearch in the 1990s and used to produce *Haematococcus* (Olaizola 2000, Fig. 1.3) and more recently used by Cellana, both in Hawaii. These reactors were used for 6–10 weeks before undergoing a deep cleaning cycle and re-inoculation. After 2–3 cycles, the plastic tubes were replaced. Similar plastic material has been used successfully to develop the Green Wall PBR by Tredici and his team (Rodolfi et al. 2008; Tredici et al. 2015). Some more complex plastic film PBRs have been developed such as the water-supported panels developed by Proviron and Solix or the hanging panels by Valcent Products (Fig. 1.6). Because of the extra manufacturing cost of these engineered panels, it is not clear whether the economics would permit replacing the panels versus investing in their cleaning.

Cleaning and sanitizing production PBRs can represent a significant fraction of the production cost, and we encourage algae producers to take this matter into account when considering different types of PBRs whether open or closed.

1.3 Photobioreactor Management

Producers are interested in learning which photobioreactor they should use for their proposed product. Based on the previous material in this chapter, there are many different factors to have in mind. Those factors affect the cost of purchasing the system and the costs associated with running the system. In the end, one might want to look for the photobioreactor that maximizes production of the desired material (e.g., tons per year) normalized to the input costs (e.g., $ per year). In this section, we present the concept that a simple production model can be fitted to data obtained from production runs and compared among different PBR platforms to help with PBR selection.

1.3.1 A Simple Production Model

A simple production model has been created that can be used in a spreadsheet for any type of photobioreactor (Fig. 1.5). The model needs the following inputs:

(a) Light energy (usually solar) impinging on a certain surface area of ground

Fig. 1.5 Engineered film panels for production of microalgae biomass. Water-supported panels (left photograph, Solix) and hanging panels (right photograph, Valcent Products)

(b) A conversion factor that transforms light energy into biochemical energy as biomass
(c) A photosynthetic efficiency representing the biochemical energy generated in a certain surface area of ground normalized to the solar energy impinging on that same surface area of ground
(d) A respiration rate which represent the fraction of biochemical energy produced that is respired by the microalgae
(e) A maximum intrinsic growth rate for the chosen organism

For a simplified approach, we have chosen 24 h intervals over which to run the calculations. Thus one measures the amount of solar energy intercepted by the installation over 24 h (kWh m^{-2} d^{-1}) and calculates the resulting gross productivity (g m^{-2} d^{-1}), the estimated respiration (g m^{-2} d^{-1}) and the net productivity (g m^{-2} d^{-1}) after making some assumptions about the system's photosynthetic efficiency, the biochemical energy contained in the biomass, the respiration rate, and the maximum intrinsic growth rate for the organism (μ_{max}).

Using these parameters and assuming a starting culture biomass concentration, one can use a simple model to produce a growth curve such as the one in Fig. 1.6 over an arbitrarily chosen 42-day production cycle (6 weeks). If the starting biomass concentration is low enough, the first part of the growth curve will show an exponential increase in standing biomass for a few days until the culture is concentrated enough that all the light energy that can be absorbed (which is a function of the biomass concentration) is absorbed. This biomass concentration is identified as B_{Imax} (the biomass concentration at which all available light is intercepted and bulk light limitation starts). At this point, and for the subsequent days, the culture is light limited and gross productivity is constant (gross productivity = energy input × photosynthetic efficiency). In this simple model, the assumption is made that a constant fraction of the standing biomass in the culture is respired. Thus, as time passes, the rate of biomass accumulation decreases (since respiration increases linearly as biomass accumulates in the culture) and, at some point, the standing

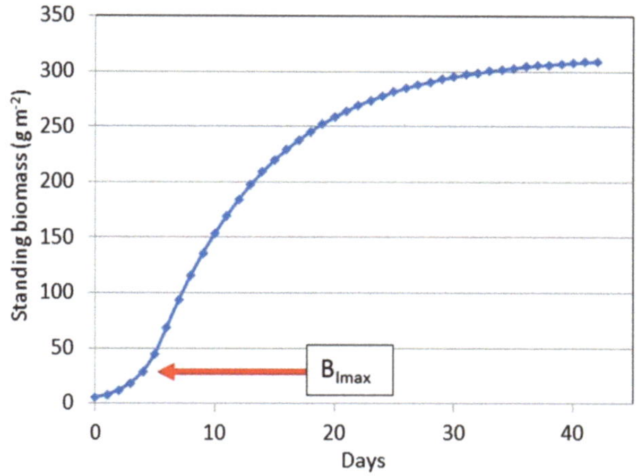

Fig. 1.6 A typical growth curve for a batch culture allowed to grow for 42 days. Assumptions: starting biomass 5 g m^{-2} (equivalent to 0.05 g L^{-1} for a 10-cm-deep open raceway, e.g.), daily solar energy input of 7 kWh m^{-2} d^{-1}, 2% photosynthetic efficiency, 10% daily respiration loss and a maximum intrinsic growth rate of 0.5 d^{-1}. B_{Imax} represents the biomass concentration at which bulk light limitation sets in

biomass values will asymptote at the point in time when gross productivity is equivalent to gross respiration (and net productivity is zero). At this point, the standing biomass of the culture represents the carrying capacity of the system.

This can be seen graphically in Fig. 1.7. The increase in gross productivity of the culture is nearly exponential until the standing biomass in the culture is such that adding more biomass to the culture via growth does not result in more light (energy) absorbed by the cells themselves (B_{Imax}). At that point, and for the subsequent days, the gross productivity of the system is constant (assuming constant solar energy and constant photosynthetic efficiency). In this simple model, it is also assumed that the daily respiration rate is 10% of the standing biomass at the beginning of the day which seems reasonable considering other published values (Torzillo et al. 1993; Doucha and Lívanský 2009). Thus, the respiration rate curve follows a similar trajectory to that of the standing biomass shown in Fig. 1.6. The difference between the gross productivity and respiration results in the net productivity. As can be seen in Fig. 1.7, maximum daily net productivity occurs relatively quickly after inoculation in a culture, and it occurs at a standing biomass concentration well below 1/3 of the carrying capacity of the system.

Daily net productivity does change significantly depending on the stage of the culture. One should keep this in mind when comparing productivity estimates from the literature based on relatively short production runs.

1.3.2 Harvest Strategies

One of the uses for this simple production model is to develop protocols that can help to guide how one manages the culture, independent of the type of PBR in use. On the assumption that one desires to maximize productivity, we can design a harvesting schedule that maximizes productivity while minimizing costs.

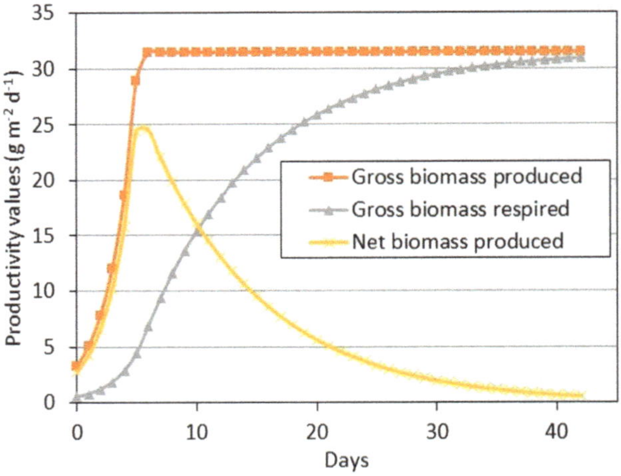

Fig. 1.7 The biomass accumulation curve in Fig. 1.6 is the result of the daily net productivity (g m^{-2} d^{-1}) which is calculated as the difference between the gross daily productivity and the daily respiratory losses

It has already been established that we can determine the standing biomass at which net productivity is maximized (see previous section and Figs. 1.6 and 1.7). In principle, one would want to harvest at this standing biomass as often as possible to maximize productivity. However, the biomass concentration is relatively low, and harvesting costs would actually be high per kg or ton of biomass harvested. We use two arbitrary examples to illustrate this point in Fig. 1.8. In one case (less frequent harvests at higher biomass concentration), we chose to trigger a harvest event when the culture's biomass reaches about 6.5 times the value of B_{Imax} and to harvest 50% of the biomass in the culture each time. The result is that a harvest is triggered about once per week. In a second case (more frequent harvests at lower biomass concentration), we chose to trigger a harvest event when the culture's biomass reaches about three times the value of B_{Imax} and to harvest 70% of the biomass in the culture each time. While a less aggressive harvest schedule likely lowers the cost of harvesting per unit mass harvested, the amount of mass harvested over the 42-day arbitrary production period is calculated as 683 g m^{-2} (which over 42 days represents a net productivity rate of 16.3 g m^{-2} d^{-1}). The more aggressive harvest schedule likely incurs higher cost, but the amount of mass harvested over the 42-day production period is 901 g m^{-2} (which over 42 days represents a net productivity rate of 21.4 g m^{-2} d^{-1}). By comparison, if the culture would have been treated as a batch culture and harvested only after the 42-day growth period, the amount produced would have been 310 g m^{-2} (average net productivity would have been 7.4 g m^{-2} d^{-1}).

By setting up simple production equations in a spreadsheet, it is possible to predict which harvest strategy may be the most appropriate for the chosen process. Of course, this is an example of a very simple model; a more sophisticated effort would include, for example, variable daily solar energy (if one can predict it), temperature effects and actual harvest costs. Recent examples of more sophisticated models have been published for specific PBR types or for specific growth factors such as those by Endres et al. (2018), Gao et al. (2018), Huesemann et al. (2016) and Ramirez et al. (2018).

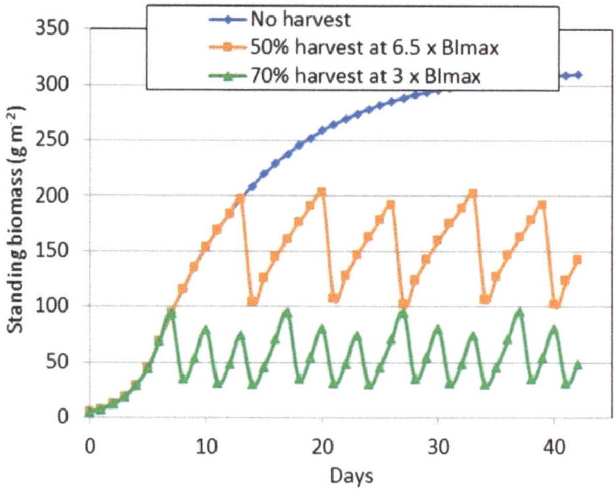

Fig. 1.8 Two examples of expected daily standing biomass values superimposed on a non-harvest growth curve. The green line represents an aggressive harvesting schedule (every 2–3 days) that strives to maintain the standing biomass of the culture closer to the maximum productivity range (see also Fig. 1.7). The orange line represents a less aggressive harvest schedule that strives to maintain the standing biomass at higher levels during harvest and, thus, decrease the cost of harvesting per kg of biomass obtained

1.3.3 Intrinsic Growth Rate, Initial Biomass and Respiration

In the previous section, it was shown how one aspect of PBR management can have a large influence on the productivity of the system. In our experience, productivity can also be affected by other factors which researches have tried to optimize.

1.3.3.1 Intrinsic Growth Rate

The intrinsic growth rate of a microalgal strain is the maximum growth rate under optimal, non-limiting growth conditions (μ_{max}, d^{-1}), and efforts have been made to find strains with high μ_{max}. However, using our simple production model, we find that the effect of a higher μ_{max} on a PBR culture can be relatively small. Higher μ_{max} does translate into a faster initial ramp-up of a culture, but once the culture reaches the standing biomass at which all the available light is effectively used (B_{Imax}), the productivity of the culture is only partially dependent on μ_{max}.

For example, using our simple production model we have calculated, for our standard 42-day production period, the total biomass produced assuming three different intrinsic growth rates (0.4, 0.6 and 1.0 d^{-1}) under the two different harvesting strategies used earlier. The results of our calculations are summarized in Table 1.4. As might be expected, the effects of μ_{max} on productivity are larger when the culture is managed in the "high frequency/low biomass harvest" strategy. As was indicated above, this strategy does result in higher harvest expenses per kg of

Table 1.4 Effects of the microalgae intrinsic growth rate on culture productivity under two different harvest strategies

Harvest strategy	Low biomass		
μ_{max} (d^{-1})	0.4	0.6	1.0
42-day production (g m^{-2})	764.7	984.9	1081.6
Daily average (g m^{-2} d^{-1})	18.2	23.4	25.7
% more production (over 0.4 d^{-1})		29	41
Harvest strategy	High biomass		
μ_{max} (d^{-1})	0.4	0.6	1.0
42-day production (g m^{-2})	636.4	692.7	718.7
Daily average (g m^{-2} d^{-1})	15.2	16.5	17.1
% more production (over 0.4 d^{-1})		9	13

To take advantage of production strains with faster growth rates, attention must be paid to the harvest strategy

biomass produced. The farm manager would need to make the determination of whether the extra harvesting expense is worthwhile (likely dependent on the value of the crop) and whether extra expenses associated with bioprospecting for (or developing) a high μ_{max} strain are worth the extra cost.

1.3.3.2 Initial Biomass

A higher initial standing biomass translates into a quicker ramp-up of the culture to the level at which harvesting can start. A higher initial standing biomass also translates into insurance against possible damage by high irradiance.

Using the simple production model presented here, we calculate that the effect of higher initial biomass over a 42-day production run is relatively small: changing the starting biomass concentration from 0.05 to 0.15 and 0.30 g L^{-1} results in production increases of 6–10% over the 42-day run. Protection against photoinhibition can result in bigger gains, however, specially is the risk of losing the whole culture is averted.

So, it does make sense to start with denser cultures outdoors. But denser initial concentrations can only be obtained with a larger inoculum production system and all its associated costs. In the end, the farm manager must decide whether the extra expenses justify the relatively small gains in productivity. With regard to insurance against photodamage, shading material can be used at, perhaps, lower cost than a larger inoculum production system.

1.3.3.3 Respiration

Microalgae respire constantly as dictated by cellular metabolism although it is mainly noticed in the dark. Respiration translates into a decrease in net photosynthetic efficiency. If it is assumed, for simplicity, that the respiration rate represents a fraction of the biomass present in the culture, the apparent drop in photosynthetic efficiency increases at higher biomass densities. Strategies have been proposed that include cooling of the culture during the night-time to lower the respiration rate (Edmundson and Huesemann 2015; Han et al. 2013; Belay 1997). However, the cultures chilled during the night need to be warmed up before exposure to high light the following day or photodamage can ensue (Vonshak et al. 2001). This approach could include the use of heat exchangers and might be easier to accomplish in closed reactors with a high surface-to-volume ratio placed in greenhouses, where temperature rises quickly after dawn. The solar collector, which has a high surface area-to-volume ratio, serves here as the heat exchanger.

1.4 New Developments in PBRs

As was indicated earlier, PBRs attempt to provide optimal growth conditions to the largest possible fraction of the microalgal population. Besides the PBRs that have already been touched upon (and have been reviewed by others), new PBR designs continue to emerge. Here we present a few examples that are less familiar but that employ somewhat different approaches to solve the problems of light utilization, gas exchange and turbulence.

In order to overcome light limitation in algal cultures caused by layer thickness, a thin layer system has been developed, called mesh ultra-thin layer (MUTL). In the MUTL, a short light path is realized to enhance the efficiency with which light energy can be converted into biomass. The principle was introduced and demonstrated to the public by Otto Pulz and colleagues at the 9th European Workshop of Microalgae, in 2012, in Potsdam, Germany, and was developed and patented by IGV GmbH. The engineering principle of the MUTL biomimics a deciduous tree with its thin leaves and homogeneous distribution of chlorophyll in three dimensions. Consequently, an algae suspension is scattered in an open PBR space, protected by a transparent cover. The algae culture is circulated by a small pump and the suspension is sprayed via nozzles (see Fig. 1.9a). Resulting drops fall by gravity and are retained by mesh elements. Subsequently the suspension is collected in a basin and recirculated. The algae are cultivated in a CO_2-rich atmosphere and are harvested in a semicontinuous mode. Optimization was carried out by changing the type and throughput of the nozzles, their distribution in the PBR, the mesh material and its spacing and the geometry of the collecting basin. Outdoor cultivation reactors with ground areas of up to 500 m^2 in scale using *Chlorella vulgaris* were tested (Schreiber et al. 2017). The main goal, from an engineering point of view, was to enhance the illuminated area per ground area and to obtain high surface-to-volume ratios. The illuminated area per ground area in tubular PBRs lies around 4–9. The surface-to-volume ratio in the MUTL reactor is about 80–120 m^{-1}, and design ratios of up to 1000 m^{-1} are possible. Using *Nannochloropsis oculata*, biomass productivities of 0.75–2.1 g L^{-1} d^{-1} have been measured, as well as biomass concentrations of up to 25 g L^{-1} (Grewe and Broneske 2015). The short light path in a falling drop results in shorter light-dark cycles for suspended algae as well as in higher mass transfer for both carbon dioxide and oxygen. At the same time, the buildup of inhibitory oxygen concentration is avoided.

Fig. 1.9 New approaches to PBR design. (**a**) The mesh ultra-thin layer (MUTL) PBR developed by Prof. Pulz and colleagues at (and patented by) IGV GmbH creates a scattered algae suspension for better light utilization and gas exchange. The close-up shows the drop-retaining elements (mesh). (**b**) 1000 m^2 greenhouse (photograph originally published in Grewe and Pulz 2012; reproduced here with permission) populated with hanging mesh members to support microalgal growth. The greenhouse was flooded with flue gas from a natural gas power plant. (**c**) Fluid movement through

(d)

(e)

Fig. 1.9 (continued) the closed PBRs developed at Synthetic Genomics advantageously alleviates the need for plug flow to create enhanced turbulence. (**d**) Photograph of the Revolving Algal Biofilm (RAB) photobioreactor provided by Dr. M Gross. (**e**) Light augmentation via artificial means can be attractive when the desired product is of high enough value or the electric rates are low enough (OmegaAlgae PBR in Iceland)

A very different design but with a similar objective of enhancing light utilization and gas exchange is shown in Fig. 1.9b. This design also originated in Prof. Pulz's laboratory, and it entails the distribution of the microalgal culture over large sheets of fabric suspended vertically. The largest installation with such an approach was installed by Greenfuels Inc at the Red Hawk power plant in Arizona in 2006–2007. The system was designed to attenuate flue gas emissions; flue gas was pumped directly from the power plant's smoke stack into a 1000 m^2 greenhouse containing the sheets of fabric. Culture medium was recirculated to maintain the fabric sheets moist. At this scale, the system suffered some engineering setbacks.

However, a smaller version (100 m^2) did prove to be quite productive (Grewe and Pulz 2012).

Other PBRs have expanded on the idea of utilizing different types of fabric to maintain productive algal biofilms and take advantage of enhanced surface area for gas exchange. One such design is the so-called twin-layer PBR, where cultivation is carried out on a vertically oriented porous substrate, covered by newsprint paper (Naumann et al. 2013). High surface-to-volume ratios, short gas diffusion pathways, no growth-limiting oxygen concentrations and low investment costs have been discussed, and area productivities of up to 50 g m^{-2} d^{-1} were reported for *Halochlorella rubescens* (Schultze et al. 2015) using a twin-layer PBR following light and CO_2 optimization. A significant advantage of this type of PBRs is the higher biomass concentration, which enhances energy efficiency per kg of biomass produced, in both up- and downstream processing.

A further development utilizing biofilm-based PBRs consists in moving the biofilm in and out of the growth medium, instead of moving the medium onto the biofilm. One example of this approach is the rotating disc PBR, which employs discs as algae growth supporting elements installed on a drive shaft, that periodically submerges the disc area into a basin containing the growth medium. At lab scale, a biomass productivity of 20 g m^{-2} d^{-1} per disc area was measured for *Chlorella sorokiniana* (Blanken et al. 2014).

Another approach to rotating the biofilm in and out of the growth medium has been developed in which continuous sheets are rotated in and out of the growth medium (Revolving Algal Biofilm—RAB). Early designs (Gross et al. 2013) resulted in productivity improvements over open pond designs. Later improvements of this design, which increased the ratio of biofilm surface area to land area occupied by the PBR (Fig. 1.9c), resulted in larger productivity gains (Gross et al. 2015).

Another approach to enclosed PBRs, in use on some outdoor studies, was recently developed at Synthetic Genomics. This approach includes a tubular PBR constructed of polyethylene film and PVC, making it inexpensive. It has design aspects such that it does not exhibit plug flow. Instead, the design promotes fluid movement through the length of the PBR providing enhanced turbulence and gas exchange (Fig. 1.9d). The PBR does not strictly require pumps to circulate the culture; a low pressure/high volume blower is used to provide air and CO_2 and to generate the characteristic fluid movement (Olaizola et al. 2019). Because the PBR is constructed with inexpensive materials, the solar collector can be replaced if it becomes too heavily encrusted with biofilm.

Another approach to better light utilization has little to do with the design of the PBR itself but with the use of artificial light. As was noted earlier, the large variability of natural sunlight represents a problem: there is too little sometimes and too much at other times to be used efficiently. Augmentation with artificial light, however, can alleviate these problems. Although the use of artificial light represents a significant cost, the production of a high value product or the availability of very inexpensive electricity makes this approach attractive. For example, OmegaAlgae in Iceland (Fig. 1.9e) augments the limited amount of sunlight available with artificial light.

Finally, PBRs of different designs may be instrumental in bringing microalgae to the city. An interesting application of this approach is the "greening" of building facades as part of urban farming (Schmidt et al. 2017), which has been realized with suspension-based PBRs before (smart Material House, International Building Exhibition Hamburg 2013, Kerner et al. 2019). If biomass of sufficient quality can be produced, these facades can be seen as agricultural area while taking up pollutants and helping manage the energy balance of a building.

1.5 Conclusions

The authors share the opinion of Borowitzka and Vonshak (2017) that production of microalgal biomass at scale is not trivial, and there is still a large gap between the potential for microalgal biotechnology and actual achievements. At commercial scale, many factors need to be controlled or, at least, managed, and different PBR designs attempt to do this. We believe that both closed PBRs and open raceways can play a role in closing the gap between potential and reality. A few years ago, during the hyped-up microalgae biofuels period, the focus was on meeting the requirements for the lowest possible production cost. Today, microalgal producers are focused on high value products, and, while still desiring to keep investments as low as possible, this new value proposition allows more flexibility regarding the production system to be chosen.

Choosing the right PBR for a specific product or service can be challenging, and here we have offered a few thoughts and considerations to keep in mind when comparing one design versus another: take short-term productivity estimates as well as extrapolations to larger ground areas with healthy skepticism, create a simple productivity model to help guide your cultivation strategy and inquire about the procedure to clean and sanitize a fouled PBR.

References

Acien Fernandez FG, Fernandez Sevilla JM, Molina Grima E (2012) Principles of photobioreactor design. In: Posten C, Walter C (eds) Microalgal biotechnology: potential and production. De Gruyter, Berlin, pp 151–180

Arbib Z, Ruiz J, Alvarez-Diaz P et al (2013) Long term outdoor operation of a tubular airlift pilot photobioreactor and a high rate algal pond as tertiary treatment of urban wastewater. Ecol Eng 52:143–153

Barclay W, Weaver C, Metz J et al (2010) Development of a docosahexaenoic acid production technology using Schizochytrium: historical perspective and update. In: Cohen Z, Ratledge C (eds) Single cell oils, 2nd edn. Elsevier, Amsterdam, pp 75–96

Belay A (1997) Mass culture of *Spirulina* outdoors – The Earthrise experience. In: Vonshak A (ed) *Spirulina platensis* (*Arthrospira*): physiology, cell biology and biotechnology. Taylor & Francis, London, pp 131–158

Blanken W, Janssen MGJ, Cuaresma M et al (2014) Biofilm growth of *Chlorella sorokiniana* in a rotating biological contactor based photobioreactor. Biotechnol Bioeng 111(12):2436–2445

Blankley WF (1973) Toxic and inhibitory materials associated with with culturing. In: Stein JR (ed) Handbook of phycological methods. Cambridge University Press, Cambridge, pp 207–229

Borowitzka MA, Vonshak A (2017) Scaling up microalgal cultures to commercial scale. Eur J Phycol 52(4):407–418

Burgess G, Fernandez-Velasio JG, Lovegrove K (2007) Materials, geometry, net energy ratio of tubular photobioreactor for microalgal hydrogen production. Int J Hydrogen Energy 32 (9):1225–1234

Chang JS, Show PL, Ling TC et al (2017) Photobioreactors. In: Larroche C, Sanroman M, Du G et al (eds) Current developments in biotechnology and bioengineering: bioprocesses, bioreactors and controls. Elsevier, Amsterdam, pp 313–352

Chisti Y (2012) Raceway-based production of algal crude oil. In: Posten C, Walter C (eds) Microalgal biotechnology: potential and production. De Gruyter, Berlin, pp 113–146

Chisti Y (2016) Large-scale production of algal biomass: raceway ponds. In: Bux F, Chity Y (eds) Algae biotechnology. Green energy and technology. Springer, Berlin, pp 21–40

De Vree JH, Bosma R, Janssen M et al (2015) Comparison of four outdoor pilot-scale photobioreactors. Biotechnol Biofuels 8:215–226

Dodge C, Peers G, McCarren J et al (2016) Method of illuminating algae for algae growth. US Patent # 9329131

Doucha J, Lívanský K (2009) Outdoor open thin-layer microalgal photobioreactor: potential productivity. J Appl Phycol 21(1):111–117

Edmundson SJ, Huesemann MH (2015) The dark side of algae cultivation: characterizing night biomass loss in three photosynthetic algae, *Chlorella sorokiniana*, *Nannochloropsis salina* and *Picochlorum sp*. Algal Res 12:470–476

Endres CH, Roth A, Bruck TB (2018) Modeling microalgae productivity in industrial scale vertical flat panel photobioreactors. Environ Sci Technol 52(9):5490–5498

Gao X, Kong B, Dennis Vigil R (2018) Simulation of algal photobioreactors: recent developments and challenges. Biotechnol Lett 40(9-10):1311–1327

Goldman JC, Porcella DB, Middlebrooks JE et al (1971) The effect of carbon on algal growth - its relationship to eutrophication. Reports. Paper 462. https://digitalcommons.usu.edu/cgi/viewcontent.cgi?article=1461&context=water rep

Grewe CB, Broneske J (2015) Development of photobioreactors, cultivation strategies and algae-based products. Algae Eur, Lisbon

Grewe CB, Pulz O (2012) The biotechnology of cyanobacteria. In: Whitton BA (ed) Ecology of cyanobacteria II: their diversity in space and time. Springer, Berlin, pp 707–739

Gross M, Henry W, Michael C et al (2013) Development of a rotating algal biofilm growth system for attached microalgae growth with *in situ* biomass harvest. Biores Technol 150:195–201

Gross M, Mascarenhas V, Wen Z (2015) Evaluating algal growth performance and water use efficiency of pilot-scale Revolving Algal Biofilm (RAB) culture systems. Biotechnol Bioeng 112(10):2040–2050

Han F, Wang W, Li Y et al (2013) Changes of biomass, lipid content and fatty acids composition under a light–dark cyclic culture of *Chlorella pyrenoidosa* in response to different temperature. Biores Technol 132:182–189

Huang Q, Jiang F, Wang L et al (2017) Design of photobioreactors for mass cultivation of photosynthetic organisms. Engineering 3:318–329

Huesemann M, Crowe B, Waller P et al (2016) A validated model to predict microalgal growth in outdoor pond cultures subjected to fluctuating light intensities and water temperatures. Algal Res 13:195–206

Hussain I, Hamid H (2004) Plastics in agriculture. In: Andrady AL (ed) Plastics and the environment. Wiley, New York, pp 185–209

Johnson TJ, Katuwal S, Anderson GA et al (2018) Photobioreactor cultivation strategies for microalgae and cyanobacteria. Biotechnol Prog 34(4):811–827

Karam AL, de los Reyes FL, Ducoste JJ (2018) Development of photochemical microsensors for evaluating photosynthetic light dose distributions in microalgal photobioreactors. Environ Sci Technol 52(21):12538–12545

Kerner M, Gebken T, Sundarrao I et al (2019) Development of a control system to cover the demand for heat in a building with algae production in a bioenergy facade. Energy and Buildings 184:65–71

Leyland B, Leu S, Boussiba S (2017) Are Traustochytrids algae? Fungal Biol. 121(10):835–840

Naumann T, Cebi S, Podola B et al (2013) Growing microalgae as aquaculture feeds on twin-layers: a novel solid-state photobioreactor. J Appl Phycol 25(5):1413–1420

Oeschger L, Posten C (2012) Construction and assessment parameters of photobioreactors. In: Posten C, Walter C (eds) Microalgal biotechnology: potential and production. De Gruyter, pp 225–236

Olaizola M (2000) Commercial production of astaxanthin from *Haematococcus pluvialis* using 25,000-liter outdoor photobioreactors. J Appl Phycol 12:499–506

Olaizola M (2003) Commercial development of microalgal biotechnology: from the test tube to the marketplace. Biomol Eng 20:459–466

Olaizola M, Brown RC, Orchard ED (2018) Present and future economic and environmental impacts of microalgal technology. In: Malcata FX, Sousa-Pinto I, Guedes AC (eds) Marine algae – features and applications. CRC, Boca Raton, pp 300–329

Olaizola M, Davenport J, Coard J et al (2019) Photobioreactor for contained microorganism cultivation. WO 2019/113116 A1

Olivieri G, Salatino P, Marzochella A (2014) Advances in photobioreactors for intensive microalgal production: configurations, operating strategies and applications. J Chem Technol Biotechnol 89(2):178–195

Perner-Nochta I, Posten C (2007) Simulations of light intensity variation in photobioreactors. J Biotechnol 131(3):276–285

Raso S, van Genugten B, Vermue MH et al (2012) Effect of oxygen concentration on the growth of *Nannochloropsis* sp. at low light intensity. J Appl Phycol 24:863–871

Rodolfi et al (2008) Microalgae for oil: strain selection, induction of lipid synthesis and outdoor mass cultivation in a low-cost photobioreactor. Biotechnol Bioeng 102(1):100–112

Schmidt T, Nguyen M-K, Lakatos M (2017) Fassadenintegrierte Bioreaktorsysteme. Fassade 2:24–26

Schreiber C, Behrendt D, Huber G et al (2017) Growth of algal biomass in laboratory and in large-scale algal photobioreactors in the temperate climate of western Germany. Bioresour Technol 234:140–149

Schultze LKP, Simon M-V, Li T et al (2015) High light and carbon dioxide optimize surface productivity in a twin-Layer biofilm photobioreactor. Algal Res 8:37–44

Singh RN, Sharma S (2012) Development of suitable photobioreactor for algae production - a review. Renew Sustain Energy Rev 16:2347–2353

Sivakaminathan S, Hankamer B, Wolf J et al (2018) High throughput optimisation of light-driven microalgal biotechnologies. Sci Rep 8:11687

Stephens E, Ross IL, Hankamer B (2013) Expanding the microalgal industry – continuing controversy or compelling case? Curr Opin Chem Biol 17(3):444–452

Torzillo G, Chini Zittelli G (2015) Tubular photobioreactors. In: Prokop A (ed) Algal biorefineries. Springer, Berlin, pp 187–212

Torzillo G et al (1993) A two-plane tubular photobioreactor for outdoor culture of Spirulina. Biotechnol Bioeng 42(7):891–898

Tredici M, Bassi N, Prussi M et al (2015) Energy balance of algal biomass production in a 1 ha "Green Wall Panel" plant: how to produce algal biomass in a closed reactor achieving a high net energy ratio. Appl Energy 154:1103–1111

Vonshak A, Torzillo G, Tomaseli L (1994) Use of chlorophyll fluorescence to estimate the effect of photoinhibition in outdoor cultures of *Spirulina platensis*. J Appl Phycol 6:31–34

Vonshak A, Torzillo G, Masojidek J et al (2001) Sub-optimal morning temperature induces photoinhibition in dense culture of the alga *Monodus subterraneus* (Eustigmatphytoa). Plant Cell Environ 24:1113–1118

Wilson MH, Groppo J, Grubbs T et al (2017) Cyclic photobioreactor and method for biofilm control. US2017/0355942 A1

Wintersteller F (2018) Think outside the pond. Lecture, Innovationsforum Algae Food, 6 Aug 2018, Magdeburg, Germany

Zeriouh O, Reinoso-Moreno JV, Lopez-Rosales L et al (2017) Biofouling in photobioreactors for marine microalgae. Crit Rev Biotechnol. https://doi.org/10.1080/07388551.2017.1299681

Chapter 2
Operational, Prophylactic, and Interdictive Technologies for Algal Crop Protection

Carolyn L. Fisher and Todd W. Lane

Abstract Both open and closed algal mass culture systems, for aquaculture, biofuel, and biotechnology applications, are subject to contamination with deleterious species such as grazers, parasites, and pathogens. In open cultivation systems, pond crashes due to infection with deleterious species can result in the loss of 30% of annualized production. More expensive closed photobioreactors are less prone to infection but nonetheless do fail and can be more difficult to decontaminate than open systems. The development of effective methodologies and technologies to interdict or prevent crashes of algal mass culture systems is necessary to increase biomass production, drive down costs, reduce the risk involved with algal cultivation, and ultimately make it more favorable for investment by entrepreneurs and biotechnology companies. A variety of chemical, biological, and technological controls have been proposed and developed for algae crop protection. We review these previously developed crop protection strategies, along with examining novel strategies and areas of ongoing research and development.

2.1 Introduction

The variety of deleterious species that can lead to biomass loss in algal production systems is extensive, diverse, and not completely understood. There are a few systematic studies of biological agents that infect production systems, and several important culprits have been identified (Fig. 2.1). Carney et al. (2016b) used second-generation sequencing to determine that rotifers (*Brachionus*), gastrotrichs (*Chaetonotus*), parasitic vampyrellids (*Vampyrella*), ciliates (*Acineta*), and free-living amoebae (*Nolandella*) were most commonly identified in crashed ponds, suggesting that these grazers played a critical role in the demise of the algal mass culture systems. Eukaryotes are not the only benefactors of algal cultures as a carbon and nutrient source. Bacteria have spent millennia co-evolving with microalgae,

C. L. Fisher · T. W. Lane (✉)
Systems Biology Department, Sandia National Laboratories, Livermore, CA, USA
e-mail: twlane@sandia.gov

A. Hallmann, P. H. Rampelotto (eds.), *Grand Challenges in Algae Biotechnology*, Grand Challenges in Biology and Biotechnology,
https://doi.org/10.1007/978-3-030-25233-5_2

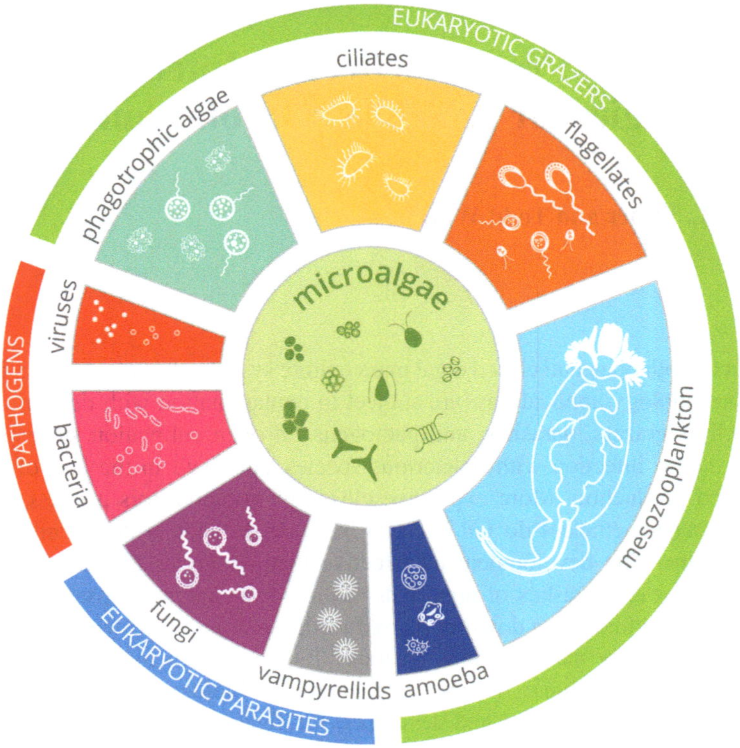

Fig. 2.1 Diverse array of microorganisms are known to feast upon algae

some in symbiotic mutualism and some as parasitic hosts of specific microalgal strains. Bacteria are, to a great degree, morphologically indistinguishable, and in some cases difficult to cultivate, and therefore can prove challenging to identify and characterize. Thus, demonstrating that a particular bacterial species is responsible for failure of algal mass culture (essentially satisfying Koch's postulates) is challenging. Although algicidal bacteria are also well known in natural systems, only two bacteria, *Vampirovibrio chlorellavorus* (Ganuza et al. 2016) and the pleomorphic bacterial strain FD111 (Lee et al. 2018), have been implicated in the demise of algal mass cultures. Thus, recent work has identified etiological agents of pond crashes that can be grouped into eukaryotic grazers, eukaryotic parasites, and bacterial pathogens.

Algae are plagued by diseases caused by eukaryotic parasites and grazers (for review, see Gachon et al. 2010), but information on how detrimental these organisms are to algal mass culture systems is limited. Major classes of eukaryotic parasites include vampyrellids and fungi (for review, see Carney and Lane 2014). Chytrid fungi are a well-established fungal parasite known to induce disease in host plants as well as freshwater (Atkinson 1909) and marine algae (Li et al. 2010). Calcofluor white was used by Rasconi et al. (2009) for the detection of chytrids in environmental samples and could easily be applied to mass culture systems. Within mass cultures of *Graesiella* sp., Ding et al. (2018) identified the epibiotic fungal parasite *Rhizophydium scenedesmi* as the biological agent of reduced growth and lipid accumulation. Karpov et al. (2018) identified novel strains of *Amoeboradix gromovi*, a chytrid-like parasite of the alga *Tribonema gayanum*. Additional pests include eukaryotic parasites, *Amoeboaphelidium protococcarum*, isolated from an outdoor algal culture of *Scenedesmus dimorphus* (Letcher et al. 2013), and *Paraphysoderma sedebokerense*, an amoeboid parasite also isolated from *S. dimorphus* in outdoor culture (Letcher et al. 2016). Microzooplanktonic grazers are also a devastating threat to algae. Rotifers, ciliates, flagellates, and amoeba are all known to cause detrimental effects to open and closed algal mass cultures (for review see Day et al. 2017). Additionally, amoeboid grazers have been identified for cultures of the cyanobacteria, *Leptolyngbya* and *Synechococcus* (Ma et al. 2016). In a closed photobioreactor system, Troschl et al. (2017) found that the ciliated protozoa *Colpoda steinii* was frequently found, despite bleaching, and would regularly clear dense cultures of the cyanobacterium *Synechocystis* sp. in 2–3 days.

In addition, algal viruses are ubiquitous in natural systems and have been implicated in the demise of algal blooms (Van Etten and Dunigan 2012). Viruses are well known for being highly abundant in marine environments with estimates reaching 10^{10} virions mL^{-1} of surface ocean water (Villain et al. 2016). Despite their staggering presence in aquatic communities, there are currently no published reports of algal viruses causing pond crashes. However, it is possible that this could be due to the difficulty of detecting or identifying viral agents. The impact of algal viruses on algae culture has been implicated in the report by Evans and Wilson (2008), showing that the dinoflagellate *Oxyrrhis marina* preferentially grazed virus-infected *Emiliania huxleyi* compared to uninfected controls. One possible explanation of this could be an evolutionarily driven preference for algal grazers to obtain higher quantities of carbon per meal, as algal viruses are some of the largest known in the world with genomes as high as 2.5 Mb and capsids as large as 1.5 μm (Villain et al. 2016). Or, perhaps virus-infected algal cells are manipulated to produce less chemical deterrents, thus promoting algal grazing and virus dispersal. Although not currently implicated in the demise of mass algal culture systems, in light of the important role that viruses play in natural systems, it does seem likely that viral agents would have a significant impact on algal production systems.

There is a long history of research and development for physical, chemical, and biological algal crop protection strategies (for review, see Wang et al. 2013; Montemezzani et al. 2015), and it is still an area of significant interest, innovation, and active research. Operational strategies, such as higher salinity and pH, have been investigated as methods for mitigating contamination by understanding the optimal environmental conditions of deleterious species and employing cultivation methods that antagonize grazer health, reproduction, and viability (Wang et al. 2016; von Alvensleben et al. 2013). Use of supportive or protective algal strains in polyculture has recently been investigated as biological methods for mitigating contamination. Several species of cyanobacteria have known biocidal properties and can be used to provide an algal defense in polycultures for production strains without an innate defensive system (see review in Renuka et al. 2018). Operational controls can sometimes overlap with prophylactic strategies meant to prevent contamination. Prophylactic strategies are applied to all production units in advance of infection, designed to prevent infection by deleterious species or reduce the impact of such infection. For example, higher salinity, pH, or CO_2 levels minimize contamination by generating an environment unsuitable for invasive species (see review Day et al. 2012a). Mechanical hydrodynamic stress and filtration have been shown to effectively kill or capture grazers and reduce pond contamination (Montemezzani et al. 2017b). In many cases, a compound or technique can be used in either a prophylactic or interdictive strategy, such as with sodium hypochlorite. It can be used in low doses to deter contamination from susceptible grazers or in higher doses to treat algal cultures with an alarming level of detrimental organisms present (Zmora et al. 2013). Although often costly, prophylactic and operational strategies can be effective measures to prevent devastating losses of algal crop and profits.

Ultimately, contamination is a serious problem in both open algal ponds and closed photobioreactor systems and thus the economic need for crop protection is substantial (Hannon et al. 2010). It has been estimated that crashes can result in the loss of up to 30% of annualized production in open systems (McBride et al. 2014). A primary key to successful crop protection in algal production systems is early detection. There are several strategies for the detection of deleterious species (see summary in Carney et al. 2016a). Early detection enables effective interdiction strategies. Interdictive strategies, targeted applications of a biocide in response to detection of a deleterious species, are common in the production of low-value products where economic demands are the highest (McBride et al. 2014). This limits the expense associated with the treatment of the production units that need it. Perhaps the most common interdictive strategy is the "rescue harvest," where the infected production unit is harvested immediately upon detection of a deleterious species to prevent total loss through a culture crash.

In this chapter, we will focus on providing an overview on the various operational, prophylactic, and interdictive methodologies that have been developed and tested for mass cultivation of algae. Additionally, we will highlight some advanced, novel methods for detection and algal crop protection. Our goal is to define the state of the art for algal crop protection in order to facilitate further innovation and development within the field to help eliminate a significant technical and economic barrier to algal biotechnology including production of biofuel, commodity chemicals, and high-value products.

2.2 Chemical and Biochemical Interventions

Some of the oldest and best studied methods for crop protection include the use of relatively simple chemical compounds to treat infected systems. The overall goal is to identify compounds that can be added to production systems at concentrations that kill, inhibit, or mitigate contamination by deleterious species. Ideally, chemical treatments will not be detrimental to algal production, adversely impact downstream use of the biomass, or add excessive additional cost. This balance can be difficult to achieve and limit the utility of these approaches. Although many chemical treatments can both be used in prophylaxis and interdiction, extensive and continued use of chemical agents can lead to desensitization and resistance in the targeted pest species. Here, we discuss the variety of chemical treatments and their effectiveness on prophylaxis and interdiction of algal cultures.

2.2.1 Copper

Heavy metals are often found in wastewater treatment facilities, for which bioremediation via algal cultivation is a tempting prospect. Pradeep et al. (2015) tested the effectiveness of copper treatment on the rotifer *Brachionus calyciflorus* in *Chlorella kessleri* cultures. The lethal concentration for 50% of the population (LC50) of Cu for isolated *B. calyciflorus* was <0.1 ppm and 1.5 ppm (about 0.2 and 24 μM, respectively) when 7 rotifers per mL were in cultivation with the algae at a starting algal biomass concentration of approximately 100 mg L^{-1} cell dry weight. This type of result is not atypical; often the effective dose is found to be higher in dense biomass cultures. Dissolved organic carbon and sunlight can also reduce the effectiveness of many agents. Thus, the effectiveness of any crop protection strategy needs to be tested under realistic cultivation conditions. The need to test under realistic conditions was also demonstrated by Wang et al. (2017b). In outdoor cultivation, 10.0 mg L^{-1} (about 157 μM) of copper in raceways eradicated all of the *Colpoda* sp. ciliates found in *Chlorella vulgaris* and the *Vorticella* sp. ciliates found in *Scenedesmus acutus* cultures but also inhibited algal growth which was not observed in 24-h flask/column tests with copper sulfate. The effects of heavy metal toxicity on algal culture were further investigated by Huang et al. (2016). In the presence of nanomolar levels of free copper II ions, the innate grazer-defense mechanism of *S. obliquus* of colony formation was inhibited along with photosynthetic efficiency and algal growth. However, this effect might only be limited to algal cultures that form colonies or *Scenedesmus* sp. cultures, and other algal species might respond significantly better or worse. These results do corroborate what was observed by Wang et al. (2017b) after treatments of micromolar to millimolar levels of copper were added to algal raceways. Lower concentrations of copper also limited ciliate populations and did not inhibit algal growth. However, these results need to be further tested on other algal cultures, raceway conditions, and grazer populations. Additionally, environmental caution and the potential for accumulation in algal biomass should be considered whenever adding heavy metals to a water system.

2.2.2 Bleach

Sodium hypochlorite (NaOCl) is a commonly utilized oxidant for the interdiction of pond crashes. Bleach, defined as 3–8% sodium hypochlorite, is readily available and commonly used in production facilities for the pretreatment of source water and decontamination or disinfection of equipment. Park et al. (2016) evaluated the effectiveness of sodium hypochlorite (bleach) against the rotifer *Brachionus calyciflorus* in co-culture with *Chlorella kessleri*. Although the 24-h LC50 for *B. calyciflorus* was determined to be 0.198 mg L^{-1} (5.59 μM) of chlorine, pond treatments of 0.45 mg L^{-1} (13 μM) of chlorine, the rotifer population was uninhibited for 96 h and ended up crashing the entire algal culture. Park et al. (2016) increased chlorine dosage to 0.45–0.6 mg L^{-1} (13–17 μM) every 2 h and were able to kill *B. calyciflorus* without harming *C. kessleri*, despite its 24-h LC50 of 0.321 mg L^{-1} (9.06 μM) to chlorine. Higher concentrations of chlorine were tested by Wang et al. (2017b) in controlling ciliate populations in algal cultures. While 10.0 mg L^{-1} (282 μM) of chlorine eliminated 95% of the *Colpoda* sp. ciliate in *Chlorella vulgaris* LRB-1201 ponds, this same chlorine dosage only eliminated 65% of *Vorticella* sp. ciliates in *Scenedesmus acutus* ponds and severely inhibited the growth of the alga. Although no dosage information is provided, Lee et al. (2018) similarly indicate that sodium hypochlorite is an effective treatment strategy for the abatement of the new pleomorphic bacterial strain, FD111, and prevents the culture crash of *Microchloropsis salina* (note: *Nannochloropsis salina*, CCMP 1776, has recently been reclassified and renamed *Microchloropsis salina*). However, timing of bleach treatment, in addition to concentration, might be key since some alga might not be able to recover after bleach treatment if their numbers are greatly reduced from grazing. When outdoor culture systems of *Microchloropsis salina* started to diminish due to grazing on day 9, Carney et al. (2016b) reported that sodium hypochlorite treatments were performed on day 10 at 2.5 ppm (0.034 mM), day 12 at 5 ppm (0.067 mM), and day 14 at 10 ppm (0.13 mM) with little success. Perhaps the addition of bleach starting at low concentrations on day 10 of the culture cycle was insufficient to rescue the pond. Weissman and Radelli (2010) took the opposite approach and found that highly concentrated "shock" treatments of sodium hypochlorite at 40–80 mg L^{-1} (0.5–1 mM) at the beginning of the culture run significantly helped control contamination. Zmora et al. (2013) recommend doses of 4–10 ppm (112–282 μM) of active chlorination in order to eliminate grazers from algal cultures. Although bleach treatment has previously succeeded as a selective biocide against some zooplankton cultures, further outdoor trials are required to determine effective dosages, biological sensitivity, and if prophylactic or interdictive measures are optimal.

2.2.3 *Other Oxidants*

Many other oxidants have been tested for their effectiveness as an algal crop protectant with limited success. Karuppasamy et al. (2018) tested a number of oxidants, along with other chemicals, for effectiveness against grazers in microalgal cultures and found that sodium perborate and sodium percarbonate, oxidizers typically found in laundry detergent, were effective against *Oxyrrhis* sp.

grazers with an LD50 of 50 ppm (0.5 and 0.3 mM, respectively) but were ineffective against *Euplotes* sp. and not tested for any negative effects on their choice of alga, *Chlorella vulgaris*. Potassium permanganate, a well-established oxidizer used in a variety of applications ranging from human medicine to biological control of aquaculture, was effective at inhibiting *Oxyrrhis*, *Euplotes*, as well as the alga *C. vulgaris* at 1 ppm (7 μM). The most promising oxidant tested was peracetic acid, which had an LD50 of 25 ppm (0.35 mM) against both grazers and 50 ppm (0.7 mM) against the algae. Peracetic acid has been previously analyzed and utilized for its antimicrobial effects. However, the US Environmental Protection Agency warns against handling concentrations of peracetic acid that exceed 1.3 ppm (0.017 mM) as it can be hazardous to human health. Additionally, the cost for routine utilization of peracetic acid was not deemed to be favorable per the technoeconomic model generated and referenced within Karuppasamy et al. (2018). These mixed results suggest that even low concentrations of oxidants could be lethal to all living organisms within pond systems. Further studies are required to determine what algal strains are tolerant of oxidants in order to optimize interdictive methods for pond contaminants.

Ozone, chlorine dioxide, and hydrogen peroxide are also effective disinfectants for aquaculture, and they have been explored as treatment options for algal mass culture systems as well. Kamaroddin et al. (2016) found that ten minute treatments with 8 mg L^{-1} (0.2 μM) of ozone reduced bacterial contamination by *Halomonas* sp. within *Dunaliella salina* cultures. Chlorine dioxide, in comparison to liquid chlorine treatment, has been found to be more effective against viruses and mesoplankton, but similarly effective toward algae (Junli et al. 1997). At 3.0 mg L^{-1} (44.5 μM) of chlorine dioxide, the algae tested were mostly unaffected, but the viruses and several animal planktons were impacted. These results suggest that additional tests could determine an appropriate interdictive dosage for chlorine dioxide for treatment of mass algal cultures.

Originally classified as an algicide (Kay et al. 1982), hydrogen peroxide treatment at 150–200 mg L^{-1} (4.4–6 mM) was lethal to ciliate populations present within *Dunaliella salina* mass culture; however, the algae was also susceptible (Moreno-Garrido and Canavate 2001). Drábková et al. (2007) found that the cyanobacterium *Microcystis aeruginosa*, the green alga *Pseudokirchneriella subcapitata*, and the marine diatom *Navicula seminulum* were all susceptible to pulsed hydrogen peroxide exposure, but at different levels. The rapid chemical decomposition and light sensitivity of hydrogen peroxide also were found to play a significant role in the growth of all the algae. For the most sensitive algae tested, cyanobacterium *M. aeruginosa*, high hydrogen peroxide concentrations inhibited photosynthetic oxygen evolution and may have inhibited growth. Although it is currently unclear why different algae are susceptible at different dosages of hydrogen peroxide, understanding the impacts of temperature, light, salinity, and pH on toxicants added to treat algal production systems is necessary. Future work to elucidate the impacts of abiotic factors on chemical efficacy will further advance available treatment options for algal ponds.

2.2.4 Natural Compounds

In addition to ozone, other naturally occurring chemicals have been investigated for their use in generating and maintaining axenic algal cultures. Xenic cultures of the freshwater *Ettlia* sp. YC001 were treated for 3 days with an antibiotic and antifungal cocktail of chloramphenicol, imipenem, carbendazim, rifampicin, and tetracycline by Lee et al. (2015). The antibiotic cocktail was initially toxic to the algae, but the cultures were recovered and confirmed to be axenic after serial dilutions, gel electrophoresis, and SEM imaging. Karuppasamy et al. (2018) tested the effect of several antimicrobial compounds on the grazers *Euplotes* and *Oxyrrhis*. Several of the quaternary amines were found to be extremely potent against grazers, but some also inhibited the *Chlorella vulgaris* algal growth. Benzalkonium chloride (BAC) was optimal in price and toxicity toward grazers and was further assessed in an outdoor culture trial where *C. vulgaris* was infected with *Euplotes*. Only the BAC-treated algal culture, and not the untreated control, successfully recovered. Although successful, the use of antibiotics and antimicrobials is broadly considered to be environmentally harmful and largely unscalable for mass algal culture systems. Antibiotics and antimicrobials remain an effective laboratory tool in order to specifically exclude bacteria and facilitate studies of algal physiology and genetic engineering.

Rotenone is a naturally occurring insecticide, piscicide, and pesticide that is found in the stems and roots of the *Fabaceae* family, such as the North American jicama plant, *Pachyrhizus erosus*. Van Ginkel et al. (2015) first tested the effectiveness of rotenone as a rotifer-killing agent and found that the freshwater *Brachionus calyciflorus* had a 24-h LC50 of 0.074 μM and marine rotifers, *Brachionus rotundiformis* and *Brachionus manjavacas*, had LC50s of 0.13 μM and 0.26 μM, respectively. The freshwater alga, *Chlorella kessleri*, and marine alga, *Tetraselmis suecica*, were uninhibited by all rotenone doses tested. Similarly, Van Ginkel et al. (2016) found that *Nannochloropsis oculata* was unaffected up to 7.6 μM of rotenone while *B. manjavacas* and *B. rotundiformis* were reproductively inhibited at 0.35 and 0.18 μM, respectively. However, treatment of *T. suecica* with rotenone had an effective concentration for 50% of the population (EC50) of 0.56 μM, indicating that algae were susceptible to rotenone toxicity. El-Sayed et al. (2018) further examined the absorptivity of rotenone by freshwater *Chlorella kessleri* in order to determine an effective dosage strategy for algal production ponds. Karuppasamy et al. (2018) found that *Euplotes* and *Oxyrrhis* sp. were also both susceptible to rotenone with LC50 values of 5 and 1.3 μM, respectively, while *Chlorella vulgaris* was uninhibited. Rotenone is biodegradable and will break down in 3–8 days depending on a variety of conditions, such as pH, temperature, and light intensity. Additionally, as a known piscicide and moderately hazardous chemical marked by the World Health Organization, concentrated stocks of rotenone can be harmful to humans and devastating to fish and insect populations if accidentally spilled into the water column and natural environment. Despite these risks, Van Ginkel et al. (2015, 2016) assert that rotenone could readily support biofuel development for $2000 per year for biweekly rotenone prophylaxis to support 1 million gallons of biodiesel production, assuming 50% lipid content of the algae. Interestingly, availability and cost were cited by Karuppasamy et al. (2018) as to reasons why rotenone is not ideal for pond treatment methods.

The antimalarial compound, quinine, is found naturally in cinchona and reijia tree bark. Xu et al. (2015) tested co-cultures of *Brachionus calyciflorus* and *Chlorella kessleri* with quinine sulfate to determine sensitivity and toxicity concentrations. Although the LC50 for *B. calyciflorus* was 17 μM, complete inhibition of the freshwater rotifers was observed at 7.7 μM when in co-culture with *C. kessleri*. In contrast, quinine sulfate concentrations up to 23 μM had no effect on the growth of *C. kessleri*. Similarly, 20.0 mg L^{-1} (26 μM) of quinine sulfate eliminated more than 85% of the contaminating *Colpoda* sp. from *Chlorella vulgaris* ponds and 90% of the *Vorticella* sp. ciliates from *Scenedesmus acutus* ponds (Wang et al. 2017b). In both cases, neither alga was inhibited after more than 24 h. Additionally, 10 mg L^{-1} (13 μM) of quinine sulfate rapidly extinguished ciliate populations contaminating outdoor *Dunaliella salina* cultures while posing no threat to algal growth (Moreno-Garrido and Canavate 2001). These data suggest that quinine sulfate is a broadly applicable interdictive treatment option that works rapidly to destroy grazers without causing harm to algal cultures.

Several naturally occurring botanical pesticides were tested for their toxicity against rotifers and algal strains. Huang et al. (2014a) tested four pesticides, celangulin, matrine, toosendanin, and azadirachtin, against the marine rotifer, *Brachionus plicatilis*. The nonalkaloidal insecticide celangulin, originally isolated from a bittersweet plant found in China, *Celastrus angulatus*, inhibited rotifers with a 24-h LC50 of 0.175 mg L^{-1} (0.259 μM). Matrine, an anticancer nematicidal alkaloid found in *Sophora* genus members, was similarly toxic to *B. plicatilis* with a 24-h LC50 of 0.061 mg L^{-1} (0.22 μM). Both celangulin and matrine were about 100 times more potent than azadirachtin, a carcinogenic, genotoxic insecticide isolated from the Indian lilac (*Meliaceae*). Although azadirachtin is active against over 200 species of insects, it was the least toxic botanical pesticide tested against *Brachionus plicatilis* with a 24-h LC50 of 18.386 mg L^{-1} (25.5 μM). Toosendanin is a triterpenoid extract from *Meliaceae* family members, such as the chinaberry. Historically used as a parasiticide and agricultural insecticide in ancient China, toosendanin has been found to induce apoptosis in many cell lines and was similarly toxic to *B. plicatilis* with a 24-h LC50 value of 0.002131 mg L^{-1} (3.7 nM). Fortunately, neither *Chlorella* nor *Nannochloropsis* sp. exhibited growth or photosynthetic inhibition after 0.001755–0.002132 mg L^{-1} (3.1–3.7 nM) toosendanin treatments. Huang et al. (2014b) further assessed the potential for botanical pesticide treatment of algal cultivation systems by testing celangulin and toosendanin together in a 1:9 ratio for toxicity on the freshwater rotifer *Brachionus calyciflorus* in co-culture with the cyanobacteria *Arthrospira platensis*. Two treatments of the 1:9 pesticide mixture over 3 days successfully inhibited the rotifer population without harming the *S. platensis* in outdoor mass cultivation systems. The physiological effects of toosendanin treatment on *Brachionus plicatilis* were further analyzed by Huang et al. (2017). Toosendanin reduced body size, population size, and the reproductive ability of the marine rotifer over the course of 48 h. Pepsase and tryptase were directly inhibited by toosendanin, suggesting a mechanism for the antifeedant effect of the pesticide on *B. plicatilis*. Yun et al. (2018) demonstrated that 6 mg L^{-1} of a biosurfactant produced by *Bacillus subtilis* C9 was effective at controlling the cladoceran grazers *Daphnia pulex* and *Moina macrocopa* within 24 h without deleterious impact on the growth or lipid content of the algal species tested.

2.2.5 Biocides

Synthetic biocides are commonly utilized for treating fungi, pests, insects, and parasites. Karuppasamy et al. (2018) screened a variety of biocides to determine which would be most effective and economical for the treatment of mass culture systems. Fungicides captan, pyraclostrobin, fluoxastrobin, and propiconazole were found to be potent inhibitors of the grazers *Euplotes* and *Oxyrrhis* and unharmful to *Chlorella vulgaris*. At dosage concentrations of 1.0–2.0 μg L^{-1} (3.4–6.8 nM), the antifungal Benomyl was effective at killing *Chytridium* sp. within cultures of *Scenedesmus obliquus* (Abeliovich and Dikbuck 1977). Woo and Kamei (2003) identified an anti-*Pythium* protein (referred to as "SAP" by the authors), isolated from a marine *Streptomyces* sp., that was an effective inhibitor of the infectious agent *Pythium porphyrae*, known for causing red rot disease in *Porphyra* sp. The mechanism of SAP was determined to be through alteration of the fungus membrane permeability. Vital dyes, such as Lugol's iodine, methylene blue, and toluidine blue, interact with the cellular membrane in order to penetrate living cells. Karuppasamy et al. (2018) found that low doses of these vital dyes were effective at inhibiting *Euplotes* and *Oxyrrhis* but were also toxic to *Chlorella vulgaris*. Despite these compelling results, the fungicides and vital dyes were determined to be too costly for the treatment of algal ponds (Karuppasamy et al. 2018).

"Pesticide" is a broad term that includes insecticides, herbicides, parasiticides, and any other chemical that inhibits pests associated with aqua- and agriculture. Karuppasamy et al. (2018) tested the effects of cypermethrin, methyl parathion, carbaryl, profenofos, and triazophos on *Euplotes*, *Oxyrrhis*, and *Chlorella vulgaris*. The authors determined that all the pesticides showed robust, selective inhibition of grazers and not the algae. Similarly, Huang et al. (2011) determined the 24-h LC50 of cypermethrin on *Brachionus calyciflorus* to be 3.38 mg L^{-1} (8.11 μM) and of deltamethrin to be 0.595 mg L^{-1} (1.18 μM). Herbicides pendimethalin, ethalfluralin, and sodium dimethyldithiocarbamate, as well as the flavonol quercetin, were also found to be quite effective against grazer populations (Karuppasamy et al. 2018). Antiparasitic compounds have been tested for use as anti-feeding agents to prevent algal culture loss. Ivermectin, a macrocyclic lactone naturally produced by *Streptomyces avermitilis*, has been identified by the World Health Organization as the most effective and safest human treatment of parasites. Garric et al. (2007) tested the toxicity of ivermectin on *Daphnia magna* and determined the 48-h LC50 to be

5.7 ng L^{-1} or 0.006 nM. This highly potent parasiticide was comparably ineffective at inhibiting cultures of *Pseudokirchneriella subcapitata*. Karuppasamy et al. (2018) observed similar potency of ivermectin toward *Euplotes* and *Oxyrrhis* although not against *Chlorella vulgaris*. Other antiparasitic compounds tested, benzyl isothiocyanate, abamectin, and niclosamide, exhibited a similar trend. The potency of the biocides often came with a high environmental, and oftentimes economical, cost due to the high toxicity, likelihood of off-target effects, and lasting ecological harm. While chemical treatment strategies are some of the oldest known contaminant mitigation techniques, many additional novel methods in development have demonstrated use in abatement procedures and are worthy of review (Table 2.1).

2.3 Physical Disruption or Removal

2.3.1 Hydrodynamic Shear

The differential velocities of different layers in a fluid impart a shear force that is referred to as hydrodynamic. Montemezzani et al. (2017a) evaluated the effectiveness of hydrodynamic shear to disrupt the rotifer *Brachionus calyciflorus* and the cladoceran *Moina tenuicornis*. The laboratory-scale system that was employed induced hydrodynamic shear by pumping cultures through perforated aluminum plates. The highest mortality rates (100% of *M. tenuicornis* and 80% of *B. calyciflorus*) were achieved by passage through a 4 mm orifice with a pressure differential across the plate of 2.5 bar. The authors described some disruption of multicellular algal aggregates and colonial forms but did not characterize the overall effect of hydrodynamic shear on algal viability.

2.3.2 Hydrodynamic Cavitation

Hydrodynamic cavitation is caused in flowing liquid that has undergone a rapid reduction of pressure causing the formation of vapor cavities (microbubbles). Upon increase in pressure, these vapor cavities implode causing an intense physical shockwave that can physically disrupt biological structures. Microalgae and grazer species are differentially sensitive to cavitation, a phenomenon that can be leveraged in crop protection applications. Kim et al. (2017) demonstrated that hydrodynamic cavitation physically disrupted the rotifer *Brachionus rotundiformis* in cultures of *Microchloropsis salina*. Four passes of the culture through the cavitation systems resulted in a 99.9% disruption of the rotifers. It was unclear if the rotifer eggs retain viability after treatment, thus periodic treatments were suggested to fully protect the cultivation system from rotifer infestation. Little apparent reduction was observed in the viability of *M. salina* by this treatment. The differential sensitivity of the algae and the rotifers to cavitation is due to the differences in cell sizes, with *M. salina* averaging 2–3 μm in length and *B. rotundiformis* averaging 90–150 μm in length.

Table 2.1 Summary of various chemical interdictive procedures for contamination abatement

Chemical additive	Tested against	Advantages	Disadvantages	References
Copper Copper sulfate	*Brachionus calyciflorus*, *Colpoda* sp., *Vorticella* sp.	High heavy metal concentrations found in many wastewater treatment facilities	Inhibits algal growth, limits photosynthetic efficiency, inhibits colony formation of *S. obliquus*, heavy metal pollutant	Pradeep et al. (2015) Wang et al. (2017a) Huang et al. (2016)
Sodium hypochlorite (active ingredient in bleach at 3–8% by weight)	*Brachionus calyciflorus*, *Colpoda* sp., *Vorticella* sp., bacterial strain FD111	No long-lasting residual, chlorine dissipates rapidly	Chlorine dissipates rapidly, requires multiple doses, selective activity against various zooplankton	Park et al. (2016) Lee et al. (2018) Carney et al. (2016b) Wang et al. (2017a)
Oxidants Sodium perborate Sodium percarbonate Peracetic acid Potassium permanganate Ozone Chlorine dioxide Hydrogen peroxide	Varying impacts on *Oxyrrhis* and *Euplotes* (see ref), ClO_2 effective against viruses and animal plankton, H_2O_2 effective against ciliates, ozone effective against bacteria and *Halomonas* sp.	Degradable, several oxidizers, known biocides, ozone microbubbles can spur algal CO_2 uptake and growth, ClO_2 more effective than liquid Cl_2	High cost for peracetic acid, EPA warns against use >1.3 ppm for peracetic acid, H_2O_2, ClO_2, and $KMnO_4$ are algicides, $NaBO_3$ is carcinogenic	Karuppasamy et al. (2018) Moreno-Garrido and Canavate (2001) Kamaroddin et al. (2016) Drábková et al. (2007) Junli et al. (1997)
Antibiotic/antifungal cocktail Carbendazim Chloramphenicol Imipenem Rifampicin Tetracycline	Bacteria and fungi associated with xenic algal culture	In 3 days can make xenic alga axenic	Environmentally harmful due to increasing antibiotic resistance, too expensive, not scalable for outdoor cultures	Lee et al. (2015)

Rotenone	*Brachionus calyciflorus*, *Brachionus rotundiformis*, *Brachionus manjavacas*, *Euplotes*, *Oxyrrhis* sp.	Rapidly degrades to nontoxic chemicals, algae are largely insensitive, effective against marine and freshwater rotifers	Rapidly degrades in sunlight, extremely toxic to insects and aquatic life, harmful to humans in concentrated doses	Van Ginkel et al. (2015) Van Ginkel et al. (2016) El-Sayed et al. (2018) Karuppasamy et al. (2018)
Quinine sulfate	*Brachionus calyciflorus*, *Colpoda* sp., *Vorticella* sp., unknown ciliate	Effective against variety of grazers without harming algae, long-lasting in water system	Expensive, toxic to human health, environmentally harmful do to developing resistance	Wang et al. (2017a) Xu et al. (2015) Moreno-Garrido and Canavate (2001)
Antimicrobial compounds Benzalkonium chloride (BAC) Benzethonium chloride (BEC) Cetyltrimethylammonium bromide Cetylpyridinium chloride Tetraethylammonium bromide Tetraethylammonium chloride Tetraethylammonium iodide Dichlorohexidine gluconate Imidazolidinyl urea Methylisothiazolinone	*Euplotes*, *Oxyrrhis* sp.	Quaternary amines potently inhibit grazers by damaging cell membrane	Environmentally harmful due to increasing antibiotic resistance, too expensive, not scalable for outdoor cultures, some antimicrobials were too potent for algae	Karuppasamy et al. (2018)

(continued)

Table 2.1 (continued)

Chemical additive	Tested against	Advantages	Disadvantages	References
Botanical pesticides Celangulin Matrine Toosendanin Azadirachtin	*Brachionus plicatilis*, *Brachionus calyciflorus*	Biodegradable, low toxicity to mammals, safe for microalgae	Broad spectrum pesticide in environment, untested in algae raceways, biodegradable and need repeat doses, toxic to humans in highly concentrated stocks	Huang et al. (2014a) Huang et al. (2014b) Huang et al. (2017)
Antifungals Captan Pyraclostrobin Fluoxastrobin Propiconazole Benomyl	*Euplotes*, *Oxyrrhis*, *Chytridium* sp.	Potent grazer inhibition	Environmentally harmful due to antifungal resistance, too expensive, not scalable for outdoor cultures	Karuppasamy et al. (2018) Abeliovich and Dikbuck (1977)
Pesticides Cypermethrin (Cyp) Deltamethrin (Del) Methyl parathion Carbaryl profenofos Triazophos	*Euplotes*, *Oxyrrhis*, *Brachionus calyciflorus*	Potent grazer inhibition	Cyp decomposition increases with exposure to sun, water, and oxygen, is highly toxic to fish, bees, and insects, Del highly toxic to aquatic life	Karuppasamy et al. (2018) Huang et al. (2011)
Vital dyes Lugol's iodine Methylene blue Toluidine blue	*Euplotes*, *Oxyrrhis*	Potent grazer inhibition	Too expensive, too toxic for algae	Karuppasamy et al. (2018)
Herbicides Pendimethalin Ethalfluralin Sodium dimethyldithiocarbamate	*Euplotes*, *Oxyrrhis*	Potent grazer inhibition	Toxic off-target environmental effects	Karuppasamy et al. (2018)
Anti-feeding agent, antiparasitic compounds Benzyl isothiocyanate Ivermectin Abamectin Niclosamide	*Euplotes*, *Oxyrrhis*, *Daphnia magna*	Potent grazer inhibition	Toxic off-target environmental effects	Karuppasamy et al. (2018) Garric et al. (2007)

This also suggests that grazer species larger than rotifers, such as *Daphnia*, will be even more susceptible to disruption by cavitation. Sawant et al. (2008) demonstrated that with a combination of shear forces and cavitation induced by their system, 80% of zooplankton (copepods, bivalve, gastropod, decapod and polychaete larvae, *Favella*, and nematodes) was destroyed in a single pass. Thus, hydrodynamic cavitation is a compelling prophylactic and interdictive method for contamination control in algal mass culture.

2.3.3 *Ultrasonication*

In ultrasonication, sound energy at frequencies in excess of 20 kHz is introduced into an aqueous system resulting in both physical and chemical effects within the sample. Physical effects include hydro-mechanical shear forces. Ultrasonication induces chemical effects, referred to as sonochemistry, through the formation and implosion of microbubbles, resulting in highly localized high-temperature regimes (Suslick 1990). The combination of these effects is known to result in the disruption of cellular membranes (sonoporation). Holm et al. (2008) characterized the effectiveness of ultrasonication on the disruption of bacteria, phytoplankton, and zooplankton in ballast water. In general, the effectiveness of the treatment in terms of energy and time required varied directly with the size of the organism, with bacteria being the least susceptible and the zooplankton being the most. The differences in susceptibility that were observed indicated that it could be possible to disrupt grazer species without causing undue harm to algal species. Wang et al. (2018) tested the effectiveness of ultrasonication on the disruption of a range of deleterious species, including the flagellate *Poterioochromonas*, in outdoor raceway cultures of *Chlorella* sp. Despite its relatively small size, *Poterioochromonas* was found to be sensitive to sonication, likely due to its lack of a cell wall. Additional contaminating species present in the pond, including an unknown fungus, *Acanthamoeba*, and ciliates, were disrupted by the treatment, but a number of species, including the flagellate *Bodomorpha* sp., the ciliate *Colpoda* sp., and the amoeba *Paradermamoeba* sp., were unaffected by ultrasonication, as was *Chlorella* sp. Ultrasonication would provide an effective interdictive method, but only toward those species that are susceptible.

2.3.4 *Pulsed Electric Fields*

Pulsed electric fields (PEF) are known to cause disruption in biological membranes (electroporation). This method works by inducing an electrical potential across the cellular membrane, resulting in a rearrangement in the structure of the lipid bilayer and resulting in the formation of a pore. It is well known that electroporation is limited by the conductivity of the medium and may not be effective for marine systems. Rego et al. (2015) demonstrated the effectiveness of PEF in the control of rotifers found in outdoor, closed, tubular photobioreactor (PBR) cultures of *Chlorella* sp. without impacting algal growth rate. The PEF system was integrated directly into the PBR so that the culture was treated as it circulated through the system. Over a 6-h test, it was calculated that each cell in the system received 36 electrical pulses (900 V/cm, 65 μs pulses of 50 Hz) and 6 h of treatment was sufficient to disrupt 85% of the rotifer population. Encouragingly, continuous treatment over a period of days resulted in 100% disruption without a deleterious effect on the algal growth rate.

2.3.5 Foam Flotation

Umar et al. (2018) tested the efficacy of foam flotation for the removal of the ciliate *Tetrahymena pyriformis* from laboratory-scale PBR cultures of *Chlorella vulgaris*. In this process, algae were cultivated in a columnar PBR sparged with air from the base (effectively a bubble column). The air was passed through a gas permeable membrane to create bubbles at the millimeter scale. The addition of sodium dodecyl sulfate (SDS) in combination with the sparging created a fine foam that differentially lifted the ciliates (*Tetrahymena pyriformis* CCAP 1630/1W) to the top of the column for removal. Efficiency of ciliate removal was directly dependent on SDS concentration, air flow rate, and number of treatment cycles. At 10 mg L^{-1} SDS and an air flow rate of 1 L min^{-1}, approximately 20% of ciliates were removed in a single treatment cycle. At 40 mg L^{-1} SDS and an air flow rate of 2 L min^{-1}, 96% of ciliates were removed after three treatment cycles. This treatment did not affect algal growth but did provide an effective way of removing the deleterious species from the cultivation system.

2.3.6 Hydrocyclone

Hydrocyclone separation technology is based on the flowing of particle-containing liquid through a vortex thus inducing a centripetal force that results in the removal of suspended particles. This technology has been applied for the removal of exotic species from ballast water (Abu-Khader et al. 2011). Although hydrocyclone technology has not been reported for the differential removal of grazers or other deleterious species from algal cultures, it may be useful to evaluate this technology for this application.

2.3.7 Filtration

Many grazer species are large enough that they can be removed by passage of the culture through a mesh or net (Borowitzka 2005); however, the author does not give any specific examples. Montemezzani et al. (2017b) reported 100% removal of the cladoceran *Moina tenuicornis* after a single passage through a 300 μm filter. Effectiveness of this technique is, of course, dependent on the relative sizes of the algae and grazers and the volume of culture to be treated.

2.4 Nutritional Control

2.4.1 Phosphate

Phosphate limitation has been shown to reduce the rate of infection of the diatom *Asterionella formosa* by the parasitic fungus *Rhizophydium planktonicum* in turbidostat cultures (Bruning 1991). However, once an infection had been established, the growth rate of the fungus was only slightly impaired, whereas the host algae displayed a reduced growth rate due to the phosphate limitation. This reduction on algal growth rate allowed the fungal infection to eventually outpace algal growth. Phosphate limitation did, however, increase the algal density required to sustain the fungal infection. A similar decrease in susceptibility to chytrid infection was observed, at high C/P ratios, in the diatom *Synedra* in pond simulating mesocosms (Flynn et al. 2017). The authors modelled the effect of phosphate limitation on the production of algae with a high C/N ratio, resulting in biomass that was an inferior food source for zooplankton. The key was to devise phosphate limitation conditions that changed the biochemical makeup of the algae without reducing their growth rate. Such a balance may be achievable in closed systems with rigorous control of the physiochemical parameters but seems unlikely to be achieved in open and/or outdoor cultivation systems.

2.4.2 Nitrogen Source: Urea and Ammonia Versus Nitrate

Addition of reduced nitrogen compounds can be an effective way to control contamination of algal cultures with deleterious species. Mendez and Uribe (2012) grew both indoor and outdoor cultures of *Arthrospira* sp. in the presence of urea and ammonium bicarbonate and at a pH of 9.5–10.5. At concentrations of 60 mg L^{-1} (1 mM) of urea or 100 mg L^{-1} (1.3 mM) of ammonium bicarbonate, *Arthrospira* growth was not affected, but both *Brachionus* sp. and *Amoeba* sp. were inhibited. Reduced nitrogen compound concentrations above this were detrimental to *Arthrospira* growth. Free ammonia in the form of ammonium hydroxide has also been used to control grazers. Lincoln et al. (1983) used ammonium hydroxide to control the zooplanktons, *Brachionus rubens* and *Diaphanosoma brachurum*, in cultures being grown on facultative lagoon effluent from swine wastewater (pH 7.8–8.0). Laboratory and field testing showed that the addition of 16–18 mg L^{-1} (0.52–0.58 mM) free NH_3-N was lethal to *Brachionus* and

20 mg L^{-1} (0.64 mM) killed *Diaphanosoma*, but algal growth was unaffected in both cases. Schluter and Groeneweg (1985) showed that free NH_3-N at concentrations below 5 mg L^{-1} (0.16 mM) controlled but did not kill *Brachionus rubens* and that the effect was reversible. At free NH_3-N concentrations above 5 mg L^{-1} (0.16 mM), the rotifers were dead after 48 h. In laboratory-scale experiments of *N. oculata* infected with *Brachionus plicatilis*, Thomas et al. (2017) demonstrated that exposure to NH_4Cl at 42–53 mg L^{-1} (1.4–1.7 mM) free NH_3-N for at least 6 hours was effective at killing the rotifers while not affecting *N. oculata* growth. Tests were done in artificial sea water medium and ASW supplemented with dairy anaerobic digester effluent. Free ammonia concentrations were manipulated by altering pH over a range from 7 to 9. Some of these studies show that algal growth was not affected by addition of ammonia compounds, but caution should be taken. Many algal species are sensitive to reduced nitrogen compounds and free NH_3-N concentrations, and growth may be reduced or inhibited (see reference Gutierrez et al. 2016).

2.4.3 Other Inorganic Salts

Interestingly, Abeliovich and Dikbuck (1977) found that *Chytridium* sp. were inhibited from infecting *Scenedesmus obliquus* cultures when Na^+ and Ca^{2+} cations were replaced with K^+ and Mg^{2+} cations at concentrations of 0.01–0.1 M within the growth medium. The authors also show that while this cation replacement inhibits chytrid infection, algal growth of *S. obliquus* is unaffected. This promising cation replacement strategy requires further trials in mass culture systems to determine the efficacy of cation replacement for large-scale algal production.

2.5 Cultivation System Operation Strategies

2.5.1 Cultivation Under Alkaline pH

Cultivation of algae at alkaline pH has been shown to be an effective method of reducing contamination by deleterious species. Moheimani and Borowitzka (2006) reported the semi-continuous outdoor cultivation of *Pleurochrysis carterae* over a period of 13 months. The ponds ranged from a pH of 8 at the end of dark phase to a maximum of pH 11 during daylight. This extreme range in pH is credited with reducing contamination by deleterious species. Wang et al. (2017b) reported that cultivation of *Scenedesmus acutus* LRB-1201 under CO_2-limiting conditions resulted in an increase of culture pH from 7.63 to 10.88. These high pH conditions resulted in an 80% decrease in *Vorticella*. Peng et al. (2015) reported that growth of *Neochloris oleoabundans* at high-sodium-bicarbonate (160 mM)-containing medium at pH 9.5 was effective in limiting the growth of unspecified protozoa.

Vadlamani et al. (2017) reported the cultivation of a high-pH-adapted strain of *Chlorella sorokiniana* SLA-04 at a pH in excess of 10 in both laboratory and outdoor open ponds. Outdoor ponds were challenged with an experimental infection with the grazer *Daphnia magna*. Under high pH conditions *D. magna* were rapidly killed, further demonstrating the effectiveness of utilizing high pH-tolerant systems and algal strains.

2.5.2 Transient pH Shifts

Transient low pH treatments either through acid addition or CO_2 treatment have been shown to control deleterious species in production systems. Ganuza et al. (2016) demonstrated that 15-min shift of *Chlorella* mass cultures to pH 3.5, through HCl addition, in the presence of 0.5 g L^{-1} acetate killed 90% of the bacteria *Vampirovibrio chlorellavorus* and was well tolerated by the algae. Similarly, Becker (1994) determined that shifting culture pH to 3 for a period of 1–2 h effectively eliminates rotifers.

2.5.3 CO_2 Asphyxiation and Anoxia

Long-term growth of pilot-scale cultures of *Chlorella sorokiniana* cultures maintained at pH 6 or pH 6.5 through CO_2 sparging (but not through acid addition) resulted in the death of *Poterioochromonas malhamensis*, and death or inhibition of the ciliates *Tetrahymena thermophila*, *Colpoda* sp., and *Vennalla* sp., and the amoeba *Sterkiella* sp. (Ma et al. 2017). Montemezzani et al. (2017a) demonstrated the effectiveness of long-term CO_2 sparging in controlling grazers in 20 L outdoor mesocosm cultures derived from wastewater treatment high-rate algal ponds. Continuous sparging with 5% and 10% CO_2 was determined to be effective in controlling multiple species of rotifers. Limiting CO_2 asphyxiation to overnight was also demonstrated to be successful in eradicating grazers in outdoor ponds (Montemezzani et al. (2017c). In an 8 m^3 high-rate algal pond, nightly 100% CO_2 injections of 1–6 L/min was successful in reducing the populations of some, but not all, zooplankton species present. Affected species included *Moina tenuicornis*, *Paracyclops fimbriatus*, and *Filinia longiseta*. Recalcitrant species included *Heterocypris incongruens*, *Asplanchna sieboldi*, *Cephalodella catellina*, and *Brachionus calyciflorus*. Despite the reported success of CO_2 asphyxiation, the protozoa *Colpoda steinii* is unaffected by this treatment as well as high ammonia (Troschl et al. 2017). However, partially anoxic conditions did inhibit growth of this protozoa in *Synechocystis* sp. cultures, suggesting that this could be a promising avenue for further investigation.

2.6 Biological Control

2.6.1 Polyculture, Natural Assemblages, and Crop Rotation

Most algal cultivation systems are predicated on the production of a single algal species with optimal characteristics (e.g., biochemical makeup or the production of a specific product). A few species have been identified with innate resistance to specific grazers which may make them better choices for mass culture. These include strains of *Chlorella* sp. that have pronounced resistance against protozoa (Zhang and Kong 2011) and *Tetreselmis* sp. (Erkelens et al. 2014). In cases where monocultures can be avoided, the cultivation of deliberate polycultures or natural assemblages can have significant advantages in crop protection. A number of research groups (Hillebrand and Cardinale 2004; Corcoran and Boeing 2012; Roelke 2017; Smith and Crews 2014; Cho et al. 2017; Godwin et al. 2018) have determined that polycultures can increase the resilience and production rate of a cultivation system in the presence of grazers. This increase in resistance to grazing may be due to an "interference effect" where prey diversity decreases effective grazing. For a given set of growth conditions, polycultures may not outperform a monoculture of the optimal species, but under grazing pressure they will have higher rates of survival and production. Under appropriate circumstances, a polyculture can be a highly effective crop protection strategy. Furthermore, crop rotation strategies have also been proposed to limit grazer or pathogen carry-over in batch strategy production systems (Kagami et al. 2007; Smith et al. 2015).

2.6.2 Trophic Control

Trophic cascades have been proposed as an alternative approach to controlling grazer populations. Such mechanisms work by altering the concentrations of organisms that prey upon or infect grazer species, thus reducing their number and impact on algal populations. Montemezzani et al. (2017b) demonstrated the trophic control by the addition of rotifer-specific predators that displayed limited grazing on microalgae. In these experiments, inoculation with 2500 individuals per liter of the cladoceran *Moina tenuicornis* or with 1000 individuals per liter of the ostracod *Heterocypris incongruens* decreased rotifer populations by 70–80% compared to the untreated controls. Thom et al. (2018) reported the trophic control of the cladoceran *Daphnia* by predators such as dragonfly larvae [Odonata: Libellulidae] and backswimmers [Hemiptera: Notonectidae] in outdoor tanks containing *Chlorella*. Predation of *Daphnia* reduced grazing losses of the algae.

2.6.3 Allelopathy and Natural Defenses of Microalgae

In nature, competition for resources and survival has evolutionarily driven the natural selection of organisms with the ability to generate secondary metabolites as "defense" molecules that chemically inhibit predators and competitors. These chemicals, often referred to as "allelochemicals" to reference their allelopathic purpose, have been considered and tested as a method of algal protection in mass culture systems (for review, see Mendes and Vermelho 2013). Relative to frequent, toxic, and costly chemical or physical treatments for pond maintenance, utilizing the natural defense systems of microalgae is a novel, innate, and natural method for bolstering algal culture protection. Polycultures including allelochemical-producing algae strains and/or genetic modification to engineer defenseless algae to generate their own chemical defense system are two applications for current work in this field. The low cost of growing algae that continually generate their own chemical defense system is worth further investigation.

Dimethylsulfoniopropionate (DMSP) is a well-known chemical signal produced by marine algae, such as *Phaeocystis* and *Emiliania huxleyi*, when they are under attack. Glycine betaine is a similar molecule to DMSP and has been shown to be negatively allelopathic. During algal grazing, oxidation of cellular membrane fatty acids generates several aldehydes, including trans,trans-2,4-decadienal. Hue et al. (2018) tested these established allelochemicals, along with the amino acid proline and the chemical analog of DMSP known as methyl 3-(methylthio)propionate (MMP), on several ciliate populations in the presence of the microalga *Chlamydomonas* sp. All chemicals inhibited ciliate populations of *Sterkiella*, *Stylonychia notophora*, *Oxytricha* sp. and two different *Paramecium* species, but this effect required very high concentrations of proline and glycine betaine (250–300 mM) compared to <10 mM concentrations needed for the other chemical treatments. Grazer populations were most sensitive to 0.13 mM of decadienal, but the *Chlamydomonas* culture was also inhibited at this concentration. DMSP and MMP were preferred for their potency (4.75 mM and 10 mM, respectively), but the authors suspected that their addition to mass culture systems might be problematic in that both are volatile compounds that could dissipate quickly in large, open, outdoor cultures. Introduction of the necessary gene cluster for innate algal synthesis of DMSP would mitigate this concern.

Differences in species and strain responses to grazers suggest the presence of allelochemicals and require further study. Yuan et al. (2017) found that both the culture suspension and 0.45-μm filtrates of *Didymogenes* sp. HN-4 were able to suppress *Rataria* sp. rotifer grazing rates. The authors postulate that the alga must be generating a chemical defense to deter rotifer grazing since control rotifers feasted upon *Chlorella* sp. normally. *Rataria* sp. rotifers were found to swim more slowly and exhibited morphological differences in the presence of *Didymogenes* sp. HN-4. Although a definitive chemical or mechanism has yet to be determined, the cultivation of microalgae capable of producing defensive allelochemicals is a novel avenue for algal crop protection strategies. Allewaert et al. (2018) found that of the 44 strains of *Haematococcus* analyzed, 10 strains were determined to have lower susceptibility to the blastocladialean fungal parasite *Paraphysoderma sedebokerense* PS1. These differences in susceptibility were not correlated with phylogeny or sampling location, but several did have flagellated morphologies that seemed to be dominant. Possibly, these select strains have inherited morphological and chemical defenses that mitigate fungal infection. Better understanding the mechanism of defense can be used for genetic modification of *Haematococcus* sp. to resist the detrimental fungal pathogen.

Plants, including microalgae, are known to trigger immunological responses as a result of activation of microbe-associated molecular patterns (MAMPs). Oligosaccharides, polysaccharides, chitin, and other microbially derived molecules are recognized by algae, and innate defense systems are activated to mitigate or inhibit the nearby pathogen (for review, see Burketova et al. 2015). In order to better understand the attraction of chytrid zoospores to their algal hosts, Scholz et al. (2017) assessed the attraction of eight carbohydrates, five fatty acids, six amino acids, and three compatible solutes, both individually and combined, in chemotaxis experiments of four chytrid species. Of these, the combined carbohydrate solution was the best attractant for chytrid pathogens. Interestingly, carbohydrates, polyunsaturated fatty acids, and aldehydes were the chemical classes present more often in the resistant diatom strains tested. In addition to chemical defenses, morphological changes can also aid in algal survival. Verschoor et al. (2009) found that *Scenedesmus obliquus* form colonies when in co-culture with the freshwater rotifer *Brachionus calyciflorus*. *S. obliquus* is known to rapidly form algal colonies in the presence of rotifers, but their disintegration was slowed under continuous light conditions and due to "lingering infochemicals of herbivory past," even after all rotifers were removed from the culture. The lengthy disintegration of *S. obliquus* colonies could be explained by the high cost of such a defensive morphological change or could point to highly sensitive putative chemical sensors of *S. obliquus* that keep the alga fully "armored" until the threat has passed. Interestingly, if a competitive algal strain is present in culture with *S. obliquus*, colony formation is inhibited. Such is the case Zhu et al. (2015) observed when *S. obliquus* cultures were grown with the grazer *Daphnia* (or in the presence of *Daphnia* filtrate) along with the competitive alga, *Microcystis aeruginosa*. The authors observed that when given the choice between colony formation to reduce grazing rates and maintaining a high competitive ability, the *S. obliquus* cultures do not form colonies. This could be because *S. obliquus* colonies do not grow and compete as well, or it is possible that chemicals from *M. aeruginosa* are inhibiting the infochemicals from grazing and *S. obliquus* does not respond normally. Regardless of mechanism, this shows that defense mechanisms are inducible and different environmental stresses result in different responses by the algae.

Various chemical treatments, nutrient limitations, and environmental changes can influence the use of defense molecules used by an alga. Abscisic acid is a plant hormone known for its role in development, growth, and stress tolerance. Pouneva (2006) discovered that both higher endogenous abscisic acid levels and ~100 μM treatments of abscisic acid prevented infection by the fungal parasite *Phlyctidium scenedesmi* on cultures of *Scenedesmus acutus*. Absence of vital nutrients can be a signal that competition is high and the algae metabolically respond by generating chemicals that might help capture the necessary nutrients to allow for survival. Bagwell et al. (2016) found that algal cultures, *Scenedesmus* sp. strain 18B and *Chlorella* sp. strain 15, in an iron-limited environment were more cytotoxic. The authors hypothesized that the secondary metabolites produced by the algae were meant to scavenge iron. What was found, however, was that the *Chlorella* sp. grown in an iron-limited environment was incredibly resistant to devastation by the *Vampirovibrio chlorellavorus* bacteria; only a slight loss (0–9%) of the iron-limited *Chlorella* sp. culture was observed compared to nutrient-replete control cultures that had 72% loss of algal biomass. In this case, nutrient depletion led to an increase in cytotoxicity and defense molecules, but environmental changes can also lead to loss of algal defense systems. Demott and McKinney (2015) observed that after 3 years of semi-continuous culture of *Oocystis* in light (i.e., 24 h of fluorescent light, exposed daily for three consecutive years) in the absence of grazers, the *Oocystis* culture lost innate defense molecules. Control *Oocystis* cultures grew more slowly but were more resistant to *Daphnia* grazing. The authors postulate that there might be a "defense versus growth" trade-off, similar to the results of colony formation inhibition by *S. obliquus* (Zhu et al. 2015). These studies indicate that algae likely have a suite of unknown chemical defenses that need to be better understood for use in algal production systems.

2.7 Advanced Methods

2.7.1 Genetic Engineering Strategies

The genetic engineering of algal production strains to resist deleterious species is an approach that, although confronted by both ecological and regulatory hurdles, is likely to be an area of active research and innovation. Larkum et al. (2012) and Qin et al. (2012) review the status of genetic engineering technology in a variety of microalgal species. A significant barrier to the deployment of genetically modified (GM) algae is the potential environmental risk if such strains were to be cultivated in open systems. In one such outdoor cultivation study by Szyjka et al. (2017), a GM strain of *Acutodesmus dimorphus* was developed to contain exogenous genes for enhanced fatty acid biosynthesis and green fluorescence protein (GFP). It was found that the strains retained the exogenous genes and their associated phenotypes for the length of a 50-day cultivation trial. When the cultivation systems were inoculated in water containing natural assemblages from the local aquatic environment, the GM strain proved incapable of outcompeting native strains.

In order to control for any potentially deleterious effects of GM algae on the natural environment, a number of different strategies have been proposed for their biocontainment. A common strategy for the biocontainment of GM microalgae is to cripple the production strain so that it is dependent on the pond environment and unable to compete in the natural environment. An example of this is to genetically knockout the gene for nitrate reductase. Most production-scale ponds utilize ammonium as a nitrogen source as it is much cheaper than nitrate salts. Nitrate reductase deficient mutants are able to grow in ammonium-fed production ponds but suffer a severe fitness deficit in the natural environment (Navarro et al. 2005). Another example is the genetic knockout of high-affinity trace metal transporters, as trace metals are usually replete in the pond environment but present at low concentrations or even limiting in the natural environment (Blaby-Haas and Merchant 2012).

GM algae strains have been developed to outcompete weed species and to create algal strains that are resistant to chemical agents that may be used for the control of grazer populations. Loera-Quezada et al. (2016) developed a GM strain of *Chlamydomonas reinhardtii* capable of using phosphite as a sole phosphorus source and thus outcompeting the growth of exogenous weed species. Bruggeman et al. (2014) developed a *C. reinhardtii* strain engineered to contain genes conferring resistance to the herbicides glyphosate (Vick 2010), oxyfluorfen, and norflurazon. Along with serving as selectable markers for genetic engineering, these genes confer resistance and thus facilitate the use of broad-spectrum herbicides to control contamination by weed species. Corcoran et al. (2018) took an alternative UV mutagenesis and selection approach to the development of strains of the green algae *Desmodesmus armatus* able to tolerate a broad spectrum fungicide.

2.7.2 Industrial Microbial Ecology

With the advent of modern methods for microbiome and metagenome analysis, there is now a focus on manipulating the microbial ecology of the production system for purposes of crop protection, growth enhancement, and control of biomass composition (e.g., lipid content) (Beyter et al. 2016; Mooij et al. 2015; Smith and McBride 2015). Algal polyculture is an example of a preliminary form of industrial ecology where the deliberate co-culture of multiple algae species can yield improved grazer and disease resistance (Shurin et al. 2014). More challenging is the control of the species composition and function of the underlying microbiome. Identification of and co-cultivation of algae with probiotic bacterial species has been shown to increase the growth and productivity of algal culture systems. Fisher et al. (2019) have shown that algae associated microbial consortia can be selected that protect algae from rotifer grazing. It is postulated that by maintaining a healthy normal microbial flora, niches that could be exploited by deleterious species would be otherwise occupied. The effective leveraging and utilization of industrial ecology to enhance stability of production systems is an area of active research where considerable work is needed and advances are possible.

2.8 Detection Methodologies

Although not a direct means of crop protection, identification and detection of deleterious species is an important aspect of any crop protection strategy. Early and accurate identification of deleterious species is important for any interdictions strategy: in many cases treatment is too expensive to apply generally as a prophylaxis and therefore is only utilized when there is evidence of infection. Light microscopy is a standard and essential method for monitoring ponds (Borowitzka 2005); however, the method has its limitations. It takes some level of training to identify grazer species, and the method is unable to discriminate between morphologically identical bacteria and viruses. Furthermore, microscopic evaluation can be labor-intensive, especially as production operations increase in scale. A number of alternative means have been developed to solve this problem, including semi-automated imaging flow cytometry and a variety of molecular techniques.

Day et al. (2012b) and Wang et al. (2017a) demonstrated the use of a FlowCAM imaging flow-cytometer to detect a wide variety of grazer species, such as *Euplotes vannus*, *Oxyrrhis marina*, *Uronema marinum*, *Brachionus calyciflorus*, *Cohnilembus reniformis*, *Paramoeba eihardii*, *Hartmannella* sp., and *Poterioocrhromonas*, in indoor and outdoor cultivation of *Chlorella* and *Nannochloropsis oculata* at sensitivities of less than 10 cells per mL. Once the image recognition system is trained, the FlowCAM is capable of operation in a semi-autonomous fashion. A number of nucleic acid-based methods, including polymerase chain reaction (PCR), and nucleic acid hybridization-based assays have been developed for the detection of deleterious species (for review, see Carney et al. 2016a). Such assays have the advantage of being potentially less time-consuming than microscopy and having the ability to detect and identify bio-contaminants, such as bacteria and viruses, that are not amenable to microscopy analysis.

2.9 Conclusion

Decades of research and innovation have generated a litany of operational, prophylactic, and interdictive procedures for mitigating and treating biological contaminants in algal mass cultivation systems. Currently proven technologies are summarized in Table 2.2. Here, we have provided an unbiased summary of both established and novel techniques (Fig. 2.2) in hopes of providing the reader with sufficient background knowledge of the field and opportunities for further investigation. Cultivation of algal cultures in open versus closed systems is one of the first decisions that algal production facilities need to answer. Both options are further implicated in downstream operational strategies, economical parameters, and possible contamination issues. While open outdoor cultures seem to be the most susceptible to contamination and culture crashes, they do remain the most inexpensive method for algal culture production and financial considerations. Closed

Table 2.2 Summary of current treatment options for biological contaminants of algal cultures

Algae	Treatment	Grazer or pathogen	References
Arthrospira sp.	Urea, ammonium bicarbonate	*Brachionus* sp. *and Amoeba* sp.	Mendez and Uribe (2012)
Chlorella sp.	pH shock	*Vampirovibrio chlorellavorus*	Ganuza et al. (2016)
Chlorella sp., *Nannochloropsis* sp.	Four botanical pesticides	*Brachionus plicatilis*	Huang et al. (2014a)
Chlorella sp., *Scenedesmus* sp.	Fe	*Vampirovibrio chlorellavorus*	Bagwell et al. (2016)
Chlorella kessleri	Chlorine	*Brachionus calyciflorus*	Park et al. (2016)
Chlorella kessleri	Copper	*Brachionus calyciflorus*	Pradeep et al. (2015)
Chlorella kessleri	Quinine sulfate	*Rotifer Brachionus calyciflorus*	Xu et al. (2015)
Chlorella kessleri, Tetraselmis suecica	Rotenone	*Brachionus calyciflorus, Brachionus rotundiformis, Brachionus manjavacas*	Van Ginkel et al. (2015)
Chlorella sorokiniana	CO_2-mediated low culture pH	*Poterioochromonas malhamensis*	Ma et al. (2017)
Chlorella sp.	Ultrasonication	*Poterioochromonas* sp.	Wang et al. (2018)
Chlorella vulgaris	Benzalkonium chloride	*Oxyrrhis, Euplotes*	Karuppasamy et al. (2018)
Chlorella vulgaris, Scenedesmus acutus	Copper, chlorine, quinine sulfate	*Vorticella* sp.	Wang et al. (2017a)
Chlorella vulgaris	Foam flotation	*Tetrahymena pyriformis*	Umar et al. (2018)
Dunaliella salina	Ozone	*Halomonas* sp.	Kamaroddin et al. (2016)
Dunaliella salina	Hydrogen peroxide	Ciliates	Moreno-Garrido and Canavate (2001)
Nannochloropsis oculata	Rotenone	*B. manjavacas, B. rotundiformis*	Van Ginkel et al. (2016)
Nannochloropsis oculata	Pesticides	*Brachionus plicatilis*	Huang et al. (2017)
Nannochloropsis oculata	Ammonia	*Brachionus plicatilis*	Thomas et al. (2017)
Microchloropsis salina	Hydrodynamic cavitation	*Brachionus rotundiformis*	Kim et al. (2017)
Natural assemblage	Ammonia hydroxide	*Brachionus rubens, Diaphanosoma brachyurum*	Lincoln et al. (1983)

(continued)

Table 2.2 (continued)

Algae	Treatment	Grazer or pathogen	References
Natural assemblage	CO_2 asphyxiation, hydrodynamic shear	*Moina tenuicornis, Brachionus calyciflorus*	Montemezzani et al. (2017a)
Natural assemblage	CO_2 asphyxiation, biological control using competitor species, filtration, and mechanical disruption using hydrodynamic shear stress	*Brachionus* spp.	Montemezzani et al. (2017b)
Natural assemblage	CO_2 asphyxiation	Microcrustaceans, rotifers, and copepods	Montemezzani et al. (2017c)
Natural assemblage	Sonication	Cladocerans and rotifers	Holm et al. (2008)
Pseudokirchneriella subcapitata	Parasiticide ivermectin	*Daphnia magna*	Garric et al. (2007)
Scenedesmus costato-granulatus	Ammonia	*Brachionus rubens*	Schluter and Groeneweg (1985)
Scenedesmus obliquus	Nutrient substitution, Benomyl	*Chytridium* sp.	Abeliovich and Dikbuck (1977)
Scenedesmus obliquus	Pyrethroid insecticides	*Brachionus calyciflorus*	Huang et al. (2011)
Spirulina platensis	Pesticides	*Brachionus calyciflorus*	Huang et al. (2014b)
Synechocystis sp.	Anoxia	*Colpoda steinii*	Troschl et al. (2017)

photobioreactors are considerably more expensive to set up, maintain, and operate but are less susceptible to infestations of detrimental species. However, if a closed system does become contaminated, it can be extremely difficult to disinfect completely, leading to additional expensive decisions to either scourge the system with highly toxic, expensive chemicals, or purchase a new one.

After the choice of an open or closed culture system is made, many operational considerations can prevent infection by deleterious species. Some algal strains with high salinity, pH, and CO_2 tolerance are less likely to be infected by microorganisms, zooplankton, parasites, and other harmful contaminants that are unable to thrive in such extreme environments. Wang et al. (2016) determined than an integrated pest management strategy in tuning the pH and salinity parameters away from the ecologically optimal values for rotifers, ciliates, and amoebae was an effective strategy in mitigating contamination in large-scale raceway ponds. Nightly CO_2 asphyxiation is another promising strategy that purges sensitive organisms intolerant of temporarily acidic conditions. While this could impact algal growth and cellular respiration, the benefits might outweigh this risk. Similarly, routine pH "shock" treatments have been shown to mitigate grazers while causing no lasting harm to algal cultures. Polycultures, crop rotation, and natural assemblages

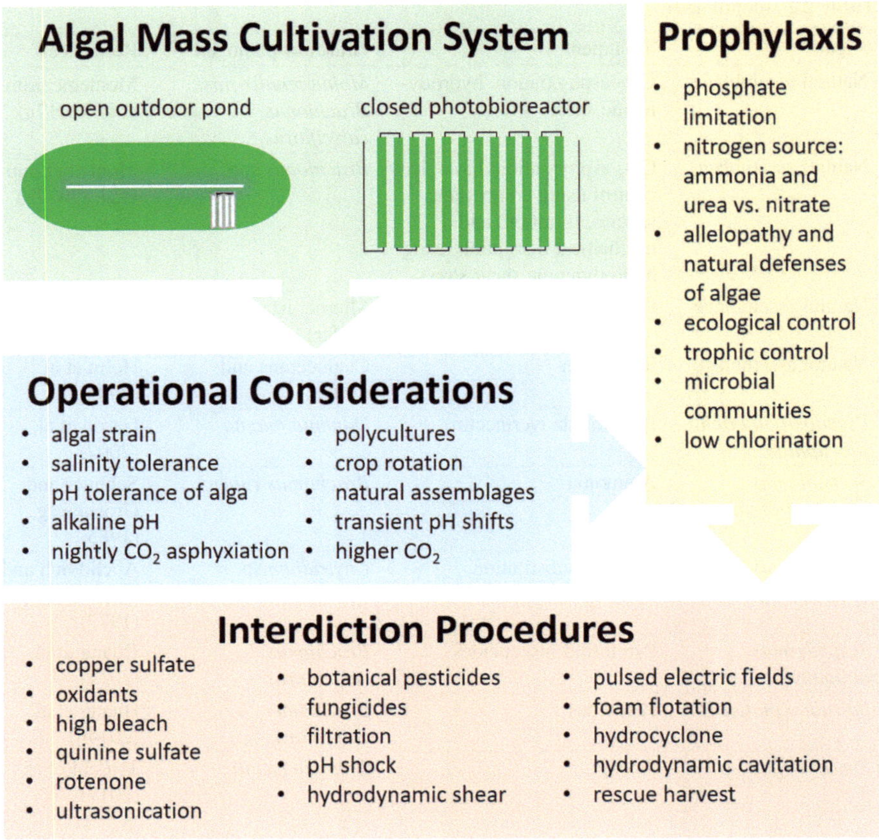

Fig. 2.2 Summary of the various operational strategies, prophylactic methods, and interdiction procedures that can aid in preventing, mitigating, and rescuing algal mass culture systems from devastating contamination and culture crashes

are biological control methods that bolster the aquaculture community and defend against invasive species, in some cases. Some algal cultures are known to generate defense molecules that could aid in their own or other community members' protection against grazing. Thus, choice of algal strain(s) and the operational setup of the mass cultures could tremendously impact the productivity and stability of algal cultivation.

Prophylactic measures can overlap with operational strategies but can also be a separate, additional layer of protection for protecting mass culture systems. Nutrient limitation or alteration is one example of how an operational strategy for culture setup can have prophylactic effects on the culture system. Many biological organisms can easily uptake phosphate and nitrate as critical P and N nutrients. Some algae, however, can also metabolize urea and ammonia as nitrogen sources and can survive on lower phosphate. However, lowering phosphate will impact algal growth and productivity, so these choices are not made without important downstream repercussions. Ammonia and urea are toxic to some algal grazers, thus high levels of these chemicals can be readily used by some algae as a nitrogen source and simultaneously deter algal contaminants. Allelopathy and natural chemical defenses of some algal strains alone or in polycultures can also act as prophylaxis for contamination. Ecological and trophic level control are biological control mechanisms that are currently breaking ground in their novelty and whole-system approach to prevent culture crashes. Monocultures of algae are more susceptible to infection and harsh changes in abiotic factors (temperature, rain, pH, light), while systems with diverse taxonomic makeups, including microbes, other algal strains, and predators of known algal grazers, are generally more stabilized and resistant to devastating culture crashes. In general, advanced planning to incorporate prophylactic measures into algal mass culture operational strategies can go a long way in mitigating grazer-induced culture crashes.

Prophylactic procedures can also overlap with interdictive strategies, to some extent. Chlorination, via sodium hypochlorite or chlorine dioxide application, can be used in low doses as prophylaxis or high doses as a treatment option for algicidal species. Similarly, pH, copper sulfate, some oxidants, quinine sulfate, and rotenone have been used as effective chemical treatment options shown to selectively impact only the algal grazers and not the algae. Several biocides have also been assessed for their impact on algae cultures, including fungicides, botanical pesticides, and parasiticides. Although most of the tested biocides were highly effective at eliminating contaminants, they were also often too expensive, environmentally harmful, or detrimental to algal strain productivity and thus are currently of limited utility. Further research is required to ascertain tolerant chemical concentrations, generate resistant genetically engineered algal strains, and conduct additional broad chemical screens against both grazer and algal species to identify new chemical treatments worth further study. Physical interdictive strategies are another category of algal pest remediation. Hydrodynamic shear and cavitation, filtration, pulsed electric fields, foam flotation, hydrocyclone, and ultrasonication are all demonstrated techniques for mitigating and eliminating grazer species without changing the chemical constituents or operational parameters of the aquaculture. Oftentimes, these physical abatement procedures come with a high energy, equipment, and economical cost, thus are often neglected as a rational prophylactic and treatment regime. Finally, the "rescue harvest" is an interdiction of last resort—should operational, prophylactic, and treatment options all fail, the culture can be immediately harvested to avoid a complete culture crash and negative profit margins. Hence, advanced detection methodologies for determining the species and concentration of contaminants present within mass cultures is crucial to culture maintenance and generating a profit. Furthermore, early detection procedures can facilitate targeted, effective treatment procedures for specific deleterious species. With adequate operational, prophylactic, and interdictive methods in place, algal pond culture systems can be productive, profitable, and effective industrial-scale operations.

Acknowledgments The authors thank Amanda Thompson, Pamela D. Lane for editing, and Sandia Creative Services for assistance in preparing this chapter. Sandia National Laboratories is a multi-mission laboratory managed and operated by National Technology and Engineering Solutions of Sandia, LLC., a wholly owned subsidiary of Honeywell International, Inc., for the US Department of Energy's National Nuclear Security Administration under contract DE-NA0003525. Research and writing of this paper was supported by the BioEnergy Technology Office, US Department of Energy under Award # NL0032356. This paper describes objective technical results and analysis. Any subjective views or opinions that might be expressed in the paper do not necessarily represent the views of the US Department of Energy or the US Government.

References

Abeliovich A, Dikbuck S (1977) Factors affecting infection of *Scenedesmus obliquus* by a *Chytridium* sp. in sewage oxidation ponds. Appl Environ Microbiol 34(6):832–836

Abu-Khader MM, Badran O, Attarakih M (2011) Ballast water treatment technologies: hydrocyclonic a viable option. Clean Technol Environ Policy 13(2):403–413. https://doi.org/10.1007/s10098-010-0325-1

Allewaert CC, Hiegle N, Strittmatter M, de Blok R, Guerra T, Gachon CMM, Vyverman W (2018) Life history determinants of the susceptibility of the blood alga Haematococcus to infection by *Paraphysoderma sedebokerense* (Blastocladiomycota). Algal Res 31:282–290. https://doi.org/10.1016/j.algal.2018.02.015

Atkinson GF (1909) Some fungus parasites of algae. Bot Gaz XLVIII (5)

Bagwell CE, Abernathy A, Barnwell R, Milliken CE, Noble PA, Dale T, Beauchesne KR, Moeller PD (2016) Discovery of bioactive metabolites in biofuel microalgae that offer protection against predatory bacteria. Front Microbiol 7:516. https://doi.org/10.3389/fmicb.2016.00516

Becker EW (1994) Microalgae: biotechnology and microbiology. Cambridge University Press, Cambridge

Beyter D, Tang PZ, Becker S, Hoang T, Bilgin D, Lim YW, Peterson TC, Mayfield S, Haerizadeh F, Shurin JB, Bafna V, McBride R (2016) Diversity, productivity, and stability of an industrial microbial ecosystem. Appl Environ Microbiol 82(8):2494–2505. https://doi.org/10.1128/AEM.03965-15

Blaby-Haas CE, Merchant SS (2012) The ins and outs of algal metal transport. Biochim Biophys Acta 1823(9):1531–1552. https://doi.org/10.1016/j.bbamcr.2012.04.010

Borowitzka MA (2005) Culturing microalgae in outdoor ponds. In: Algal culturing techniques, pp 205–218

Bruggeman AJ, Kuehler D, Weeks DP (2014) Evaluation of three herbicide resistance genes for use in genetic transformations and for potential crop protection in algae production. Plant Biotechnol J 12(7):894–902. https://doi.org/10.1111/pbi.12192

Bruning K (1991) Effects of phosphorus limitation on the epidemiology of a chytrid phytoplankton parasite. Freshwater Biol 25:409–417

Burketova L, Trda L, Ott PG, Valentova O (2015) Bio-based resistance inducers for sustainable plant protection against pathogens. Biotechnol Adv 33(6 Pt2):994–1004. https://doi.org/10.1016/j.biotechadv.2015.01.004

Carney LT, Lane TW (2014) Parasites in algae mass culture. Front Microbiol. https://doi.org/10.3389/fmicb.2014.00278

Carney LT, McBride RC, Smith VH, Lane TW (2016a) Molecular diagnostic solutions in algal cultivation systems. Micro-algal production for biomass and high-value products. CRC, Boca Raton

Carney LT, Wilkenfeld JS, Lane PD, Solberg OD, Fuqua ZB, Cornelius NG, Gillespie S, Williams KP, Samocha TM, Lane TW (2016b) Pond crash forensics: presumptive identification of pond crash agents by next generation sequencing in replicate raceway mass cultures of *Nannochloropsis salina*. Algal Res 17:341–347. https://doi.org/10.1016/j.algal.2016.05.011

Cho DH, Choi JW, Kang Z, Kim BH, Oh HM, Kim HS, Ramanan R (2017) Microalgal diversity fosters stable biomass productivity in open ponds treating wastewater. Sci Rep 7(1):1979. https://doi.org/10.1038/s41598-017-02139-8

Corcoran AA, Boeing WJ (2012) Biodiversity increases the productivity and stability of phytoplankton communities. PLoS One 7(11):e49397. https://doi.org/10.1371/journal.pone.0049397

Corcoran AA, Saunders MA, Hanley AP, Lee PA, Lopez S, Ryan R, Yohn CB (2018) Iterative screening of an evolutionary engineered Desmodesmus generates robust field strains with pesticide tolerance. Algal Res 31:443–453. https://doi.org/10.1016/j.algal.2018.02.026

Day JG, Slocombe SP, Stanley MS (2012a) Overcoming biological constraints to enable the exploitation of microalgae for biofuels. Bioresour Technol 109:245–251. https://doi.org/10.1016/j.biortech.2011.05.033

Day JG, Thomas NJ, Achilles-Day UEM, Leakey RJG (2012b) Early detection of protozoan grazers in algal biofuel cultures. Bioresour Technol 114:715–719

Day JG, Gong Y, Hu Q (2017) Microzooplanktonic grazers – A potentially devastating threat to the commercial success of microalgal mass culture. Algal Res 27:356–365. https://doi.org/10.1016/j.algal.2017.08.024

Demott WR, McKinney EN (2015) Use it or lose it? Loss of grazing defenses during laboratory culture of the digestion-resistant green alga Oocystis. J Plankton Res 37(2):399–408. https://doi.org/10.1093/plankt/fbv013

Ding Y, Peng X, Wang Z, Wen X, Geng Y, Zhang D, Li Y (2018) Occurrence and characterization of an epibiotic parasite in cultures of oleaginous microalga *Graesiella* sp. WBG-1. J Appl Phycol 30(2):819–830. https://doi.org/10.1007/s10811-017-1302-4

Drábková M, Admiraal W, Maršálek B (2007) Combined exposure to hydrogen peroxide and light selective effects on cyanobacteria, green algae, and diatoms. Environ Sci Technol 41 (1):309–314. https://doi.org/10.1021/es060746i

El-Sayed WMM, Van Ginkel SW, Igou T, Ibrahim HA, Abdul-Raouf UM, Chen Y (2018) Environmental influence on rotenone performance as an algal crop protective agent to prevent pond crashes for biofuel production. Algal Res 33:277–283. https://doi.org/10.1016/j.algal.2018.05.015

Erkelens M, Ball AS, Lewis DM (2014) The influence of protozoa with a filtered and non-filtered seawater culture of *Tetraselmis* sp., and effects to the bacterial and algal communities over 10 days. Bioresour Technol 173:361–366. https://doi.org/10.1016/j.biortech.2014.09.115

Evans C, Wilson WH (2008) Preferential grazing of *Oxyrrhis marina* on virus-infected *Emiliania huxleyi*. Limnol Oceanog 53(5):2035–2040

Fisher CL, Ward CW, Lane PD, Kimbrel JA, Sale KL, Stuart RK, Mayali X, Lane TW (2019) Bacterial communities protect the alga *Microchloropsis salina* from grazing by the rotifer *Brachionus plicatilis*. Algal Res (in press)

Flynn KJ, Kenny P, Mitra A (2017) Minimising losses to predation during microalgae cultivation. J Appl Phycol 29(4):1829–1840. https://doi.org/10.1007/s10811-017-1112-8

Gachon CM, Sime-Ngando T, Strittmatter M, Chambouvet A, Kim GH (2010) Algal diseases: spotlight on a black box. Trends Plant Sci 15(11):633–640. https://doi.org/10.1016/j.tplants.2010.08.005

Ganuza E, Sellers CE, Bennett BW, Lyons EM, Carney LT (2016) A novel treatment protects chlorella at commercial scale from the predatory bacterium *Vampirovibrio chlorellavorus*. Front Microbiol 7:848. https://doi.org/10.3389/fmicb.2016.00848

Garric J, Vollat B, Duis K, Pery A, Junker T, Ramil M, Fink G, Ternes T (2007) Effects of the parasiticide ivermectin on the cladoceran *Daphnia magna* and the green alga *Pseudokirchneriella subcapitata*. Chemosphere 96(2007):903–910. https://doi.org/10.1016/j.chemosphere.2007.05.070

Godwin CM, Lashaway AR, Hietala DC, Savage PE, Cardinale BJ (2018) Biodiversity improves the ecological design of sustainable biofuel systems. GCB Bioenergy 10(10):752–765. https://doi.org/10.1111/gcbb.12524

Gutierrez J, Kwan TA, Zimmerman JB, Peccia J (2016) Ammonia inhibition in oleaginous microalgae. Algal Res 19:123–127. https://doi.org/10.1016/j.algal.2016.07.016

Hannon M, Gimpel J, Tran M, Rasla B, Mayfield S (2010) Biofuels from algae: challenges and potential. Biofuels 1(5):763–784

Hao JC, Lian B, Zhou RB, Liu HF, Shi LJ, Liu CQ (2016) Microbial flocculant combined ferric trichloride facilitates floating aggregation of *Microcystis aeruginosa* for efficient removal. Desalin Water Treat 57(43):20483–20493. https://doi.org/10.1080/19443994.2015.1108236

Hillebrand H, Cardinale BJ (2004) Consumer effects decline with prey diversity. Ecol Lett 7:192–201

Holm ER, Stamper DM, Brizzolara RA, Barnes L, Deamer N, Burkholder JM (2008) Sonication of bacteria, phytoplankton and zooplankton: application to treatment of ballast water. Mar Pollut Bull 56(6):1201–1208. https://doi.org/10.1016/j.marpolbul.2008.02.007

Huang L, Liu C, Wei C, Li J (2011) Acute toxicity of two pyrethroids insecticides on *Brachionus calyciflorus*. Plant Dis Pests 2(3):62–64, 68. https://doi.org/10.19579/j.cnki.plant-d.p.2011.03.018

Huang Y, Li L, Liu J, Lin W (2014a) Botanical pesticides as potential rotifer-control agents in microalgal mass culture. Algal Res 4:62–69. https://doi.org/10.1016/j.algal.2013.08.001

Huang Y, Liu J, Wang H, Gao Z (2014b) Treatment potential of a synergistic botanical pesticide combination for rotifer extermination during outdoor mass cultivation of *Spirulina platensis*. Algal Res 6:139–144. https://doi.org/10.1016/j.algal.2014.11.003

Huang Y, Nan H, Zhu X, Li B, Zhang Z, Yang Z (2016) Waterborne copper impairs grazer-induced colony formation and photosynthetic efficiency in *Scenedesmus obliquus*. Limnol Oceanogr 61 (2):625–634. https://doi.org/10.1002/lno.10236

Huang Y, Liu J, Pang T, Li L (2017) Growth inhibitory and antifeedant effects of sublethal concentrations of toosendanin on the rotifer *Brachionus plicatilis*. Biomass Bioenergy 99:31–37. https://doi.org/10.1016/j.biombioe.2017.02.013

Hue NT, Deruyck B, Decaestecker E, Vandamme D, Muylaert K (2018) Natural chemicals produced by marine microalgae as predator deterrents can be used to control ciliates contamination in microalgal cultures. Algal Res 29:297–303. https://doi.org/10.1016/j.algal.2017.11.036

Junli H, Li W, Nenqi R, Liu XL, Sun RF, Guanle Y (1997) Disinfection effect of chlorine dioxide on viruses, algae and animal planktons in water. Water Res 31(3):455–460

Kagami M, von Elert E, Ibelings BW, de Bruin A, Van Donk E (2007) The parasitic chytrid, Zygorhizidium, facilitates the growth of the cladoceran zooplankter, Daphnia, in cultures of the inedible alga, Asterionella. Proc R Soc Lond B Biol Sci 274(1617):1561–1566. https://doi.org/10.1098/rspb.2007.0425

Kamaroddin MF, Hanotu J, Gilmour DJ, Zimmerman WB (2016) In-situ disinfection and a new downstream processing scheme from algal harvesting to lipid extraction using ozone-rich microbubbles for biofuel production. Algal Res 17:217–226. https://doi.org/10.1016/j.algal.2016.05.006

Karpov SA, Lopez-Garcia P, Mamkaeva MA, Klimov VI, Vishnyakov AE, Tcvetkova VS, Moreira D (2018) The chytrid-like parasites of algae *Amoeboradix gromovi* gen. et sp. nov. and *Sanchytrium tribonematis* belong to a new fungal lineage. Protist 169(1):122–140. https://doi.org/10.1016/j.protis.2017.11.002

Karuppasamy S, Musale AS, Soni B, Bhadra B, Gujarathi N, Sundaram M, Sapre A, Dasgupta S, Kumar C (2018) Integrated grazer management mediated by chemicals for sustainable cultivation of algae in open ponds. Algal Res 35:439–448. https://doi.org/10.1016/j.algal.2018.09.017

Kay SH, Quimby PC, Ouzts JD (1982) H_2O_2: a potential algicide for aquaculture. In: Proceedings of the 35th annual meeting of the Southern Weed Science Society. Southern Weed Science Society, Atlanta, GA, pp 275–289

Kim D, Kim EK, Koh HG, Kim K, Han J-I, Chang YK (2017) Selective removal of rotifers in microalgae cultivation using hydrodynamic cavitation. Algal Res 28:24–29. https://doi.org/10.1016/j.algal.2017.09.026

Larkum AW, Ross IL, Kruse O, Hankamer B (2012) Selection, breeding and engineering of microalgae for bioenergy and biofuel production. Trends Biotechnol 30(4):198–205. https://doi.org/10.1016/j.tibtech.2011.11.003

Lee PA (2014) Crop protection in algal cultures grown for oil production. Phytopathology 104 (5):7–7

Lee H-G, Shin S-Y, Jin L, Yoo C, Srivastava A, La H-J, Ahn C-Y, Kim H-S, Oh H-M (2015) Establishment and maintenance of an axenic culture of *Ettlia* sp. using a species-specific approach. Biotechnol Bioprocess Eng 20(6):1056–1063. https://doi.org/10.1007/s12257-015-0289-4

Lee PA, Martinez KJL, Letcher PM, Corcoran AA, Ryan RA (2018) A novel predatory bacterium infecting the eukaryotic alga Nannochloropsis. Algal Res 32:314–320. https://doi.org/10.1016/j.algal.2018.04.003

Letcher PM, Lopez S, Schmieder R, Lee PA, Behnke C, Powell MJ, McBride RC (2013) Characterization of *Amoeboaphelidium protococcarum*, an algal parasite new to the cryptomycota isolated from an outdoor algal pond used for the production of biofuel. PLoS One 8(2):e56232. https://doi.org/10.1371/journal.pone.0056232

Letcher PM, Lee PA, Lopez S, Burnett M, McBride RC, Powell MJ (2016) An ultrastructural study of *Paraphysoderma sedebokerense* (Blastocladiomycota), an epibiotic parasite of microalgae. Fungal Biol 120(3):324–337. https://doi.org/10.1016/j.funbio.2015.11.003

Lincoln EP, Hall TW, Koopman B (1983) Zooplankton control in mass algal cultures. Aquaculture 32:331–337

Li W, Zhang T, Tang X, Wang B (2010) Oomycetes and fungi: important parasites on marine algae. Acta Oceanol Sin 29(5):74–81

Loera-Quezada MM, Leyva-Gonzalez MA, Velazquez-Juarez G, Sanchez-Calderon L, Do Nascimento M, Lopez-Arredondo D, Herrera-Estrella L (2016) A novel genetic engineering platform for the effective management of biological contaminants for the production of microalgae. Plant Biotechnol J 14(10):2066–2076. https://doi.org/10.1111/pbi.12564

Ma AT, Daniels EF, Gulizia N, Brahamsha B (2016) Isolation of diverse amoebal grazers of freshwater cyanobacteria for the development of model systems to study predator-prey interactions. Algal Res Biomass Biofuels Bioproducts 13:85–93. https://doi.org/10.1016/j.algal.2015.11.010

Ma M, Yuan D, He Y, Park M, Gong Y, Hu Q (2017) Effective control of *Poterioochromonas malhamensis* in pilot-scale culture of *Chlorella sorokiniana* GT-1 by maintaining CO_2-mediated low culture pH. Algal Res 26:436–444. https://doi.org/10.1016/j.algal.2017.06.023

McBride RC, Lopez S, Meenach C, Burnett M, Lee PA, Nohilly F, Behnke C (2014) Contamination management in low cost open algae ponds for biofuels production. Ind Biotechnol 10 (3):221–227. https://doi.org/10.1089/ind.2013.0036

Mendes LBB, Vermelho AB (2013) Allelopathy as a potential strategy to improve microalgae cultivation. Biotechnol Biofuels 6:152

Mendez C, Uribe E (2012) Control of *Branchionus* sp. and *Amoeba* sp. in cultures of *Arthrospira* sp. Lat Am M Aquat Res 40(3):553–561

Moheimani NR, Borowitzka MA (2006) The long-term culture of the coccolithophore *Pleurochrysis carterae* (Haptophyta) in outdoor raceway ponds. J Appl Phycol 18:703–712. https://doi.org/10.1007/s10811-006-9075-1

Montemezzani V, Duggan IC, Hogg ID, Craggs RJ (2015) A review of potential methods for zooplankton control in wastewater treatment high rate algal ponds and algal production raceways. Algal Res 11:211–226. https://doi.org/10.1016/j.algal.2015.06.024

Montemezzani V, Duggan IC, Hogg ID, Craggs RJ (2017a) Screening of potential zooplankton control technologies for wastewater treatment high rate algal ponds. Algal Res 22:1–13. https://doi.org/10.1016/j.algal.2016.11.022

Montemezzani V, Duggan IC, Hogg ID, Craggs RJ (2017b) Assessment of potential zooplankton control treatments for wastewater treatment high rate algal ponds. Algal Res 24:40–63. https://doi.org/10.1016/j.algal.2017.03.013

Montemezzani V, Duggan IC, Hogg ID, Craggs RJ (2017c) Control of zooplankton populations in a wastewater treatment high rate algal pond using overnight CO_2 asphyxiation. Algal Res 26:250–264. https://doi.org/10.1016/j.algal.2017.08.004

Mooij PR, Stouten GR, van Loosdrecht MC, Kleerebezem R (2015) Ecology-based selective environments as solution to contamination in microalgal cultivation. Curr Opin Biotechnol 33:46–51. https://doi.org/10.1016/j.copbio.2014.11.001

Moreno-Garrido I, Canavate JP (2001) Assessing chemical compounds for controlling predator ciliates in outdoor mass cultures of the green algae *Dunaliella salina*. Aquac Eng 24:107–114

Navarro MT, Mariscal V, Macías MI, Fernández E, Galván A (2005) *Chlamydomonas reinhardtii* strains expressing nitrate reductase under control of the cabII-1 promoter: isolation of chlorate resistant mutants and identification of new loci for nitrate assimilation. Photosynth Res 83 (2):151–161. https://doi.org/10.1007/s11120-004-9297-5

Park S, Van Ginkel SW, Pradeep P, Igou T, Yi C, Snell T, Chen YS (2016) The selective use of hypochlorite to prevent pond crashes for algae-biofuel production. Water Environ Res 88 (1):70–78. https://doi.org/10.2175/106143015x14362865227670

Peng L, Lan CQ, Zhang Z, Sarch C, Laporte M (2015) Control of protozoa contamination and lipid accumulation in *Neochloris oleoabundans* culture: effects of pH and dissolved inorganic carbon. Bioresour Technol 197:143–151. https://doi.org/10.1016/j.biortech.2015.07.101

Pouneva ID (2006) Effect of abscisic acid and ontogenic phases of the host alga on the infection process in the pathosystem *Scenedesmus acutus* - *Phlyctidium scenenedesmi*. Acta Physiol Plant 28(5):395–400

Pradeep V, Van Ginkel SW, Park S, Igou T, Yi C, Fu H, Johnston R, Snell T, Chen Y (2015) Use of copper to selectively inhibit *Brachionus calyciflorus* (predator) growth in *Chlorella kessleri* (prey) mass cultures for algae biodiesel production. Int J Mol Sci 16(9):20674–20684. https://doi.org/10.3390/ijms160920674

Qin S, Lin H, Jiang P (2012) Advances in genetic engineering of marine algae. Biotechnol Adv 30 (6):1602–1613. https://doi.org/10.1016/j.biotechadv.2012.05.004

Rasconi S, Jobard M, Jouve L, Sime-Ngando T (2009) Use of calcofluor white for detection, identification, and quantification of phytoplanktonic fungal parasites. Appl Environ Microbiol 75(8):2545–2553. https://doi.org/10.1128/AEM.02211-08

Rego D, Redondo LM, Geraldes V, Costa L, Navalho J, Pereira MT (2015) Control of predators in industrial scale microalgae cultures with pulsed electric fields. Bioelectrochemistry 103:60–64. https://doi.org/10.1016/j.bioelechem.2014.08.004

Renuka N, Guldhe A, Prasanna R, Singh P, Bux F (2018) Microalgae as multi-functional options in modern agriculture: current trends, prospects and challenges. Biotechnol Adv 36(4):1255–1273. https://doi.org/10.1016/j.biotechadv.2018.04.004

Roelke DL (2017) Applying principles of resource competition theory to microalgae biomass production: a more refined relationship between species richness and productivity. Algal Res 25:431–438. https://doi.org/10.1016/j.algal.2017.06.002

Sawant SS, Anil AC, Krishnamurthy V, Gaonkar C, Kolwalkar J, Khandeparker L, Desai D, Mahulkar AV, Ranade VV, Pandit AB (2008) Effect of hydrodynamic cavitation on zooplankton: a tool for disinfection. Biochem Eng J 42(3):320–328. https://doi.org/10.1016/j.bej.2008.08.001

Schluter M, Groeneweg J (1985) The inhibition by ammonia of population growth of the rotifer, *Brachionus rubens*, in continuous culture. Aquaculture 46:215–220

Scholz B, Kupper FC, Vyverman W, Olafsson HG, Karsten U (2017) Chytridiomycosis of marine diatoms - the role of stress physiology and resistance in parasite-host recognition and accumulation of defense molecules. Mar Drugs 15(2). https://doi.org/10.3390/md15020026

Shurin JB, Abbott RL, Deal MS, Kwan GT, Litchman E, McBride RC, Mandal S, Smith VH (2014) Industrial-strength ecology: trade-offs and opportunities in algal biofuel production. Ecol Lett 16(11):1393–1404. https://doi.org/10.1111/ele.12176

Smith VH, Crews T (2014) Applying ecological principles of crop cultivation in large-scale algal biomass production. Algal Res 4:23–34. https://doi.org/10.1016/j.algal.2013.11.005

Smith VH, McBride RC (2015) Key ecological challenges in sustainable algal biofuels production. J Plankton Res 37(4):671–682. https://doi.org/10.1093/plankt/fbv053

Smith VH, McBride RC, Shurin JB, Bever JD, Crews TE, Tilman GD (2015) Crop diversification can contribute to disease risk control in sustainable biofuels production. Front Ecol Environ 13 (10):561–567. https://doi.org/10.1890/150094

Suslick KS (1990) Sonochemistry. Science 247(4949):1439–1445. https://doi.org/10.1126/science.247.4949.1439

Szyjka SJ, Mandal S, Schoepp NG, Tyler BM, Yohn CB, Poon YS, Villareal S, Burkart MD, Shurin JB, Mayfield SP (2017) Evaluation of phenotype stability and ecological risk of a genetically engineered alga in open pond production. Algal Res Biomass Biofuels Bioproducts 24:378–386. https://doi.org/10.1016/j.algal.2017.04.006

Tchokouaha Yamthe LR, Appiah-Opong R, Tsouh Fokou PV, Tsabang N, Fekam Boyom F, Nyarko AK, Wilson MD (2017) Marine algae as source of novel antileishmanial drugs: a review. Mar Drugs 15(11). https://doi.org/10.3390/md15110323

Thom AE, Bhattacharjee M, Siemann E (2018) Predaceous insects may limit algal grazers in high-nutrient environments: implications for wastewater remediation in open bioreactors. Aquat Ecol 52(1):107–118. https://doi.org/10.1007/s10452-018-9648-5

Thomas PK, Dunn GP, Passero M, Feris KP (2017) Free ammonia offers algal crop protection from predators in dairy wastewater and ammonium-rich media. Bioresour Technol 243:724–730. https://doi.org/10.1016/j.biortech.2017.07.008

Troschl C, Fritz I, Sodnikar K, Drosg B (2017) Contaminations in mass cultivation of cyanobacteria: highly resilient *Colpoda steinii* leads to rapid crash of *Synechocystis* sp. cultures and is inhibited by partially anoxic conditions. Algal Res 28:229–234. https://doi.org/10.1016/j.algal.2017.11.002

Umar A, Caldwell GS, Lee JGM (2018) Foam flotation can remove and eradicate ciliates contaminating algae culture systems. Algal Res 29:337–342. https://doi.org/10.1016/j.algal.2017.12.002

Vadlamani A, Viamajala S, Pendyala B, Varanasi S (2017) Cultivation of microalgae at extreme alkaline pH conditions: a novel approach for biofuel production. ACS Sustain Chem Eng 5 (8):7284–7294. https://doi.org/10.1021/acssuschemeng.7b01534

Van Etten JL, Dunigan DD (2012) Chloroviruses: not your everyday plant virus. Trends Plant Sci 17(1):1–8. https://doi.org/10.1016/j.tplants.2011.10.005

Van Ginkel SW, Igou T, Hu Z, Narode A, Cheruvu S, Doi S, Johnston R, Snell T, Chen Y (2015) Taking advantage of rotifer sensitivity to rotenone to prevent pond crashes for algal-biofuel production. Algal Res 10:100–103

Van Ginkel SW, Bidwell M, Igou T, Gijon-Felix R, Ribeiro Salvi EJNRS, De Oliveira SHR, Duarte LHK, Steiner D, Hu Z, Johnston R, Snell T, Chen Y (2016) The prevention of saltwater algal pond contamination using the electron transport chain disruptor, rotenone. Algal Res 18:209–212

Verschoor AM, Bekmezci OK, Van Donk E, Vijverberg J (2009) The ghost of herbivory past: slow defence relaxation in the chlorophyte *Scenedesmus obliquus*. J Limnol 68(2):327–335. https://doi.org/10.3274/JL09-68-2-15

Vick, B (2010) Glyphosate applications in aquaculture. US Patent No. 201000022393A1. United States Patent Office

Villain A, Gallot-Lavallee L, Blanc G, Maumus F (2016) Giant viruses at the core of microscopic wars with global impacts. Curr Opin Virol 17:130–137. https://doi.org/10.1016/j.coviro.2016.03.007

von Alvensleben N, Stookey K, Magnusson M, Heimann K (2013) Salinity tolerance of *Picochlorum atomus* and the use of salinity for contamination control by the freshwater cyanobacterium *Pseudanabaena limnetica*. PLoS One 8(5):e63569. https://doi.org/10.1371/journal.pone.0063569

Wang H, Zhang W, Chen L, Wang J, Liu T (2013) The contamination and control of biological pollutants in mass cultivation of microalgae. Bioresour Technol 128:745–750. https://doi.org/10.1016/j.biortech.2012.10.158

Wang L, Yuan D, Li Y, Ma M, Hu Q, Gong Y (2016) Contaminating microzooplankton in outdoor microalgal mass culture systems: an ecological viewpoint. Algal Res 20:258–266. https://doi.org/10.1016/j.algal.2016.10.013

Wang Y, Eustance E, Castillo-Keller M, Sommerfeld M (2017a) Evaluation of chemical treatments for control of ciliate grazers in algae cultures: a site study. J Appl Phycol 29(6):2761–2770. https://doi.org/10.1007/s10811-017-1150-2

Wang YF, Castillo-Keller M, Eustance E, Sommerfeld M (2017b) Early detection and quantification of zooplankton grazers in algal cultures by FlowCAM. Algal Res Biomass Biofuels Bioproducts 21:98–102. https://doi.org/10.1016/j.algal.2016.11.012

Wang Y, Gong Y, Dai L, Sommerfeld M, Zhang C, Hu Q (2018) Identification of harmful protozoa in outdoor cultivation of Chlorella and the use of ultrasonication to control contamination. Algal Res 31:298–310. https://doi.org/10.1016/j.algal.2018.02.002

Weissman J, Radelli G (2010) Aurora biofuels, assignee. Systems and methods for maintaining the dominance of Nannochloropsis in an algae cultivation system. US Patent No. 20100183744. United States Patent Office

Woo JH, Kamei Y (2003) Antifungal mechanism of an anti-Pythium protein (SAP) from the marine bacterium *Streptomyces* sp. strain AP77 is specific for *Pythium porphyrae*, a causative agent of red rot disease in *Porphyra* spp. Appl Microbiol Biotechnol 62(4):407–413. https://doi.org/10.1007/s00253-003-1235-5

Xu C, Wu K, Van Ginkel SW, Igou T, Lee HJ, Bhargava A, Johnston R, Snell T, Chen Y (2015) The use of the schizonticidal agent Quinie sulfate to prevent pond crashes for algal-biofuel production. Int J Mol Sci 16:27450–27456

Yuan C, Zheng Y-L, Zhang W-L, He R, Fan Y, Hu G-R, Li F-L (2017) Lipid accumulation and anti-rotifer robustness of microalgal strains isolated from Eastern China. J Appl Phycol 29 (6):2789–2800. https://doi.org/10.1007/s10811-017-1167-6

Yun JH, Cho DH, Lee B, Kim HS, Chang YK (2018) Application of biosurfactant from *Bacillus subtilis* C9 for controlling cladoceran grazers in algal cultivation systems. Sci Rep 8(1):5365. https://doi.org/10.1038/s41598-018-23535-8

Zhang Z, Kong F (2011) Enhanced growth and esterase activity of *Chlorella pyrenoidosa* (Chlorophyta) in response to short-term direct grazing and grazing-associated infochemicals from *Daphnia carinata*. J Freshw Ecol 26(4):553–561. https://doi.org/10.1080/02705060.2011.577973

Zhu X, Wang J, Lu Y, Chen Q, Yang Z (2015) Grazer-induced morphological defense in *Scenedesmus obliquus* is affected by competition against *Microcystis aeruginosa*. Sci Rep 5:12743. https://doi.org/10.1038/srep12743

Zmora O, Grosse DJ, Zou N, Samocha TM (2013) Microalga for aquaculture: practical implications. In: Handbook of microalgal culture: applied phycology and biotechnology. Wiley, New York, pp 628–652

Chapter 3
Heterotrophic Growth of Microalgae

Michele Carone, Amélie Corato, Thomas Dauvrin, Tung Le Thanh, Lorenzo Durante, Bernard Joris, Fabrice Franck, and Claire Remacle

Abstract Microalgae constitute a phylogenetically diverse group of eukaryotic photosynthetic microorganisms that thrive in various environments. They are the subject of increasing interest from both the scientific community and the industrial world due, on the one hand, to their ecological significance at the global scale and, on the other hand, to their large potential as sources of economically important products, from biofuels to new pharmaceutical drugs. Some species (such as *Chlorella* sp., *Haematococcus pluvialis* or *Dunaliella salina*) are already cultivated on large scales, mainly in the photoautotrophic mode. Due to the production cost of microalgal biomass, relatively high-value compounds are targeted for economically viable processes. Among them, fatty acids and pigments are common products for which relevant species and cultivation processes are studied. Antibacterial substances appear as promising but less studied targets. This chapter makes a review about the production of fatty acids, pigments and antibacterial/antifungal compounds, with a special emphasis on their production in the heterotrophic mode. Compared to phototrophic growth, heterotrophic growth may result in higher biomass productivity by avoiding light limitations. In addition, it decreases area footprint and installation costs. The heterotrophic metabolism of some model microalgae is also described as well as the productivities that could be encountered in heterotrophy.

Michele Carone, Amélie Corato and Thomas Dauvrin contributed equally with all other contributors.

M. Carone · L. Durante · C. Remacle (✉)
Genetics and Physiology of Microalgae, UR InBios/Phytosystems, University of Liège, Liège, Belgium
e-mail: c.remacle@uliege.be

A. Corato · T. Le Thanh · F. Franck
Laboratory of Bioenergetics, UR InBios/Phytosystems, University of Liège, Liège, Belgium

T. Dauvrin · B. Joris
Genetics and Physiology of Bacteria, UR InBios/CIP, University of Liège, Liège, Belgium

A. Hallmann, P. H. Rampelotto (eds.), *Grand Challenges in Algae Biotechnology*, Grand Challenges in Biology and Biotechnology,
https://doi.org/10.1007/978-3-030-25233-5_3

3.1 Eukaryotic Microalgae as Cell Factories for the Production of Valuable Biomolecules

3.1.1 Heterotrophic Production of Fatty Acids and Lipids

Among the valuable compounds produced by microalgae, lipids and fatty acids are the most studied and exploited ones. Industrially, they are mostly used in biofuel refineries, as food additives and as health-promoting supplements (Chisti 2007; Koller et al. 2014; Mata et al. 2010). Microalgae can produce a wide range and large amounts of lipids, i.e. contents from 4.8 to 60% of dry weight (DW) depending on the algal strain and culture conditions (Liu et al. 2014b; Sajjadi et al. 2018). Among all the range of lipids, neutral lipids are the most interesting ones, in particular, triacylglycerol (TAG) which may account for 80% of total neutral lipids (Morales-Sánchez et al. 2014). Indeed, as biofuels are composed of fatty acid methyl esters (FAMEs), made by transesterification of TAGs, rich-oil microalgae may be valorized as third-generation biofuels (Liu et al. 2011, 2014a, b). Moreover, in some microalgal strains, TAGs contain a high quantity of polyunsaturated fatty acids (PUFAs) which have an increasing commercial interest due to their nutritive values for economically important marine animals and for human health (Hu and Gao 2003). The most important ω-3 PUFAs are eicosapentaenoic acid (EPA C20:5n-3) and docosapentaenoic acid (DHA C22:6n-3) (Sajjadi et al. 2018). Nowadays, the common source of EPA and DHA is fish oil which faces challenges in the food industry because of its significant taste and odour, heavy-metal contamination and because of some stability problems (Barclay et al. 1994). Thus, alternative sources of natural EPA and DHA are studied, and microalgae that contain large amounts of ω-3 PUFAs seem to be a promising source, especially under heterotrophic conditions (Perez-Garcia et al. 2011).

In microalgae, the lipid accumulation strongly depends on culture conditions. Indeed, several studies have performed comparison between autotrophic and heterotrophic growths and showed that numerous microalgal species accumulate more biomass and lipids and have a higher productivity under heterotrophic conditions (Perez-Garcia et al. 2011; Lowrey et al. 2015). For example, Xu et al. (2006) showed that *Chlorella protothecoides* has a lipid content four times higher under heterotrophic conditions (55.2% of lipids) than under phototrophic ones (14.6%) (Xu et al. 2006). Moreover, *Chlorella zofingiensis* produces two times more lipids when grown heterotrophically compared with autotrophic growth (respectively, 51.1% and 25.8% of lipids) (Liu et al. 2011). It was also shown that the diatom *Nitzschia laevis* has a higher ω-3 PUFA productivity in heterotrophic conditions (6.37 mg/L/day, in presence of 5 g/L glucose) than in autotrophic ones (3.39 mg/L/day) (Wen and Chen 2000). This could be explained by the fact that exogenous organic carbon sources provided during heterotrophic growth (glucose, fructose, etc.) offer an excess of chemical energy that need to be stored as lipid by the microalgae (Ratledge 2004; Wan et al. 2012). In addition, microalgae can metabolize different carbon sources such as glucose, acetate, sucrose, pure glycerol, molasses and also crude glycerol, which is a by-product from the biodiesel refinery and is available at a very low price (Chen and Walker 2011; Chi et al. 2007). These carbon sources may be used at different concentrations which were found to impact the lipid distribution. Indeed, when sugar concentration increases, the proportion of neutral lipids and TAGs tends to increase and, in contrast, the content of membrane lipids (phospholipids and glycolipids) tends to decrease (Liu et al. 2014b). Moreover, saturation of the fatty acids is directly dependent on the amount of excess sugars: as the concentration of the sugars increases, the fatty acids become more saturated (Wood et al. 1999). Although high sugar concentrations increase the desired lipids content, too high sugar concentrations lead to inhibition of growth and lipid production. Thus, for large-scale heterotrophic cultures, the use of fed-batch cultures is considered to increase cell density and lipid content and avoid substrate inhibition (Xiong et al. 2008).

It is widely known that lipid content and fatty acid composition can be also greatly influenced by a variety of medium nutrients and environmental factors (Liu et al. 2014a, b). On the one hand, nutrient (such as N, P, S and Si) limitation conditions have been intensively studied (Perez-Garcia et al. 2011; Sakarika and Kornaros 2017). Some studies show that low-nitrogen medium induced lipid accumulation up to 70% within 20 days in most microalgae (Sajjadi et al. 2018). Otherwise, in diatoms, the accumulation of lipids is related to depletion of silicates because of their dependence on silica for growth (Wilhelm et al. 2006). In heterotrophy, it is the higher C:(N or Si) ratios (corresponding to higher carbon content when the initial nitrogen (silicate) is fixed or lower nitrogen (silicate) when the carbon is fixed) that significantly enhance lipid content and yield, in particular, the neutral lipids (Perez-Garcia et al. 2011; Wang et al. 2017). On the other hand, environmental factors, such as temperature, pH, salinity or dissolved oxygen, play an important role in influencing the lipid profile and content in microalgae (Liu et al. 2014a, b). For example, it was shown that when temperature is below the optimal growth temperature, microalgae produce more unsaturated fatty acids with no change in the total fatty acid content (Tatsuzawa and Takizawa 1995).

3.1.1.1 Examples of Total Lipid Production in Heterotrophy

The green microalgae of the genus *Chlorella* are extensively studied because they are appropriate for heterotrophic cultivation and show great potential for lipid production. Indeed, under heterotrophic conditions, some *Chlorella* species may accumulate lipid at levels ranging from about 10 to 55% on a DW basis, and their fatty acid composition is between C14:0 and C20:0 (Isleen-Hosoglu et al. 2012). Sakarika and Kornaros (2017) studied the heterotrophic accumulation of lipids in *Chlorella vulgaris* (SAG211-11b) under macro-nutriment limitation (N, P or S) (Sakarika and Kornaros 2017). They found that the total lipid content and productivity were higher under sulphur limitation than under N or P limitations (Table 3.1). Moreover, they determined that, in all the tested conditions, the fatty acids (FAs)

Table 3.1 Heterotrophic production of lipids by green microalgae

Microalgae		Culture mode	Carbon source	Nitrogen source	Accumulation condition	Total lipid content (% g/g DW)	Total lipid productivity (mg/L/day)	References
Chlorella vulgaris	UTEX #259	250 mL, flasks, batch	Glucose—1%	$NaNO_3$—0.25 g/L	–	23 ± 2	35 ± 6	Liang et al. (2009)
			Acetate—1%		Nitrogen 0 g/L	36 ± 1	29 ± 4	
			Glycerol—2%		/	34 ± 4	31 ± 4	
	SAG 211-11b	0.8 L, flasks, batch	Glucose—10 g/L	$NaNO_3$—1.5 g/L	Sulphur limitation	53.43 ± 3.93	119.87 ± 9.56	Sakarika and Kornaros (2017)
			Glucose—40 g/L		Nitrogen limitation	21.48 ± 1.77	98.93 ± 8.89	
			Glucose—27 g/L		Phosphorous limitation	17.42 ± 6.42	96.47 ± 16.47	
Chlorella protothecoides	University of Texas	5 L, fermentor, fed-batch	Glucose—2–24 g/L	Yeast extract—4 g/L	–	50.3	–	Xiong et al. (2008)
	IOCAS038F	3 L, fermentor, fed-batch	Glucose—10–20 g/L	$NaNO_3$—2.5 g/L	Nitrogen 0 g/L and hyperosmotic stress (1200 mOsm/kg)	39.2	4255.2	Wang et al. (2017)
	UTEX 255	Flasks, batch	Glucose:Glycerol (9:1)—30 g/L	Yeast extract—4 g/L	–	–	2352	O'Grady and Morgan (2011)
			Glucose:Crude Glycerol (9:1)—30 g/L	Yeast extract—4 g/L	–	–	2496	
			Glucose 40 g/L	KNO_3—2.4 g/L	–	–	654 ± 3	Shen et al. (2010)
	UTEX 256	200 mL, flasks, batch	Glucose 30 g/L	Yeast extract—4 g/L	–	0.5	1530	Chen and Walker (2011)
			Crude glycerol 30 g/L		–	0.62	2460	
		5.5 L, bioreactor, fed-batch	Glucose 30 g/L		–	0.533	3040	
			Crude glycerol 30 g/L		–	0.542	2990	

Chlorella sorokiniana	UTEX 1602	0.8 L, flasks, batch	Glucose 10 g/L	$NaNO_3$	Nitrogen starvation (C/N = 35.4)	29.1	–	Chen and Johns (1991)
	M209220	Flasks	Glucose 10 g/L	$(NH_4)_2SO_4$	Natural nitrogen starvation	53 ± 3	–	Wan et al. (2012)
Chlorella sp.	LAM-H	100 mL, flasks, batch	Glucose 26.2 g/L	$NaNO_3$	–	–	247.16	Xie et al. (2012)
	PKU AC102	7.5 L, fer-mentor, batch	Glucose 12.37 g/L	KNO_3	Natural Nitrogen starvation	–	220	Jia et al. (2014)
Neochloris oleoabundans	UTEX 1185	5 L, biore-actor, batch	Glucose 50 g/L	$NaNO_3$	C/N ratio = 278	51.7 ± 1.7	–	Morales-Sánchez et al. (2013)
		40 L, bio-reactor fed-batch			Nitrogen limitation and glucose excess	44.6	1020	Morales-Sánchez et al. (2014)
Scenedesmus obliquus	NIES-2280	1 L, flasks, batch	Acetate 6 g/L	$NaNO_3$	Nitrogen starvation	47	56	Shen et al. (2015)

were mainly composed of monounsaturated fatty acids (MUFAs) with the dominant one being oleic acid (C18:1), and its content was constant between the conditions. They conclude that the optimal limitation condition for biofuel production using *C. vulgaris* is sulphur limitation (Sakarika and Kornaros 2017). Liang and collaborators explored the lipid production of *C. vulgaris* UTEX259 under autotrophic and heterotrophic conditions using different sugars as carbon source (Liang et al. 2009). The highest lipid content was found under autotrophic growth (38% DW), but due to a lower biomass concentration, the lipid productivity stayed low (4 mg/L/day) compared to the heterotrophic one (35 mg/L/day). They showed that the total lipid content was higher using acetate 1% and glycerol 2% (36% DW and 34% DW respectively) than using glucose 1% (23% DW), but the total lipid productivities were about the same in all conditions (Table 3.1). Several studies focussed on *Chlorella protothecoides*, which could reach lipid contents up to 55.2% DW in presence of a low-price carbon source, which was slightly more than in the presence of 10 g/L glucose (54.7% DW) (Miao and Wu 2006; Xu et al. 2006). The same group cultivated *C. protothecoides* in a 11,000 L stirred tank bioreactor in the presence of 10 g/L glucose. In these conditions, without any nutrient limitation applied, *C. protothecoides* was able to produce 43% DW of total lipids (Li et al. 2007). In addition, several studies on *C. protothecoides* were made using crude glycerol and batch or fed-batch culture modes (Table 3.1). The data in Table 1 show that the fed-batch culture did not allow to obtain higher lipid content than the batch condition (Chen and Walker 2011), but the application of nitrogen starvation during a fed-batch culture (i.e. C:N ratio is high) led to a higher total lipid productivity (Wang et al. 2017). Other *Chlorella* strains were also studied, such as *C. sorokiniana* (Table 3.1), *Chlorella* sp. (Table 1) and *C. saccharophila* (Tan and Johns 1991; Isleen-Hosoglu et al. 2012), but their lipid productivities were not as high as for *Chlorella protothecoides*, which makes them less interesting.

More recently, *Neochloris oleoabundans* has gained some interest as it can metabolize glucose at high concentration and produce large amount of lipids, up to 51.7% DW (Table 3.1). Moreover, another green microalga, *Scenedesmus obliquus*, is also studied for lipid production (Table 3.1). Shen et al. (2015) showed that in the presence of a high amount of acetate (6 g/L) and under nitrogen starvation, *S. obliquus* may produce 47% of total lipids (Shen et al. 2015). Thus, the researches for finding a strain which could produce large amount of biomass and lipids under heterotrophic conditions, and which may be competitive with current large-scale lipid productions, are still ongoing.

3.1.1.2 Examples of ω-3 Polyunsaturated Fatty Acid Production in Heterotrophy

Ω-3 PUFAs are mainly produced by marine microalgae from the genus *Schizochytrium* (Thraustochytriales), *Crypthecodinium* (Dinoflagellata) and *Nitzschia* (Bacillariophyceae). Indeed, in heterotrophic conditions, *Schizochytrium* sp. can yield about 40% DW of DHA from its total fatty acids and *Crypthecodinium cohnii* may produce DHA content up to 65% DW (de Swaaf et al. 1999, 2003; Sijtsma et al. 2010). Heterotrophic growths in lab- or large-scale cultures in order to enhance ω-3 PUFA content in these strains have been performed, and some are summarized in Table 3.2. Ethier et al. (2011) studied the lipid and PUFA productions by the microalga *Schizochytrium limacinum* SR21 ATCC MYA-1381 in the presence of crude glycerol, the major by-product of biodiesel, in continuous cultures (Ethier et al. 2011). These experiments (Table 3.2) showed that the DHA content and productivity in continuous cultivation (0.52 ± 0.03 g/L/day and 148.2 ± 2.9 mg/g, respectively) were similar to those in batch cultures (0.51 ± 0.04 g/L/day and 170.45 ± 11.27 mg/g, respectively) (Chi et al. 2007). Experiments performed on the diatom *Cyclotella cryptica* in batch cultures with glucose and different nitrogen sources and stresses applied showed no differences in the PUFA content between all the different conditions (Pahl et al. 2010, 2012). Moreover, *C. cryptica* produces lower amounts of DHA than *S. limacinum* (Table 3.2). Wen and Chen (2001) analysed the nitrogen source as the most suitable for EPA accumulation (Wen and Chen 2001). They found that nitrate and urea were the preferred sources for cell growth and EPA content, whereas tryptone and yeast extract were found to enhance EPA production, as shown in Table 3.2.

Nowadays, some companies that produce ω-3 PUFAs already exist: Omega Tech Inc. and Martek Biosciences (Barclay et al. 1994). The latter uses the heterotrophic production of PUFAs by *Nitzschia* and *Crypthecodinium cohnii*. They reported that *Nitzschia* species may reach an oil content up to 50% DW with an EPA productivity of 0.25 g/L/day (Kyle and Gladue 1993). Furthermore, by using *Crypthecodinium cohnii*, oil content up to 15–30% DW could be reached, with a DHA productivity of 1.2 g/L/day (Kyle et al. 1992). On the other hand, Omega Tech Inc. uses heterotrophic cultures of thraustochytrids, such as *Schizochytrium* sp. with which they may obtain ω-3 PUFA productivity of 1 g/L/day (Barclay 1992).

3.1.2 Pigments

Pigments are compounds that absorb light in the visible spectral region. In microalgae, three major groups are found: chlorophylls, carotenoids and phycobilins. They are acting as light energy absorbers in photosynthetic systems and some as protective agents against the production of reactive oxygen species (ROS) (Demmig-Adams and Adams 2002; Gong and Bassi 2016). Commercially, they are generally used as food and feed additives and as health-promoting supplements (Spolaore et al. 2006). Among these pigments, some are considered as algal products of high potential for commercial success, such as β-carotene, lutein, cantaxanthin, astaxanthin, phycocyanin and fucoxanthin (Gong and Bassi 2016; Hu et al. 2018). Although their large-scale production conditions are currently mostly photoautotrophic ones, heterotrophic cultivations have gained interest and are more and more developed (Bumbak et al. 2011). However, the main restriction to the use of heterotrophic conditions to produce pigments is that many biosynthetic

Table 3.2 Heterotrophic production of ω-3 polyunsaturated fatty acids by microalgae

Microalgae		Culture mode	Carbon source	Nitrogen source	Accumulation condition	Total lipid content (%g/g DW)	Total FA content (mg/g DW)	ω-3 Long-chain PUFA (mg/g)	References
Schizochytrium limacinum	SR21 (ATCC MYA-1381)	50 mL, flasks, batch	Crude glycerol 75 g/L	Yeast extract + peptone	–	45	505.69 ± 45.14	170.45 ± 11.27 (DHA)	Chi et al. (2007)
		4.5 L, bio-reactor continuous	Crude glycerol 90 g/L		–	–	502.5 ± 6.57	148.2 ± 2.9 (DHA)	Ethier et al. (2011)
Cyclotella cryptica	UTEX 1269	100 mL, flasks, batch	Glucose 10 g/L	Yeast extract, $NaNO_3$	Salt stress (11.2 psu)	7.4	–	13.5 ± 1.2 (EPA) 1.7 ± 0.1 (DHA)	Pahl et al. (2010)
					Temperature (25 °C)	5.67	–	10.5 ± 0.5 (EPA) 2.5 ± 0.1 (DHA)	
			Glucose 10 g/L	Urea	Nitrogen starvation	–	61.0 ± 3.4	12.5	Pahl et al. (2012)
Nitzschia laevis	UTEX 2047	125 mL flasks, batch	Glucose 10 g/L	Yeast extract + peptone	–	39.3	–	23.2 (EPA)	Tan and Johns (1996)
		100 mL, flasks, batch	Glucose 5 g/L	$NaNO_3$ + tryptone	Salt stress 20 g/L	–	–	21.0 ± 0.2 (EPA)	Chen et al. (2008)
Crypthecodinium cohnii	ATCC 30772	500 mL, flasks, batch	Glucose 25 g/L	Yeast extract	–	15	–	4.6	de Swaaf et al. (2003)

pathways are light-dependent. Indeed, in many algal species, chlorophyll and primary carotenoid production is controlled by light availability (Im et al. 2006) and, thus, the pigment content is often lower in complete darkness. To overcome this issue, it is possible to generate mutants that are able to increase the pigment content in the dark. Indeed, Ishikawa et al. generated chlorophyll mutants in which the light control of pigment biosynthesis is weakened (Ishikawa et al. 2004).

Otherwise, when the biosynthetic pathway is not light-dependent (as for secondary carotenoids), the pigment accumulation may proceed under heterotrophic conditions and may be increased by the application of stress conditions. On the one hand, some studies show that simply by using heterotrophic (or mixotrophic) culture conditions, we may enhance pigment productivity on a volumetric basis. It was shown that a high carbon to nitrogen (C/N) medium (presence of acetate or glucose) can induce astaxanthin biosynthesis (Ip and Chen 2005a). Indeed, glucose might be essential for providing the carbon skeleton for the formation of secondary carotenoids, and thus using heterotrophic growth may enhance pigment content (Ip and Chen 2005a). On the other hand, using stress conditions can lead to an efficient increase in pigment content. These stresses are often applied after a period of heterotrophic growth in optimal conditions and may be created in the dark by high/low temperature and pH exposure, by nutrient starvation (N or P), by high salinity or by oxidative stress (H_2O_2 and/or NaClO in presence of Fe^{2+}) (Wei et al. 2008; Markou and Nerantzis 2013; Gong and Bassi 2016; Rajesh et al. 2017). For example, nitrate starvation alters the normal cellular metabolism (such as protein synthesis) and, by doing so, affects indirectly pigment accumulation (Ip and Chen 2005b). Moreover, the addition of chemicals produces ROS that enhance carotenogenesis (Kobayashi 2003; Markou and Nerantzis 2013).

The major heterotrophic pigment production processes, in lab- or large-scale conditions, are for phycobilins or carotenoids (mainly lutein and astaxanthin). The red microalgae *Galdieria sulphuraria* and the cyanobacteria *Spirulina platensis* are mostly used for phycobilin production, whereas microalgae from the genus *Chlorella* are used for carotenoids. Examples of heterotrophic production of these pigments are discussed below.

3.1.2.1 Lutein

Lutein ((3R, 3′R, 6′R)-β, ε-carotene-3, 3′-diol) is a primary xanthophyll. Its industrial and commercial supply is nowadays depending on marigold flower petals from the genus *Tagetes* (Lin et al. 2015). Nevertheless, the use of marigold flowers presents a lot of disadvantages, such as their low and variable lutein content: from 0.3 up to 50 mg/g (Piccaglia et al. 1998; Hu et al. 2018). Thus, alternative sources need to be considered, and microalgae appear as a promising source of lutein.

Heterotrophic microalgae cultures for lutein production have been investigated (Table 3.3). Among all the species, green microalgae of the genus *Chlorella* are the most studied ones. Indeed, they can grow easily and rapidly in heterotrophy with a high productivity (Wei et al. 2008). Depending on the strain used, the choice of the

Table 3.3 Heterotrophic cultivation of green microalgae for production of lutein

Microalgae species	Cultivation mode	Carbon source	Accumulation condition	Content (mg/g)	Yield (mg/L)	Productivity (mg/L/day)	References
Chlorella protothecoides CS-41	3.7 L, fermentor, fed-batch	Glucose (40 g/L)	/	4.65	216.92	39.12 ± 19.64	Shi and Chen (2002)
Chlorella protothecoides CS-41	30 L, fermentor, fed-batch	Glucose (40 g/L)	N starvation	5.35	209.08	49.18 ± 19.18	Shi and Chen (2002)
Chlorella protothecoides CS-41	150 mL, flask, batch	Glucose (40 g/L)	pH = 6.6	4.75	77.92	/	Shi et al. (2006)
Chlorella protothecoides CS-41	150 mL, flask, batch	Glucose (40 g/L)	Late exponential phase	4.43	/	/	Shi and Chen (1999)
Chlorella protothecoides CS-41	3.7 L, fermentor, fed-batch	Glucose (40 g/L)	/	4.58	83.81	/	Shi et al. (2000)
Chlorella protothecoides UTEX29	100 mL, flask, batch	Glucose (36 g/L)	0.01 mM H_2O_2 + 0.5 mM NaClO	1.98 ± 0.02	31.40 ± 0.35	/	Wei et al. (2008)
Chlorella protothecoides CS-41	30 L, fermentor, fed-batch	Glucose (36 g/L)	/	4.9	66.3	/	Shi et al. (1997)
Chlorella protothecoides CS-41	3.7 L, fermentor, fed-batch	Glucose (40 g/L)	/	4.44	76.56	/	Shi et al. (1999)
Chlorella pyrenoidosa 15-2070	19 L, fermentor, fed-batch	Glucose (10 g/L), fed-batch: glucose 400 g/L	/	/	178	/	Wu et al. (2007)

carbon source is crucial. Indeed, in the case of *C. protothecoides* and *C. pyrenoidosa*, to enhance the lutein production, glucose is preferable in heterotrophy, while acetate should be used for mixotrophy (Theriault 1965). For example, growing *Chlorella protothecoides* CS-41 on glucose 40 g/L and urea can lead to a lutein yield of 83.81 mg/L (Shi et al. 2000). In some cases, when microalgae are submitted to stress conditions, their lutein content can increase. Indeed, Shi et al. showed that after increasing the temperature (from 25 to 35 °C) and modifying the pH (6.6) of heterotrophic cultures of *C. protothecoides* CS-41, the cellular lutein content is increased but the lutein yield is decreased (Table 3.3) (Shi et al. 2006). Moreover, with the same strain, higher lutein yields may be obtained in N-limited and fed-batch cultures, and, in the case of *C. protothecoides* UTEX 29, higher lutein content can be achieved by addition of chemicals (H_2O_2 and NaClO in presence of Fe^{2+}) (Table 3.3) (Shi and Chen 2002; Wei et al. 2008). These chemicals, in the presence of Fe^{2+}, affect the redox state and generate ROS, which activate the production of carotenoids such as lutein (Guedes et al. 2011). On the other hand, since lutein is a typical primary xanthophyll, its accumulation can also suffer from stress conditions, and thus, a too stressful condition can lead to a decrease in lutein content (Gong and Bassi 2016).

3.1.2.2 Astaxanthin

Astaxanthin (3,3′-dihydroxy-β, β-1-carotene-4,4′-dione) is a secondary xanthophyll mainly produced under oxidative stresses (Sarada et al. 2002; Markou and Nerantzis 2013; Rajesh et al. 2017). Nowadays, the market is dominated by synthetic astaxanthin produced from petrochemical derivatives (>95%) (Bernhard 1989; Lim et al. 2018). However, the high cost of synthetic pigment production and the growing demand for natural ones stimulate the search for natural pigment sources such as microalgae (Higuera-Ciapara et al. 2006). Moreover, for an economically competitive astaxanthin production process, heterotrophic growth and adapted stress conditions should be considered (Hu et al. 2018).

Although heterotrophic production of astaxanthin by *Haematococcus pluvialis* has been reported, its content was confined to 9 mg/L even after stress application, which is actually low compared to phototrophically grown cells (Kobayashi et al. 1997) (Table 3.4). Therefore, alternative fast-growing, astaxanthin-producing microalgae and optimal culture conditions are desirable (Orosa et al. 2001). *Chlorella zofingiensis*, *Chlorella protothecoides* and *Chlorococcum* sp. have been considered as alternative strains (Ma and Chen 2001; Liu et al. 2014a; Minhas et al. 2016). These strains are largely studied although their astaxanthin content is lower than in *H. pluvialis*: 1 mg/g for heterotrophic *C. zofingiensis* versus 40 mg/g for phototrophic *H. pluvialis* (Sun et al. 2008). In heterotrophic conditions, *C. zofingiensis* was found to have an astaxanthin yield of 8.4–10.7 mg/L in the presence of 100 mM pyruvate (Guedes et al. 2011). In the presence of 50 g/L glucose, its astaxanthin content was about 1.01 mg/g (Ip and Chen 2005a) and about 1 mg/g also in the presence of 30 g/L pretreated molasses (Liu et al. 2012)

Table 3.4 Heterotrophic cultivation of green microalgae for production of astaxanthin

Microalgae species	Cultivation mode	Carbon source	Accumulation condition	Content (mg/g)	Yield (mg/L)	Productivity (mg/L/day)	References
Chlorella zofingiensis	Heterotrophy 100 mL, flask, batch	Glucose 50 g/L	High C/N report	1.01 ± 0.04	10.29 ± 0.4	/	Ip and Chen (2005a)
Haematococcus pluvialis Flotow NIES-144	Heterotrophy 100 mL, flask, fed-batch	Acetate 15 mM + 45 mM NaAc	Salt stress (NaCl 0.1%)	/	9	/	Kobayashi et al. (1997)
Chlorococcum sp. strain MA-1	Heterotrophy 2 L, fermentor, fed-batch	Glucose 40 mM	Addition of 80 mM acetate, 1 mM Fe2+ and 10 μM H_2O_2	0.54	9.8	0.98	Zhang and Lee (2001)
			Light stress (300 μmol. $m^{-2}.s^{-1}$)	1.77	32.2	3.22	
Chlorococcum sp.	Heterotrophy 250 mL, culture, batch	Glucose 44 g/L	/	1.034 ± 0.08	/	/	Ma and Chen (2001)
			H_2O_2 0.1 mM	1.782 ± 0.12	/	/	
Chlorella zofingiensis ATCC 30412	Heterotrophy 100 mL, culture, batch	Glucose 50 g/L	/	1.02 ± 0.05	10.51 ± 0.2	/	Sun et al. (2008)
	Heterotrophy 2 L, fermentor, fed-batch		/	0.69 ± 0.06	32.4	/	
Chlorella zofingiensis ATCC 30412 + A11: H11	Heterotrophy 100 mL, culture, batch	Glucose 30 g/L	10 mM H_2O_2 + 0.5 mM NaClO	/	12.58	/	Ip and Chen (2005b)
Chlorella zofingiensis ATCC 30412	Heterotrophy 100 mL, culture, batch	Pretreated molasses 30 g/L	/	1	/	1.7	Liu et al. (2012)
Chlorella zofingiensis Mutant E17	Heterotrophy 50 mL, culture, batch	Glucose 20 g/L	/	1.21	/	1.95	Liu et al. (2013)

(Table 3.4). Thus, although it is clear that the carbon source has a high impact on heterotrophic batch growth and accumulation of biomass, it does not impact that much the astaxanthin content. Regarding stress conditions with a culture under heterotrophic conditions, Zhang and Lee (Table 3.4) studied the impact of a chemical stress on *Chlorococcum* sp. astaxanthin production and showed that astaxanthin yield improved up to 9.8 mg/L (Zhang and Lee 2001). Moreover, some studies show that different C/N ratios, namely, nitrogen limitation coupled with an excess of organic carbon substrates, may increase astaxanthin production (Ip and Chen 2005a; Cuellar-Bermudez et al. 2015). Finally, as it was shown by Liu et al. (Table 3.4), other lines of research focus on increasing astaxanthin content by the selection of overproducing mutants (Liu et al. 2013).

3.1.2.3 Phycocyanin

Phycocyanin is a blue, fluorescent pigment which is only produced by some microalgae, belonging to Rhodophyta or Cryptophyta, and by cyanobacteria (Stadnichuk et al. 2000; Kuddus et al. 2013). Currently phycocyanin is recovered from outdoor cultures of the cyanobacterium *Spirulina* (*Arthrospira*) *platensis* using sunlight as the energy source (Schmidt et al. 2005; Eriksen 2008). However, in the recent years more attention was gradually transferred to Rhodophyta, and in particular to *G. sulphuraria*. Indeed, its phycocyanin can tolerate temperatures as high as 73 °C while phycocyanin from *Spirulina* deactivated in temperatures higher than 46 °C (Moon et al. 2014). The physiology and ecology of this red alga are described in Sect. 2.1.3.

Although *Galdieria sulphuraria* heterotrophic growth allows to obtain high biomass concentration, up to 100 g/L (Schmidt et al. 2005), it has been shown that the cells display a yellow-green colour, due to the arrest of pigment synthesis (Rhie and Beale 1994; Stadnichuk et al. 1998, 2000; Eriksen 2008). However, Gross and Schnarrenberger isolated a mutated strain—named 074G—which maintains high specific pigment concentrations when grown heterotrophically in darkness (Gross and Schnarrenberger 1995). Sloth et al. studied this particular *G. sulphuraria* strain in heterotrophic conditions, using carbon (mainly glucose)- or nitrogen (ammonium)-limited cultures (Sloth et al. 2006). They demonstrated that carbon limitation, with an excess of nitrogen, results in a relatively high phycocyanin content (up to 10–20 mg/g DW), while nitrogen starvation results in less than 1.3 mg/g DW (Sloth et al. 2006). Moreover, Schmidt et al. analysed phycocyanin production in the presence of different sugars and showed that for a fed-batch culture containing glucose or molasses, the phycocyanin concentration is between 250 and 400 mg/L and the phycocyanin productivity can reach up to 142 mg/L/day (Schmidt et al. 2005). Finally, Wan and collaborators have applied a novel "sequential heterotrophy-dilution-photoinduction" (SHDP) cultivation mode to increase the phycocyanin content produced by *G. sulphuraria* (Wan et al. 2016). This culture mode is based on a two-step strategy: a heterotrophic growth step in a short period of time followed by a photoinduction step after dilution for enhanced phycocyanin accumulation (Wan et al. 2016). Using these conditions, they obtained a final phycocyanin content of 13.2% DW, which is comparable to autotrophic production by *Spirulina*, and a global phycocyanin productivity of 2209.2 mg/L/day which is 15 times more than the one observed by Schmidt and collaborators (Schmidt et al. 2005; Wan et al. 2016).

3.1.3 Antibacterial and Antifungal Activities in Heterotrophy

Since the discovery of chlorellin (a mixture of fatty acids showing an antibacterial activity) extracted from *Chlorella* species in 1944 (Pratt et al. 1944), studies aiming at the identification of new antimicrobial compounds produced by microalgae have boomed. Some of these studies have resulted in the identification of the active molecules, and these often belong to the same class and include pigments, lipids, fatty acids, carbohydrates, flavonoid and phenolic compounds (for review, see Falaise et al. 2016).

All the screening programs have been carried out in autotrophic conditions. Heterotrophy is one of the routes that need to be further explored to find new antibacterial compounds in microalgae: it changes the microalgae metabolism, and these other metabolic pathways may lead to the production of unknown antimicrobial compounds. Mashhadinejad et al. showed that heterotrophy allows to obtain a stronger antibacterial activity of *Chlorella vulgaris*. These authors have found that the growth in heterotrophic conditions increases the antimicrobial activity against Gram-negative bacteria (*Escherichia coli* and *Pseudomonas aeruginosa*) and yeast (*Candida albicans*), while the mixotrophic conditions increase the antibacterial activity against Gram-positive bacteria (*Bacillus subtilis* and *Staphylococcus aureus*) compared to autotrophy. They also showed that mixotrophy and heterotrophy increase the total lipid content and the lipid productivity compared to autotrophic conditions (Mashhadinejad et al. 2016). To our knowledge, this is the only study that has assessed the antimicrobial potential of microalgae in heterotrophy. Heterotrophic cultivation of microalgae is a large unexplored field to search for new antimicrobial metabolites.

3.1.4 Bioactive Products of Cyanobacteria

Cyanobacteria are photosynthetic prokaryotic microorganisms, phylogenetically related to microalgae of the Archaeplastida lineage since these eukaryotic microalgae acquired their chloroplasts by primary endosymbiosis of cyanobacteria between 2.1 and 1.5 billion years ago (McFadden 2014). Cyanobacteria are well known for the production of bioactive secondary metabolites which are synthesized by specific pathways known as the polyketide (PK), non-ribosomal polypeptide (NRP) or polyketide-NRP biosynthetic pathways (Newman and Cragg 2007), also found in heterotrophic bacteria and fungi (Wang et al. 2014). Briefly, the NRP pathway relies on the assembly of non-ribosomal peptides catalysed by non-ribosomal peptide synthetases; the polyketide pathway relies on a sequential set of condensation reactions of acyl-CoA units catalysed by polyketide synthases; the hybrid PK-NRP biosynthetic pathway contains both polyketide and peptide modules. The bioactive compounds produced by cyanobacteria using these pathways include, for example, the antibiotic vancomycin, the immunosuppressive agent cyclosporine and the anticancer agent bleomycin (reviewed in Tan 2007). Until now, these bioactive molecules have been found in marine cyanobacteria (Tan 2007), and to our knowledge production of these compounds in heterotrophic conditions has not been investigated yet.

3.2 Basal Carbon Metabolism of Microalgae Under Heterotrophy

3.2.1 Molecular Analysis of Heterotrophic Metabolism in Model Microalgal Species

3.2.1.1 *Chlamydomonas reinhardtii*

C. reinhardtii is a green freshwater microalga 10 μM long, which has been known for decades as a model system for photosynthetic studies (Rochaix 2002), and contains a unique chloroplast typical of primary endosymbiosis. It belongs to the Chlorophytes which diverged from land plants over 1 billion years ago (Merchant et al. 2007). The three genomes are sequenced (Vahrenholz et al. 1993; Maul et al. 2002; Merchant et al. 2007), and many molecular tools for genome manipulation have been developed (Jinkerson and Jonikas 2015). *C. reinhardtii* is a facultative phototroph, able to assimilate acetate either mixotrophically when combined with CO_2 assimilation in the light or heterotrophically in darkness as the sole carbon source. *C. reinhardtii* is accumulating starch as carbon storage (Hicks et al. 2001), whose metabolism allows continuous growth in the dark, provided that acetate is supplied in the medium. Heterotrophic growth of *Chlamydomonas* has not usually been considered for biotechnological applications. However, this mode of growth has recently also drawn attention as a new cost-effective mode of cultivation for enhancing lipid accumulation (Fan and Zheng 2017), considering that the detailed knowledge of the metabolism allows the modelling and metabolic flux analysis of carbon storage accumulation (Boyle et al. 2017; Shene et al. 2017).

Acetate Assimilation and Heterotrophic Growth

Acetate is the only organic carbon source that *Chlamydomonas* is able to assimilate. In order to allow glucose assimilation, a hexose symporter of *Chlorella* has been expressed in *C. reinhardtii*, and the impact on H_2 production in dark anoxic conditions has been analysed (Doebbe et al. 2007). The utilization of acetate as well as the characterization of the metabolism in dark oxic and anoxic conditions has been reviewed in Yang et al. 2015. We thus provide here with an update based on the most recent papers. Once inside the cell, acetate is metabolized into acetyl-CoA by acetyl-CoA synthase. Three acetyl-CoA synthase isoforms are present: ACS1 is located in the cytosol (Lauersen et al. 2016), ACS2 in the chloroplast (Terashima et al. 2010) and ACS3 in the peroxisomes (Lauersen et al. 2016). ACS3 could also be located in mitochondria (Atteia et al. 2009). Acetyl-CoA can be used in the glyoxylate cycle to fuel gluconeogenesis and other anabolic pathways such as amino acid synthesis or in the Krebs cycle to produce NADH and $FADH_2$ that are used by the respiratory chain for ATP production. These two biochemical pathways are necessary for growth in the dark oxic conditions since mutants of the respiratory chain (Salinas et al. 2014) and of the glyoxylate cycle (Plancke et al. 2014) are unable to grow in the dark. We have recently clarified the localization of the enzymes of the glyoxylate (Lauersen et al. 2016); five of the six enzymes are located in the peroxisomes, malate dehydrogenase (MDH1), citrate synthase (CIS2), acetyl-CoA synthase (ACS3), the lone form of malate synthase (MAS1) and of aconitase (ACH1), which defines a central role to peroxisomes in acetate assimilation. Only isocitrate lyase (ICL1) is localized in the cytosol. Metabolites such glyoxylate or acetate should shuttle between the two compartments. This is made possible by channel-forming proteins whose characterization is under way in our laboratory (Durante et al. 2019).

The oxidative phosphorylation (OXPHOS) mitochondrial electron transfer chain is composed of four major multiprotein complexes: NADH:ubiquinone oxidoreductase (complex I), succinate dehydrogenase (complex II), ubiquinone:cytochrome *c* oxidoreductase (complex III) and cytochrome *c* oxidase (complex IV). Three out of these four complexes (complexes I, III and IV) pump protons from the matrix to the intermembrane space (IMS) and build a gradient which is used by a fifth complex, the ATP synthase, to produce ATP. In *Chlamydomonas*, like in other eukaryotes, proteins forming these complexes have a dual genetic origin. Only five subunits of complex I (ND1, 2, 4, 5 and 6) and one subunit of complex III (COB) and complex IV (COX1), highly hydrophobic, are produced directly inside the mitochondria by expression of the mitochondrial genome, while the vast majority, encoded in the nucleus, is imported from the cytoplasm. This dual origin accounts for the general complexity of the assembly processes as many chaperones or other assembly factors are needed to coordinate all those proteins (reviewed in Massoz et al. 2017).

Survival Mechanisms in Darkness Under Aerobic, Hypoxic and Anaerobic Conditions

C. reinhardtii is able to survive in transient hypoxic and anaerobic conditions and exhibits a complex fermentative metabolism. In case of anoxia, the respiratory chain is not functioning, and ATP production relies on glycolysis whose end product is pyruvate. NADH, which is a co-product of glycolysis, should be reoxidized to sustain glycolysis. Pyruvate is metabolized into a variety of fermentative products such as formate, acetate, lactate, succinate, glycerol, ethanol and H_2 (Yang et al.

2014, reviewed in Yang et al. 2015). Two main enzymes catalyse the metabolization of pyruvate in *C. reinhardtii*: pyruvate formate lyase (PFL1) and pyruvate ferredoxin oxidoreductase (PFR1). PFL1, dually located in mitochondria and chloroplasts, catalyses the conversion of pyruvate and CoASH into acetyl-CoA and formate. PFR1 located exclusively in chloroplasts catalyses the conversion of pyruvate and CoASH into acetyl-CoA, CO_2 and reduced ferredoxin. The reduced ferredoxin can be reoxidized by hydrogenases, while the acetyl-CoA is either reduced into ethanol by alcohol aldehyde dehydrogenase (ADH1) or converted to acetate by a pair of acetate kinase and phosphotransacetylase (ACK-PAT) present in mitochondria and chloroplasts.

3.2.1.2 *Euglena gracilis*

Euglenids are a group of free-living, single-celled flagellates that thrive predominantly in aquatic environments. *Euglena gracilis*, a model organism among Euglenids, is a secondary photosynthetic unicellular eukaryote that arises from an endosymbiotic event between a green alga and an ancient phagotroph euglenozoan species (Gibbs 1981; Turmel et al. 2009). This freshwater protist possesses a remarkable adaptability to various environmental conditions. It can grow photoautotrophically, heterotrophically and photoheterotrophically (Buetow 1989; Ogbonna et al. 2002), and it is able to use various exogenous carbon sources such as glucose, glutamate, malate, pyruvate, lactate and ethanol (Tani and Tsumura 1989; Moreno-Sánchez and Jasso-Chávez 2003; Jasso-Chávez et al. 2005; Rodríguez-Zavala et al. 2006). Since the aim of this chapter is a description of heterotrophic metabolism in microalgae, we will focus on the main metabolites of interest that can be obtained in high amounts during both aerobic and anaerobic heterotrophic growth of *E. gracilis*.

The Storage Polysaccharide Paramylon

E. gracilis produces and accumulates paramylon as a storage polysaccharide. Paramylon is a linear glucan made only of D-glucose units (up to 700 units) linked by β-1,3-glucosidic bonds. It is accumulated in the form of discoidal granules characterized by high crystallinity (over 90%, Marchessault and Deslandes 1979; Buetow 1989). Paramylon belongs to a group of naturally occurring polysaccharides such as lentinan and fungal glucans, generally recognized as bioactive molecules (Šantek et al. 2009). In addition, paramylon can also be used as starter for production of novel biomaterials, as discussed in a recent review by (Krajčovič et al. 2015). Contrary to the starch metabolism which is well detailed in *C. reinhardtii* (Ball et al. 2011), the entire mechanism of paramylon biosynthesis in *E. gracilis* still needs to be clarified. It has been shown that paramylon synthase is a complex enzyme with a molecular mass of 670 kDa, which includes two UDP-glucose-binding peptides of 37 and 54 kDa (Bäumer et al. 2001; Tanaka et al. 2017) Enzymological analyses revealed that paramylon synthase is a uridine diphosphate (UDP) glucose-β-1,3-glucan β-3-glucosyltransferase, utilizing UDP-glucose specifically as a substrate

(Goldemberg and Marechal 1963; Marechal and Goldemberg 1964). Tanaka and collaborators recently identified a key enzyme for paramylon production [glucan synthase-like 2, EgGSL2 (Tanaka et al. 2017)] and suggested the presence of a similar mechanism for the biosynthesis of paramylon as the one for callose synthesis in *Arabidopsis thaliana* (for review see Hong et al. 2001). The control of growth conditions is fundamental for paramylon production, and it is generally recognized that the levels of paramylon in dark-grown cells are higher (9.7 g L^{-1}) than those in cells grown in photoautotrophic conditions (0.78 g L^{-1}) (Rodríguez-Zavala et al. 2010; Grimm et al. 2015). Glucose and fructose usually induce the highest heterotrophic paramylon production (Barsanti et al. 2001; Ivušić and Šantek 2015), which can reach up to 90% of total dry weight in the presence of excess carbon sources in the culture medium (Buetow 1989). In addition, many studies (for review see Šantek et al. 2009; Rodríguez-Zavala et al. 2010; Grimm et al. 2015; Ivušić and Šantek 2015) aim at replacing the synthetic medium according to Hutner (Hutner et al. 1956), which is too complex to conduct medium optimization. As a matter of fact, Šantek and co-workers showed that *E. gracilis* could be cultivated on liquid media based on potato liquor, a waste of considerable impact on the environment, with only a supplementation of glucose and some vitamins for paramylon production [reaching values about 80% of total biomass (Šantek et al. 2010, 2012)]. In 2015 Ivušić and Šantek also showed that the use of corn steep solid as a nitrogen and growth factor source, coupled with glucose as a carbon source, produces the highest bioprocess efficiency parameters during the cultivation of *E. gracilis* in a stirred tank bioreactor. It is worth to mention that Wang and collaborators have recently renewed the interest for phototrophic growth by showing that *E. gracilis* can accumulate up to $\approx$ 50% of the cell biomass as paramylon during photoautotrophic cultivation, values much higher than previously reported (Wang et al. 2018). Therefore, in view of the vast heterogeneity of the parameters and conditions tested in the aforementioned studies, further efforts are needed to optimize the culture conditions and strategies when the cultivation of *E. gracilis* is tailored to the production of paramylon as a specific target product.

Vitamin E: α-Tocopherol

Tocopherols are lipophilic antioxidants widely used as a food complement and food preservative due to their action as free-radical scavengers (McCay and King 1980). They belong to the vitamin E family, which comprises a mixture of α-, β-, γ- and δ-tocopherols and α-, β-, γ- and δ-tocotrienols. Of the four isoforms, α-tocopherol exhibits the highest biological activity thanks to its chemical structure (Kaiser et al. 1990; Kamal-Eldin and Appelqvist 1996), and interestingly, α-tocopherol represents 97% of total tocopherols produced in *E. gracilis* (Shigeoka et al. 1986). Furthermore, it has been described that *E. gracilis* was the most effective tocopherol producer among 285 microorganisms tested (Tani and Tsumura 1989). The optimization of α-tocopherol production in *E. gracilis* has been intensively investigated (Ogbonna et al. 1998, 1999; Fujita et al. 2008; Rodríguez-Zavala et al. 2010; Grimm et al. 2015). Grimm and co-authors showed that the productivity of α-tocopherol was higher in photoautotrophic cultures with respect to heterotrophic and photoheterotrophic conditions (8.6, 6.0 and 5.3 mg L^{-1}, respectively). However, the authors suggested that an economical production of α-tocopherol would be achievable in all the three cultivation modes, due to the higher biomass concentrations reached in heterotrophic and photoheterotrophic conditions. Indeed, heterotrophic cultivations of *E. gracilis* have been already reported and shown to be a suitable source for α-tocopherol production (Ogbonna et al. 1998; Rodríguez-Zavala et al. 2010). Ethanol as a carbon source in the medium led to higher tocopherol values (almost 4 mg/g DW) than those previously reported for heterotrophic cultivation of *E. gracilis* by Fujita et al. 2008 (1.2 mg/g DW). Ethanol was chosen to subject *E. gracilis* cells to a manageable oxidative stress in order to increase the production of antioxidant metabolites. Ethanol is a good non-carbohydrate carbon source for Euglenoids (Nakazawa 2017), and its concentration was reported to be 177 mmol L^{-1} with higher concentrations inhibiting cell growth (Rodríguez-Zavala et al. 2006). *E. gracilis* rapidly converts ethanol into acetaldehyde and then acetate through the sequential reactions of alcohol dehydrogenase and acetaldehyde dehydrogenase. Therefore, the ATP-dependent reaction catalysed by the enzyme acetyl-CoA synthetase transforms acetate into acetyl-CoA which enters the glyoxylate cycle. For a detailed description of the C2 metabolism and the peculiarities of the glyoxylate pathway of *E. gracilis*, we refer the reader to the recently published review of Nakazawa (2017).

Dark Anaerobic Metabolism: Wax Esters and Succinate

E. gracilis cells possess facultatively anaerobic mitochondria (Fig. 3.1). When grown in aerobic conditions, *E. gracilis* mitochondria act like mammalian mitochondria: the pyruvate obtained from the glycolytic pathway enters in the mitochondrion and is transformed in acetyl-CoA through oxidative decarboxylation reaction via pyruvate dehydrogenase (PDH—Hoffmeister et al. 2004). Then, acetyl-CoA enters in a slightly modified tricarboxylic acid cycle (TCA) (Buetow 1989) where α-ketoglutarate decarboxylase (αKGDC) and succinate semialdehyde dehydrogenase (SSDH) replace the α-ketoglutarate dehydrogenase of the classical TCA cycle. The electrons from the reducing equivalents (NADH, $FADH_2$) are transferred to oxygen through the mitochondrial respiratory chain releasing energy which is used to produce adenosine triphosphate (ATP) via oxidative phosphorylation (Fig. 3.1). On the contrary under dark oxygen-limited conditions, *Euglena* expresses the oxygen-sensitive enzyme pyruvate:NADP+ oxidoreductase (PNO) (Inui et al. 1985, 1991; Nakazawa et al. 2000), which decarboxylates pyruvate to acetyl-CoA. The latter is used as the terminal electron acceptor, and wax esters represent the end products of the anaerobic metabolism (Fig. 3.1) (Inui et al. 1982, 1983, 1984; Buetow 1989; Tucci et al. 2010; Müller et al. 2012; Ogawa et al. 2015). The wax ester synthesis has been designated as "wax ester fermentation" and allows a net gain of ATP synthesis from paramylon breakdown, thanks to a malonyl-CoA independent fatty acid synthesis. Indeed, the mitochondrial fatty acid synthetic system is acetyl-CoA dependent, meaning that acetyl-CoA is used directly as C2 donor for the fatty acid synthesis, bypassing the ATP-dependent reaction converting acetyl-CoA to malonyl-CoA (Inui et al. 1982, 1984; Schneider and Betz 1985). Acetyl-CoA acts

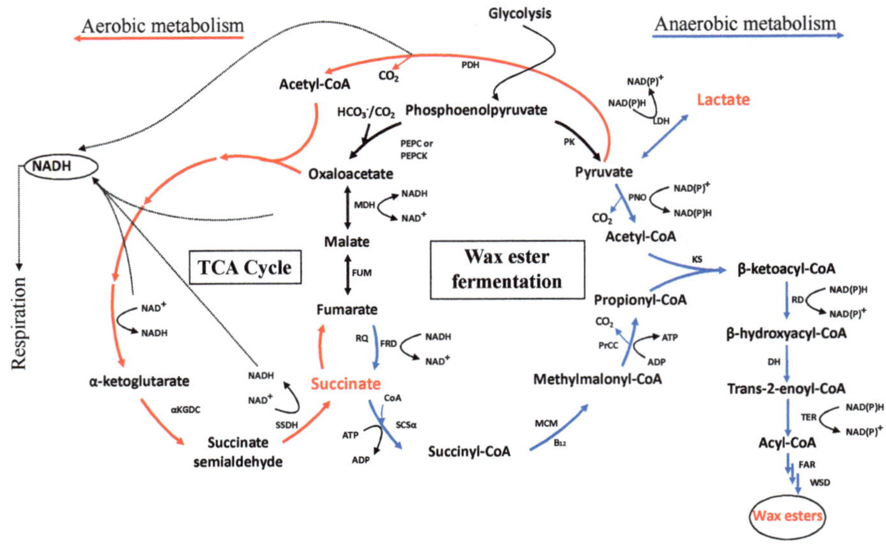

Fig. 3.1 Working model for the energy metabolism in the facultatively anaerobic mitochondrion of *E. gracilis*. This metabolic map is adapted from Tucci et al. (2010) and modified according to Müller et al. (2012) and Tomita et al. (2016). *α-KGDC* α-ketoglutarate decarboxylase (Buetow 1989), *SSDH* succinate semialdehyde dehydrogenase (Buetow 1989), *MDH* malate dehydrogenase, *FUM* fumarase, *FRD* fumarate reductase, *RQ* rhodoquinone (Hoffmeister et al. 2004), *SCSα* succinyl-CoA α-synthetase, *MCM* methylmalonyl mutase, *PrCC* propionyl-CoA carboxylase, *PDH* pyruvate dehydrogenase (Hoffmeister et al. 2004), *PNO* pyruvate:NADP1oxidoreductase (Rotte et al. 2001), *LDH* lactate dehydrogenase, *PEPC* phosphoenolpyruvate carboxylase, *PEPCK* phosphoenolpyruvate carboxykinase, *PK* pyruvate kinase, *KS* β-ketosynthase, *RD* β-ketoreductase, *DH* β-hydroxyacyl-dehydrogenase, *TER* trans-2-enoyl-CoA reductase (Hoffmeister et al. 2005), *FAR* acyl-CoA reductase, *WSD* wax ester synthase/diacylglycerol acyltransferase

both as a primer and C2 donor using NAD(P)H as an electron donor by the reverse of the β-oxidation reaction except that trans-2-enoyl-CoA reductase (TER) is used instead of acyl-CoA dehydrogenase (Fig. 3.1) (Hoffmeister et al. 2005; Inui et al. 2017). Also, propionyl-CoA—produced via the same methylmalonyl-CoA route as the one found in animal mitochondria (Schneider and Betz 1985)—is used as a starter for the synthesis of odd-numbered chain length fatty acids. A portion of the resultant acyl-CoA(s) is exported to the endoplasmic reticulum and reduced to fatty alcohol via acyl-CoA reductase (FAR—Kolattukudy 1970; Khan and Kolattukudy 1973). Finally, a wax ester synthase/diacylglycerol acyltransferase (WSD) catalyses the esterification of fatty acyl-CoA and fatty alcohol to form waxes, which are deposited into the cytosol (Teerawanichpan and Qiu 2010). Recently, Tomiyama and collaborators have characterized the WSD isoenzymes as effectively involved in the final step of wax ester formation in vivo under anaerobic conditions, identifying EgWSD2 and EgWSD5 as predominantly wax ester synthases along six possible WSD orthologs predicted by BLASTX search (Tomiyama et al. 2017). The wax ester fermentation is quite unique and an unusual phenomenon because—in contrast to classical fermentations—waxes produced from paramylon degradation are high molecular weight molecules retained in the cytosol in an insoluble crystalline form, and they are rapidly converted back to paramylon in the presence of oxygen (Inui et al. 1982, 1992; Tucci et al. 2010; Ogawa et al. 2015). A detailed description of all the biosynthetic routes to wax esters in *Euglena* has been recently published (Inui et al. 2017). Nowadays the main interest in the anaerobic metabolism of *E. gracilis* focuses on the application of wax esters for biofuel production. Tucci and co-authors have reported that some *E. gracilis* strains accumulate wax esters up to 57% of total dry weight (Tucci et al. 2010). *Euglena* wax ester consists of saturated fatty acids and alcohols with carbon chain lengths ranging from 10 to 18. The main ester is myristyl myristate (C28), composed of myristic acid (C14:0) and myristyl alcohol (C14:0), which represent the 44% of total wax ester content (Inui et al. 1983, 2017). In contrast to the most common medium-length algal fatty acids, like palmitic (C16:0) and stearic acid (C18:0), myristic acid (C14:0) possesses a low freezing point and a high cetane number (66.2)(Yanowitz et al. 2014; Tomiyama et al. 2017; Inui et al. 2017), being thus suitable as a drop-in jet fuel.

Furthermore, recently Tomita and collaborators have reported that *E. gracilis* excretes organic acids, such as succinate and lactate, when transferred from light aerobic conditions to dark anaerobic conditions (Tomita et al. 2016). Succinate levels were particularly high (up to 869.6 mg L^{-1}) when a commercial strain of *E. gracilis*—commercially cultivated by the Euglena Co., Ltd. (Yamada et al. 2016)—was cultivated under nitrogen-starved conditions prior to the incubation under dark anaerobic conditions. Succinate and lactate are typical fermentative products excreted under dark anaerobic conditions to oxidize NAD(P)H for substrate-level phosphorylation (Fig. 3.1) and are used as bulk chemicals for bioplastic production, such as the widely used plastic polybutylene succinate (Tomita et al. 2016).

3.2.1.3 *Galdieria sulphuraria*

Galdieria sulphuraria is a unicellular extremophile organism belonging to the order of Cyanidiales. This order represents a small monophyletic group—within the phylum Rhodophyta—of evolutionary anciently diverged, unicellular microalgae, located at the root of the secondary endosymbiosis (Yoon et al. 2002, 2004, 2006). Being red algae, all the Cyanidiales are characterized by the presence of only chlorophyll a as major photosynthetic pigment and of phycobiliproteins as accessory pigments. In particular, phycocyanin represents the main accessory pigment of all the Cyanidiales, which thus exhibits a blue-green colour. The members of the order Cyanidiales are known to be thermoacidophilic microorganisms inhabiting hot sulphur springs, volcanic areas and hostile environments deriving from human activities such as strips or opencast mining scattered all around the world. Those environments are characterized by temperatures up to 56 °C (Seckbach 1999); therefore, a value close to the upper limit for eukaryotic life (Rothschild and Mancinelli 2001; Pinto et al. 2003), and low pH values ranging from 0 to 4 due to the high presence of H_2S (hydrogen sulphide) that is progressively oxidized to sulphur and H_2SO_4 (Doemel and Brock 1971; Brock 1978; Bottone et al. 2018).

The members of the genus *Galdieria* are known to be the only organisms within the Cyanidiales to grow both autotrophically, performing photosynthesis, and heterotrophically in complete darkness. It has been reported that *G. sulphuraria* is able to use more than 50 different carbon sources when grown in the dark such as sugars, polyols, disaccharides, amino acids and organic acids (Rigano et al. 1976; Gross and Schnarrenberger 1995; Gross 1999; Oesterhelt et al. 1999). This metabolic versatility is not conferred by the presence of genes encoding essential enzymes for carbohydrate metabolism, since they can also be found in the genome of the obligate autotroph *Cyanidioschyzon merolae*, but by a multitude of genes encoding metabolite transporters, which allow the uptake of compounds from the environment into the cells (Weber et al. 2004; Barbier et al. 2005; Oesterhelt et al. 1999; Oesterhelt and Gross 2002). Oesterhelt and collaborators observed that the uptake system of *G. sulphuraria* for sugars and polyols is not induced by the absence of light but by the presence of a metabolizable substrate (Oesterhelt et al. 1999), and thus the general heterotrophic metabolism cannot be considered the key for the induction of the uptake system. Following the publication of the highly compact genome of G. *sulphuraria* in 2013, Schönknecht and co-authors found that it possesses a higher number of genes (5.2% of *G. sulphuraria* genes) encoding membrane transporters than in most of other eukaryotes, in agreement with previous observations (Schönknecht et al. 2013). Furthermore, the study revealed that the high metabolic flexibility and tolerance to harsh conditions of *G. sulphuraria* were facilitated by the horizontal acquisition of numerous critical genes from extremophilic bacteria. Since *G. sulphuraria* grows mainly in cryptoendolithic cell mats poor in organic compounds (Gross and Oesterhelt 1999), its extraordinary metabolic plasticity is regarded as a survival strategy that enables the cells—when light is insufficient to perform photosynthesis—to use as substrates for heterotrophic growth the compounds released by surrounding dying organisms within this mat.

From a biotechnological point of view, *G. sulphuraria* is thus interesting by two aspects, linked to its physiology and to its bioactive components content. Indeed, its ability to grow in acidic conditions and to reach high cell densities heterotrophically on a wide range of carbon sources (Gross and Schnarrenberger 1995), including by-products such as glycerol, molasses from the sugar industry or food waste hydrolysates (Schmidt et al. 2005; Graziani et al. 2013; Sloth et al. 2017), makes *G. sulphuraria* an interesting organism for mass cultivation with a significant reduction of the risk of contamination. Furthermore, the red alga is able to produce phycocyanin (see above, point 1.2.3), floridoside (2-O-α-D-galactopyranosylglycerol) and a highly branched glycogen (Martinez-Garcia and van der Maarel 2016; Martinez-Garcia et al. 2016, 2017). In the above-mentioned studies, the authors showed that *G. sulphuraria* grown heterotrophically can accumulate constitutively high amount of glycogen (up to 50% of cell dry weight) under conditions that do not limit cell growth, without differentiating between a biomass accumulating phase and a nutrient-deprived, glycogen accumulation phase. The authors also characterized this glycogen which contains the highest number of ramifications [18% of α-(1 → 6) linkages] found to date. In addition, this glycogen is also different from the one found in other organisms by its composition, consisting only in short chains, its small molecular weight and particle size (Martinez-Garcia et al. 2016). As suggested by the authors, thanks to these features, this highly branched glycogen could represent a valid alternative to the so-called highly branched glucose polymers derived from the acid and/or enzymatic treatment of starch, which found applications as osmotic agents for peritoneal dialysis or as slowly digestible carbohydrates in sport drinks (Backer and Saniez 2005; Fuertes et al. 2009). For a detailed description of the highly branched glycogen of *G. sulphuraria* and its potential applications, we refer the reader to the above-mentioned studies of Martinez-Garcia and collaborators (2016, 2017). Moreover, the same authors have reported in another study the potential of *G. sulphuraria* for production of the compatible solute floridoside (2-O-α-D-galactopyranosylglycerol) (Martinez-Garcia and van der Maarel 2016), a molecule that has been shown to possess interesting antifouling and therapeutic properties (Hellio et al. 2004; Kim et al. 2013; Ryu et al. 2015). Compatible solutes are small organic molecules synthesized by almost all red algae under high osmotic stress conditions that can be accumulated at high intracellular concentrations and do not interfere with the normal functioning of the metabolism (Ekman et al. 1991; Da Costa et al. 1998; Roberts 2005; Hagemann and Pade 2015). This glycoside also constitutes the major soluble pool of carbon fixed by photosynthesis and is a precursor for cell wall polysaccharides (Li et al. 2001, 2002).

3.2.1.4 Heterotrophic Metabolism of Microalgae Belonging to Other Phyla

Knowledge on the heterotrophic metabolism of microalgae presenting interesting valuable compounds and belonging to other phyla than those presented above is still scarce. Diatoms such as *Nitzschia alba* and *Cyclotella cryptica* which are interesting for aquaculture needs have been presented for their capabilities to grow in fermentors (Gladue and Maxey 1994), but to our knowledge, there is no report about the description of their heterotrophic metabolism. A glucose transporter has been expressed in the diatom *Phaeodactylum tricornutum* together with a Δ5-elongase to allow the heterotrophic growth and Ω-3 long-chain polyunsaturated fatty acid (Ω-3 LC-PUFAs) production. However, the highest Ω-3 LC-PUFA accumulation was observed in cultures grown under mixotrophic conditions in the presence of 1% glucose and not under heterotrophy. Nevertheless, this study demonstrates the potential for *P. tricornutum* to be developed as a viable commercial strain for Ω-3 LC-PUFAs production under mixo- and heterotrophic conditions (Hamilton et al. 2016) and is in favour for a better knowledge of heterotrophic metabolism in diatoms.

3.3 Productivities of Heterotrophic Microalgal Cultures

For selected species, volumetric productivities of microalgal heterotrophic cultures are well known for being much higher than those of phototrophic cultures. While the productivity of well-conducted phototrophic cultures is no more than a few grams

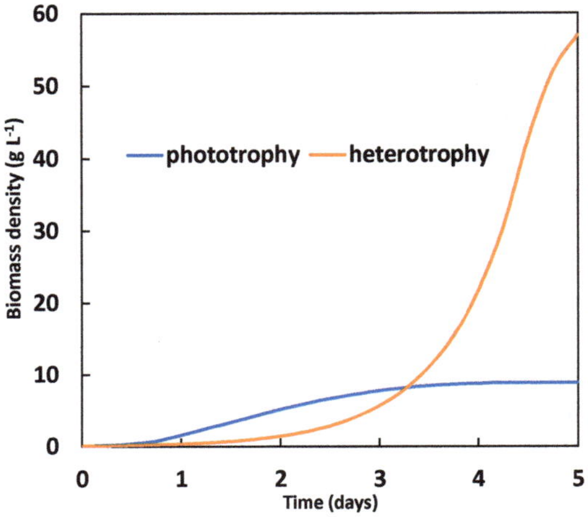

Fig. 3.2 Schematic representation of heterotrophic compared to phototrophic microalgal growth curves. The exponential growth phase is slower under heterotrophy but can be maintained for a longer time, leading to superior productivity

(DW) L^{-1} d^{-1}, productivities as high as 80 g (DW) L^{-1} d^{-1} were reported with fed-batch heterotrophic cultures (Doucha and Livansky 2012 for *Chlorella vulgaris*). The superior productivity of heterotrophic cultures is explained by the fact that the exponential growth phase can be maintained at a much higher biomass density than for phototrophic cultures (Fig. 3.2). This is because phototrophic growth becomes light-limited already at low biomass densities even in short light-path cultures [generally around 0.5 g (DW) L^{-1}, see de Marchin et al. (2015), for instance]. In contrast, the exponential phase can be maintained over a broader range of biomass densities under heterotrophy as long as proper nutrient balance is maintained and no autoinhibitory substance is produced. This feature outweighs the generally reported low maximal specific growth rates (μ_{max}) of the heterotrophic growth (on most commonly used carbon substrates such as glucose or acetate), compared to the maximum specific growth rates of phototrophic cultures. μ_{max} values of the order of 0.1 h^{-1} have been frequently reported for phototrophic cultures, whereas lower values are found under heterotrophy (Lee 2004). However, there are cases of quite high μ_{max} values for heterotrophic cultures, such as 0.11 h^{-1} for *Chlorella protothecoides* on glucose (Guidossi et al. 2017) or 0.28 h^{-1} for *Chlorella regularis* on acetate (Endo et al. 1974). Such values are of the same order than those commonly found for yeast (Ejiofor et al. 1996).

The most frequently used carbon substrate for heterotrophic growth has been glucose which, in many species, is tolerated at concentrations as high as 20 g L^{-1}. With well-balanced media, a biomass yield on glucose around 0.5 g (DW) g^{-1} has repeatedly been found (reviewed in Bumbak et al. 2011); therefore, biomass concentrations around 10 g L^{-1} are easily obtained in batch cultures. Inhibitory effects of very high glucose concentrations have been reported (Sakarika and Kornaros 2017 for *Chlorella vulgaris*). This is species-dependent; therefore, the maximal glucose concentration to be used for a particular strain has to be first determined. In order to maintain growth and reach very high biomass densities ($\approx$ 100 g L^{-1}), heterotrophic cultures can be operated in fed-batch mode (most often intermittent fed-batch). The mode of feeding control can be based on the monitoring of glucose concentration (Shi et al. 2002 for *Chlorella protothecoides*; Doucha and Livansky 2012 for *Chlorella vulgaris*) or of dissolved oxygen concentration (Schmidt et al. 2005 for *Galdieria sulphuraria*), which increases when glucose concentration becomes low.

Using acetate as a carbon source is less easy because progressive inhibitory effects were reported already at moderate concentrations (Chen and Johns 1994 for *Chlamydomonas reinhardtii*) and because acetate consumption leads to alkalinization and accumulation of Na^+ ions. Variable values of biomass yield on acetate were reported from 0.13 to 0.6 g (DW) g^{-1} (De Swaaf et al. 2003; Bumbak et al. 2011). Fed-batch experiments were performed using pH as a controlling parameter for acetic acid feeding. With this approach, high biomass concentrations were obtained (Endo et al. 1974 for *Chlorella regularis*; De Swaaf et al. 2003 for *Crypthecodinium cohnii*).

The nitrogen source used in heterotrophy is an important issue for the stability of the culture and for obtaining dense cultures. In a fed-batch mode, it has to be present at balanced concentrations in the feeding medium. Although nitrate is frequently used, its consumption leads to alkalinization because it is reduced to ammonium in the cells. Ammonium consumption leads to the opposite effect. Urea consumption entails less pH drifts and can be used in some species (mostly *Chlorella* species) [see Perez-Garcia et al. (2011) for a detailed discussion on N-source effects]. Bacterial peptone (Isleen-Hosoglu et al. 2012 for *Chlorella saccharophila*) or yeast extracts (Guidossi et al. 2017 for *Chlorella protothecoides*) were also used.

Autoinhibitory substances may be secreted and may therefore accumulate at high biomass concentrations in fed-batch mode. This was already reported in early works on *Chlorella* sp. for which a 'chlorellin' inhibitory factor was hypothesized and later ascribed to free fatty acids. One way to avoid this problem is to run perfusion cultures, in which the algal cells are retained in the culture vessel while the culture medium is replaced by fresh medium (Wen and Chen 2003).

Productive heterotrophic cultures of high biomass densities are also strongly conditioned by the efficiency of the oxygen supply (Smith et al. 2015 for *Micractinium inermum*; Bouyam et al. 2017 for *Chlorella* sp.). Respiration rates of heterotrophically grown cells are known as being much higher than for phototrophically grown ones, and this is dependent on the carbon source used (Villarejo et al. 1995 for *Chlorella vulgaris* with glucose; Gérin et al. 2014 for *Chlamydomonas reinhardtii* with acetate). On glucose, for instance, a biomass yield on glucose around 0.5 g (DW) g^{-1} implies that a significant part of the imported glucose is oxidized to CO_2 through respiration, which is likely needed to generate the ATP consumed in biosynthetic processes. Oligo- and polysaccharide synthesis from glucose, for instance, generates an important ATP demand, which can be met by glycolysis and oxidative phosphorylation. When using acetate as a substrate, ATP is also needed in the first place to generate acetyl-CoA (Perez-Garcia et al. 2011). Oxygen supply to heterotrophic cultures generally necessitates to combine strong air bubbling and mechanical stirring for efficient oxygen transfer. However, the resistance of algal cells to shear forces is strongly species-dependent depending on cell wall structure and composition (Wang and Lang 2008), and robust strains as well as appropriate mixing devices must be selected. Whereas this question has been relatively well tackled for phototrophic cultures (Tredici 2010), it remains to be assessed with appropriate analytical methods for heterotrophic cultures, considering the algal oxygen demand and the particular geometries and mixing devices of bioreactors.

Besides the above-discussed eukaryotic microalga, cyanobacteria were also considered for their capabilities of heterotrophic growth. These unicellular or filamentous photosynthetic prokaryotes were long considered as obligate phototrophs (Holm-Hansen 1968), but several species were shown to grow slowly in the dark using various carbon substrates (glucose, fructose, saccharose, etc.). A non-exhaustive list of such species with their respective carbon substrates can be found in Stebegg et al. (2019).

Specific growth rates and productivities of heterotrophic cultures of cyanobacteria are often low. For instance, *Arthrospira platensis*, the popular phototrophic phycocyanin producer, showed a μ_{max} value of only 0.008 day^{-1} and a productivity of less than 1 $g.L^{-1}$ after more than 10 days on glucose in darkness (Marquez et al. 1993). As another example of slow heterotrophic growth, *Nostoc* sp. was found to grow on glucose with a μ_{max} value of 0.009 (Tredici et al. 1988). A similar value was reported in Raboy et al. (1976) for *Plectonema boryanum* on glucose. It is noteworthy that for several species, a relatively long acclimation time (several generations) in the presence of an organic carbon substrate (in light or dark) is needed before significant growth in the dark is observed. This is well documented in the case of *Cyanothece* sp. for which a relatively fast-growing strain ($\mu_{max} = 0.030\ h^{-1}$) could be obtained on glycerol after several months (Schneegurt et al. 1997). A special case is encountered with *Synechocystis* sp. PCC 6803, which can only grow on glucose in darkness ($\mu_{max} = 0.019\ h^{-1}$) if it is exposed daily to a short pulse of light (Anderson and McIntosh 1991).

In contrast to progress already made on (eukaryotic) microalgal processes, research aimed at optimizing and upscaling heterotrophic production of cyanobacterial biomass is only starting, probably due to the modest growth results of flask-scale growth experiments. In a recent study, it was possible to increase growth performances of an *Aphanothece* sp. strain in well-controlled bioreactor experiments using starch-containing media (dos Santos et al. 2017).

3.4 Conclusions

The ability of many microalgal species to grow in complete darkness on a variety of organic carbon substrates was recognized already in the 1950s. Suitable substrates include simple sugars (mainly glucose), some small organic acids (acetate, butyrate), small polyalcohols (glycerol, mannitol) and, in some cases, ethanol. Assimilation of complex sugars (saccharose) was only occasionally reported, probably due to the absence of suitable hydrolases in microalgae. The practical interest of growing microalgae heterotrophically was recognized in the 1990s when limitations of outdoor phototrophic mass cultures became apparent.

For improvement of the system, additional screening for algal strains able to grow heterotrophically is needed. Ideal strains should present the following characteristics:

- High specific rate constant during exponential phase (low generation time)
- The capacity to reach high biomass density
  The capacity to rapidly activate heterotrophic metabolism in the presence of a suitable organic carbon substrate of low cost
- An active biosynthesis of valuable bioproducts under heterotrophic conditions
- The easiness of cultivation (absence of fouling, robustness) and harvest (for instance, by flocculation).

In addition, a good knowledge of the heterotrophic carbon metabolism of microalgae is necessary to understand adaptive responses to the absence of photosynthetic carbon assimilation in the presence of a given organic carbon substrate. Fermentors originally designed for bacterial growth should be adapted to algal growth, and upscaling to (semi)-industrial levels should be performed in order to obtain life cycle assessment and techno-economic analyses of microalgae-based compounds produced in heterotrophy.

Acknowledgements The research is funded through the ARC grant (DARKMET proposal) for Concerted Research Actions (2017), financed by the French Community of Belgium (Wallonia-Brussels Federation). CR acknowledges Fonds National de Recherche Scientifique (FNRS, Belgium, CDR J.0265.17). LD is a recipient of the Fund for Research Training in Industry and Agriculture (FRIA). FF is a research director of FNRS.

References

Anderson SL, McIntosh L (1991) Light-activated heterotrophic growth of the cyanobacterium *Synechocystis* sp. strain PCC 6803: a blue-light-requiring process. J Bacteriol 173:2761–2767

Atteia A, Adrait A, Brugière S et al (2009) A proteomic survey of *Chlamydomonas reinhardtii* mitochondria sheds new light on the metabolic plasticity of the organelle and on the nature of the alpha-proteobacterial mitochondrial ancestor. Mol Biol Evol 26(7):1533–1548

Backer D, Saniez MH (2005) Soluble highly branched glucose polymers and their method of production. US6861519 B2

Ball S, Colleoni C, Cenci U, Raj JN, Tirtiaux C (2011) The evolution of glycogen and starch metabolism in eukaryotes gives molecular clues to understand the establishment of plastid endosymbiosis. J Exp Bot 62(6):1775–1801. https://doi.org/10.1093/jxb/erq411

Barbier G, Oesterhelt C, Larson MD, Halgren RG, Wilkerson C, Garavito RM, Benning C, APM W (2005) Comparative genomics of two closely related unicellular thermo-acidophilic red algae, *Galdieria sulphuraria* and *Cyanidioschyzon merolae*, reveals the molecular basis of the metabolic flexibility of *Galdieria sulphuriaria* and significant differences in carbohydrate metabolism of both algae. Plant Physiol 137:460–474. https://doi.org/10.1104/pp.104.051169.460

Barclay WR (1992) Process for the heterotrophic production of microbial products with high concentrations of omega-3 highly unsaturated fatty acids. Adv Biores 1:40–45

Barclay WR, Meager KM, Abril JR (1994) Heterotrophic production of long chain omega-3 fatty acids utilizing algae and algae-like microorganisms. J Appl Phycol 6:123–129. https://doi.org/10.1007/BF02186066

Barsanti L, Vismara R, Passarelli V, Gualtieri PJ (2001) Paramylon (β-1,3-glucan) content in wild type and WZSL mutant of *Euglena gracilis*: effects of growth conditions. J Appl Phycol 13:59–65

Bäumer D, Preisfeld A, Ruppel HG (2001) Isolation and characterization of paramylon synthase from *Euglena gracilis* (Euglenophyceae). J Phycol 37:38–46

Bernhard K (1989) Synthetic astaxanthin. The route of a carotenoid from research to commercialisation. In: Krinsky NI, Mathews-Roth MM, Taylor RF (eds) Carotenoids: chemistry and biology. Springer, Boston, MA, pp 337–363. https://doi.org/10.1007/978-1-4613-0849-2_24

Bottone C, Camerlingo R, Miceli R, Salbitani G, Sessa G, Pirozzi G, Carfagna S (2018) Antioxidant and anti-proliferative properties of extracts from heterotrophic cultures of *Galdieria sulphuraria*. Nat Prod Res 6419:1–5. https://doi.org/10.1080/14786419.2018.1425853

Bouyam S, Choorit W, Sirisansaneeyakul S, Chisty Y (2017) Heterotrophic production of *Chlorella* sp. TISTR 8990 – biomass growth and composition under various production conditions. Biotechnol Prog 33:1589–1600

Boyle NR, Sengupta N, Morgan JA (2017) Metabolic flux analysis of heterotrophic growth in *Chlamydomonas reinhardtii*. PLoS One 12(5):e0177292. https://doi.org/10.1371/journal.pone.0177292

Brock TD (1978) The genus Cyanidium. In: Brock TD (ed) Thermophilic microorganisms and life at high temperatures. Springer, New York, pp 255–302

Buetow DE (1989) The mitochondrion. In: Buetow DE (ed) The biology of Euglena, vol IV. Academic, New York and London, pp 247–314

Bumbak F, Cook S, Zachleder V, Hauser S, Kovar K (2011) Best practices in heterotrophic high-cell-density microalgal processes: achievements, potential and possible limitations. Appl Microbiol Biotechnol 91:31–46

Chen F, Johns MR (1991) Effect of C/N ratio and aeration on the fatty acid composition of heterotrophic *Chlorella sorokiniana*. J Appl Phycol 3:203–209. https://doi.org/10.1007/BF00003578

Chen F, Johns MR (1994) Substrate inhibition of *Chlamydomonas reinhardtii* by acetate in heterotrophic culture. Process Biochem 29:245–252

Chen Y-H, Walker TH (2011) Biomass and lipid production of heterotrophic microalgae *Chlorella protothecoides* by using biodiesel-derived crude glycerol. Biotechnol Lett 33:1973. https://doi.org/10.1007/s10529-011-0672-y

Chen G-Q, Jiang Y, Chen F (2008) Salt-induced alterations in lipid composition of diatom *Nitzschia laevis* (bacillariophyceae) under heterotrophic culture condition. J Phycol 44:1309–1314. https://doi.org/10.1111/j.1529-8817.2008.00565.x

Chi Z, Pyle D, Wen Z, Frear C, Chen S (2007) A laboratory study of producing docosahexaenoic acid from biodiesel-waste glycerol by microalgal fermentation. Process Biochem 42:1537–1545. https://doi.org/10.1016/j.procbio.2007.08.008

Chisti Y (2007) Biodiesel from microalgae. Biotechnol Adv 25:294–306

Cuellar-Bermudez SP, Aguilar-Hernandez I, Cardenas-Chavez DL, Ornelas-Soto N, Romero-Ogawa MA, Parra-Saldivar R (2015) Extraction and purification of high-value metabolites from microalgae: essential lipids, astaxanthin and phycobiliproteins. Microb Biotechnol 8:190–209. https://doi.org/10.1111/1751-7915.12167

Da Costa MS, Santos H, Galinski EA (1998) An overview of the role and diversity of compatible solutes in Bacteria and Archaea. Adv Biochem Eng Biotechnol 61:117–153

de Marchin T, Erpicum M, Franck F (2015) Photosynthesis of *Scenedesmus obliquus* in outdoor open thin-layer cascade system in high and low CO_2 in Belgium. J Biotechnol 215:2–12

De Swaaf ME, de Rijk TC, Eggink G, Sijtsma L (1999) Optimisation of docosahexaenoic acid production in batch cultivations by *Crypthecodinium cohnii*. In: Osinga R, Tramper J, Burgess JG, Wijffels RH (eds) Marine bioprocess engineering. Progress in industrial microbiology. Elsevier, The Netherlands, pp 185–192. https://doi.org/10.1016/S0079-6352(99)80111-8

De Swaaf ME, Sijtsma L, Pronk JT (2003) High-cell-density fed-batch cultivation of the docosahexaenoic acid producing marine alga *Crypthecodinium cohnii*. Biotechnol Bioeng 81:666–672

Demmig-Adams B, Adams WW (2002) Antioxidants in photosynthesis and human nutrition. Science 298(5601):2149–2153. https://doi.org/10.1126/science.1078002

Doebbe A, Rupprecht J, Beckmann J, Mussgnug JH, Hallmann A, Hankamer B, Kruse O (2007) Functional integration of the HUP1 hexose symporter gene into the genome of *C. reinhardtii*: impacts on biological $H(_2)$ production. J Biotechnol 131:27–33

Doemel WN, Brock TD (1971) The physiological ecology of *Cyanidium caldarium*. J Gen Microbiol 67:17–32

dos Santos AM, Vieira KR, Sartori RB, dos Santos AM, Queiros MI, Zepka LQ, Jacob-Lopez E (2017) Heterotrophic cultivation of cyanobacteria: study of effect of exogenous sources of organic carbon, absolute amount of nutrients, and stirring speed on biomass and lipid productivity. Frontiers Bioeng Biotech 5:12

Doucha J, Livansky K (2012) Production of high-density *Chlorella* culture grown in fermenters. J Appl Phycol 24:35–43

Durante L, Hübner W, Lauersen KJ, Remacle C (2019) Characterization of the GPR1/FUN34/YaaH protein family in the green microalga Chlamydomonas suggests their role as intracellular membrane acetate channels. Plant Direct 3(6):e00148. https://doi.org/10.1002/pld3.148

Ejiofor AO, Chisty Y, Moo-Young M (1996) Culture of *Saccharomyces cerevisiae* on hydrolyzed waste cassava starch for production of baking-quality yeast. Enzym Microb Technol 18:519–525

Ekman P, Yu S, Pedersen M (1991) Effects of altered salinity, darkness and algal nutrient status on floridoside and starch content, α-galactosidase activity and agar yield of cultivated *Gracilaria sordida*. Br Phycol J 26:123–131

Endo H, Nakajima K, Chino R, Shirota M (1974) Growth characteristics and cellular components of *Chlorella regularis*, heterotrophic fast growing strain. Agr Biol Chem 38:9–18

Eriksen NT (2008) Production of phycocyanin—a pigment with applications in biology, biotechnology, foods and medicine. Appl Microbiol Biotechnol 80:1–14. https://doi.org/10.1007/s00253-008-1542-y

Ethier S, Woisard K, Vaughan D, Wen Z (2011) Continuous culture of the microalgae *Schizochytrium limacinum* on biodiesel-derived crude glycerol for producing docosahexaenoic acid. Bioresour Technol 102:88–93. https://doi.org/10.1016/j.biortech.2010.05.021

Falaise C, François C, Travers MA et al (2016) Antimicrobial compounds from eukaryotic microalgae against human pathogens and diseases in aquaculture. Mar Drugs 14(9):1–27

Fan J, Zheng L (2017) Acclimation to NaCl and light stress of heterotrophic *Chlamydomonas reinhardtii* for lipid accumulation. J Biosci Bioeng 124(3):302–308. https://doi.org/10.1016/j.jbiosc.2017.04.009

Fuertes P, Roturier JM, Petitjean C (2009) Soluble highly branched glucose polymers. US7612198 B2

Fujita T, Aoyagi H, Ogbonna JC, Tanaka H (2008) Effect of mixed organic substrate on α-tocopherol production by *Euglena gracilis* in photoheterotrophic culture. Appl Microbiol Biotechnol 79:371–378

Gérin S, Mathy G, Franck F (2014) Modeling the dependence of respiration and photosynthesis upon light, acetate, carbon dioxide, nitrate and ammonium in *Chlamydomonas reinhardtii* using design of experiments and multiple regression. BMC Systems Biol 8:96

Gibbs SP (1981) The chloroplasts of some algal groups may have evolved from endosymbiotic eukaryotic algae. Ann N Y Acad Sci 81:193–208

Gladue R, Maxey JE (1994) Microalgal feeds for aquaculture. J Appl Phycol 6:131–141

Goldemberg SH, Marechal LR (1963) Biosynthesis of paramylon in *Euglena gracilis*. Biochim Biophys Acta 71:743–744

Gong M, Bassi A (2016) Carotenoids from microalgae: a review of recent developments. Biotechnol Adv 34:1396–1412. https://doi.org/10.1016/J.BIOTECHADV.2016.10.005

Graziani G, Schiavo S, Nicolai MA, Buono S, Fogliano V, Pinto G, Pollio A (2013) Microalgae as human food: chemical and nutritional characteristics of the thermo-acidophilic microalga *Galdieria sulphuraria*. Food Funct 4(1):144–152. https://doi.org/10.1039/c2fo30198a

Grimm P, Risse JM, Cholewa D et al (2015) Applicability of *Euglena gracilis* for biorefineries demonstrated by the production of α-tocopherol and paramylon followed by anaerobic digestion. J Biotechnol 215:72–79

Gross W (1999) Revision of comparative traits for the acid- and thermo- philic red algae *Cyanidium* and *Galdieria*. In: Seckbach J (ed) Enigmatic microorganisms and life in extreme environments. Kluwer Academic Publishers, London, pp 439–446

Gross W, Oesterhelt C (1999) Ecophysiological studies on the red alga *Galdieria sulphuraria* isolated from Southwest Iceland. Plant Biol 1:694–700

Gross W, Schnarrenberger C (1995) Heterotrophic growth of two strains of the acido-thermophilic red alga *Galdieria sulphuraria*. Plant Cell Physiol 36(4):633–638. https://doi.org/10.1093/oxfordjournals.pcp.a078803

Guedes AC, Amaro HM, Malcata FX (2011) Microalgae as sources of carotenoids. Mar Drugs 9(4):625–644. https://doi.org/10.3390/md9040625

Guidossi T, Marison I, Devery R, Gaffney D, Forde C (2017) Characterization and optimization of a fermentation process for the production of high cell densities and lipids using heterotrophic cultivation of *Chlorella protothecoides*. Ind Biotechnol 13:253–259

Hagemann M, Pade N (2015) Heterosides – compatible solutes occurring in prokaryotic and eukaryotic phototrophs. Plant Biol 17:927–934

Hamilton ML, Powers S, Napier JA, Sayanova O (2016) Heterotrophic production of omega-3 long-chain polyunsaturated fatty acids by trophically converted marine diatom *Phaeodactylum tricornutum*. Mar Drugs 14:53. https://doi.org/10.3390/md14030053

Hellio C, Simon-Colin C, Clare A, Deslandes E (2004) Isethionic acid and floridoside isolated from the red alga *Grateloupia turuturu* inhibit settlement of *Balanus amphitrite* crypid larvae. Biofouling 20:139–145

Hicks GR, Hironaka CM, Dauvillee D, Funke RP, D'Hulst C, Waffenschmidt S, Ball SG (2001) When simpler is better. Unicellular green algae for discovering new genes and functions in carbohydrate metabolism. Plant Physiol 127(4):1334–1338

Higuera-Ciapara I, Félix-Valenzuela L, Goycoolea FM (2006) Astaxanthin: a review of its chemistry and applications. Crit Rev Food Sci Nutr 46:185–196. https://doi.org/10.1080/10408690590957188

Hoffmeister M, van der Klei A, Rotte C et al (2004) *Euglena gracilis* rhodoquinone:ubiquinone ratio and mitochondrial proteome differ under aerobic and anaerobic conditions. J Biol Chem 279:22422–22429

Hoffmeister M, Piotrowski M, Nowitzki U, Martin W (2005) Mitochondrial trans-2-enoyl-CoA reductase of wax ester fermentation from *Euglena gracilis* defines a new family of enzymes involved in lipid synthesis. J Biol Chem 280:4329–4338

Holm-Hansen O (1968) Ecology, physiology and biochemistry of blue-green algae. Ann Rev Microbiol 22:47–70

Hong Z, Zhang Z, Olson JM, Verma DP (2001) A novel UDP-glucose transferase is part of the callose synthase complex and interacts with phragmoplastin at the forming cell plate. Plant Cell 13:769–779

Hu H, Gao K (2003) Optimization of growth and fatty acid composition of a unicellular marine picoplankton, *Nannochloropsis* sp., with enriched carbon sources. Biotechnol Lett 25:421–425. https://doi.org/10.1023/A:1022489108980

Hu J, Nagarajan D, Zhang Q, Chang JS, Lee DJ (2018) Heterotrophic cultivation of microalgae for pigment production: a review. Biotechnol Adv 36:54–67. https://doi.org/10.1016/J.BIOTECHADV.2017.09.009

Hutner SH, Bach MK, Ross GTM (1956) A sugar- containing basal medium for vitamin B-12 assay with *Euglena*: application to body fluids. J Protozool 3:101–105

Im C-S, Eberhard S, Huang K, Beck CF, Grossman AR (2006) Phototropin involvement in the expression of genes encoding chlorophyll and carotenoid biosynthesis enzymes and LHC apoproteins in *Chlamydomonas reinhardtii*. Plant J 48:1–16. https://doi.org/10.1111/j.1365-313X.2006.02852.x

Inui H, Miyatake K, Nakano Y, Kitaoka S (1982) Wax ester fermentation in *Euglena gracilis*. FEBS Lett 150:89–93

Inui H, Miyatake K, Nakano Y, Kitaoka S (1983) Production and composition of wax esters by fermentation of *Euglena gracilis*. Agric Biol Chem 47:2669–2671

Inui H, Miyatake K, Nakano Y, Kitaoka S (1984) Fatty acid synthesis in mitochondria of *Euglena gracilis*. Eur J Biochem 142:121–126

Inui H, Miyatake K, Nakano Y, Kitaoka S (1985) The physiological role of oxygen-sensitive pyruvate dehydrogenase in mitochondrial fatty acid synthesis in *Euglena gracilis*. Arch Biochem Biophys 237:423–429

Inui H, Yamaji R, Saidoh H, Miyatake K, Kitaoka S (1991) Pyruvate:NADP oxidoreductase from *Euglena gracilis*: limited proteolysis of the enzyme with trypsin. Arch Biochem Biophys 286:270–276

Inui H, Miyatake K, Nakano Y, Kitaoka S (1992) Synthesis of reserved polysaccharide from wax esters accumulated as the result of anaerobic energy generation in *Euglena gracilis* returned from anaerobic to aerobic conditions. Int J Biochem 24:799–803

Inui H, Ishikawa T, Tamoi M (2017) Wax esters fermentation and its application for biofuel production. In: Schwartzbach S, Shingeoka S (eds) Euglena: biochemistry, cell and molecular biology. Springer, Cham, pp 269–283

Ip PF, Chen F (2005a) Production of astaxanthin by the green microalga *Chlorella zofingiensis* in the dark. Process Biochem 40:733–738. https://doi.org/10.1016/j.procbio.2004.01.039

Ip PF, Chen F (2005b) Employment of reactive oxygen species to enhance astaxanthin formation in *Chlorella zofingiensis* in heterotrophic culture. Process Biochem 40:3491–3496. https://doi.org/10.1016/J.PROCBIO.2005.02.014

Ishikawa E, Sansawa H, Abe H (2004) Isolation and characterization of a *Chlorella* mutant producing high amounts of chlorophyll and carotenoids. J Appl Phycol 16:385–393. https://doi.org/10.1023/B:JAPH.0000047949.31288.33

Isleen-Hosoglu M, Gultepe I, Elibol M (2012) Optimization of carbon and nitrogen sources for biomass and lipid production by *Chlorella saccharophila* under heterotrophic conditions and development of Nile red fluorescence based method for quantification of its neutral lipid content. Biochem Eng J 61:11–19

Ivušić F, Šantek B (2015) Optimization of complex medium composition for heterotrophic cultivation of *Euglena gracilis* and paramylon production. Bioprocess Biosyst Eng 38(6):1103–1112. https://doi.org/10.1007/s00449-015-1353-3

Jasso-Chávez R, Garcáa-Cano I, Marin-Hernández A, Mendoza-Cózatl D, Rendon JL, Moreno-Sánchez R (2005) The bacterial-like lactate shuttle components from heterotrophic *Euglena gracilis*. Biochim Biophys Acta 1709:181–190

Jia Z, Liu Y, Daroch M, Geng S, Cheng JJ (2014) Screening, growth medium optimisation and heterotrophic cultivation of microalgae for biodiesel production. Appl Biochem Biotechnol 173:1667–1679. https://doi.org/10.1007/s12010-014-0954-7

Jinkerson RE, Jonikas MC (2015) Molecular techniques to interrogate and edit the *Chlamydomonas* nuclear genome. Plant J 82(3):393–412

Kaiser S, Di Mascio P, Murphy ME (1990) Physical and chemical scavenging of singlet molecular oxygen by tocopherols. Arch Biochem Biophys 277:101–108

Kamal-Eldin A, Appelqvist LA (1996) The chemistry and antioxidant properties of tocopherols and tocotrienols. Lipids 31:671–701

Khan AA, Kolattukudy PE (1973) Control of synthesis and distribution of acyl moieties in etiolated *Euglena gracilis*. Biochemistry 12:1939–1948

Kim MJ, Li YX, Dewapriya P, Ryu BM, Kim SK (2013) Floridoside suppresses pro-inflamatory reponses by blocking MAPK signaling in activated microglia. BMB Rep 46:398–403

Kobayashi M (2003) Astaxanthin biosynthesis enhanced by reactive oxygen species in the green alga *Haematococcus pluvialis*. Biotechnol Bioprocess Eng 8:322. https://doi.org/10.1007/BF02949275

Kobayashi M, Kurimura Y, Tsuji Y (1997) Light-independent, astaxanthin production by the green microalga *Haematococcus pluvialis* under salt stress. Biotechnol Lett 19:507–509. https://doi.org/10.1023/A:1018372900649

Kolattukudy PE (1970) Reduction of fatty acids to alcohols by cell-free preparations of *Euglena gracilis*. Biochemistry 9:1095–1102

Koller M, Muhr A, Braunegg G (2014) Microalgae as versatile cellular factories for valued products. Algal Res 6:52–63. https://doi.org/10.1016/J.ALGAL.2014.09.002

Krajčovič J, Vesteg M, Schwartzbach SD (2015) Euglenoid flagellates: a multi- faceted biotechnology platform. J Biotechnol 202:135–145

Kuddus M, Singh P, Thomas G, Al-Hazimi A (2013) Recent developments in production and biotechnological applications of C-phycocyanin. Biomed Res Int 2013:491–512

Kyle DJ, Gladue R (1993) Eicosapentaenoic acids and methods for their production. US Patent 5,244,921

Kyle DJ, Boswell KDB, Gladue RM, Reeb SE (1992) Designer oils from microalgae as nutritional supplements. In: Bills DD, Kung S-D (eds) Biotechnology and nutrition. Newnes, Boston, pp 451–468. https://doi.org/10.1016/B978-0-7506-9259-5.50029-7

Lauersen KJ, Willamme R, Coosemans N, Joris M, Kruse O, Remacle C (2016) Peroxisomal microbodies are at the crossroads of acetate assimilation in the green microalga *Chlamydomonas reinhardtii*. Algal Res 16:266–274

Lee YK (2004) Heterotrophic carbon nutrition. In: Richmond A (ed) Handbook of microalgal culture. Blackwell, Oxford, pp 116–124

Li SY, Lellouche JP, Shabtai Y, Arad S (2001) Fixed carbon partitioning in the red microalga *Porphyridium sp.* (Rhodophyta). J Phycol 37:289–297

Li SY, Shabtai Y, Arad S (2002) Floridoside as a carbon precursor for the synthesis of cell wall polysaccharide in the red microalga *Porphyridium sp.* (Rhodophyta). J Phycol 38:931–938

Li X, Xu H, Wu Q (2007) Large-scale biodiesel production from microalga *Chlorella protothecoides* through heterotrophic cultivation in bioreactors. Biotechnol Bioeng 98:764–771. https://doi.org/10.1002/bit.21489

Liang Y, Sarkany N, Cui Y (2009) Biomass and lipid productivities of *Chlorella vulgaris* under autotrophic, heterotrophic and mixotrophic growth conditions. Biotechnol Lett 31:1043–1049. https://doi.org/10.1007/s10529-009-9975-7

Lim KC, Yusoff FM, Shariff M, Kamarudin MS (2018) Astaxanthin as feed supplement in aquatic animals. Rev Aquac 10:738–773. https://doi.org/10.1111/raq.12200

Lin JH, Lee DJ, Chang JS (2015) Lutein production from biomass: marigold flowers versus microalgae. Bioresour Technol 184:421–428. https://doi.org/10.1016/j.biortech.2014.09.099

Liu J, Huang J, Sun Z, Zhong Y, Jiang Y, Chen F (2011) Differential lipid and fatty acid profiles of photoautotrophic and heterotrophic *Chlorella zofingiensis*: assessment of algal oils for biodiesel production. Bioresour Technol 102:106–110. https://doi.org/10.1016/j.biortech.2010.06.017

Liu J, Huang J, Jiang Y, Chen F (2012) Molasses-based growth and production of oil and astaxanthin by *Chlorella zofingiensis*. Bioresour Technol 107:393–398. https://doi.org/10.1016/J.BIORTECH.2011.12.047

Liu J, Sun Z, Zhong Y, Gerken H, Huang J, Chen F (2013) Utilization of cane molasses towards cost-saving astaxanthin production by a *Chlorella zofingiensis* mutant. J Appl Phycol 25:1447–1456. https://doi.org/10.1007/s10811-013-9974-x

Liu J, Sun Z, Gerken H, Liu Z, Jiang Y, Chen F (2014a) *Chlorella zofingiensis* as an alternative microalgal producer of astaxanthin: biology and industrial potential. Mar Drugs 12:3487–3515. https://doi.org/10.3390/md12063487

Liu J, Sun Z, Chen F (2014b) Chapter 6 - heterotrophic production of algal oils. In: Pandey A, Lee D-J, Chisti Y, Soccol CR (eds) Biofuels from algae. Elsevier, Amsterdam, pp 111–142. https://doi.org/10.1016/B978-0-444-59558-4.00006-1

Lowrey J, Brooks MS, McGinn PJ (2015) Heterotrophic and mixotrophic cultivation of microalgae for biodiesel production in agricultural wastewaters and associated challenges—a critical review. J Appl Phycol 27:1485–1498. https://doi.org/10.1007/s10811-014-0459-3

Ma RYN, Chen F (2001) Enhanced production of free trans-astaxanthin by oxidative stress in the cultures of the green microalga *Chlorococcum sp.* Process Biochem 36:1175–1179. https://doi.org/10.1016/S0032-9592(01)00157-1

Marchessault RH, Deslandes Y (1979) Fine structure of (1→3)-β-d-glucans: curdlan and paramylon. Carbohydr Res 75:231–242

Marechal LR, Goldemberg SH (1964) Uridine diphosphate glucose-beta-1,3-glucan beta-3-glucosyltransferase from *Euglena gracilis*. J Biol Chem 239:3163–3167

Markou G, Nerantzis E (2013) Microalgae for high-value compounds and biofuels production: a review with focus on cultivation under stress conditions. Biotechnol Adv 31:1532–1542. https://doi.org/10.1016/J.BIOTECHADV.2013.07.011

Marquez FJ, Sasaki KS, Kakizono T, Nishio N, Nagai S (1993) Growth characteristics of *Spirulina platensis* in mixotrophic and heterotrophic conditions. J Ferment Bioeng 76:408–410

Martinez-Garcia M, van der Maarel MJEC (2016) Floridoside production by the red microalga *Galdieria sulphuraria* under different conditions of growth and osmotic stress. AMB Expr 6 (1):1–7. https://doi.org/10.1186/s13568-016-0244-6

Martinez-Garcia M, Stuart MCA, van der Maarel MJEC (2016) Characterization of the highly branched glycogen from the thermoacidophilic red microalga *Galdieria sulphuraria* and comparison with other glycogens. Int J Biol Macromol 89:12–18. https://doi.org/10.1016/j.ijbiomac.2016.04.051

Martinez-Garcia M, Kormpa A, van der Maarel MJEC (2017) The glycogen of *Galdieria sulphuraria* as alternative to starch for the production of slowly digestible and resistant glucose polymers. Carbohydr Polym 169:75–82. https://doi.org/10.1016/j.carbpol.2017.04.004

Mashhadinejad A, Zamani H, Sarmad J (2016) Effect of growth conditions and extraction solvents on enhancement of antimicrobial activity of the microalgae *Chlorella vulgaris*. Pharm Biomed Res 2(4):65–73

Massoz M, Cardol P, Gonzalez-Halphen D, Remacle C (2017) Mitochondrial bioenergetics pathways in *Chlamydomonas*. In: Hippler M (ed) *Chlamydomonas*: molecular genetics and physiology, microbiology monographs. Springer, Cham

Mata TM, Martins AA, Caetano NS (2010) Microalgae for biodiesel production and other applications: a review. Renew Sust Energ Rev 14:217–232. https://doi.org/10.1016/j.rser.2009.07.020

Maul JE, Lilly JW, Cui L, dePamphilis CW, Miller W, Harris EH, Stern DB (2002) The *Chlamydomonas reinhardtii* plastid chromosome: islands of genes in a sea of repeats. Plant Cell 14(11):2659–2679

McCay PB, King MM (1980) Vitamin E: its role as biologic free radical scavenger and its relationship to the microsomal mixed-function oxidase system. In: Machin LJ (ed) Vitamin E: a comprehensive treatise. Marcel Dekker, New York, pp 289–318

McFadden GI (2014) Origin and evolution of plastids and photosynthesis in eukaryotes. Cold Spring Harb Perspect Biol 6:a016105

Merchant SS, Prochnik SE, Vallon O et al (2007) The *Chlamydomonas* genome reveals the evolution of key animal and plant functions. Science 318:245–252. https://doi.org/10.1126/science.1143609

Miao X, Wu Q (2006) Biodiesel production from heterotrophic microalgal oil. Bioresour Technol 97:841–846. https://doi.org/10.1016/j.biortech.2005.04.008

Minhas AK, Hodgson P, Barrow CJ, Adholeya A (2016) A review on the assessment of stress conditions for simultaneous production of microalgal lipids and carotenoids. Front Microbiol 7:546

Moon M, Mishra SK, Kim CW, Suh WI, Min SP, Yang JW (2014) Isolation and characterization of thermostable phycocyanin from *Galdieria sulphuraria*. Korean J Chem Eng 31:490–495

Morales-Sánchez D, Tinoco-Valencia R, Kyndt J, Martinez A (2013) Heterotrophic growth of *Neochloris oleoabundans* using glucose as a carbon source. Biotechnol Biofuels 6:100. https://doi.org/10.1186/1754-6834-6-100

Morales-Sánchez D, Tinoco-Valencia R, Caro-Bermúdez MA, Martinez A (2014) Culturing *Neochloris oleoabundans* microalga in a nitrogen-limited, heterotrophic fed-batch system to enhance lipid and carbohydrate accumulation. Algal Res 5:61–69. https://doi.org/10.1016/j.algal.2014.05.006

Moreno-Sánchez R, Jasso-Chávez R (2003) Cytosol – mitochondria transfer of reducing equivalents by a lactate shuttle in heterotrophic *Euglena*. Eur J Biochem 270(24):4942–4951. https://doi.org/10.1046/j.1432-1033.2003.03896.x

Müller M, Mentel M, Van Hellemond JJ et al (2012) Biochemistry and evolution of anaerobic energy metabolism in eukaryotes. Microbiol Mol Biol Rev 76(2):444–495. https://doi.org/10.1128/MMBR.05024-11

Nakazawa M (2017) C2 metabolism in Euglena. In: Schwartzbach S, Shingeoka S (eds) Euglena: biochemistry, cell and molecular biology. Springer, Cham, pp 39–45

Nakazawa M, Inui H, Yamaji R, Yamamoto T, Takenaka S, Ueda M, Nakano Y, Miyatake K (2000) The origin of pyruvate:NADP+ oxidoreductase in mitochondria of *Euglena gracilis*. FEBS Lett 479:155–156

Newman DJ, Cragg GM (2007) Natural products as sources of new drugs over the last 25 years. J Nat Prod 70:461–477

O'Grady J, Morgan JA (2011) Heterotrophic growth and lipid production of *Chlorella protothecoides* on glycerol. Bioprocess Biosyst Eng 34:121–125. https://doi.org/10.1007/s00449-010-0474-y

Oesterhelt C, Gross W (2002) Different sugar kinases are involved in the sugar sensing of *Galdieria sulphuraria*. Plant Physiol 128(1):291–299. https://doi.org/10.1104/pp.010553

Oesterhelt C, Schnarrenberger C, Gross W (1999) Characterization of a sugar/polyol uptake system in the red alga *Galdieria sulphuraria*. Eur J Phycol 34:271–277

Ogawa T, Kimura A, Sakuyama H, Tamoi M, Ishikawa T, Shigeoka S (2015) Identification and characterization of cytosolic fructose-1,6-bisphosphatase in *Euglena gracilis*. Biosci Biotechnol Biochem 79:1957–1964

Ogbonna JC, Tomiyama S, Tanaka H (1998) Heterotrophic cultivation of *Euglena gracilis Z* for efficient production of α-tocopherol. J Appl Phycol 10:67–74

Ogbonna JC, Tomiyama S, Tanaka H (1999) Production of α-tocopherol by sequential heterotrophic–photoautotrophic cultivation of *Euglena gracilis*. J Biotechnol 70:213–221

Ogbonna JC, Ichige E, Tanaka H (2002) Interactions between photoautotrophic and heterotrophic metabolism in photoheterotrophic cultures of *Euglena gracilis*. Appl Microbiol Biotechnol 58:532–538

Orosa M, Valero JF, Herrero C, Abalde J (2001) Comparison of the accumulation of astaxanthin in *Haematococcus pluvialis* and other green microalgae under N-starvation and high light conditions. Biotechnol Lett 23:1079–1085. https://doi.org/10.1023/A:1010510508384

Pahl SL, Lewis DM, Chen F, King KD (2010) Heterotrophic growth and nutritional aspects of the diatom *Cyclotella cryptica* (Bacillariophyceae): effect of some environmental factors. J Biosci Bioeng 109:235–239. https://doi.org/10.1016/j.jbiosc.2009.08.480

Pahl SL, Lewis DM, King KD, Chen F (2012) Heterotrophic growth and nutritional aspects of the diatom *Cyclotella cryptica* (Bacillariophyceae): effect of nitrogen source and concentration. J Appl Phycol 24:301–307. https://doi.org/10.1007/s10811-011-9680-5

Perez-Garcia O, Escalante FM, de-Bashan LE, Bashan Y (2011) Heterotrophic culture of microalgae: metabolism and potential products. Water Res 45(1):11–36

Piccaglia R, Marotti M, Grandi S (1998) Lutein and lutein ester content in different types of *Tagetes patula* and *T. erecta*. Ind Crop Prod 8:45–51. https://doi.org/10.1016/S0926-6690(97)10005-X

Pinto G, Albertano P, Ciniglia C, Cozzolino S, Pollio A, Yoon HS, Bhattacharya D (2003) Comparative approaches to the taxonomy of the genus *Galdieria Merola* (Cyanidiales, Rhodophyta). Cryptogamie Algol 24:13–32

Plancke C, Vigeolas H, Höhner R et al (2014) Lack of isocitrate lyase in *Chlamydomonas* leads to changes in carbon metabolism and in the response to oxidative stress under mixotrophic growth. Plant J 77(3):404–417

Pratt R, Daniels TC, Eiler JJ et al (1944) Chlorellin, an antibacterial substance from *Chlorella*. Science 99(2574):351–352

Raboy B, Padan E, Shilo M (1976) Heterotrophic capacities of *Plectonema boryanum*. Arch Microbiol 110:77–85

Rajesh K, Rohit MV, Venkata Mohan S (2017) Microalgae-based carotenoids production. Algal Green Chem. pp 139–147. https://doi.org/10.1016/B978-0-444-63784-0.00007-2

Ratledge C (2004) Fatty acid biosynthesis in microorganisms being used for single cell oil production. Biochimie 86:807–815. https://doi.org/10.1016/j.biochi.2004.09.017

Rhie G, Beale SI (1994) Regulation of heme oxygenase activity in *Cyanidium caldarium* by light, glucose, and phycobilin precursors. J Biol Chem 269:9620–9626

Rigano C, Fuggi A, Rigano VDM, Aliotta G (1976) Studies on utilization of 2-ketoglutarate, glutamate and other amino acids by the unicellular alga *Cyanidium caldarium*. Arch Microbiol 107:133–138

Roberts MF (2005) Organic compatible solutes of halotolerant and halophilic microorganisms. Saline Syst 1:5

Rochaix JD (2002) *Chlamydomonas*, a model system for studying the assembly and dynamics of photosynthetic complexes. FEBS Lett 529:34–38

Rodríguez-Zavala JS, Ortiz-Cruz MA, Moreno-Sánchez R (2006) Characterization of an aldehyde dehydrogenase from *Euglena gracilis*. J Eukaryot Microbiol 53:36–42

Rodríguez-Zavala JS, Ortiz-Cruz MA, Mendoza-Hernández G, Moreno-Sánchez R (2010) Increased synthesis of α-tocopherol, paramylon and tyrosine by *Euglena gracilis* under conditions of high biomass production. J Appl Microbiol 109:2160–2171

Rothschild LJ, Mancinelli RL (2001) Life in extreme environments. Nature 409:1092–1101

Rotte C, Stejskal F, Zhu G, Keithly JS, Martin W (2001) Pyruvate:NADP+ oxidoreductase from the mitochondrion of *Euglena gracilis* and from the apicomplexan *Cryptosporidium parvum*: a biochemical relic linking pyruvate metabolism in mitochondriate and amitochondriate protists. Mol Biol Evol 18:710–720

Ryu BM, Li YX, Kang KH, Kim SK, Kim DG (2015) Floridoside from *Laurencia undulata* promotes osteogenic differentiation in murine bone marrow mesenchymal cells. J Funct Foods 19:505–511

Sajjadi B, Chen W-Y, Raman AAA, Ibrahim S (2018) Microalgae lipid and biomass for biofuel production: a comprehensive review on lipid enhancement strategies and their effects on fatty acid composition. Renew Sust Energ Rev 97:200–232. https://doi.org/10.1016/j.rser.2018.07.050

Sakarika M, Kornaros M (2017) Kinetics of growth and lipids accumulation in *Chlorella vulgaris* during batch heterotrophic cultivation: effect of different nutrient limitation strategies. Bioresour Technol 243:356–365

Salinas T, Larosa V, Cardol P, Maréchal-Drouard L, Remacle C (2014) Respiratory-deficient mutants of the unicellular green alga *Chlamydomonas*: a review. Biochimie 100:207–218

Šantek B, Felski M, Friehs K, Lotz M, Flaschel E (2009) Production of paramylon, a β-1,3-glucan, by heterotrophic cultivation of *Euglena gracilis* on a synthetic medium. Eng Life Sci 9:23–28

Šantek B, Felski M, Friehs K, Lotz M, Flaschel E (2010) Production of paramylon, a β-1,3-glucan, by heterotrophic cultivation of *Euglena gracilis* on potato liquor. Eng Life Sci 10:165–170

Šantek B, Friehs K, Lotz M, Flaschel E (2012) Production of paramylon, a β-1,3-glucan, by heterotrophic growth of *Euglena gracilis* on potato liquor in fed-batch and repeated-batch mode of cultivation. Eng Life Sci 12:89–94

Sarada R, Tripathi U, Ravishankar G (2002) Influence of stress on astaxanthin production in *Haematococcus pluvialis* grown under different culture conditions. Process Biochem 37:623–627. https://doi.org/10.1016/S0032-9592(01)00246-1

Schmidt RA, Wiebe MG, Eriksen NT (2005) Heterotrophic high cell-density fed-batch cultures of the phycocyanin-producing red alga *Galdieria sulphuraria*. Biotechnol Bioeng 90:77–84

Schneegurt MA, Sherman DM, Sherman LA (1997) Growth, physiology and ultrastructure of a diazotrophic cyanobacterium, *Cyanothece* sp. strain ATCC 51142 in mixotrophic and chemoheterotrophic cultures. J Phycol 33:632–642

Schneider T, Betz A (1985) Wax ester fermentation in *Euglena gracilis* T. Factors favouring the synthesis of odd-numbered fatty acids and alcohols. Planta 166:67–73

Schönknecht G, Chen WH, Ternes CM et al (2013) Gene transfer from bacteria and archaea facilitated evolution of an extremophilic eukaryote. Science 339:1207–1209

Seckbach J (1999) The Cyanidiophyceae: hot spring acidophilic algae. In: Seckbach J (ed) Enigmatic microorganisms and life in extreme environments. Springer, Netherlands, pp 425–435

Shen Y, Yuan W, Pei Z, Mao E (2010) Heterotrophic culture of *Chlorella protothecoides* in various nitrogen sources for lipid production. Appl Biochem Biotechnol 160:1674–1684. https://doi.org/10.1007/s12010-009-8659-z

Shen XF, Liu JJ, Chu FF, Lam PKS, Zeng RJ (2015) Enhancement of FAME productivity of *Scenedesmus obliquus* by combining nitrogen deficiency with sufficient phosphorus supply in heterotrophic cultivation. Appl Energy 158:348–354. https://doi.org/10.1016/j.apenergy.2015.08.057

Shene C, Asenjo JA, Chisti Y (2017) Metabolic modelling and simulation of the light and dark metabolism of *Chlamydomonas reinhardtii*. Plant J 96(5):1076–1088. https://doi.org/10.1111/tpj.14078

Shi XM, Chen F (1999) Production and rapid extraction of lutein and the other lipid-soluble pigments from *Chlorella protothecoides* grown under heterotrophic and mixotrophic conditions. Mol Nutr Food Res 43(2):109–113. https://doi.org/10.1002/(SICI)1521-3803(19990301)43:2<109::AID-FOOD109>3.0.CO;2-K

Shi XM, Chen F (2002) High-yield production of lutein by the green microalga *Chlorella protothecoides* in heterotrophic fed-batch culture. Biotechnol Prog 18:723–727

Shi XM, Chen F, Yuan JP, Chen H (1997) Heterotrophic production of lutein by selected *Chlorella* strains. J Appl Phycol 9:445–450. https://doi.org/10.1023/A:1007938215655

Shi X-M, Liu H-J, Zhang X-W, Chen F (1999) Production of biomass and lutein by *Chlorella protothecoides* at various glucose concentrations in heterotrophic cultures. Process Biochem 34:341–347. https://doi.org/10.1016/S0032-9592(98)00101-0

Shi XM, Zhang XW, Chen F (2000) Heterotrophic production of biomass and lutein by *Chlorella protothecoides* on various nitrogen sources. Enzym Microb Technol 27:312–318. https://doi.org/10.1016/S0141-0229(00)00208-8

Shi XM, Jiang Y, Chen F (2002) High-yield production of lutein by the green microalga *Chlorella protothecoides* in heterotrophic fed-batch culture. Biotechnol Prog 18:723–727

Shi X, Wu Z, Chen F (2006) Kinetic modeling of lutein production by heterotrophic *Chlorella* at various pH and temperatures. Mol Nutr Food Res 50:763–768. https://doi.org/10.1002/mnfr.200600037

Shigeoka S, Onishi T, Nakano Y, Kitaoka S (1986) The contents and subcellular distribution of tocopherols in *Euglena gracilis*. Agric Biol Chem 50:1063–1065

Sijtsma L, Anderson AJ, Ratledge C (2010) Alternative carbon sources for heterotrophic production of Docosahexaenoic acid by the marine alga *Crypthecodinium cohnii*. In: Cohen Z, Ratledge C (eds) Single cell oils, 2nd edn. AOCS, Champaign, pp 131–149. https://doi.org/10.1016/B978-1-893997-73-8.50011-6

Sloth JK, Wiebe MG, Eriksen NT (2006) Accumulation of phycocyanin in heterotrophic and mixotrophic cultures of the acidophilic red alga *Galdieria sulphuraria*. Enzym Microb Technol 38(1–2):168–175. https://doi.org/10.1016/j.enzmictec.2005.05.010

Sloth JK, Jensen HC, Pleissner D, Eriksen NT (2017) Growth and phycocyanin synthesis in the heterotrophic microalga *Galdieria sulphuraria* on substrates made of food waste from restaurants and bakeries. Bioresour Technol 238:296–305. https://doi.org/10.1016/j.biortech.2017.04.043

Smith RT, Bangert K, Wilkinson SJ, Gilmour DJ (2015) Synergistic carbon metabolism in a fast growing mixotrophic freshwater microalgal species *Micractinium inermum*. Biomass Bioenergy 82:73–86

Spolaore P, Joannis-Cassan C, Duran E, Isambert A (2006) Commercial applications of microalgae. J Biosci Bioeng 101:87–96. https://doi.org/10.1263/JBB.101.87

Stadnichuk IN, Rakhimberdieva MG, Bolychevtseva YV, Yurina NP, Karapetyan NV, Selyakh IO (1998) Inhibition by glucose of chlorophyll a and phycocyanobilin biosynthesis in the unicellular red alga *Galdieria partita* at the stage of coproporphyrinogen III formation. Plant Sci 136:11–23

Stadnichuk IN, Rakhimberdieva MG, Boichenko VA, Karapetyan NV, Selyakh IO, Bolychevtseva YV (2000) Glucose-induced inhibition of the photosynthetic pigment apparatus in heterotrophically-grown *Galdieria partita*. Russ J Plant Physiol 47:585–592

Stebegg R, Schmetterer G, Rompel A (2019) Transport of organic substances through the cytoplasmic membrane of cyanobacteria. Phytochemistry 157:206–218

Sun N, Wang Y, Li YT, Huang JC, Chen F (2008) Sugar-based growth, astaxanthin accumulation and carotenogenic transcription of heterotrophic *Chlorella zofingiensis* (Chlorophyta). Process Biochem 43:1288–1292. https://doi.org/10.1016/J.PROCBIO.2008.07.014

Tan LT (2007) Bioactive natural products from marine cyanobacteria for drug discovery. Phytochemistry 68:954–979

Tan CK, Johns MR (1991) Fatty acid production by heterotrophic *Chlorella saccharophila*. Hydrobiologia 215:13–19. https://doi.org/10.1007/BF00005896

Tan CK, Johns MR (1996) Screening of diatoms for heterotrophic eicosapentaenoic acid production. J Appl Phycol 8:59–64. https://doi.org/10.1007/BF02186223

Tanaka Y, Ogawa T, Maruta T, Yoshida Y, Arakawa K, Ishikawa T (2017) Glucan synthase-like 2 is indispensable for paramylon synthesis in *Euglena gracilis*. FEBS Lett 591(10):1360–1370. https://doi.org/10.1002/1873-3468.12659

Tani Y, Tsumura H (1989) Screening for tocopherol-producing microorganisms and α-tocopherol production by *Euglena gracilis* Z. Agric Biol Chem 53:305–312

Tatsuzawa H, Takizawa E (1995) Changes in lipid and fatty acid composition of *Pavlova lutheri*. Phytochemistry 40:397–400. https://doi.org/10.1016/0031-9422(95)00327-4

Teerawanichpan P, Qiu X (2010) Fatty acyl-CoA reductase and wax synthase from *Euglena gracilis* in the biosynthesis of medium-chain wax esters. Lipids 45:263–273

Terashima M, Specht M, Naumann B, Hippler M (2010) Characterizing the anaerobic response of *Chlamydomonas reinhardtii* by quantitative proteomics. Mol Cell Proteomics 9(7):1514–1532

Theriault RJ (1965) Heterotrophic growth and production of xanthophylls by *Chlorella pyrenoidosa*. Appl Microbiol 13:402–416

Tomita Y, Yoshioka K, Iijima H et al (2016) Succinate and lactate production from *Euglena gracilis* during dark, anaerobic conditions. Front Microbiol 7:2050

Tomiyama T, Kurihara K, Ogawa T et al (2017) Wax ester synthase/diacylglycerol acyltransferase isoenzymes play a pivotal role in wax ester biosynthesis in *Euglena gracilis*. Sci Rep 7(1):1–13. https://doi.org/10.1038/s41598-017-14077-6

Tredici MR (2010) Photobiology of microalgae mass cultures: understanding the tools for the next green revolution. Biofuels 1:143–162

Tredici MR, Margheri MC, Giovannetti L, De Philippos R, Vincenzini M (1988) Heterotrophic metabolism and diazotrophic growth of *Nostoc* sp. from *Cycas circinalis*. Plant Soil 110:199–206

Tucci S, Vacula R, Krajčovič J, Proksch P, Martin W (2010) Variability of wax-ester fermentation in natural and bleached *Euglena gracilis* strains in response to oxygen and the elongase inhibitor flufenacet. J Eukaryot Microbiol 57:63–69

Turmel M, Gagnon MC, O'Kelly CJ, Otis C, Lemieux C (2009) The chloroplast genomes of the green algae *Pyramimonas*, *Monomastix*, and *Pycnococcus* shed new light on the evolutionary history of prasinophytes and the origin of the secondary chloroplasts of euglenids. Mol Biol Evol 26:631–648

Vahrenholz C, Riemen G, Pratje E, Dujon B, Michaelis G (1993) Mitochondrial DNA of *Chlamydomonas reinhardtii*: the structure of the ends of the linear 15.8-kb genome suggests mechanisms for DNA replication. Curr Genet 24(3):241–247

Villarejo A, Orus MI, Martinez F (1995) Coordination of photosynthetic and respiratory metabolism in *Chlorella vulgaris* UAM 101 in the light. Physiol Plant 94:680–686

Wan M-X, Wang R-M, Xia J-L, Rosenberg JN, Nie Z-Y, Kobayashi N, Oyler GA, Betenbaugh MJ (2012) Physiological evaluation of a new *Chlorella sorokiniana* isolate for its biomass production and lipid accumulation in photoautotrophic and heterotrophic cultures. Biotechnol Bioeng 109:1958–1964. https://doi.org/10.1002/bit.24477

Wan M, Wang Z, Zhang Z, Wang J, Li S, Yu A, Li Y (2016) A novel paradigm for the high-efficient production of phycocyanin from *Galdieria sulphuraria*. Bioresour Technol 218:272–278. https://doi.org/10.1016/j.biortech.2016.06.045

Wang C, Lang CQ (2008) Effects of shear stress on microalgae – a review. Biotechnol Advances 36:986–1002

Wang H, Fewer DP, Holm L, Rouhiainen L, Sivonen K (2014) Atlas of nonribosomal peptide and polyketide biosynthetic pathways reveals common occurrence of nonmodular enzymes. Proc Natl Acad Sci U S A 111:9259–9264

Wang T, Tian X, Liu T, Wang Z, Guan W, Guo M, Chu J, Zhuang Y (2017) A two-stage fed-batch heterotrophic culture of *Chlorella protothecoides* that combined nitrogen depletion with hyperosmotic stress strategy enhanced lipid yield and productivity. Process Biochem 60:74–83. https://doi.org/10.1016/j.procbio.2017.05.027

Wang Y, Seppänen-Laakso T, Rischer H, Wiebe MG (2018) *Euglena gracilis* growth and cell composition under different temperature, light and trophic conditions. PLoS One 13(4): e0195329. https://doi.org/10.1371/journal.pone.0195329

Weber A, Oesterhelt C, Gross W et al (2004) EST-analysis of the thermo-acidophilic red microalga *Galdieria sulphuraria* reveals potential for lipid a biosynthesis and unveils the pathway of carbon export from rhodoplasts. Plant Mol Biol 55:17–32

Wei D, Chen F, Chen G, Zhang X, Liu L, Zhang H (2008) Enhanced production of lutein in heterotrophic *Chlorella protothecoides* by oxidative stress. Sci China Ser C Life Sci 51:1088–1093. https://doi.org/10.1007/s11427-008-0145-2

Wen Z-Y, Chen F (2000) Production potential of eicosapentaenoic acid by the diatom *Nitzschia laevis*. Biotechnol Lett 22:727–733. https://doi.org/10.1023/A:1005666219163

Wen Z-Y, Chen F (2001) Optimization of nitrogen sources for heterotrophic production of eicosapentaenoic acid by the diatom *Nitzschia laevis*. Enzym Microb Technol 29:341–347. https://doi.org/10.1016/S0141-0229(01)00385-4

Wen ZY, Chen F (2003) Heterotrophic production of eicosapentoic acid by microalgae. Biotechnol Adv 21:273–294

Wilhelm C, Büchel C, Fisahn J, Goss R, Jakob T, LaRoche J, Lavaud J, Lohr M, Riebesell U, Stehfest K, Valentin K, Kroth PG (2006) The regulation of carbon and nutrient assimilation in diatoms is significantly different from green algae. Protist 157:91–124. https://doi.org/10.1016/j.protis.2006.02.003

Wood BJB, Grimson PHK, German JB, Turner M (1999) Photoheterotrophy in the production of phytoplankton organisms. J Biotechnol 70:175–183. https://doi.org/10.1016/S0168-1656(99)00070-X

Wu Z, Wu S, Shi X (2007) Supercritical fluid extraction and determination of lutein in heterotrophically cultivated *Chlorella pyrenoidosa*. J Food Process Eng 30:174–185. https://doi.org/10.1111/j.1745-4530.2007.00102.x

Xie T, Sun Y, Du K, Liang B, Cheng R, Zhang Y (2012) Optimization of heterotrophic cultivation of *Chlorella* sp. for oil production. Bioresour Technol 118:235–242. https://doi.org/10.1016/j.biortech.2012.05.004

Xiong W, Li X, Xiang J, Wu Q (2008) High-density fermentation of microalga *Chlorella protothecoides* in bioreactor for microbio-diesel production. Appl Microbiol Biotechnol 78:29–36. https://doi.org/10.1007/s00253-007-1285-1

Xu H, Miao X, Wu Q (2006) High quality biodiesel production from a microalga *Chlorella protothecoides* by heterotrophic growth in fermenters. J Biotechnol 126:499–507. https://doi.org/10.1016/j.jbiotec.2006.05.002

Yamada K, Suzuki H, Takeuchi T et al (2016) Efficient selective breeding of live oil-rich *Euglena gracilis* with fluorescence-activated cell sorting. Sci Rep 6:26327

Yang W, Catalanotti C, D'Adamo S et al (2014) Alternative acetate production pathways in *Chlamydomonas reinhardtii* during dark anoxia and the dominant role of chloroplasts in fermentative acetate production. Plant Cell 26(11):4499–4518

Yang W, Catalanotti C, Wittkopp TM, Posewitz MC, Grossman AR (2015) Algae after dark: mechanisms to cope with anoxic/hypoxic conditions. Plant J 82(3):481–503

Yanowitz J, Ratcliff MA, McCormick RL, Taylor JD, Murphy MJ (2014) Compendium of experimental cetane numbers. Technical report. National Renewable Energy Laboratory, Golden, CO

Yoon HS, Hackett JD, Pinto G, Bhattacharya D (2002) The single, ancient origin of chromist plastids. Proc Natl Acad Sci U S A 99:15507–15512

Yoon HS, Hackett JD, Ciniglia C, Pinto G, Bhattacharya D (2004) A molecular timeline for the origin of photosynthetic eukaryotes. Mol Biol Evol 21:809–818

Yoon HS, Muller KM, Sheath RG, Ott FD, Bhattacharya D (2006) Defining the major lineages of red algae (Rhodophyta). J Phycol 42:482–492

Zhang DH, Lee YK (2001) Two-step process for ketocarotenoid production by a green alga, *Chlorococcum sp.* strain MA-1. Appl Microbiol Biotechnol 55:537–540. https://doi.org/10.1007/s002530000526

Chapter 4
Agronomic Practices for Photoautotrophic Production of Algae Biomass

Philip A. Lee and Rebecca L. White

Abstract Phototrophic production of algae, either in open raceway ponds or closed systems such as PBRs or covered raceways, is increasingly considered an agricultural enterprise, albeit one not requiring soil. Particular focus on "algae as agriculture" in the last 10 years has been the result of the critical integration of traditional agricultural techniques with decades of algae cultivation knowledge to form a set of ever-growing agronomic practices for algae biomass production. There are two parts to agronomic practices: ensuring the health of the crop by providing optimum growth conditions and preventing and/or treating pests to protect the crop. The critical advancements in algae cultivation in recent years have been in crop protection, particularly the translation of integrated pest management practices from traditional agriculture. This chapter summarizes how agronomic practices can be applied to each stage of algal production which should drive future optimization.

4.1 Introduction

Algae (micro- and macro-) have long been of interest for production of renewable fuel, food, and other materials, including, but not limited to, soil amendments, pharmaceuticals, plastics, cosmetics, and more. This chapter focuses on cultivation of microalgae. Multiple product streams can be produced from one algae source (e.g., protein as well as oils) (Masojídek and Torzillo 2014). Algae can be grown virtually anywhere with the right energy source and can be grown in many different ways. Algae are naturally photosynthetic organisms but can be grown with or

---

P. A. Lee
Department of Biology, Midland College, Midland, TX, USA

Pebble Labs, Inc., Los Alamos, NM, USA

R. L. White (✉)
Qualitas Health, Inc, Houston, TX, USA

Pebble Labs, Inc., Los Alamos, NM, USA

A. Hallmann, P. H. Rampelotto (eds.), *Grand Challenges in Algae Biotechnology*, Grand Challenges in Biology and Biotechnology,
https://doi.org/10.1007/978-3-030-25233-5_4

without light and with varying degrees of complexity from open ponds mixed only with wind to highly controlled fermentation systems (Masojídek and Torzillo 2014).

Since microalgae are microorganisms, the large-scale production of microalgae is, on its face, industrial microbiology. Industrial microbiology or microbial biotechnology is a branch of applied microbiology that comprises the application of scientific and engineering principles to the commercial exploitation of microorganisms (including algae) or cells from multicellular organisms to create useful products or processes (http://www.simbhq.org/careers/career-information/). While all algae production systems are technically industrial microbiology, the authors here propose that photoautotrophic production of algae more closely resembles agriculture, given the extensive cross-over of disciplines between agronomy and photoautotrophic production of algae.

Agriculture is the science and technology of farming—that is, using land for food production, including plant crops and livestock rearing. Agronomy is the plant-focused portion of agriculture and encompasses the science and technology of producing and using plants for food, fuel, fiber, and land reclamation, as well as the commercial aspects, development, and practical management of plant production systems. Agronomy requires work in plant genetics, plant physiology, meteorology, and soil science and the application of biology, chemistry, economics, ecology, earth science, and genetics (Benton Jones 2002) (https://educalingo.com/en/dic-en/agronomist).

The idea of "algae as agriculture" is not new—indeed, algae production facilities have been called "farms" for several decades (Becker 1994), although the dedicated practice of transitioning agronomic practices to photoautotrophic algae production is only found in the literature for the last ~10 years, with a few publications focusing specifically on agronomic practices for algae (White and Ryan 2015; Knoshaug et al. 2018). Calls for policy changes to include algae in agriculture support programs in the United States were first outlined in 2014 (Trentacoste et al. 2015). The 2018 Farm Bill included these changes, expanding and formalizing support for algae agriculture and identifying algae as a crop in US policy.

4.1.1 Photoautotrophic Growth Systems

Even within photoautotrophic production of algae, a considerable diversity of growth platforms exist, from closed systems such as those used for the production of astaxanthin from *Haematococcus*, or for the production of ethanol, to covered or open paddle wheel-driven "Oswald"-style ponds for growth of *Arthrospira* (commonly referred to as "Spirulina"), *Nannochloropsis*, and other strains and to very large open ponds for *Dunaliella* production and green algae as feed for shrimp production (Fig. 4.1.)

There are a number of recent publications comparing photobioreactors to open pond systems (Chew et al. 2018; Jerney and Spilling 2018). Comparison tables often include criteria around contamination control and risk, water loss, light utilization, capital and operating costs, and others. However, those publications are generally focused on a single end product and do not include location of the facility as a factor.

Fig. 4.1 Examples of photoautotrophic growth systems for algae. Top left: Open pond production of Spirulina and *Haematococcus* at Cyanotech in Hawaii (photo courtesy of Cyantotech). Top center and top right: Open pond production of *Nannochloropsis* at Green Stream Farms/Qualitas Health facility in Columbus, New Mexico (photos courtesy of Qualitas Health, Inc.). Lower left: Open pond production of *Dunaliella* at BASF's Hutt Lagoon Farm in Western Australia (Photo courtesy of BASF). Lower left center: Covered pond system for *Spirulina* production at AlgaeMor in Israel (photo courtesy of AlgaeMor). Top right: Lower right center: Outdoor PBRs for *Haematococcus* production at Algatech in Israel (photo courtesy of Algatech). Lower left: PBRs under greenhouses for cultivation of various microalgae at AlgEternal Technologies in La Grange, Texas (photo courtesy of AlgEternal Technologies)

However, it is these two factors that, commercially, are the main drivers of growth system selection. End product considerations are the primary driver, e.g., pharmaceuticals versus biofuels.

Location heavily influences things like evaporation rate/water use, growth period/season, pest pressure, operational costs, and other factors (Oswald 1988; Becker 1994). There are examples of companies producing the same or similar products in different locales, with the location setting the requirement for the growth system. For example, astaxanthin is produced in closed photobioreactors in Israel by the company Algatech and in open ponds in Hawaii by Cyanotech. Evaporation rates are significantly different between the two locations, making the use of open ponds in one location untenable but acceptable in the other. Also, the production of *Nannochloropsis* for omega-3 oils is accomplished in open ponds in the southwestern United States by Qualitas Health, Inc., but Algaennovation Ltd. in Iceland uses closed photobioreactors (Isaac Berzin, personal communication). The major differences are due to the location of the facilities—one functioning as a "farm," using sunlight and the temperate climate, and one requiring light, geothermal heat, and cover for year-round production. A very brief introduction to the types of photoautotrophic growth systems is below. For a more thorough review, see Becker (1994). The majority of the information in this chapter will focus on pond systems, rather than PBRs, since pond systems are the most common commercial production systems (Borowitzka and Vonshak 2017).

4.1.1.1 Photobioreactors

Photobioreactors or PBRs are closed or semi-closed systems and range in size from a few liters to over 150,000 l. PBRs are made of transparent materials, have large surface area-to-volume ratios, and can be placed outside to use sunlight for illumination or in greenhouses with artificial light. When compared with open systems, PBRs are less susceptible to water loss, CO_2 loss, and contamination and have been reported to have higher productivity. However, PBRs are more difficult to clean, are susceptible to biofilm formation, and must be temperature controlled (especially cooled), and buildup of oxygen must be degassed properly. Additionally, the material the PBR is made of may block light penetration and impact growth and therefore must be selected carefully. Operationally, PBRs require up to 30 times more energy than open ponds (Nogueira Junior et al. 2018) and have higher labor requirements. Furthermore, the cost of construction is about one order of magnitude higher than that of open ponds (Mata et al. 2010; Masojídek and Torzillo 2014; Chew et al. 2018; Jerney and Spilling 2018). There are a variety of types of PBRs, each with unique benefits and drawbacks (Tredici 2007; Chew et al. 2018).

4.1.1.2 Open Ponds

The simplest way to grow algae is in an open vessel. Open ponds are a common growth system in commercial production of algae biomass (Borowitzka and Vonshak 2017), likely owing to their lower capital and operational costs and relatively large scale and despite the reported issues with biotic and abiotic contamination and control of culture conditions (Mata et al. 2010; Chew et al. 2018; Jerney and Spilling 2018). Open pond systems range in size from a few hundred liters (usually used for testing) to multiple millions of liters in volume (White and Ryan 2015).

There are three main types of open pond systems: unstirred pond, circular pond, and the extensively used Oswald-style raceway pond (Chew et al. 2018). All are subject to the drawbacks of open systems, such as seasonal variation and foreign material contamination from wind and rain depositions.

4.1.1.3 Covered Ponds

Covered ponds have similar benefits and issues as PBRs but with some of the lower operational and capital costs of open pond systems. Typically, ponds are not covered year-round, and instead removable covers are used to extend the growing season of warm-water algae (Richmond and Becker 1986). Covered ponds are in use for the commercial production of Spirulina (Fig. 4.1).

4.2 Agronomic Practices for Algae Production

To produce commercial-scale quantities of photoautotrophic algal biomass reliably, an agricultural mind-set is needed. Borowitzka and Vonshak described, in detail, the issues with scaling to commercial levels (Borowitzka and Vonshak 2017). In their analysis, they state that "the maintenance of long-term, stable, high-productivity, large scale cultures, usually under prevailing outdoor conditions of variable irradiance, temperature and rainfall, presents additional challenges, most of which are not seen…in the laboratory." These challenges, or very similar ones, have been studied and mitigated for hundreds of years in agriculture. In fact, the goal of agronomic practices is to increase crop yield quality and quantity while also minimizing inputs and environmental impacts. This is accomplished by both ensuring optimum growth conditions for the crop to give it an advantage over invasive species and pests and preventing and/or treating pests to protect crop yield.

There are a number of clear transition points for traditional agronomic practices to the algae farm, such as assuring proper balance of major ions for high yield and quality, pH control, fertilizer formulation and application(s), micronutrient applications, harvesting, integrated pest management, and what to test for and when to ensure high yield and quality (Kinsey and Walters 2009). These principles, along with new technology that resulted in high-yielding crop varieties, new chemicals for pest control, increased use of fertilizers and other chemicals, improved use of irrigation, and new cultivation strategies including increased use of mechanization, led to the "Green Revolution" of the 1950s and 1960s (Evenson and Gollin 2003). The result of this revolution was dramatically increased agricultural productivity, and the algae industry should be learning from this period to generate the next green revolution, this time with algae.

In traditional agriculture, agronomic practices should, in the long run, improve soil quality, enhance water use, reduce fertilizer and pesticide application, and be better for the environment. Working on the principle that "the whole is greater than the sum of its parts," the result will be from iterative combinations of many factors. There is thus a manifest difficulty in researching such methods, as with many approaches in agriculture and aquaculture, due to the massive combinatorial effects of the list of variables. Figure 4.2 is a short summary of some of these factors, specifically for agronomic practices in algae production. Long-term analysis of multi-season, multi-year, and multi-site data on crop yield is required to build hypotheses as well as to test them. The algae industry and academic groups studying cultivation are not growing algae at the scale required for this level of analysis and testing, but recent efforts have made a step in the right direction and shown the power of this approach (McGowen et al. 2017).

The larger the scale of production, the more difficult research becomes, as the effect of uncontrolled variables grows and costs increase. The pressure to produce versus the need to complete research at scale is a difficult balance, and the tension between the two can be counterproductive if the organization is not sufficiently strategic as to what is tested at small scale versus commercial scale. This highlights

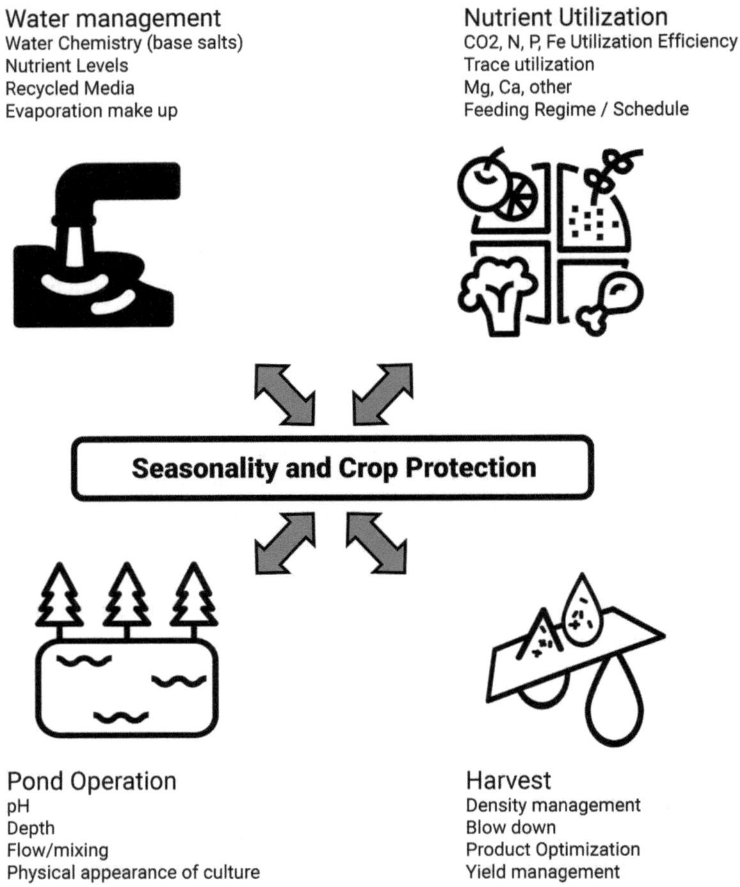

Fig. 4.2 Agronomic practices for algal cultivation (credit: Rebecca White and Otavio Santiago; courtesy of Qualitas Health, Inc.)

the need for a robust lab to field translation in any research, where variables can be tested at small scale in more controlled environments before successful ideas are tested at larger-scale outdoors. It is important to fail early (and often) at small scale.

Field trials are critically essential to successful commercial production. Testing the selected strain outside, year-round (preferably for multiple years) before scaling it commercially allows producers to understand annualized productivity and yield potential, pest pressures and mitigations, and other operational challenges before learning them at a scale that can be financially disastrous. In addition to testing the strain, field trials should also be used to vet new fertilizers, other operational parameters (e.g., different pH ranges or salinity), crop protection methods, monitoring equipment, etc. However, it is also important to properly test changes in operational parameters at large scale. Thinking of the cultivation system as another unit of operation ("unit op"), rather than the complex biological system it is, can help production staff understand the need for testing changes to operational parameters at commercial scale and can help focus the research program around these operational parameters. All other unit ops require commissioning and optimization, and the same is true of the cultivation system.

Field to lab translation is equally important in this process. Recapitulation of a field observation in a laboratory setting is essential to research, and many variables may not be isolated in less controlled systems. This is highlighted most prominently by research into pests. If the infection process observed in the field cannot be repeated at lab scale, it would be very difficult and expensive to research the pest life cycle or potential mitigation strategies.

The first five subsections below describe the factors to consider ensuring optimal crop health, while the sixth section covers crop protection. It is important to note that while the crop protection section is substantially longer than those sections before it, this focus on crop protection is not unique to algae as a crop. Recent estimates show that the burden of pathogens and pests is responsible for 20–40% of yield loss for major food crops worldwide and indicates that prioritization of reducing these losses is critical (Savary et al. 2019). The same is true for algae, especially in open pond systems.

4.2.1 Balance of Manual Monitoring Versus Automation

In designing the cultivation system, economic and practical considerations will determine the level of automation of monitoring and parameter control that is acceptable. Most sites opt to monitor pH via PLC and use that to then control the pH by injection of CO_2 into the culture medium, providing both pH control and carbon supply. The pH typically rises during photosynthetic growth and the addition of CO_2 lowers pH. Media composition, particularly around nitrogen source, can affect this pH control (Scherholz and Curtis 2013). Additional potential variables that can be controlled by automation are mixing (for open pond systems, e.g., paddle wheel power and speed); water volume/depth (both monitoring and autofill to control volume); optical density monitoring; temperature monitoring; oxidation reduction potential (ORP); dissolved oxygen; and others.

Smaller research scale ponds may have a high degree of automation, since it is necessary to control and isolate variables in service of ensuring that only successful ideas are moved to larger scales where there is less control. Large production systems mean that distances, in particular, dramatically increase; automated systems for things such as nutrient addition or crop protection chemical dosing become capex intensive and are unlikely to be economically feasible. Additionally, because equipment and sensors are exposed to the elements and weather, they need to be significantly more robust and/or protected from those elements, further increasing costs.

4.2.2 *Cultivation Management*

4.2.2.1 Year-Round Cultivation Versus Seasonal Crop Rotation

In agriculture, crop rotation aims to reduce soil erosion, increase soil fertility, and improve yield. One benefit of crop rotation in traditional agriculture is in maintaining soil condition by preventing the leaching of nutrients by a single crop over long periods, since different plants have different nutrient use rates. This benefit seems unlikely to transfer to algal production since nutrients can easily be measured and replenished in what is essentially a closed system. Conversely to this reasoning, crop rotation in algal culture could be beneficial due to buildup of undesirable chemistry in the water that prevents growth of one crop species but not another. Changing crop species is unlikely to benefit producers of specific products but may work in more strain agnostic production of biomass for other products, e.g., biofuels.

Seasonal crop rotation allows cultivation of multiple strains in 1 year at the same location. There are potential benefits to this approach if it could be achieved economically. Scale-up of a new strain from initial inoculation to a single acre of production would take more than a month, if you could double biomass on a daily basis. After that point, depending on the time of year, a new pond may be inoculated every 5–7 days, meaning that a 50-acre farm could take more than another month before crop rotation would be complete. This simple calculation does not even consider the declining growth rate of the previous season's crop or the fact that scale-up may be happening during the new crops "off" season. Thus, in a perfect scenario where biomass is doubling rapidly at all scales, it should be expected to take at least 2 months' time to change a crop.

A more realistic timeline for scale-up from inoculation to 50 acres is 3–4 months, given differences in growth rate, weather, and logistics. Additionally, the trade-off between the portion of a facility being used for scale and the portion being used for production must be considered. Any portion of the facility that is not actively in production for product reduces the overall yield for the facility on a per annum basis and therefore may have considerable impact on the economics of production.

A considerable amount of investment in infrastructure is required to have the capacity to scale up culture. The equipment for this scale-up would then be unused unless seed culture was continually produced to reduce sunk costs in unused equipment and to provide a continuous supply of clean culture. This is an approach successfully used by some growers but typically restricted to processes that require a two-stage process for product synthesis (Huntley and Redalje 2007; Narala et al. 2016). Ponds can be successfully inoculated from harvested biomass, meaning that the inoculum stream does not necessarily need to be colocated with the production facility (Bart Reid, personal communication). Thus, toll producers at smaller-scale sites may be able to provide inoculum more quickly than could be produced in house, resulting in lower capital and operational expenditure.

Another major reason for crop rotation is to prevent pest population buildup. There is a potential benefit to this approach in algal cultivation as some pests are specific to a given algal species, and pest pressure observed with a new strain is often lower in the first year of cultivation. However, there are also many pests that are not unique to one species, and it remains to be seen if stopping cultivation of a given strain and then later restarting would yield the same "honeymoon period" of cultivation that is sometimes observed. In light of the length of time for scale-up of a second crop and the likely increased expense, if a rotation is shown to be beneficial, it may be practical on a multi-year cycle rather than seasonally.

A major difference to agriculture is in the likely genetic drift of algal crop over time. A plant is planted and grown from a seed of known genetics, and this genotype will remain with the plant. Any changes to that plant will not be propagated through the field since most crop harvesting destroys the plant, and new plants will be grown from further supplies of seed. Since an algae culture is a homogeneous mix of a single-celled organism, any genetic changes in a single cell may propagate through the system over time. Many of these changes will be insignificant to productivity, and some may even be beneficial by creating a field-adapted strain.

4.2.3 Water Chemistry

4.2.3.1 Fertilizer Selection

Since plants require similar basic nutrients as algae, namely, nitrogen and phosphorus, many fertilizers available to traditional agriculture are applicable to use in algal cultivation. Numerous forms of nitrogen are readily available including nitrate, ammonia, and urea, as well as mixtures such as urea–ammonium nitrate (UAN). Various species of algae have been successfully cultivated in media utilizing components from agricultural suppliers (Ashraf et al. 2011; Nabris 2012; Camacho-Rodríguez et al. 2013). Generally, unless otherwise indicated for a specific final product, the most economical option available locally for traditional agriculture can be used for algal culture.

Algae can utilize multiple sources of nitrogen but show preference between forms, for example, ammonia is preferentially metabolized over nitrate. One cause for this is that the different forms of nitrogen have different cellular energy requirements to initiate import or metabolism. Nitrate is first converted to nitrite and then ammonia within the cell which requires energy (Sanz-Luque et al. 2015); thus, while both ammonia and nitrate use require energy, the energy cost of nitrate is higher. Utilization efficiency should be considered (tested for) when choosing a nitrogen source, including such factors as nutrient availability due to media pH (e.g., off-gassing of ammonia).

For pond systems, it is advisable to monitor all nitrogen forms as well as total nitrogen, since the local microbiome will use and convert fertilizer and atmospheric nitrogen to other forms within the established pond ecosystem. For example, nitrite, which algae can utilize, is produced from ammonia by nitrification.

Due to the increased lipid phenotype associated with N starvation, nitrogen is the most investigated nutrient. Reported optimal phosphorus levels also vary considerably and standard lab media recipes may vary by as much as sixfold. As with other media components, the optimal level will depend on the growth environment, strain choice, and goal of cultivation. Kanaga et al. (2016) report an optimum of 8.9 mg/L phosphorus for biomass production but 10.8 mg/L for lipid production in *Chlorella*. Yang et al. (2014) calculated an optimum level of 3.1 mg/L for *Scenedesmus* culture, whereas Hakalin et al. (2014) report that 49 mg/L was optimum for the same genus. In our experience, 6.5 mg/L phosphorus would be a good starting point for optimization of open pond cultivation of most species of green algae.

The ratio of N:P is often also used as a benchmark for media composition. However, the effect of the N:P ratio on growth is likely only important when one nutrient is limited, which would not be optimal if biomass production is the goal. The solution to media optimization is likely a method that can predict combinatorial effects such as design of experiment (DOE) or response surface methodology (Li et al. 2011; Yang et al. 2014; Kanaga et al. 2016) in combination with experimental confirmation under the specific conditions that would be used at production scale.

Since the pond is a liquid environment, liquid fertilizers and other chemical additions to ponds (such as crop protection treatments) are advantageous over powdered forms. Liquids can more easily be added slowly to the culture to prevent large chemical concentration gradients and to avoid localized areas reaching toxic levels. Powdered chemicals can also be problematic when applied during windy conditions which would not affect liquid additions made directly into the culture. Powdered chemicals also may cause operational issues if workers are required to lift, carry, and empty sacks of fertilizers, which can lead to a high incidence of back strain or other injury. Preventing aerosolization of liquid additions or airborne particles of powdered chemicals is obviously important for operator safety.

Any chemical additions made to the cultivation medium can result in the buildup of chemical constituents over time due to the counterions that are added at the same time. For example, the use of sodium nitrate adds a useful nitrogen source in the form of the nitrate ion but will also increase the concentration of sodium ions in the media. This may confer an advantage to certain fertilizers that do not contain counterions. For example, ammonium nitrate as a nitrogen source and ammonium phosphate as a phosphate (and nitrogen) source provide only the nutrients required for algal growth, without significant amounts of other ions.

An additional layer of complexity in water chemistry management is post-harvest water recycle. The ability to recycle water removed from algae culture during the harvesting process back into the production system is unique to algae. Water in traditional crops is lost, both to the soil and to evaporation. In open algal cultivation systems, the largest water loss is to evaporation. If the open system has a clay liner, rather than a plastic liner, there will also be small losses to the soil, but these are minor in comparison with evaporative losses (Rebecca White, unpublished data). Because evaporative loss occurs year-round in non-tropical climates, recycling water post-harvest will require careful management of total dissolved solids (TDS) through a blowdown stream or other discharge method, along with the addition of makeup water to maintain total system volume. Managing TDS is more complicated when the optimal water chemistry is very close in composition to the source water. When this is the case, substantial blowdown streams are required to keep the water chemistry within optimal ranges, leading to higher water usage (Guieyss et al. 2013; Talent et al. 2014; White and Ryan 2015).

4.2.3.2 Trace Elements

Trace elements provided in the growth medium include anything required for normal growth and function outside of N, P, Fe, and major salts/ions. The level and composition of trace element additions that are required will depend on the levels found in the water source used to make media. Most can be monitored over time for both replenishment and to prevent accumulation to toxic levels. These additions represent relatively little cost but may be important for product yield or growth (Kropat et al. 2011; Malakootian et al. 2015).

4.2.3.3 Other Components

Alkalinity affects buffering capacity of media and can be crop protective against a number of known and unknown pests. Maintenance of alkalinity is a tool used in the cultivation of Spirulina and also is crop protective in cultivation of *Desmodesmus* sp. (unpublished). The decision to use bicarbonate in this way must be weighed against the desire to use large pH swings as a crop protection response, since the increased buffering capacity delivered by bicarbonate buffer system directly affects these techniques.

Total dissolved solids and salinity are other media components that can be manipulated and have traditionally been used for propagation of crop protective conditions. Aside from salinity (sodium chloride), the major ions that make up the total dissolved solids are magnesium, calcium, potassium, and sulfate; the amounts of these in the media are strain and growth system dependent.

4.2.3.4 Carbon Dioxide

Carbon is required to build biomass and CO_2 is necessary for photosynthesis. Exogenous CO_2 is typically supplied to improve biomass production and also to reduce pH to maintain optimum cultivation conditions. Cost of CO_2 has significant impact on the overall cost of algal biomass production, and CO_2 use efficiency is a critical component to control for cost-efficient production (Davis et al. 2012).

4.2.4 Fertilizer Use Efficiency and Feeding Management

Management of fertilizer application concerns both the amount applied and the frequency. Reported optimal nutrient levels vary widely and common media recipes may differ by an order of magnitude in the level of nitrogen they contain (Wong 2017). Even within the same genus, different reports can vary significantly in their optimal concentration. For example, Li et al. (2015) showed the highest biomass productivity for *Chlorella sorokiniana* at 210 mg/l N (although this was the highest level tested), whereas Li et al. (2011) reported a value of 2600 mg/l N for *Chlorella minutissima*. Likewise, for Spirulina cultivation, the media recipes often contain over 400 mg/l N, but Vieira Costa et al. (2001) show 420 mg/l is optimal over 140 or 700, while Colla et al. (2007) report that the nitrogen level can be reduced to 103 mg/ml without any loss of biomass productivity. The reason for these large differences probably stems from differences in cultivation method, scale, and environmental conditions between studies.

Often, reports modifying media are aimed at utilizing nitrogen starvation to improve lipid synthesis and thus are biased toward lower levels of nitrogen, where peak biomass productivity is shown at the highest nitrogen level tested (e.g., Converti et al. 2009; Li et al. 2015). For *Desmodesmus* sp., an obvious nitrogen starvation phenotype may be observed visually as a color change or in pond data when nitrogen levels fall below approximately 50 mg/l N (unpublished observation), and experiments optimizing for lipid production often report N levels in this range being ideal (Li et al. 2015). Physiological changes (cell/micro-level phenotypic changes) are likely to occur at the cellular level prior to the visual changes in pond phenotype (macro-level phenotypic changes).

Nitrogen level for biomass productivity optimization is typically higher than that for lipid production (Li et al. 2011; Kanaga et al. 2016). When biomass production is the goal in optimization experiments, the amount seems to usually be above 200 mg/l N for multiple genera such as *Nannochloropsis* (Wan et al. 2013), *Chlorella* (Kanaga et al. 2016), or Spirulina (Vieira Costa et al. 2001). The reported differences in optimal N suggest that algae can tolerate a wide range of nitrogen levels, and avoiding the lower end of the range would be advisable when biomass production is the goal. For new media formulations, starting with a concentration around 200 mg/l and optimizing for use in the growth system would seem a reasonable starting point.

Reported optimal phosphorus levels also vary considerably. The solution to media optimization is likely a method that can predict combinatorial effects such as design of experiment (DOE) or response surface methodology (Li et al. 2011; Yang et al. 2014; Kanaga et al. 2016) in combination with experimental confirmation under the specific conditions that would be used at production scale.

If optimal levels fall within a small range of concentrations, automatic feeding of nutrients to maintain in-pond nutrient levels or very frequent manual feeding will be necessary. There are obvious operational cost drawbacks to the latter approach and capex cost considerations for the former, which must be balanced. As volumetric growth increases, use of fertilizer will obviously increase, so ponds operated at lower depths or high-density growth systems will have higher fertilizer use rates. Peak growing season in these systems may necessitate automated feeding methods to maintain levels within optimal ranges. The nutrient use efficiency should also be taken into consideration when determining target levels in the growth system. This is calculated as the percentage of the nutrient placed in the pond that is recovered in the biomass at harvest. If use efficiency is low, then using concentrations in the lower end of the acceptable range may be more economical, so long as that does not affect productivity or final product yield.

4.2.5 Harvest Management and Impacts

The primary goal of harvest is to separate the algal biomass from culture media for downstream processing. It also serves to maintain the density range of the culture at a desired level. The optimal density to maximize biomass productivity may differ to that required for optimizing product yield if, for example, your product is a metabolite whose synthesis is affected by cell density. However, maximum biomass productivity remains the goal of most systems. There are times where sub-optimal biomass concentration for productivity would be beneficial for overall yield production, when taking into consideration factors other than just the algae growth and cell division. As with all aspects of algal cultivation, there are economical, operational, biological, and chemical considerations which may not all pull in the same direction.

Harvest will remove the crop from the culture media, but it also removes other organisms and constituents of the culture. Thus, it should be looked upon as a method for realizing yield as well as maintaining future yield. The total of agronomic practices should aim to design and maintain a cultural environment that favors growth of the crop over that of pest organisms. Removal of a percentage of both populations by harvest should then favor the crop over time. At times, the crop may benefit from being grown outside of the optimal ranges as determined in optimization studies. For example, maintaining the culture at higher than optimal density for growth may benefit the culture in the presence of pests. Denser cultures may compete better against invading weed algae (Richmond et al. 1990). This may also be applicable to grazing pests such as rotifers where higher cell densities can affect rotifer population growth and hatching. For these reasons, it may be preferable to sacrifice a small amount of growth rate in order to increase the likelihood of achieving higher harvested yield and quality in the long run. This could be achieved by simply harvesting down to optimal density, rather than aiming to maintain culture average density at the optimum. Harvesting at higher cell densities also means that less water is moved per unit of harvested biomass, reducing operational costs.

The timing of harvest in traditional agriculture can be critical to optimize yield since the plant may need to mature, ripen, defoliate, etc. before harvest can begin. For small-scale algal production, or for products that have multi-stage cultivation that includes a maturation phase, such as astaxanthin production from

Haematococcus, the timing of harvest may be important for final product yield. However, for large-scale production of algal biomass, this is unlikely to be a practical approach since harvest will likely be a 24-h operation at scale. Although, in such an operation during periods of slower growth such as winter, harvest timing is a variable that could still be considered as a means to improve productivity or otherwise reduce operational costs.

In agriculture, the aerial surface area is used to calculate fertilizer and crop protection chemical application rates. Similar calculations for algal cultivation must take into account the culture depth since this will affect the final concentration of the chemical added to the pond. Similarly, although productivity is also calculated on a per area basis (e.g., g of algae produced per m^2), the actual concentration of algae in the media is dependent on the culture volume. Light is attenuated quickly as you move deeper from the surface of an algal culture, and at a pond density of 0.5 g/l, little to no light is reaching below the top inch of the culture (Richmond and Zou 1999; Borowitzka 2016).

Light exposure is one of the limiting factors for algal productivity, and as density increases, self-shading reduces light per cell. Average light on the culture can be changed by adjusting mixing or mixing (frequency that each individual cell is exposed to light) or adding or removing water to adjust depth (the former makes the culture more dilute, with a higher percentage of culture volume that is exposed to light; the latter, which can be accomplished by allowing evaporation to make the culture more dense, decreases the percentage of culture volume that is exposed to light).

At any set of culture conditions, an optimum cell density exists that will yield optimum culture aerial productivity. At lower depths or higher mixing rates, this density is increased. The specific growth rate of the culture will also be higher at lower culture depth, when more light is available on average to the algal cells (Richmond and Zou 1999). This set of observations show that culture depth, density, and mixing rate can be controlled to maximize and maintain productivity. At different times of the year, it may be preferential to maintain ponds at different depths to advantage the algae over other organisms (i.e., increase specific growth rate of the algae to compete with a pest) or to aid in culture temperature control (shallower ponds will heat and cool more quickly).

The density of the culture can also impact the level of biomass lost to respiration at night. Lower-density cultures that have more light exposure during the day lose a higher percentage of biomass at night than cultures at higher densities (Edmundson and Huesemann 2015). One cost reduction method for outdoor ponds is to reduce the energy used to mix ponds during nighttime hours (Eustance et al. 2015) which may also reduce the biomass loss due to respiration (Ogbonna and Tanaka 1996).

Crop protection chemicals can be a significant cost factor in production. These are typically administered at a given target concentration. Treating a pond that is at a lower depth (and thus volume) results in a reduction in chemical cost per dose which could represent a significant annual saving in production costs. This principle could be applied to ponds without changing normal cultivation depth by treating at harvest, prior to culture volume being replenished. There are logistical considerations to this approach of course, and it would also rely on the water being used for volume replacement being free of pests.

The use of a harvest system that is less than 100% efficient increases the risk of selecting for organisms that are less likely to be removed by that technology. For example, if harvesting by screen filter, small contaminating algal cells (weeds) will be returned to the pond in the recycled media (Vonshak et al. 1983). Harvest systems using sub-micron level filtration should prevent these issues but have yet to be proven at scale (Drexler and Yeh 2014; Mo et al. 2015).

4.2.6 Crop Protection

Pesticide use increased dramatically after the Second World War as many broad-spectrum pesticides were developed and brought to market. Extensive use of such chemicals resulted in vast quantities of chemical released into the environment. Their low cost meant pesticides were used even in the absence of pest detection (Beers and Brunner 1993; Weddle et al. 2009). With this high use came the development of resistance of many pests to the pesticides, as well as significant impacts to populations of non-target organisms (Weddle et al. 2009). Some of these organisms included natural predators of pests as well as other beneficial organisms such as pollinators. Integrated pest management (IPM) concepts were introduced to stem the use of chemicals, reduce the prevalence of pest resistance, and limit the toxic effects on non-pests. IPM uses a variety of pest control tactics and aims to use pesticides as a last resort rather than the first tool in response to pest pressure.

IPM is difficult to implement but is a sound framework for crop protection that has shown success (Ehler 2006; Stenberg 2017). For example, mite control in apple orchards of the Pacific Northwest became difficult in the 1950s–1960s due to overuse of insecticides and the development of resistance to miticides. However, with the selective uses of insecticides that allowed biological control of mites by predatory mite populations as part of an IPM strategy, many orchards have not used miticides for over 50 years. There are now over 65 definitions of IPM (Prokopy and Kogan 2009). However, there are a number of key attributes to an IPM program, many of which can be directly translated or adapted for use in algal production. These include management of multiple pests simultaneously, monitoring for pests, use of preventative measures ahead of pesticides, and the use of economic treatment thresholds for deciding when to use pesticides (Ehler 2006). An example of how IPM can be applied to algae is summarized in Fig. 4.3.

Key to an IPM program is identification of pests and developing an understanding of their life cycle. This includes how the life cycle is affected by climate, season, and interactions with other organisms including the host and natural predators of the pest. Once pests are identified, their populations must be monitored accurately enough to enable treatment decisions to be made and to determine the efficacy of any treatment program. Use of integrated preventative or suppressive cultural controls along with encouraging biological control is also a key component of an IPM program. For

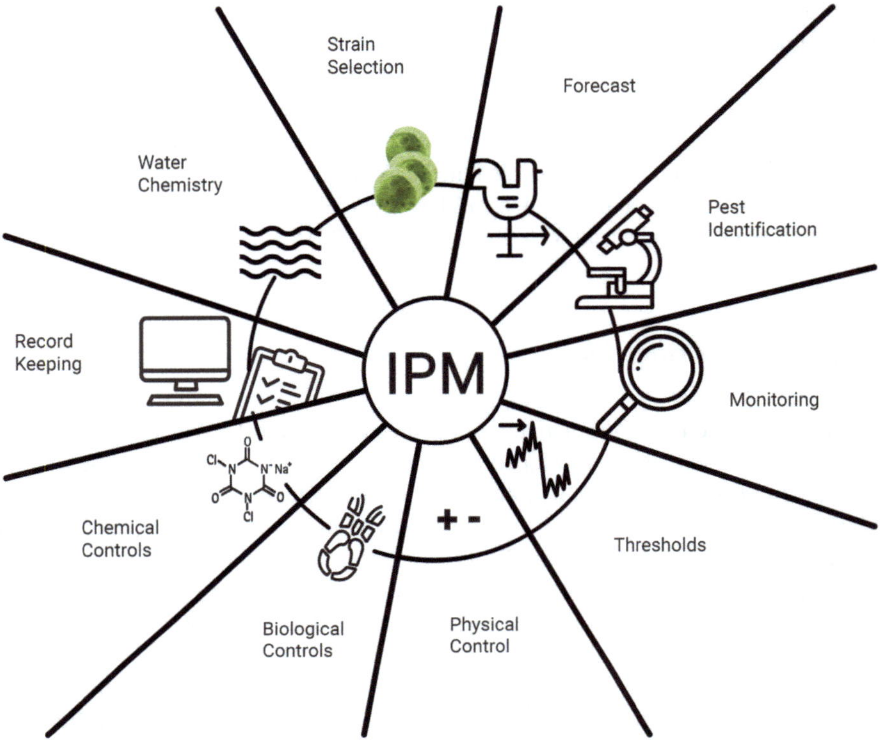

Fig. 4.3 Integrated Pest Management (IPM) for algal cultivation (credit: Rebecca White and Otavio Santiago; courtesy of Qualitas Health, Inc.)

example, timing of pruning and fertilization, and the use of cover crops to suppress weeds and provide habit for beneficial organisms, and good sanitation techniques can all reduce pest pressure and reduce the need for pesticides (Beers and Brunner 1993).

Strain selection is probably the underlying key upon which an IPM program can be built. A robust strain that can withstand a wide range in various operational parameters (e.g., salinity, pH) allows for more options for crop protection. Selecting the correct variety of crop for the location should yield the healthiest and most productive crop. The development of pest- or pesticide-resistant cultivars can further reduce the risk of crop failure and decrease dependence on reactive crop protection measures.

All steps of the cultivation process affect crop protection. The cultivation system design, media composition, depth, harvest system, target densities, etc. can all affect the productivity of the crop as well as growth of pests. The components of an IPM program as discussed in this chapter should move toward a broader, more proactive philosophy of "yield protection" from a traditional, reactive "crop protection" mind-set. A healthy and productive crop will require less crop protection.

4.2.6.1 Monitoring

Monitoring for pests is an essential component of crop protection and integral to an IPM program. Since the algal production system is essentially a homogeneous system, this aspect of IPM may be more easily applied to algae production than others. Monitoring techniques will include those for specific pest populations as well as for the algal phenotypic response to pests. Algae have a set of generalized responses to pest organisms including clumping/flocking, loss of fluorescence, etc. that can be relatively easy to detect (Schlüter et al. 1987; Hessen and Donk 1993; Juneau et al. 2003; Lurling and Beekman 2006; McBride et al. 2014). The first appearance of a new pest will be detected by generalized measures of algal health or by simple observation of the pond. Once a pest has been identified, then monitoring techniques specific to that organism can be developed. As such, the monitoring tool kit will include a group of generalized assays of algal "health" that would be deployed frequently (including daily) along with a set of pest-specific assays that are deployed during times of potential pest pressure (Fig. 4.4).

Reactions to these different types of assays will differ. A specific response should be triggered by an assay for a given pest population once a predetermined threshold is met, whereas the response to a general measure of algal health would be to trigger other actions such as secondary assays, microscopy, or other observations of the data to determine the best course of action. As understanding of pest life cycles improves, the goal of a monitoring program should be to move from the observation that something has happened toward an assay that is predictive of future yield loss, enabling the farmer time to respond prior to yield being significantly impacted. The more sensitive and thus more predictive a monitoring technique can become, the higher realized yield should be.

The first level of monitoring should begin at the pond scale. The initial sign of a new pest phenotype is typically observed in the coloration or other macroscopic observations of the culture, in the cultivation setting itself. The most common observation caused by an unidentified pest is a color change, typically from a lush green to a brown color, or from a very deep green to a green with yellow undertones, which can happen in less than 24 h (McBride et al. 2014; Lee et al. 2018). Other macroscopic signs of pest infection that could be observed at the pond include foaming, flocculation, flotation, and other color changes. This highlights the need for eyes on the pond on a daily basis, especially in seasons with high pest pressure, and for training operators to be familiar with these symptoms, their implications, and the urgency to report them to the agronomist or a crop protection specialist on staff. Presently, this may be a physical person observing each pond during sample collection, but with the advent of cheap access to technology such as drones, the "big picture" look at the pond may be achievable remotely with some automation and quantification. Pond imaging devices are also in development that can collect data to correlate with culture growth and health in real time (Reichardt et al. 2014).

Strong record keeping and communication by operational staff that see the ponds daily is essential to build patterns of observations and to build correlations of these

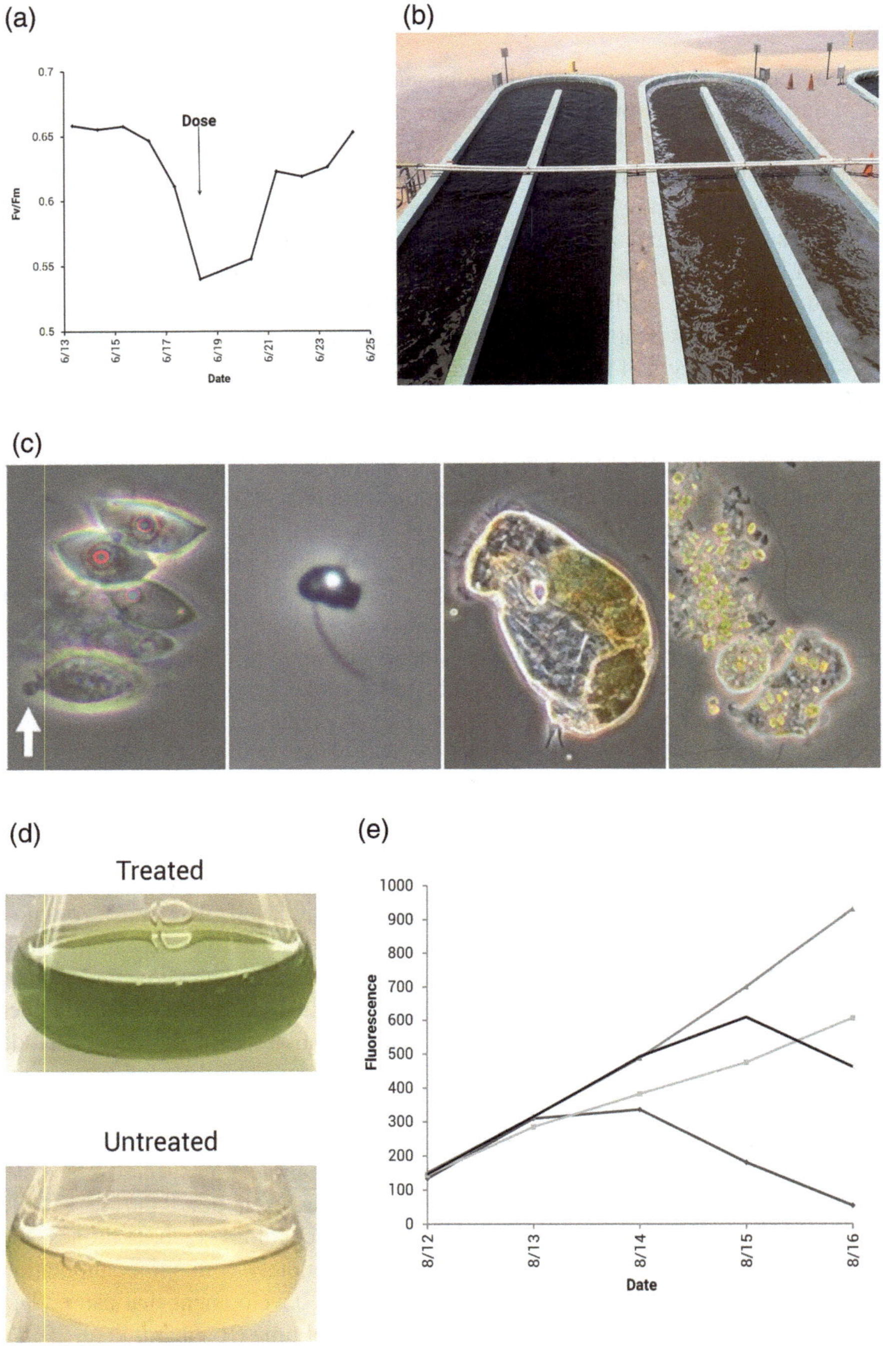

Fig. 4.4 Examples of monitoring techniques for algal cultivation. Panel **a**: Graph of F_v/F_m over time for a raceway pond. Arrow indicates when a crop protection chemical dose was applied in response to F_v/F_m value decline. Panel **b**: Observation at the pond scale. Ponds are different colors and show different levels of foam. Panel **c**: Microscope images of various pest phenotypes. From left to right: attached sporangia (arrow), free swimming zoospore, rotifer, amoeba. Panel **d**:

with season, temperature, operational activities, etc. This should help determine which observations could be considered normal and which require a response. As with these relatively simple observations, all components of an IPM system should build on previous year's observations, to predict or aid in planning future activities. All observations of differences to established normal ranges should be recorded and acted upon. When operational ranges have been established, the response to finding a parameter out of this range must also be established. Depending on the observation, these responses may be as simple as logging information or informing others, through to extra sample collection or storage of samples for long-term research (e.g., in the case of potential pest presence).

An experienced observer can detect small changes in coloration and other macroscopic phenotypes between samples, or over time, that can be indicative of pest presence, nutritional status, settling rate, foreign material presence, and pond density. Close observation of sample bottle contents and selection of sample bottle (i.e., to allow visualization of the contents) is important. Photographs of pond samples over time can, in hindsight, provide clues to when an unidentified problem may have started and may aid in the development of a predictive assay for future analysis. They also provide a quick snapshot of general state of the culture that can be easily communicated and stored. Software could be designed to analyze photographic images of samples which would remove some of the subjectivity of this process as well as increasing automation while decreasing labor.

Microscopy is time-consuming but essential for pest monitoring in the research setting and early on in production scale-up. Some level of routine microscopy is probably also essential at production scale. Readily available books can be helpful in identifying the myriad organisms observable under the microscope (e.g., Patterson 2003). In an open system, other organisms will always be present so determining which are potential pests is important to avoid unnecessary monitoring or treatment. Many organisms observed in the pond have little impact on yield, and some are likely beneficial (Cole 1982; Kazamia et al. 2012; Lian et al. 2018; Yao et al. 2019). Even those organisms that can consume algae may be beneficial in some cases due to their consumption of pest organisms or control of other unwanted biomass (Kagami et al. 2004; Johnke et al. 2017). An algal-specific pest handbook would be a useful addition to the industry and would be an area for potential collaboration across groups.

Typical magnification ranges for light microscopy are 100–400X, sufficient to identify many pest phenotypes and especially the algal phenotypic response to pests at the cellular level. However, some pest populations would be at catastrophic levels before being detected by microscopy of a typical sample volume of 5–10 μl. Once a pest has been identified, the faster a more sensitive assay can be developed the better.

Fig. 4.4 (continued) Qualitative satellite flask culture. Flasks of pond culture grown with and without treatment in lab conditions. Panel **e**: Quantitative satellite culture. Graph shows fluorescence over time for pond culture samples grown in laboratory condition with and without different treatment options that can be applied in the field

Magnification levels offered by stereoscopes can be more useful for organisms such as flagellates, amoebae, ciliates, and rotifers where milliliters of culture can be observed at once, potentially increasing detection sensitivity by orders of magnitude.

Microscopy can be made quantitative by using either hemocytometers, Sedgewick Rafter slides, or other types of counting slides or chambers. An approximate number of pest organisms per given volume are often enough to trigger a crop protective action at the farm scale, but more rigorous quantification would be required when studying pests and potential treatments in a research setting. Microscopy can also be automated although most systems are not inexpensive, and current technology makes it difficult to set these systems up effectively, mainly due to the need to manually create initial screening filters and the low resolution of resulting images. Many organisms are difficult to observe in light microscopy, and the use of phase contrast is often necessary to enable visualization of pest organisms.

Pulse amplitude modulated fluorometry (PAM) is used to measure the quantum yield of photosystem II (F_v/F_m), the minimal or maximal fluorescence (F_o or F_m), as well as other photosynthetic parameters. All three of these measurements can correlate with culture "health" and broadly speaking can be indicators of pest presence or other detrimental effects on cultures (Juneau et al. 2003; Dong et al. 2013; McBride et al. 2014). A rapid drop in F_v/F_m is typically observed prior to a pond crash due to a parasitic pest infection (e.g., fungal/chytrid-like or bacterial infections) and, depending on the pest and time of year, may allow time to respond with a crop protective treatment or process prior to culture loss (Fig. 4.4, panel a).

Other factors such as temperature, culture density (i.e., harvest or dilution), and light can also affect F_v/F_m measurements (as well as time of the day of sample collection and assay methodology) (Alboresi et al. 2016). Thus, in most cases, data from PAM should be used as a trigger for further investigation, rather than for a specific action. During most pest infections, chlorophyll fluorescence will change prior to changes being detected in biomass measurements such as OD and dry weight. Because of this, monitoring of fluorescence either as part of PAM assays or directly via a spectrofluorometer can give advanced warning of potential problems over other measurements of biomass. Certain pigments may be absent from crop or weed strains, and so the monitoring of fluorescence at specific wavelengths can be a useful tool for detection of contaminating algal strains (Beutler et al. 2002; Winckelmann et al. 2016). For example, monitoring for phycocyanin fluorescence can show presence of certain cyanobacteria, and chlorophyll *b* fluorescence can be used to monitor weed algae in *Nannochloropsis* culture since this alga only contains chlorophyll *a*.

In terms of monitoring, a major advantage that algal farming has over traditional agriculture is that the farm can essentially be miniaturized in a controlled growth chamber where variables can be manipulated. This is mostly driven by the fact that the liquid pond culture is an essentially homogeneous entity (excluding biofilm and surface interfaces). This ability to culture small volumes of pond samples referred to as "satellite cultures" in a laboratory setting is essential to pest research as well as providing a predictive monitoring tool for pond crashes in the production setting (Fig. 4.4).

If a sample of pond culture brought into the laboratory crashes or grows at a reduced rate, this suggests that a significant population of pest or other inhibitory factor existed in the original sample and given the correct set of conditions could reduce yield. More importantly, if the crash can then be successfully passaged to clean culture, it shows that the pest crash can be studied in the lab setting (and one of Koch's postulates has been met) (McBride et al. 2014). If satellite cultures are cultivated in warm conditions where pest populations should propagate quickly, they offer a prediction of how the pond culture may act in the near future, without intervention. Satellite culture can thus be more predictive than other general monitoring techniques. Coupled with selective additions to the pond sample (such as available crop protection treatments, fertilizer, etc.), this technique can also be used to predict which treatments may prevent the crash or improve algal growth with direct feedback to cultivation practices, often within 24–48 h of sample collection (Fig. 4.4, panel e). This type of assay can utilize any growth system from 96-well plates to miniponds or even production-sized ponds if resources allow.

Some systems will not work for all algal crops or pests and offer different levels of control along with differing costs. In our experience, all pests studied thus far have worked well in traditional Erlenmeyer flasks shaken at 100–150 rpm in a lighted box with CO_2 supply (McBride et al. 2014), but higher throughput can be gained by utilizing smaller culture volumes in plates or tubes (Lee et al. 2018). Capital expenditure for a simple growth system is relatively small, but utilizing it can be labor-intensive depending on the level of data collection and number of cultures desired. These systems also require trained lab staff to operate them and to collect and interpret the data, although this can be accomplished by careful coordination with an offsite agronomist and sufficiently trained operations staff.

Useful and actionable data can be gained from simply observing the presence or absence of a crash in satellite cultures set up once or twice per week with and without the addition of crop protection chemicals (Fig. 4.4, panel d). The level of data collection beyond that can be determined by resource availability. The amount of time between an observed crash in a satellite culture and the parent pond crashing will need to be determined for each season and pest. Knowing how predictive the system is is obviously vital to the successful use of this tool.

Some organisms can be seen by the naked eye within small culture samples, for example, daphnia, water fleas, and rotifers in 5–50 ml sample tubes. Larger organisms such as water fleas may settle to the bottom of conical bottom tubes or be collected by low-speed centrifugation without concentrating the algae. Rotifers can be observed at the top of culture tubes that have been left to settle under a light source since they are phototaxic (Cornillac et al. 1983). An experienced observer can detect rotifers in this manner at levels below 1/ml which is a reasonable threshold to consider deploying crop protection treatments or monitoring more closely. With the use of readily available and affordable equipment such as stereoscopes, organisms of rotifer size (75–250 μm) can easily be enumerated in pond samples, or samples concentrated via filtration, providing a quantitative assay for the population. Stains or dyes can aid observation of rotifers and similar-sized organisms such as Lugol's iodine or neutral red (Crippen and Perrier 1974). Quantification for this type of pest can also be automated with the use of equipment such as Flowcam or other similar imaging equipment (Day et al. 2012).

Genetic sequences unique to a pest organism can be identified from mixed samples or more easily after pest isolation. Typically, rDNA is targeted for identification purposes (e.g., 16S, 18S, ITS1&2), while others may be used for tracking purposes (Carney et al. 2016). Once a unique sequence identifier is generated, a number of techniques may be used for monitoring purposes. The method of quantitative polymerase chain reaction (qPCR) has been utilized in many applications, for example, detection of pathogenic bacteria in various water sources (Collins et al. 2015) or marine organisms (Bott et al. 2010). It is a common method for algal pest detection (Letcher et al. 2013; Fulbright et al. 2014; Carney et al. 2016). For assays of this nature, the sample size is the limiting factor for detection limit. The simplest sample preparation method is boiling in a 96-well plate format after addition of a lysis buffer with sample sizes in the range of 50–150 μl (Letcher et al. 2013). To increase detection limits, additional sample preparation steps would be required to concentrate the pest DNA which add time and cost, as well as potentially increasing the background level of DNA and other contaminants from non-pest organisms.

Next generation sequencing is a useful tool for monitoring both prokaryotes and eukaryotes in the microbiome. This data can also be used to identify pests and correlate population changes to pond crashes (Carney et al. 2016). Relative levels of organisms can also be determined using these sequencing techniques and thus can be useful if cultivating mixed cultures, transitioning between strains with similar phenotypes, or identifying weed species with similar morphologies to the crop.

The detection of one pest organism should not be sufficient to trigger a crop protective action. Thresholds for reaction to detected pests must take into account the economics of treatment versus the risk of no treatment. Economic injury levels (EIL) are defined as the population level of pests that can cause the amount of damage equal to the cost of treatment (Stern et al. 1959; Pedigo and Higley 1992). Pest numbers above the economic injury level thus impact profit. The goal of a crop protection program is to maintain pest populations below this level. Treatment thresholds (i.e., the pest population that triggers a response) may be lower than the EIL since the goal is to prevent the population growing to that size.

Prior to the development of specific pest assays, monitoring techniques will rely heavily on labor-intensive methods such as microscopy. Over time, as new assays are developed, the labor cost should be reduced. These techniques should be developed concomitantly with improvements in other areas of agronomic practices that ultimately reduce the need for monitoring. For example, scheduled dosing and maintenance of conditions that prevent pest crashes as discussed below should improve yield while also reducing the reliance on pest monitoring over the long run.

4.2.6.2 Proactive Versus Reactive Treatments

Adding a chemical to the pond in response to the detection of a pest is probably the most common action associated with crop protection. Once a pest is detected at a predetermined threshold level, the appropriate crop protective action is administered to the pond(s) which should render the pest undetectable within a reasonable timeframe (McBride et al. 2014). The type of crop protective treatments used in this reactive approach includes general biocides such as hypochlorite (Zmora and Richmond 2007; Park et al. 2016; Wang et al. 2017), benzalkonium (Steichen and Brown 2018), as well as agricultural pesticides (Huang et al. 2014; McBride et al. 2014).

Though affordable in most cases, the use of cidal agents comes with inherent drawbacks and risks. Most, if not all, chemicals reportedly used in this way have some level of toxicity to the algae. When determining dose concentration, the balance between efficacy against the pest and toxicity to the crop (and therefore loss of productivity) must be considered. Safety of the operators should also be a concern which may add further costs. Many consumer groups that would likely be key targets for algal-based products may also be against the use of chemicals such as pesticides. Pesticide use is also highly regulated, and laws prevent the use of products for purposes other than defined on the label. The cost of testing for relabeling a formula or developing and testing a new formula for use in algae production is likely prohibitive. Industry advocate groups or companies working together may be the only way to change regulations in the favor of the algae farmer.

The trigger for reactive action is highly dependent on the detection threshold of the assay used to measure pest pressure. Typical assays will use μl to ml sample volumes, many orders of magnitude smaller than an acre-sized production pond. The pest population must have been present in the pond prior to meeting the assay detection threshold and therefore must have been affecting productivity prior to that point. To get around this, two solutions seem obvious—improve the detection limit of the assays or take a proactive rather than reactive approach to crop protection. The best solution is likely a combination of both.

For certain pests that are present annually when conditions prevail, the approach of scheduling doses ahead of time may reduce the total amount of chemical treatment used over time as well as increasing crop yield (due to less productivity loss from pest consumption/infection prior to detection and lower toxicity from fewer chemical additions). When taking this proactive approach, monitoring is still important to confirm efficacy of treatment, but the frequency of reactive responses should be reduced. Adjustments to the schedule would be made upon detection of pests (indicating the current schedule is insufficient) or during environmental conditions that have been linked to increased pest pressure such as weather events. Pest pressure is often greater after wind, rain, or when temperatures differ from seasonal norms.

Monitoring and predictions of pest pressure are combined in agriculture for the practice of degree-day models to predict pest development times and to allow the timing of crop protective measures to coincide with particular life cycle of the pest or to result in maximum control with the minimum dose. Degree-day models have been successfully used for bacterial, fungal, and insect pests in agriculture (Pruess 1983; Moschini and Fortugno 1996; Fukui et al.1999). Building degree-day models relies on the fact that the life cycle time for non-warm-blooded organisms is highly dependent on temperature. A minimum temperature, below which growth and development is inhibited, must be known for a given pest, and a starting point or day 0 must be defined to begin accumulating degree days (this could be annual, seasonal, or upon triggering of detection threshold or other measurement). For most organisms there is an approximately linear relationship with temperature and population growth, above the minimum temperature, until a maximum temperature is reached where growth is again inhibited. Thus, the length of time that the pest spends between these minimum and maximum temperatures will determine the time required for the population growth to reach economic injury level (i.e., the point at which product yield has been impacted).

A degree day is 1 day where the average temperature is above this minimum temperature by one degree, and they accumulate from the defined day 0. A treatment threshold can be set at a defined number of cumulative degree days since day 0. Since many pests of algal cultures show seasonality, it is likely that this kind of an approach to predicting treatment thresholds would be applicable. However, in order to apply this approach, at least the minimum temperature for pest growth must be known, and further details about the life cycle and monitoring techniques are also important. Unlike most agricultural crops, algae are harvested regularly and this will remove a portion of the pest population. To develop a degree-day model for algal farming, this dynamic population will have to be factored into any calculations.

One potential downside to a scheduled dosing approach is that it may prevent identification of new pests, if they are also sensitive to the treatments used. The use of satellite cultures with close inspection of any crashed cultures to confirm causal agents can help with this. As it is true that pests were present in the pond culture prior to reaching detection levels, it is also true that after treatment it is highly unlikely that the entire pest population will be removed. With use of monitoring and proactive treatments, the pest population should remain below either detection or economic thresholds and allow production to continue. Steichen and Brown (2018) showed that the application of benzalkonium chloride at 4-day intervals was enough to prevent propagation of *Vampirovibrio chlorellavorus* to high enough levels to cause culture crashes, though it remained detectable by qPCR. Growth cycle was extended, and overall productivity increased even though some algal toxicity was observed due to the treatment itself and the pest was still present.

Hundreds or more chemicals exist that may be useful for crop protection. The list is drastically shortened when you consider cost, algal toxicity, safety, consumer acceptance, and regulatory issues. Toxicity of potential chemical treatments against the algal crop is experimentally simple to determine. If pests can be cultivated in the laboratory, chemicals can also be tested for efficacy in a simple lab cultivation system. Successful candidates can then be used in outdoor cultures to confirm efficacy while looking for any potential off effects.

Some treatments are more effective in laboratory conditions than outdoors so lab to field translation is not guaranteed. Light intensity and UV light are major differences between indoor and outdoor cultivation systems that may affect stability or efficacy of chemicals. The presence of greater amounts of other biological or chemical components that chemicals could interact with may also reduce efficacy outside. This is evident in the use of chlorine which is less effective outside than in the laboratory (Zmora and Richmond 2007). As new strains, media, or other cultivation conditions are developed, and new pests are identified, toxicity and efficacy studies should be repeated. This activity is ideal for winter months when pest pressure is lower with the aim of developing the crop protection armamentarium for the following year. This process was recently illustrated in Karuppasamy et al. (2018) where known grazing pest models were tested against a large list of potential crop protection measures, with successful candidates demonstrated at field scale.

IPM strategies were introduced partly to reduce the reliance on pesticides. With the limited options for chemical treatments available for use on algal crops, there is a reliance on less specific treatments such as chlorine (Zmora and Richmond 2007; Park et al. 2016; Wang et al. 2017). Presumably such chemicals affect much of the pond ecology other than the target pest. Overuse of this type of treatment may result in the loss of potential beneficial organisms present in the pond microbiome, including natural predators to the pest organisms (Cole 1982; Johnke et al. 2017). Indeed, seeding or manipulation of the microbiome may be a potential future crop protection method to improve yield (Lian et al. 2018; Yao et al. 2019).

In order to reduce the use of pesticides in agriculture, the development of pest-resistant plants was important. These traits can be gained by breeding or genetic modification. Breeding of algae is not well established for production strains at this time, and though genetic tools are available for many algae, the path to cultivation of GM algae outdoors is yet to be determined (Szyjka et al. 2017). However, since most microalgae are single celled, directed evolution after mutagenesis and selection for desired traits is simpler than in plants. The development of cell lines resistant to chemicals or pests has been demonstrated in algae or related organisms (Simkovsky et al. 2012; Corcoran et al. 2018). The ability to study and cultivate pests in the laboratory setting is essential to such projects aimed at pest resistance. Robust secondary screens and a lab to field translation pipeline are also required to select against off-types and improve the likelihood of success.

In order to move away from the use of chemicals, another essential component of an IPM strategy is the use of biological control. That is allowing natural predators of the algal pests to persist and reduce pest population. This approach is impossible without a good understanding of the pest, its life cycle, and its interactions with other organisms in the local area. This principle has been demonstrated for control of weed algae in multiple algal crops. Rotifers discriminate food source by size, and thus, small algae are consumed preferentially and more prolifically over larger cells (Rothhaupt 1990). This can be used as an advantage as a biological control mechanism if the weed algae are smaller than the crop and the crop algae is too large to be consumed. Biological control using rotifers has been successfully used for removal of *Chlorella* contamination in Spirulina cultures (Mitchell and Richmond 1987) as well as for small algae contamination of *Desmodesmus* cultivation (unpublished). Some algae will clump together in the presence of rotifers, presumably to increase their relative size, so this approach may be suitable to algal strains that demonstrate this phenotype even if the individual cell size would be suitable for rotifer consumption. Since bacteria are a food source for many protists, some may be potential candidates for biological control of predatory bacteria (Johnke et al. 2017). To further expand the use of biological control, more must be understood about pond ecology and a move away from reliance on general biocides is necessary (Smith and Crews 2014).

Polycultures have been shown in multiple plant systems to offer potential increases in crop yield (e.g., Tooker and Frank 2012). The most obvious way this could be beneficial in algae is through pest resistance and prevention of pest propagation throughout a genetically and phenotypically diverse population (Shurin et al. 2013; McBride et al. 2016). This may be less obvious in a mixed homogeneous algal growth system where a pest is likely to meet its host randomly, but the frequency of this interaction should be reduced in a mixed population.

There may be barriers to use of mixed algal culture when the goal is to make a specific product, but for applications such as biofuel production, polyculture would seem a sound option (Newby et al. 2016). Increased biomass production and resilience to grazers has been demonstrated for cultures with increased algal species richness (Hillebrand and Cardinale 2004; Corcoran and Boeing 2012; Shurin et al. 2013). In our experience, cultivating mixed cultures in the presence of multiple pests, one single strain tends to dominate any culture that survives (unpublished observation). However, these observations only apply to short-term experiments at small scale with small consortia, and whether species richness would reestablish is unknown. Stability and productivity of high-rate wastewater ponds is an example of successful long-term use of polyculture which could likely be applied to biofuel production. Some of the obvious drawbacks to polyculture in traditional agriculture (such as different maturation rates, plant heights, etc. that may impede the use of mechanical equipment) are absent from an algal growth system where a harvest system is ideally strain agnostic.

4.2.6.3 Cultural Practices for Pest Prevention

Cultural practices in IPM include any physical activities around crop management that may be modified in order to confer an advantage to the crop or a disadvantage to a pest species. In traditional agriculture this could be activities such as timing of pruning, cleaning of site or equipment, etc. Sanitation is an area that is directly translatable to algal culture. Cleaning of ponds and equipment at regular intervals may be necessary, especially if pests have been a problem in recent months (Zmora and Richmond 2007). The method of sanitation will depend on the growth system and whether it contains a solid surface, but removal of built-up material should be a primary goal. Other cultural practices are discussed below.

Culture movement can be manipulated for crop protection. The flow rate of Oswald-style raceway ponds is typically kept between 20 and 30 cm/s. This flow rate permits the adequate mixing of the culture such that each algal cell is exposed to light at a frequent rate. Mixing with paddle wheels also produces aerosol that permits off-gassing of volatile gases (such as oxygen) and heat loss through evaporative cooling. The ideal mixing rate should be determined for yield but may be manipulated temporarily to provide crop protection. Some algal strains can withstand long periods without mixing, with no significant impact to productivity. Larger weed species that settle easily will be confined to the dark areas of the culture if flow rates are reduced, impeding their growth. Cost savings may also be possible if paddle wheels are turned off at night, since photosynthesis is only occurring at any significant rate during daylight, and when culture temperature is amenable (Moheimani and Borowitzka 2007; Borowitzka and Vonshak 2017). Allowing culture to stand briefly in pre-harvest containers or large ponds may also remove contaminants with faster settling rates from the harvest stream.

Pests may be removed or destroyed by physical means by differentiation on size or strength. Rotifers can be destroyed by cavitation (Mason et al. 2003; Kim et al. 2017) which can be introduced in several ways. Presumably shear forces within pumps used to move culture could be designed or selected to harm pests but not the algae. In-pond filtration devices can remove large organisms such as flies and larvae (Belay et al. 1994; Ravikumar 2014). Plankton nets used to sample zooplankton in the ocean, often dragged behind boats, could be adapted for sampling or removal of organisms greater than 50–100 um, for example, rotifers. Large quantities of fluids are filtered in other industries such as food and wastewater treatment so there may be off-the-shelf technology available that can be applied to algal culture.

4.2.6.4 Maintaining an Environment Versus Temporary Changes

Permanently modifying culture conditions can be crop protective by preventing propagation of pests or reducing pathogenicity. Several well-known cultivation methods use this approach such as alkalinity/high pH in Spirulina cultivation and high salt in *Dunaliella* cultures (Lee 2001; Borowitzka and Moheimani 2013). In these examples the somewhat extreme condition favors or advantages the algae while also inhibiting or preventing the growth or pathogenicity of the pest(s). However, many commercially interesting algal species do not grow in such extreme conditions, but maintenance of certain constituents of media can still be crop protective if a concentration that favors the crop over the pest can be experimentally determined.

There are only a limited number of additions in a typical algal media, and thus testing ranges for each component is relatively simple, but combinatorial effects can be complicated to assess. Exactly what concentration limits are viable will depend on economics as well as the chemistry of the water source available for cultivation (adding is much easier than taking away). Pests that infect algae such as bacteria and fungal-like organisms require a binding or recognition event for host attachment. This is likely mediated by proteins, and thus the pH, temperature, salt level, etc. may be manipulated to disrupt this interaction. For example, it was shown that an ion concentration of 20 mM and the inclusion of calcium are required for the infection of *Haematococcus pluvialis* by *Paraphysoderma sedebokerensis*, and host recognition could also be prevented by the addition of lectins that presumably block binding to cell surface carbohydrates (Gutman et al. 2011).

Phosphate levels have been shown to affect pathogenicity of many bacteria in other host systems. A phosphate level of below 0.2 mM was sufficient to induce virulence phenotypes in *Pseudomonas aeruginosa* (Bains et al. 2012), and *P. aeruginosa* PAO1 lethality to *Caenorhabditis elegans* required phosphate levels to be below 0.2–0.5 mM (Zaborin et al. 2009). Increased phosphate levels can protect some algae against a presumed bacterial pest infection (unpublished observation).

Alkalinity levels in the media, far below those used for Spirulina cultivation, can also be crop protective for certain algal species (unpublished observation). These and other observations show that extreme conditions are not always necessary to provide a crop protective environment. Media development typically focuses on biomass productivity and product synthesis, but the observations above show that crop protection should also be a consideration at this stage.

Algae appear to tolerate large temporary changes in some culture conditions, and this has been used as an advantage against some pest organisms. For example, the pH of the culture can be temporarily lowered or raised to protect against bacteria, rotifers, and flagellates (Becker 1994; Zmora and Richmond 2007; Ganuza et al. 2016; Ma et al. 2017). If the pH shift is accomplished with the use of CO_2 (to lower pH), or the growth of the algae itself (to raise pH), this is another treatment that may have limited long-term effects on culture chemistry since no residual chemical is left behind and requires little to no extra equipment. However, precipitation of some media components may be of concern if they do not dissolve after the culture returns to normal pH.

Unlike in traditional agriculture where run-off and percolation are an issue, anything that is put into a lined pond will remain there unless it is consumed (by the crop or other organisms), is volatile, or is removed by harvest. For example, repeated use of hypochlorite will result in the buildup of its conjugate salt (sodium or calcium) in the water over time. Since oxidation is a commonly used treatment for algal cultures, this would suggest an advantage to oxidizing treatments such as ozone and peroxide, which break down to oxygen and water and thus are unlikely to cause chemical buildup over time.

Bacterial pathogens are common algal pests found in cultivation systems (Ganuza et al. 2016; Lee et al. 2018) or in natural algal populations (Mayali and Azam 2004). Ions such as copper and zinc are known to reduce the growth of some organisms including bacteria, fungi, and algae and are often components of antifouling paints and other biocides (Babich and Stotzky 1978; Borkow and Gabbay 2004). However, resistance to both elements in bacterial populations is reported and would be a concern if considering their use (Cervantes and Gutierrez-Corona 1994; Choudhury and Srivastava 2001).

4.2.6.5 Management over Changing Seasons

All organisms have a range of conditions conducive to optimal growth. This is of course true for pests as well as the crop. Thus, there is seasonality to pest pressure observed in algal cultures. The majority of observed pests are most prevalent during the warmer months with limited detection of pests in winter in climates that have distinct seasons. However, some pests will be detected only in spring or fall, while others may be present as soon as culture temperatures reach a certain point and remain active until cooler temperatures return.

Once the seasonality of pests for the crop species is determined, culture management practices can be developed and may be modified by season. For example, if the ideal operating density for maximum algal productivity is 0.5 g/l, the maximum yield will be gained by maintaining culture density as close as possible to this value. Automation or small-volume cultivation systems may enable this, but in larger systems, a density range is inevitable as algae are pulled from the system only at harvest intervals. When to harvest and what culture density range to target is driven by operational and economic constraints as well as the cellular productivity. Crop protection should also be a consideration. For example, light and temperature are the major factors affecting rotifer hatching from resting cysts, and rotifer populations may propagate less well in denser algal cultures. Thus, lower losses to grazing may occur at higher culture densities in summer, when rotifer presence is highest. The slight reduction in maximum aerial productivity from growing at higher densities will result in higher actualized productivity due to more stable production and lower losses to grazing.

As with traditional agriculture where degree-day models can predict pest pressure, the same should be true for algal pests. Seasonality is clearly evident, but as more is understood of the life cycle of pests and more cultivation data is gathered year over year, such models should be developed for all relevant algal pests. Recent efforts from the multi-site ATP3 project should be a step toward developing this kind of methodology and making it available to the industry (McGowen et al. 2017; Knoshaug et al. 2018). One advantage of a liquid growth system over traditional agriculture is that although the weather is a major driver of culture temperature, there are ways to affect extremes of culture temperature such as modifying culture depth and changing paddle wheel speeds/schedules (to reduce/increase evaporative cooling). If economics allow, heating or cooling systems may be employed for temperature control.

4.2.6.6 Identifying New Pests

A pest is an organism that can reduce harvested productivity by any means. Broadly, they can be divided into three categories: those that consume the algae such as multicellular grazers; those that infect and use the algal cell to produce offspring such as viruses, bacteria, and fungi; and those that compete with the crop strain either by consuming nutrients, shading, or producing a growth-inhibitory factor such as weed organisms.

Algal pests should be considered ubiquitous. The quote "Everything is everywhere, but, the environment selects," although debated (O'Malley 2008), is a good mantra to live by with respect to algal pests. Pests should be considered present in the local area or will arrive and establish soon after cultivation begins (i.e., after provision of a suitable environment).

Bacteria, viruses, and small invertebrates can be transported to long distances by wind. For example, rotifers can be collected in vessels containing algae placed on top of a four-story building in the desert southwest (Elizabeth Walsh, personal communication). In our experience pest presence in open production ponds is often correlated with weather events such as wind storms or rain. *Vampirovibrio chlorellavorus* levels in *Chlorella* cultures increased after rain events, suggesting it was present in aerosols in the atmosphere (Steichen and Brown 2018).

The presence of bacteria in atmospheric aerosols has been shown for many bacterial species (Burrows et al. 2009). Bacteria and fungi contribute a significant portion of the sub-micron particles in atmospheric aerosols (DeLeon-Rodriguez et al. 2013), and bacterial concentrations in ambient air may average at least 1×10^4 cells m^{-3} over land (Bauer et al. 2002). The volume and movement of particles in the air is demonstrated by the intercontinental transport of dust. As much as 800 million metric tons of soil dust may be transported from North Africa each year (Huneeus et al. 2011), and much of that ends up crossing the Atlantic Ocean and can be measured in the air in North and South America (Prospero and Mayol-Bracero 2013). Park et al. (2018) observed the host alga *Chlorella* and parasite *Vampirovibrio chlorellavorus* in soil samples from nearby the cultivation location, showing that algae and specific pests can be found in natural reservoirs in the local environment (Park et al. 2018).

Grazers are probably the most common problem in algal cultures (Day et al. 2017). The rotifer genus *Brachionus* is the most recognizable example of a ubiquitous grazing pest for microalgae and is distributed globally (Mills et al. 2007). Many algae and organisms that consume or infect algae have a long persisting life stage that may survive in the environment for decades or longer. The resting cyst stage of the rotifer life cycle can be found in natural soil and sediment samples, referred to as egg banks, and may contain up to 2.9–13.6×10^4 eggs per m^2 (Hairston 1996). Any site or location that has propagated rotifers intentionally or otherwise will likely house egg banks in any local sediment. This is one observation that suggests regular cleaning of ponds would be beneficial especially if rotifers have ever been detected in previous seasons.

Given the correct range of temperature and light, resting eggs of rotifers will hatch within 24–48 h when water is present (Hagiwara and Hino 1989). After rain events rotifers can be found in standing water where they will consume as much biomass as possible, reproduce, and then produce resting cysts again when food source or water runs out or the population density reaches a critical level (Walsh et al. 2014). Adding local soil samples to cultivation media will result in identification of potential weed species and algal pests already present in the local area although this approach will identify numerous organisms that will have no impact on algae cultivation. Resting eggs from rotifers and other invertebrates such as *Daphnia* can be isolated and identified relatively easily from local soil or sediment samples (Briski et al. 2013).

Other algal pests with more specific host ranges have been identified in multiple global locations. *Paraphysoderma sedebokerense* identified from cultures of *Scenedesmus* in Las Cruces, New Mexico, USA (Letcher et al. 2016), is considered conspecific (from rDNA sequencing) with *Paraphysoderma* identified in cultures of *Haematococcus pluvialis* in Israel (Hoffman et al. 2008). This is particularly interesting since it also shows a broad host range for this pest which attaches to (and thus presumably identifies in some way) the algal cell in order to initiate infection (Gutman et al. 2011). Previous attempts at infecting other species had suggested that certain species of *Scenedesmus* would not be sensitive to this pest, possibly showing the effect of intraspecies differences or the effect of cultivation media and conditions (Gutman et al. 2009).

Electron microscopy images of bacterial infection of green algae (Lee et al. 2018) bear a striking resemblance to images of samples from Germany in the 1970s (Schnepf et al. 1974). Anyone who has used DNA sequence identified from an algal culture in a BLAST search will be familiar with the top hits being "uncultured organism" from "environmental DNA" from all over the world. The use of next generation sequencing techniques on environmental samples has highlighted the distribution of most genera.

As early as possible when beginning to cultivate algae at a new site, it can be beneficial to perform the "goat on a stick" experiment to assess local populations of pest organisms. That is to say that algae can be used as bait to attract pests. Algal culture placed outside without crop protection application will eventually contain and succumb to pests. The sooner the chosen algal species is grown outside and at as large a scale as possible, the sooner novel pests can be identified and researched before the site is fully operational. As cultivation area increases, the likelihood of novel pest arrival and identification increases (Smith and Crews 2014). As much upfront pest research as possible can be done during scale-up, before resources will be needed in production including targeted environmental sampling in combination with cultivation of unprotected culture.

Viruses are ubiquitous and studied in bloom forming as well as other algal species but are not yet widely reported as being detrimental to commercially grown algal cultures (Coy et al. 2018). The giant viruses that infect *Chlorella* are distributed globally and can typically be found in natural freshwaters at concentrations of 1–100 plaque forming units per milliliter (Quispe et al. 2016).

Many general microbiological techniques can be used to study and isolate pest organisms. Since the algae are growing in a liquid medium, most pests can be co-cultivated in liquid media with the host, and this is typically enough to allow further study. Many strains of algal pests can also be isolated in either liquid or solid media. It is typically easier to cultivate organisms with their host as devising culture media for axenic growth can be difficult and time consuming, and many organisms are likely to be obligate pathogens.

Amoebae can be grown in isolation or on algal or bacterial hosts and will crawl across the surface of solid media consuming food organisms inoculated on the surface. This method can facilitate isolation of amoebae or for use in screening potential resistant algal strains (Simkovsky et al. 2012). A common technique to isolate viruses is using plaque plates, where a dilution of viral sample is layered over a lawn of the host organism. Small holes, or plaques, form where the virus has propagated and destroyed the host. Plaque plating can be used for fungal/fungal-like pests (Letcher et al. 2013) as well as for viruses of algae (VAN Etten et al. 1983) and could likely be applied to other parasitic pests with algal hosts.

Most algal parasites are smaller than the host alga and/or are translucent and thus can be difficult to observe via basic light microscopy. The use of phase contrast is essential to observe many of these organisms and should be considered for routine microscopy of all cultures. Electron microscopy can be valuable for connecting a pest phenotype to a culture crash. Without electron microscopic observation of the bacterial predator of *Nannochloropsis* FD111, the BALO-like life cycle would not have been observed and a connection between the bacterial phenotypes observed via light microscopy in crashed samples and the life cycle stage internal to the host would have been difficult to make (Lee et al. 2018). Similarly, electron microscopy in conjunction with rDNA sequencing of fungal pests of *Scenedesmus* revealed similarities with previously published organisms (Hoffman et al. 2008; Letcher et al. 2016). These observations highlight the utility of interdisciplinary research and collaboration, surely an essential component of any algal pest research.

Since pests are to be considered ubiquitous and pest presence is determined by assay detection limits, all ponds should be treated as routine upon detection in any one culture—if the pest is in one pond, it is in all of them.

4.2.6.7 Identifying and Responding to Potential Crop Failure

At the research or pilot scale, where new algal strains and culture conditions, etc. are being trialed, multiple new pest phenotypes should be expected annually. For a given individual strain, new pests will likely appear for several years after a strain is initially cultivated outdoors. The first year of cultivation of a new strain may show relatively little pest pressure, presumably until pest populations in the local area are established to high enough levels. This "honeymoon period" can give a false sense of pest resilience for a newly cultivated strain, highlighting the need for trials that span multiple seasons, years, and locations. Conversely, if a new strain succumbs to pests very early in a field trial, it may not mean that the strain itself is highly susceptible to pests, but that the pests that the strain is susceptible to are already established in the vicinity. In our experience, the first pests to appear for most strains are grazers such as rotifers and parasitic pests. Once you can keep the crop from being eaten or infected, weed algae are then the more common pest.

Reacting quickly to a potential crop failure is obviously important. As with all areas of crop management, determining a set of triggers and the immediate action required when those triggers are met will help prevent missed opportunities. Once triggered the goal of pest identification activities should be to confirm that a pest is indeed the cause of the observed phenotype and then to characterize the organism and develop treatment and tracking methods. Determining the trigger requires careful consideration and analysis of historical data. An overly sensitive trigger will result in many false positives and excess work, while an insensitive trigger could yield missed opportunities and risk crop or experiment failure.

Prior to identifying a pest, there is a heavy reliance on general measures of algal health and growth, that are crop and pest agnostic, to serve as a signal for potential crop failure. Types of data that may be used as a primary trigger for investigative action may include changes to culture color, drops in F_v/F_m or other measured parameters, slower than expected growth, etc. Some may be generally applied, but others will need to be determined for each site, cultivation method, and strain. The most obvious trigger that requires an immediate response is a crashed pond, which typically will manifest as an overnight color change from green to brown, with severe drop in F_v/F_m values and possibly biomass loss. If satellite cultures are being used as part of the monitoring program, the crash should appear in these laboratory cultures prior to occurring in the field, hopefully giving time to respond in the field prior to pond failure. A flow diagram of activities for pest discovery work is shown in Fig. 4.5.

The initial reaction to a health trigger should be to establish if the observed phenotype has been caused by a known pest, in which case the appropriate response can be taken. All pond parameters and recent additions and harvests should be investigated and confirmed to be normal or correct. If no obvious reason for the observed phenotype can be found, pest identification procedures should be initiated. The primary goal should be to gather data on the phenotype, store samples for later use, and recapitulate the phenotype in the lab. The breadth of response will be governed by resource availability, but gathering as much data as possible in a short timeframe should be the goal with initial activities being completed the same day. Samples should be collected as soon as possible, and satellite culture from this sample should be observed in the laboratory. Samples of the culture should also be stored frozen using multiple cryopreservation techniques (e.g., 10% DMSO, 25% glycerol). The availability of infection sources greatly improves the success of later research. Once pest propagation and the optimum cryopreservation technique are established, further samples should be stored and monitored for viability at regular intervals.

In the early stages of identifying a new pest, it is not always clear what the pest phenotype is or the algal response, and thus gathering as much data as possible is important. Retrospective analysis of this data when more details have been determined can help identify characteristics that may initially be missed. Collecting many random microscopic images using different magnification or styles (e.g., phase contrast versus light or dark field) is often a useful first activity. Taking pictures at random can help prevent natural bias toward phenotypes that are present in the sample but later turn out to be coincidental to the crash.

The next level of actions is aimed at characterizing and identifying the pest. Determining if the observed reduction in productivity or culture health was caused by biotic factors is an important step in pest identification which may be included in primary actions or later depending on resources and initial observations. If the pond culture has crashed, this can be started immediately. Alternatively, if a crash is observed in satellite culture within a reasonable timeframe (e.g., 1–5 days), this

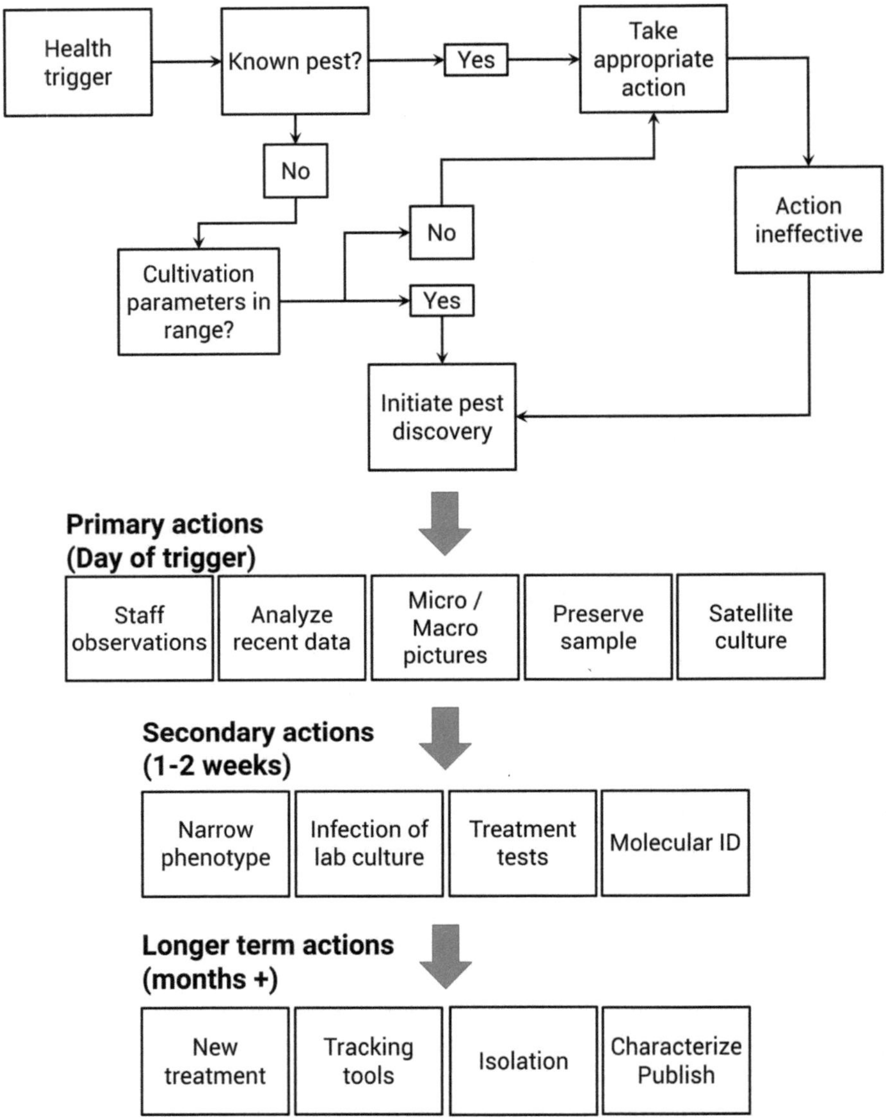

Fig. 4.5 Response to potential pond failure. Flow diagram of potential actions in response to an algal health decline due to potential pest organism

sample can be used. A culture crash could be indicative of a pest infection (i.e., biotic factors have caused the crash), or alternatively, the medium could either contain a toxin or be lacking in some essential component. To help distinguish between these and other possible causes, a recently crashed culture can be sub-cultured into clean laboratory culture in fresh medium (typically at a dilution rate of 0.1–1.0% v/v). To be successful this sub-culture or re-infection experiment should be completed within a day of the initial crash. Some pests may propagate well after longer periods, but typically the infection is most active during a crash or shortly after the crash is observed. If a secondary crash is observed in this new culture, that is suggestive of infection from a biotic agent since the media chemistry should be close to optimum at this point. A secondary crash is also important since it demonstrates that the pest can be propagated under laboratory conditions, and thus it can be studied (and one of Koch's postulates has been met). From this secondary crash, molecular identification through sequencing of the rDNA can be completed via various means.

To further characterize the causal agent, a number of relatively simple experiments can narrow down the likely cause and aid in future efforts for pest identification. If a recently crashed culture is used as an infection source, it can be treated by various means prior to infection of clean culture, to help determine the causal agent. If the causal agent is biotic, heat denaturation (e.g., by boiling) should prevent culture crash. To determine the size of the potential pest, the infection source may be filtered using various pore sizes. A 0.22 μm filter should eliminate all bacteria but not all viruses, while 0.45 μm would remove most bacterial species. Larger pore sizes such as 5 μm or 50 μm remove pests such as grazers and other eukaryotes. Certain pests may attach to the host algae by recognition of certain cell surface carbohydrates (Gutman et al. 2011) so selection of filter material may be important in these experiments since some organisms may bind to those containing cellulose (a component of some algal cell walls), irrespective of pore size. Addition of antibiotics with known modes of action may be added to infection experiments to help determine the type of pest organism present while also yielding information that may be useful later when attempting pest isolation and clean-up.

If an unknown pest is suspected in a production system, the priority is to keep the culture alive and productive and preventing other nearby cultures from also deteriorating. Using all available treatment options in infection experiments, or in multiple satellite cultures of the initial pond sample, will indicate which treatments may be successful in the field. If no suitable treatments are discovered this way, longer-term systematic research into treatment options will be required once a reliable pest model is established (Karuppasamy et al. 2018). Other longer-term activities include steps to further characterize and isolate the pest organism. Sequencing and taxonomic analysis can lead to the development of tracking tools such qPCR as well as directing research efforts based on literature around similar organisms.

Crop protection is an area where even rival companies ought to be able to find common ground to collaborate and is also an area of research where the academic labs should work heavily. For little upfront cost (especially compared to cultivation system costs), novel pest organisms can be studied at lab scale generating publishable content very quickly. They are also ideal experiments for academic collaboration since there are many laboratories worldwide studying the physiology and taxonomy of algal pests or organisms with highly similar life cycles such as amoebas, rotifers, bacteria, viruses, etc. These researchers routinely use techniques useful to algal crop protection research but are currently applied to other host organisms. There is a wealth of methodology ready to be transferred for use in algal crop protection laboratories.

One significant issue with crop protection methodologies is that it is very difficult to be prescriptive. The interaction between strain, location, water chemistry, fertilizer selection, feeding regime and timing, harvest management, water recycle, seasonality, etc. means that what works for one strain at one location may or may not work for another strain at another location. For example, Qualitas Health operates two farms, one in West Texas and one in Southern New Mexico; the same strain is cultivated on both farms. The farms are close enough geographically to share similar weather and seasonal profiles, but their water sources have very different chemistry profiles, and their major pests in summer are different. While the farms are managed as similarly as possible, some crop protection techniques are specific to each site, and the chief agronomist is responsible for both the overarching management of both facilities and the specialization of any methods or crop protection responses and crop protection and yield management in general (Rebecca White, unpublished data).

4.2.7 The Role of the Agronomist on an Algae Farm

An agronomist, in the traditional definition, is a scientist whose job is to ensure the health and productivity of a crop. In practicality, this person often reviews data regarding the current crop, directs sampling and testing to interrogate issues with the current crop, and reviews new scientific developments for that crop, often liaising between crop researchers and farmers, in order to provide recommendations to the farmers to maximize yield and quality (https://www.sokanu.com/careers/agronomist/). Algae producers often have personnel in this role, but rarely do they use this title (although they should).

Ideally, an agronomist would have available to him or her a large historical set of statistically meaningful data from multi-season field trails, from geographically separated locations. Along with this, they would have ability to do highly replicated and controlled experiments at varying scales smaller than the production system. Of course, few in the algae industry have this level of historical data or have the resources to experiment in this detail at a production facility while remaining (or becoming) profitable. Once at production scale, the ability to run well-controlled experiments becomes more difficult and the experience and insight of the staff becomes very important. As with the medical profession where replication and controls are challenging, "a strong intuition is much more powerful than a weak test" (Mukherjee 2015). The power of observations from operations and lab staff who interact with the algae on a daily basis *cannot be understated*, and if acted upon, these observations can improve cultivation practices over time. The authors agree with the statement by Borowitzka and Vonshak "A common feature of all large-scale operations known to us is that, over time (years), both productivity and reliability of the cultures improve as the operators gather experience in managing their cultures." (Borowitzka and Vonshak 2017).

The main measurement of success or failure for a farm is yield. Crop yield is typically measured in units per area, for example, bushels per acre. Productivity for algal production is typically defined as in pond growth in grams of algal dry weight per unit area over time, for example, g/m^2/day. This is a useful measure at small scale and in a research setting, but what is more important at scale is total product yield, meaning what is sent for downstream processing from harvest, expressed as g/m^2/day (where m^2 is the area of the actively cultivated portion of the farm). Relying on productivity data for calculations obscures that fact that many downstream processes are not 100% efficient, and thus what is grown in the pond does not all end up getting sold as a product. It is the agronomist's role to marry both productivity and yield and to manage the farm in such a way as to minimize difference between the two measures. Williams and Laurens (2010) provide an excellent outline of the major factors that impact productivity—light, temperature, mixing, and CO_2—which is where any algae agronomist should start when optimizing the parameters on his or her farm.

4.3 Remaining Challenges

Algae are grown productively and profitably in many places around the world yet still hold great promise to be used for even more purposes. However, there are some key limitations in the state of current technology and market acceptance that remain as barriers for widespread algae production.

4.3.1 Infrastructure

Capital costs for building algae farms are high (Davis et al. 2011) compared to traditional farming, and there is a lack of standard technology for algal biomass production. Although numerous consulting companies and technology companies do exist, there is still a need for self-design of equipment and processes, of which many will be covered by invention protection. As the industry has matured, more off-the-shelf technology is becoming available, but thus far this is restricted to technology-heavy areas (e.g., PBRs) or those that have translated directly from other industries (e.g., centrifuges).

Unlike a field of wheat, a field of algae requires much more connectivity to the rest of the world, due to the year-round need for inputs and the constant need for electricity. Although much land and water are available and applicable for algal cultivation, the infrastructure may not be in place to allow constant access and power, etc. Additionally, access to CO_2 is a barrier; CO_2 use via colocalization increases land cost, but where land is inexpensive, CO_2 must be transported over long distances, increasing the cost and decreasing algae's value as a CO_2 sink. Development of commercial direct air capture (DAC) technology for CO_2 delivery is required, along with a tax credit structure that would allow CO_2 emitters to operator DAC units at algae farms distant from the original omission source while still receiving credit for the carbon capture and use.

Lack of efficient and scalable harvest technology continues to be an issue for the industry. Methods for harvesting algae have been reviewed extensively, comparing their energy use, operational expenses, capital costs, achievable solids concentrations, impacts on products, as well as other characteristics (Barros et al. 2015; Mathimani and Mallick 2018; Sing and Patidar 2018; Show et al. 2019). Technology borrowed from other industries, such as centrifugation, is too costly and too energy inefficient, and others such as DAF are not efficient enough in terms of biomass capture and involve adding chemical which ends up both in the product and in the recycled water stream. Newer technologies, admittedly still borrowing from other industries, such as hollow fiber membrane filtration hold great promise but are still unproven at any relevant scale. A combination of different harvesting methods may be required (a primary dewatering step combined with a secondary dewatering step using the more expensive centrifugation) in order to achieve the necessary solids concentration while maintaining the integrity of the product for downstream processing (Borowitzka 1992).

4.3.2 Crop Protection Options

Most of the algae that have been successfully cultivated at scale to date rely on culture conditions that are inhospitable to most pests (e.g., *Dunaliella* or Spirulina). Alternatively, for other species some form of batch culture is used (*Haematococcus*) to minimize contamination and product loss. Both of these approaches reduce the requirement for crop protection processes. In order to cultivate other desirable strains that require less saline conditions and more neutral pH ranges, more options for crop protection will be required. Currently, cultivation of common alga such as *Scenedesmus*, *Chlorella*, *Desmodesmus*, and *Nannochloropsis* in any published media recipe will result in culture crash without significant intervention.

Most reactive treatment options published so far are general cidal agents which affect many organisms other than the target. Because of this, these broad-spectrum biocides actually go against the principle of IPM that we should be striving for. Specific treatments would be more beneficial to prevent off-target effects on culture ecology and may be less likely to be toxic to the crop itself. Growers of land crops often have multiple active ingredient, formulation, and brand options of pesticide for any given crop and pest combination. Algal crop protection is far more limited in scope. The size of the industry means investment in pesticide development by chemical producers is unlikely to happen. Pesticide use is also unattractive to the end consumer of many algal products. The use of agronomic practices and general biocides typically used in swimming pools and home or industrial sanitation has enabled the large-scale cultivation of a number of algal species, but more options are required for new algal crop species and as new pest organisms continue to be identified. There is enough diversity in strains, production methods, extraction protocols, final products, etc. that the industry ought to be able to work together or share crop protection information. Lobbying as a group may be necessary to gain regulatory approval for use of certain treatments.

4.3.3 *Other*

The National Institute for Food and Agriculture (NIFA) recently celebrated the 100-year anniversary of the extension service. Through this service land grant colleges and universities bring vital, practical information to agricultural producers. They provide services such as technology transfer and advice and translate science for practical application in areas from cultivation to marketing and economics. They also provide education to children through the 4-H program and training to beginning farmers through the Beginning Farmer and Rancher Development Program (https://nifa.usda.gov/). The extensive service does not currently have any algae-related information or activities, as far as the authors are aware, as this is a critical gap to bringing large-scale algae cultivation into the agricultural community at large and solving some of the critical challenges the industry faces.

In spite of the lack of activity from the extension service, from publication and presentation records, it is clear that most, if not all, companies in the algae industry are willing to collaborate and share information, at least at some level. The national labs of the United States also receive governmental funding for algal research, and many researchers at these labs have published with industry co-authors. The Algae Foundation's Algae Technology Educational Consortium (ATEC) (www.algaefoundationatec.org) consists of people from universities, community colleges, national labs, and industry, and ATEC is currently developing a 2-year associates degree course in Algae Biology, Technology, and Cultivation for workforce development. The Algae Foundation also has a K-12 STEM education program, the Algae Academy, that reaches schools in nearly half the states in the United States each year (personal communication, Marissa Nalley). The Arizona Center for Algae Technology and Innovation (AzCATI) also provides various services relating to algae including equipment testing, biomass supply, and education and training. These types of service should provide the workforce for the future algae industry. It is clear that the industry has recognized the need for collaboration, awareness, and education; with the passage of the 2018 Farm Bill, the authors hope that the USDA will begin to include this base of established activities into the extension service.

4.4 Conclusions

Algae as agriculture is growing—more farms, more products, in more areas of the world. The most critical advancements, particularly for open pond production, have been in the adaptation of integrated pest management practices. With a view toward the complexity of the growth system and how each factor impacts the other, and in particular crop protection, it has become clear that the major focus of agronomic practices should be maximizing the health of the culture while minimizing the opportunities or ability of pests to propagate. A strong program of agronomics and a philosophy of preventative protection of the crop are critical to continued growth of this sector of the industry.

References

Alboresi A, Perin G, Vitulo N et al (2016) Light remodels lipid biosynthesis in *Nannochloropsis gaditana* by modulating carbon partitioning between organelles. Plant Physiol 171 (4):2468–2482. https://doi.org/10.1104/pp.16.00599

Ashraf M (University of V and AS), Javaid M, Fish SH, Rashid T (Fisheries R and TI et al.) (2011) Replacement of expensive pure nutritive media with low cost commercial fertilizers for mass culture of freshwater algae, *Chlorella vulgaris*. Int J Agric Biol (Pakistan) 13:484–490

Babich H, Stotzky G (1978) Toxicity of zinc to fungi, bacteria, and coliphages: influence of chloride ions. Appl Environ Microbiol 36:906–914

Bains M, Fernández L, Hancock REW (2012) Phosphate starvation promotes swarming motility and cytotoxicity of *Pseudomonas aeruginosa*. Appl Environ Microbiol 78:6762–6768

Barros AI, Goncalves AL, Simoes M, Pires JCM (2015) Harvesting techniques applied to microalgae: a review. Renew Sustain Energy Rev 41:1489–1500

Bauer H, Kasper-Giebl A, Löflund M et al (2002) The contribution of bacteria and fungal spores to the organic carbon content of cloud water, precipitation and aerosols. Atmos Res 64:109–119

Becker EW (1994) Microalgae: biotechnology: biotechnology and microbiology, 1st edn. Cambridge University Press, Cambridge

Beers EH, Brunner JF (1993) Orchard pest management online. Washington State University, Wenatchee, WA

Belay A, Ota Y, Miyakawa K et al (1994) Production of high quality *Spirulina* at Earthrise farms. In: Vonshak A (ed) Algal biotechnology in the Asia-pacific region. Taylor and Frances, London, pp 131–158

Benton Jones J (2002) Agronomic handbook: management of crops, soils and their fertility. CRC, Boca Raton, FL

Beutler M, Wiltshire KH, Meyer B et al (2002) A fluorometric method for the differentiation of algal populations in vivo and in situ. Photosynth Res 72:39–53

Borkow G, Gabbay J (2004) Putting copper into action: copper-impregnated products with potent biocidal activities. FASEB J 18:1728–1730

Borowitzka MA (1992) Algal biotechnology products and processes—matching science and economics. J Appl Phycol 4:267–279

Borowitzka MA (2016) Algal physiology and large-scale outdoor cultures of microalgae. In: Borowitzka MA, Beardall J, Raven JA (eds) The physiology of microalgae. Springer, Cham, pp 601–652

Borowitzka MA, Moheimani NR (2013) Open pond culture systems. In: Borowitzka MA, Moheimani NR (eds) Algae for biofuels and energy, Developments in applied phycology. Springer Netherlands, Dordrecht, pp 133–152. https://doi.org/10.1007/978-94-007-5479-9_8

Borowitzka MA, Vonshak A (2017) Scaling up microalgal cultures to commercial scale. Eur J Phycol 52:407–418

Bott NJ, Ophel-Keller KM, Sierp MT et al (2010) Toward routine, DNA-based detection methods for marine pests. Biotechnol Adv 28:706–714

Briski E, Bailey SA, MacIsaac HJ (2013) Separation strategies for invertebrate dormant stages contained in sediment. Aquat Biol 18:209–215

Burrows SM, Elbert W, Lawrence MG et al (2009) Bacteria in the global atmosphere – Part 1: review and synthesis of literature data for different ecosystems. Atmos Chem Phys 9:9263–9280

Camacho-Rodríguez J, Cerón-García MC, González-López CV et al (2013) A low-cost culture medium for the production of *Nannochloropsis gaditana* biomass optimized for aquaculture. Bioresour Technol 144:57–66

Carney LT, Wilkenfeld JS, Lane PD et al (2016) Pond crash forensics: presumptive identification of pond crash agents by next generation sequencing in replicate raceway mass cultures of *Nannochloropsis salina*. Algal Res 17:341–347

Cervantes C, Gutierrez-Corona F (1994) Copper resistance mechanisms in bacteria and fungi. FEMS Microbiol Rev 14:121–137

Chew KW, Chia SR, Show PL et al (2018) Effects of water culture medium, cultivation systems and growth modes for microalgae cultivation: a review. J Taiwan Inst Chem Eng 91:332–344

Choudhury R, Srivastava S (2001) Zinc resistance mechanisms in bacteria. Curr Sci 81:768–775

Cole JJ (1982) Interactions between bacteria and algae in aquatic ecosystems. Annu Rev Ecol Syst 13:291–314

Colla LM, Oliveira Reinehr C, Reichert C et al (2007) Production of biomass and nutraceutical compounds by *Spirulina platensis* under different temperature and nitrogen regimes. Bioresour Technol 98:1489–1493

Collins S, Jorgensen F, Willis C et al (2015) Real-time PCR to supplement gold-standard culture-based detection of *Legionella* in environmental samples. J Appl Microbiol 119:1158–1169

Converti A, Casazza AA, Ortiz EY et al (2009) Effect of temperature and nitrogen concentration on the growth and lipid content of *Nannochloropsis oculata* and *Chlorella vulgaris* for biodiesel production. Chem Eng Process Process Intensif 48:1146–1151

Corcoran AA, Boeing WJ (2012) Biodiversity increases the productivity and stability of phytoplankton communities. PLoS One 7:e49397

Corcoran AA, Saunders MA, Hanley AP et al (2018) Iterative screening of an evolutionary engineered *Desmodesmus* generates robust field strains with pesticide tolerance. Algal Res 31:443–453

Cornillac A, Wurdak E, Clément P (1983) Phototaxis in monochromatic light and microspectrophotometry of the cerebral eye of the rotifer *Brachionus calyciflorus*. Hydrobiologia 104:191–196

Coy SR, Gann ER, Pound HL et al (2018) Viruses of eukaryotic algae: diversity, methods for detection, and future directions. Viruses 10:487

Crippen RW, Perrier JL (1974) The use of neutral red and Evans blue for live-dead determinations of marine plankton (with comments on the use of rotenone for inhibition of grazing). Stain Technol 49:97–104

Davis R, Aden A, Pienkos PT (2011) Techno-economic analysis of autotrophic microalgae for fuel production. Appl Energy 88:3524–3531

Davis R, Fishman D, Frank ED et al (2012) Renewable diesel from algal lipids: an integrated baseline for cost, emissions, and resource potential from a harmonized model. National Renewable Energy Lab. (NREL), Golden, CO

Day JG, Thomas NJ, Achilles-Day UEM et al (2012) Early detection of protozoan grazers in algal biofuel cultures. Bioresour Technol 114:715–719

Day JG, Gong Y, Hu Q (2017) Microzooplanktonic grazers – a potentially devastating threat to the commercial success of microalgal mass culture. Algal Res 27:356–365

DeLeon-Rodriguez N, Lathem TL, Rodriguez-R LM et al (2013) Microbiome of the upper troposphere: species composition and prevalence, effects of tropical storms, and atmospheric implications. Proc Natl Acad Sci U S A 110:2575–2580

Dong H-P, Williams E, Wang D-Z et al (2013) Responses of *Nannochloropsis oceanica* IMET1 to long-term nitrogen starvation and recovery. Plant Physiol 162:1110–1126

Drexler ILC, Yeh DH (2014) Membrane applications for microalgae cultivation and harvesting: a review. Rev Environ Sci Biotechnol 13:487–504

Edmundson SJ, Huesemann MH (2015) The dark side of algae cultivation: characterizing night biomass loss in three photosynthetic algae, *Chlorella sorokiniana*, *Nannochloropsis salina* and *Picochlorum sp*. Algal Res 12:470–476

Ehler LE (2006) Integrated pest management (IPM): definition, historical development and implementation, and the other IPM. Pest Manag Sci 62:787–789

Eustance E, Wray JT, Badvipour S et al (2015) The effects of limiting nighttime aeration on productivity and lipid accumulation in *Scenedesmus dimorphous*. Algal Res 10:33–40

Evenson RE, Gollin D (2003) Assessing the impact of the green revolution, 1960 to 2000. Science 300:758–762

Fukui R, Fukui H, Alvarez AM (1999) Effect of temperature on the incubation period and leaf colonization in bacterial blight of *Anthurium*. Phytopathology 89:1007–1014

Fulbright SP, Dean MK, Wardle G et al (2014) Molecular diagnostics for monitoring contaminants in algal cultivation. Algal Res 4:41–51

Ganuza E, Sellers CE, Bennett BW et al (2016) A novel treatment protects *Chlorella* at commercial scale from the predatory bacterium *Vampirovibrio chlorellavorus*. Front Microbiol 7:848

Guieysse B, Bechet Q, Shilton A (2013) Variability and uncertainty in water demand and water footprint assessments of fresh algae cultivation based on case studies from five climatic regions. Bioresour Technol 128:317–323

Gutman J, Zarka A, Boussiba S (2009) The host-range of *Paraphysoderma sedebokerensis*, a chytrid that infects *Haematococcus pluvialis*. Eur J Phycol 44:509–514

Gutman J, Zarka A, Boussiba S (2011) Evidence for the involvement of surface carbohydrates in the recognition of *Haematococcus pluvialis* by the parasitic blastoclad *Paraphysoderma sedebokerensis*. Fungal Biol 115:803–811

Hagiwara A, Hino A (1989) Effect of incubation and preservation on resting egg hatching and mixis in the derived clones of the rotifer *Brachionus plicatilis*. In: Ricci C, Snell TW, King CE (eds) *Rotifer symposium V*. Springer, Netherlands, pp 415–421

Hairston NG (1996) Zooplankton egg banks as biotic reservoirs in changing environments. Limnol Oceanogr 41:1087–1092

Hakalin N, Paz A, Aranda D, Moraes L (2014) Enhancement of cell growth and lipid content of a freshwater microalga *Scenedesmus* sp. by optimizing nitrogen, phosphorus and vitamin concentrations for biodiesel production. Nat Sci 6:1044–1054. https://doi.org/10.4236/ns.2014.612095

Hessen DO, van Donk E (1993) Morphological changes in *Scenedesmus* induced by substances released from *Daphnia*. Arch Hydrobiol 127:129–140

Hillebrand H, Cardinale BJ (2004) Consumer effects decline with prey diversity. Ecol Lett 7:192–201

Hoffman Y, Aflalo C, Zarka A et al (2008) Isolation and characterization of a novel chytrid species (phylum *Blastocladiomycota*), parasitic on the green alga *Haematococcus*. Mycol Res 112:70–81

Huang Y, Li L, Liu J et al (2014) Botanical pesticides as potential rotifer-control agents in microalgal mass culture. Algal Res 4:62–69

Huneeus N, Schulz M, Balkanski Y et al (2011) Global dust model intercomparison in AeroCom phase I. Atmos Chem Phys 11:7781–7816

Huntley ME, Redalje DG (2007) CO_2 mitigation and renewable oil from photosynthetic microbes: a new appraisal. Mitig Adapt Strat Glob Change 12:573–608

Jerney J, Spilling K (2018) Large scale cultivation of microalgae: open and closed systems. Methods Mol Biol. https://doi.org/10.1007/7651_2018_130

Johnke J, Boenigk J, Harms H et al (2017) Killing the killer: predation between protists and predatory bacteria. FEMS Microbiol Lett 364. https://doi.org/10.1093/femsle/fnx089

Juneau P, Lawrence JE, Suttle CA et al (2003) Effects of viral infection on photosynthetic processes in the bloom-forming alga *Heterosigma akashiwo*. Aquat Microb Ecol 31:9–17

Kagami M, Donk EV, de Bruin A et al (2004) *Daphnia* can protect diatoms from fungal parasitism. Limnol Oceanogr 49:680–685

Kanaga K, Pandey A, Kumar S et al (2016) Multi-objective optimization of media nutrients for enhanced production of algae biomass and fatty acid biosynthesis from *Chlorella pyrenoidosa* NCIM 2738. Bioresour Technol 200:940–950

Karuppasamy S, Musale AS, Soni B et al (2018) Integrated grazer management mediated by chemicals for sustainable cultivation of algae in open ponds. Algal Res 35:439–448
Kazamia E, Czesnick H, Nguyen TTV et al (2012) Mutualistic interactions between vitamin B12 -dependent algae and heterotrophic bacteria exhibit regulation. Environ Microbiol 14:1466–1476
Kim D, Kim EK, Koh HG et al (2017) Selective removal of rotifers in microalgae cultivation using hydrodynamic cavitation. Algal Res 28:24–29
Kinsey N, Walters C (2009) Hands-on agronomy. Acres USA, Austin
Knoshaug EP, Wolfrum E, Laurens LML et al (2018) Unified field studies of the algae testbed public-private partnership as the benchmark for algae agronomics. Sci Data 5:180267
Kropat J, Hong-Hermesdorf A, Casero D et al (2011) A revised mineral nutrient supplement increases biomass and growth rate in *Chlamydomonas reinhardtii*: a revised mineral nutrient supplement for *Chlamydomonas*. Plant J 66:770–780
Lee YK (2001) Microalgal mass culture systems and methods: their limitation and potential. J Appl Phycol 13:307–315. https://doi.org/10.1023/A:1017560006941
Lee PA, Martinez KJL, Letcher PM et al (2018) A novel predatory bacterium infecting the eukaryotic alga *Nannochloropsis*. Algal Res 32:314–320
Letcher PM, Lopez S, Schmieder R et al (2013) Characterization of *Amoeboaphelidium protococcarum*, an algal parasite new to the Cryptomycota isolated from an outdoor algal pond used for the production of biofuel. PLoS One 8:e56232
Letcher PM, Lee PA, Lopez S et al (2016) An ultrastructural study of *Paraphysoderma sedebokerense* (*Blastocladiomycota*), an epibiotic parasite of microalgae. Fungal Biol 120:324–337
Li Z, Yuan H, Yang J et al (2011) Optimization of the biomass production of oil algae *Chlorella minutissima* UTEX2341. Bioresour Technol 102:9128–9134
Li YX, Zhao FJ, Yu DD et al (2015) Effect of nitrogen limitation on cell growth, lipid accumulation and gene expression in *Chlorella sorokiniana*. Braz Arch Biol Technol 58:462–467
Lian J, Wijffels RH, Smidt H et al (2018) The effect of the algal microbiome on industrial production of microalgae. Microb Biotechnol 11:806–818
Lurling M, Beekman W (2006) Palmelloids formation in *Chlamydomonas reinhardtii*: defence against rotifer predators? Ann Limnol - Int J Lim 42:65–72
Ma M, Yuan D, He Y et al (2017) Effective control of *Poterioochromonas malhamensis* in pilot-scale culture of *Chlorella sorokiniana* GT-1 by maintaining CO_2-mediated low culture pH. Algal Res 26:436–444
Malakootian M, Hatami B, Dowlatshahi S et al (2015) Optimization of culture media for lipid production by *Nannochloropsis oculata* for biodiesel production. Environ Health Eng Manag J 2:141–147
Masojídek J, Torzillo G (2014) Mass cultivation of freshwater microalgae. Reference module in earth systems and environmental sciences. Elsevier, USA
Mason TJ, Joyce E, Phull SS et al (2003) Potential uses of ultrasound in the biological decontamination of water. Ultrason Sonochem 10:319–323
Mata TM, Martins AA, Caetano NS (2010) Microalgae for biodiesel production and other applications: a review. Renew Sust Energ Rev 14:217–232
Mathimani T, Mallick N (2018) A comprehensive review on harvesting of microalgae for biodiesel - key challenges and future directions. Renew Sust Energ Rev 91:1103–1120
Mayali X, Azam F (2004) Algicidal bacteria in the sea and their impact on algal blooms. J Eukaryot Microbiol 51:139–144
McBride RC, Lopez S, Meenach C et al (2014) Contamination management in low cost open algae ponds for biofuels production. Ind Biotechnol 10:221–227
McBride RC, Smith VH, Carney LT et al (2016) Crop protection in open ponds. In: Microalgal production for biomass and high-value products. CRC, Boca Raton, FL, pp 165–182

McGowen J, Knoshaug EP, Laurens LML et al (2017) The algae testbed public-private partnership (ATP3) framework; establishment of a national network of testbed sites to support sustainable algae production. Algal Res 25:168–177

Mills S, Lunt DH, Gómez A (2007) Global isolation by distance despite strong regional phylogeography in a small metazoan. BMC Evol Biol 7:225

Mitchell SA, Richmond A (1987) The use of rotifers for the maintenance of monoalgal mass cultures of *Spirulina*. Biotechnol Bioeng 30:164–168

Mo W, Soh L, Werber JR et al (2015) Application of membrane dewatering for algal biofuel. Algal Res 11:1–12

Moheimani NR, Borowitzka MA (2007) Limits to productivity of the alga *Pleurochrysis carterae* (*Haptophyta*) grown in outdoor raceway ponds. Biotechnol Bioeng 96:27–36

Moschini RC, Fortugno C (1996) Predicting wheat head blight incidence using models based on meteorological factors in Pergamino, Argentina. Eur J Plant Pathol 102:211–218

Mukherjee S (2015) The Laws of medicine: field notes from an uncertain science, 1st edn. Simon & Schuster/TED, New York

Nabris KJ-AE (2012) Development of cheap and simple culture medium for the microalgae *Nannochloropsis* sp. based on agricultural grade fertilizers available in the local market of gaza strip (Palestine). J Al Azhar University-Gaza (Natural Sciences) 14:61–76

Narala RR, Garg S, Sharma KK et al (2016) Comparison of microalgae cultivation in photobioreactor, open raceway pond, and a two-stage hybrid system. Front Energy Res 4. https://doi.org/10.3389/fenrg.2016.00029

Newby DT, Mathews TJ, Pate RC et al (2016) Assessing the potential of polyculture to accelerate algal biofuel production. Algal Res 19:264–277

Nogueira Junior E, Kumar M, Pankratz S et al (2018) Development of life cycle water footprints for the production of fuels and chemicals from algae biomass. Water Res 140:311–322

O'Malley MA (2008) 'Everything is everywhere: but the environment selects': ubiquitous distribution and ecological determinism in microbial biogeography. Stud Hist Philos Sci Part C 39:314–325

Ogbonna JC, Tanaka H (1996) Night biomass loss and changes in biochemical composition of cells during light/dark cyclic culture of *Chlorella pyrenoidosa*. J Ferment Bioeng 82:558–564

Oswald WJ (1988) Large-scale algal culture systems (engineering aspects). In: Micro-algal biotechnology. Cambridge University Press, Cambridge, pp 357–394

Park S, Van Ginkel SW, Pradeep P et al (2016) The selective use of hypochlorite to prevent pond crashes for algae-biofuel production. Water Environ Res 88:70–78

Park S-H, Steichen SA, Li X et al (2018) Association of *Vampirovibrio chlorellavorus* with decline and death of *Chlorella sorokiniana* in outdoor reactors. J Appl Phycol. https://doi.org/10.1007/s10811-018-1633-9

Patterson DJ (2003) Free-living freshwater Protozoa: a color guide, 1st edn. ASM Press, Washington

Pedigo LP, Higley LG (1992) The economic injury level concept and environmental quality: a new perspective. Am Entomol 38:12–21

Prokopy R, Kogan M (2009) Chapter 139 - Integrated pest management. In: Resh VH, Cardé RT (eds) Encyclopedia of insects, 2nd edn. Academic, San Diego, pp 523–528

Prospero JM, Mayol-Bracero OL (2013) Understanding the transport and impact of African dust on the Caribbean Basin. Bull Amer Meteor Soc 94:1329–1337

Pruess KP (1983) Day-degree methods for pest management. Environ Entomol 12:613–619

Quispe CF, Sonderman O, Seng A et al (2016) Three-year survey of abundance, prevalence and genetic diversity of chlorovirus populations in a small urban lake. Arch Virol 161:1839–1847

Ravikumar R (2014) Microalgae in open raceways. In: Bajpai R, Prokop A, Zappi M (eds) Algal biorefineries, vol 1: Cultivation of cells and products. Springer, Dordrecht, pp 127–146

Reichardt TA, Collins AM, McBride RC et al (2014) Spectroradiometric monitoring for open outdoor culturing of algae and cyanobacteria. Appl Opt 53:F31–F45

Richmond A, Becker EW (1986) Technological aspects of mass cultivation — a general outline. Handbook of microalgal mass culture. CRC, Boca Raton

Richmond A, Zou N (1999) Efficient utilisation of high photon irradiance for mass production of photoautotrophic micro-organisms. In: Kain JM, Brown MT, Lahaye M (eds) Sixteenth international seaweed symposium. Springer, Cham, pp 637–641
Richmond A, Lichtenberg E, Stahl B et al (1990) Quantitative assessment of the major limitations on productivity of *Spirulina platensis* in open raceways. J Appl Phycol 2:195–206
Rothhaupt KO (1990) Population growth rates of two closely related rotifer species: effects of food quantity, particle size, and nutritional quality. Freshw Biol 23:561–570
Sanz-Luque E, Chamizo-Ampudia A, Llamas A et al (2015) Understanding nitrate assimilation and its regulation in microalgae. Front Plant Sci 6. https://doi.org/10.3389/fpls.2015.00899
Savary S, Willocquet L, Pethybridge SJ, Esker P, McRoberts N, Nelson A (2019) The global burende of pathogens and pests on major food crops. Nat Ecol Evol 3:430–439
Scherholz ML, Curtis WR (2013) Achieving pH control in microalgal cultures through fed-batch addition of stoichiometrically-balanced growth media. BMC Biotechnol 13:39
Schlüter M, Groeneweg J, Soeder CJ (1987) Impact of rotifer grazing on population dynamics of green microalgae in high-rate ponds. Water Res 21:1293–1297
Schnepf E, Hegewald E, Soeder CJ (1974) Elektronenmikroskopische beobachtungen an parasiten aus *Scenedesmus* massenkulturen. Arch Microbiol 98:133–145
Show K, Yan Y, Lee D (2019) Algal biomass harvesting and drying. In: Panday A, Chnag J, Soccol CR, Lee D, Chisti Y (eds) Biofuels from algae: sustainable platform for fuels, chemicals and remeditation, 2nd edn. Elsevier, Amsterdam, pp 135–166. https://doi.org/10.1016/B978-0-444-64192-2.00007-X
Shurin JB, Abbott RL, Deal MS et al (2013) Industrial-strength ecology: trade-offs and opportunities in algal biofuel production. Ecol Lett 16:1393–1404
Simkovsky R, Daniels EF, Tang K et al (2012) Impairment of O-antigen production confers resistance to grazing in a model amoeba–cyanobacterium predator–prey system. PNAS 109:16678–16683
Sing G, Patidar SK (2018) Microalgae harvesting techniques: a review. J Environ Manag 217:499–508
Smith VH, Crews T (2014) Applying ecological principles of crop cultivation in large-scale algal biomass production. Algal Res 4:23–34
Steichen SA, Brown JK (2018) Real-time quantitative detection of *Vampirovibrio chlorellavorus*, an obligate bacterial pathogen of *Chlorella sorokiniana*. J Appl Phycol:1–13. https://doi.org/10.1007/s10811-018-1659-z
Stenberg JA (2017) A conceptual framework for integrated pest management. Trends Plant Sci 22:759–769
Stern V, Smith R, van den Bosch R et al (1959) The integration of chemical and biological control of the spotted alfalfa aphid: the integrated control concept. Hilgardia 29:81–101
Szyjka SJ, Mandal S, Schoepp NG et al (2017) Evaluation of phenotype stability and ecological risk of a genetically engineered alga in open pond production. Algal Res 24:378–386
Talent M, Burgess G, Fernandez-Velasco JG (2014) Protocol to compensate net evaporation and net precipitation in open-pond microalgal massive cultures and permit maximal steady-state productivities. Biomass Bioenergy 64:81–90
Tooker JF, Frank SD (2012) Genotypically diverse cultivar mixtures for insect pest management and increased crop yields. J Appl Ecol 49:974–985
Tredici MR (2007) Mass production of microalgae: photobioreactors. Handbook of microalgal culture. Wiley, West Sussex, UK, pp 178–214
Trentacoste EM, Martinez AM, Zenk T (2015) The place of algae in agriculture: policies for algal biomass production. Photosynth Res 123:305–315
VAN Etten JL, Burbank DE, Kuczmarski D et al (1983) Virus infection of culturable chlorella-like algae and development of a plaque assay. Science 219:994–996
Vieira Costa JA, Cozza KL, Oliveira L et al (2001) Different nitrogen sources and growth responses of *Spirulina platensis* in microenvironments. World J Microbiol Biotechnol 17:439–442

Vonshak A, Boussiba S, Abeliovich A et al (1983) Production of *Spirulina* biomass: maintenance of monoalgal culture outdoors. Biotechnol Bioeng 25:341–349

Walsh EJ, Smith HA, Wallace RL (2014) Rotifers of temporary waters. Int Rev Hydrobiol 99:3–19

Wan C, Bai F-W, Zhao X-Q (2013) Effects of nitrogen concentration and media replacement on cell growth and lipid production of oleaginous marine microalga *Nannochloropsis oceanica* DUT01. Biochem Eng J 78:32–38

Wang Y, Eustance E, Castillo-Keller M et al (2017) Evaluation of chemical treatments for control of ciliate grazers in algae cultures: a site study. J Appl Phycol 29:2761–2770

Weddle PW, Welter SC, Thomson D (2009) History of IPM in California pears - 50 years of pesticide use and the transition to biologically intensive IPM. Pest Manag Sci 65:1287–1292

White RL, Ryan RA (2015) Long-term cultivation of algae in open-raceway ponds: lessons from the field. Ind Biotechnol 11:213–220

Williams PJB, Laurens LML (2010) Microalgae as biodiesel & biomass feedstocks: review & analysis of the biochemistry, energetics & economics. Energy Environ Sci 3:554–590. https://doi.org/10.1039/B924978H

Winckelmann D, Bleeke F, Bergmann P et al (2016) Detection of weed algae in open pond cultures of *Cyanobacterium aponinum* using PAM. Int Aquat Res 8:81–90

Wong Y (2017) Growth medium screening for *Chlorella vulgaris* growth and lipid production. J Aquaculture Marine Biol 6. https://doi.org/10.15406/jamb.2017.06.00143

Yang F, Long L, Sun X et al (2014) Optimization of medium using response surface methodology for lipid production by *Scenedesmus sp.* Mar Drugs 12:1245–1257

Yao S, Lyu S, An Y et al (2019) Microalgae-bacteria symbiosis in microalgal growth and biofuel production: a review. J Appl Microbiol 126:359–368

Zaborin A, Romanowski K, Gerdes S et al (2009) Red death in *Caenorhabditis elegans* caused by *Pseudomonas aeruginosa* PAO1. PNAS 106:6327–6332

Zmora O, Richmond A (2007) Microalgae for aquaculture: microalgae production for aquaculture. Handbook of microalgal culture. Wiley, West Sussex, UK, pp 365–379

Part II
Genetic and Metabolic Engineering

Chapter 5
Advances in Genetic Engineering of Microalgae

Armin Hallmann

Abstract Microalgae biotechnology gains more and more commercial importance in quite different fields such as food, pharmaceutical, nutraceutical, cosmetic, animal feed, and energy industries. Genetic modification of microalgae promises enhanced biotechnological exploitation possibilities in nearly all contexts where unaltered microalgae are involved. Furthermore, genetic engineering makes it possible to produce compounds or to add traits that are not normally present in algae. Recent developments in genetic engineering of microalgae have been supported by rapid advances in omics technologies and a sharp increase of key elements for construction of vectors, transformation, and selection. Even genome editing by application of the CRISPR/Cas system became available. Both the omics groundwork and the key elements of genetic engineering are described in detail. The chapter also discusses a series of applications for transgenic microalgae including the production of high-value compounds like antibodies or vaccines, food additives, biofuels, and even optogenetic tools for neuroscience.

5.1 Introduction

Algae are eukaryotic photosynthetic organisms that are present in almost all terrestrial, aquatic, and even atmospheric habitats. The aquatic habitats, where the majority of them lives, include freshwater, marine, and intermediate levels of salinity. The importance of algae for ecosystems around the world is underlined by the fact that algae achieve a net primary production of ~52,000,000,000 tons of organic carbon per year, which is ~50% of the total organic carbon produced on earth each year (Field et al. 1998). Algae form an extremely large group with estimated 350,000 species, but only a few species are currently domesticated (Bux and Chisti 2016). However, the group of algae constitutes a polyphyletic group that describes a life form, not a systematic unit; this is one reason why algae exhibit extremely diverse

A. Hallmann (✉)
Department of Cellular and Developmental Biology of Plants, University of Bielefeld, Bielefeld, Germany

A. Hallmann, P. H. Rampelotto (eds.), *Grand Challenges in Algae Biotechnology*, Grand Challenges in Biology and Biotechnology,
https://doi.org/10.1007/978-3-030-25233-5_5

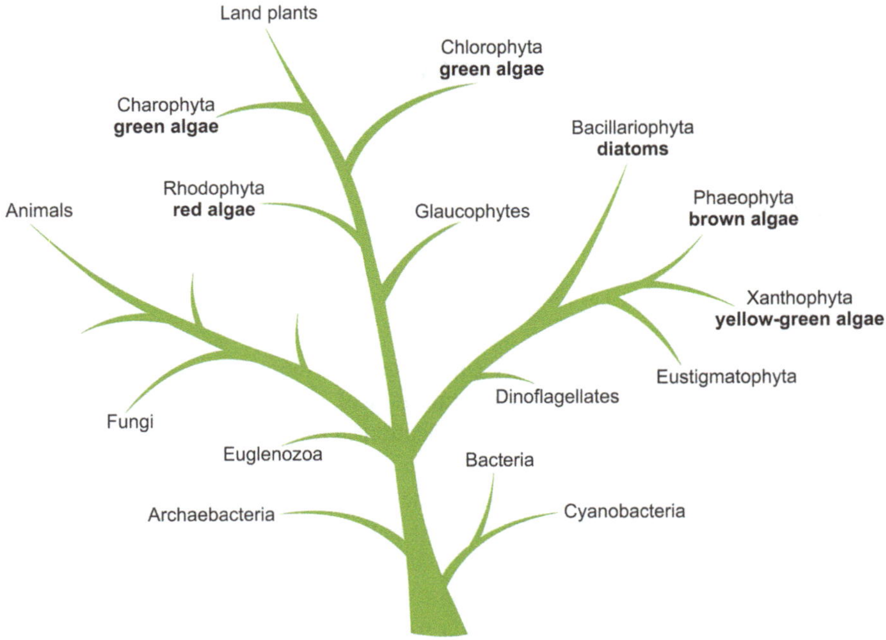

Fig. 5.1 Simplified tree of life with emphasis on algae. The tree was extracted from earlier phylogenetic analyses (Baldauf 2003; Hallmann 2011; Kianianmomeni and Hallmann 2014; Kranz et al. 1995; Maddison and Schulz 2007; Prochnik et al. 2010)

morphologies. A simplified tree of life with emphasis on algae in Fig. 5.1 gives an overview of the phylogenetic relationships among algae branches.

For orientation, the tree also contains well-known groups such as animals, land plants, fungi, archaebacteria, and bacteria. The types of algae include green algae (Chlorophyta and Charophyta), red algae (Rhodophyta), brown algae (Phaeophyta), yellow-green algae (Xanthophyta), diatoms (Bacillariophyta), glaucophytes, Eustigmatophyta, Euglenozoa, and dinoflagellates. The (eukaryotic) algae, together with land plants and the prokaryotic Cyanobacteria, have the remarkable ability to convert light energy into chemical energy by photosynthesis.

This chapter deals with small-sized (eukaryotic) algae, the microalgae, which gain increasingly more commercial importance in quite different fields of biotechnology such as food, pharmaceutical, nutraceutical, cosmetic, animal feed, and energy industries. Genetic engineering of microalgae promises enhanced biotechnological exploitation possibilities.

As is the case with algae in general, the phenotypes of microalgae are quite diverse. For illustration, phenotypes of some genetically modifiable microalgae species are shown in Fig. 5.2.

Microalgae are normally easy to grow, and they belong to the fastest-growing photosynthetic organisms. Cultivation of algae can be performed even on non-arable land as long as there is enough water available. Moreover, broadly diversified traits

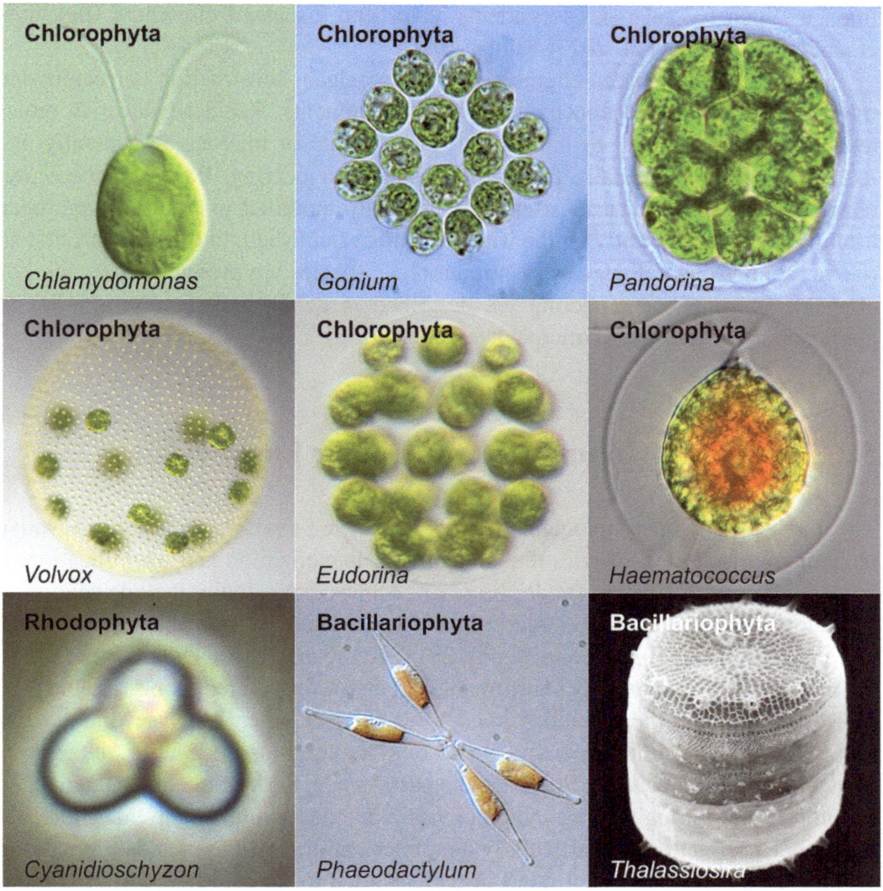

Fig. 5.2 Phenotypes of some genetically modifiable microalgae species. The complete list of transformed species and the corresponding references are shown in Table 5.4. The following species are shown: *Chlamydomonas reinhardtii*, *Gonium pectorale*, *Pandorina morum*, *Volvox carteri*, *Eudorina elegans*, *Haematococcus pluvialis*, *Cyanidioschyzon merolae*, *Phaeodactylum tricornutum*, and *Thalassiosira pseudonana*. Photos by Alessandra de Martino and Chris Bowler (*Phaeodactylum*), Nils Kröger (*Thalassiosira*), Wiedehopf20 (*Haematococcus*) and own work

and living conditions of microalgae species make a couple of them extremely attractive for commercial utilization, particularly if their traits can be further improved by genetic engineering. Moreover, it is important to recognize that significant advances in the development of genetic manipulation tools have recently been achieved with microalgal model systems. Due to these outstanding framework conditions, transgenic microalgae have the potential to impact diverse businesses such as human and animal nutraceuticals, pharmaceuticals, cosmetics, exquisite chemicals, and energy (Rasala et al. 2014). However, any manufacturing processes involving genetically modified algae or any other organisms require governmental regulations that closely scrutinize environmental, economic, and human health impacts. Such federal risk assessments have been discussed elsewhere (Beacham et al. 2017; Henley et al. 2013; Kumar 2015).

In the course of genetic engineering of microalgae, construction of vectors and transformation and selection methods are key elements. Rapid advances in omics technologies are helping to unlock the full potential of microalgae generally and more particularly by greatly facilitating genetic engineering. Based on these preconditions, several industrial uses for genetically modified microalgae have been established or are currently on the way of getting established. In this chapter, first all relevant omics technologies used in microalgae species are presented, later the key elements of genetic engineering are described in detail, and, finally, a series of applications for transgenic microalgae are discussed.

5.2 The Omics Groundwork for Genetic Engineering

The English-language neologism "omics" stands for a whole range of innovative technologies that all deal with the collective characterization and quantification of large pools of biological molecules (Fig. 5.3). Omics technologies include genomics, epigenomics, metagenomics, transcriptomics, proteomics, lipidomics, glycomics, and metabolomics (Horgan and Kenny 2011; Karahalil 2016). The different omics technologies are not mutually substitutable but deal with different pools of biological molecules and offer the possibility to complement each other. When several of these omics are applied simultaneously and results are combined, the term "integrative omics" is used (Dihazi et al. 2018). In the past decade, the volume of data generated by omics high-throughput technologies has expanded exponentially, and the dramatic increase continues. Particularly genomics and transcriptomics data stand out. The generated overwhelming flood of data also led to the rapid growth of computational biology and bioinformatics (Schneider and Orchard 2011).

Advances in molecular-biological and biotechnological techniques were essential for the emergence and rapid evolution of the omics era. However, this is not a one-way street because there is a multidirectional flow of information, a constant exchange of know-how, and experience between all kinds of involved disciplines (Fig. 5.3).

Biotechnology, particularly algae biotechnology, profits tremendously from the results and comprehensive data repositories provided by omics technologies.

The omics technologies also constitute fundamental building blocks of systems biology, which is an interdisciplinary approach that integrates experimental data with computational and theoretical methods (Karahalil 2016). The aim of systems biology is to develop a quantitative understanding of complex biological systems in terms of their components and interactions. While traditional scientific investigations are generally hypothesis-driven, both systems biology and all omics applications are holistic approaches where no hypothesis is generated at the beginning. Instead, large-scale data acquisition needs to be completed first, then the data are analyzed, and a hypothesis is generated, which then can be further tested.

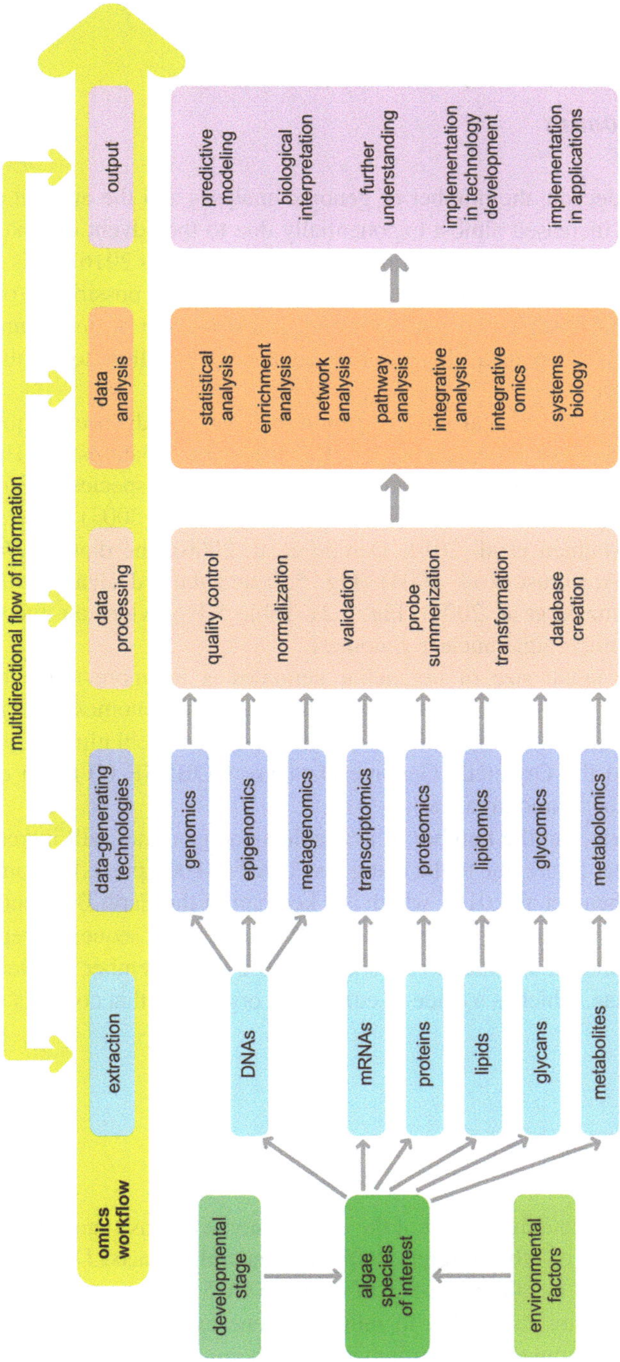

Fig. 5.3 Workflow of microalgae omics technologies

Relevant omics technologies that lay the groundwork for genetic engineering in microalgae are discussed below.

5.2.1 *Genomics*

Over the last decade, the number of genome analyses and the amount of genomic sequence data increased almost exponentially due to the advent of next-generation sequencing (NGS) technologies (Hui 2014; Levy and Myers 2016). This applies also to the group of microalgae. Genome sequencing makes it possible to obtain information about the architecture, organization, and evolution of genes and genomes. Together with mRNA data, the genome sequences also allow to display genome-wide expression.

More than 30 microalgal nuclear genomes have already been sequenced. The genome projects mainly concern green algae, red algae, diatoms and Haptophytes. The most advanced projects are those for the green algal species *Chlamydomonas reinhardtii* (Fig. 5.2) (Merchant et al. 2007; Shrager et al. 2003) and *Ostreococcus tauri* (Blanc-Mathieu et al. 2014; Derelle et al. 2006), the diatom *Thalassiosira pseudonana* (Armbrust et al. 2004) (Fig. 5.2) and the red alga *Cyanidioschyzon merolae* (Matsuzaki et al. 2004) (Fig. 5.2). Table 5.1 provides an overview of the 34 sequenced microalgal nuclear genomes.

Due to the small size of organellar genomes, a lot more mitochondrial and chloroplast genomes have been sequenced than nuclear genomes. Many sequences are available at the NCBI organelle database (http://www.ncbi.nlm.nih.gov/genome/organelle/) and the Organelle Genome Database GOBASE (O'Brien et al. 2009) (http://gobase.bcm.umontreal.ca/).

It seems clear that the number of available algae genomes will increase dramatically in the near future due to the instigated 10KP (10,000 plants) genome sequencing project (Cheng et al. 2018), which is a key part of the Earth BioGenome Project (EBP). Within the next 5 years, the 10KP project will sequence representative genomes to build an annotated reference genome for a member of every genus of the Viridiplantae, which is a clade of eukaryotic organisms that contains many green algae genera.

5.2.2 *Epigenomics*

Epigenetic modifications are part of the cell's genome defense against viruses, transposable elements, or other foreign DNA or unnaturally placed DNA (Cerutti et al. 1997b; van Dijk et al. 2005; Wu-Scharf et al. 2000). Moreover, epigenetic modifications give organisms the ability to influence gene expression in heritable manner depending on the immediate environment (Kronholm et al. 2017). The simultaneous study of all epigenetic modifications on the genetic material of a cell is called

Table 5.1 Sequenced microalgae nuclear genomes

Species	Lineage	References
	Green algae	
Chlamydomonas reinhardtii	Chlorophyta; Chlorophyceae; Volvocales; Chlamydomonadaceae	Merchant et al. (2007), Shrager et al. (2003)
Volvox carteri f. *nagariensis*	Chlorophyta; Chlorophyceae; Volvocales; Volvocaceae	Prochnik et al. (2010)
Dunaliella salina	Chlorophyta; Chlorophyceae; Volvocales; Dunaliellaceae	Polle et al. (2017)
Coccomyxa subellipsoidea	Chlorophyta; Chlorophyceae; Chlorococcales; Coccomyxaceae	Blanc et al. (2012)
Monoraphidium neglectum	Chlorophyta; Chlorophyceae; Sphaeropleales; Selenastraceae	Bogen et al. (2013)
Scenedesmus obliquus	Chlorophyta; Chlorophyceae; Sphaeropleales; Scenedesmaceae	Starkenburg et al. (2017)
Chlorella variabilis	Chlorophyta; Trebouxiophyceae; Chlorellales; Chlorellaceae	Blanc et al. (2010)
Chlorella sorokiniana	Chlorophyta; Trebouxiophyceae; Chlorellales; Chlorellaceae	Arriola et al. (2018)
Chlorella vulgaris	Chlorophyta; Trebouxiophyceae; Chlorellales; Chlorellaceae	GenBank Ac.No. LDKB00000000
Auxenochlorella protothecoides	Chlorophyta; Trebouxiophyceae; Chlorellales; Chlorellaceae	Gao et al. (2014)
Picochlorum soloecismus	Chlorophyta; Trebouxiophyceae; Chlorellales; incertae sedis	Foflonker et al. (2015), Gonzalez-Esquer et al. (2018)
Botryococcus braunii	Chlorophyta; Trebouxiophyceae; Trebouxiales; Botryococcaceae	Browne et al. (2017)
Ostreococcus tauri	Chlorophyta; Prasinophyceae; Mamiellales; Mamiellaceae	Blanc-Mathieu et al. (2014), Derelle et al. (2006)
Ostreococcus lucimarinus	Chlorophyta; Prasinophyceae; Mamiellales; Mamiellaceae	Palenik et al. (2007)
Micromonas pusilla	Chlorophyta; Prasinophyceae; Mamiellales; Mamiellaceae	Worden et al. (2009)
Bathycoccus prasinos	Chlorophyta; Prasinophyceae; Mamiellales; Mamiellaceae	Moreau et al. (2012)
	Red algae	
Cyanidioschyzon merolae	Rhodophyta; Cyanidiophyceae; Cyanidiales; Cyanidiaceae	Matsuzaki et al. (2004)
Galdieria sulphuraria	Rhodophyta; Cyanidiophyceae; Galdieriaceae; Cyanidiaceae	Barbier et al. (2005)
Porphyridium purpureum	Rhodophyta; Porphyridiophyceae; Porphyridiales; Porphyridiaceae	Bhattacharya et al. (2013)
	Diatoms	
Phaeodactylum tricornutum	Heterokontophyta; Bacillariophyceae; Naviculales; Phaeodactylaceae	Bowler et al. (2008)
Fragilariopsis cylindrus	Heterokontophyta; Bacillariophyceae; Bacillariales; Bacillariaceae	Mock et al. (2017)
Pseudo-nitzschia multiseries	Heterokontophyta; Bacillariophyceae; Bacillariales; Bacillariaceae	JGI project id: 16870

(continued)

Table 5.1 (continued)

Species	Lineage	References
Thalassiosira pseudonana	Heterokontophyta; Coscinodiscophyceae; Thalassiosirales; Thalassiosiraceae	Armbrust et al. (2004), Bowler et al. (2008)
Thalassiosira oceanica	Heterokontophyta; Coscinodiscophyceae; Thalassiosirales; Thalassiosiraceae	Lommer et al. (2012)
	Eustigmatophyta	
Nannochloropsis gaditana	Heterokontophyta; Eustigmatophyceae; Eustigmatales; Monodopsidaceae	Corteggiani Carpinelli et al. (2014), Radakovits et al. (2012)
Nannochloropsis oceanica	Heterokontophyta; Eustigmatophyceae; Eustigmatales; Monodopsidaceae	Vieler et al. (2012)
	Pelagophyta	
Aureococcus anophagefferens	Heterokontophyta; Pelagophyceae; Pelagomonadales; Pelagomonadaceae	Gobler et al. (2011)
	Haptophytes	
Emiliania huxleyi	Haptophyta; Prymnesiophyceae; Isochrysidales; Noelaerhabdaceae	Read et al. (2013)
Chrysochromulina tobin	Haptophyta; Prymnesiophyceae; Prymnesiales; Prymnesiaceae	Hovde et al. (2015)
Chrysochromulina parva	Haptophyta; Prymnesiophyceae; Prymnesiales; Prymnesiaceae	GenBank Ac.No. PJAB00000000
	Cryptomonads	
Guillardia theta	Cryptophyta; Cryptophyceae; Pyrenomonadales; Geminigeraceae	Curtis et al. (2012), Douglas et al. (2001)
	Chlorarachniophytes	
Bigelowiella natans	Cercozoa; Chlorarachniophyceae; Chlorarachniales; Chlorarachniaceae	Curtis et al. (2012)
	Glaucophytes	
Cyanophora paradoxa	Glaucophyta; Glaucophyceae; Glaucocystales; Glaucocystaceae	Price et al. (2012)
	Dinoflagellates	
Symbiodinium minutum	Alveolata; Dinophyceae; Suessiales; Symbiodiniaceae	Shoguchi et al. (2013)

epigenomics. One of the best-characterized epigenetic modifications is DNA methylation, which is highly conserved among eukaryotes and can impact gene expression (Lee et al. 2010). The majority of DNA methylation occurs on cytosines that precede a guanine nucleotide, the CpG islands, and the methyl group is added to the 5 position of cytosine by a DNA methyltransferase to form N^5-methyldeoxycytosine (5mC). Another important conserved modification is N^6-methyldeoxyadenosine (6 mA). All epigenetic modifications are systematically missed by conventional DNA sequence-based genomic analyses, even though the identification of DNA methyltransferase genes suggests that epigenetic modifications are to be expected. An experimental verification of cytosine methylation was accomplished for a couple of microalgae including Chlorophyceae (*Chlamydomonas reinhardtii, Volvox carteri* (Fig. 5.2), *Chlorella ellipsoidea,* and *Monoraphidium minutum*), Bacillariophyceae (*Phaeodactylum tricornutum* (Fig. 5.2), *Cyclotella cryptica, Navicula saprophila,* and *Nitzschia pusilla*), Trebouxiophyceae (*Stichococcus* sp.), Dinophyceae (*Crypthecodinium cohnii*), and Chlorodendrophyceae (*Tetraselmis suecica*) (Jarvis et al. 1992; Veluchamy et al. 2013; Zemach et al. 2010). However, genome-wide epigenetic analyses have only been performed in *Chlamydomonas reinhardtii* (Lopez et al. 2015), *Volvox carteri* (Fig. 5.2) (Zemach et al. 2010), *Chlorella* sp. (Zemach et al. 2010), and *Phaeodactylum tricornutum* (Veluchamy et al. 2013) (Fig. 5.2). In *Chlamydomonas,* 6 mA was shown to be present in 84% of all genes, and it marks active transcription start sites (Fu et al. 2015). Genome-wide epigenetic analyses can be done by ChIP sequencing, MeDIP sequencing, whole-genome bisulfite sequencing (WGBS), and reduced representation bisulfite sequencing (RRBS) (Kim et al. 2014).

For algal molecular biologists and biotechnologists, epigenetic causes frequently have a negative impact on experimental results. One of the reasons for this is that most restriction enzymes are sensitive to the DNA methylation state of their recognition site where cleavage can be blocked or impaired if genomic DNA is used. Another even more serious reason is that epigenetic modifications can cause unforeseen low gene-expression levels of transgenes in genetically modified organisms. Transgenes may become actively silenced by DNA methylation (Babinger et al. 2001, 2007; Kim et al. 2015; Kurniasih et al. 2016) or by enzymes that place specific histone modifications onto nucleosomes at the transgene loci to trigger chromatin compaction (Bannister and Kouzarides 2011; Casas-Mollano et al. 2008; Kim et al. 2015).

5.2.3 Metagenomics

As indicated by the name, metagenomics has its roots in conventional genomics. Metagenomics is defined as the direct genetic analysis of genomes contained with an environmental sample (Toulza et al. 2012). The method allows to determine the number of species and their abundance distribution across a certain environment. Also new and unculturable species can be identified in metagenomic analyses. The DNA typically is extracted from a sample with many species and mechanically sheared into fragments, and the fragments are subject to shotgun sequencing without heterologous cloning. Thus, this technique bypasses the requirement for culturing species and for obtaining pure cultures with only a single species per sample. As a consequence, the assembly of sequences is even more complicated compared to conventional genomics and special-purpose assembly algorithms are required (Ghurye et al. 2016; Vollmers et al. 2017). For example, metagenomics has been used to analyze biocenoses of algae, bacteria, and any other organisms in water samples or in biofilms (Krohn-Molt et al. 2013; Yergeau et al. 2017).

Due to the bioinformatic possibilities offered by metagenomics and the extremely high number of sequences produced by modern sequencing machines, the situation with several or even many genomes in a single sample is also artificially generated with increasing frequency. This kind of sequencing after pooling of different genomes is called multiplex sequencing (Anand et al. 2016; Wong et al. 2013). However, in contrast to metagenomic analyses of multi-species environmental samples, samples for multiplexing can be tagged with an individual "barcode" sequence during library preparation before pooling of the genomes. After multiplex sequencing, the obtained sequences can be easily assigned to the correct genome (de-multiplexed) due to the barcode tag.

5.2.4 Transcriptomics

For interpreting the functional elements of the genome, it is essential to know the complete set of transcripts, the transcriptome. The transcriptome includes all RNA molecules from protein coding mRNA to noncoding RNA, including rRNA, tRNA, lncRNA, pri-miRNA, and others.

Transcriptomics technologies include expressed sequence tags (ESTs) (Marra et al. 1998), serial and cap analysis of gene expression (SAGE/CAGE) (Shiraki et al. 2003; Velculescu et al. 1995), microarrays (Romanov et al. 2014), and RNA-Seq (Ozsolak and Milos 2011). The last two of these are key contemporary techniques in the field. Microarrays are used to measure the abundances of a predetermined set of transcripts via their hybridization to an array of complementary probes (Lowe et al. 2017; Pozhitkov et al. 2007; Romanov et al. 2014). RNA-Seq, which currently is the most dominant technology, refers to the sequencing of transcript cDNAs, in which abundance is derived from the number of counts from each transcript (Lowe et al. 2017; Ozsolak and Milos 2011).

As with nuclear genomic sequences, there is also a continuous, dramatic increase in transcriptome sequence information of microalgae. EST data from many microalgae species are available at the Taxonomically Broad EST Database (TBestDB, http://tbestdb.bcm.umontreal.ca) (O'Brien et al. 2007) and the EST sequence databases of the National Center for Biotechnology Information (NCBI) (https://www.ncbi.nlm.nih.gov/nucest/). The NCBI also provides pre-assembled UniGenes (mRNA contigs) of microalgae in the Transcriptome Shotgun Assembly (TSA) sequence database (https://www.ncbi.nlm.nih.gov/genbank/tsa/). However, the largest contribution to transcriptome resources of algae has been made by the 1000 Plants Initiative (1KP, www.onekp.com) (Matasci et al. 2014). This international consortium has made transcriptome assemblies available for 1341 plant species with exemplars for all of the major lineages across the Viridiplantae including 241 algae species.

The number of available algae transcriptomes will further increase in the near future due to the 10KP Genome Sequencing Project (Cheng et al. 2018), which will not only sequence genomes but also transcriptomes.

5.2.5 Proteomics

Proteomics refers to the large-scale experimental analysis of all proteins of a cell, tissue, organ, or organism including protein sequences, structures, and functions (Aslam et al. 2017). The proteome can differ from cell to cell depending on the cell's interactions with its own cellular components, with other cells of the same organism, or with the environment. In addition, proteins can have posttranslational modifications, which are not easily deducible from the underlying transcriptome.

The challenge of proteomics is to fractionate complex peptide or protein mixtures, use mass spectrometry to obtain raw mass spectral data, and bioinformatically analyze and assemble the mass spectrometry data by using different kinds of databases to identify individual proteins. Frequently, proteolytic peptide mixtures are used as a starting material (bottom-up proteomics), but longer peptides and intact proteins allow for a more complete characterization of protein isoforms and post-translational modifications (middle and top-down proteomics) (Han et al. 2008; Thompson 2017).

Proteolytically digested proteins are frequently fractionated by liquid chromatography techniques such as ion exchange chromatography (IEC) (Jungbauer and Hahn 2009), size exclusion chromatography (SEC) (Voedisch and Thie 2010), and affinity chromatography (AC) (Hage et al. 2012), while intact proteins are separated by gel electrophoresis such as sodium dodecyl sulfate-polyacrylamide gel electrophoresis (SDS-PAGE) (Dunn 1986), two-dimensional gel electrophoresis (2DE) (Issaq and Veenstra 2008), and two-dimensional differential gel electrophoresis (2D-DIGE) (Marouga et al. 2005). For mass spectrometric measures, four types of mass analyzers are mainly used: quadrupole (Q), ion trap (quadrupole ion trap, QIT; linear ion trap, LIT; linear trap quadrupole, LTQ), time-of-flight (TOF) mass analyzer, and Fourier-transform ion cyclotron resonance (FTICR) mass analyzer (Han et al. 2008; Thompson 2017). In addition, there are hybrid mass spectrometers with combined capabilities including Q-Q-Q, Q-Q-LIT, Q-TOF, TOF-TOF, and LTQ-FTICR. The use of isotope labeling strategies further expands the capabilities of mass spectrometry because it allows for measuring and quantifying dynamic changes in protein expression, interaction, and modification (Han et al. 2008). This strategy makes use of stable isotopes like ^{2}H, ^{13}C, ^{15}N, and ^{18}O, which are incorporated by metabolic labeling during protein synthesis.

An alternative approach to mass spectrometric measures is offered by antibody proteomics, which is defined as the systematic generation and use of protein-specific antibodies to functionally explore the proteome (Solier and Langen 2014; Uhlen and Ponten 2005; Wingren 2016). The generated antibodies are used for both tissue profiling and protein assays (ELISAs, protein arrays). In addition, these antibodies can be used as capture ("pull-down") reagents for purification of specific (native) proteins and their associated complexes for structural and biochemical analyses (Uhlen and Ponten 2005). The specific capture of the target proteins by antibodies can also be combined with mass spectrometry (antibody-enriched selected reaction monitoring) to allow for specific and sensitive detection and even quantification of proteins (Whiteaker et al. 2010).

Looking specifically at microalgae, proteomics has been used in basic research to better understand protein functions in all kinds of cellular mechanisms and in response to environmental conditions. Previous research was mainly focused on model organisms such as *Chlamydomonas reinhardtii* (Fig. 5.2) (Mastrobuoni et al. 2012; Schmidt et al. 2006; Winck et al. 2012; Zhan et al. 2018), *Thalassiosira pseudonana* (Carvalho and Lettieri 2011; Dong et al. 2016; Dyhrman et al. 2012) (Fig. 5.2), *Phaeodactylum tricornutum* (Bai et al. 2016; Xie et al. 2015) (Fig. 5.2), and *Emiliania huxleyi* (Jones et al. 2011, 2013; McKew et al. 2013). Biotechnologically oriented proteomic approaches in microalgae were driven by the desire to find ideal algae or at least ideal growth conditions for the production of biodiesel, bioethanol, and biogas (Anand et al. 2017) or high-value compounds like astaxanthin (Tran et al. 2009).

5.2.6 Lipidomics

The lipidome corresponds to the totality of lipids in a biological system, and lipidomics deals with the holistic analysis of these lipids. Lipidomics is an emerging sector particularly with regard to microalgae. It is also a key sector because lipids play many essential roles in cellular functions, including cellular barriers, membrane matrices, signaling, and energy depots (Yang and Han 2016).

Typically, the lipidome analysis comprises a lipid extraction using appropriate internal standards followed by lipid profiling with mass spectrometry techniques. In order to make use of the hydrophobic nature of lipids, the most commonly utilized lipid extraction procedures include the solvents chloroform, methyl *tert*-butyl ether (MTBE), butanol and butanol-methanol (BUME), or acetic acid combined with isopropanol and hexane (Tumanov and Kamphorst 2017). When necessary, extracted lipids are derivatized. The mass spectrometry analyses are attributed either to shotgun lipidomics or to chromatography-based lipidomics. The techniques for mass spectrometry ionization include electrospray ionization (ESI), matrix-assisted laser desorption/ionization (MALDI), atmospheric pressure chemical ionization (APCI), atmospheric pressure photoionization (APPI), secondary ion mass spectrometry (SIMS), and desorption ESI (DESI) (Yang and Han 2016). Following ionization, full mass spectrometry analyses (MS), tandem mass spectrometry (MS/MS) analyses, or both can be applied. The utilized MS/MS techniques include product ion scan, precursor ion scan (PIS), neutral loss scan (NLS) and selected/multiple reaction monitoring (SRM/MRM) (Yang and Han 2016). The acquired MS data are then processed for deisotoping, followed by identification and quantification of individual lipid species.

Previous lipidome analysis of microalgae species already generated important insights into their lipid metabolism (Li et al. 2014b; Martin et al. 2014; Matich et al. 2018; Siaut et al. 2011; Willette et al. 2018; Yang et al. 2015). The lipid metabolism of microalgae is of particular interest for biotechnologists because many green microalgae species have been referred to as oleaginous microorganisms and they therefore can be utilized as feedstocks for biofuel production (Boyle et al. 2012; Campos et al. 2014; Griffiths et al. 2014; MacDougall et al. 2011; Matich et al. 2018; Yang et al. 2015). Moreover, accumulation of lipids usually even can be induced under nutrient stress conditions. This makes it possible to first cultivate the microalgae under optimal conditions for fast growth, and once there is a substantial amount of biomass, accumulation of lipids can be induced. In this way it is less problematic that the growth rate decreases during accumulation of lipids.

5.2.7 Glycomics

Glycosylation is a major posttranslational modification in which glycan (polysaccharide) molecules are covalently bonded to proteins or other organic molecules such as lipids. It was estimated that at least 50% of all proteins are glycosylated (Haltiwanger and Lowe 2004). Glycomics refers to the large-scale experimental analysis of all glycans in a cell, tissue, organ, or organism. Unfortunately, glycans are a quite difficult class of biological molecules to study by large-scale experimental analysis due to the chemical similarities between the constituent monosaccharide building blocks, the branched or isomeric status of many glycans, the template-less biosynthesis, and the lack of clearly identifiable consensus sequences for the glycan modification of cohorts of glycoproteins (Rakus and Mahal 2011).

The glycome analysis starts with an extraction and, in case of glycoconjugates, the enzymatic or chemical release of the bonded glycans. The obtained glycans are either derivatized or analyzed directly. Derivatized glycans are separated by HPLC and other approaches and further analyzed by mass spectrometry or NMR (Cummings and Pierce 2014; North et al. 2009; Zaia 2010). In large-scale experimental analyses (shotgun glycomics), separated glycans also can be printed to generate glycan microarrays. Such glycan libraries on microarrays can be systematically processed to identify the specificities of glycan-binding proteins to enable investigations into their biological roles (Rillahan and Paulson 2011). In another approach, glycan structures and functions are also studied by lectin microarrays, in which the carbohydrate-binding lectins are immobilized on the microarray and the glycosylation status of a biological sample is investigated (Hirabayashi 2008; Krishnamoorthy and Mahal 2009). Insight into the functional roles of glycans is also gained by synthesis of small molecules that inhibit key glycan-binding proteins or key biosynthetic enzymes (Kiessling and Splain 2010).

Looking at the microalgae, it becomes clear that glycomics currently lags behind other omics technologies probably due to the complexity and technical effort of glycome analyses. However, glycans *N*-linked to proteins have been investigated in the algal model organisms *Chlamydomonas reinhardtii* (Mathieu-Rivet et al. 2013) and *Phaeodactylum tricornutum* (Baiet et al. 2011) (Fig. 5.2). In a couple of other microalgae species, thus far only the genes coding for proteins involved in *N*-glycan maturation have been identified in the corresponding genomes (Mathieu-Rivet et al. 2014).

5.2.8 Metabolomics

The metabolome of a cell or organism includes the complete set of metabolites, the nongenetically encoded substances, intermediates, and products of metabolic pathways. Among the major metabolite compound classes are carbohydrates, amino acids, peptides, nucleotides, lipids, fatty acids, phenols, terpenoids, flavonoids, alkaloids, polyols, carotenoids, vitamins, and phytohormones. Metabolomics, also termed metabolic profiling or metabolomic analysis, makes it possible to simultaneously investigate metabolite pool sizes, fluxes, and mechanisms of metabolic regulation in the course of normal development or in response to chemicals, mutations, changing environment parameters, or environmental stresses (Zamboni et al. 2015). Metabolomic analyses are also used to analyze genetically modified organisms to better understand the intended and unintended consequences of genetic engineering and, finally, to improve productivity for biotechnological applications. Metabolomics is also a key technology for systems biology because the metabolome of a cell provides information about its molecular phenotype, physiological condition, and interaction with its environment. The diverse functional applications of metabolomic technologies include chemical inventorying, chemotaxonomy, enzyme annotation, biochemical phenotyping, generation of metabolomic fingerprints, in situ enzymology, pathway analysis, and for a greater understanding of global system biology (Aldridge and Rhee 2014). The central analytical method in metabolomics is mass spectrometry, which is commonly combined with chromatographic procedures (e.g., LC-MS). In addition, NMR metabolomics frequently complements mass spectrometry.

Metabolomic analyses have been performed in brown algae, red algae, green algae, and diatoms including both microalgae and macroalgae from freshwater, marine, or saline habitats. There are large-scale investigations using a great number of species (Belghit et al. 2017; Bromke et al. 2015; Lang et al. 2011; Sun et al. 2018) or investigations that focus just on a single species (Chen et al. 2017; Lamers et al. 2008; Ritter et al. 2014; Wang et al. 2017; Willamme et al. 2015; Willette et al. 2018; Wördenweber et al. 2018).

Genetic engineering of microalgae is greatly benefiting from the metabolic network reconstruction in *Chlamydomonas reinhardtii*, *Ostreococcus lucimarinus*, *Ostreococcus tauri*, *Botryococcus braunii*, and *Synechocystis* sp. (Boyle and Morgan 2009; Chang et al. 2011; Dal'Molin et al. 2011; Krumholz et al. 2012; Molnar et al. 2012; Yoshikawa et al. 2011). Based on this extensive knowledge base, flux balance analyses (e.g., using ^{13}C-labeled substrates) are being applied to identify bottlenecks and new targets to metabolically engineer microalgae (Boyle and Morgan 2009; Dal'Molin et al. 2011; Wu et al. 2015; Xiong et al. 2010).

5.3 Key Elements of Genetic Engineering

For a given target species, the availability of omics data sets is a great advantage for its genetic engineering. Particularly, genomics and transcriptomics data are more than welcome. However, the existence of such data is no mandatory prerequisite. By contrast, there are a number of key elements and steps that are imperative for genetic engineering. Fundamental prerequisites for genetic engineering are that the target species can be grown in the lab under standardized conditions and that it exists as an axenic culture. For later transformation and selection, it is necessary that the target species can be re-grown from single cells. The mandatory construction of transformation vectors requires several DNA elements, such as selectable marker genes, reporter genes, genes of interest, regulatory sequences, introns, and untranslated regions (UTRs). DNA elements can be cloned from total DNA or RNA of various source organisms or synthesized de novo based on available sequence data. The codon usage of the target species must be considered, and helpful restriction site or sequence tags can be added. The constructed transformation vector needs to be delivered into the cell using transformation methods like particle bombardment, agitation with glass beads, or electroporation. Usually, only a low number of target cells are transformed among countless unchanged cells, which requires efficient selection of the transformed cells. Aside from the selection marker gene, the transformed cells must be inspected for appropriate expression of the integrated reporter genes or other genes of interest. A generalized workflow of genetic engineering of microalgae is shown in Fig. 5.4. Essential DNA elements and key steps are discussed in more detail below.

5.3.1 Selectable Marker Genes

Stable transformation of algae requires the use of selectable marker genes that confer selective advantage because only a very small percentage of treated organisms is successfully transformed in a transformation experiment. The expression of selectable marker genes and the application of selective conditions allow for artificial selection of few successfully transformed organisms. Both nuclear and chloroplast genomes are subject to genetic transformation. Frequently, selectable marker genes are antibiotic resistance genes, which are dominant markers because they confer a new trait to any transformed target species. There are also herbicide resistance genes and metabolic marker genes that can be useful selectable marker genes. However, the latter require isolation of auxotrophic mutants, which can be time-consuming and tedious.

By far the highest number of selectable marker genes for nuclear transformation, i.e., more than 20, has been established for the unicellular model alga *Chlamydomonas reinhardtii* (Table 5.2): The R100.1 plasmid/bacteriophage T4/synthetic aminoglycoside adenyltransferase gene *aadA* confers resistance to spectinomycin and streptomycin (Cerutti et al. 1997b), the *Streptoalloteichus hindustanus ble* gene to zeomycin and phleomycin (Stevens et al. 1996), the Transposon *Tn9* chloramphenicol acetyltransferase *cat* gene to chloramphenicol (Tang et al. 1995), the *Streptomyces rimosus* aminoglycoside phosphotransferase *aphVIII* (*aphH*) gene to paromomycin (Sizova et al. 2001), both the *Streptomyces hygroscopicus* aminoglycoside phosphotransferase *aph7″* gene and the *E. coli* hygromycin phosphotransferase *hpt*

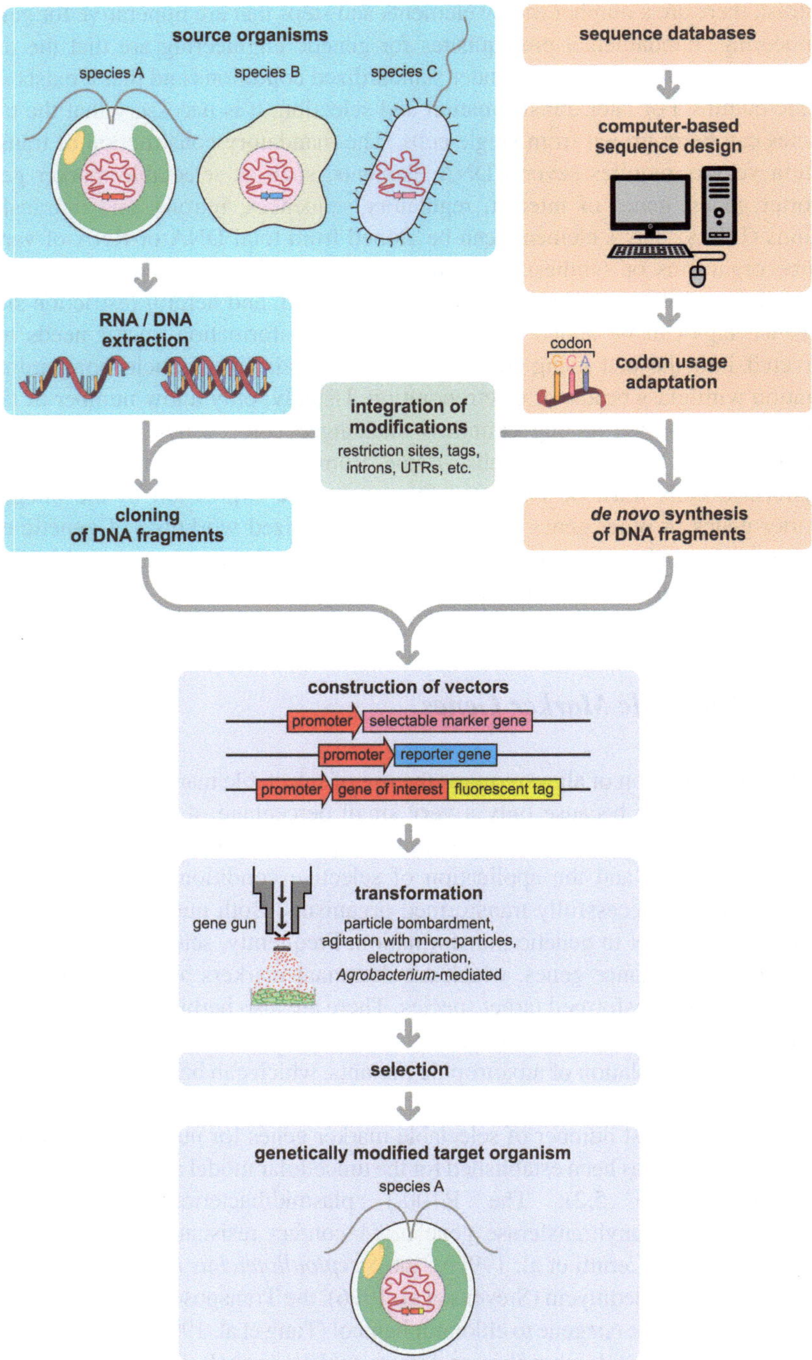

Fig. 5.4 Workflow of genetic engineering of microalgae

Table 5.2 Microalgae species with available selectable marker genes for nuclear transformation

Transformed species	Lineage	Available selectable marker genes	References
	Green algae		
Chlamydomonas reinhardtii	Chlorophyta; Chlorophyceae; Volvocales; Chlamydomonadaceae	*aadA, ble, cat, aphVIII (aphH), aph7", hpt, CRY1–1, nptII, TETX, ALS, PPX (PPO), GAT, PDS, NIT1, NIT2, ARG7, ARG9, ATPC, AC29 (ALB3.1), THI10, NIC7, PSY* and *OEE1*	Berthold et al. (2002), Brueggeman et al. (2014), Butanaev (1994), Cerutti et al. (1997b), Debuchy et al. (1989), Garcia-Echauri and Cardineau (2015), Hall et al. (1993), Kindle et al. (1989), Kovar et al. (2002), Nelson et al. (1994), Randolph-Anderson et al. (1998), Remacle et al. (2009), Schnell and Lefebvre (1993), Sizova et al. (2001), Stevens et al. (1996), Tang et al. (1995), Ferris (1995), Mayfield and Kindle (1990), McCarthy et al. (2004), Smart and Selman (1993)
Volvox carteri f. *nagariensis*	Chlorophyta; Chlorophyceae; Volvocales; Volvocaceae	*ble, aphVIII (aphH), nitA*	Hallmann and Rappel (1999), Hallmann and Wodniok (2006), Jakobiak et al. (2004), Schiedlmeier et al. (1994)
Gonium pectorale	Chlorophyta; Chlorophyceae; Volvocales; Volvocaceae	*aphVIII (aphH)*	Lerche and Hallmann (2009)
Pandorina morum	Chlorophyta; Chlorophyceae; Volvocales; Volvocaceae	*aphVIII (aphH)*	Lerche and Hallmann (2014)
Eudorina elegans	Chlorophyta; Chlorophyceae; Volvocales; Volvocaceae	*aphVIII (aphH)*	Lerche and Hallmann (2013)
Haematococcus pluvialis	Chlorophyta; Chlorophyceae; Volvocales; Haematococcaceae	*aph7, PDS*	Kathiresan et al. (2015), Sharon-Gojman et al. (2015), Steinbrenner and Sandmann (2006)

(continued)

Table 5.2 (continued)

Transformed species	Lineage	Available selectable marker genes	References
Dunaliella sp.	Chlorophyta; Chlorophyceae; Volvocales; Dunaliellaceae	*Ble, cat, NIT1*	Geng et al. (2004), Sun et al. (2006), Sun et al. (2005)
Scenedesmus obliquus	Chlorophyta; Chlorophyceae; Sphaeropleales; Scenedesmaceae	*Ble*	Guo et al. (2013)
Chlorella sp.	Chlorophyta; Trebouxiophyceae; Chlorellales; Chlorellaceae	*Ble, aph7, nptII, PDS, NIT1*	Chow and Tung (1999), Dawson et al. (1997), Hawkins and Nakamura (1999), Huang et al. (2008), Kim et al. (2002)
	Diatoms		
Phaeodactylum tricornutum	Heterokontophyta; Bacillariophyceae; Naviculales; Phaeodactylaceae	*Ble, cat, nptII*	Apt et al. (1996), Falciatore et al. (1999), Niu et al. (2012), Zaslavskaia et al. (2000)
Navicula saprophila	Heterokontophyta; Bacillariophyceae; Naviculales; Naviculaceae	*nptII*	Dunahay et al. (1995)
Cylindrotheca fusiformis	Heterokontophyta; Bacillariophyceae; Bacillariales; Bacillariaceae	*Ble*	Fischer et al. (1999), Poulsen and Kröger (2005)
Cyclotella cryptica	Heterokontophyta; Coscinodiscophyceae; Thalassiosirales; Stephanodiscaceae	*nptII*	Dunahay et al. (1995)
	Eustigmatophyta		
Nannochloropsis sp.	Heterokontophyta; Eustigmatophyceae; Eustigmatales; Monodopsidaceae	*Ble*	Kilian et al. (2011), Li et al. (2014a)
	Dinoflagellates		
Symbiodinium microadriaticum	Alveolata; Dinophyceae; Suessiales; Symbiodiniaceae	*nptII*	ten Lohuis and Miller (1998)
Amphidinium sp.	Alveolata; Dinophyceae; Gymnodiniales; Gymnodiniaceae	*nptII*	ten Lohuis and Miller (1998)

gene to hygromycin B (Berthold et al. 2002; Butanaev 1994), the mutated version of the *Chlamydomonas reinhardtii* ribosomal protein gene *S14*, *CRY1–1*, to emetine and cryptopleurine (Nelson et al. 1994), the neomycin phosphotransferase *nptII* gene to neomycin (Hall et al. 1993), the *Enterobacteriaceae bacterium* NADP-requiring oxidoreductase *TETX* gene to tetracycline (Garcia-Echauri and Cardineau 2015), the mutated *Chlamydomonas reinhardtii* acetolactate synthase *ALS* gene to sulfonylurea herbicides (Kovar et al. 2002), the mutated *Chlamydomonas reinhardtii* protoporphyrinogen oxidase *PPX* (*PPO*) gene to oxyfluorfen (Brueggeman et al. 2014; Randolph-Anderson et al. 1998), the glyphosate aminotransferase *GAT* gene to glyphosate (Brueggeman et al. 2014), and the phytoene desaturase *PDS* gene to norflurazone (Brueggeman et al. 2014). Working metabolic markers for *Chlamydomonas reinhardtii* are the nitrate reductase *NIT1* gene (Kindle et al. 1989), the nitrate reductase *NIT2* gene (Schnell and Lefebvre 1993), the argininosuccinate lyase *ARG7* gene (Debuchy et al. 1989), the *Arabidopsis thaliana/C. reinhardtii* N-acetyl ornithine aminotransferase *ARG9* gene (Remacle et al. 2009), the ATP synthase subunit *ATPC* gene (Smart and Selman 1993), the *C. reinhardtii AC29* (*ALB3.1*) gene (Ferris 1995), the hydroxyethylthiazole kinase *THI10* gene (Ferris 1995), the quinolinate synthetase A *NIC7* gene (Ferris 1995), the phytoene synthase *PSY* gene (McCarthy et al. 2004), and, finally, the oxygen-evolving enhancer *OEE1* gene (Mayfield and Kindle 1990). Selectable marker genes also have been established for nuclear transformation of quite a few other microalgae species, which are listed in Table 5.2. In addition, chloroplast transformation has been established for *Chlamydomonas reinhardtii* and several other microalgal species using antibiotic resistance genes, herbicide resistance genes, and metabolic marker genes (Doron et al. 2016; Esland et al. 2018).

The remarkable repertoire of selectable marker genes, which has been shown to work in different groups of microalgae, is an excellent basis for all those researchers who intend to transform other algal species.

5.3.2 Reporter Genes

Like selectable marker genes, reporter genes are also basic essentials in the molecular toolbox of a genetic engineer. Reporter genes frequently code for enzymes that convert a substrate into a colored product or result in light emission, or the reporter gene product is a fluorescent protein. The coding sequences of reporter genes are either placed under the control of the regulatory sequences of a gene of interest or they are attached to the coding sequence of the gene of interest, which is under control of its regulatory sequences, to produce a fusion protein. These approaches allow for obtaining dynamic spatial and temporal information about the gene of interest's protein expression. Quite a few reporter genes have been established for *Chlamydomonas reinhardtii* and some other green algal and diatom species (Table 5.3). Because most reporter genes come from heterologous sources, the genes frequently have been adapted to the codon usage of the target microalgae.

Table 5.3 Microalgae species with available reporter genes

Transformed species	Lineage	Available reporter genes	References
	Green algae		
Chlamydomonas reinhardtii	Chlorophyta; Chlorophyceae; Volvocales; Chlamydomonadaceae	*ARS, GFP, BFP, CFP, OFP, RFP, YFP, GLUC, RLUC, LUCCP, LUXCT, xyn1*	Davies et al. (1992), Fuhrmann et al. (1999, 2004), Lauersen et al. (2015), Matsuo et al. (2006), Mayfield and Schultz (2004), Rasala et al. (2013, 2012), Shao and Bock (2008)
Volvox carteri f. *nagariensis*	Chlorophyta; Chlorophyceae; Volvocales; Volvocaceae	*ARS, GFP, HUP1, GLUC*	Ender et al. (2002), Hallmann and Sumper (1994,1996), von der Heyde et al. (2015)
Gonium pectorale	Chlorophyta; Chlorophyceae; Volvocales; Volvocaceae	*GLUC*	Lerche and Hallmann (2009)
Pandorina morum	Chlorophyta; Chlorophyceae; Volvocales; Volvocaceae	*GLUC*	Lerche and Hallmann (2014)
Eudorina elegans	Chlorophyta; Chlorophyceae; Volvocales; Volvocaceae	*GLUC*	Lerche and Hallmann (2013)
Haematococcus pluvialis	Chlorophyta; Chlorophyceae; Volvocales; Haematococcaceae	*lacZ, PDS*	Steinbrenner and Sandmann (2006), Teng et al. (2002)
Dunaliella salina	Chlorophyta; Chlorophyceae; Volvocales; Dunaliellaceae	*GUS*	Tan et al. (2005)
Chlorella sp.	Chlorophyta; Trebouxiophyceae; Chlorellales; Chlorellaceae	*GUS, LUC*	Chow and Tung (1999), El-Sheekh (1999), Jarvis and Brown (1991)
	Diatoms		
Phaeodactylum tricornutum	Heterokontophyta; Bacillariophyceae; Naviculales; Phaeodactylaceae	*LUC, GFP, HUP1, glut1*	Falciatore et al. (1999), Zaslavskaia et al. (2000)
Cylindrotheca fusiformis	Heterokontophyta; Bacillariophyceae; Bacillariales; Bacillariaceae	*GFP*	Poulsen and Kröger (2005)

(continued)

Table 5.3 (continued)

Transformed species	Lineage	Available reporter genes	References
Thalassiosira weissflogii	Heterokontophyta; Coscinodiscophyceae; Thalassiosirales; Thalassiosiraceae	*GUS*	Falciatore et al. (1999)

The most successful reporters seem to be the green fluorescent protein (GFP) and its blue, cyan, orange, red, and yellow variants BFP, CFP, OFP, RFP, and YFP because they allow for colorful in vivo fluorescence labeling of transgenic organisms (Day and Davidson 2009; Rasala et al. 2013). Furthermore, it turned out that luciferase enzymes are suitable reporters because they allow for luminometer quantification of emitted chemiluminescence. The utilized luciferase sequences come from the marine copepod *Gaussia princeps* (*GLUC*) (Shao and Bock 2008), the soft coral *Renilla reniformis* (*RLUC*) (Fuhrmann et al. 2004), the firefly (beetle) *Photinus pyralis* (*LUCCP*) (Matsuo et al. 2006), or the bacterium *Vibrio harveyi* (*LUXCT*) (Mayfield and Schultz 2004). Further enzymatic reporters like arylsulfatase convert chromogenic substrates and allow for spectrophotometric quantification (Davies et al. 1992; Hallmann and Sumper 1996).

5.3.3 Regulatory Sequences

The regulatory sequences include the promoter region together with enhancer or silencer elements. Whenever genes from heterologous sources, like many selectable marker or reporter genes, are used, endogenous regulatory sequences from the target species are required because in most cases heterologous regulatory sequences would result in little or no expression. Even for expression of endogenous genes, researchers frequently replace regulatory sequences to achieve a difference in the expression pattern or an increased expression rate. Also constitutive regulatory sequences are useful, for example, to permanently express selectable marker genes. There is also a need for inducible regulatory sequences, which are regulated by chemicals or light, to switch genes of interest on and off at any time. When regulatory sequences are utilized for artificial gene constructs, the exact position of the essential DNA elements is frequently unknown, and, therefore, larger "regulatory regions" or "promoter regions" are cloned.

Quite a few constitutive and inducible, endogenous regulatory sequences of microalgae have been shown to drive the desired (heterologous) genes. In particular, there are more than 30 working regulatory sequences (promoter regions) for expression in *C. reinhardtii* (Mussgnug 2015) and about 6 for *V. carteri* (Hallmann 2015). Working regulatory sequences come from genes of the following proteins: small chain of ribulose bisphosphate carboxylase (RBCS2) (Stevens et al. 1996), heat shock protein 70A (HSP70A) (Schroda et al. 2000), HSP70A/RBCS2 fusion (Schroda et al. 2000), HSP70/RBCS3 fusion (Jakobiak et al. 2004), HSP70A/β2-tubulin (β2TUB) fusion (Schroda et al. 2000), β-tubulin (Blankenship and Kindle 1992; Hallmann and Sumper 1996), α-tubulin (tubA1) (Kozminski et al. 1993), arylsulfatase (ARS) (Hallmann and Sumper 1994), abundant protein of photosystem I complex (psaD) (Fischer and Rochaix 2001), cytochrome c6 (CYC6) (Quinn et al. 2003), coproporphyrinogen III oxidase (CPX1) (Quinn et al. 2003), plastocyanin (PCY1/petE) (Quinn and Merchant 1995), γ-subunit of the chloroplast ATPase (AtpC) (Quinn and Merchant 1995), nopaline synthase (NOS) (Diaz-Santos et al. 2013), NAD(P)H nitrate reductase (nia1) (Llamas et al. 2002), light-harvesting chlorophyll a/b-binding protein of photosystem II (cabII-1) (Blankenship and Kindle 1992), carbonic anhydrase (ca1/ca2) (Villand et al. 1997), and *Chlorella* virus adenine methyltransferase (amt) (Mitra and Higgins 1994; Mitra et al. 1994). The regulatory sequences of the fucoxanthin-chlorophyll a/c-binding protein (FCP) (Falciatore et al. 1999) were used in diatoms (i.e., *Phaeodactylum*). Besides, regulatory sequences of *chlL*, *petD*, *rbcL*, *atpA*, *atpB*, and *rrn16* genes were suitable for chloroplast transformations (Kim 2015). Even quite a few synthetic regulatory sequences are capable of driving robust (nuclear) gene expression (Scranton et al. 2016).

Moreover, computational approaches allowed for the identification of tens of thousands of potential cis-regulatory elements by using the novel algorithm MERCED ("modeling evolution rate across species for cis-regulatory element discovery") (Ding et al. 2012).

5.3.4 UTRs

Both 5′ and 3′ untranslated regions (UTRs) can influence the expression of transgenes. Expression is supported if 3′ UTRs contain an effective polyadenylation signal, like UGUAA in *C. reinhardtii* (Bell et al. 2016). 5′ UTRs might contain introns and these intron sequences can include regulatory elements that regulate the transcription. Therefore, for expression studies of a gene of interest, it might be not sufficient to place just a reporter gene behind the promoter region of the gene of interest. Using both UTRs of the gene of interest might also be required if the expression of the reporter gene construct should most closely approximate the expression pattern of the gene of interest. Regulatory sequences are frequently cloned as larger promoter regions together with the 5′ UTRs of the genes of interest because both the actual 5′ end of the mRNA and the exact position of regulatory elements are frequently uncertain. If it is about 3′ UTRs, the use of some specific UTRs has been preferred, like those of *RBCS2* or *PSAD* (Cerutti et al. 1997b; Fischer and Rochaix 2001; Lumbreras et al. 1998). However, it has not yet been systematically proved that these 3′ UTRs generally have a positive impact on expression.

After transcription, processing and nuclear export of an mRNA, the cap structure at the 5′ UTR recruits the ribosomal complex for initiation of translation and usually sequence-dependent secondary and tertiary structures within the 5′ UTR regulate translation. The 3′ UTRs frequently also play a role in posttranscriptional regulation and can contain regulatory elements that attract translational enhancers. These are all reasons to carefully select the desired UTRs.

5.3.5 Introns

Spliceosomal introns are present in all characterized eukaryotes. Even if the number of introns per gene varies greatly, most of all microalgal genes contain at least one intron. Since introns can contain regulatory elements that influence transcription, they can be important for adequate expression of a gene of interest. Moreover, there are several publications, which demonstrate that expression of transgenes in microalgae is improved if appropriate introns are present (Baier et al. 2018; Eichler-Stahlberg et al. 2009; Gruber et al. 1996; Hallmann and Rappel 1999; Kovar et al. 2002; Lumbreras et al. 1998). For expression of homologous genes, genomic sequences with introns can be cloned. The genes then contain their natural introns at natural positions. However, enhancement of expression is also achievable by artificially integrating introns of the target species at unnatural positions into heterologous or homologous cDNAs (Hallmann and Rappel 1999).

5.3.6 Codon Usage

The degeneracy of the genetic code implies that most amino acids are encoded by a family of synonymous codons. The frequency of use of alternative, synonymous codons, the codon usage, varies among all species (Grantham et al. 1981; Ikemura 1985). In each species, more frequent (preferred) codons are those that are recognized by more abundant tRNA molecules. Especially in the case of highly expressed genes, the preferred codons of the given species can be found. The codon usage and availability of tRNAs influence the speed of mRNA translation elongation and affect also the kinetics of in vivo folding (Angov 2011). Especially when foreign genes are going to be expressed in a certain species, the codon usages of donor and recipient species need to be compared. In more closely related species, the codon usage bias might be low and tolerable. However, if there is a high codon usage bias, de novo DNA synthesis (see below) with an optimized codon usage is indispensable. Special algorithms for codon optimizing have been developed (Liu et al. 2012). Codon optimization improves protein expression because it counteracts the differences in codon usage between the coding DNA of the donor and the recipient species or between weakly and highly expressed genes (Angov 2011). Aside from codon usage, a similar G + C content between transgene and the genome of the recipient species was suggested to be important for proper transgene expression. However, the

G + C content of a coding sequence is actually just a consequence of the codon usage (Barahimipour et al. 2016; Barahimipour et al. 2015).

5.3.7 De Novo DNA Synthesis

De novo DNA synthesis refers to chemically synthesizing a gene or any piece of DNA in vitro, giving researchers the freedom of generating any DNA sequence in the absence of a DNA or RNA template strand, simply based on electronically available or designed sequences. Main factors for a de novo DNA synthesis are a high codon usage bias between donor and recipient species (see above), the non-availability of a template DNA, or intended massive modifications of an existing sequence. Moreover, unsuitable restriction sites can be removed, suitable sites added, potential cryptic splice sites removed, and polymerase slippage sites avoided. Simultaneously, it is possible to optimize the mRNA secondary structure and the folding free energy of native mRNA. Routinely, synthetic DNA is manufactured in the form of single-stranded oligonucleotides using the nucleoside phosphoramidite method (Beaucage and Caruthers 1981), which currently has its upper limit in length at 200–300 nucleotides. As a consequence, all longer DNA molecules have to be synthesized from a set of overlapping single-stranded oligonucleotides. However, the synthesis of longer DNA sequences needs to be planned and customized using DNA design software. The individually designed DNA is divided into chemically synthesizable pieces, so-called synthons, which have a length of up to 1500 bp (Hughes and Ellington 2017). Designed and synthesized oligonucleotides are then assembled together into the designed synthons using different gene synthesis techniques (Hughes and Ellington 2017). If required, several synthons can be assembled together into larger DNAs. The assembled DNAs are cloned into vectors and sequenced. Minor sequence errors can be repaired or otherwise the process is repeated with an improved design strategy. Recently, a new DNA synthesis technique was presented, which promises to revolutionize DNA synthesis with a faster, cheaper, and more accurate approach (Palluk et al. 2018). The new technology could lead to "DNA printers."

Already in 2010, de novo DNA synthesis of the complete genome of the bacterium *Mycoplasma mycoides* could be achieved (Gibson et al. 2010). Its genome consists of 1.1 million base pairs. Currently, artificial synthesis of the first eukaryotic genome is in progress, and it should be completed soon (Burgess 2017; Richardson et al. 2017). It is a redesigned genome of the yeast *Saccharomyces cerevisiae*, which is constructed in a way that allows to extensively rearrange the genome on demand (Foo and Chang 2018). The *S. cerevisiae* genome comprises 16 chromosomes totalling approximately 12 million base pairs. When considering that the genome of the green microalga *Ostreococcus tauri* (Prasinophyceae) has a very similar size, i.e., it has only 12.6 million base pairs (Derelle et al. 2006), de novo synthesis of complete microalgal genomes is within reach.

5.3.8 Vector Construction

Vectors for transformation of microalgae frequently include not only the above-discussed elements for expression in eukaryotic microalgae but also the backbone of bacterial standard vectors (e.g., pBluescript), which makes it possible to perform the vector construction in *Escherichia coli* strains (e.g., DH5α). The minimal requirement for propagation in *E. coli* is the presence of an origin of replication (e.g., pUC ori), an antibiotic resistance gene (e.g., AmpR), and, for easy cloning, a multiple cloning site. To facilitate artificial gene construction, quite a few special vectors and construction strategies have been developed (Hallmann and Rappel 1999; Hallmann and Wodniok 2006; Heitzer and Zschoernig 2007; Oey et al. 2014; Song et al. 2013; Xie et al. 2014). One example is a simplification of the vector construction by using the λ-phage based gateway system, which makes use of recombination sequences that allow the transfer of DNA fragments between vectors in vitro (Hartley et al. 2000; Oey et al. 2014). Another strategy, the cassette multiplication technique, allows for easy production of vectors with multiple gene copies, which can affect a higher gene dosage and, as a consequence, an increased expression rate (Hallmann and Rappel 1999; Hallmann and Wodniok 2006). For *Chlamydomonas reinhardtii*, a modular cloning toolkit has recently been developed based on Golden Gate cloning with standard syntax (Crozet et al. 2018). The toolkit contains 119 openly distributed genetic parts including regulatory sequences, UTRs, terminators, tags, reporters, antibiotic resistance genes, and introns cloned in various positions to allow maximum modularity.

In several cases, it might be difficult to pack all required eukaryotic and bacterial elements onto a single plasmid due to the large vector size and unsuitable restriction sites. In such cases, the eukaryotic selectable marker gene can be placed on one plasmid (including a bacterial vector backbone), and the unselectable gene of interest or reporter gene is placed on a second plasmid (including a bacterial vector backbone). Then, both plasmids are co-transformed. Fortunately, the reported co-transformation rates are quite high: The co-transformation rate was about 80% in *Chlamydomonas reinhardtii* (Stevens et al. 1996), 30–50% in *Gonium pectorale* (Lerche and Hallmann 2009) (Fig. 5.2), 80% in *Pandorina morum* (Lerche and Hallmann 2014) (Fig. 5.2), and 50–100% in *Eudorina elegans* (Lerche and Hallmann 2013) (Fig. 5.2). For *V. carteri* (Fig. 5.2), co-transformation rates of 10–60% (Hallmann and Rappel 1999) or 40–80% (Schiedlmeier et al. 1994) have been reported. Co-transformation has the advantage that always the same selectable marker plasmid can be used for all transformation experiments in addition to an always changing unselectable plasmid containing the gene of interest.

Vector construction requires also considerations about the desired cellular localization of the expressed protein. Among others, localization sequences for targeting expressed proteins to the nucleus, mitochondria, endoplasmic reticulum, chloroplast, plasma membrane, or outward are known. Without use of a localization sequence, the default localization is the cytoplasm. Difficulties in obtaining a high level of transgene expression in the cytoplasm frequently urged researchers to target the proteins of interest to the chloroplast. Sophisticated transformation vectors that direct engineered gene products to different specific subcellular locations have been constructed (Rasala et al. 2014).

In addition, vectors that lead to efficient secretion of recombinant proteins have been made for *C. reinhardtii* (Molino et al. 2018) and *Phaeodactylum tricornutum* (Vanier et al. 2018). One great advantage of secreted proteins is that purification becomes much easier.

Both circular (super-coiled) and linearized plasmids are used for transformation. However, the open or closed form of the plasmid does not seem to significantly change the efficiency of transformation (Coll 2006). Since it has been shown that the efficiency of DNA transformation was increased when vector sequences were omitted from nuclear transformation setups (Meslet-Cladiere and Vallon 2011), researchers should consider to cut away unnecessary DNA regions before transformation.

5.3.9 Transformation Methods

Transformation of microalgae encompasses methods for inducing the uptake of exogenous DNA into a cell and, finally, into the nucleus. While the DNA entry requires temporary permeabilization of the cell membrane, the entrance of the DNA into the nucleus and, if desired, its integration into the genome occurs without outside assistance. Any DNA integrations normally occur by illegitimate recombination, resulting in ectopic integration of the introduced DNA and, therefore, in stable genetic transformation (Dent et al. 2015; Tam and Lefebvre 1993). The integration of exogenous genes into the genome typically occurs at random sites because the illegitimate recombination is neither homology nor sequence dependent, although regions of micro-homology may be involved. Depending on the transformation method, also random joining and integration of several copies of the exogenous gene are possible (Flickinger 2013). Depending on the researcher's goal, these conditions can be an advantage or disadvantage for genetic engineering.

Contrary to illegitimate recombination, targeted genomic integration of exogenous DNA by homologous recombination is rare in microalgae and other eukaryotes. Only for a few microalgae species, it was shown that nuclear homologous recombination is basically possible but difficult to achieve and rather time-consuming (Gumpel et al. 1994; Hallmann et al. 1997; Minoda et al. 2004; Nelson and Lefebvre 1995; Sodeinde and Kindle 1993; Zorin et al. 2005). Efficient nuclear gene targeting and removal of exogenous DNA by homologous recombination were shown for just three microalgae species, *Nannochloropsis* (Kilian et al. 2011; Weeks 2011), *Ostreococcus* (Lozano et al. 2014), and *Cyanidioschyzon merolae* (Fig. 5.2) (Takemura et al. 2018). The situation looks different in plastid genomes of microalgae, which frequently integrate exogenous DNA by homologous recombination due to a natural homologous recombination machinery (Day and Goldschmidt-Clermont 2011; Verma and Daniell 2007).

There are a number of working methods that describe the introduction of DNA into microalgae cells, transformation of the cells, and recovery of viable transformants. The challenging task is to permeabilize cell membranes in order to introduce DNA without killing the cells with this life-threatening membrane damage and DNA invasion. The cells also must be able to resume cell division after this genomic intervention.

One of the most popular transformation methods in microalgae is particle bombardment, also referred to as micro-particle bombardment, micro-projectile bombardment, particle gun transformation, gene gun transformation, or biolistic transformation. This method uses DNA-coated micro-projectiles made of gold or other high-density metals. Particle bombardment allows for transformation of almost any type of cell, regardless of the thickness or rigidity of the cell wall or extracellular matrix. In addition, particle bombardment is often used for transformation of plastids. Particle bombardment using gold micro-projectiles was successfully applied in the volvocalean microalgae *Chlamydomonas reinhardtii* (Kindle et al. 1989), *Volvox carteri* (Hallmann and Sumper 1994; Schiedlmeier et al. 1994) (Fig. 5.2), *Gonium pectorale* (Lerche and Hallmann 2009), *Pandorina morum* (Lerche and Hallmann 2014), *Eudorina elegans* (Lerche and Hallmann 2013), *Haematococcus pluvialis* (Steinbrenner and Sandmann 2006) (Fig. 5.2), and *Dunaliella salina* (Tan et al. 2005). It is also possible in the diatoms *Phaeodactylum tricornutum* (Apt et al. 1996), *Cylindrotheca fusiformis* (Fischer et al. 1999), *Navicula saprophila*, (Dunahay et al. 1995), and *Cyclotella cryptica* (Dunahay et al. 1995). Moreover, *Euglena gracilis* (Doetsch et al. 2001), *Porphyridium* sp. (Lapidot et al. 2002), '*Chlorella*' *kessleri* (El-Sheekh 1999), and *Chlorella sorokiniana* (Dawson et al. 1997) can be transformed in this way.

Another less complex and less expensive transformation method is the agitation of cells in the presence of DNA and micro-particles. Frequently, polyethylene glycol is also added. Several researchers used silicon carbide fibers as micro-particles. These hard and rigid fibers are a ceramic compound of silicon and carbon. They even allowed for transformation of cells with intact cell walls including *Chlamydomonas reinhardtii* (Dunahay 1993), *Dunaliella salina* (Feng et al. 2009), *Symbiodinium microadriaticum* (ten Lohuis and Miller 1998), and *Amphidinium* sp. (ten Lohuis and Miller 1998). However, the method seems to be more appropriate for cell wall-reduced algae strains. In *Chlamydomonas reinhardtii*, cell wall-reduced mutants are transformed by agitation in the presence of glass beads (0.4–0.5 mm in diameter), polyethylene glycol, and DNA (Kindle 1990). This is the default mode for transformation of *Chlamydomonas*, even though other methods also work. Cell wall-free protoplasts of the green alga '*Chlorella*' *ellipsoidea* even can be transformed without any micro-particles only by agitating the cells in the presence of DNA and polyethylene glycol (Jarvis and Brown 1991).

Transformation of some microalgae species can also be done by electroporation. An important precondition is the availability of naked cells, protoplasts, cell wall-reduced mutants, or at least cells with thin cell walls. With this method, specially designed electrodes affect voltage across the plasma membrane that exceeds its dielectric strength. This large electronic pulse temporarily disturbs the phospholipid bilayer of the cell membrane, allowing molecules like DNA to pass. The microalgae *Chlamydomonas reinhardtii* (Brown et al. 1991; Shimogawara et al. 1998), *Dunaliella salina* (Geng et al. 2003), *Cyanidioschyzon merolae* (Minoda et al. 2004) (Fig. 5.2), *Nannochloropsis* sp. (Kilian et al. 2011), *Chlorella vulgaris* (Chow and Tung 1999), *Chlorella ellipsoidea* (Chen et al. 2001), and *Chlorella saccharophila* (Maruyama et al. 1994) have been transformed by electroporation. A multi-pulse electroporation variant was successfully applied in *Phaeodactylum tricornutum* (Miyahara et al. 2013).

Like in higher plants, the transformation of microalgae was also achieved through the use of *Agrobacterium tumefaciens*, which possesses a tumor-inducing (Ti) plasmid that integrates semi-randomly into the genome of infected cells. The infection with *Agrobacterium* causes tumors in dicots and some monocots, and most surprisingly, even some algae become infected, but, of course, they do not develop tumors. *Agrobacterium*-mediated transformation was shown to work in the microalgae *Chlamydomonas reinhardtii* (Kumar et al. 2004), *Haematococcus pluvialis* (Kathiresan et al. 2009; Kathiresan and Sarada 2009) (Fig. 5.2), *Dunaliella bardawil* (Anila et al. 2011), and three clades of *Symbiodinium* spp. (Ortiz-Matamoros et al. 2015).

5.3.10 Selection

Any algae transformation usually produces a mixture of relatively few transformed cells and an abundance of non-transformed cells. The reported transformation efficiencies vary greatly between species and used methods (Coll 2006). Due to the very diverse basic conditions, the efficiencies are not really comparable with each other. However, even the highest reported transformation efficiencies produce only 1–3 transformants per 1000 treated cells (Coll 2006; Shimogawara et al. 1998). Thus, selection of transformed cells using the above-described selectable markers is a significant step of any microalgae transformation procedure. An important precondition is that the recipient species can be re-grown from single cells. Ideally, re-growth of transformants can be done on agar plates under selective pressure, which allows to obtain over 100 independent transformant clones from a single plate. If re-growth of transformants on plates is not working, treated cells need to be grown in many separated volumes of liquid medium under selective pressure (Lerche and Hallmann 2009, 2013, 2014). To verify genomic integration, genomic DNA of potential transformants has to be isolated, and unique fragments of the introduced DNA have to be amplified by PCR and sequenced. If desired, transformants can also be analyzed by both Southern blot and real-time PCR to determine the number of integrated copies.

Successful transformation does not guarantee satisfactory expression of an introduced, non-selectable gene of interest. Moreover, expression rates are quite different among simultaneously produced transformants. If production of numerous transformants is possible, easy identification of strongly expressing transformants with high expression can be achieved by fusing the gene of interest via a special linker sequence to a selectable antibiotic resistance gene (Rasala et al. 2012, 2014). By selecting those transformant strains that withstand high levels of the antibiotic, efficient expression of the fusion protein is ensured. After protein synthesis, the selectable marker protein is cleaved off from the protein of interest by a self-cleavage peptide that is encoded within the linker sequence (Rasala et al. 2012, 2014).

5.3.11 Genetically Transformable Microalgae Species

Until now, successful genetic transformation has been demonstrated for more than 30 species of microalgae (Table 5.4). The wild-type phenotypes of several transformable species are shown in Fig. 5.2. Most of the corresponding transformation reports deal with nuclear transformation experiments and just a few examine chloroplast transformations. Many recipient species belong to the green algae lineage: 15 species of green microalgae have been transformed, and stable nuclear transformation has been shown for 13 of them. Within the green algae, the order Volvocales stands out with nine transformed species (Table 5.4, Fig. 5.2). The most prominent members are *Chlamydomonas*, *Volvox*, *Dunaliella*, and *Haematococcus*. The unicellular alga *Chlamydomonas* probably is the best-studied microalga also because of its genetic accessibility and the high number of available molecular tools and methods. *Chlamydomonas* has been used to study fundamental biological processes such as the photosynthetic apparatus, microtubule assembly, flagella movement, mineral nutrition, and other aspects. *Chlamydomonas* and its multicellular relative *Volvox* as well as *Gonium*, *Pandorina*, and *Eudorina* (Table 5.4, Fig. 5.2) belong to the closely related subgroup of volvocine algae, which span the full range of organizational complexity from unicellular and colonial genera to multicellular genera with a full germ-soma division of labor and male-female dichotomy. The existing possibility to transform five different species of volvocine algae with different grades of complexity makes this group a transformable model lineage for addressing fundamental issues related to the evolutionary transition to multicellularity with cellular differentiation and for discovering and testing universal rules that characterize this transition (Hallmann 2011; Kirk 1998).

Transformation has also been shown for seven species of diatoms, and stable transformation has been shown for six of them (Table 5.4). Diatoms are ecologically the most important group of photosynthetic eukaryotes and can be found in nearly every freshwater and marine habitat. Diatoms contribute about 20% of global carbon fixation and 40% of marine primary productivity. There are also approaches to use diatoms for biotechnological applications (Bozarth et al. 2009; Levitan et al. 2014). For all of these reasons, it is advantageous or even indispensable that several diatoms can be transformed.

Reports about successful transformation also relate to several other microalgae species that phylogenetically belong to the red algae, Charophyta, Eustigmatophyta, euglenids, Chlorarachniophyceae, and dinoflagellates (Table 5.4).

Table 5.4 Genetically transformed microalgae species

Species	Lineage	Transformation	References
	Green algae		
Chlamydomonas reinhardtii	Chlorophyta; Chlorophyceae; Volvocales; Chlamydomonadaceae	Stable	Debuchy et al. (1989), Dunahay (1993), Kindle (1990, 1989), Kumar et al. (2004)
Volvox carteri	Chlorophyta; Chlorophyceae; Volvocales; Volvocaceae	Stable	Hallmann and Sumper (1994), Schiedlmeier et al. (1994)
Eudorina elegans	Chlorophyta; Chlorophyceae; Volvocales; Volvocaceae	Stable	Lerche and Hallmann (2013)
Pandorina morum	Chlorophyta; Chlorophyceae; Volvocales; Volvocaceae	Stable	Lerche and Hallmann (2014)
Platymonas subcordiformis	Chlorophyta; Chlorophyceae; Volvocales; Volvocaceae	Stable	Cui et al. (2014)
Gonium pectorale	Chlorophyta; Chlorophyceae; Volvocales; Goniaceae	Stable	Lerche and Hallmann (2009)
Dunaliella salina	Chlorophyta; Chlorophyceae; Volvocales; Dunaliellaceae	Stable	Geng et al. (2003, 2004), Tan et al. (2005)
Dunaliella viridis	Chlorophyta; Chlorophyceae; Volvocales; Dunaliellaceae	Stable	Sun et al. (2006)
Haematococcus pluvialis	Chlorophyta; Chlorophyceae; Volvocales; Haematococcaceae	Stable	Steinbrenner and Sandmann (2006)
Lobosphaera incisa (Parietochloris incisa)	Chlorophyta; Trebouxiophyceae; Trebouxiales; Trebouxiaceae	Stable	Zorin et al. (2014)
Chlorella sorokiniana	Chlorophyta; Trebouxiophyceae; Chlorellales; Chlorellaceae	Stable	Dawson et al. (1997)
'Chlorella' kessleri (Parachlorella kessleri)	Chlorophyta; Trebouxiophyceae; Chlorellales; Chlorellales incertae sedis	Stable	El-Sheekh (1999)
'Chlorella' ellipsoidea	Chlorophyta; Trebouxiophyceae; Trebouxiophyceae incertae sedis	Stable	Chen et al. (2001), Jarvis and Brown (1991), Liu et al. (2013)
Chlorella vulgaris	Chlorophyta; Trebouxiophyceae; Chlorellales; Chlorellaceae	Transient	Chow and Tung (1999)

(continued)

Table 5.4 (continued)

Species	Lineage	Transformation	References
Chlorella saccharophila	Chlorophyta; Trebouxiophyceae; Chlorellales; Chlorellaceae	Transient	Maruyama et al. (1994)
	Charophyta		
Closterium peracerosum-strigosum-littorale	Charophyta; Zygnematophyceae; Desmidiales; Closteriaceae	Stable	Abe et al. (2011)
	Red algae		
Cyanidioschyzon merolae	Rhodophyta; Bangiophyceae; Cyanidiales; Cyanidiaceae	Stable	Minoda et al. (2004)
Porphyridium sp.	Rhodophyta; Bangiophyceae; Porphyridiales; Porphyridiaceae	Stable	Lapidot et al. (2002)
	Diatoms		
Phaeodactylum tricornutum	Bacillariophyta; Bacillariophyceae; Bacillariophycidae; Naviculales; Phaeodactylaceae	Stable	Apt et al. (1996), Falciatore et al. (1999, 2000), Xie et al. (2014), Zaslavskaia et al. (2000, 2001)
Navicula saprophila (=Fistulifera saprophila)	Bacillariophyta; Bacillariophyceae; Bacillariophycidae; Naviculales; Naviculaceae	Stable	Dunahay et al. (1995)
Cylindrotheca fusiformis	Bacillariophyta; Bacillariophyceae; Bacillariophycidae; Bacillariales; Bacillariaceae	Stable	Fischer et al. (1999)
Cyclotella cryptica	Bacillariophyta; Coscinodiscophyceae; Thalassiosirophycidae; Thalassiosirales; Thalassiosiraceae	Stable	Dunahay et al. (1995)
Thalassiosira pseudonana	Bacillariophyta; Coscinodiscophyceae; Thalassiosirophycidae; Thalassiosirales; Thalassiosiraceae	Stable	Poulsen et al. (2006)
Thalassiosira weissflogii	Bacillariophyta; Coscinodiscophyceae; Thalassiosirophycidae; Thalassiosirales; Thalassiosiraceae	Transient	Falciatore et al. (1999)
Chaetoceros gracilis	Bacillariophyta; Mediophyceae; Chaetocerotophycidae; Chaetocerotales; Chaetocerotaceae	Stable	Ifuku et al. (2015)

(continued)

Table 5.4 (continued)

Species	Lineage	Transformation	References
	Eustigmatophyta		
Nannochloropsis oculata	Heterokontophyta; Eustigmatophyceae; Eustigmatales; Monodopsidaceae	Stable	Chen et al. (2008)
Nannochloropsis sp.	Heterokontophyta; Eustigmatophyceae; Eustigmatales; Monodopsidaceae	Stable	Kilian et al. (2011), Weeks (2011)
Nannochloropsis oceanica	Heterokontophyta; Eustigmatophyceae; Eustigmatales; Monodopsidaceae	Stable	Vieler et al. (2012)
Nannochloropsis gaditana	Heterokontophyta; Eustigmatophyceae; Eustigmatales; Monodopsidaceae	Stable	Radakovits et al. (2012)
	Euglenozoa		
Euglena gracilis	Euglenozoa; Euglenida; Euglenales	Stable	Doetsch et al. (2001)
	Haptophytes		
Pleurochrysis carterae	Haptophyta; Prymnesiophyceae; Isochrysidales; Pleurochrysidae	Stable	Endo et al. (2016)
	Chlorarachniophyceae		
Lotharella amoebiformis	Cercozoa; Chlorarachniophyceae; Chlorarachniales; Chlorarachniaceae	Transient	Hirakawa et al. (2008)
	Dinoflagellates		
Amphidinium sp.	Alveolata; Dinophyceae; Gymnodiniales; Gymnodiniaceae	Stable	ten Lohuis and Miller (1998)
Symbiodinium microadriaticum	Alveolata; Dinophyceae; Suessiales; Symbiodiniaceae	Stable	ten Lohuis and Miller (1998)

5.3.12 Gene Silencing, Gene Knockout, and Genome Editing

In most reported transformation experiments in microalgae, DNA has been inserted randomly into a host genome. This means that DNA is added at an unforeseeable position to the genome. However, when a researcher intends to investigate the function of an unknown gene by reducing its expression, a precisely defined gene locus in the host genome needs to be affected. Such a sequence-specific gene silencing or gene knockdown can be achieved in microalgae by introduction or expression of artificial microRNAs or by application of RNA interference (RNAi) techniques (Cerutti et al. 2011; Kim and Cerutti 2009; Molnár et al. 2009). The RNA-mediated silencing is either only transient if exogenously synthesized nucleic acids are introduced or the effect is more or less stable if genome-integrated transgenes are expressed. However, RNA-mediated techniques never lead to a complete switch-off of the target gene, and the degree of reduction of expression varies greatly between transformants, and it frequently changes over time in one and the same transformant.

Furthermore, true gene knockouts have been generated in microalgae when homologous recombination could be achieved (Day and Goldschmidt-Clermont 2011; Gumpel et al. 1994; Hallmann et al. 1997; Kilian et al. 2011; Lozano et al. 2014; Minoda et al. 2004; Nelson and Lefebvre 1995; Sodeinde and Kindle 1993; Verma and Daniell 2007; Weeks 2011; Zorin et al. 2005). Gene knockouts were also possible by using engineered zinc-finger nucleases (ZFNs) and transcription activator-like effector nucleases (TALENs) (Daboussi et al. 2014; Sizova et al. 2013). It should be noted that all these three approaches for obtaining knockout strains are very laborious and error-prone.

The true step-change regarding knockouts arose through the discovery and application of the CRISPR/Cas system, which belongs to the bacterial adaptive immune system (Bolotin et al. 2005; Cong et al. 2013) and which is applied increasingly also in microalgae (Jeon et al. 2017; Ng et al. 2017). This system requires a certain endonuclease (mostly Cas9 but also Cpf1) or its gene for expression and a customized small piece of RNA with a short guide sequence that specifically binds both to the gene sequence to be edited and to the endonuclease (Cas9 or Cpf1) enzyme. The endonuclease then introduces double-strand breaks into the DNA, and the DNA repair machinery of the cell adds or deletes pieces of DNA. The first CRISPR-based genome editing in microalgae was reported for the unicellular green alga *C. reinhardtii* in 2014 (Jiang et al. 2014). In subsequent years, further reports on genome editing applications followed, targeting not only *C. reinhardtii* (Baek et al. 2016; Ferenczi et al. 2017; Greiner et al. 2017) but also the multicellular green alga *V. carteri* (Ortega-Escalante et al. 2019), the diatoms *Phaeodactylum tricornutum* (Nymark et al. 2016) and *Thalassiosira pseudonana* (Hopes et al. 2016), and the Eustigmatophyte *Nannochloropsis oceanica* (Wang et al. 2016). Due to its success in microalgae and many other species within a quite short period of time, the CRISPR/Cas system is likely to become the standard procedure for genome editing.

5.3.13 Unwanted Silencing of Transgenes

While it is demanding to achieve intended gene silencing of a certain target gene at its natural position, unwanted silencing of transgenes that are randomly integrated into the genome frequently occurs spontaneously and leads to significant reduction or even complete loss of expression (Blankenship and Kindle 1992; Cerutti et al. 1997a; Stevens et al. 1996). The cause for unwanted silencing frequently seems to be the DNA methyltransferase (MET I), which generates high-level DNA methylation of transgenes (Babinger et al. 2001, 2007; Kim et al. 2015; Kurniasih et al. 2016). Further problems are caused by enzymes that place specific histone modifications onto nucleosomes at the transgene loci to trigger chromatin compaction (Bannister and Kouzarides 2011; Casas-Mollano et al. 2008; Kim et al. 2015). The silencing might happen immediately or within weeks, months, or even years, which means after countless microalgae generations (Babinger et al. 2001). Even if there are only few reports about unwanted gene silencing in microalgae, it frequently happens. As for intended silencing, the degree of reduction of expression varies greatly between transformants. Placing matrix attachment regions (MARs) at the sides of the transgene could reduce transgene silencing (Allen et al. 2000). If heterologous, intronless sequences are utilized, the integration of homologous intron sequences can minimize the problem (Dong et al. 2017). Also codon optimization and the use of homologous regulatory sequences and 5′ and 3′ UTRs are helpful. Eventually, the extent of gene silencing is hard to predict for a given transgene. Therefore, the best strategy seems to be to produce several suitable transformants and keep them all in parallel as a backup.

5.4 The Application of Genetic Engineering

The scientific communities' primary driving force for the ongoing improvement of the molecular toolkit for genetic engineering of microalgae is the substantial potential of biotechnological applications (Fig. 5.5). A comprehensive assortment of molecular tools is already utilized in basic research and some industrial applications. One of the major advantages of microalgae is the fact that they can be cultivated in a cheap and easy manner. Moreover, they are free from toxins and viral agents, grow fast, and can reach high cell densities. Genetic engineering of microalgae frequently aims to increase the production of a desired product that is otherwise produced at lower concentrations. Another important aim is to enable the production of compounds or to add traits that are not normally present in algae. With these aims in mind, many different recombinant proteins have already been produced in microalgae (Barrera and Mayfield 2013; Fletcher et al. 2007; Franklin and Mayfield 2004; Griesbeck et al. 2006; Hallmann 2007, 2015; Hempel et al. 2011; Ma et al. 2005; Rasala and Mayfield 2011; Rasala et al. 2010; Rosales-Mendoza 2016; Specht et al. 2010; Stoger et al. 2014; Yan et al. 2016). Biotechnological purposes with participation of microalgae include the production of compounds that are valuable to medicine and pharmaceutics. Therefore, these compounds are also called biopharmaceuticals, and the biotechnological approach is called molecular farming (or molecular pharming, biopharming, or gene pharming) (Franklin and Mayfield 2004; Griesbeck et al. 2006; Hallmann 2007; 2015; Ma et al. 2005; Rasala and

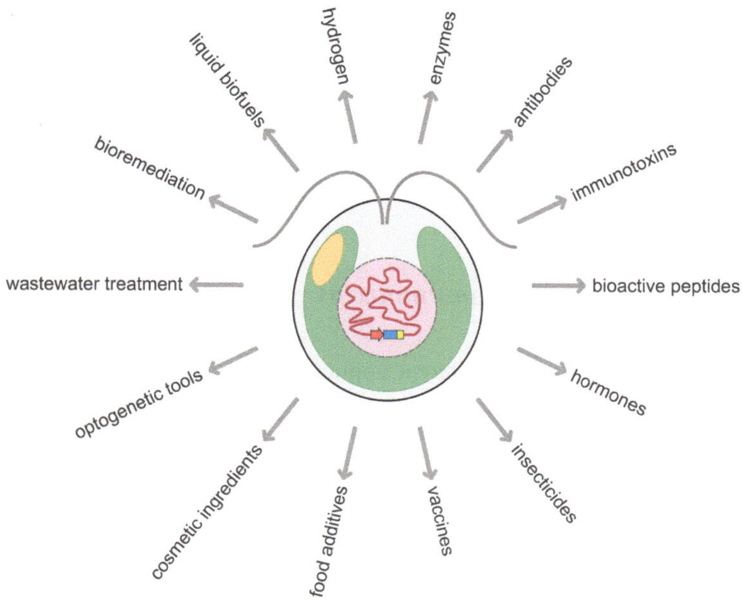

Fig. 5.5 Possible applications of genetic engineering in microalgae

Mayfield 2011; Rosales-Mendoza 2016; Stoger et al. 2014; Yan et al. 2016). Biopharmaceuticals include enzymes, antibodies, immunotoxins, bioactive peptides, hormones, and vaccines (Fig. 5.5). In principle, algae-made biopharmaceuticals can be applicable for humans and animals. Numerous therapeutic proteins have already been produced successfully in the *C. reinhardtii* chloroplast (Dyo and Purton 2018).

Genetically modified microalgae can also be relevant for the production of food additives, cosmetics, insecticides, and biofuels. Moreover, they can be employed for cleaning polluted water or soil sources. Light-dependent ion channels of microalgae are applied in neuroscience to switch neurons on and off.

Current applications of genetic engineering for biotechnological exploitation of microalgae are discussed in more detail below.

5.4.1 Enzymes

Several foreign enzymes have been expressed in microalgae (Brasil et al. 2017). Enzyme production often involved the correct subcellular localization, folding, assembly, and posttranslational modification. The numerous enzymes that have been successfully expressed in microalgae comprise cellulases, amylolytic enzymes, galactosidases, xylanases, mannases, proteases, lipases, sulfatases, phosphatases, sugar transporters, phytases, laccases, antioxidant enzymes (superoxide dismutase), and enzymes involved in carbohydrate accumulation (Brasil et al. 2017; Hallmann 2007; Hallmann and Rappel 1999). For example, five different classes of recombinant enzymes have been produced in the chloroplasts of *Chlamydomonas reinhardtii* and *Dunaliella tertiolecta* (Georgianna et al. 2013). The enzymes included xylanase, α-galactosidase, phytase, phosphatase, and β-mannanase. Enzymes can be used in a wide range of industrial processes.

5.4.2 Antibodies and Immunotoxins

Production of antibodies for different applications using mammalian cell culture systems causes very high costs. An alternative could be expression of antibodies in microalgal systems, which was basically demonstrated on a small scale in *C. reinhardtii* by expression of a recombinant human monoclonal IgA antibody (Mayfield et al. 2003). Moreover, a human IgG antibody and also the respective antigen have been successfully expressed in the diatom *Phaeodactylum tricornutum* (Fig. 5.2) (Hempel et al. 2011).

Antibody fragments can be fused to toxic proteins and become immunotoxins for targeted cancer therapy. If the antibody part targets a specific cell type of a tumor tissue, the coupled toxin part can be concentrated in the tumor and specifically destroy the tumor tissue. An immunotoxin-containing antibody domain that targets CD22, a B-cell surface epitope, fused to the enzymatic domain of exotoxin A from the bacterium *Pseudomonas aeruginosa* has been successfully produced in chloroplasts of transgenic *Chlamydomonas reinhardtii* algae (Tran et al. 2013b). This immunotoxin protein significantly prolonged the survival of mice with implanted human B-cell tumors (Tran et al. 2013b). In a similar approach, the anti-CD22 domain has been fused to the toxin gelonin from the plant *Suregada multiflora* (*Gelonium multiflorum*, false lime) and produced also in transgenic *Chlamydomonas reinhardtii* algae (Tran et al. 2013a).

5.4.3 Bioactive Peptides and Hormones

One type of bioactive peptides are defensins, which can be found both in vertebrates and invertebrates and which act mainly by disrupting the structure of bacterial cell membranes. Thus, defensins can be used as promising alternatives to other antibiotics. In a proof-of-concept experiment, one of these defensins, a mutated rabbit neutrophil peptide 1 (NP-1), has been successfully produced at high levels in genetically modified *Chlorella ellipsoidea* (Bai et al. 2013). NP-1 has a broad antimicrobial activity that is effective against Gram-negative and Gram-positive bacteria, certain viruses, and pathogenic fungi.

Researchers also investigated the expression of hormones in microalgae. Specifically, the flounder growth hormone (fGH) was produced at high levels in transgenic

Chlorella ellipsoidea (Kim et al. 2002). When flounders were fed with these transgenic algae, the size of the fish increased by 25% after 30 days of feeding (Kim et al. 2002). Moreover, human growth hormone (hGH) has been made in the *C. reinhardtii* chloroplast, and its biological activity has been confirmed by a proliferation assay (Wannathong et al. 2016).

5.4.4 *Insecticides*

Expression of a toxin again insects, an insecticide, has been successfully achieved in the microalga *Chlorella*, which is one possible food for mosquito larvae. The heterologously expressed insecticidal protein was the mosquito hormone trypsin-modulating oostatic factor (TMOF) (Borovsky 2003). TMOF causes the termination of trypsin biosynthesis in the mosquito gut. After feeding mosquito larvae with TMOF expressing *Chlorella* cells, the larvae died within 3 days (Borovsky 2003). Mosquitoes are transmitting severe diseases such as malaria, dengue, and West Nile fever, and mosquito abatement is a costly requirement in quite a few tropical countries. Creating a strategy by using transgenic algae might be a cost-efficient alternative.

5.4.5 *Vaccines*

Expressing antigens in microalgae is particularly attractive, because many wild-type microalgae species can be used as a healthy and safe food supplement. Therefore, there might be no need for purification of the antigen, and the intact algae could be used to deliver a vaccine. The latter advantage makes the algae expression system most attractive to low- and middle-income countries with region-specific infectious diseases.

In preliminary experiments, transgenic microalgae have been shown to be suitable for synthesizing vaccines for human use (Gregory and Mayfield 2014; Specht and Mayfield 2014). For example, stable expression of the hepatitis B virus surface antigen gene has been shown in *Dunaliella salina* (Geng et al. 2003; Sayre et al. 2001; Sun et al. 2003). Hepatitis B is a serious infection that attacks the liver and can cause both acute and chronic disease. In another approach, a transgene encoding the E7 oncoprotein (E7GGG protein) of human papillomavirus (HPV) type 16 was expressed in *Chlamydomonas reinhardtii* (Demurtas et al. 2013). HPV 16 is a high-risk type known to significantly increase the risk of cervical, vaginal, and vulvar cancer in women and other female mammals. In mice, subcutaneous injections of the algae-derived E7GGG protein yielded protection in 60% of the experimental animals (Demurtas et al. 2013).

It can also be useful to make vaccines against animal pathogens from genetically engineered microalgae. To get this idea started, the foot-and-mouth disease virus

(FMDV) VP1 antigen protein fused with the cholera toxin B subunit has been successfully produced in chloroplasts of transgenic *Chlamydomonas reinhardtii* algae (Sun et al. 2003). The foot-and-mouth disease is an important disease of livestock. Binding assays showed that the produced fusion protein can bind to the intestinal membrane GM1-ganglioside receptor via the cholera toxin B part. However, the produced fusion protein was not examined any further. In a similar approach, the antigen for vaccination against classical swine fever virus (CSFV) was successfully produced in *Chlamydomonas* chloroplasts. Subsequently, subcutaneous injection was shown to produce serum antibodies against CSFV in mice (He et al. 2007).

Also actions against another animal virus, the white spot syndrome virus (WSSV), yielded some promising results. The main viral envelop protein VP28 of the WSS virus was expressed both in *Dunaliella salina* and *Chlamydomonas reinhardtii* (Feng et al. 2014; Surzycki et al. 2009). The WSS virus can infect almost all species of shrimps and other crustaceans, and outbreaks of this disease have previously wiped out entire populations of many shrimp farms. By oral delivery of microalgal VP28, the mortality rate of WSSV-infected crayfish could be reduced from 100 to 41% (Feng et al. 2014). In the same way, researchers aim to use an alga-produced antigen to vaccinate fish against the infectious hematopoietic necrosis virus (IHNV), which causes an infectious disease that kills 30% of the US trout population each year. The goal is to realize vaccination simply by feeding the fish with transgenic microalgae (Banicki 2004).

Aside from viruses, efforts also target at potentially harmful bacteria. For it, the D2 fibronectin-binding domain of the bacterium *Staphylococcus aureus* fused with the cholera toxin B subunit has been stably expressed in the chloroplast of *Chlamydomonas* (Dreesen et al. 2010). *S. aureus* bacteria are one of the most common causes of bacteremia, infective endocarditis, and various skin and soft-tissue infections in humans and animals. Oral vaccination of mice using transgenic *Chlamydomonas* microalgae that express the above D2-fusion protein protected 80% of them against otherwise lethal doses of *S. aureus* (Dreesen et al. 2010).

Further antigens for vaccination have been successfully expressed in microalgae such as the glutamic acid decarboxylase-65 (GAD65), which is a major autoantigen for human type 1 diabetes (Wang et al. 2008); malaria antigens (GBSS-pfMSP$_{1-19}$, GBSS-pfAMA$_{1\text{-}C}$) that can protect against lethal *Plasmodium berghei* infection (Dauvillee et al. 2010); angiotensin II, which raises blood pressure, fused with a nucleocapsid antigen of hepatitis B virus (Soria-Guerra et al. 2014); the p210 epitope of ApoB100 (fused with the cholera toxin B subunit) implicated in atherosclerosis immunotherapies (Beltran-Lopez et al. 2016); and the HIV antigen P24 (Barahimipour et al. 2016).

5.4.6 Food Additives and Cosmetic Ingredients

So far only preliminary experiments have been conducted to prepare the ground for the enhanced production of food additives and cosmetic ingredients in genetically modified microalgae. Such commercially interesting compounds are, for example, β-carotene (Varela et al. 2015), astaxanthin, and long-chain polyunsaturated fatty acids (Borowitzka 2013; Sharon-Gojman et al. 2015). Genetic engineering mainly aims at the overexpression of enzymes of the corresponding pathways or at the knockout of enzymes to direct the metabolic flow into the desired direction.

In a probably commercially less relevant experiment, a human selenoprotein, Sep15, has been produced in *Chlamydomonas reinhardtii* as a potential selenium source against selenium deficiency (Hou et al. 2013).

5.4.7 Optogenetic Tools for Neuroscience

Optogenetics is a quite new field of applied biotechnology that combines genetic engineering, optics, and light-sensitive molecules to achieve precise control of cell activity (Deisseroth 2015; Fenno et al. 2011; Hegemann and Sigrist 2013; Kianianmomeni and Hallmann 2015; Tischer and Weiner 2014). The goal of optogenetics is to use molecular light sensors to produce light-switchable biochemical tools. The utilization of optogenetics is particularly interesting for neuroscientists who acquired an ability to control neurons by light (Kim et al. 2017). Optogenetics has its origins in microalgae because the first light-dependent ion channels, the channelrhodopsins 1 and 2, have been identified in *Chlamydomonas reinhardtii* (Nagel et al. 2002, 2003), and from then on, a rapid growth of the new field began. The channelrhodopsins are localized at the eyespot, a primitive "eye" of *Chlamydomonas* and many related green algae. They are directly light-gated cation channels that function as primary photoreceptors. Blue light initiates a fast inward-directed photocurrent, and after amplification of this electrical signal by voltage-gated secondary channels, it is transmitted to the flagella, which in turn change their beating behavior (i.e., the beating plane, beating pattern, and frequency). The membrane-spanning domain of channelrhodopsins with its seven transmembrane helices was shown to drive photocurrents. Moreover, it was possible to express this domain in other organisms to depolarize their cells with light (Nagel et al. 2003). Channelrhodopsins have been heterologously expressed in cells of quite different origin such as *Xenopus* oocytes, the retina of blind mice, hippocampal neurons, spine of living chicken embryos, PC12 cells, mouse brain slices, and worm cells (Bi et al. 2006; Boyden et al. 2005; Ishizuka et al. 2006; Li et al. 2005; Nagel et al. 2005).

Also due to the success of channelrhodopsins, further photoreceptors have been searched and found in microalgae (Kianianmomeni and Hallmann 2014). It became clear that microalgae evolved quite a few different photoreceptors to constantly sense light of different intensities and wavelengths in order to adapt their physiology in response to environmental changes (Foster and Smyth 1980). Such light-sensitive proteins also regulate developmental processes, photosynthesis, and photoorientation, and they interact with the circadian clock. In the last 15 years, several of the identified photoreceptors have been characterized in some detail (Hegemann 2008; Kianianmomeni and Hallmann 2014; Tian et al. 2018). The photoreceptors are classified as LOV domain-containing photoreceptors (phototropins, aureochromes, and neochromes), cryptochromes, phytochromes, and rhodopsins (Hegemann 2008; Kianianmomeni and Hallmann 2014). Photoreceptors can also be classified by the kind of induced reaction. There are not only light-controlled ion channels but also photoswitchable enzymes. Even ATP-dependent and light-inhibited guanylyl cyclases have been identified that can change the level of the signaling molecule cGMP in a light-dependent manner (Tian et al. 2018).

Due to the ongoing increase in sequenced microalgal genomes and transcriptomes, more proteins with light-sensitive domains are being discovered. New photoreceptors not only can provide new insights into light-regulated cellular processes but can enrich the molecular toolbox for optogenetics, especially if they are triggered by different wavelengths. Quite a few photoswitchable tools have already been designed and engineered to control cellular and physiological activities in organisms of different complexity (Hegemann and Nagel 2013; Hegemann and Sigrist 2013; Kianianmomeni and Hallmann 2014, 2015, 2016). Target cells, particularly neurons, can even be genetically modified with different photoswitchable tools simultaneously. If these photoswitches not only trigger different reactions but are also sensitive for different wavelengths, the transgenic target cells can be urged to change their cellular behavior in response to the color of light (Kianianmomeni and Hallmann 2015, 2016).

5.4.8 Wastewater Treatment and Bioremediation

A progressive way to return polluted water and soil to its original condition is to use living microalgae or other microorganisms for absorption of toxic contaminants, a process known as bioremediation. Some of the most critical contaminants in the environment are heavy metals and the most frequent water- and soil-polluting heavy metals are lead, cadmium, and mercury. Industrial processes, including plastic manufacturing, electroplating, Ni-Cd battery production, mining, and smelting industries, continuously release substantial amounts of such heavy metals into the environment. Microalgae readily take up such heavy metals from their environment, but the uptake has negative consequences for them. At lower concentrations, a heavy metal stress response is induced, which includes production of heavy metal binding factors and proteins. However, higher heavy metal concentrations obstruct main cellular processes (e.g., photosynthesis, growth) and finally kill the cells. At this point genetic engineering comes into play aiming at an increased heavy metal tolerance of the target cells. Genetically modified *Chlamydomonas* strains that heterologously express the moth bean *P5CS* gene have been shown to grow in the presence of much higher heavy metal concentrations than the parent *Chlamydomonas* wild-type strain. The P5CS protein catalyzes the first step in proline synthesis, and proline is known to increase the heavy metal tolerance because it detoxifies free radicals produced as a result of heavy metal poisoning. Expression of the *P5CS* gene in transgenic cells therefore results in an 80% higher free proline level and a fourfold increase in cadmium-binding capacity relative to wild-type cells (Siripornadulsil et al. 2002). In addition, expression of this gene results in rapid growth at otherwise deadly cadmium concentrations (Siripornadulsil et al. 2002).

In phytoremediation studies with higher plants, heterologous expression of a bacterial mercuric ion reductase (*merA*) and a bacterial organomercurial lyase (*merB*) significantly improved growth on soil contaminated with the heavy metal mercury. With regard to microalgae, transgenic *Chlorella* strains heterologously expressing the *merA* gene withstood up to 40 mM $HgCl_2$ (Huang et al. 2006). Unfortunately, mercuric ion reductase cannot protect against the more toxic and environmentally relevant organomercurials.

In principle, the previous transgenetic approaches seem to be transferable to other microalgae species. An appropriate downstream processing of heavy metal-loaded microalgae needs to be established particularly in combination with biomass or bioenery concepts. In addition, any approaches must take the environmental impact of genetically modified organisms into account. Wastewater treatment and bioremediation with transgenic microalgae is still in its early stages of development, while the use of wild-type microalgae or microalgal-bacterial consortia is already quite advanced (Munoz and Guieysse 2006; Olguín 2012; Ramanan et al. 2016; Rawat et al. 2011; Subashchandrabose et al. 2011; Unnithan et al. 2014).

5.4.9 Liquid Biofuels and Hydrogen

Efforts have long been directed at the industrial use of transgenic microalgal biomass for production of liquid biofuels and hydrogen as energy carriers, although the relatively high cost and low yield are of concern.

In order to use microalgae for biodiesel production, several attempts were made to enhance their lipid production and lipid content by genetic engineering. With this objective in mind, genetic engineering targets at key genes of the anabolism and catabolism of lipids (Chu 2017; Singh et al. 2016). To stimulate the lipid anabolism, overexpression of a key gene is a usual approach. In *Phaeodactylum tricornutum*, the malic enzyme, which is involved in pyruvate metabolism and carbon fixation, has been overexpressed and caused a 2.5-fold increase of the total lipid content in transgenic algae (Xue et al. 2015). Another strategy is to overexpress not only single enzymes but suitable transcription factors in lipid anabolism that can control downstream expression of numerous genes and can have considerable multiplication effects. For this purpose, the gene of a transcription factor in lipid synthesis of soybean (*Glycine max*) has been heterologously expressed in *Chlorella ellipsoidea*

(Zhang et al. 2014). This expression enhanced the lipid content in transformants by up to 53% without affecting the growth rate.

Not only stimulation of the anabolism but also inhibition of the catabolism can increase the lipid content. To that end, a multifunctional lipase/phospholipase/acyltransferase has been silenced through RNAi in the diatom *Thalassiosira pseudonana* (Trentacoste et al. 2013). The corresponding antisense-expressing knockdown strains showed an up to 3.3-fold increase of the total lipid content during exponential growth without negatively affecting the cell growth rate (Trentacoste et al. 2013).

Aside from direct interventions in lipid metabolism, maximization of the light capture efficiency of photosynthesis can be a promising approach because light capture stands at the beginning in all biofuel production processes (Larkum et al. 2012). Photosynthetic efficiency can be enhanced, for example, by extending the absorption range of the photosynthetically active pigments (Larkum et al. 2012) or by reducing the total antenna size in the photosystem of microalgae to lower the losses of captured energy due to non-photochemical quenching (Chu 2017).

Apart from biodiesel, hydrogen gas can be another relevant energy carrier. Several microalgae species have been shown to produce hydrogen using (Fe–Fe)-hydrogenases (Meuser et al. 2009). A couple of projects have been started to enhance this hydrogen production. The applied strategies include depletion and repletion of cultures with sulfur, screens for mutants with increased hydrogen production, genetic modifications of the light-harvesting antennae complexes, forced overproduction of protons and electrons, and optimization of the essential hydrogenase enzyme (Kruse et al. 2005; Melis et al. 2000; Prince and Kheshgi 2005). Unfortunately, the hydrogenases require anoxic conditions for their function and hydrogen cannot be generated during photosynthesis. Attempts to genetically engineer (Fe–Fe)-hydrogenases with increased oxygen tolerance (Dubini and Ghirardi 2015) or to heterologously express hydrogenases with better oxygen tolerance have not been successful. Another strategy was to increase the enzyme concentration by overexpressing the (Fe–Fe)-hydrogenase genes, which actually led to an up to tenfold increased hydrogen production (Chien et al. 2012). It was also possible to inactivate photosystem II by disturbing the sulfate metabolism through an RNAi knockdown of a sulfate transporter (Chen et al. 2005). Likewise, an increase in hydrogen was obtained by mutating D1 of photosysystem II, which led to another tenfold increase of hydrogen (Torzillo et al. 2009). There were also attempts to capture oxygen by expressing a soybean leghemoglobin gene (Wu et al. 2011). Although there has been progress in hydrogen production using microalgae, large-scale production is still not feasible because of low biomass concentrations, expensive downstreaming processes, and other complications (Sharma and Arya 2017).

5.5 Conclusions

Genetic engineering of microalgae promises enhanced exploitation possibilities in nearly all contexts where otherwise unmodified microalgae are involved. Even the production of compounds or the addition of traits that are not normally present in algae is possible. The emergence and rapid evolution of omics technologies in microalgae has already formed an excellent basis for genetic engineering. Particularly, the quantity of genomics and transcriptomics data is still exponentially increasing. Several decades of work in model algae like *Chlamydomonas* have clarified gene expression, biochemical processes, and metabolic pathways. Moreover, significant advances in the development of key elements of genetic engineering already have been achieved with several microalgal model systems.

Applied algae biotechnology profits tremendously both from the massive databases created by means of omics technologies and availability of countless molecular tools and methods. Also the acceptance and importance of microalgae in sustainability networks is constantly increasing. However, still more basic research needs to be performed before biotechnology with genetically engineered microalgae reaches a capacity and profitability to compete with other systems.

References

Abe J, Hori S, Tsuchikane Y, Kitao N, Kato M, Sekimoto H (2011) Stable nuclear transformation of the *Closterium peracerosum-strigosum-littorale* complex. Plant Cell Physiol 52:1676–1685

Aldridge BB, Rhee KY (2014) Microbial metabolomics: innovation, application, insight. Curr Opin Microbiol 19:90–96

Allen GC, Spiker S, Thompson WF (2000) Use of matrix attachment regions (MARs) to minimize transgene silencing. Plant Mol Biol 43:361–376

Anand S, Mangano E, Barizzone N, Bordoni R, Sorosina M, Clarelli F, Corrado L, Martinelli Boneschi F, D'Alfonso S, De Bellis G (2016) Next generation sequencing of pooled samples: guideline for variants' filtering. Sci Rep 6:33735

Anand V, Singh PK, Banerjee C, Shukla P (2017) Proteomic approaches in microalgae: perspectives and applications. 3 Biotech 7:197

Angov E (2011) Codon usage: nature's roadmap to expression and folding of proteins. Biotechnol J 6:650–659

Anila N, Chandrashekar A, Ravishankar GA, Sarada R (2011) Establishment of *Agrobacterium tumefaciens*-mediated genetic transformation in *Dunaliella bardawil*. Eur J Phycol 46:36–44

Apt KE, Kroth-Pancic PG, Grossman AR (1996) Stable nuclear transformation of the diatom *Phaeodactylum tricornutum*. Mol Gen Genet 252:572–579

Armbrust EV, Berges JA, Bowler C, Green BR, Martinez D, Putnam NH, Zhou S, Allen AE, Apt KE, Bechner M et al (2004) The genome of the diatom *Thalassiosira pseudonana*: ecology, evolution, and metabolism. Science 306:79–86

Arriola MB, Velmurugan N, Zhang Y, Plunkett MH, Hondzo H, Barney BM (2018) Genome sequences of *Chlorella sorokiniana* UTEX 1602 and *Micractinium conductrix* SAG 241.80: implications to maltose excretion by a green alga. Plant J 93:566–586

Aslam B, Basit M, Nisar MA, Khurshid M, Rasool MH (2017) Proteomics: technologies and their applications. J Chromatogr Sci 55:182–196

Babinger P, Kobl I, Mages W, Schmitt R (2001) A link between DNA methylation and epigenetic silencing in transgenic *Volvox carteri*. Nucleic Acids Res 29:1261–1271

Babinger P, Völkl R, Cakstina I, Maftei A, Schmitt R (2007) Maintenance DNA methyltransferase (Met1) and silencing of CpG-methylated foreign DNA in *Volvox carteri*. Plant Mol Biol 63:325–336

Baek K, Kim DH, Jeong J, Sim SJ, Melis A, Kim JS, Jin E, Bae S (2016) DNA-free two-gene knockout in *Chlamydomonas reinhardtii* via CRISPR-Cas9 ribonucleoproteins. Sci Rep 6:30620

Bai LL, Yin WB, Chen YH, Niu LL, Sun YR, Zhao SM, Yang FQ, Wang RR, Wu Q, Zhang XQ et al (2013) A new strategy to produce a defensin: stable production of mutated NP-1 in nitrate reductase-deficient *Chlorella ellipsoidea*. PLoS One 8:e54966

Bai X, Song H, Lavoie M, Zhu K, Su Y, Ye H, Chen S, Fu Z, Qian H (2016) Proteomic analyses bring new insights into the effect of a dark stress on lipid biosynthesis in *Phaeodactylum tricornutum*. Sci Rep 6:25494

Baier T, Wichmann J, Kruse O, Lauersen KJ (2018) Intron-containing algal transgenes mediate efficient recombinant gene expression in the green microalga *Chlamydomonas reinhardtii*. Nucleic Acids Res 46:6909–6919

Baiet B, Burel C, Saint-Jean B, Louvet R, Menu-Bouaouiche L, Kiefer-Meyer MC, Mathieu-Rivet E, Lefebvre T, Castel H, Carlier A et al (2011) N-glycans of *Phaeodactylum tricornutum* diatom and functional characterization of its N-acetylglucosaminyltransferase I enzyme. J Biol Chem 286:6152–6164

Baldauf SL (2003) The deep roots of eukaryotes. Science 300:1703–1706

Banicki JJ (2004) An alga a day keeps the doctor away. Engineered algae as a new means to vaccinate fish. Twine Line 26:1–5

Bannister AJ, Kouzarides T (2011) Regulation of chromatin by histone modifications. Cell Res 21:381–395

Barahimipour R, Strenkert D, Neupert J, Schroda M, Merchant SS, Bock R (2015) Dissecting the contributions of GC content and codon usage to gene expression in the model alga *Chlamydomonas reinhardtii*. Plant J 84:704–717

Barahimipour R, Neupert J, Bock R (2016) Efficient expression of nuclear transgenes in the green alga *Chlamydomonas*: synthesis of an HIV antigen and development of a new selectable marker. Plant Mol Biol 90:403–418

Barbier G, Oesterhelt C, Larson MD, Halgren RG, Wilkerson C, Garavito RM, Benning C, Weber AP (2005) Comparative genomics of two closely related unicellular thermo-acidophilic red algae, *Galdieria sulphuraria* and *Cyanidioschyzon merolae*, reveals the molecular basis of the metabolic flexibility of *Galdieria sulphuraria* and significant differences in carbohydrate metabolism of both algae. Plant Physiol 137:460–474

Barrera DJ, Mayfield SP (2013) High-value recombinant protein production in microalgae. In: Richmond A, Hu Q (eds) Handbook of microalgal culture: applied phycology and biotechnology, 2nd edn. Wiley, Chichester, pp 532–544

Beacham TA, Sweet JB, Allen MJ (2017) Large scale cultivation of genetically modified microalgae: a new era for environmental risk assessment. Algal Res 25:90–100

Beaucage SL, Caruthers MH (1981) Deoxynucleoside phosphoramidites - a new class of key intermediates for deoxypolynucleotide synthesis. Tetrahedron Lett 22:1859–1862

Belghit I, Rasinger JD, Heesch S, Biancarosa I, Liland N, Torstensen B, Waagbo R, Lock EJ, Bruckner CG (2017) In-depth metabolic profiling of marine macroalgae confirms strong biochemical differences between brown, red and green algae. Algal Res 26:240–249

Bell SA, Shen C, Brown A, Hunt AG (2016) Experimental genome-wide determination of RNA polyadenylation in *Chlamydomonas reinhardtii*. PLoS One 11:e0146107

Beltran-Lopez JI, Romero-Maldonado A, Monreal-Escalante E, Banuelos-Hernandez B, Paz-Maldonado LM, Rosales-Mendoza S (2016) *Chlamydomonas reinhardtii* chloroplasts express an orally immunogenic protein targeting the p210 epitope implicated in atherosclerosis immunotherapies. Plant Cell Rep 35:1133–1141

Berthold P, Schmitt R, Mages W (2002) An engineered *Streptomyces hygroscopicus aph 7″* gene mediates dominant resistance against hygromycin B in *Chlamydomonas reinhardtii.* Protist 153:401–412

Bhattacharya D, Price DC, Chan CX, Qiu H, Rose N, Ball S, Weber AP, Arias MC, Henrissat B, Coutinho PM et al (2013) Genome of the red alga *Porphyridium purpureum.* Nat Commun 4:1941

Bi A, Cui J, Ma YP, Olshevskaya E, Pu M, Dizhoor AM, Pan ZH (2006) Ectopic expression of a microbial-type rhodopsin restores visual responses in mice with photoreceptor degeneration. Neuron 50:23–33

Blanc G, Duncan G, Agarkova I, Borodovsky M, Gurnon J, Kuo A, Lindquist E, Lucas S, Pangilinan J, Polle J et al (2010) The *Chlorella variabilis* NC64A genome reveals adaptation to photosymbiosis, coevolution with viruses, and cryptic sex. Plant Cell 22:2943–2955

Blanc G, Agarkova I, Grimwood J, Kuo A, Brueggeman A, Dunigan DD, Gurnon J, Ladunga I, Lindquist E, Lucas S et al (2012) The genome of the polar eukaryotic microalga *Coccomyxa subellipsoidea* reveals traits of cold adaptation. Genome Biol 13:R39

Blanc-Mathieu R, Verhelst B, Derelle E, Rombauts S, Bouget FY, Carre I, Chateau A, Eyre-Walker A, Grimsley N, Moreau H et al (2014) An improved genome of the model marine alga *Ostreococcus tauri* unfolds by assessing Illumina de novo assemblies. BMC Genomics 15:1103

Blankenship JE, Kindle KL (1992) Expression of chimeric genes by the light-regulated *cabII*-1 promoter in *Chlamydomonas reinhardtii*: a *cabII*-1/*nit1* gene functions as a dominant selectable marker in a *nit1*$^{-}$ *nit2*$^{-}$ strain. Mol Cell Biol 12:5268–5279

Bogen C, Al-Dilaimi A, Albersmeier A, Wichmann J, Grundmann M, Rupp O, Lauersen KJ, Blifernez-Klassen O, Kalinowski J, Goesmann A et al (2013) Reconstruction of the lipid metabolism for the microalga *Monoraphidium neglectum* from its genome sequence reveals characteristics suitable for biofuel production. BMC Genomics 14:926

Bolotin A, Quinquis B, Sorokin A, Ehrlich SD (2005) Clustered regularly interspaced short palindrome repeats (CRISPRs) have spacers of extrachromosomal origin. Microbiology 151:2551–2561

Borovsky D (2003) Trypsin-modulating oostatic factor: a potential new larvicide for mosquito control. J Exp Biol 206:3869–3875

Borowitzka MA (2013) High-value products from microalgae - their development and commercialisation. J Appl Phycol 25:743–756

Bowler C, Allen AE, Badger JH, Grimwood J, Jabbari K, Kuo A, Maheswari U, Martens C, Maumus F, Otillar RP et al (2008) The *Phaeodactylum* genome reveals the evolutionary history of diatom genomes. Nature 456:239–244

Boyden ES, Zhang F, Bamberg E, Nagel G, Deisseroth K (2005) Millisecond-timescale, genetically targeted optical control of neural activity. Nat Neurosci 8:1263–1268

Boyle NR, Morgan JA (2009) Flux balance analysis of primary metabolism in *Chlamydomonas reinhardtii.* BMC Syst Biol 3:4

Boyle NR, Page MD, Liu B, Blaby IK, Casero D, Kropat J, Cokus SJ, Hong-Hermesdorf A, Shaw J, Karpowicz SJ et al (2012) Three acyltransferases and nitrogen-responsive regulator are implicated in nitrogen starvation-induced triacylglycerol accumulation in *Chlamydomonas.* J Biol Chem 287:15811–15825

Bozarth A, Maier UG, Zauner S (2009) Diatoms in biotechnology: modern tools and applications. Appl Microbiol Biotechnol 82:195–201

Brasil BDSAF, Siqueira FG, Salum TFC, Zanette CM, Spier MR (2017) Microalgae and cyanobacteria as enzyme biofactories. Algal Res 25:76–89

Bromke MA, Sabir JS, Alfassi FA, Hajarah NH, Kabli SA, Al-Malki AL, Ashworth MP, Meret M, Jansen RK, Willmitzer L (2015) Metabolomic profiling of 13 diatom cultures and their adaptation to nitrate-limited growth conditions. PLoS One 10:e0138965

Brown LE, Sprecher SL, Keller LR (1991) Introduction of exogenous DNA into *Chlamydomonas reinhardtii* by electroporation. Mol Cell Biol 11:2328–2332

Browne DR, Jenkins J, Schmutz J, Shu S, Barry K, Grimwood J, Chiniquy J, Sharma A, Niehaus TD, Weiss TL et al (2017) Draft nuclear genome sequence of the liquid hydrocarbon-accumulating green microalga *Botryococcus braunii* Race B (Showa). Genome Announc 5: e00215-17

Brueggeman AJ, Kuehler D, Weeks DP (2014) Evaluation of three herbicide resistance genes for use in genetic transformations and for potential crop protection in algae production. Plant Biotechnol J 12:894–902

Burgess DJ (2017) Building a custom eukaryotic genome de novo. Nat Rev Genet 18:274–274

Butanaev AM (1994) Use of the hygromycin phosphotransferase gene as the dominant selective marker for *Chlamydomonas reinhardtii* transformation. Mol Biol (Mosk) 28:1061–1068

Bux F, Chisti Y (2016) Algae biotechnology: products and processes (Green energy and technology). Springer, Cham

Campos H, Boeing WJ, Dungan BN, Schaub T (2014) Cultivating the marine microalga *Nannochloropsis salina* under various nitrogen sources: effect on biovolume yields, lipid content and composition, and invasive organisms. Biomass Bioenergy 66:301–307

Carvalho RN, Lettieri T (2011) Proteomic analysis of the marine diatom *Thalassiosira pseudonana* upon exposure to benzo(a)pyrene. BMC Genomics 12:159

Casas-Mollano JA, Jeong BR, Xu J, Moriyama H, Cerutti H (2008) The MUT9p kinase phosphorylates histone H3 threonine 3 and is necessary for heritable epigenetic silencing in *Chlamydomonas*. Proc Natl Acad Sci U S A 105:6486–6491

Cerutti H, Johnson AM, Gillham NW, Boynton JE (1997a) Epigenetic silencing of a foreign gene in nuclear transformants of *Chlamydomonas*. Plant Cell 9:925–945

Cerutti H, Johnson AM, Gillham NW, Boynton JE (1997b) A eubacterial gene conferring spectinomycin resistance on *Chlamydomonas reinhardtii*: integration into the nuclear genome and gene expression. Genetics 145:97–110

Cerutti H, Ma X, Msanne J, Repas T (2011) RNA-mediated silencing in algae: biological roles and tools for analysis of gene function. Eukaryot Cell 10:1164–1172

Chang RL, Ghamsari L, Manichaikul A, Hom EF, Balaji S, Fu W, Shen Y, Hao T, Palsson BO, Salehi-Ashtiani K et al (2011) Metabolic network reconstruction of *Chlamydomonas* offers insight into light-driven algal metabolism. Mol Syst Biol 7:518

Chen Y, Wang Y, Sun Y, Zhang L, Li W (2001) Highly efficient expression of rabbit neutrophil peptide-1 gene in *Chlorella ellipsoidea* cells. Curr Genet 39:365–370

Chen HC, Newton AJ, Melis A (2005) Role of SulP, a nuclear-encoded chloroplast sulfate permease, in sulfate transport and H2 evolution in *Chlamydomonas reinhardtii*. Photosynth Res 84:289–296

Chen HL, Li SS, Huang R, Tsai HJ (2008) Conditional production of a functional fish growth hormone in the transgenic line of *Nannochloropsis oculata* (Eustigmatophyceae). J Phycol 44:768–776

Chen H, Zheng Y, Zhan J, He C, Wang Q (2017) Comparative metabolic profiling of the lipid-producing green microalga *Chlorella* reveals that nitrogen and carbon metabolic pathways contribute to lipid metabolism. Biotechnol Biofuels 10:153

Cheng S, Melkonian M, Smith SA, Brockington S, Archibald JM, Delaux PM, Li FW, Melkonian B, Mavrodiev EV, Sun W et al (2018) 10KP: A phylodiverse genome sequencing plan. Gigascience 7:1–9

Chien LF, Kuo TT, Liu BH, Lin HD, Feng TY, Huang CC (2012) Solar-to-bioH_2 production enhanced by homologous overexpression of hydrogenase in green alga *Chlorella* sp. DT. Int J Hydrog Energy 37:17738–17748

Chow K-C, Tung WL (1999) Electrotransformation of *Chlorella vulgaris*. Plant Cell Rep 18:778–780

Chu W-L (2017) Strategies to enhance production of microalgal biomass and lipids for biofuel feedstock. Eur J Phycol 52:419–437

Coll JM (2006) Methodologies for transferring DNA into eukaryotic microalgae. Span J Agric Res 4:316–330

Cong L, Ran FA, Cox D, Lin S, Barretto R, Habib N, Hsu PD, Wu X, Jiang W, Marraffini LA et al (2013) Multiplex genome engineering using CRISPR/Cas systems. Science 339:819–823

Corteggiani Carpinelli E, Telatin A, Vitulo N, Forcato C, D'Angelo M, Schiavon R, Vezzi A, Giacometti GM, Morosinotto T, Valle G (2014) Chromosome scale genome assembly and transcriptome profiling of *Nannochloropsis gaditana* in nitrogen depletion. Mol Plant 7:323–335

Crozet P, Navarro FJ, Willmund F, Mehrshahi P, Bakowski K, Lauersen KJ, Perez-Perez ME, Auroy P, Gorchs Rovira A, Sauret-Gueto S et al (2018) Birth of a photosynthetic chassis: a MoClo toolkit enabling synthetic biology in the microalga *Chlamydomonas reinhardtii*. ACS Synth Biol 7:2074–2086

Cui Y, Qin S, Jiang P (2014) Chloroplast transformation of *Platymonas* (*Tetraselmis*) *subcordiformis* with the bar gene as selectable marker. PLoS One 9:e98607

Cummings RD, Pierce JM (2014) The challenge and promise of glycomics. Chem Biol 21:1–15

Curtis BA, Tanifuji G, Burki F, Gruber A, Irimia M, Maruyama S, Arias MC, Ball SG, Gile GH, Hirakawa Y et al (2012) Algal genomes reveal evolutionary mosaicism and the fate of nucleomorphs. Nature 492:59–65

Daboussi F, Leduc S, Marechal A, Dubois G, Guyot V, Perez-Michaut C, Amato A, Falciatore A, Juillerat A, Beurdeley M et al (2014) Genome engineering empowers the diatom *Phaeodactylum tricornutum* for biotechnology. Nat Commun 5:3831

Dal'Molin CG, Quek LE, Palfreyman RW, Nielsen LK (2011) AlgaGEM - a genome-scale metabolic reconstruction of algae based on the *Chlamydomonas reinhardtii* genome. BMC Genomics 12(Suppl 4):S5

Dauvillee D, Delhaye S, Gruyer S, Slomianny C, Moretz SE, d'Hulst C, Long CA, Ball SG, Tomavo S (2010) Engineering the chloroplast targeted malarial vaccine antigens in *Chlamydomonas* starch granules. PLoS One 5:e15424

Davies JP, Weeks DP, Grossman AR (1992) Expression of the arylsulfatase gene from the β2-tubulin promoter in *Chlamydomonas reinhardtii*. Nucleic Acids Res 20:2959–2965

Dawson HN, Burlingame R, Cannons AC (1997) Stable transformation of *Chlorella*: rescue of nitrate reductase-deficient mutants with the nitrate reductase gene. Curr Microbiol 35:356–362

Day RN, Davidson MW (2009) The fluorescent protein palette: tools for cellular imaging. Chem Soc Rev 38:2887–2921

Day A, Goldschmidt-Clermont M (2011) The chloroplast transformation toolbox: selectable markers and marker removal. Plant Biotechnol J 9:540–553

Debuchy R, Purton S, Rochaix JD (1989) The argininosuccinate lyase gene of *Chlamydomonas reinhardtii*: an important tool for nuclear transformation and for correlating the genetic and molecular maps of the ARG7 locus. EMBO J 8:2803–2809

Deisseroth K (2015) Optogenetics: 10 years of microbial opsins in neuroscience. Nat Neurosci 18:1213–1225

Demurtas OC, Massa S, Ferrante P, Venuti A, Franconi R, Giuliano G (2013) A *Chlamydomonas*-derived human papillomavirus 16 E7 vaccine induces specific tumor protection. PLoS One 8: e61473

Dent RM, Sharifi MN, Malnoe A, Haglund C, Calderon RH, Wakao S, Niyogi KK (2015) Large-scale insertional mutagenesis of *Chlamydomonas* supports phylogenomic functional prediction of photosynthetic genes and analysis of classical acetate-requiring mutants. Plant J 82:337–351

Derelle E, Ferraz C, Rombauts S, Rouzé P, Worden AZ, Robbens S, Partensky F, Degroeve S, Echeynié S, Cooke R et al (2006) Genome analysis of the smallest free-living eukaryote *Ostreococcus tauri* unveils many unique features. Proc Natl Acad Sci U S A 103:11647–11652

Diaz-Santos E, de la Vega M, Vila M, Vigara J, Leon R (2013) Efficiency of different heterologous promoters in the unicellular microalga *Chlamydomonas reinhardtii*. Biotechnol Prog 29:319–328

Dihazi H, Asif AR, Beissbarth T, Bohrer R, Feussner K, Feussner I, Jahn O, Lenz C, Majcherczyk A, Schmidt B et al (2018) Integrative omics - from data to biology. Expert Rev Proteomics 15:463–466

Ding J, Li X, Hu H (2012) Systematic prediction of cis-regulatory elements in the *Chlamydomonas reinhardtii* genome using comparative genomics. Plant Physiol 160:613–623

Doetsch NA, Favreau MR, Kuscuoglu N, Thompson MD, Hallick RB (2001) Chloroplast transformation in *Euglena gracilis*: splicing of a group III twintron transcribed from a transgenic *psbK* operon. Curr Genet 39:49–60

Dong HP, Dong YL, Cui L, Balamurugan S, Gao J, Lu SH, Jiang T (2016) High light stress triggers distinct proteomic responses in the marine diatom *Thalassiosira pseudonana*. BMC Genomics 17:994

Dong B, Hu HH, Li ZF, Cheng RQ, Meng DM, Wang J, Fan ZC (2017) A novel bicistronic expression system composed of the intraflagellar transport protein gene *ift25* and FMDV 2A sequence directs robust nuclear gene expression in *Chlamydomonas reinhardtii*. Appl Microbiol Biotechnol 101:4227–4245

Doron L, Segal N, Shapira M (2016) Transgene expression in microalgae - from tools to applications. Front Plant Sci 7:505

Douglas S, Zauner S, Fraunholz M, Beaton M, Penny S, Deng LT, Wu X, Reith M, Cavalier-Smith T, Maier UG (2001) The highly reduced genome of an enslaved algal nucleus. Nature 410:1091–1096

Dreesen IA, Charpin-El Hamri G, Fussenegger M (2010) Heat-stable oral alga-based vaccine protects mice from *Staphylococcus aureus* infection. J Biotechnol 145:273–280

Dubini A, Ghirardi ML (2015) Engineering photosynthetic organisms for the production of biohydrogen. Photosynth Res 123:241–253

Dunahay TG (1993) Transformation of *Chlamydomonas reinhardtii* with silicon carbide whiskers. BioTechniques 15:452–455, 457–458, 460

Dunahay TG, Jarvis EE, Roessler PG (1995) Genetic transformation of the diatoms *Cyclotella cryptica* and *Navicula saprophila*. J Phycol 31:1004–1012

Dunn MJ (1986) Gel electrophoresis of proteins. Wright, Bristol

Dyhrman ST, Jenkins BD, Rynearson TA, Saito MA, Mercier ML, Alexander H, Whitney LP, Drzewianowski A, Bulygin VV, Bertrand EM et al (2012) The transcriptome and proteome of the diatom *Thalassiosira pseudonana* reveal a diverse phosphorus stress response. PLoS One 7: e33768

Dyo YM, Purton S (2018) The algal chloroplast as a synthetic biology platform for production of therapeutic proteins. Microbiology 164:113–121

Eichler-Stahlberg A, Weisheit W, Ruecker O, Heitzer M (2009) Strategies to facilitate transgene expression in *Chlamydomonas reinhardtii*. Planta 229:873–883

El-Sheekh MM (1999) Stable transformation of the intact cells of *Chlorella kessleri* with high velocity microprojectiles *Biol*. Plantarum 42:209–216

Ender F, Godl K, Wenzl S, Sumper M (2002) Evidence for autocatalytic cross-linking of hydroxyproline-rich glycoproteins during extracellular matrix assembly in *Volvox*. Plant Cell 14:1147–1160

Endo H, Yoshida M, Uji T, Saga N, Inoue K, Nagasawa H (2016) Stable nuclear transformation system for the coccolithophorid alga *Pleurochrysis carterae*. Sci Rep 6:22252

Esland L, Larrea-Alvarez M, Purton S (2018) Selectable markers and reporter genes for engineering the chloroplast of *Chlamydomonas reinhardtii*. Biology (Basel) 7:46

Falciatore A, Casotti R, Leblanc C, Abrescia C, Bowler C (1999) Transformation of nonselectable reporter genes in marine diatoms. Mar Biotechnol 1:239–251

Falciatore A, d'Alcalà MR, Croot P, Bowler C (2000) Perception of environmental signals by a marine diatom. Science 288:2363–2366

Feng S, Xue L, Liu H, Lu P (2009) Improvement of efficiency of genetic transformation for *Dunaliella salina* by glass beads method. Mol Biol Rep 36:1433–1439

Feng S, Feng W, Zhao L, Gu H, Li Q, Shi K, Guo S, Zhang N (2014) Preparation of transgenic *Dunaliella salina* for immunization against white spot syndrome virus in crayfish. Arch Virol 159:519–525

Fenno L, Yizhar O, Deisseroth K (2011) The development and application of optogenetics. Annu Rev Neurosci 34:389–412

Ferenczi A, Pyott DE, Xipnitou A, Molnar A (2017) Efficient targeted DNA editing and replacement in *Chlamydomonas reinhardtii* using Cpf1 ribonucleoproteins and single-stranded DNA. Proc Natl Acad Sci U S A 114:13567–13572

Ferris PJ (1995) Localization of the *nic-7*, *ac-29* and *thi-10* genes within the mating-type locus of *Chlamydomonas reinhardtii*. Genetics 141:543–549

Field CB, Behrenfeld MJ, Randerson JT, Falkowski P (1998) Primary production of the biosphere: integrating terrestrial and oceanic components. Science 281:237–240

Fischer N, Rochaix JD (2001) The flanking regions of *PsaD* drive efficient gene expression in the nucleus of the green alga *Chlamydomonas reinhardtii*. Mol Gen Genomics 265:888–894

Fischer H, Robl I, Sumper M, Kröger N (1999) Targeting and covalent modification of cell wall and membrane proteins heterologously expressed in the diatom *Cylindrotheca fusiformis*. J Phycol 35:113–120

Fletcher SP, Muto M, Mayfield SP (2007) Optimization of recombinant protein expression in the chloroplasts of green algae. Adv Exp Med Biol 616:90–98

Flickinger MC (2013) Upstream industrial biotechnology. Wiley, Hoboken, NJ

Foflonker F, Price DC, Qiu H, Palenik B, Wang S, Bhattacharya D (2015) Genome of the halotolerant green alga *Picochlorum* sp. reveals strategies for thriving under fluctuating environmental conditions. Environ Microbiol 17:412–426

Foo JL, Chang MW (2018) Synthetic yeast genome reveals its versatility. Nature 557:647–648

Foster KW, Smyth RD (1980) Light antennas in phototactic algae. Microbiol Rev 44:572–630

Franklin SE, Mayfield SP (2004) Prospects for molecular farming in the green alga *Chlamydomonas*. Curr Opin Plant Biol 7:159–165

Fu Y, Luo GZ, Chen K, Deng X, Yu M, Han D, Hao Z, Liu J, Lu X, Dore LC et al (2015) N6-methyldeoxyadenosine marks active transcription start sites in *Chlamydomonas*. Cell 161:879–892

Fuhrmann M, Oertel W, Hegemann P (1999) A synthetic gene coding for the green fluorescent protein (GFP) is a versatile reporter in *Chlamydomonas reinhardtii*. Plant J 19:353–361

Fuhrmann M, Hausherr A, Ferbitz L, Schödl T, Heitzer M, Hegemann P (2004) Monitoring dynamic expression of nuclear genes in *Chlamydomonas reinhardtii* by using a synthetic luciferase reporter gene. Plant Mol Biol 55:869–881

Gao C, Wang Y, Shen Y, Yan D, He X, Dai J, Wu Q (2014) Oil accumulation mechanisms of the oleaginous microalga *Chlorella protothecoides* revealed through its genome, transcriptomes, and proteomes. BMC Genomics 15:582

Garcia-Echauri SA, Cardineau GA (2015) TETX: a novel nuclear selection marker for *Chlamydomonas reinhardtii* transformation. Plant Methods 11:27

Geng D, Wang Y, Wang P, Li W, Sun Y (2003) Stable expression of hepatitis B surface antigen gene in *Dunaliella salina* (Chlorophyta). J Appl Phycol 15:451–456

Geng DG, Han Y, Wang YQ, Wang P, Zhang LM, Li WB, Sun YR (2004) Construction of a system for the stable expression of foreign genes in *Dunaliella salina*. Acta Bot Sin 46:342–346

Georgianna DR, Hannon MJ, Marcuschi M, Wu S, Botsch K, Lewis AJ, Hyun J, Mendez M, Mayfield SP (2013) Production of recombinant enzymes in the marine alga *Dunaliella tertiolecta*. Algal Res 2:2–9

Ghurye JS, Cepeda-Espinoza V, Pop M (2016) Metagenomic assembly: overview, challenges and applications. Yale J Biol Med 89:353–362

Gibson DG, Glass JI, Lartigue C, Noskov VN, Chuang RY, Algire MA, Benders GA, Montague MG, Ma L, Moodie MM et al (2010) Creation of a bacterial cell controlled by a chemically synthesized genome. Science 329:52–56

Gobler CJ, Berry DL, Dyhrman ST, Wilhelm SW, Salamov A, Lobanov AV, Zhang Y, Collier JL, Wurch LL, Kustka AB et al (2011) Niche of harmful alga *Aureococcus anophagefferens* revealed through ecogenomics. Proc Natl Acad Sci U S A 108:4352–4357

Gonzalez-Esquer CR, Twary SN, Hovde BT, Starkenburg SR (2018) Nuclear, chloroplast, and mitochondrial genome sequences of the prospective microalgal biofuel strain *Picochlorum soloecismus*. Genome Announc 6:e01498-17

Grantham R, Gautier C, Gouy M, Jacobzone M, Mercier R (1981) Codon catalog usage is a genome strategy modulated for gene expressivity. Nucleic Acids Res 9:r43–r74

Gregory JA, Mayfield SP (2014) Developing inexpensive malaria vaccines from plants and algae. Appl Microbiol Biotechnol 98:1983–1990

Greiner A, Kelterborn S, Evers H, Kreimer G, Sizova I, Hegemann P (2017) Targeting of photoreceptor genes in *Chlamydomonas reinhardtii* via zinc-finger nucleases and CRISPR/Cas9. Plant Cell 29:2498–2518

Griesbeck C, Kobl I, Heitzer M (2006) *Chlamydomonas reinhardtii*: a protein expression system for pharmaceutical and biotechnological proteins. Mol Biotechnol 34:213–223

Griffiths MJ, van Hille RP, Harrison ST (2014) The effect of degree and timing of nitrogen limitation on lipid productivity in *Chlorella vulgaris*. Appl Microbiol Biotechnol 98:6147–6159

Gruber H, Kirzinger SH, Schmitt R (1996) Expression of the *Volvox* gene encoding nitrate reductase: mutation-dependent activation of cryptic splice sites and intron-enhanced gene expression from a cDNA. Plant Mol Biol 31:1–12

Gumpel NJ, Rochaix JD, Purton S (1994) Studies on homologous recombination in the green alga *Chlamydomonas reinhardtii*. Curr Genet 26:438–442

Guo SL, Zhao XQ, Tang Y, Wan C, Alam MA, Ho SH, Bai FW, Chang JS (2013) Establishment of an efficient genetic transformation system in *Scenedesmus obliquus*. J Biotechnol 163:61–68

Hage DS, Anguizola JA, Bi C, Li R, Matsuda R, Papastavros E, Pfaunmiller E, Vargas J, Zheng X (2012) Pharmaceutical and biomedical applications of affinity chromatography: recent trends and developments. J Pharm Biomed Anal 69:93–105

Hall LM, Taylor KB, Jones DD (1993) Expression of a foreign gene in *Chlamydomonas reinhardtii*. Gene 124:75–81

Hallmann A (2007) Algal transgenics and biotechnology. Transgenic Plant J 1:81–98

Hallmann A (2011) Evolution of reproductive development in the volvocine algae. Sex Plant Reprod 24:97–112

Hallmann A (2015) Algae biotechnology – green cell-factories on the rise. Curr Biotechnol 4:389–415

Hallmann A, Rappel A (1999) Genetic engineering of the multicellular green alga *Volvox*: a modified and multiplied bacterial antibiotic resistance gene as a dominant selectable marker. Plant J 17:99–109

Hallmann A, Sumper M (1994) Reporter genes and highly regulated promoters as tools for transformation experiments in *Volvox carteri*. Proc Natl Acad Sci U S A 91:11562–11566

Hallmann A, Sumper M (1996) The *Chlorella* hexose/H^+ symporter is a useful selectable marker and biochemical reagent when expressed in *Volvox*. Proc Natl Acad Sci U S A 93:669–673

Hallmann A, Wodniok S (2006) Swapped green algal promoters: *aphVIII*-based gene constructs with *Chlamydomonas* flanking sequences work as dominant selectable markers in *Volvox* and vice versa. Plant Cell Rep 25:582–591

Hallmann A, Rappel A, Sumper M (1997) Gene replacement by homologous recombination in the multicellular green alga *Volvox carteri*. Proc Natl Acad Sci U S A 94:7469–7474

Haltiwanger RS, Lowe JB (2004) Role of glycosylation in development. Annu Rev Biochem 73:491–537

Han X, Aslanian A, Yates JR 3rd. (2008) Mass spectrometry for proteomics. Curr Opin Chem Biol 12:483–490

Hartley JL, Temple GF, Brasch MA (2000) DNA cloning using in vitro site-specific recombination. Genome Res 10:1788–1795

Hawkins RL, Nakamura M (1999) Expression of human growth hormone by the eukaryotic alga, *Chlorella*. Curr Microbiol 38:335–341

He DM, Qian KX, Shen GF, Zhang ZF, Li YN, Su ZL, Shao HB (2007) Recombination and expression of classical swine fever virus (CSFV) structural protein E2 gene in *Chlamydomonas reinhardtii* chroloplasts. Colloids Surf B Biointerfaces 55:26–30

Hegemann P (2008) Algal sensory photoreceptors. Annu Rev Plant Biol 59:167–189

Hegemann P, Nagel G (2013) From channelrhodopsins to optogenetics. EMBO Mol Med 5:173–176

Hegemann P, Sigrist S (2013) Optogenetics. De Gruyter, Berlin

Heitzer M, Zschoernig B (2007) Construction of modular tandem expression vectors for the green alga *Chlamydomonas reinhardtii* using the Cre/lox-system. BioTechniques 43:324, 326, 328 passim

Hempel F, Lau J, Klingl A, Maier UG (2011) Algae as protein factories: expression of a human antibody and the respective antigen in the diatom *Phaeodactylum tricornutum*. PLoS One 6: e28424

Henley WJ, Litaker RW, Novoveská L, Duke CS, Quemada HD, Sayre RT (2013) Initial risk assessment of genetically modified (GM) microalgae for commodity-scale biofuel cultivation. Algal Res 2:66–77

Hirabayashi J (2008) Concept, strategy and realization of lectin-based glycan profiling. J Biochem 144:139–147

Hirakawa Y, Kofuji R, Ishida K (2008) Transient transformation of a chlorarachniophyte alga, *Lotharella amoebiformis* (Chlorarachniophyceae), with *uidA* and *egfp* reporter genes. J Phycol 44:814–820

Hopes A, Nekrasov V, Kamoun S, Mock T (2016) Editing of the urease gene by CRISPR-Cas in the diatom *Thalassiosira pseudonana*. Plant Methods 12:49

Horgan RP, Kenny LC (2011) 'Omic' technologies: genomics, transcriptomics, proteomics and metabolomics. Obstet Gynaecol 13:189–195

Hou Q, Qiu S, Liu Q, Tian J, Hu Z, Ni J (2013) Selenoprotein-transgenic *Chlamydomonas reinhardtii*. Nutrients 5:624–636

Hovde BT, Deodato CR, Hunsperger HM, Ryken SA, Yost W, Jha RK, Patterson J, Monnat RJ Jr, Barlow SB, Starkenburg SR et al (2015) Genome sequence and transcriptome analyses of *Chrysochromulina tobin*: metabolic tools for enhanced algal fitness in the prominent order Prymnesiales (Haptophyceae). PLoS Genet 11:e1005469

Huang CC, Chen MW, Hsieh JL, Lin WH, Chen PC, Chien LF (2006) Expression of mercuric reductase from *Bacillus megaterium* MB1 in eukaryotic microalga *Chlorella* sp. DT: an approach for mercury phytoremediation. Appl Microbiol Biotechnol 72:197–205

Huang J, Liu J, Li Y, Chen F (2008) Isolation and characterization of the phytoene desaturase gene as a potential selective marker for genetic engineering of the astaxanthin-producing green alga *Chlorella zofingiensis* (Chlorophyta). J Phycol 44:684–690

Hughes RA, Ellington AD (2017) Synthetic DNA synthesis and assembly: putting the synthetic in synthetic biology. CSH Perspect Biol 9:a023812

Hui P (2014) Next generation sequencing: chemistry, technology and applications. Top Curr Chem 336:1–18

Ifuku K, Yan D, Miyahara M, Inoue-Kashino N, Yamamoto YY, Kashino Y (2015) A stable and efficient nuclear transformation system for the diatom *Chaetoceros gracilis*. Photosynth Res 123:203–211

Ikemura T (1985) Codon usage and tRNA content in unicellular and multicellular organisms. Mol Biol Evol 2:13–34

Ishizuka T, Kakuda M, Araki R, Yawo H (2006) Kinetic evaluation of photosensitivity in genetically engineered neurons expressing green algae light-gated channels. Neurosci Res 54:85–94

Issaq H, Veenstra T (2008) Two-dimensional polyacrylamide gel electrophoresis (2D-PAGE): advances and perspectives. BioTechniques 44:697–698, 700

Jakobiak T, Mages W, Scharf B, Babinger P, Stark K, Schmitt R (2004) The bacterial paromomycin resistance gene, *aphH*, as a dominant selectable marker in *Volvox carteri*. Protist 155:381–393

Jarvis EE, Brown LM (1991) Transient expression of firefly luciferase in protoplasts of the green alga *Chlorella ellipsoidea*. Curr Genet 19:317–321

Jarvis EE, Dunahay TG, Brown LM (1992) DNA nucleoside composition and methylation in several species of microalgae. J Phycol 28:356–362

Jeon S, Lim JM, Lee HG, Shin SE, Kang NK, Park YI, Oh HM, Jeong WJ, Jeong BR, Chang YK (2017) Current status and perspectives of genome editing technology for microalgae. Biotechnol Biofuels 10:267

Jiang W, Brueggeman AJ, Horken KM, Plucinak TM, Weeks DP (2014) Successful transient expression of Cas9 and single guide RNA genes in *Chlamydomonas reinhardtii*. Eukaryot Cell 13:1465–1469

Jones BM, Edwards RJ, Skipp PJ, O'Connor CD, Iglesias-Rodriguez MD (2011) Shotgun proteomic analysis of *Emiliania huxleyi*, a marine phytoplankton species of major biogeochemical importance. Mar Biotechnol (NY) 13:496–504

Jones BM, Iglesias-Rodriguez MD, Skipp PJ, Edwards RJ, Greaves MJ, Young JR, Elderfield H, O'Connor CD (2013) Responses of the *Emiliania huxleyi* proteome to ocean acidification. PLoS One 8:e61868

Jungbauer A, Hahn R (2009) Ion-exchange chromatography. Methods Enzymol 463:349–371

Karahalil B (2016) Overview of systems biology and omics technologies. Curr Med Chem 23:4221–4230

Kathiresan S, Sarada R (2009) Towards genetic improvement of commercially important microalga *Haematococcus pluvialis* for biotech applications. J Appl Phycol 21:553–558

Kathiresan S, Chandrashekar A, Ravishankar GA, Sarada R (2009) *Agrobacterium*-mediated transformation in the green alga *Haematococcus pluvialis* (Chlorophyceae, Volvocales). J Phycol 45:642–649

Kathiresan S, Chandrashekar A, Ravishankar GA, Sarada R (2015) Regulation of astaxanthin and its intermediates through cloning and genetic transformation of β-carotene ketolase in *Haematococcus pluvialis*. J Biotechnol 196–197:33–41

Kianianmomeni A, Hallmann A (2014) Algal photoreceptors: in vivo functions and potential applications. Planta 239:1–26

Kianianmomeni A, Hallmann A (2015) Spotlighted brains: Optogenetic activation and silencing of neurons. Trends Biochem Sci 40:624–627

Kianianmomeni A, Hallmann A (2016) Algal photobiology: A rich source of unusual light sensitive proteins for synthetic biology and optogenetics. Methods Mol Biol 1408:37–54

Kiessling LL, Splain RA (2010) Chemical approaches to glycobiology. Annu Rev Biochem 79:619–653

Kilian O, Benemann CS, Niyogi KK, Vick B (2011) High-efficiency homologous recombination in the oil-producing alga *Nannochloropsis* sp. Proc Natl Acad Sci U S A 108:21265–21269

Kim S-K (2015) Handbook of marine microalgae: biotechnology advances. Academic/Elsevier Science Publishing, Waltham, MA

Kim EJ, Cerutti H (2009) Targeted gene silencing by RNA interference in *Chlamydomonas*. Methods Cell Biol 93:99–110

Kim DH, Kim YT, Cho JJ, Bae JH, Hur SB, Hwang I, Choi TJ (2002) Stable integration and functional expression of flounder growth hormone gene in transformed microalga, *Chlorella ellipsoidea*. Mar Biotechnol 4:63–73

Kim KM, Park JH, Bhattacharya D, Yoon HS (2014) Applications of next-generation sequencing to unravelling the evolutionary history of algae. Int J Syst Evol Microbiol 64:333–345

Kim EJ, Ma X, Cerutti H (2015) Gene silencing in microalgae: mechanisms and biological roles. Bioresour Technol 184:23–32

Kim CK, Adhikari A, Deisseroth K (2017) Integration of optogenetics with complementary methodologies in systems neuroscience. Nat Rev Neurosci 18:222–235

Kindle KL (1990) High-frequency nuclear transformation of *Chlamydomonas reinhardtii*. Proc Natl Acad Sci U S A 87:1228–1232

Kindle KL, Schnell RA, Fernandez E, Lefebvre PA (1989) Stable nuclear transformation of *Chlamydomonas* using the *Chlamydomonas* gene for nitrate reductase. J Cell Biol 109:2589–2601

Kirk DL (1998) *Volvox*: molecular-genetic origins of multicellularity and cellular differentiation. Cambridge University Press, Cambridge

Kovar JL, Zhang J, Funke RP, Weeks DP (2002) Molecular analysis of the acetolactate synthase gene of *Chlamydomonas reinhardtii* and development of a genetically engineered gene as a dominant selectable marker for genetic transformation. Plant J 29:109–117

Kozminski KG, Diener DR, Rosenbaum JL (1993) High level expression of nonacetylatable α-tubulin in *Chlamydomonas reinhardtii*. Cell Motil Cytoskeleton 25:158–170

Kranz HD, Miks D, Siegler ML, Capesius I, Sensen CW, Huss VA (1995) The origin of land plants: phylogenetic relationships among charophytes, bryophytes, and vascular plants inferred from complete small-subunit ribosomal RNA gene sequences. J Mol Evol 41:74–84

Krishnamoorthy L, Mahal LK (2009) Glycomic analysis: an array of technologies. ACS Chem Biol 4:715–732

Krohn-Molt I, Wemheuer B, Alawi M, Poehlein A, Güllert S, Schmeisser C, Pommerening-Röser A, Grundhoff A, Daniel R, Hanelt D et al (2013) Metagenome survey of a multispecies and alga-associated biofilm revealed key elements of bacterial-algal interactions in photobioreactors. Appl Environ Microbiol 79:6196–6206

Kronholm I, Bassett A, Baulcombe D, Collins S (2017) Epigenetic and genetic contributions to adaptation in *Chlamydomonas*. Mol Biol Evol 34:2285–2306

Krumholz EW, Yang H, Weisenhorn P, Henry CS, Libourel IG (2012) Genome-wide metabolic network reconstruction of the picoalga *Ostreococcus*. J Exp Bot 63:2353–2362

Kruse O, Rupprecht J, Mussgnug JH, Dismukes GC, Hankamer B (2005) Photosynthesis: a blueprint for solar energy capture and biohydrogen production technologies. Photochem Photobiol Sci 4:957–970

Kumar S (2015) GM algae for biofuel production: biosafety and risk assessment. Collect Biosaf Rev 9:52–75

Kumar SV, Misquitta RW, Reddy VS, Rao BJ, Rajam MV (2004) Genetic transformation of the green alga *Chlamydomonas reinhardtii* by *Agrobacterium tumefaciens*. Plant Sci 166:731–738

Kurniasih SD, Yamasaki T, Kong F, Okada S, Widyaningrum D, Ohama T (2016) UV-mediated *Chlamydomonas* mutants with enhanced nuclear transgene expression by disruption of DNA methylation-dependent and independent silencing systems. Plant Mol Biol 92:629–641

Lamers PP, Janssen M, De Vos RC, Bino RJ, Wijffels RH (2008) Exploring and exploiting carotenoid accumulation in *Dunaliella salina* for cell-factory applications. Trends Biotechnol 26:631–638

Lang I, Hodac L, Friedl T, Feussner I (2011) Fatty acid profiles and their distribution patterns in microalgae: a comprehensive analysis of more than 2000 strains from the SAG culture collection. BMC Plant Biol 11:124

Lapidot M, Raveh D, Sivan A, Arad SM, Shapira M (2002) Stable chloroplast transformation of the unicellular red alga *Porphyridium* species. Plant Physiol 129:7–12

Larkum AW, Ross IL, Kruse O, Hankamer B (2012) Selection, breeding and engineering of microalgae for bioenergy and biofuel production. Trends Biotechnol 30:198–205

Lauersen KJ, Kruse O, Mussgnug JH (2015) Targeted expression of nuclear transgenes in *Chlamydomonas reinhardtii* with a versatile, modular vector toolkit. Appl Microbiol Biotechnol 99:3491–3503

Lee TF, Zhai J, Meyers BC (2010) Conservation and divergence in eukaryotic DNA methylation. Proc Natl Acad Sci U S A 107:9027–9028

Lerche K, Hallmann A (2009) Stable nuclear transformation of *Gonium pectorale*. BMC Biotechnol 9:64

Lerche K, Hallmann A (2013) Stable nuclear transformation of *Eudorina elegans*. BMC Biotechnol 13:11

Lerche K, Hallmann A (2014) Stable nuclear transformation of *Pandorina morum*. BMC Biotechnol 14:65

Levitan O, Dinamarca J, Hochman G, Falkowski PG (2014) Diatoms: a fossil fuel of the future. Trends Biotechnol 32:117–124

Levy SE, Myers RM (2016) Advancements in next-generation sequencing. Annu Rev Genomics Hum Genet 17:95–115

Li X, Gutierrez DV, Hanson MG, Han J, Mark MD, Chiel H, Hegemann P, Landmesser LT, Herlitze S (2005) Fast noninvasive activation and inhibition of neural and network activity by vertebrate rhodopsin and green algae channelrhodopsin. Proc Natl Acad Sci U S A 102:17816–17821

Li F, Gao D, Hu H (2014a) High-efficiency nuclear transformation of the oleaginous marine *Nannochloropsis* species using PCR product. Biosci Biotechnol Biochem 78:812–817

Li J, Han D, Wang D, Ning K, Jia J, Wei L, Jing X, Huang S, Chen J, Li Y et al (2014b) Choreography of transcriptomes and lipidomes of *Nannochloropsis* reveals the mechanisms of oil synthesis in microalgae. Plant Cell 26:1645–1665

Liu X, He Y, Wang Z, Wang C, Liu Z, Xia T, Wang L, Zhang FL (2012) Method of sequence optimization for improved recombinant protein expression using a particle swarm optimization algorithm. United States Patent 8326547

Liu L, Wang Y, Zhang Y, Chen X, Zhang P, Ma S (2013) Development of a new method for genetic transformation of the green alga *Chlorella ellipsoidea*. Mol Biotechnol 54:211–219

Llamas A, Igeño MI, Galván A, Fernández E (2002) Nitrate signalling on the nitrate reductase gene promoter depends directly on the activity of the nitrate transport systems in *Chlamydomonas*. Plant J 30:261–271

Lommer M, Specht M, Roy AS, Kraemer L, Andreson R, Gutowska MA, Wolf J, Bergner SV, Schilhabel MB, Klostermeier UC et al (2012) Genome and low-iron response of an oceanic diatom adapted to chronic iron limitation. Genome Biol 13:R66

Lopez D, Hamaji T, Kropat J, De Hoff P, Morselli M, Rubbi L, Fitz-Gibbon S, Gallaher SD, Merchant SS, Umen J et al (2015) Dynamic changes in the transcriptome and methylome of *Chlamydomonas reinhardtii* throughout its life cycle. Plant Physiol 169:2730–2743

Lowe R, Shirley N, Bleackley M, Dolan S, Shafee T (2017) Transcriptomics technologies. PLoS Comput Biol 13:e1005457

Lozano JC, Schatt P, Botebol H, Verge V, Lesuisse E, Blain S, Carre IA, Bouget FY (2014) Efficient gene targeting and removal of foreign DNA by homologous recombination in the picoeukaryote *Ostreococcus*. Plant J 78:1073–1083

Lumbreras V, Stevens DL, Purton S (1998) Efficient foreign gene expression in *Chlamydomonas reinhardtii* mediated by an endogenous intron. Plant J 14:441–447

Ma JK, Barros E, Bock R, Christou P, Dale PJ, Dix PJ, Fischer R, Irwin J, Mahoney R, Pezzotti M et al (2005) Molecular farming for new drugs and vaccines. Current perspectives on the production of pharmaceuticals in transgenic plants. EMBO Rep 6:593–599

MacDougall KM, McNichol J, McGinn PJ, O'Leary SJ, Melanson JE (2011) Triacylglycerol profiling of microalgae strains for biofuel feedstock by liquid chromatography-high-resolution mass spectrometry. Anal Bioanal Chem 401:2609–2616

Maddison DR, Schulz K-S (2007) The tree of life web project

Marouga R, David S, Hawkins E (2005) The development of the DIGE system: 2D fluorescence difference gel analysis technology. Anal Bioanal Chem 382:669–678

Marra MA, Hillier L, Waterston RH (1998) Expressed sequence tags - ESTablishing bridges between genomes. Trends Genet 14:4–7

Martin GJ, Hill DR, Olmstead IL, Bergamin A, Shears MJ, Dias DA, Kentish SE, Scales PJ, Botte CY, Callahan DL (2014) Lipid profile remodeling in response to nitrogen deprivation in the microalgae *Chlorella* sp. (Trebouxiophyceae) and *Nannochloropsis* sp. (Eustigmatophyceae). PloS One 9:e103389

Maruyama M, Horáková I, Honda H, Xing X, Shiragami N, Unno H (1994) Introduction of foreign DNA into *Chlorella saccharophila* by electroporation. Biotechnol Tech 8:821–826

Mastrobuoni G, Irgang S, Pietzke M, Assmus HE, Wenzel M, Schulze WX, Kempa S (2012) Proteome dynamics and early salt stress response of the photosynthetic organism *Chlamydomonas reinhardtii*. BMC Genomics 13:215

Matasci N, Hung LH, Yan Z, Carpenter EJ, Wickett NJ, Mirarab S, Nguyen N, Warnow T, Ayyampalayam S, Barker M et al (2014) Data access for the 1,000 plants (1KP) project. Gigascience 3:17

Mathieu-Rivet E, Scholz M, Arias C, Dardelle F, Schulze S, Le Mauff F, Teo G, Hochmal AK, Blanco-Rivero A, Loutelier-Bourhis C et al (2013) Exploring the N-glycosylation pathway in *Chlamydomonas reinhardtii* unravels novel complex structures. Mol Cell Proteomics 12:3160–3183

Mathieu-Rivet E, Kiefer-Meyer MC, Vanier G, Ovide C, Burel C, Lerouge P, Bardor M (2014) Protein N-glycosylation in eukaryotic microalgae and its impact on the production of nuclear expressed biopharmaceuticals. Front Plant Sci 5:359

Matich EK, Ghafari M, Camgoz E, Caliskan E, Pfeifer BA, Haznedaroglu BZ, Atilla-Gokcumen GE (2018) Time-series lipidomic analysis of the oleaginous green microalga species *Ettlia oleoabundans* under nutrient stress. Biotechnol Biofuels 11:29

Matsuo T, Onai K, Okamoto K, Minagawa J, Ishiura M (2006) Real-time monitoring of chloroplast gene expression by a luciferase reporter: evidence for nuclear regulation of chloroplast circadian period. Mol Cell Biol 26:863–870

Matsuzaki M, Misumi O, Shin IT, Maruyama S, Takahara M, Miyagishima SY, Mori T, Nishida K, Yagisawa F, Nishida K et al (2004) Genome sequence of the ultrasmall unicellular red alga *Cyanidioschyzon merolae* 10D. Nature 428:653–657

Mayfield SP, Kindle KL (1990) Stable nuclear transformation of *Chlamydomonas reinhardtii* by using a *C. reinhardtii* gene as the selectable marker. Proc Natl Acad Sci U S A 87:2087–2091

Mayfield SP, Schultz J (2004) Development of a luciferase reporter gene, *luxCt*, for *Chlamydomonas reinhardtii* chloroplast. Plant J 37:449–458

Mayfield SP, Franklin SE, Lerner RA (2003) Expression and assembly of a fully active antibody in algae. Proc Natl Acad Sci U S A 100:438–442

McCarthy SS, Kobayashi MC, Niyogi KK (2004) White mutants of *Chlamydomonas reinhardtii* are defective in phytoene synthase. Genetics 168:1249–1257

McKew BA, Lefebvre SC, Achterberg EP, Metodieva G, Raines CA, Metodiev MV, Geider RJ (2013) Plasticity in the proteome of *Emiliania huxleyi* CCMP 1516 to extremes of light is highly targeted. New Phytol 200:61–73

Melis A, Zhang L, Forestier M, Ghirardi ML, Seibert M (2000) Sustained photobiological hydrogen gas production upon reversible inactivation of oxygen evolution in the green alga *Chlamydomonas reinhardtii*. Plant Physiol 122:127–136

Merchant SS, Prochnik SE, Vallon O, Harris EH, Karpowicz SJ, Witman GB, Terry A, Salamov A, Fritz-Laylin LK, Marechal-Drouard L et al (2007) The *Chlamydomonas* genome reveals the evolution of key animal and plant functions. Science 318:245–250

Meslet-Cladiere L, Vallon O (2011) Novel shuttle markers for nuclear transformation of the green alga *Chlamydomonas reinhardtii*. Eukaryot Cell 10:1670–1678

Meuser JE, Ananyev G, Wittig LE, Kosourov S, Ghirardi ML, Seibert M, Dismukes GC, Posewitz MC (2009) Phenotypic diversity of hydrogen production in chlorophycean algae reflects distinct anaerobic metabolisms. J Biotechnol 142:21–30

Minoda A, Sakagami R, Yagisawa F, Kuroiwa T, Tanaka K (2004) Improvement of culture conditions and evidence for nuclear transformation by homologous recombination in a red alga, *Cyanidioschyzon merolae* 10D. Plant Cell Physiol 45:667–671

Mitra A, Higgins DW (1994) The *Chlorella* virus adenine methyltransferase gene promoter is a strong promoter in plants. Plant Mol Biol 26:85–93

Mitra A, Higgins DW, Rohe NJ (1994) A *Chlorella* virus gene promoter functions as a strong promoter both in plants and bacteria. Biochem Biophys Res Commun 204:187–194

Miyahara M, Aoi M, Inoue-Kashino N, Kashino Y, Ifuku K (2013) Highly efficient transformation of the diatom *Phaeodactylum tricornutum* by multi-pulse electroporation. Biosci Biotechnol Biochem 77:874–876

Mock T, Otillar RP, Strauss J, McMullan M, Paajanen P, Schmutz J, Salamov A, Sanges R, Toseland A, Ward BJ et al (2017) Evolutionary genomics of the cold-adapted diatom *Fragilariopsis cylindrus*. Nature 541:536–540

Molino JVD, de Carvalho JCM, Mayfield SP (2018) Comparison of secretory signal peptides for heterologous protein expression in microalgae: expanding the secretion portfolio for *Chlamydomonas reinhardtii*. PLoS One 13:e0192433

Molnár A, Bassett A, Thuenemann E, Schwach F, Karkare S, Ossowski S, Weigel D, Baulcombe D (2009) Highly specific gene silencing by artificial microRNAs in the unicellular alga *Chlamydomonas reinhardtii*. Plant J 58:165–174

Molnar I, Lopez D, Wisecaver JH, Devarenne TP, Weiss TL, Pellegrini M, Hackett JD (2012) Bio-crude transcriptomics: gene discovery and metabolic network reconstruction for the biosynthesis of the terpenome of the hydrocarbon oil-producing green alga, *Botryococcus braunii* race B (Showa). BMC Genomics 13:576

Moreau H, Verhelst B, Couloux A, Derelle E, Rombauts S, Grimsley N, Van Bel M, Poulain J, Katinka M, Hohmann-Marriott MF et al (2012) Gene functionalities and genome structure in *Bathycoccus prasinos* reflect cellular specializations at the base of the green lineage. Genome Biol 13:R74

Munoz R, Guieysse B (2006) Algal-bacterial processes for the treatment of hazardous contaminants: a review. Water Res 40:2799–2815

Mussgnug JH (2015) Genetic tools and techniques for *Chlamydomonas reinhardtii*. Appl Microbiol Biotechnol 99:5407–5418

Nagel G, Ollig D, Fuhrmann M, Kateriya S, Musti AM, Bamberg E, Hegemann P (2002) Channelrhodopsin-1: a light-gated proton channel in green algae. Science 296:2395–2398

Nagel G, Szellas T, Huhn W, Kateriya S, Adeishvili N, Berthold P, Ollig D, Hegemann P, Bamberg E (2003) Channelrhodopsin-2, a directly light-gated cation-selective membrane channel. Proc Natl Acad Sci U S A 100:13940–13945

Nagel G, Szellas T, Kateriya S, Adeishvili N, Hegemann P, Bamberg E (2005) Channelrhodopsins: directly light-gated cation channels. Biochem Soc Trans 33:863–866

Nelson JA, Lefebvre PA (1995) Targeted disruption of the *NIT8* gene in *Chlamydomonas reinhardtii*. Mol Cell Biol 15:5762–5769

Nelson JA, Savereide PB, Lefebvre PA (1994) The *CRY1* gene in *Chlamydomonas reinhardtii*: structure and use as a dominant selectable marker for nuclear transformation. Mol Cell Biol 14:4011–4019

Ng IS, Tan SI, Kao PH, Chang YK, Chang JS (2017) Recent developments on genetic engineering of microalgae for biofuels and bio-based chemicals. Biotechnol J 12:1600644

Niu YF, Yang ZK, Zhang MH, Zhu CC, Yang WD, Liu JS, Li HY (2012) Transformation of diatom *Phaeodactylum tricornutum* by electroporation and establishment of inducible selection marker. BioTechniques 52:1–3

North SJ, Hitchen PG, Haslam SM, Dell A (2009) Mass spectrometry in the analysis of N-linked and O-linked glycans. Curr Opin Struct Biol 19:498–506

Nymark M, Sharma AK, Sparstad T, Bones AM, Winge P (2016) A CRISPR/Cas9 system adapted for gene editing in marine algae. Sci Rep 6:24951

O'Brien EA, Koski LB, Zhang Y, Yang L, Wang E, Gray MW, Burger G, Lang BF (2007) TBestDB: a taxonomically broad database of expressed sequence tags (ESTs). Nucleic Acids Res 35:D445–D451

O'Brien EA, Zhang Y, Wang E, Marie V, Badejoko W, Lang BF, Burger G (2009) GOBASE: an organelle genome database. Nucleic Acids Res 37:D946–D950

Oey M, Ross IL, Hankamer B (2014) Gateway-assisted vector construction to facilitate expression of foreign proteins in the chloroplast of single celled algae. PLoS One 9:e86841

Olguín EJ (2012) Dual purpose microalgae-bacteria-based systems that treat wastewater and produce biodiesel and chemical products within a biorefinery. Biotechnol Adv 30:1031–1046

Ortega-Escalante JA, Jasper R, Miller SM (2019) CRISPR/Cas9 mutagenesis in *Volvox carteri*. Plant J 97:661–672

Ortiz-Matamoros MF, Islas-Flores T, Voigt B, Menzel D, Baluska F, Villanueva MA (2015) Heterologous DNA uptake in cultured *Symbiodinium* spp aided by *Agrobacterium tumefaciens*. PloS one 10:e0132693

Ozsolak F, Milos PM (2011) RNA sequencing: advances, challenges and opportunities. Nat Rev Genet 12:87–98

Palenik B, Grimwood J, Aerts A, Rouzé P, Salamov A, Putnam N, Dupont C, Jorgensen R, Derelle E, Rombauts S et al (2007) The tiny eukaryote *Ostreococcus* provides genomic insights into the paradox of plankton speciation. Proc Natl Acad Sci U S A 104:7705–7710

Palluk S, Arlow DH, de Rond T, Barthel S, Kang JS, Bector R, Baghdassarian HM, Truong AN, Kim PW, Singh AK et al (2018) De novo DNA synthesis using polymerase-nucleotide conjugates. Nat Biotechnol 36:645–650

Polle JEW, Barry K, Cushman J, Schmutz J, Tran D, Hathwaik LT, Yim WC, Jenkins J, McKie-Krisberg Z, Prochnik S et al (2017) Draft nuclear genome sequence of the halophilic and beta-carotene-accumulating green alga *Dunaliella salina* strain CCAP19/18. Genome Announc 5: e01105-17

Poulsen N, Kröger N (2005) A new molecular tool for transgenic diatoms: control of mRNA and protein biosynthesis by an inducible promoter-terminator cassette. FEBS J 272:3413–3423

Poulsen N, Chesley PM, Kröger N (2006) Molecular genetic manipulation of the diatom *Thalassiosira pseudonana* (Bacillariophyceae). J Phycol 42:1059–1065

Pozhitkov AE, Tautz D, Noble PA (2007) Oligonucleotide microarrays: widely applied - poorly understood. Brief Funct Genomic Proteomic 6:141–148

Price DC, Chan CX, Yoon HS, Yang EC, Qiu H, Weber AP, Schwacke R, Gross J, Blouin NA, Lane C et al (2012) *Cyanophora paradoxa* genome elucidates origin of photosynthesis in algae and plants. Science 335:843–847

Prince RC, Kheshgi HS (2005) The photobiological production of hydrogen: potential efficiency and effectiveness as a renewable fuel. Crit Rev Microbiol 31:19–31

Prochnik SE, Umen J, Nedelcu AM, Hallmann A, Miller SM, Nishii I, Ferris P, Kuo A, Mitros T, Fritz-Laylin LK et al (2010) Genomic analysis of organismal complexity in the multicellular green alga *Volvox carteri*. Science 329:223–226

Quinn JM, Merchant S (1995) Two copper-responsive elements associated with the *Chlamydomonas Cyc6* gene function as targets for transcriptional activators. Plant Cell 7:623–628

Quinn JM, Kropat J, Merchant S (2003) Copper response element and Crr1-dependent Ni^{2+}-responsive promoter for induced, reversible gene expression in *Chlamydomonas reinhardtii*. Eukaryot Cell 2:995–1002

Radakovits R, Jinkerson RE, Fuerstenberg SI, Tae H, Settlage RE, Boore JL, Posewitz MC (2012) Draft genome sequence and genetic transformation of the oleaginous alga *Nannochloropsis gaditana*. Nat Commun 3:686

Rakus JF, Mahal LK (2011) New technologies for glycomic analysis: toward a systematic understanding of the glycome. Annu Rev Anal Chem (Palo Alto, Calif) 4:367–392

Ramanan R, Kim BH, Cho DH, Oh HM, Kim HS (2016) Algae-bacteria interactions: evolution, ecology and emerging applications. Biotechnol Adv 34:14–29

Randolph-Anderson BL, Sato R, Johnson AM, Harris EH, Hauser CR, Oeda K, Ishige F, Nishio S, Gillham NW, Boynton JE (1998) Isolation and characterization of a mutant protoporphyrinogen oxidase gene from *Chlamydomonas reinhardtii* conferring resistance to porphyric herbicides. Plant Mol Biol 38:839–859

Rasala BA, Mayfield SP (2011) The microalga *Chlamydomonas reinhardtii* as a platform for the production of human protein therapeutics. Bioeng Bugs 2:50–54

Rasala BA, Muto M, Lee PA, Jager M, Cardoso RM, Behnke CA, Kirk P, Hokanson CA, Crea R, Mendez M et al (2010) Production of therapeutic proteins in algae, analysis of expression of seven human proteins in the chloroplast of *Chlamydomonas reinhardtii*. Plant Biotechnol J 8:719–733

Rasala BA, Lee PA, Shen Z, Briggs SP, Mendez M, Mayfield SP (2012) Robust expression and secretion of Xylanase1 in *Chlamydomonas reinhardtii* by fusion to a selection gene and processing with the FMDV 2A peptide. PLoS One 7:e43349

Rasala BA, Barrera DJ, Ng J, Plucinak TM, Rosenberg JN, Weeks DP, Oyler GA, Peterson TC, Haerizadeh F, Mayfield SP (2013) Expanding the spectral palette of fluorescent proteins for the green microalga *Chlamydomonas reinhardtii*. Plant J 74:545–556

Rasala BA, Chao SS, Pier M, Barrera DJ, Mayfield SP (2014) Enhanced genetic tools for engineering multigene traits into green algae. PLoS One 9:e94028

Rawat I, Ranjith Kumar R, Mutanda T, Bux F (2011) Dual role of microalgae: Phycoremediation of domestic wastewater and biomass production for sustainable biofuels production. Appl Energy 88:3411–3424

Read BA, Kegel J, Klute MJ, Kuo A, Lefebvre SC, Maumus F, Mayer C, Miller J, Monier A, Salamov A et al (2013) Pan genome of the phytoplankton *Emiliania* underpins its global distribution. Nature 499:209–213

Remacle C, Cline S, Boutaffala L, Gabilly S, Larosa V, Barbieri MR, Coosemans N, Hamel PP (2009) The *ARG9* gene encodes the plastid-resident N-acetyl ornithine aminotransferase in the green alga *Chlamydomonas reinhardtii*. Eukaryot Cell 8:1460–1463

Richardson SM, Mitchell LA, Stracquadanio G, Yang K, Dymond JS, DiCarlo JE, Lee D, Huang CL, Chandrasegaran S, Cai Y et al (2017) Design of a synthetic yeast genome. Science 355:1040–1044

Rillahan CD, Paulson JC (2011) Glycan microarrays for decoding the glycome. Annu Rev Biochem 80:797–823

Ritter A, Dittami SM, Goulitquer S, Correa JA, Boyen C, Potin P, Tonon T (2014) Transcriptomic and metabolomic analysis of copper stress acclimation in *Ectocarpus siliculosus* highlights signaling and tolerance mechanisms in brown algae. BMC Plant Biol 14:116

Romanov V, Davidoff SN, Miles AR, Grainger DW, Gale BK, Brooks BD (2014) A critical comparison of protein microarray fabrication technologies. Analyst 139:1303–1326

Rosales-Mendoza S (2016) Algae-based biopharmaceuticals. Springer, Cham

Sayre RT, Wagner RE, Sirporanadulsil S, Farias C (2001) Transgenic algae for delivery antigens to animals. In Int. Patent Number W.O. 01/98335 A2

Schiedlmeier B, Schmitt R, Müller W, Kirk MM, Gruber H, Mages W, Kirk DL (1994) Nuclear transformation of *Volvox carteri*. Proc Natl Acad Sci U S A 91:5080–5084

Schmidt M, Gessner G, Luff M, Heiland I, Wagner V, Kaminski M, Geimer S, Eitzinger N, Reissenweber T, Voytsekh O et al (2006) Proteomic analysis of the eyespot of *Chlamydomonas reinhardtii* provides novel insights into its components and tactic movements. Plant Cell 18:1908–1930

Schneider MV, Orchard S (2011) Omics technologies, data and bioinformatics principles. Methods Mol Biol 719:3–30

Schnell RA, Lefebvre PA (1993) Isolation of the *Chlamydomonas* regulatory gene *NIT2* by transposon tagging. Genetics 134:737–747

Schroda M, Blöcker D, Beck CF (2000) The *HSP70A* promoter as a tool for the improved expression of transgenes in *Chlamydomonas*. Plant J 21:121–131

Scranton MA, Ostrand JT, Georgianna DR, Lofgren SM, Li D, Ellis RC, Carruthers DN, Drager A, Masica DL, Mayfield SP (2016) Synthetic promoters capable of driving robust nuclear gene expression in the green alga *Chlamydomonas reinhardtii*. Algal Res 15:135–142

Shao N, Bock R (2008) A codon-optimized luciferase from *Gaussia princeps* facilitates the in vivo monitoring of gene expression in the model alga *Chlamydomonas reinhardtii*. Curr Genet 53:381–388

Sharma A, Arya SK (2017) Hydrogen from algal biomass: a review of production process. Biotechnol Rep 15:63–69

Sharon-Gojman R, Maimon E, Leu S, Zarka A, Boussiba S (2015) Advanced methods for genetic engineering of *Haematococcus pluvialis* (Chlorophyceae, Volvocales). Algal Res 10:8–15

Shimogawara K, Fujiwara S, Grossman A, Usuda H (1998) High-efficiency transformation of *Chlamydomonas reinhardtii* by electroporation. Genetics 148:1821–1828

Shiraki T, Kondo S, Katayama S, Waki K, Kasukawa T, Kawaji H, Kodzius R, Watahiki A, Nakamura M, Arakawa T et al (2003) Cap analysis gene expression for high-throughput analysis of transcriptional starting point and identification of promoter usage. Proc Natl Acad Sci U S A 100:15776–15781

Shoguchi E, Shinzato C, Kawashima T, Gyoja F, Mungpakdee S, Koyanagi R, Takeuchi T, Hisata K, Tanaka M, Fujiwara M et al (2013) Draft assembly of the *Symbiodinium minutum* nuclear genome reveals dinoflagellate gene structure. Curr Biol 23:1399–1408

Shrager J, Hauser C, Chang CW, Harris EH, Davies J, McDermott J, Tamse R, Zhang Z, Grossman AR (2003) *Chlamydomonas reinhardtii* genome project. A guide to the generation and use of the cDNA information. Plant Physiol 131:401–408

Siaut M, Cuine S, Cagnon C, Fessler B, Nguyen M, Carrier P, Beyly A, Beisson F, Triantaphylides C, Li-Beisson Y et al (2011) Oil accumulation in the model green alga *Chlamydomonas reinhardtii*: characterization, variability between common laboratory strains and relationship with starch reserves. BMC Biotechnol 11:7

Singh P, Kumari S, Guldhe A, Misra R, Rawat I, Bux F (2016) Trends and novel strategies for enhancing lipid accumulation and quality in microalgae. Renew Sust Energ Rev 55:1–16

Siripornadulsil S, Traina S, Verma DP, Sayre RT (2002) Molecular mechanisms of proline-mediated tolerance to toxic heavy metals in transgenic microalgae. Plant Cell 14:2837–2847

Sizova I, Fuhrmann M, Hegemann P (2001) A *Streptomyces rimosus aphVIII* gene coding for a new type phosphotransferase provides stable antibiotic resistance to *Chlamydomonas reinhardtii*. Gene 277:221–229

Sizova I, Greiner A, Awasthi M, Kateriya S, Hegemann P (2013) Nuclear gene targeting in *Chlamydomonas* using engineered zinc-finger nucleases. Plant J 73:873–882

Smart EJ, Selman BR (1993) Complementation of a *Chlamydomonas reinhardtii* mutant defective in the nuclear gene encoding the chloroplast coupling factor 1 (CF1) γ-subunit (*atpC*). J Bioenerg Biomembr 25:275–284

Sodeinde OA, Kindle KL (1993) Homologous recombination in the nuclear genome of *Chlamydomonas reinhardtii*. Proc Natl Acad Sci U S A 90:9199–9203

Solier C, Langen H (2014) Antibody-based proteomics and biomarker research - current status and limitations. Proteomics 14:774–783

Song D, Li J, Hu X, Xi B (2013) Construction of a shuttle vector for heterologous gene expression in *Escherichia coli* and microalgae *Anabaena*. Engineering 5:540–544

Soria-Guerra RE, Ramírez-Alonso JI, Ibáñez-Salazar A, Govea-Alonso DO, Paz-Maldonado LMT, BañuelosHernández B, Korban SS, Rosales-Mendoza S (2014) Expression of an HBcAg-based antigen carrying angiotensin II in *Chlamydomonas reinhardtii* as a candidate hypertension vaccine. Plant Cell Tissue Organ Cult 116:133–139

Specht EA, Mayfield SP (2014) Algae-based oral recombinant vaccines. Front Microbiol 5:60

Specht E, Miyake-Stoner S, Mayfield S (2010) Micro-algae come of age as a platform for recombinant protein production. Biotechnol Lett 32:1373–1383

Starkenburg SR, Polle JEW, Hovde B, Daligault HE, Davenport KW, Huang A, Neofotis P, McKie-Krisberg Z (2017) Draft nuclear genome, complete chloroplast genome, and complete mitochondrial genome for the biofuel/bioproduct feedstock species *Scenedesmus obliquus* strain DOE0152z. Genome Announc 5:e00617-17

Steinbrenner J, Sandmann G (2006) Transformation of the green alga *Haematococcus pluvialis* with a phytoene desaturase for accelerated astaxanthin biosynthesis. Appl Environ Microbiol 72:7477–7484

Stevens DR, Rochaix JD, Purton S (1996) The bacterial phleomycin resistance gene *ble* as a dominant selectable marker in *Chlamydomonas*. Mol Gen Genet 251:23–30

Stoger E, Fischer R, Moloney M, Ma JK (2014) Plant molecular pharming for the treatment of chronic and infectious diseases. Annu Rev Plant Biol 65:743–768

Subashchandrabose SR, Ramakrishnan B, Megharaj M, Venkateswarlu K, Naidu R (2011) Consortia of cyanobacteria/microalgae and bacteria: biotechnological potential. Biotechnol Adv 29:896–907

Sun M, Qian K, Su N, Chang H, Liu J, Shen G (2003) Foot-and-mouth disease virus VP1 protein fused with cholera toxin B subunit expressed in *Chlamydomonas reinhardtii* chloroplast. Biotechnol Lett 25:1087–1092

Sun Y, Yang Z, Gao X, Li Q, Zhang Q, Xu Z (2005) Expression of foreign genes in *Dunaliella* by electroporation. Mol Biotechnol 30:185–192

Sun Y, Gao X, Li Q, Zhang Q, Xu Z (2006) Functional complementation of a nitrate reductase defective mutant of a green alga *Dunaliella viridis* by introducing the nitrate reductase gene. Gene 377:140–149

Sun M, Yang Z, Wawrik B (2018) Metabolomic fingerprints of individual algal cells using the single-probe mass spectrometry technique. Front Plant Sci 9:571

Surzycki R, Greenham K, Kitayama K, Dibal F, Wagner R, Rochaix JD, Ajam T, Surzycki S (2009) Factors effecting expression of vaccines in microalgae. Biologicals 37:133–138

Takemura T, Imamura S, Kobayashi Y, Tanaka K (2018) Construction of a selectable marker recycling system and the use in epitope tagging of multiple nuclear genes in the unicellular red alga *Cyanidioschyzon merolae*. Plant Cell Physiol 59:2308–2316

Tam LW, Lefebvre PA (1993) Cloning of flagellar genes in *Chlamydomonas reinhardtii* by DNA insertional mutagenesis. Genetics 135:375–384

Tan C, Qin S, Zhang Q, Jiang P, Zhao F (2005) Establishment of a micro-particle bombardment transformation system for *Dunaliella salina*. J Microbiol 43:361–365

Tang DK, Qiao SY, Wu M (1995) Insertion mutagenesis of *Chlamydomonas reinhardtii* by electroporation and heterologous DNA. Biochem Mol Biol Int 36:1025–1035

ten Lohuis MR, Miller DJ (1998) Genetic transformation of dinoflagellates (*Amphidinium* and *Symbiodinium*): expression of GUS in microalgae using heterologous promoter constructs. Plant J 13:427–435

Teng CY, Qin S, Liu JG, Yu DZ, Liang CW, Tseng CK (2002) Transient expression of lacZ in bombarded unicellular green alga *Haematococcus pluvialis*. J Appl Phycol 14:497–500

Thompson JM (2017) Mass spectrometry. Pan Stanford, Singapore

Tian Y, Gao S, von der Heyde EL, Hallmann A, Nagel G (2018) Two-component cyclase opsins of green algae are ATP-dependent and light-inhibited guanylyl cyclases. BMC Biol 16:144

Tischer D, Weiner OD (2014) Illuminating cell signalling with optogenetic tools. Nat Rev Mol Cell Biol 15:551–558

Torzillo G, Scoma A, Faraloni C, Ena A, Johanningmeier U (2009) Increased hydrogen photoproduction by means of a sulfur-deprived *Chlamydomonas reinhardtii* D1 protein mutant. Int J Hydrogen Energy 34:1–8

Toulza E, Blanc-Mathieu R, Gourbiere S, Piganeau G (2012) Environmental and evolutionary genomics of microbial algae: power and challenges of metagenomics. Adv Bot Res 64:383–427

Tran NP, Park JK, Lee CG (2009) Proteomics analysis of proteins in green alga *Haematococcus lacustris* (Chlorophyceae) expressed under combined stress of nitrogen starvation and high irradiance. Enzym Microb Technol 45:241–246

Tran M, Henry RE, Siefker D, Van C, Newkirk G, Kim J, Bui J, Mayfield SP (2013a) Production of anti-cancer immunotoxins in algae: ribosome inactivating proteins as fusion partners. Biotechnol Bioeng 110:2826–2835

Tran M, Van C, Barrera DJ, Pettersson PL, Peinado CD, Bui J, Mayfield SP (2013b) Production of unique immunotoxin cancer therapeutics in algal chloroplasts. Proc Natl Acad Sci U S A 110: E15–E22

Trentacoste EM, Shrestha RP, Smith SR, Gle C, Hartmann AC, Hildebrand M, Gerwick WH (2013) Metabolic engineering of lipid catabolism increases microalgal lipid accumulation without compromising growth. Proc Natl Acad Sci U S A 110:19748–19753

Tumanov S, Kamphorst JJ (2017) Recent advances in expanding the coverage of the lipidome. Curr Opin Biotechnol 43:127–133

Uhlen M, Ponten F (2005) Antibody-based proteomics for human tissue profiling. Mol Cell Proteomics 4:384–393

Unnithan VV, Unc A, Smith GB (2014) Mini-review: a priori considerations for bacteria–algae interactions in algal biofuel systems receiving municipal wastewaters. Algal Res 4:35–40

van Dijk K, Xu H, Cerutti H (2005) Epigenetic silencing of transposons in the green alga *Chlamydomonas reinhardtii*. In: Small RNAs, vol 17. Springer, Berlin, Heidelberg, pp 159–178

Vanier G, Stelter S, Vanier J, Hempel F, Maier UG, Lerouge P, Ma J, Bardor M (2018) Alga-made anti-hepatitis B antibody binds to human Fcγ receptors. Biotechnol J 13:e1700496

Varela JC, Pereira H, Vila M, Leon R (2015) Production of carotenoids by microalgae: achievements and challenges. Photosynth Res 125:423–436

Velculescu VE, Zhang L, Vogelstein B, Kinzler KW (1995) Serial analysis of gene expression. Science 270:484–487

Veluchamy A, Lin X, Maumus F, Rivarola M, Bhavsar J, Creasy T, O'Brien K, Sengamalay NA, Tallon LJ, Smith AD et al (2013) Insights into the role of DNA methylation in diatoms by genome-wide profiling in *Phaeodactylum tricornutum*. Nat Commun 4:2091

Verma D, Daniell H (2007) Chloroplast vector systems for biotechnology applications. Plant Physiol 145:1129–1143

Vieler A, Wu G, Tsai CH, Bullard B, Cornish AJ, Harvey C, Reca IB, Thornburg C, Achawanantakun R, Buehl CJ et al (2012) Genome, functional gene annotation, and nuclear transformation of the heterokont oleaginous alga *Nannochloropsis oceanica* CCMP1779. PLoS Genet 8:e1003064

Villand P, Eriksson M, Samuelsson G (1997) Carbon dioxide and light regulation of promoters controlling the expression of mitochondrial carbonic anhydrase in *Chlamydomonas reinhardtii*. Biochem J 327(Pt 1):51–57

Voedisch B, Thie H (2010) Size exclusion chromatography. In: Antibody Engineering. Springer, Berlin, Heidelberg, pp 607–612

Vollmers J, Wiegand S, Kaster AK (2017) Comparing and evaluating metagenome assembly tools from a microbiologist's perspective - not only size matters! PLoS One 12:e0169662

von der Heyde EL, Klein B, Abram L, Hallmann A (2015) The inducible *nitA* promoter provides a powerful molecular switch for transgene expression in *Volvox carteri*. BMC Biotechnol 15:5

Wang X, Brandsma M, Tremblay R, Maxwell D, Jevnikar AM, Huner N, Ma S (2008) A novel expression platform for the production of diabetes-associated autoantigen human glutamic acid decarboxylase (hGAD65). BMC Biotechnol 8:87

Wang Q, Lu Y, Xin Y, Wei L, Huang S, Xu J (2016) Genome editing of model oleaginous microalgae *Nannochloropsis* spp. by CRISPR/Cas9. Plant J 88:1071–1081

Wang H, Gao L, Shao H, Zhou W, Liu T (2017) Lipid accumulation and metabolic analysis based on transcriptome sequencing of filamentous oleaginous microalgae *Tribonema minus* at different growth phases. Bioprocess Biosyst Eng 40:1327–1335

Wannathong T, Waterhouse JC, Young RE, Economou CK, Purton S (2016) New tools for chloroplast genetic engineering allow the synthesis of human growth hormone in the green alga *Chlamydomonas reinhardtii*. Appl Microbiol Biotechnol 100:5467–5477

Weeks DP (2011) Homologous recombination in *Nannochloropsis*: a powerful tool in an industrially relevant alga. Proc Natl Acad Sci U S A 108:20859–20860

Whiteaker JR, Zhao L, Anderson L, Paulovich AG (2010) An automated and multiplexed method for high throughput peptide immunoaffinity enrichment and multiple reaction monitoring mass spectrometry-based quantification of protein biomarkers. Mol Cell Proteomics 9:184–196

Willamme R, Alsafra Z, Arumugam R, Eppe G, Remacle F, Levine RD, Remacle C (2015) Metabolomic analysis of the green microalga *Chlamydomonas reinhardtii* cultivated under day/night conditions. J Biotechnol 215:20–26

Willette S, Gill SS, Dungan B, Schaub TM, Jarvis JM, St Hilaire R, Holguin FO (2018) Alterations in lipidome and metabolome profiles of *Nannochloropsis salina* in response to reduced culture temperature during sinusoidal temperature and light. Algal Res 32:79–92

Winck FV, Riano-Pachon DM, Sommer F, Rupprecht J, Mueller-Roeber B (2012) The nuclear proteome of the green alga *Chlamydomonas reinhardtii*. Proteomics 12:95–100

Wingren C (2016) Antibody-based proteomics. Adv Exp Med Biol 926:163–179

Wong KH, Jin Y, Moqtaderi Z (2013) Multiplex Illumina sequencing using DNA barcoding. Curr Protoc Mol Biol Chapter 7, Unit 7 11

Worden AZ, Lee JH, Mock T, Rouze P, Simmons MP, Aerts AL, Allen AE, Cuvelier ML, Derelle E, Everett MV et al (2009) Green evolution and dynamic adaptations revealed by genomes of the marine picoeukaryotes *Micromonas*. Science 324:268–272

Wördenweber R, Rokitta SD, Heidenreich E, Corona K, Kirschhöfer F, Fahl K, Klocke J, Kottke T, Brenner-Weiß G, Rost B et al (2018) Phosphorus and nitrogen starvation reveal life-cycle specific responses in the metabolome of *Emiliania huxleyi* (Haptophyta). Limnol Oceanogr 63:203–226

Wu S, Xu L, Huang R, Wang Q (2011) Improved biohydrogen production with an expression of codon-optimized hemH and lba genes in the chloroplast of *Chlamydomonas reinhardtii*. Bioresour Technol 102:2610–2616

Wu C, Xiong W, Dai J, Wu Q (2015) Genome-based metabolic mapping and ^{13}C flux analysis reveal systematic properties of an oleaginous microalga *Chlorella protothecoides*. Plant Physiol 167:586–599

Wu-Scharf D, Jeong B, Zhang C, Cerutti H (2000) Transgene and transposon silencing in *Chlamydomonas reinhardtii* by a DEAH-box RNA helicase. Science 290:1159–1162

Xie WH, Zhu CC, Zhang NS, Li DW, Yang WD, Liu JS, Sathishkumar R, Li HY (2014) Construction of novel chloroplast expression vector and development of an efficient transformation system for the diatom *Phaeodactylum tricornutum*. Mar Biotechnol (NY) 16:538–546

Xie J, Bai X, Lavoie M, Lu H, Fan X, Pan X, Fu Z, Qian H (2015) Analysis of the proteome of the marine diatom *Phaeodactylum tricornutum* exposed to aluminum providing insights into aluminum toxicity mechanisms. Environ Sci Technol 49:11182–11190

Xiong W, Liu L, Wu C, Yang C, Wu Q (2010) ^{13}C-tracer and gas chromatography-mass spectrometry analyses reveal metabolic flux distribution in the oleaginous microalga *Chlorella protothecoides*. Plant Physiol 154:1001–1011

Xue J, Niu YF, Huang T, Yang WD, Liu JS, Li HY (2015) Genetic improvement of the microalga *Phaeodactylum tricornutum* for boosting neutral lipid accumulation. Metab Eng 27:1–9

Yan N, Fan C, Chen Y, Hu Z (2016) The potential for microalgae as bioreactors to produce pharmaceuticals. Int J Mol Sci 17:962

Yang K, Han X (2016) Lipidomics: techniques, applications, and outcomes related to biomedical sciences. Trends Biochem Sci 41:954–969

Yang D, Song D, Kind T, Ma Y, Hoefkens J, Fiehn O (2015) Lipidomic analysis of *Chlamydomonas reinhardtii* under nitrogen and sulfur deprivation. PLoS One 10:e0137948

Yergeau E, Michel C, Tremblay J, Niemi A, King TL, Wyglinski J, Lee K, Greer CW (2017) Metagenomic survey of the taxonomic and functional microbial communities of seawater and sea ice from the Canadian arctic. Sci Rep 7:42242

Yoshikawa K, Kojima Y, Nakajima T, Furusawa C, Hirasawa T, Shimizu H (2011) Reconstruction and verification of a genome-scale metabolic model for *Synechocystis* sp. PCC6803. Appl Microbiol Biotechnol 92:347–358

Zaia J (2010) Mass spectrometry and glycomics. OMICS 14:401–418

Zamboni N, Saghatelian A, Patti GJ (2015) Defining the metabolome: size, flux, and regulation. Mol Cell 58:699–706

Zaslavskaia LA, Lippmeier JC, Kroth PG, Grossman AR, Apt KE (2000) Transformation of the diatom *Phaeodactylum tricornutum* (Bacillariophyceae) with a variety of selectable marker and reporter genes. J Phycol 36:379–386

Zaslavskaia LA, Lippmeier JC, Shih C, Ehrhardt D, Grossman AR, Apt KE (2001) Trophic conversion of an obligate photoautotrophic organism through metabolic engineering. Science 292:2073–2075

Zemach A, McDaniel IE, Silva P, Zilberman D (2010) Genome-wide evolutionary analysis of eukaryotic DNA methylation. Science 328:916–919

Zhan Y, Marchand CH, Maes A, Mauries A, Sun Y, Dhaliwal JS, Uniacke J, Arragain S, Jiang H, Gold ND et al (2018) Pyrenoid functions revealed by proteomics in *Chlamydomonas reinhardtii*. PLoS One 13:e0185039

Zhang J, Hao Q, Bai L, Xu J, Yin W, Song L, Xu L, Guo X, Fan C, Chen Y et al (2014) Overexpression of the soybean transcription factor GmDof4 significantly enhances the lipid content of *Chlorella ellipsoidea*. Biotechnol Biofuels 7:128

Zorin B, Hegemann P, Sizova I (2005) Nuclear-gene targeting by using single-stranded DNA avoids illegitimate DNA integration in *Chlamydomonas reinhardtii*. Eukaryot Cell 4:1264–1272

Zorin B, Grundman O, Khozin-Goldberg I, Leu S, Shapira M, Kaye Y, Tourasse N, Vallon O, Boussiba S (2014) Development of a nuclear transformation system for oleaginous green alga *Lobosphaera* (*Parietochloris*) *incisa* and genetic complementation of a mutant strain, deficient in arachidonic acid biosynthesis. PLoS One 9:e105223

Chapter 6
Optimization of Microalgae Photosynthetic Metabolism to Close the Gap with Potential Productivity

Giorgio Perin and Tomas Morosinotto

Abstract Microalgae metabolism is powered only by sustainable energy and carbon sources, representing a valuable alternative to develop clean industrial processes. Moreover, this group of unicellular photosynthetic microorganisms shows high versatility, including species from different ecological niches which evolved a variety of pathways to synthesize a wide spectrum of bioactive compounds. However, sophisticated industrial cultivation systems are needed to control the stability of the production process during intensive cultivation. This artificial environment is far different from the ecological niches that shaped these organisms, limiting photon-to-biomass conversion efficiency (PBCE) to values far below those achieved at the lab scale. Moreover, large-scale cultivation has high energetic and operational costs due to initial investment and maintenance, that current PBCE values cannot compensate for, preventing commercial feasibility. Tuning microalgae photosynthetic metabolism represents an unavoidable challenge to improve PBCE and meet the theoretical potential of these organisms.

6.1 Introduction

Current global economy development trends rely on industrial processes causing the release of greenhouse gases (GHG) and other pollutants in the atmosphere (Gies 2017). If future economy continues on current trends, available models foresee a 30% increase in anthropogenic GHG emissions from 2010 to 2100 and a consequent rise in the average global temperature up to 3 °C (Walsh et al. 2015), with a massive negative effects on food and water security, as well as on functional biodiversity. In order to curb this scenario, an unavoidable challenge for our society is to identify

G. Perin
Department of Biology, University of Padova, Padova, Italy

Department of Life Sciences, Imperial College London, London, UK

T. Morosinotto (✉)
Department of Biology, University of Padova, Padova, Italy
e-mail: tomas.morosinotto@unipd.it

A. Hallmann, P. H. Rampelotto (eds.), *Grand Challenges in Algae Biotechnology*, Grand Challenges in Biology and Biotechnology,
https://doi.org/10.1007/978-3-030-25233-5_6

environmentally sustainable solutions to current needs for energy, food and materials (Rockström et al. 2017). Plants biomass is a potential source for many interesting molecules and it might represent a promising alternative to feed current industrial processes. However, such feedstock is in competition with food productions for agricultural land use, risking the inflation of food prices (Rulli et al. 2016). Furthermore the land-use change (LUC) derived from the conversion of existing forests and grasslands to new croplands would release stored carbon (C) as CO_2 in the atmosphere, leading to an increased carbon debt (Fargione et al. 2008). Only developing plant biomass cultivation from marginal lands could avoid this scenario (Robertson et al. 2017) and exploitation of unconventional sources of biomass like microalgae can provide a seminal contribution.

6.2 The Untapped Potential of Microalgae

6.2.1 Microalgae as Feedstock for a Sustainable Global Economy

Microalgae are emerging as promising alternative as source of biomass to complement crops cultivation. Microalgae are also photosynthetic microorganisms and thus their metabolism is fuelled by sustainable energy/carbon sources, i.e. sunlight and atmospheric CO_2. Microalgae photosynthesis could indeed represent a cheap and efficient carbon sequestering technology, thanks to the fact that some species can withstand high CO_2 concentrations [i.e. up to 90% in open ponds (Salih 2011)], enabling their cultivation in connection to industrial sites. Moreover, such organisms are able to extract nutrients from civil and industrial wastewaters and are particularly efficient in removing nitrates and phosphates, making them highly promising for the development of a sustainable bioeconomy (Ramos Tercero et al. 2014; Santos and Pires 2018).

Microalgae include different species that adapted to various ecological niches, shaping a large metabolic diversity that makes them potential platforms to synthesize a wide spectrum of bioactive compounds (Bule et al. 2018). At present, many molecules commercialized by pharmaceutical and chemical industries are indeed obtained through environmentally unsustainable processes and the development of sustainable enzymatic alternatives could be seminal to reduce the environmental impact of such practices.

Microalgae biomass is also a valuable feedstock of food, feed additives, pigments (i.e. carotenoids and keto-carotenoids) and fatty acids (i.e. omega-3 fatty acids), showing great potential in nutraceutics and pharmaceutics market (Bilal et al. 2017; Koutra et al. 2018; Maeda et al. 2018; Ng et al. 2017; Renuka et al. 2018). They also represent a valuable feedstock for energy production, where the production of biofuels can be combined with wastewater treatment and nutrients recycling to develop environmentally/energetically sustainable processes.

The theoretical potential of microalgae to replace current industrial processes with sustainable solutions is well recognized, but still largely untapped because implementing this technology on a global scale is not commercially feasible yet and research efforts are still needed to concretize such technology into an industrial reality (Acién Fernández et al. 2003; Lam et al. 2018). Several critical issues still need to be addressed, including the development of efficient biomass-harvesting and product-extraction technologies. Among those challenges, one major barrier to reach economic competitiveness is the maximization of algae efficiency in converting sunlight into biomass, which has been identified as the major factor affecting algae competitiveness (Ruiz et al. 2016).

Microalgae have a higher photon-to-biomass conversion efficiency (PBCE) than plants, up to 3% in the lab with respect to 0.2–1% in crops (De Vree et al. 2015; Melis 2009), a value still far from the theoretical maximum of natural photosynthesis, estimated to be around 12% (Ooms et al. 2016). This difference is mainly due to the fact that the whole microalgae biomass is photosynthetically active, while in plants only leaves perform photosynthesis. Microalgae have the potential to maintain photosynthetic activity all year, at difference with crops which are photosynthetically active only for a few months in temperate climates, compromising the total biomass productivity per year. On the other hand, algae industrial cultivation systems require larger initial investments and have higher energetic and operational costs with respect to agriculture. Since energy fuelling the biomass growth is coming from sunlight, higher PBCE implies smaller cultivations systems and lower costs (Simionato et al. 2013) meaning that this parameter has strategic role to achieve commercial feasibility (Carneiro et al. 2017; Park and Lee 2016).

In order to make microalgae production process competitive, it is thus seminal to optimize PBCE by reproducing at industrial scale the performances observed in the lab and by closing the gap with the maximal theoretical photosynthetic efficiency (Melis 2009; Ooms et al. 2016). The investigation and optimization of algae photosynthetic metabolism is thus a strategic target to close the gap between current achievements and theoretical potential and achieve commercial competitiveness.

6.2.2 Microalgae Photosynthetic Metabolism

6.2.2.1 Light Reactions

Out of the total sunlight reaching our planet, only the 46% is photosynthetic active radiation (PAR, between 400 and 700 nm) and can thus be exploited by oxygenic photosynthesis (Wobbe and Remacle 2014). This energy is harvested through pigments binding proteins called antenna or light-harvesting complexes (LHC), which serve the biological functions of harvesting photons and transfer their energy, in the form of electronic excitation, to the reaction centers (RC) of photosystems (PS) (Büchel 2015). Light absorption drives the generation of chlorophyll (Chl) excited forms, i.e. singlet chlorophyll ($^1Chl^*$), that fuels photochemical reactions, leading to the synthesis of ATP and NADPH. Microalgae evolved large antenna systems to maximize their light-harvesting ability and compete with other photosynthetic organisms for available energy (Kirk 1994).

If absorbed light is saturating, however, this is not all exploitable by photochemical reactions, generating an excess of $^{1}Chl^{*}$ that eventually leads to intersystem crossing and formation of Chl triplets ($^{3}Chl^{*}$). $^{3}Chl^{*}$ are stable enough to react with oxygen and generate singlet oxygen ($^{1}O_2$), which is a highly reactive molecule that oxidizes pigments, proteins and lipids, leading to damage and photo-inhibition.

In order to limit the negative consequences of saturating light, photosynthetic organisms evolved mechanisms to repair photosynthetic components damaged by reactive oxygen species. As an example, $^{1}O_2$ damage targets in particular photosystem II (PSII) and its subunit D1 is constantly re-synthesized under strong illumination to restore photochemical activity (Kato et al. 2012). Another strategy to reduce light damage involves mechanisms to dissipate excess energy as heat [i.e. non-photochemical quenching (NPQ) (Melis 2009)], driving thus to a safe de-excitation of excess $^{1}Chl^{*}$.

In algae, a peculiar class of LHC, known as LHCSR or LHCX (stress-related light-harvesting complex) modulate the response to light excess, triggering NPQ activation (Büchel 2015; Ruiz-Sola and Petroutsos 2018). Upon exposition to excess illumination, protons accumulate in the thylakoid lumen, inducing the protonation of LHCSR and triggering activation of NPQ.

A second protection mechanism also activated by the decrease in luminal pH involves the activation of the enzyme violaxanthin de-epoxidase (VDE) that catalyzes the conversion of carotenoid violaxanthin into zeaxanthin. Once cells return in light limiting conditions, zeaxanthin is converted back to violaxanthin, thanks to the activity of the enzyme zeaxanthin epoxidase (ZE). Zeaxanthin is a carotenoid highly effective in scavenging ROS (Kuczynska et al. 2015) but also enhancing NPQ.

On a longer term, algae modulate the accumulation of photosynthetic pigments/proteins to optimize the composition of photosynthetic apparatus to the light environment. This phenomenon, called photo-acclimation, requires de-novo synthesis or degradation of pigments (i.e. Chl, Car and xanthophylls) and proteins (Meneghesso et al. 2016) and it is thus activated in a time-scale of days.

6.2.2.2 Carbon Fixation

Light energy fuels the photosynthetic electrons transport chain, finally leading to the generation of reducing power in form of NADPH. During electrons transport, protons are pumped into the thylakoids lumen, and the consequent proton motive force (*pmf*) is used by ATP synthase to store chemical energy in form of ATP. Reducing power and ATP generated during photosynthesis are consequently exploited to assimilate CO_2 through the Calvin–Benson cycle, starting with the carboxylation of ribulose-1,5-bisphosphate (RuBP) in two molecules of 3-phosphoglycerate (3-PGA), catalysed by Ribulose-1,5-bisphosphate carboxylase (RuBisCO). CO_2

availability and Calvin–Benson cycle efficiency thus have a strong impact on photosynthetic productivity and efficiency in converting sunlight into biomass.

In aquatic photosynthetic organisms, CO_2 availability is often limited by its slow diffusion in water as well as by the presence of hydration/dehydration reactions which interconvert this substrate in other forms of dissolved inorganic carbon (HCO_3^-), not accessible to RuBisCO (Riebesell et al. 1993). As a consequence, these organisms evolved carbon concentrating mechanisms (CCM), to increase CO_2 availability inside the cell (Tomar et al. 2017), up to 1000-fold with respect to the external environment (Wang et al. 2015). These mechanisms generally include transporters to translocate bicarbonate ions inside the cell and carbonic anhydrases to convert it into CO_2 (Gee and Niyogi 2017; Giordano et al. 2005).

RuBisCO is well known to be a rather inefficient enzyme with a maximal rate at least one order of magnitude lower than the average of the central metabolism enzymes (Bar-Even et al. 2011). Moreover, it shows a low specificity for CO_2 and it can also react with oxygen (Savir et al. 2010), leading to the generation of one molecule of 3-PGA and one molecule of 2-phosphoglycolate (2-PG). The latter is toxic and needs to be recycled by the photorespiration pathway with a consumption of 2 and 3.25 mol of NADPH and ATP, respectively. Oxygenation reaction of RuBisCO thus has a large negative effect on productivity causing a net loss of reducing power, ATP and a release of CO_2 (Wingler et al. 2000).

6.2.3 How Microalgae Photosynthetic Metabolism Responds to Intensive Cultivation

Microalgae cultivation for industrial purposes takes place in photobioreactor (PBR) or ponds, respectively closed or open systems allowing a controlled cultivation environment (De Vree et al. 2015). Here microalgae cultures are normally supplemented with CO_2-enriched gas, bubbled or pumped into the system (Brennan and Owende 2010). The number of inlets in the reactor or the efficiency in the culture mixing has a major influence on CO_2 distribution in the culture (Cheah et al. 2015). When CO_2 is limiting, Calvin–Benson cycle cannot re-generate photosynthesis substrates (i.e. ADP, Pi, $NADP^+$) at a sufficient rate to enable the processing of the available radiation, risking light saturation and lowering light-use efficiency (Sect. 6.2.2.2).

Algae evolved in ecological niches where light is often limiting for growth and therefore have large antenna systems (Kirk 1994), including hundreds of chlorophyll per reaction center in most eukaryotic algae, to maximize the chance of harvesting photons and compete with other photosynthetic organisms in a natural environment. In PBR, microalgae are generally cultivated at high concentrations that, combined with the high pigment content per cell, causes large optical densities of the cultures, preventing a homogeneous light distribution (Simionato et al. 2013). Because of the optical density of the culture, cells at the surface absorb most of the incident

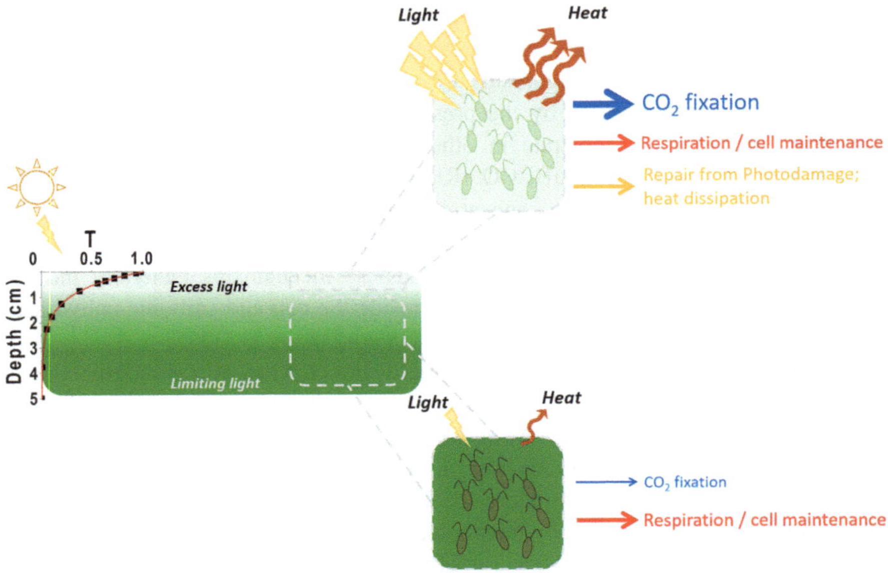

Fig. 6.1 Consequences on photosynthesis of the inhomogeneous light distribution in PBR. An inhomogeneous light distribution in industrial microalgae cultures is due to their high optical densities. The latter is responsible for the light attenuation profile here depicted, expressed as transmittance (T) (i.e. the ratio between transmitted and incident light intensities) depending on the culture depth (in centimetres, cm), calculated from experimental data collected for *Nannochloropsis* cultures at 1 g/l concentration in cylindrical PBR with 5 cm diameter (Perin et al. 2017b). As a consequence, cells close to the light source experience excess light (paler color), while inner layers are in limiting light (darker color). The former perform CO_2 fixation at a high rate but also waste part of the absorbed energy to repair damaged photosystems as well as in the form of heat to reduce oxidative stress. Cells in limiting light show a reduced CO_2 fixation, sometimes going below the compensation point where respiration and photosynthesis rates are equivalent. Figure adapted from (Perin et al. 2018)

radiation. In the example shown in Fig. 6.1, 60% of available radiation is absorbed by the first cm of the culture. This leaves inner layers in light limitation with not enough energy to fuel a sustained growth. If light intensity is too low, it can go below the compensation point where energy spent for cells maintenance is bigger than the one available for carbon fixation (Fig. 6.1).

Cells in more exposed PBR layers perform CO_2 fixation at their maximal capacity, but because of their high light-harvesting ability, they also often experience light saturation, activating repair and photoprotection mechanisms (Fig. 6.1). NPQ dissipates part of the absorbed energy as heat (Melis 2009), counteracting the light-induced damage but also strongly reducing the amount of energy available for photochemistry.

It is worth noting that during industrial cultivation, environmental conditions are not stable and cells are exposed to fluctuating light intensities, because of changing weather conditions experienced daily and seasonally. Furthermore, the mass culture is not steady, but continuously mixed to ensure optimal distribution of CO_2. Because of this mixing, cells can move continuously between culture layers, thus experiencing sudden changes to their exposition to sunlight with dark/light cycles that are normally faster in closed PBR (e.g. millisecond time-scale) than in open ponds (i.e. seconds–minutes time-scale) (Carvalho et al. 2011; Molina et al. 2001).

Kinetics of photoprotection processes are often slower than the mixing kinetics, and, as a consequence, cells inefficiently spend energy to counteract the environmental changes (Eberhard et al. 2008). As an example, in most microalgae, full NPQ expression in vivo requires the presence of zeaxanthin, which is synthesized in a time-scale of minutes (Murchie and Niyogi 2011; Nilkens et al. 2010). Moreover, PSII repair mechanism has been estimated to work within tens of minutes [D1 turnover in illuminated cells is around 30 min (Barber and Andersson 1992)]. These time-scales are averagely two orders of magnitude larger than closed PBR mixing kinetics, meaning that cells being exposed to full irradiation do not have time to fully activate photoprotection mechanisms. On the other hand, when cells switch from oversaturating to light limiting conditions, photoprotection relaxes thanks to the conversion of zeaxanthin to violaxanthin (Sect. 6.2.2.1). The reaction occurs within tens of minutes to hours (Goss and Jakob 2010), meaning that when cells switch to inner layers, they still have photoprotection mechanisms activated, leading to dissipation of available light, decreasing even further the overall efficiency.

6.3 Genetic Engineering of Photosynthetic Metabolism

The maximization of PBCE is clearly dependent on the optimization of the whole cultivation process to maximize productivity. These efforts should however be complemented by the metabolic engineering of microalgae photosynthesis for its domestication to artificial conditions.

6.3.1 Improvement of Light Reactions

6.3.1.1 Engineering Light-Harvesting to Increase Light Homogeneity

As discussed above, inhomogeneous light distribution in PBR has negative impact on productivity. Moreover, since cultures in PBR normally have high density, microalgae cells are mostly low-light adapted. Therefore, it has been suggested that reduction of photosystems antenna size would allow a more homogeneous light distribution with a positive impact on productivity (Fig. 6.2). As a matter of fact, only about 50 and 90 out of, respectively, 350 and 300 Chl molecules found in each PSII and PSI unit in *C. reinhardtii* are strictly required for the correct assembly of the reaction centers and photochemical activity (Melis 1991), while the remaining are not necessary for electron transport reactions. Therefore, antenna size reduction

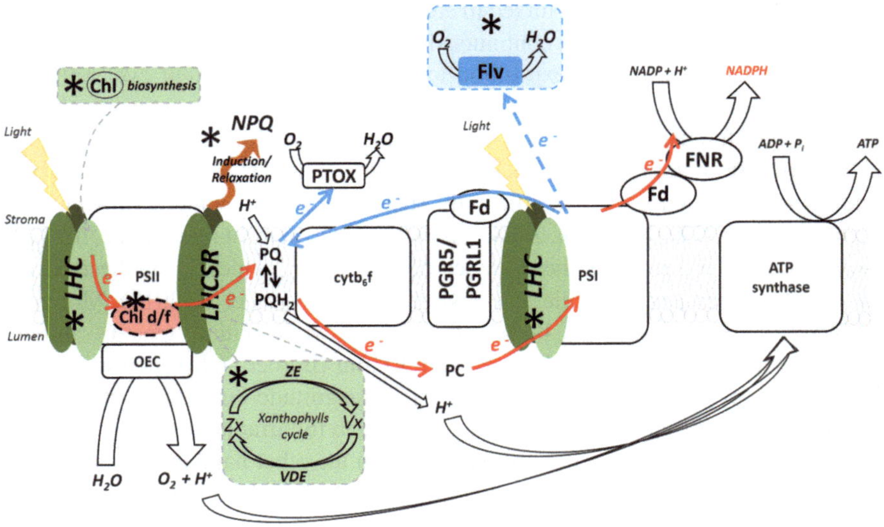

Fig. 6.2 Potential targets to improve microalgae photosynthetic efficiency. Photosynthesis components are schematized as white boxes (core complexes) and as colored ellipses [light-harvesting complexes (LHC)]. In algae, NPQ is triggered by light-harvesting complexes stress related (LHCSR), responsible for heat dissipation of excess energy (red wave). Xanthophylls cycle is indicated in a green dashed box and its kinetics control NPQ induction/relaxation. Linear electron flow (red solid arrows) from water to $NADP^+$ generates reducing power and also triggers protons pumping into the thylakoid lumen, used to drive ATP synthesis. Alternative electron pathways (i.e. PGR5/PGRL1 cyclic electron flow around PSI and PTOX-mediated water to water cycle) are indicated by blue solid arrows. Pseudo-cyclic electron flow, redirecting electrons downstream PSI to water by non-native Flv proteins (dashed blue box), is indicated by the dashed blue arrow. Reactions are not balanced. Targets for algae photosynthesis improvement are indicated by asterisks and are: Chl content per cell, accumulation of LHC proteins, expansion of the usable light spectrum by introducing red-forms of Chl (Chld/f), NPQ capacity, modulation of xanthophyll cycle/NPQ photoprotection induction/relaxation kinetics, optimization of electron flows. *cytb$_6$f* cytochrome b_6f, *Fd* ferredoxin, *Flv* flavodiiron protein, *FNR* ferredoxin:$NADP^+$ reductase, *OEC* oxygen evolving complex, *PC* plastocyanin, *PQ* plastoquinone, *PGH_2* plastoquinol, *PGR5* proton gradient regulation 5, *PGRL1* PGR5-like protein 1, *PSI* photosystem I, *PSII* photosystem II, *PTOX* plastid terminal oxidase

is a theoretically viable strategy to reduce culture optical density without affecting the photosystems ability to perform photochemistry. If this is achieved, cells exposed to excess light should minimize the energy lost by heat dissipation and saturate photosynthesis at higher irradiances, while inner cells should receive more energy to run photochemistry (Simionato et al. 2013).

Microalgae strains with reduced antenna size have been widely isolated in the past, especially in the model genus *Chlamydomonas* (Bonente et al. 2011; Nakajima et al. 2001; Nakajima and Itayama 2003; Nakajima and Ueda 1997, 2000; Polle et al. 2003), but also in other species with a more promising industrial potential (Cazzaniga et al. 2014; Melis et al. 1998; Perin et al. 2015; Sharon-Gojman et al. 2017; Shin et al. 2016; Verruto et al. 2018).

In such studies, microalgae antenna size was mainly reduced by directly targeting LHC encoding genes, exploiting insertional mutagenesis and RNAi approaches. Also indirect strategies proved their feasibility in controlling LHC abundance (Formighieri et al. 2012), with co-translational and post-translational molecular regulators of antenna proteins abundance [e.g. NAB1 repressor (Beckmann et al. 2009; Sharon-Gojman et al. 2017; Wobbe et al. 2009)] as major targets. Other indirect approaches have been directed to control genes involved in Chl biosynthetic pathway (Polle et al. 2000), chloroplast import and protein assembly (Kirst et al. 2012a, b; Kirst and Melis 2014; Jeong et al. 2018)—all showing a successful reduction in antenna size. The development of CRISPR/CAS-based editing methods (Banerjee et al. 2018; Verruto et al. 2018) is opening the possibility to elucidate the complex molecular network regulating antenna proteins accumulation also in species more interesting for industrial applications, enabling to identify new and more promising candidates (Jeong et al. 2018).

Strains with a reduced antenna size show an increased photosynthetic activity on a Chl basis and generally the saturation of light reactions at higher irradiances than parental strains (Beckmann et al. 2009; Cazzaniga et al. 2014; Perin et al. 2015), consequently pushing growth in small-scale experiments.

6.3.1.2 Engineering Light-Harvesting to Increase Exploitable Radiation

Another approach to improve light-use efficiency during industrial cultivation is to engineer microalgae light-harvesting apparatus to increase the amount of exploitable radiation, including regions of the solar light spectrum beyond PAR (Cardona et al. 2018). It was indeed estimated that just extending absorption from 700 to 750 nm would result in a 19% increase in available photons (Blankenship et al. 2011; Chen and Blankenship 2011). Some cyanobacteria are able to expand their absorption in the near infra-red, synthetizing chlorophylls d and f that have a red-shifted absorption (Ho et al. 2016), a strategy that could potentially be exported also in other organisms (Fig. 6.2).

However, in cyanobacteria species containing Chl d/f, such pigments are bound to specific PS components that are co-expressed, and the synthesis of the pigments alone is thus not expected to be sufficient to expand the exploitable radiation (Ort et al. 2015). Red-shifted PS complexes are however expected to be more susceptible to light damage (Cotton et al. 2015). Furthermore, expressing heterologous photosystem complexes raises also the challenge of their integration with the endogenous photosynthetic metabolism of the host, a task complicated by the need to coordinately manipulate both nuclear and chloroplast genomes. Such radical approaches, although highly interesting and promising, are presently only theoretical without experimental proof of concepts.

6.3.1.3 Reprogramming Photo-Protection Mechanisms

As discussed above, upon saturating irradiances, microalgae activate NPQ to dissipate the excess harvested energy and protect the photosynthetic metabolism from photoinhibition (see Sect. 6.2.2.1). NPQ can dissipate up to 80% of harvested light (Melis 2009), consequently lowering the amount of energy exploitable by photochemistry for growth. Reducing NPQ activation has thus been suggested as promising strategy to curb the amount of energy loss (Fig. 6.2). Its activation requires LHCSR/LHCX proteins to undergo conformational changes and bind photoprotective carotenoids (see Sect. 6.2.2.1). Reduction of maximal NPQ activation in *C. reinhardtii* showed some positive effects on cells growth under constant illumination in a lab-scale environment (Berteotti et al. 2016). However, during industrial cultivation, cells are exposed to highly dynamic irradiances, and a reduction in NPQ can cause decreased ability to withstand intense illumination in the most exposed layers, eliding the advantages in efficiency of the deeper layers (Perin et al. 2017b). The complete inactivation of xanthophyll cycle, in fact, causes a strong decrease in productivity in algae cultures under strong illumination even if the culture density is high (Bellan and Morosinotto, unpublished).

One alternative strategy could be to alter NPQ kinetics rather than its levels. NPQ relaxation in particular is slow, since it takes several minutes for zeaxanthin to be converted back to violaxanthin. When cells in PBR move from full sunlight to inner layers, they are still locked in a photo-protective state and thus use light less efficiently (see Sect. 6.2.3). A faster NPQ relaxation would enable cells to respond more promptly to light intensity changes, improving photosynthetic efficiency while keeping intact their photoprotection ability. Given NPQ activation in algae is dependent from zeaxanthin accumulation (see Sect. 6.2.2.1), rewiring xanthophylls cycle kinetics might influence photoprotection relaxation kinetics (Fig. 6.2). Recently, the alteration of xanthophylls cycle kinetics by overexpressing its enzymes was proven to be effective in enhancing photosynthetic productivity and biomass accumulation in tobacco plants in the field (Kromdijk et al. 2016).

Addressing light-harvesting constrains has been the major target to improve microalgae light-use efficiency during industrial cultivation so far. However, also electron transport reactions play a major role in it, with alternative electron transports (PTOX-mediated water to water cycle and cyclic and pseudo-cyclic around PSI) showing a seminal role to keep light-use homeostasis in response to light dynamics and particularly to fluctuating light environments (Cardol et al. 2011).

In this context, highly promising might also be the introduction of Flavodiiron (Flv) proteins in some eukaryotic microalgae (within brown and red algae) that lost them during evolution (Fig. 6.2). Such proteins are in fact ubiquitous in all cyanobacteria species, and are present in many plants including mosses (Gerotto et al. 2016) but were lost in the ancestor to all flowering plants (Alboresi et al. 2018) and in some eukaryotic photosynthetic microorganisms, therefore leaving room for improvement in some microalgae species. Flv avoid over-reduction of the electron transport chain in fluctuating light conditions, redirecting electrons from NADPH to oxygen and preventing PSI to be photodamaged (Rochaix 2014), which, differently from PSII, doesn't have efficient photoprotective mechanisms when electron transport chain is oversaturated (Larosa et al. 2018; Tiwari et al. 2016). Flv expression in *Arabidopsis* and tobacco plants was indeed proven to speed up the recovery of photochemistry in fluctuating light conditions (Gómez et al. 2018; Yamamoto et al. 2016). Despite the promising results achieved with higher plants, such approaches has not been attempted in microalgae so far. However, the elucidation of the regulation of photoprotective mechanisms in a wide variety of species (Goss and Lepetit 2015) and the development of efficient genome editing tools also for several microalgae species (Banerjee et al. 2018) is opening the possibility to transfer such approaches to photosynthetic microorganisms with a larger industrial potential.

6.3.2 Improvement of Carbon Fixation Rate

As discussed above, engineering light reactions can improve photosynthesis performances. However, such strategies lead to an improved light-use efficiency only if light reactions are not limited by the availability of photosynthesis substrates (i.e. $NADP^+$, ADP). Carbon fixation rate controls the regeneration of such molecules and its efficiency thus directly influences light conversion. Even if CO_2 can be provided in excess to algae PBRs, carbon fixation efficiency strongly affects microalgae photosynthetic productivity, making it a second fundamental target for improvement efforts.

6.3.2.1 Engineering Rubisco and Substrates Availability

RuBisCO catalyses RuBP carboxylation and represents the entry point of CO_2 in the Calvin–Benson cycle, contributing to control the carbon fixation rate by the cell. RuBisCO has been also identified as rate-limiting enzyme and besides it solely belongs to such pathway, making it a promising target to engineer in order to push carbon fixation, without affecting other steps of the central metabolism (Fig. 6.3b) (Hudson et al. 1992; Rosgaard et al. 2012). Heterologous overexpression of RuBisCO was proven to increase carbon assimilation and consequent biomass growth in some cyanobacteria species (Iwaki et al. 2006; Liang and Lindblad 2017; Ruffing 2014). However, such strategy is not universal and its applicability is indeed challenged by the complexity of the maturation process of RuBisCO, showing several players involved and often not compatible with different hosts (Hauser et al. 2015; Hayer-Hartl 2017).

The potential positive impact of increasing RuBisCO accumulation can also be reduced by the enzyme intrinsic low specificity for CO_2. To overcome such limitations, a second strategy to push carbon fixation is thus to engineer RuBisCO to increase its specificity for CO_2 and consequently its carboxylation efficiency. However, several attempts in this direction in general yielded unsatisfactory results with

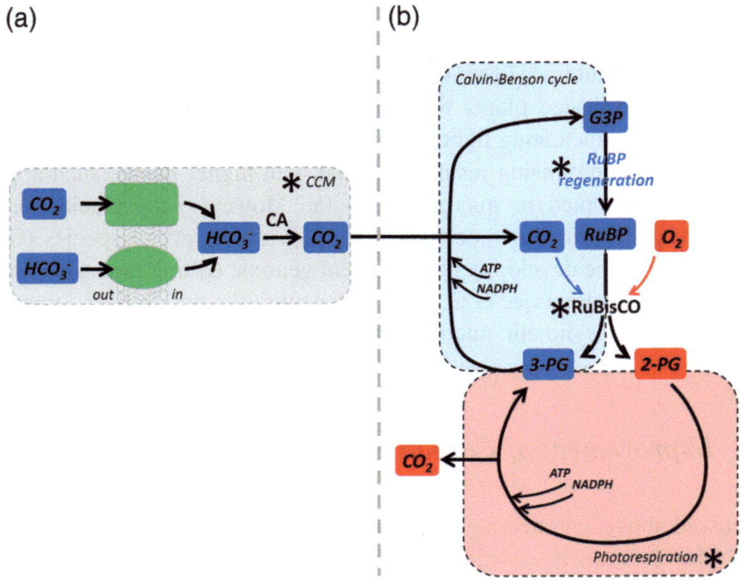

Fig. 6.3 Potential targets to improve microalgae carbon fixation efficiency. Basic operational components for microalgae CCM, carbon fixation (Calvin–Benson cycle) and photorespiration pathways are here depicted. (**a**) CCM in microalgae (in grey) includes active CO_2, HCO_3^- uptake systems, here indicated by two transmembrane transporters (i.e. green polygon and ellipse, respectively), and carbonic anhydrases (CA) to catalyse the conversion of HCO_3^- to CO_2. (**b**) CO_2 consequently enters the Calvin–Benson cycle (in blue) and is used to convert RuBP into 3-PG by RuBisCO. 3-PG is then converted in G3P to regenerate RuBP. RuBisCO can also process O_2, converting RuBP in 2-PG, a toxic metabolite which needs to be recycled through the photorespiratory pathway (in red). Both pathways consume ATP and NADPH, with photorespiration also releasing CO_2, directly counteracting RuBisCO activity. Reactions are not balanced. Targets for microalgae carbon fixation improvement are indicated by asterisks and are: introduction of more efficient CCM to increase CO_2 availability, increase RuBisCO affinity for CO_2 ad lower that for O_2, enhance RuBP regeneration rate, replace photorespiration with metabolic by-pass to curb carbon loss. *3-PG* 3-phosphoglycerate, *2-PG* 2-phosphoglycolate, *CA* carbonic anhydrase, *CCM* carbon concentration mechanism, *G3P* glyceraldehyde 3-phosphate, *RuBisCO* ribulose-1,5-bisphosphate carboxylase, *RuBP* ribulose-1,5-bisphosphate

even sometimes negative consequences on the homeostasis of the whole Calvin–Benson cycle (Carmo-Silva et al. 2015; Durall and Lindblad 2015; Parry et al. 2007). The most reasonable explanation for this lack of success is that this enzyme has already been strongly selected over billions of years of evolution (Badger and Bek 2008); therefore, it is unlikely that direct modifications may result in an increased activity.

An alternative strategy is thus to substitute RuBisCO with another carboxylating enzyme, with better kinetic properties and specificity for CO_2. However, finding good targets is challenged by the limitation to the small pool of currently characterized carboxylases and among them only to those processing the same substrates as RuBisCO, in order to allow an efficient integration with the natural cycle (Bar-Even

2018). The choice should also fall on enzymes that do not generate potential toxic metabolic intermediates, which should eventually be depleted by introducing accessory pathways, with the chance to develop an unaffordable metabolic burden. The identification process led so far to enzymes showing higher activity rates but lower specificity for CO_2 than RuBisCO, preventing their applicability (Cotton et al. 2018).

Overall efforts to improve RuBisCO carboxylation efficiency or replacement with more efficient alternative enzymes have shown many challenges and limited success. One potential alternative strategy is to improve the availability of its substrates (i.e. CO_2 and RuBP), an approach that might represent a viable alternative to indirectly push carboxylation efficiency (Fig. 6.3b). CO_2 availability might be increased by carbon concentrating mechanisms in the cell. Such approach could indeed make the difference during microalgae industrial cultivation, where cells may experience CO_2 limitations (see Sect. 6.2.3). While the introduction of microalgae-derived CCM into C3-plants chloroplasts was successfully proven to increase carbon fixation rate and biomass yield (Nölke et al. 2018), the possibility of increasing CO_2 availability in microalgae and cyanobacteria is still to be demonstrated (Rosgaard et al. 2012), because they already evolved efficient CCM to cope with the low solubility of CO_2 in the aquatic environment. However, a recent work showed that a *Chlorella* mutant with a constitutively active CCM achieves higher photosynthetic efficiency and faster growth than the parental strain, upon light saturating conditions, suggesting CCM optimization presents a valuable potential for improvement of photosynthetic microorganisms (Hwangbo et al. 2018).

RuBP availability is controlled by a series of enzymatic steps in the Calvin–Benson cycle, responsible for its regeneration after conversion of 3-PG in G3P (Fig. 6.3b). Among them, fructose-1,6 bisphosphatase (FBPase) and sedoheptulose-1,7-bisphosphatase (SBPase) in higher plants and fructose-1,6 bisphosphatase/sedoheptulos-1,7-bisphosphatase (BiBPase) in cyanobacteria have been identified as rate-limiting enzymes in such process (Ding et al. 2016; Feng et al. 2009). Speeding up RuBP regeneration rate might thus represent a second strategy to indirectly push carbon fixation (Fig. 6.3b). Overexpression of cyanobacterial FBPase in higher plants improved carbon assimilation and biomass yield (Ichikawa et al. 2010) while the same strategy in *C. reinhardtii* did not show any positive effect (Dejtisakdi and Miller 2016). On the contrary, heterologous overexpression of SBPase and BiBPase in microalgae and cyanobacteria resulted in improved photosynthesis and biomass yield (De Porcellinis et al. 2018; Fang et al. 2012; Liang and Lindblad 2016; Ogawa et al. 2015). These results highlight that indirect approaches to push RuBisCO carboxylation efficiency may be impactful, opening new opportunities for efficiently engineering the central carbon metabolism for directing carbon flux towards biomass accumulation.

6.3.2.2 Engineering Photorespiration

Photorespiration allows the Calvin–Benson cycle to operate in an oxygenic environment by recycling 2-PG eventually produced by RuBisCO oxygenase reaction with loss of chemical energy and production of CO_2 (Sect. 6.2.2.2). Photorespiration is however playing a major biological role, and in higher plants attempts to inactivate it resulted in the impossibility of growing at ambient CO_2 concentrations (Somerville 2001). Even in the presence of efficient CCM in C4-metabolism that lowers RuBisCO oxygenation, photorespiration is still not fully dispensable since it is required for the detoxification from 2-PG, i.e. an inhibitor of glycolysis (it inhibits triose phosphate isomerase) and of multiple enzymatic steps in the Calvin–Benson cycle itself (Flügel et al. 2017; Levey et al. 2018). A simple downregulation of the photorespiratory pathway is thus expected to prevent 2-PG detoxification, leading to strong growth defects as a consequence of the inhibition of carbon fixation (Levey et al. 2018).

However, rewiring photorespiration to minimize energy/reducing power consumption and carbon loss may represent a promising strategy to enhance carbon fixation efficiency, while preserving the ability to recycle 2-PG (Fig. 6.3b). This task is however challenged by the complexity of the pathway, involving enzymatic reactions spanning multiple cell compartments [chloroplast, peroxisome and mitochondrion in plant cells (Hagemann and Bauwe 2016)]. The introduction of metabolic by-pass, such as the bacterial glycerate pathway, in plants resulted in a reduced photorespiration rate and greater biomass accumulation (Dalal et al. 2015; Kebeish et al. 2007; Nölke et al. 2014), even though doubts have been raised about the full functionality of the exogenous pathway in the host (Bar-Even 2018; Hagemann and Bauwe 2016).

6.3.2.3 Synthetic Pathways for Carbon Fixation

Another strategy to push PBCE, might be to reduce energy/reducing power demand for carbon fixation. In terms of exploitation of cellular resources, the Calvin–Benson cycle is indeed the least efficient process out of the six known carbon fixing pathways in autotrophic organisms to date (Bar-Even 2018). It consumes ATP not only to activate thermodynamically unfavourable reactions but also to increase metabolic flux control. Theoretically it would be possible to operate the cycle replacing such reactions with no energy requiring alternatives, without compromising the carbon fixation efficiency while lowering its energetic cost (Bar-Even 2018). To do so, the whole Calvin–Benson cycle might be replaced with alternative pathways. Out of six known natural carbon-fixing pathways, only two are oxygen tolerant, but their compatibility with the host is challenged by the overlapping with the fatty-acid biosynthesis (Bar-Even 2018). Synthetic pathways may represent a solution to limit the disruption of native pathways, despite their design is a very complex task and is yet to be implemented. While the exploitation of synthetic pathways relying on existing enzymes is limited to only those able to process substrates available in the host, synthetic pathways employing new enzymes might represent preferable candidates to select novel and more efficient catalytic capacities, readily compatible with the target organism (Erb et al. 2017; Trudeau et al. 2018). However, further scientific efforts in both new enzymes design as well as in metabolic flux analyses in different hosts are seminal to achieve such goal.

6.4 In Silico Approaches to Drive Genetic Engineering of Photosynthetic Metabolism

6.4.1 Photosynthesis Metabolic Engineering and Industrial Cultivation

Genetic engineering approaches to rewire the photosynthetic metabolism have been extensively pursued in the last decades to domesticate microalgae to the PBR artificial environment (see Sects. 6.3.1 and 6.3.2). Overall, genetic engineering attempts performed so far proved that it is indeed feasible to get improvements in the photosynthetic metabolism performances by targeting both light and dark reactions of photosynthesis (see Sects. 6.3.1 and 6.3.2), but those strategies have never been attempted simultaneously so far, potentially limiting the achievable improvements in PBCE. This approach has been strongly curbed by the availability of laborious engineering methods, but the recent development of genome editing might contribute to open similar possibilities. Given light-use efficiency is the most important parameter to address to push microalgae phototrophic growth in industrial systems (Ruiz et al. 2016), major efforts have been put on tuning light reactions of photosynthesis so far (see Sects. 6.3.1.1 and 6.3.1.3), while engineering carbon fixation showed major limitations due to the complex set of reactions, often highly integrated with the central metabolism, and to the intrinsic low efficiency of crucial enzymes (e.g. RuBisCO), difficult to overcome while preserving overall carbon fixation homeostasis (see Sects. 6.3.2.1 and 6.3.2.2).

The strains isolated so far demonstrated interesting potential in small-scale trials, but when cultivated in intensive conditions simulating the PBR artificial environment, they led to contrasting results (Cazzaniga et al. 2014; De Mooij et al. 2014). Selected mutations on light reactions can indeed lead to intrinsic negative consequences on cells physiological, potentially hindering the benefits of a more uniform light distribution.

As discussed in Sect. 6.2.2.1, LHC proteins, the main target pursued to tune light reactions so far, play a complex biological role and are not purely responsible for light-harvesting but are also involved in NPQ photo-regulation (Büchel 2015). Therefore, an indiscriminate reduction in light-harvesting complexes might also affect photo-protection capability, leading to increased photoinhibition when cells are exposed to saturating irradiances. Such negative consequences of antenna reduction might however be curbed by balancing the benefits of lowering light-harvesting with the drawbacks of reducing photo-protection (Bonente et al. 2011; Simionato et al. 2013). The development of effective genome editing approaches is opening the way to direct modifications only to LHC proteins specifically involved in light harvesting, also balancing their reduction between the two PS, consequently avoiding undesired photo-regulatory side effects (Banerjee et al. 2018; Ferenczi et al. 2017; Jeong et al. 2018; Nymark et al. 2016; Poliner et al. 2018; Verruto et al. 2018).

Targeted engineering approaches are however only the first step since they cannot guarantee the successful improvement in biomass productivities upon large-scale cultivation in PBR. Generated mutants have been so far tested in small-scale experiments, while their performances in larger-scale conditions are still to be addressed. The cultivation environment was indeed demonstrated to strongly influence biological properties and performances of genetic engineered microalgae strains, pointing out how industrial operational conditions may enhance or reduce the advantages of modifications isolated in the lab (Perin et al. 2017b). Moreover, this picture is expected to be complicated even further when such modifications will be implemented with a rewired carbon fixation. Identifying the right genetic targets in PBR is clearly non-trivial, as it depends on many environmental parameters. As well as the environmental conditions, parameters related to the PBR design and operation such as light path, culture density, mixing rate, temperature and pH might have a strong influence because controlling light, CO_2 and nutrients availability. It is therefore impossible to extrapolate the impact of a genetic modification in the PBR environment from simple lab-scale evaluations. Given the complexity due to the interaction between genotype and productivity, it is also highly likely that there will be no superior genetically modified strain with superior performance in all circumstances, but instead, the genetic modification ought to be tailored to the local environment and PBR characteristics.

6.4.2 Mathematical Models to Direct Metabolic Engineering of Photosynthesis

Addressing the effect of PBR variables on microalgae performances via trial-and-error approaches is both time-consuming and expensive, limiting the currently available knowledge to drive effective metabolic engineering strategies in a significant environment of cultivation. For this reason, all tools capable of predicting the effect of PBR parameters on microalgae physiology are extremely valuable to direct genetic engineering a priori towards those traits with the largest impact on productivity in an industrial cultivation environment. This objective however hinges on our ability to develop computational platforms able to quantitatively describe the effect of multiple operational parameters as well as their mutual interactions on specific biological functions. Computer-based approaches represent valuable tools for this goal. They indeed allow to describe biological mechanisms with equations and parameters, thereby generating a correlation between quantitative information (i.e., change in light/CO_2 availability) and specific biological functions (i.e., what are the consequences on photosynthesis/carbon fixation?). Moreover, sensitivity analyses (Saltelli et al. 2007) applied to a multi-scale modelling approach could evaluate the effect of such parameters on key process outputs (e.g. biomass productivity) to assess whether a specific intervention line is deemed probably successful or not. This strategy finds direct application in metabolic engineering, aiming to prevent major mistakes in the prioritization of genetic interventions (e.g. when aim for antenna size reduction or rather rewire quenching mechanisms?; when rewire photorespiration?).

Given photosynthetic activity is the main driver of microalgae growth in PBR (Ruiz et al. 2016), modelling the effect of light availability on photosynthetic performances has been the prioritized target so far to reliably describe microalgae growth during industrial cultivation. Models describing photosynthesis rate as a function of the optimal irradiation have been extensively developed in the past (Eilers and Peeters 1988; Rubio et al. 2003) and further implemented to predict microalgae growth (Béchet et al. 2015; Bernardi et al. 2014). However, modelling photoproduction (i.e. O_2 evolution) in sub-optimal light environments (i.e. limiting or saturating conditions) is complicated by the need to include biological phenomena (i.e. photoregulation) spanning numerous time-scales, from milliseconds to days. Several mathematical models have thus been implemented to also include photoinhibitory and photoregulatory phenomena (Bernardi et al. 2016; Nikolaou et al. 2015), thus expanding models prediction capability to multiple steady environmental conditions.

However, without a reliable simulator than can account for scale-up effects, it may not be possible to reliably predict genetic modifications consequences on a large-scale culture system. In fact, this has been a common issue with several existing simple kinetic models of mutants which have been used to predict growth under large-scale culture conditions (Kirst et al. 2014; Nakajima and Itayama 2003; Ort and Melis 2011) but failed to describe the experimental observations (De Mooij et al. 2014). In such environment, the overall growth depends from the contribution of cells populating different culture layers, each experiencing different irradiances, as a consequence of light attenuation when culture depth increases (Fig. 6.1) (Yun and Park 2001).

Therefore, models which describe photosynthetic productivity as a function of light intensity, also taking into account the light attenuation profile in PBR, are the preferable tools to develop, in order to effectively predict microalgae growth in PBR (Perin et al. 2017a). Such models account for a mathematical description of the major parameters controlling light-use efficiency, leaving room to the possibility to rewire them in silico in order to predict the effect of the targets listed in Fig. 6.2. This way, models can contribute to determine which combinations of modifications are likely to lead to improved performances at minimum cost, and how to design/operate a PBR in order to get the most out of the mutants (Perin et al. 2017a, 2018).

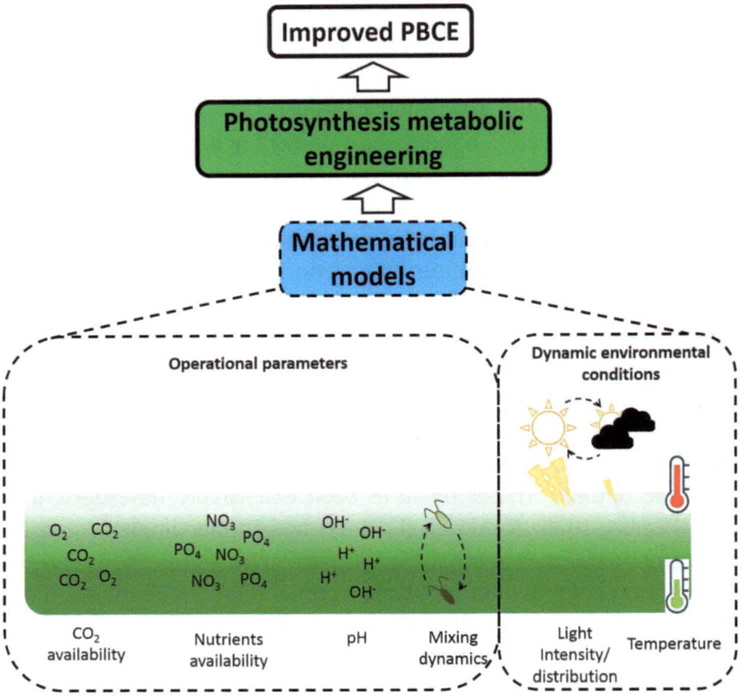

Fig. 6.4 Mathematical models for effectively engineering photosynthetic metabolism. Several genetic engineering strategies have already been successful to improve microalgae photosynthesis in lab-scale conditions. Taking into account the complexity of the final cultivation environment is, however, seminal to make such modifications effective also in industrial plants. Here, photosynthetic metabolism dynamics are affected by operational parameters (i.e. CO_2, nutrients availability, pH and mixing) and dynamic environmental conditions (i.e. light intensity/distribution and temperature) whose contribution on growth is difficult to predict, especially when mutants come into play. Mathematical models allow evaluating in silico how photosynthetic metabolism responds to such parameters, thus enabling to modify the photosynthesis targets predicted to have the major impact on PBCE

The application of such tools is however still limited by their comprehensiveness (i.e. the inclusion of all physical-chemical and biological phenomena controlling growth). The majority of currently available models indeed assume that growth strictly depends on light availability, limiting their prediction potential only to those conditions where CO_2 and nutrients are provided in excess.

In order to expand their prediction capability, models will need to be implemented to account for the biological consequences of the operational parameters adopted in PBR (Fig. 6.4), where changing CO_2/nutrients/light availabilities and pH/mixing dynamics are expected to affect growth. Moreover, outdoor PBR are also exposed to dynamic environmental conditions, leading cells to face further changes in temperature and light supply (Fig. 6.4), with a strong impact on cells physiology, as it was clearly demonstrated for both cyanobacteria and microalgae (Bamba et al. 2015; Kirst et al. 2014; Page et al. 2012; Perin et al. 2017b).

Understanding the impact that such parameters have on algae physiology will be highly strategic to speed up the development of more reliable models. Microscale devices are being developed for such aim, allowing to investigate the effect of multiple parameters simultaneously saving time and costs (Kim et al. 2014; Luke et al. 2016; Perin et al. 2016).

In this context, the effectiveness of such modelling approaches also relies on how easily they could be applied to different species (i.e. transferability). The regulatory network of microalgae photosynthetic metabolism is likely to be species-specific, complicating the development of universal models. However, models using general equations describing photosynthetic traits shared among several microalgae, allow for an easy re-parametrization with new experimental data, enabling the transfer to other species.

6.5 Conclusions

Microalgae represent a promising sustainable source of biomass with many potential future industrial applications. This potential is however still largely untapped because of many technological barriers, with photon-to-biomass conversion inefficiency being one of the main challenges to address. Metabolic engineering of photosynthesis is a critical objective in this regard, and the current state of the art demonstrates that significant improvements are achievable but not yet realized. Genetic engineering strategies targeting both light-harvesting and carbon fixation efficiency are likely to be more successful and yield more robust solutions in real conditions. Considering the metabolic complexity of the photosynthetic conversion of light energy into biomass and the number of parameters involved, mathematical models are emerging as strategic tools to address this challenge, improving effectiveness of metabolic engineering efforts.

References

Acién Fernández FG, Alías CB, García-Malea López MC, Fernández Sevilla JM, Ibáñez González MJ, Gómez RN, Molina Grima E (2003) Assessment of the production of 13C labeled compounds from phototrophic microalgae at laboratory scale. Biomol Eng 20:149–162

Alboresi A, Storti M, Morosinotto T (2018) Balancing protection and efficiency in the regulation of photosynthetic electron transport across plant evolution. New Phytol 1859:676–683

Badger MR, Bek EJ (2008) Multiple Rubisco forms in proteobacteria: their functional significance in relation to CO_2 acquisition by the CBB cycle. J Exp Bot 59:1525–1541

Bamba BSB, Lozano P, Adjé F, Ouattara A, Vian MA, Tranchant C, Lozano Y (2015) Effects of temperature and other operational parameters on *Chlorella vulgaris* mass cultivation in a simple and low-cost column photobioreactor. Appl Biochem Biotechnol 177:389–406

Banerjee A, Banerjee C, Negi S, Chang J-S, Shukla P (2018) Improvements in algal lipid production: a systems biology and gene editing approach. Crit Rev Biotechnol 38:369–385

Barber J, Andersson B (1992) Too much of a good thing: light can be bad for photosynthesis. Trends Biochem Sci 17:61–66

Bar-Even A (2018) Daring metabolic designs for enhanced plant carbon fixation. Plant Sci 273:71–83

Bar-Even A, Noor E, Savir Y, Liebermeister W, Davidi D, Tawfik DS, Milo R (2011) The moderately efficient enzyme: evolutionary and physicochemical trends shaping enzyme parameters. Biochemistry 50:19

Béchet Q, Chambonnière P, Shilton A, Guizard G, Guieysse B (2015) Algal productivity modeling: a step toward accurate assessments of full-scale algal cultivation. Biotechnol Bioeng 112:987–996

Beckmann J, Lehr F, Finazzi G, Hankamer B, Posten C, Wobbe L, Kruse O (2009) Improvement of light to biomass conversion by de-regulation of light-harvesting protein translation in *Chlamydomonas reinhardtii*. J Biotechnol 142:70–77

Bernardi A, Perin G, Sforza E, Galvanin F, Morosinotto T, Bezzo F (2014) An identifiable state model to describe light intensity influence on microalgae growth. Ind Eng Chem Res 53:6738–6749

Bernardi A, Meneghesso A, Morosinotto T, Bezzo F (2016) A model-based investigation of genetically modified microalgae strains. In: Computer aided chemical engineering

Berteotti S, Ballottari M, Bassi R (2016) Increased biomass productivity in green algae by tuning non-photochemical quenching. Sci Rep 6:21339

Bilal M, Rasheed T, Ahmed I, Iqbal HMN, Sada EG (2017) High-value compounds from microalgae with industrial exploitability – a review. Front Biosci 9:319–342

Blankenship RE, Tiede DM, Barber J, Brudvig GW, Fleming G, Ghirardi M, Gunner MR, Junge W, Kramer DM, Melis A, Moore TA, Moser CC, Nocera DG, Nozik AJ, Ort DR, Parson WW, Prince RC, Sayre RT (2011) Comparing photosynthetic and photovoltaic efficiencies and recognizing the potential for improvement. Science 332:805–809

Bonente G, Formighieri C, Mantelli M, Catalanotti C, Giuliano G, Morosinotto T, Bassi R (2011) Mutagenesis and phenotypic selection as a strategy toward domestication of *Chlamydomonas reinhardtii* strains for improved performance in photobioreactors. Photosynth Res 108:107–120

Brennan L, Owende P (2010) Biofuels from microalgae—a review of technologies for production, processing, and extractions of biofuels and co-products. Renew Sust Energ Rev 14:557–577

Büchel C (2015) Evolution and function of light harvesting proteins. J Plant Physiol 172:62–75

Bule MH, Ahmed I, Maqbool F, Bilal M, Iqbal HMN (2018) Microalgae as a source of high-value bioactive compounds. Front Biosci (Schol Ed) 10:197–216

Cardol P, Forti G, Finazzi G (2011) Regulation of electron transport in microalgae. Biochim Biophys Acta 1807:912–918

Cardona T, Shao S, Nixon PJ (2018) Enhancing photosynthesis in plants: the light reactions. Essays Biochem 62:85–94

Carmo-Silva E, Scales JC, Madgwick PJ, Parry MAJ (2015) Optimizing Rubisco and its regulation for greater resource use efficiency. Plant Cell Environ 38:1817–1832

Carneiro MLNM, Pradelle F, Braga SL, Gomes MSP, Martins ARFA, Turkovics F, Pradelle RNC (2017) Potential of biofuels from algae: comparison with fossil fuels, ethanol and biodiesel in Europe and Brazil through life cycle assessment (LCA). Renew Sust Energ Rev 73:632–653

Carvalho AP, Silva SO, Baptista JM, Malcata FX (2011) Light requirements in microalgal photobioreactors: an overview of biophotonic aspects. Appl Microbiol Biotechnol 89:1275–1288

Cazzaniga S, Dall'Osto L, Szaub J, Scibilia L, Ballottari M, Purton S, Bassi R (2014) Domestication of the green alga *Chlorella sorokiniana*: reduction of antenna size improves light-use efficiency in a photobioreactor. Biotechnol Biofuels 7:157

Cheah WY, Show PL, Chang J-S, Ling TC, Juan JC (2015) Biosequestration of atmospheric CO_2 and flue gas-containing CO_2 by microalgae. Bioresour Technol 184:190–201

Chen M, Blankenship RE (2011) Expanding the solar spectrum used by photosynthesis. Trends Plant Sci 16:427–431

Cotton CAR, Douglass JS, De Causmaecker S, Brinkert K, Cardona T, Fantuzzi A, Rutherford AW, Murray JW (2015) Photosynthetic constraints on fuel from microbes. Front Bioeng Biotechnol 3:36

Cotton CA, Edlich-Muth C, Bar-Even A (2018) Reinforcing carbon fixation: CO_2 reduction replacing and supporting carboxylation. Curr Opin Biotechnol 49:49–56

Dalal J, Lopez H, Vasani NB, Hu Z, Swift JE, Yalamanchili R, Dvora M, Lin X, Xie D, Qu R, Sederoff HW (2015) A photorespiratory bypass increases plant growth and seed yield in biofuel crop *Camelina sativa*. Biotechnol Biofuels 8:175

De Mooij T, Janssen M, Cerezo-Chinarro O, Mussgnug JH, Kruse O, Ballottari M, Bassi R, Bujaldon S, Wollman F-A, Wijffels RH (2014) Antenna size reduction as a strategy to increase biomass productivity: a great potential not yet realized. J Appl Phycol 27:1063–1077

De Porcellinis AJ, Nørgaard H, Brey LMF, Erstad SM, Jones PR, Heazlewood JL, Sakuragi Y (2018) Overexpression of bifunctional fructose-1,6-bisphosphatase/sedoheptulose-1,7-bisphosphatase leads to enhanced photosynthesis and global reprogramming of carbon metabolism in Synechococcus sp. PCC 7002. Metab Eng 47:170–183

De Vree JH, Bosma R, Janssen M, Barbosa MJ, Wijffels RH (2015) Comparison of four outdoor pilot-scale photobioreactors. Biotechnol Biofuels 8:1–12

Dejtisakdi W, Miller SM (2016) Overexpression of Calvin cycle enzyme fructose 1,6-bisphosphatase in *Chlamydomonas reinhardtii* has a detrimental effect on growth. Algal Res 14:116–126

Ding F, Wang M, Zhang S, Ai X (2016) Changes in SBPase activity influence photosynthetic capacity, growth, and tolerance to chilling stress in transgenic tomato plants. Sci Rep 6:32741

Durall C, Lindblad P (2015) Mechanisms of carbon fixation and engineering for increased carbon fixation in cyanobacteria. Algal Res 11:263–270

Eberhard S, Finazzi G, Wollman F-A (2008) The dynamics of photosynthesis. Annu Rev Genet 42:463–515

Eilers P, Peeters J (1988) A Model for the reletionship between light-intensity and the rate of photosynthesis in phytoplankton. Ecol Model 42:199–215

Erb TJ, Jones PR, Bar-Even A (2017) Synthetic metabolism: metabolic engineering meets enzyme design. Curr Opin Chem Biol 37:56–62

Fang L, Lin HX, Low CS, Wu MH, Chow Y, Lee YK (2012) Expression of the *Chlamydomonas reinhardtii* sedoheptulose-1,7-bisphosphatase in *Dunaliella bardawil* leads to enhanced photosynthesis and increased glycerol production. Plant Biotechnol J 10:1129–1135

Fargione J, Hill J, Tilman D, Polasky S, Hawthorne P (2008) Land clearing and the biofuel carbon debt. Science 319:1235–1238

Feng L, Li H, Jiao J, Li D, Zhou L, Wan J, Li Y (2009) Reduction in SBPase activity by antisense RNA in transgenic rice plants: effect on photosynthesis, growth, and biomass allocation at different nitrogen levels. J Plant Biol 52:382–394

Ferenczi A, Euan D, Xipnitou A, Molnar A (2017) Efficient targeted DNA editing and replacement in *Chlamydomonas reinhardtii* using Cpf1 ribonucleoproteins and single-stranded DNA. PNAS 114:13567–13572

Flügel F, Timm S, Arrivault S, Florian A, Stitt M, Fernie AR, Bauwe H (2017) The photorespiratory metabolite 2-phosphoglycolate regulates photosynthesis and starch accumulation in arabidopsis. Plant Cell 29:2537–2551

Formighieri C, Franck F, Bassi R (2012) Regulation of the pigment optical density of an algal cell: filling the gap between photosynthetic productivity in the laboratory and in mass culture. J Biotechnol 162:115–123

Gee CW, Niyogi KK (2017) The carbonic anhydrase CAH1 is an essential component of the carbon-concentrating mechanism in *Nannochloropsis oceanica*. Proc Natl Acad Sci 114:4537–4542

Gerotto C, Alboresi A, Meneghesso A, Jokel M, Suorsa M, Aro E-MEM, Morosinotto T (2016) Flavodiiron proteins act as safety valve for electrons in *Physcomitrella patens*. Proc Natl Acad Sci USA 113:12322–12327

Gies E (2017) The real cost of energy. Nature 551:S145–S147
Giordano M, Beardall J, Raven JA (2005) CO_2 concentrating mechanisms in algae: mechanisms, environmental modulation, and evolution. Annu Rev Plant Biol 56:99–131
Gómez R, Carrillo N, Morelli MP, Tula S (2018) Faster photosynthetic induction in tobacco by expressing cyanobacterial flavodiiron proteins in chloroplasts. Photosynth Res 136:129–138
Goss R, Jakob T (2010) Regulation and function of xanthophyll cycle-dependent photoprotection in algae. Photosynth Res 106:103–122
Goss R, Lepetit B (2015) Biodiversity of NPQ. J Plant Physiol 172C:13–32
Hagemann M, Bauwe H (2016) Photorespiration and the potential to improve photosynthesis. Curr Opin Chem Biol 35:109–116
Hauser T, Popilka L, Hartl FU, Hayer-Hartl M (2015) Role of auxiliary proteins in Rubisco biogenesis and function. Nat Plants 1:15065
Hayer-Hartl M (2017) From chaperonins to Rubisco assembly and metabolic repair. Protein Sci 26:2324–2333
Ho M-Y, Shen G, Canniffe DP, Zhao C, Bryant DA (2016) Light-dependent chlorophyll f synthase is a highly divergent paralog of PsbA of photosystem II. Science 353:aaf9178–aaf9178
Hudson GS, Evans JR, von Caemmerer S, Arvidsson YB, Andrews TJ (1992) Reduction of ribulose-1,5-bisphosphate carboxylase/oxygenase content by antisense RNA reduces photosynthesis in transgenic tobacco plants. Plant Physiol 98:294–302
Hwangbo K, Lim J-M, Jeong S-W, Vikramathithan J, Park Y-I, Jeong W-J (2018) Elevated inorganic carbon concentrating mechanism confers tolerance to high light in an arctic Chlorella sp. ArM0029B. Front Plant Sci 9:590
Ichikawa Y, Tamoi M, Sakuyama H, Maruta T, Ashida H, Yokota A, Shigeoka S (2010) Generation of transplastomic lettuce with enhanced growth and high yield. GM Crops 1:322–326
Iwaki T, Haranoh K, Inoue N, Kojima K, Satoh R, Nishino T, Wada S, Ihara H, Tsuyama S, Kobayashi H, Wadano A (2006) Expression of foreign type I ribulose-1,5-bisphosphate carboxylase/oxygenase (EC 4.1.1.39) stimulates photosynthesis in cyanobacterium Synechococcus PCC7942 cells. Photosynth Res 88:287–297
Jeong J, Baek K, Yu J, Kirst H, Betterle N, Shin W (2018) Deletion of the chloroplast LTD protein impedes LHCI import and PSI – LHCI assembly in *Chlamydomonas reinhardtii*. J Exp Bot 69:1147–1158
Kato Y, Sun X, Zhang L, Sakamoto W (2012) Cooperative D1 degradation in the photosystem II repair mediated by chloroplastic proteases in Arabidopsis. Plant Physiol 159:1428–1439
Kebeish R, Niessen M, Thiruveedhi K, Bari R, Hirsch H-J, Rosenkranz R, Stäbler N, Schönfeld B, Kreuzaler F, Peterhänsel C (2007) Chloroplastic photorespiratory bypass increases photosynthesis and biomass production in *Arabidopsis thaliana*. Nat Biotechnol 25:593–599
Kim HS, Weiss TL, Thapa HR, Devarenne TP, Han A (2014) A microfluidic photobioreactor array demonstrating high-throughput screening for microalgal oil production. Lab Chip 14:1415–1425
Kirk J (1994) Light and photosynthesis in aquatic ecosystems, 2nd edn. Cambridge University Press, Cambridge
Kirst H, Melis A (2014) The chloroplast signal recognition particle (CpSRP) pathway as a tool to minimize chlorophyll antenna size and maximize photosynthetic productivity. Biotechnol Adv 32:66–72
Kirst H, García-Cerdán JG, Zurbriggen A, Melis A (2012a) Assembly of the light-harvesting chlorophyll antenna in the green alga *Chlamydomonas reinhardtii* requires expression of the TLA2-CpFTSY gene. Plant Physiol 158:930–945
Kirst H, Garcia-Cerdan JG, Zurbriggen A, Ruehle T, Melis A (2012b) Truncated photosystem chlorophyll antenna size in the green microalga *Chlamydomonas reinhardtii* upon deletion of the TLA3-CpSRP43 gene. Plant Physiol 160:2251–2260
Kirst H, Formighieri C, Melis A (2014) Maximizing photosynthetic efficiency and culture productivity in cyanobacteria upon minimizing the phycobilisome light-harvesting antenna size. Biochim Biophys Acta 1837:1653–1664

Koutra E, Economou CN, Tsafrakidou P, Kornaros M (2018) Bio-based products from microalgae cultivated in digestates. Trends Biotechnol 36:819–833
Kromdijk J, Głowacka K, Leonelli L, Gabilly ST, Iwai M, Niyogi KK, Long SP (2016) Improving photosynthesis and crop productivity by accelerating recovery from photoprotection. Science 354:857–861
Kuczynska P, Jemiola-Rzeminska M, Strzalka K (2015) Photosynthetic pigments in diatoms. Mar Drugs 13:5847–5881
Lam GP, Vermuë MH, Eppink MHM, Wijffels RH, Van Den Berg C (2018) Multi-product microalgae biorefineries: from concept towards reality. Trends Biotechnol 36:216–227
Larosa V, Meneghesso A, La Rocca N, Steinbeck J, Hippler M, Szabò I, Morosinotto T (2018) Mitochondria affect photosynthetic electron transport and photosensitivity in a green alga. Plant Physiol 176:2305–2314
Levey M, Timm S, Mettler-Altmann T, Luca Borghi G, Koczor M, Arrivault S, Pm Weber A, Bauwe H, Gowik U, Westhoff P (2018) Efficient 2-phosphoglycolate degradation is required to maintain carbon assimilation and allocation in the C4 plant *Flaveria bidentis*. J Exp Bot 70:575–587
Liang F, Lindblad P (2016) Effects of overexpressing photosynthetic carbon flux control enzymes in the cyanobacterium Synechocystis PCC 6803. Metab Eng 38:56–64
Liang F, Lindblad P (2017) Synechocystis PCC 6803 overexpressing RuBisCO grow faster with increased photosynthesis. Metab Eng Commun 4:29–36
Luke CS, Selimkhanov J, Baumgart L, Cohen SE, Golden SS, Cookson NA, Hasty J (2016) A microfluidic platform for long-term monitoring of algae in a dynamic environment. ACS Synth Biol 5:8–14
Maeda Y, Yoshino T, Matsunaga T, Matsumoto M, Tanaka T (2018) ScienceDirect Marine microalgae for production of biofuels and chemicals. Curr Opin Biotechnol 50:111–120
Melis A (1991) Dynamics of photosynthetic membrane composition and function. BBA-Bioenergetics 1058:87–106
Melis A (2009) Solar energy conversion efficiencies in photosynthesis: minimizing the chlorophyll antennae to maximize efficiency. Plant Sci 177:272–280
Melis A, Neidhardt J, Benemann JR (1998) *Dunaliella salina* (Chlorophyta) with small chlorophyll antenna sizes exhibit higher photosynthetic productivities and photon use efficiencies than normally pigmented cells. J Appl Phycol 10:515–525
Meneghesso A, Simionato D, Gerotto C, La Rocca N, Finazzi G, Morosinotto T (2016) Photoacclimation of photosynthesis in the Eustigmatophycean *Nannochloropsis gaditana*. Photosynth Res 129:291–305
Molina E, Ferna J, Acie FG, Chisti Y (2001) Tubular photobioreactor design for algal cultures. J Biotechnol 92:113–131
Murchie EH, Niyogi KK (2011) Manipulation of photoprotection to improve plant photosynthesis. Plant Physiol 155:86–92
Nakajima Y, Itayama T (2003) Analysis of photosynthetic productivity of microalgal mass cultures. J Appl Phycol 15:497–505
Nakajima Y, Ueda R (1997) Improvement of photosynthesis in dense microalgal suspension by reduction of light harvesting pigments. J Appl Phycol 9:503–510
Nakajima Y, Ueda R (2000) The effect of reducing light-harvesting pigment on marine microalgal productivity. J Appl Phycol 12:285–290
Nakajima Y, Tsuzuki M, Ueda R (2001) Improved productivity by reduction of the content of light-harvesting pigment in *Chlamydomonas perigranulata*. J Appl Phycol 13:95–101
Ng I, Tan S, Kao P, Chang Y, Chang J (2017) Recent developments on genetic engineering of microalgae for biofuels and bio-based chemicals. Biotechnol J 1600644:1–13
Nikolaou A, Bernardi A, Meneghesso A, Bezzo F, Morosinotto T, Chachuat B (2015) A model of chlorophyll fluorescence in microalgae integrating photoproduction, photoinhibition and photoregulation. J Biotechnol 194:91–99

Nilkens M, Kress E, Lambrev P, Miloslavina Y, Müller M, Holzwarth AR, Jahns P (2010) Identification of a slowly inducible zeaxanthin-dependent component of non-photochemical quenching of chlorophyll fluorescence generated under steady-state conditions in Arabidopsis. Biochim Biophys Acta 1797:466–475

Nölke G, Houdelet M, Kreuzaler F, Peterhänsel C, Schillberg S (2014) The expression of a recombinant glycolate dehydrogenase polyprotein in potato (*Solanum tuberosum*) plastids strongly enhances photosynthesis and tuber yield. Plant Biotechnol J 12:734–742

Nölke G, Barsoum M, Houdelet M, Arcalís E, Kreuzaler F, Fischer R, Schillberg S (2018) The integration of algal carbon concentration mechanism components into tobacco chloroplasts increases photosynthetic efficiency and biomass. Biotechnol J 14:e1800170

Nymark M, Sharma AK, Sparstad T, Bones AM, Winge P (2016) A CRISPR/Cas9 system adapted for gene editing in marine algae. Sci Rep 6:24951

Ogawa T, Tamoi M, Kimura A, Mine A, Sakuyama H, Yoshida E, Maruta T, Suzuki K, Ishikawa T, Shigeoka S (2015) Enhancement of photosynthetic capacity in *Euglena gracilis* by expression of cyanobacterial fructose-1,6−/sedoheptulose-1,7-bisphosphatase leads to increases in biomass and wax ester production. Biotechnol Biofuels 8:80

Ooms MD, Dinh CT, Sargent EH, Sinton D (2016) Photon management for augmented photosynthesis. Nat Commun 7:12699

Ort DR, Melis A (2011) Optimizing antenna size to maximize photosynthetic efficiency. Plant Physiol 155:79–85

Ort DR, Merchant SS, Alric J, Barkan A, Blankenship RE, Bock R, Croce R, Hanson MR, Hibberd JM, Long SP, Moore TA, Moroney J, Niyogi KK, Parry MAJ, Peralta-Yahya PP, Prince RC, Redding KE, Spalding MH, van Wijk KJ, Vermaas WFJ, von Caemmerer S, Weber APM, Yeates TO, Yuan JS, Zhu XG (2015) Redesigning photosynthesis to sustainably meet global food and bioenergy demand. Proc Natl Acad Sci USA 112:8529–8536

Page LE, Liberton M, Pakrasi HB (2012) Reduction of photoautotrophic productivity in the cyanobacterium Synechocystis sp. strain PCC 6803 by phycobilisome antenna truncation. Appl Environ Microbiol 78:6349–6351

Park H, Lee C (2016) Theoretical calculations on the feasibility of microalgal biofuels: utilization of marine resources could help realizing the potential of microalgae. Biotechnol J 11:1461–1470

Parry MAJ, Madgwick PJ, Carvalho JFC, Andralojc PJ (2007) Prospects for increasing photosynthesis by overcoming the limitations of Rubisco. J Agric Sci 145:31

Perin G, Bellan A, Segalla A, Meneghesso A, Alboresi A, Morosinotto T (2015) Generation of random mutants to improve light-use efficiency of *Nannochloropsis gaditana* cultures for biofuel production. Biotechnol Biofuels 8:161

Perin G, Cimetta E, Monetti F, Morosinotto T, Bezzo F (2016) Novel micro-photobioreactor design and monitoring method for assessing microalgae response to light intensity. Algal Res 19:69–76

Perin G, Bernardi A, Bellan A, Bezzo F, Morosinotto T (2017a) A mathematical model to guide genetic engineering of photosynthetic metabolism. Metab Eng 44:337–347

Perin G, Simionato D, Bellan A, Carone M, Occhipinti A, Maffei ME, Morosinotto T (2017b) Cultivation in industrially relevant conditions has a strong influence on biological properties and performances of *Nannochloropsis gaditana* genetically modified strains. Algal Res 28:88–99

Perin G, Bellan A, Bernardi A, Bezzo F, Morosinotto T (2018) The potential of quantitative models to improve microalgae photosynthetic efficiency. Physiol Plant 166:380–391

Poliner E, Takeuchi T, Du ZY, Benning C, Farré EM (2018) Nontransgenic marker-free gene disruption by an episomal CRISPR system in the oleaginous microalga, *Nannochloropsis oceanica* CCMP1779. ACS Synth Biol 7:962–968

Polle JEW, Benemann JR, Tanaka A, Melis A (2000) Photosynthetic apparatus organization and function in the wild type and a chlorophyll b-less mutant of *Chlamydomonas reinhardtii*. Dependence on carbon source. Planta 211:335–344

Polle JEW, Kanakagiri S-DD, Melis A (2003) tla1, a DNA insertional transformant of the green alga *Chlamydomonas reinhardtii* with a truncated light-harvesting chlorophyll antenna size. Planta 217:49–59

Ramos Tercero EA, Sforza E, Morandini M, Bertucco A (2014) Cultivation of *Chlorella protothecoides* with urban wastewater in continuous photobioreactor: biomass productivity and nutrient removal. Appl Biochem Biotechnol 172:1470–1485
Renuka N, Guldhe A, Prasanna R, Singh P, Bux F (2018) Microalgae as multi-functional options in modern agriculture: current trends, prospects and challenges. Biotechnol Adv 36:1255–1273
Riebesell U, Wolf-Gladrow DA, Smetacek V (1993) Carbon dioxide limitation of marine phytoplankton growth rates. Nature 361:249–251
Robertson GP, Hamilton SK, Barham BL, Dale BE, Izaurralde RC, Jackson RD, Landis DA, Swinton SM, Thelen KD, Tiedje JM (2017) Cellulosic biofuel contributions to a sustainable energy future: choices and outcomes. Science 356:eaal2324
Rochaix J-D (2014) Regulation and dynamics of the light-harvesting system. Annu Rev Plant Biol 65:287–309
Rockström J, Gaffney O, Rogelj J, Meinshausen M, Nakicenovic N, Schellnhuber HJ (2017) A roadmap for rapid decarbonization. Science 355:1269–1271
Rosgaard L, de Porcellinis AJ, Jacobsen JH, Frigaard N-U, Sakuragi Y (2012) Bioengineering of carbon fixation, biofuels, and biochemicals in cyanobacteria and plants. J Biotechnol 162:134–147
Rubio FC, Camacho FG, Sevilla JMF, Chisti Y, Grima EM (2003) A mechanistic model of photosynthesis in microalgae. Biotechnol Bioeng 81:459–473
Ruffing AM (2014) Improved free fatty acid production in cyanobacteria with Synechococcus sp. PCC 7002 as host. Front Bioeng Biotechnol 2:17
Ruiz J, Olivieri G, de Vree J, Bosma R, Willems P, Reith JH, Eppink MHM, Kleinegris DMM, Wijffels RH, Barbosa MJ (2016) Towards industrial products from microalgae. Energy Environ Sci 9:3036–3043
Ruiz-Sola MÁ, Petroutsos D (2018) A toolkit for the characterization of the photoprotective capacity of green algae. Methods Mol Biol (Clifton, NJ) 1829:315–323
Rulli MC, Bellomi D, Cazzoli A, De Carolis G, Odorico PD (2016) The water-land-food nexus of first-generation biofuels. Sci Rep 6:22521
Salih FM (2011) Microalgae tolerance to high concentrations of carbon dioxide: a review. J Environ Prot (Irvine, Calif) 02:648–654
Saltelli A, Ratto M, Andres T, Campolongo F, Cariboni J, Gatelli D, Saisana M, Tarantola S (2007) Global sensitivity analysis. The primer, global sensitivity analysis. The primer. Wiley, Chichester
Santos FM, Pires JCM (2018) Nutrient recovery from wastewaters by microalgae and its potential application as bio-char. Bioresour Technol 267:725–731
Savir Y, Noor E, Milo R, Tlusty T (2010) Cross-species analysis traces adaptation of Rubisco toward optimality in a low-dimensional landscape. Proc Natl Acad Sci 107:3475–3480
Sharon-Gojman R, Leu S, Zarka A (2017) Antenna size reduction and altered division cycles in self-cloned, marker-free genetically modified strains of *Haematococcus pluvialis*. Algal Res 28:172–183
Shin W-S, Lee B, Jeong B, Chang YK, Kwon J-H (2016) Truncated light-harvesting chlorophyll antenna size in *Chlorella vulgaris* improves biomass productivity. J Appl Phycol 28:3193–3202
Simionato D, Basso S, Giacometti GMGM, Morosinotto T (2013) Optimization of light use efficiency for biofuel production in algae. Biophys Chem 182:71–78
Somerville CR (2001) An early Arabidopsis demonstration. Resolving a few issues concerning photorespiration. Plant Physiol 125:20–24
Tiwari A, Mamedov F, Grieco M, Suorsa M, Jajoo A, Styring S, Tikkanen M, Aro E-M (2016) Photodamage of iron–sulphur clusters in photosystem I induces non-photochemical energy dissipation. Nat Plants 2:16035
Tomar V, Sidhu GK, Nogia P, Mehrotra R, Mehrotra S (2017) Regulatory components of carbon concentrating mechanisms in aquatic unicellular photosynthetic organisms. Plant Cell Rep 36:1671–1688

Trudeau DL, Edlich-Muth C, Zarzycki J, Scheffen M, Goldsmith M, Khersonsky O, Avizemer Z, Fleishman SJ, Cotton CAR, Erb TJ, Tawfik DS, Bar-Even A (2018) Design and in vitro realization of carbon-conserving photorespiration. Proc Natl Acad Sci 115:E11455–E11464

Verruto J, Francis K, Wang Y, Low MC, Greiner J, Tacke S, Kuzminov F, Lambert W, McCarren J, Ajjawi I, Bauman N, Kalb R, Hannum G, Moellering ER (2018) Unrestrained markerless trait stacking in *Nannochloropsis gaditana* through combined genome editing and marker recycling technologies. Proc Natl Acad Sci USA 115:E7015–E7022

Walsh BJ, Rydzak F, Palazzo A, Kraxner F, Herrero M, Schenk PM, Ciais P, Janssens IA, Peñuelas J, Niederl-Schmidinger A, Obersteiner M (2015) New feed sources key to ambitious climate targets. Carbon Balance Manag 10:26

Wang Y, Stessman DJ, Spalding MH (2015) The CO_2 concentrating mechanism and photosynthetic carbon assimilation in limiting CO_2: how Chlamydomonas works against the gradient. Plant J 82:429–448

Wingler A, Lea PJ, Quick WP, Leegood RC (2000) Photorespiration: metabolic pathways and their role in stress protection. Philos Trans R Soc B Biol Sci 355:1517–1529

Wobbe L, Remacle C (2014) Improving the sunlight-to-biomass conversion efficiency in microalgal biofactories. J Biotechnol 201:28–42

Wobbe L, Blifernez O, Schwarz C, Mussgnug JH, Nickelsen J, Kruse O (2009) Cysteine modification of a specific repressor protein controls the translational status of nucleus-encoded LHCII mRNAs in Chlamydomonas. Proc Natl Acad Sci USA 106:13290–13295

Yamamoto H, Takahashi S, Badger MR, Shikanai T (2016) Artificial remodelling of alternative electron flow by flavodiiron proteins in Arabidopsis. Nat Plants 2:16012

Yun YS, Park JM (2001) Attenuation of monochromatic and polychromatic lights in *Chlorella vulgaris* suspensions. Appl Microbiol Biotechnol 55:765–770

Chapter 7
Metabolic Engineering and Synthetic Biology Approaches to Enhancing Production of Long-Chain Polyunsaturated Fatty Acids in Microalgae

Inna Khozin-Goldberg and Olga Sayanova

Abstract Interest in microalgal biotechnology for the production of bioactive and nutritional ingredients has increased tremendously with the urgent need for developing renewable bioresources. Microalgae represent a huge and still insufficiently tapped resource of LC-PUFA for human nutrition and health-related applications. Photosynthetic omega-3 and omega-6 LC-PUFA-producing microalgae are potent organisms and synthetic biology chassis for the production of high-value constituents for both health and aquaculture sectors. This chapter covers the diversity of eukaryotic microalgae in relation to their LC-PUFA production and outlines the fundamental role of microalgae as primary producers of LC-PUFA. We provide mechanistic insights into the biosynthesis of LC-PUFA in photosynthetic microalgae and highlight their biotechnological applications. The ongoing research aims to understand the molecular mechanisms by which microalgae of different evolutionary groups synthesize, relocate and incorporate LC-PUFA into complex lipids in the context of their multifaceted cellular organization. This research is supported by the availability of genomic information and 'omics' studies in a growing number of species. Recent progress in microalgal genetic transformation, genome editing and metabolic engineering has enabled the manipulation of LC-PUFA biosynthesis. We discuss the prospects for LC-PUFA manufacturing by microalgal biotechnology as a renewable and sustainable alternative to the finite resources of LC-PUFA.

I. Khozin-Goldberg (✉)
Microalgal Biotechnology Laboratory, The French Associates Institute for Agriculture and Biotechnology for Drylands, J. Blaustein Institutes for Desert Research, Ben-Gurion University of the Negev, Midreshet Ben-Gurion, Israel
e-mail: khozin@bgu.ac.il

O. Sayanova
Department of Biological Chemistry and Crop Protection, Rothamsted Research, Harpenden, UK
e-mail: olga.sayanova@rothamsted.ac.uk

A. Hallmann, P. H. Rampelotto (eds.), *Grand Challenges in Algae Biotechnology*,
Grand Challenges in Biology and Biotechnology,
https://doi.org/10.1007/978-3-030-25233-5_7

7.1 Introduction

Microalgae constitute a tremendously diverse group of mainly aquatic unicellular organisms that are affiliated to various taxonomic groups of eukaryotic organisms. Due to their complex evolutional history, microalgae are extremely diverse and comprise myriads of families and species that are major contributors to the global carbon and nitrogen cycles and are responsible for primary productivity in aquatic environments. Microalgae have been recognized as a potent bioresource for a variety of nutraceuticals and pharmaceuticals; among them are long-chain polyunsaturated fatty acids (LC-PUFA), known for their importance for nutrition and health. LC-PUFA contain three or more *cis*-double bonds on a fatty acid chain of 20 or 22 carbon atoms and are essential components of healthy nutrition. As primary producers of LC-PUFA, marine microalgae play a pivotal role in supplying these essential ingredients throughout the food webs. The capacity of microalgae to synthetize LC-PUFA traces back evolutionarily to the primary and secondary endosymbiotic events. Indeed, LC-PUFA are abundant components of the membrane acyl lipids in the cells of microalgae of numerous taxonomic groups but are generally rare within the green microalgae lineage. C18 PUFA and C16 PUFA are components of the chloroplast membrane glycerolipids in higher plants, as well as in the closely related green microalgae. In contrast to lower plants and microalgae, the flowering plants lack the enzymes for chain elongation of C18-PUFA and do not produce LC-PUFA.

In the following sections, we give a short basic introduction to the structural features of fatty acids. Fatty acids are principal components of complex glycerolipids, which are major constituents of structural membranes and reserve lipids in the cells. PUFAs are unsaturated fatty acids containing two and more double bonds; LC-PUFA contain 20 or 22 carbon atoms and more than three double bonds. The latter are often referred to as very long-chain PUFA to distinguish their carbon chain length. Most acyl (fatty) groups in plant and algal glycerolipids have double bonds in a *cis* configuration, which are separated by at least a single methylene group –C=C–C–C=C– (methylene-interrupted fatty acids). Non-methylene-interrupted fatty acids (NMIFAs) occur in some rare plant oils, marine invertebrates and some microalgae.

The position of the C=C double bond can be denoted by the Δ nomenclature, whereby the location of a double bond is counted from the first carbon atom located at the carboxylic group (designated α). Two major families of PUFA—the so-called ω-3 (*n*-3) or ω-6 (*n*-6) families—are distinguished by the position of the last double bond from the methyl terminus. C18 PUFA, linoleic acid (LA, $18{:}2^{\Delta 9,12}$, *n*-6) and α-linolenic acid (ALA, $18{:}3^{\Delta 9,12,15}$, *n*-3) are essential fatty acids because humans and most animals and marine fish cannot synthetize them de novo and must obtain them with a diet of plant or microalgal origin. C20 and C22 PUFA, containing three to six double bonds, are formed from C18PUFA by a 2C chain elongation. These include the LC-PUFA of the *n*-3 and *n*-6 families, such as arachidonic acid (ARA, $20{:}4^{\Delta 5,8,11,14}$, *n*-6), eicosapentaenoic acid (EPA, $20{:}5^{\Delta 5,8,11,14,17}$, *n*-3) and docosahexaenoic acid (DHA, $22{:}6^{\Delta 4,7,10,13,16,19}$, *n*-3).

7.2 LC-PUFA in Nutrition and Health

LC-PUFA are the major components of mammalian and other vertebrates' membrane phospholipids and play important roles in cellular and tissue metabolism, regulating membrane fluidity and thermal adaptation. ARA and DHA play important roles in neonatal health and development. They are naturally present in human milk and are recommended as supplemental ingredients in infant formulas all over the world (Horrocks and Yeo 1999; Innis 2000; Agostoni 2008; Mazzocchi et al. 2018).

LC-PUFA are central to the biosynthesis of the hormone-like bioactive compounds called eicosanoids, which regulate key physiological functions such as inflammatory responses, blood clotting, neurotransmission and cholesterol metabolism (Funk 2001). Eicosanoids derived from omega-6 ARA are generally considered as pro-inflammatory, whereas those derived from omega-3 EPA ($20{:}5^{\Delta 5,8,11,14,17}$) have anti-inflammatory effects (Simopoulos 2002, 2008). The balance of omega-3 and omega-6 LC-PUFA in the body is dependent on the levels of the two dietary essential fatty acids (EFA), linoleic acid (LA; $18{:}2^{\Delta 9,12}$) and α-linolenic (ALA; $18{:}3^{\Delta 9,12,15}$) that serve as the precursors for omega-6 and omega-3 LC-PUFA, respectively (Das 2006). However, all vertebrates, including humans and the majority of marine fish, have very limited capacities to synthesize EFA de novo as they lack Δ12- and Δ15-desaturase activities, which convert oleic acid (OA, $18{:}1^{\Delta 9}$) to LA and ALA, respectively. The usual Western diet contains excessive levels of omega-6 fatty acids derived from common vegetable oils (sunflower, corn) and processed foods and is under-represented in terms of omega-3 fatty acids; therefore, currently, the ratio of omega-6/omega-3 fatty acids is about 25:1 in the Western diet, whereas it is 2:1 in the ancestral human diet, which indicates a deficiency in the level of omega-3 fatty acids (Simopoulos 2002).

It is well established that the presence of omega-3 LC-PUFA in the human diet has a therapeutic effect on conditions such as eczema, inflammatory bowel disease, rheumatoid arthritis, metabolic syndrome and related diseases, such as obesity and type-2 diabetes (Calder 2015). The imbalance between the intakes of omega-3 and omega-6 LC-PUFA has been associated with a range of pathological conditions, including cardiovascular disease (CVD) (Aarsetoey et al. 2011), obesity (Nugent 2004), cancer (Gerber 2012) and various mental illnesses (Simopoulos 2008). Inflammation is a major burden in numerous human diseases and a unifying component of numerous health conditions, including cardiovascular, proliferative and neurological diseases. The anti-inflammatory activity of omega-3 LC-PUFA is well established and best studied in vitro and in vivo in the numerous disease models; however, large-scale epidemiological studies are still debating their essentiality as human dietary supplements. The resolution of the acute phase of inflammation is an active process orchestrated by structurally distinct families of LC-PUFA-derived molecules: lipoxins (derived from ARA), resolvins, protectins and maresins (derived from omega-3 LC-PUFA), collectively termed specialized pro-resolving lipid mediators (SPM) (Serhan 2014). DHA and its precursor docosapentaenoic acid (DPA, 22:5 *n*-6) are both precursors for resolvins and protectins, playing a role in regulating and resolving inflammation (Boller and Felix 2009; Recchiuti and Serhan 2012). A role for dietary EPA in cardioprotection relates to its content in the membrane lipids of the myocardium and to its role as a metabolic precursor in the synthesis of anti-inflammatory eicosanoid molecules, such as 3-series prostaglandins. Although body can synthesize both EPA and DHA from EFA, LA and ALA via desaturation and elongation steps, the conversion of ALA is very inefficient and deteriorates with age and disease. Therefore, these fatty acids should be considered essential and must be obtained from the diet. Currently, the major dietary sources of omega-3 LC-PUFA are marine fish and other seafood.

Though this chapter deals with the biotechnological potential of photosynthetic microalgae as producers of LC-PUFA, we note the current industrial producers of DHA, which are mainly the heterotrophic unicellular protists. These are frequently and incorrectly referred to as microalgae due to their positioning within the clade Stramenopiles (Leyland et al. 2017), which also comprises heterokont microalgae (haptophytes and diatoms). We shall refer to these organisms as they are currently the major single-cell oleaginous organisms fermented for omega-3 LC-PUFA production and have been successfully developed for the commercial production of DHA oils. The thraustochytrids (Labyrinthulomycetes) *Aurantiochytrium* spp. (Taoka et al. 2009), *Thraustochytrium* spp. (Barclay et al. 1994; Burja et al. 2006), *Schizochytrium* spp. (Yokochi et al. 1998), Ulkenia (Lee Chang et al. 2014) and the dinoflagellate *Crypthecodinium cohnii* (Kyle 2001; Ratledge et al. 2001) (only the latter is a heterotrophic dinoflagellate) have the ability to produce high levels of lipids and accumulate DHA or/and EPA mainly as triacylglycerols and phospholipids (reviewed in Perez-Garcia et al. 2011). *Thraustochytrium* strains contain from 30 to 40% DHA of total fatty acids (Burja et al. 2006), whereas some strains of *Schizochytrium* and *Ulkenia* are able to accumulate up to 50–56% DHA. Fermentation of *Schizochytrium* spp. is used in industrial-scale omega-3 LC-PUFA production as they can reach a DHA content over 40% (cell dry weight) [for ref., see Martins et al. (2013)]. DHA produced by *Schizochytrium* species is commercially available as a dietary supplement and is also used in animal feed and aquaculture. The DHA-rich oil of *C. cohnii* and the ARA-rich oil of the oleaginous fungus *Mortierella alpina* are included in infant formula in many countries. Several strains were developed via heterotrophic cultivation for the commercial production of DHA (mainly for infant formula) (Martins et al. 2013; Sprague et al. 2017).

During the last two decades, important advances have been made in developing alternative sources of LC-PUFA by genetic engineering, in oil seed plants and in oleaginous yeast. Successful modification of plant oil composition via metabolic engineering demonstrated the potential of making EPA and DHA in transgenic oilseeds [see Qi et al. (2004), Wu et al. (2005), Ruiz-Lopez et al. (2015) among others]. Recent studies have demonstrated that the oilseed crop *Camelina sativa* represents a new platform for the terrestrial production of omega-3 LC-PUFA (Usher et al. 2017). Oil from transgenic seeds accumulating up to 20% of non-native EPA and DHA was shown to be an effective replacement for fish oils in salmon feeding trials (Betancor et al. 2015, 2016). Moreover, the first field-based evaluation of a terrestrial source of EPA and DHA, synthesized in the seeds of transgenic *C. sativa* plants via the heterologous reconstitution of the omega-3 LC-PUFA's biosynthetic pathway, demonstrates the robust nature of this novel trait and the feasibility of making fish oils in genetically modified crops (Usher et al. 2017).

The oleaginous yeast *Yarrowia lipolytica* was engineered to produce high levels of EPA. The first successful example of metabolic engineering in *Y. lipolytica* for the production of omega-3 LC-PUFA through fermentation of sugars was reported by Xue et al. (2013). The heterologous expression of multiple copies of PUFA/LC-PUFA biosynthetic genes from various microorganisms and microalgae resulted in the production of a Gen III HP strain capable of accumulating EPA at 25% of total fatty acids (TFA) and lipids up to 50% of dry cell weight (DCW) (Xie et al. 2015). This tailored oil with a high EPA content is now commercially available.

Photosynthetic microalgae that utilize photosynthesis and nutrients for growth hold promise as a production platform for the supply of LC-PUFA for human food and nutrition, as well for fish farming (Draaisma et al. 2013; Leu and Boussiba 2014; Chauton et al. 2015). Marine microalgae are the primary producers of omega-3 LC-PUFA in the aquatic food chain, and EPA and DHA enter our diet through the consumption of marine fish. Microalgae can be grown on alternative water sources (brackish, sea water, aquacultural systems, waste waters and other side streams) and on marginal land without competing with agricultural crops for freshwater use; in addition, they are expected to be more productive per unit surface area of land than any cultivated oilseed crops (Georgianna and Mayfield 2012). However, the commercial mass production of omega-3 LC-PUFA from photosynthetic microalgae is currently more expensive than fermentation-based technologies. At present, cost-effective phototrophic cultivation and biorefinery for LC-PUFA production represent a substantial technological challenge (Vigani et al. 2015). Recent technological improvements in production systems and developments in the field of microalgal biotechnology have renewed interest in microalgae as a sustainable source of food/feed commodities and high-value ingredients. Elucidating and gaining a deep understanding of the LC-PUFA biosynthesis pathways in photosynthetic microalgae and the means to manipulate them at the industrial scale lie at the frontier of microalgal biotechnology. As an alternative approach, metabolic engineering of oleaginous microalgae to enhance LC-PUFA production may offer many advantages. Therefore, the metabolic engineering of microalgae and the genetic modification of algal strains are some of the most promising and fast-developing strategies for producing sustainable omega-3 lipids and oils.

7.3 Omega-3 Fatty Acid Production by Microalgae

EPA and DHA production driven by photosynthesis has become a promising target for microalgal photo-biotechnology. The ability of diverse marine microalgae to produce significant levels of EPA and/or DHA has been previously reviewed (Harwood and Jones 1989; Guschina and Harwood 2006; Khozin-Goldberg et al. 2011, 2016; Chauton et al. 2015) and investigated in numerous studies (Table 7.1).

In the following, we provide a brief overview focusing on the main species for which major progress was achieved during the last several decades. EPA-producing photosynthetic microalgae with industrial potential include the eustigmatophytes of the *Nannochloropsis* genus (*N. oculata*, *N. salina*, *N. oceanica* and *N. gaditana*), *Monodus subterr*aneus, diatoms *Phaeodactylum tricornutum*, *Nitzschia* spp., *Odontella aurita* and red microalgae, for example, *Porphyridium cruentum* (*purpureum*). Membrane glycerolipids of *Nannochloropsis* microalgae are rich in EPA (Sukenik 1991; Pal et al. 2013). Based on this characteristic, *Nannochloropsis* microalgae are exploited as a source of EPA for the human health product market and aquaculture. Interest in *Nannochloropsis* cultivation is driven by high EPA levels and the absence of DHA, making these microalgae a valuable source of a single ω-3 LC-PUFA for dietary purposes. The ω-6 LC-PUFA, ARA, accounts for only a small percentage of TFA in this alga; other major FAs are the C16 palmitic acid (16:0) and palmitoleic acid (16:1); the latter fatty acid holds promise for the amelioration or prevention of insulin resistance and diabetes (Frigolet and Gutiérrez-Aguilar 2017). The production of EPA by freshwater species, such as *N. limnetica*, has also been considered (Krienitz and Wirth 2006; Freire et al. 2016). Freire et al. reported (Freire et al. 2016) that *N. gaditana* and *N. limnetica* displayed almost similar fatty acid profile, EPA proportions and productivity in semi-continuous laboratory cultures grown in saline and freshwater medium, respectively. Of note, a remarkable plasticity of *N. oceanica* towards external salinity level was suggested to be governed by the structural properties of cell walls and cellular metabolic adjustments (Pal et al. 2013). Furthermore, *N. oceanica* displayed the higher contents of EPA in the nutrient-replete cultures grown in freshwater medium compared to saline media (Pal et al. 2013).

Cultivation of *Nannochloropsis* for EPA production was pioneered in the 1980s and 1990s and is currently performed with different strains and at various scales, both indoors and outdoors, and in different types and sizes of photobioreactors (Boussiba et al. 1987; Zittelli et al. 1999; Sukenik et al. 2009; Safafar et al. 2016; Hulatt et al. 2017). Despite fluctuations in temperature and natural irradiance and variations in the FA profile, EPA levels remained relatively stable (ca. 4–5% of DW) (Zittelli et al. 1999). More recent studies have reported similar values: 4.2–4.9% EPA under nutrient-replete conditions (Hulatt et al. 2017). Notably, the variations in salinity did not show drastic effects on the EPA content in the biomass, although brackish and even freshwater salinities are more favourable for EPA content (Pal et al. 2013). However, the combination of high irradiance, salinity and nitrogen deprivation appeared to be deleterious for EPA and lipid productivities (Pal et al. 2011). Under N starvation, the proportion of EPA out of total fatty acids decreased, but the EPA content of DW changed moderately, due to its redistribution within cellular lipids and partitioning to storage lipids. Numerous studies have addressed the optimization of biomass and EPA productivity by *Nannochloropsis* microalgae (Pal et al. 2011; Meng et al. 2015; Camacho-Rodríguez et al. 2015; Safafar et al. 2016), among others. EPA levels may vary from 25% to 50% of TFA, depending on growth stage, irradiance conditions and nitrogen availability (Hodgson et al. 1991; Pal et al. 2011). Large-scale production of EPA from *Nannochloropsis* has been

Table 7.1 Relative proportions of EPA and DHA in the fatty acid profile of some microalgae, indicated as mol% or weight % of total fatty acids

Phylum/Class[a]	Species	EPA	DHA	References
Bacillariophyta				
Mediophyceae	*Chaetoceros muelleri*	12.8	0.8	Zhukova and Aizdaicher (1995)
Mediophyceae	*Cyclotella* sp.	19.2		Li et al. (2017a)
Bacillariophyceae	*Fistulifera* sp.	15.0		Liang et al. (2013)
Bacillariophyceae	*Nitzschia laevis*	19.1		Wen and Chen (2003)
Mediophyceae	*Odontella aurita*	23.3	2.8	Pasquet et al. (2014)
Bacillariophyceae	*Phaeodactylum tricornutum*	34.5	<1	Yongmanitchai and Ward (1991)
Mediophyceae	*Skeletonema costatum*	29.2	3.1	Zhukova and Aizdaicher (1995)
Mediophyceae	*Thalassiosira pseudonana*	12.2	3.4	Tonon et al. (2002)
Chlorophyta				
Chlorodendrophyceae	*Tetraselmis* sp.	4–11	<1	Ryckebosch et al. (2012)
Trebouxiophyceae	*Lobosphaera incisa*	ARA >50%		Bigogno et al. (2002c)
Cryptophyta				
Cryptophyceae	*Cryptomonas* sp.	3–25	2.5–10	Ryckebosch et al. (2012)
Cryptophyceae	*Rhodomonas* sp.	4.6–10.0	1.6–4.0	Huerlimann et al. (2010)
Ochrophyta				
Pinguiophyceae	*Glossomastix chrysoplasta*	30		Hsiao and Blanch (2005)
Eustigmatophyceae				
	Monodus subterraneus	37.1		Khozin-Goldberg et al. (2002b)
	Monodus subterraneus UTEX 151	31–34		Liu and Lin (2005)
	Nannochloropsis oculata	29.7		Zhukova and Aizdaicher (1995)
	Nannochloropsis oceanica	25		Pal et al. (2011)
	Nannochloropsis (Microchloropsis) salina	>44.2		Safafar et al. (2016)
	Nannochloropsis sp.	7.6–15.3	–	Huerlimann et al. (2010)
Haptophyta				
Coccolithophyceae	*Isochrysis* sp.	0.3–0.8	8.2–14.1	Huerlimann et al. (2010)
Pavlovophyceae	*Pavlova salina*	19.1	1.5	Zhukova and Aizdaicher (1995)
	Pavlova lutheri	11.6	22–29	Guihéneuf et al. (2009)
Rhodophyta				
Porphyridiophyceae	*Porphyridium cruentum*	41		Khozin-Goldberg et al. (2000)

[a]Classification based on Guiry, M.D. and Guiry, G.M. 2019. AlgaeBase. World-wide electronic publication, National University of Ireland, Galway. http://www.algaebase.org; searched on 15 April 2019

performed in open ponds and photobioreactors [reviewed in Ma et al. (2016)]. As the proportion of EPA as total fatty acids is subject to cultivation conditions, it should be counterbalanced by substantial biomass productivities to achieve feasible EPA productivities. For example, lowering growth temperatures increases the levels of EPA, but biomass production rates decrease (Aussant et al. 2018). When *Nannochloropsis salina* was cultivated under a constant suboptimal temperature (<20 °C), it exhibited the highest quantity of EPA; however, the growth rate and photosynthetic activity were significantly reduced compared to under the optimal cultivation temperature. Interestingly, in the cultures maintained at 15 °C, EPA was found to be enriched in some triacylglycerol (TAG) molecular species (Gill et al. 2018).

EPA and DHA production by diatoms has been recently reviewed (Sayanova et al. 2017). The marine pennate diatom *Phaeodactylum tricornutum* is considered a good source for the industrial production of EPA as it accumulates up to 30% EPA of total fatty acids; however, the levels of DHA are generally low (Yongmanitchai and Ward 1991; Hamilton et al. 2014). In a recent study, EPA contents, averaging 4.4, 3.2 and 3.1% of DW, were reported for three strains of *P. tricornutum* cultivated in western Norway outdoors in flat panel bioreactors (Steinrücken et al. 2018). The annual potential yield of *P. tricornutum* grown outdoors in the green-wall panel GWP® reactors was estimated as 730 kg of EPA per hectare (Rodolfi et al. 2017). Furthermore, *P. tricornutum* and its lipid fractions were shown to exhibit a wide range of biological activities, some associated with EPA production (Andrianasolo et al. 2008; Desbois et al. 2009). A recent study discovered that under oxidative stress, *P. tricornutum* produces bioactive oxylipins—isoprostanoids—non-enzymatically (Lupette et al. 2018). An oleaginous diatom *Fistulifera solaris*, for which numerous tools have been rapidly developed post-isolation, hold promise for either biofuel or EPA production (Liang et al. 2014). A prominent example of a commercialized diatom is *Odontella aurita, whose* EPA levels reach 27–28% of total lipids (Guihéneuf et al. 2010). It is cultivated in outdoor raceway ponds of *O. aurita*, yielding an EPA-rich algal biomass. *Odontella aurita* is approved for use as a dietary supplement for human nutrition with proven beneficial effects on dyslipidaemia, platelet function and oxidative stress in high-fat-fed rats (Haimeur et al. 2012).

While EPA is abundant in diatoms and eustigmatophytes, the levels of DHA are generally low. A marine haptophyte *Isochrysis galbana* (Prymnesiophyceae) is rich in DHA (to 12% of TFA). It is used in hatcheries to feed fish larvae and clams (Hubert et al. 2017). Two strains are mainly used in aquaculture: T-Iso (Tahiti isolate) and Parke. Recently, Bendif et al. (2013) proposed reclassifying T-Iso to a new genus *Tisochrysis lutea*, referring to its original name and yellow cell pigmentation. The lipid metabolism of *T. lutea* has been intensively investigated by comparative proteomics and transcriptomics of the wild-type strain and a high lipid mutant (Carrier et al. 2014).

ARA does not account for more than a small percentage of fatty acids in LC-PUFA-producing microalgae as it generally serves as an intermediate of EPA. Exceptionally, the ARA-accumulating oleaginous microalga *Lobosphaera incisa* represents a potent source for photosynthesis-driven production and accumulation of the omega-6 LC-PUFA in TAG (Bigogno et al. 2002b; Khozin-Goldberg et al. 2002a; Solovchenko et al. 2008; Pal-Nath et al. 2017).

7.3.1 LC-PUFA-Producing Microalgae in Aquaculture

The origin of LC-PUFA lies at the base of the food chain in marine microalgae and heterotrophic organisms. Fish are rich in ω-3 LC-PUFA, as they obtain them from marine microalgae (Martins et al. 2013). The abundance of LC-PUFA in the marine environment has remarkably modulated the ability for fatty acid desaturation in extinct marine teleost fish, though some fish hold genes for LC-PUFA biosynthesis (Lopes-Marques et al. 2018). Natural marine sources of LC-PUFA are continuously diminishing due to over-fishing as fish catches reach a plateau and, in some cases, even exceed maximal sustainable levels (Shepherd and Jackson 2013; Sprague et al. 2017). At present, the most available dietary sources of LC-PUFA are oily marine fish and seafood. However, the increasing demand for fish and fish oils is putting enormous pressure on marine ecosystems. The depletion of fish stocks and pollution of the marine environment indicate an urgent need for an alternative and sustainable source of omega-3 LC-PUFA. Fish farming now supplies the most proportion of fish for human consumption: in 2014, 51% of fish consumed by humans was produced by fish farming (Fao 2016). However, farmed fish require feeds containing fish meal and fish oil as they supply the EPA and DHA essential for the optimal nutrition and stress tolerance of marine fish (Tocher and Glencross 2015). Aquaculture's consumption of fish oil has more than doubled over the past decade and the demand is rapidly increasing: about 75% of available fish oil is used for aquafeeds, reducing the EPA and DHA levels and the omega-6/omega-3 ratio, and thus lowering nutritional value for human consumption (Tocher 2015). A number of marine LC-PUFA-producing photosynthetic microalgae, namely *Isochrysis galbana*, *Chaetoceros calcitrans*, different species of *Nannochloropsis*, *Phaeodactylum tricornutum* and *Pavlova*, are routinely cultivated in aquaculture, for the enrichment of zooplankton and fish juvenile stages. Considering the importance of balanced nutrition in the aquaculture species with regard to their dietary requirements, generally a mixture of different microalgal strains is provided (Shields and Lupatsch 2012). An optimal supply of EPA and DHA, as well as an adequate provision of ARA, is essential for farmed fish (Bell and Sargent 2003), as the latter improves fish survival in the larval stage and under stress and affects immune function (Koven et al. 2001; Nayak et al. 2017; Ma et al. 2018). Substantial modifications in immune function and improvements in fish survival were demonstrated in guppies and zebrafish fed with the diets enriched with either the lipid extracts or biomass of ARA-accumulating *L. incisa* (Dagar et al. 2010; Nath et al. 2012; Nayak et al. 2018). Several farmed fish species and shrimp were shown to accept the algae-based feeds (Kiron et al. 2012). In addition, microalgal co-products, such as defatted biomass after lipid extraction, may serve as potent alternatives to fishmeal for aquaculture feed manufacturing. Alternatives are also being sought to replace unsustainable feedstocks for proteins, such as fishmeal, in aquaculture feeds. For example, the defatted co-product of *N. oculata* was evaluated in an 84-day nutritional feeding experiment in tilapia (Sarker et al. 2018). Results showed that a 33% diet replacement with microalgal products yielded fish growth, feed conversion and survival rates that were similar to the reference diet. Furthermore, due to the remaining EPA in the co-product, all applied diets yielded an omega-3/omega-6 ratio in tilapia fillet that is favourable for human consumption.

Microalgae-based supplements have been tested in mammals. A freeze-dried *N. oceanica* biomass showed promise as a natural dietary source of EPA for ruminants, based on better protection from the ruminal biohydrogenation compared to free EPA, spray-dried *N. oceanica* and *P. tricornutum* biomass (Alves et al. 2018). While studying the nutritional value and digestibility aspects of three species comprising *N. oceanica*, *P. tricornutum* and *I. galbana*, the best lipid and protein digestibility was observed for the feed that included *P. tricornutum* biomass (Skrede et al. 2011). The potency and safety of three microalgae, including *Nannochloropsis* and *Phaeodactylum*, as dietary nutrient resources were evaluated (Neumann et al. 2018). High liver EPA levels were measured following feeding with *N. oceanica*, in line with its highest EPA enrichment. Neither histological nor serum analyses revealed any heart, kidney or liver toxicity induced by any of the algal diets. Algae-rich diets were thus well accepted, well tolerated and suitable for the maintenance of body weight and normal organ function. No toxicological effects were observed (Neumann et al. 2018).

In the past decade, a number of microalgae that are used in human dietary supplements, including Spirulina, Haematococcus, *P. tricornutum* and some microalgal products, have been classified as GRAS (Generally Recognized As Safe) by the FDA (Food and Drug Administration, USA) (García et al. 2017). Also, permission was granted by Food Standards Australia New Zealand (FSANZ) for using the dried marine protist Schizochytrium, which is rich in DHA, as a novel food ingredient in a limited range of foods (García et al. 2017).

7.3.2 LC-PUFA Biosynthesis

In recent years, the biosynthetic pathways of LC-PUFA have been described from a range of different organisms, including animals, fungi, marine invertebrates and microalgae (summarized in Fig. 7.1). In recent decades, numerous microalgal genes for LC-PUFA biosynthesis were cloned and functionally characterized in heterologous organisms. The most common host has been a baker's yeast *Saccharomyces cerevisiae*, which presents a simple eukaryotic expression system for the functional characterization of eukaryotic desaturases and elongases, acting either on the acyl groups bound to the membrane phospholipids or on acyl-CoA. In this way, numerous desaturases and elongases that are engaged in LC-PUFA biosynthesis were expressed, and their substrate preference range was evaluated following fatty acid supplementation. These efforts were driven by the goal to identify the most

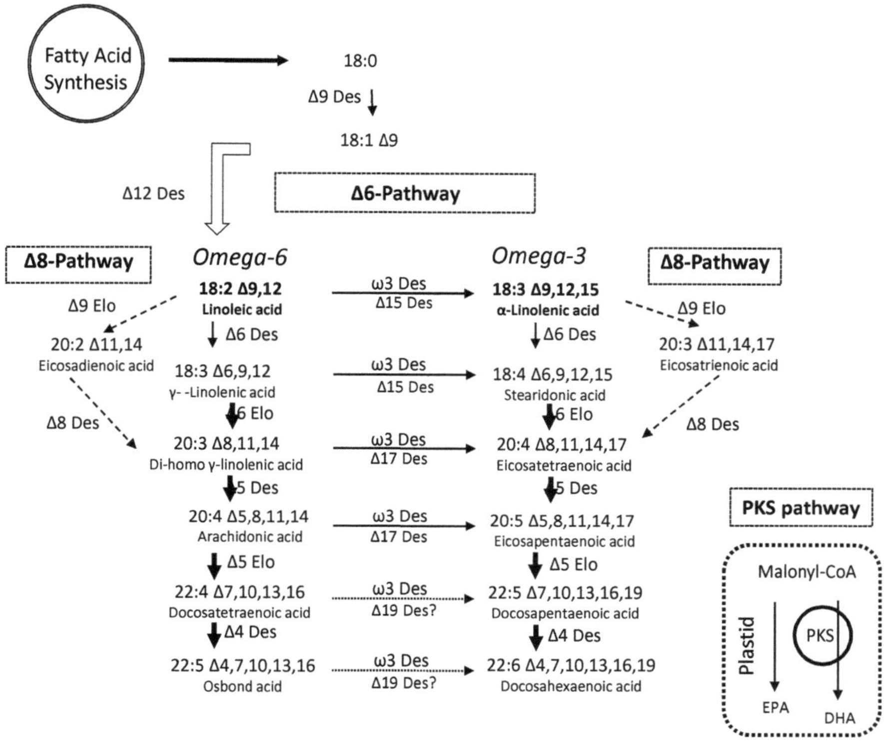

Fig. 7.1 Schematic representation of aerobic (Δ6-conventional and Δ9 alternative) and anaerobic (PKS) pathways of LC-PUFA biosynthesis. Linoleic and α-linolenic precursors are shown in bold. Bold arrows indicate a microbial Δ4-pathway

promising genes for the reconstitution of LC-PUFA biosynthesis in oilseed plants. Indeed, thanks to the dedicated efforts of several research groups and commercial companies, fish-oil-like quantities of EPA and DHA have been achieved by the synthetic redesign of plant oilseed metabolism (Haslam et al. 2016). At the same time, these studies revealed novel activities and provided many valuable insights into the operation of LC-PUFA biosynthetic pathways. The primary activities involved in LC-PUFA biosynthesis in microalgae (elongases and desaturases) have been identified and functionally characterized in different species, reviewed in Khozin-Goldberg et al. (2011, 2016) and Sayanova et al. (2017). Within the primary producers of omega-3 LC-PUFA, two independent metabolic pathways have been characterized: the aerobic desaturase/elongase pathway (Sayanova et al. 2017) and the anaerobic polyketide synthase (PKS) pathway (Metz et al. 2001).

7.3.2.1 Aerobic Pathway

The mechanisms of lipid production and fatty acid biosynthesis in microalgae remain to be studied in detail. Fatty acid and glycerolipid biosynthesis have been described in detail in the model plant *Arabidopsis thaliana* (Li-Beisson et al. 2013). Based on the sequence homology of genes identified in algal genomes, it is assumed that algal fatty acid biosynthesis resembles that of higher plants and begins with de novo fatty acid synthesis in the chloroplast (Ohlrogge and Browse 1995). However, there are indications based on genomic information that in some microalgae, fatty acid synthesis may occur in the cytoplasm (Vieler et al. 2012). The first committed step is the formation of malonyl-CoA following the ATP-dependent carboxylation of acetyl-CoA, catalysed by acetyl-CoA carboxylase (ACCase), a multi-subunit enzyme found in the chloroplasts of plants and algae. The fatty acid synthase (FAS) multi-subunit complex catalyses fatty acid elongation by condensing malonyl-CoAs and acetyl-CoA. These condensing reactions are initiated by a β-ketoacyl-ACP synthase (KAS). Each FAS cycle of reactions leads to the extension of the fatty acid chain by two carbon atoms, usually ending with the synthesis of saturated C16:0- and C18:0-ACPs. Mature fatty acids produced de novo are then either retained in the chloroplast where they are transferred to glycerol-3-phosphate for incorporation into membrane lipids (the so-called prokaryotic pathway) or are hydrolysed from ACP by acyl-ACP thioesterases (FATs), exit the plastid and are esterified to coenzyme A (CoA) to form acyl-CoAs. The saturated C16:0 and C18:0 attached to acyl carrier proteins (ACPs) can undergo further desaturation by soluble ACP-dependent desaturases and subsequent release by FATs. Desaturation of C18:0 can be catalysed either by the soluble stearoyl-ACP Δ9-desaturase (SAD) of the chloroplast stroma (Shanklin et al. 2009) or by an extra-plastidic ER-bound acyl-CoA Δ9-desaturase (ADS) (Fukuchi-Mizutani et al. 1998) generating oleic acid (OA, 18:1Δ9). OA becomes a substrate for further desaturation by Δ12-desaturase, resulting in the production of linoleic acid (LA; 18:2Δ9,12), which subsequently may be converted into α-linolenic (ALA; 18:3Δ9,12,15) by the action of Δ15-desaturase (Fig. 7.1).

The biosynthesis of LC-PUFA from LA and ALA may follow two converging routes: the predominant or 'conventional' Δ6-pathway operating in most PUFA-synthesizing organisms and an 'alternative' Δ8-pathway. To produce LC-PUFA, a series of alternating desaturation and elongation steps are required. The 'conventional' pathway starts with Δ6-desaturation of both LA and ALA, resulting in the synthesis of γ-linolenic acid (GLA, $18{:}3^{\Delta 6,9,12}$) and stearidonic acid (SDA, $18{:}4^{\Delta 6,9,12,15}$), respectively. These two Δ6-desaturated fatty acids are then chain-elongated by two carbons via a Δ6-specific elongase to yield di-homo γ-linolenic acid (DGLA, $20{:}3^{\Delta 8,11,14}$) and eicosatetraenoic acid (ETA; $20{:}4^{\Delta 8,11,14,17}$). A second desaturation, via the Δ5-desaturase, then produces arachidonic acid (ARA; $20{:}4^{\Delta 5,8,11,14}$) and EPA.

Several microalgae use an 'alternative' Δ8-pathway in which LA and ALA are elongated by the action of a Δ9-specific elongase, yielding eicosadienoic acid (EDA;

20:2 Δ11,14) and eicosatrienoic acid (ETrA; 20:3$^{\Delta 11,14,17}$), prior to two sequential desaturations by a Δ8- and a Δ5-desaturase to generate ARA and EPA, similar to the conventional pathway. This has been found in the haptophytes *Isochrysis galbana*, *Pavlova salina*, the coccolithophore *Emiliania huxleyi* (Qi et al. 2003; Zhou et al. 2007; Sayanova et al. 2011) and in the euglenoid *Euglena gracilis*, which appears to lack Δ6-desaturase activity (Wallis and Browse 1999).

The enzymatic machinery of the LC-PUFA biosynthetic pathway consists of fatty acid desaturases and elongases. Several desaturation steps of LC-PUFA's biosynthesis are carried out by microsomal 'front-end' PUFA desaturases (Δ6, Δ5, Δ8, Δ4); these all belong to the N-terminal cytochrome *b*5 fusion superfamily and introduce a new double bond between the pre-existing double bond and the carboxyl end of the fatty acid (Sayanova et al. 1997) as opposed to the Δ12- and Δ15-desaturases from plants, algae and some fungi, which lack this cytochrome *b*5 domain and insert a double bond between the pre-existing double bond and the methyl-end (Meesapyodsuk and Qiu 2012). In DHA-producing microalgae, the conversion of EPA to DHA is carried out by the linear desaturation-elongation pathway, involving C2 elongation of EPA by a specific Δ5-elongase to docosapentaenoic acid (DPA; 22:5$^{\Delta 7,10,13,16,19}$), which is then desaturated by a Δ4-specific desaturase to yield DHA (Fig. 7.1). A gene (*IgFAD4-2*), encoding a C22-Δ4 polyunsaturated fatty acid-specific desaturase, for example, has been isolated and characterized from *I. galbana* (Haptophyceae) (Qi et al. 2004; Shi et al. 2012).

All the corresponding genes encoding putative desaturases and elongases have been identified in microalgae, including diatoms, eustigmatophytes and prasinophytes. The presence of the highly active 'conventional' Δ6-pathway has been demonstrated or predicted for several diatoms such as *P. tricornutum*, *T. pseudonana*, *Fragilariopsis cylindrus*, *Fistulifera solaris* (Domergue et al. 2003; Tonon et al. 2005a; Vaezi et al. 2013; Jiang et al. 2014; Tanaka et al. 2015) and *Nitzschia* sp. (Cheng et al. 2014). The functional characterization of candidate activities involved in EPA and DHA biosynthesis (i.e. Δ9-, Δ12-, Δ6-, Δ5-, ω3-, Δ4 desaturases and Δ6-elongase) indicates that some diatoms (*P. tricornutum* and *T. pseudonana*) have the capacity to utilize both pathways to synthesize EPA and DHA (Arao and Yamada 1994; Domergue et al. 2002; Tonon et al. 2005b). The Δ6-pathway was shown to be the major operating route in the rhodophyte *P. cruentum* (Khozin et al. 1997), the freshwater chlorophyte *Parietochloris* (currently *Lobosphaera) incisa* (Bigogno et al. 2002a; Iskandarov et al. 2009, 2010) and two eustigmatophytes, the freshwater *Monodus subterraneus* (Khozin-Goldberg et al. 2002b) and the marine *Nannochloropsis* sp. (Schneider and Roessler 1994).

Some algal front-end Δ6- and Δ5-desaturases act on PUFA esterified to CoA, similar to mammalian front-end desaturases. Examples include Δ6-desaturases from the Prasinophytes: *Ostreococcus tauri* (Domergue et al. 2005), *O. lucimarinus* (Petrie et al. 2010a) and *M. pusilla* (Petrie et al. 2010b), and Δ6- and Δ5-desaturases from *Mantoniella squamata* (Hoffmann et al. 2008). The acyl-CoA-dependent enzymes insert a double bond into CoA-activated fatty acid from the acyl-CoA pool. This mechanism is distinct from phospholipid-linked desaturases, which require a fatty acid to be attached to PC or phosphoethanolamine (PE). Moreover, some acyl-CoA-dependent desaturases are highly specific for their omega-3 substrates. The use of the acyl-CoA-dependent enzymes with the correct substrate specificities may eliminate the requirement for the rate-limiting acyl-exchange with PC and thus avoid 'substrate dichotomy' for lipid-linked desaturases and acyl-CoA-dependent elongases [for further information refer to Sayanova et al. (2012)]. Microalgae have the capacity to convert omega-6 PUFA into their omega-3 counterparts by the action of ω3-desaturases. Conversion of C18, C20 or C22 omega-6 fatty acids to omega-3 fatty acids was suggested to be catalysed by the activities of ω-3 desaturases acting as Δ15-desaturase (Δ15Des), Δ17-desaturase (Δ17Des) and possibly by a hypothetical Δ19-desaturase (Δ19 Des), respectively (Fig. 7.1). Since many microalgal genomes appear to have only one isoform of ω-3 desaturase, it is thus possible that these activities are accomplished by a single enzyme acting on different available substrates (Nguyen et al. 2013; Zorin et al. 2017). In the EPA-producing *Nannochloropsis* sp., *M. subterraneus* and *P. cruentum*, it is predicted that an ω3 desaturase acts on ARA to produce EPA (Schneider and Roessler 1994; Khozin et al. 1997; Khozin-Goldberg et al. 2002b). Recently, a set of desaturases and several elongases have been identified in *Nannochloropsis* sp. (Kilian and Vick 2013). A functional characterization of the ω-3 desaturase involved in EPA synthesis has demonstrated high substrate preferences towards ARA and LA. Overexpression of the ω3-desaturase in *Nannochloropsis* sp. resulted in a large increase of the EPA/ARA ratio and a small increase of ALA (Kilian and Vick 2013).

A few algal species accumulate unusual levels of octapentaenoic acid (OPA; 18:5 *n*-3). OPA is present at high levels in *Emiliania huxleyi* (Bell and Pond 1996), microalgae that are considered to be the most productive calcifying organism on earth. The biosynthetic route for OPA is currently unknown and could be generated by either (1) desaturation of SDA or (2) via β-oxidation of EPA (Sayanova et al. 2011). The synthesis of OPA could occur in plastids, utilizing sulfolipid-linked SDA as a substrate.

7.3.2.2 Anaerobic Pathway (PKS)

Most marine organisms utilize the aerobic fatty acid desaturation/elongation pathway to produce DHA. However, some lower marine microorganisms are able to produce EPA, DPA and DHA in anaerobic conditions from malonyl-CoA by an enzymatic complex similar to bacterial polyketide synthase (PKS) (Metz et al. 2001). The anaerobic or PKS pathway represents an alternative route to EPA and DHA and was identified in both prokaryotic and eukaryotic microorganisms (Metz et al. 2001). De novo biosynthesis of LC-PUFA via PKS is similar to the fatty acid synthase (FAS) pathway and involves several rounds of reduction, dehydration, reduction and condensation with the acyl chain growing by two carbon units with each round. The complex consists of eight different PKS protein domains, including malonyl-CoA:ACP acyltransferase (MAT), acyl carrier protein (ACP), 3-ketoacyl synthase (KS), 3-ketoacyl-ACP reductase (KR), acyltransferase (AT), chain length factor (CLF), enoyl reductase (ER) and 3-hydroxyacyl-ACP dehydratase/isomerase (DH/IS) (Kaulmann and Hertweck 2002). This PKS-like pathway produces EPA and DHA as free fatty acids, which would require activation to acyl-CoAs prior to incorporation of these products into lipids (Metz et al. 2009). Interestingly, aerobic and anaerobic pathways co-exist in some heterotrophic marine LC-PUFA eukaryotes such as *Schizochytrium* and *Thraustochytrium aureum* (Qiu 2003; Matsuda et al. 2012). PKS-encoding sequences were identified in the sequenced genomes of phototrophic microalgae, but their functional significance remains to be investigated. There have been attempts to engineer the prokaryotic photosynthetic cells with PKS-coding genes. Cyanobacteria were also used as host organisms for the recombinant production of EPA by the polyketide biosynthesis pathway. For example, the EPA synthesis cluster of the *Shewanella* strain SCRC-2738 was expressed in a marine cyanobacterium *Synechococcus* sp., resulting in the production of EPA and ETA (Yu et al. 2000). The production yields of LC-PUFA have been found to be greatly affected by cultivation temperature: the highest levels of EPA in transgenic cyanobacterial strains were produced at 18 °C and relatively low light intensity.

7.3.3 Intracellular Compartmentation and Partitioning of LC-PUFA in Microalgae

In the microalgal cells, different fatty acids show distinct patterns of enrichment within the lipid classes. LC-PUFA are generally enriched in the membrane lipids, while storage lipids' TAG rarely shows accumulation of LC-PUFA. Nevertheless, some microalgae are able to deposit LC-PUFA in their TAG under certain cultivation conditions (Bigogno et al. 2002c; Tonon et al. 2002; Guihéneuf and Stengel 2013; Řezanka et al. 2018). Partitioning of omega-3 LC-PUFA to storage lipids was found in *N. oculata*, *P. tricornutum*, *T. pseudonana* and *P. lutheri* (Tonon et al. 2002). Differences with regard to the growth stage and incorporation of LC-PUFA into TAG were demonstrated across these microalgae. *T. pseudonana* and *P. lutheri* incorporated EPA and DHA into TAG in the stationary phase, while *N. oculata* translocated more EPA to TAG compared with *P. tricornutum*. A recent study investigated the translocation of EPA from membrane lipids to TAG in *N. gaditana* during N starvation using ^{13}C fatty acid labelling (Janssen et al. 2019). It was suggested that EPA was not only partitioned from membrane lipids to TAG but was also produced de novo in N-starved *N. gaditana*. A prominent and, so far, unique oleaginous microalga that produces de novo and accumulates omega-6 LC-PUFA in TAG up to 50% under N starvation is *L. incisa* (formerly *Parietochloris*) (Khozin-Goldberg et al. 2002a). ARA-biosynthesis genes are co-ordinately upregulated on the transcriptional level in this microalga (Iskandarov et al. 2010, 2011).

7.4 Recent Advances in the Metabolic Engineering of Microalgae to Enhance Production of LC-PUFA

The productivity of photosynthetic marine microalgae producing LC-PUFA is lower than that of heterotrophic microalgae producing LC-PUFA and is hampered by the relatively low productivity and efficiency of photo-assimilated carbon to biomass conversion. Other key parameters necessary for significant productivity improvements inevitably include resilience to environmental and mechanical stressors, to invading microorganisms and to a range of cultivation conditions. Regarding LC-PUFA production on a large scale, the suitable strain should combine the ability to accumulate large amounts of LC-PUFA with high biomass productivity. A promising approach is to use metabolic engineering to optimize natural LC-PUFA producer strains or to introduce LC-PUFA biosynthesis pathways in oleaginous microalgae.

To achieve this goal, the development of efficient transformation techniques for biotechnologically important species plays a critical role in strain improvement and sustainable delivery of enhanced production of omega-3 LC-PUFA. Nuclear transformation of various microalgae is now established, chloroplast transformation has been achieved for green, red and euglenoid algae, and organelle transformation (plastid and mitochondrial) is in progress (see Tirichine and Bowler 2011; Brodie et al. 2017). Various genetic transformation systems have been developed in diatoms and several other LC-PUFA-synthesizing species of industrial interest. Different diatoms have been successfully transformed by microparticle bombardment [*Chaetoceros* sp. (Miyagawa-Yamaguchi et al. 2011), *Cyclotella cryptica* and *Navicula saprophila* (Dunahay et al. 1995), *Cylindrotheca fusiformis* (Fischer et al. 1999; Poulsen and Kröger 2005), *Fistulifera* sp. (Muto et al. 2013), *P. tricornutum* (Apt et al. 1996; Zaslavskaia et al. 2000), *T. pseudonana* (Poulsen and Kröger 2005)] and electroporation [*P. tricornutum* (Miyahara et al. 2013) and *Chaetoceros gracilis* (Ifuku et al. 2015)].

Eustigmatophytes, representing another key group of EPA-producing microalgae with high industrial potential, are amendable for genetic engineering. The transformation protocols and tools have been established for the different species of *Nannochloropsis*. Electroporation (Kilian et al. 2011) was used to transform *N. oceanica* (Vieler et al. 2012), *N. gaditana* (Jinkerson et al. 2013) and *N. salina* (Koh et al. 2018). Recently, the genome sequence (Bhattacharya et al. 2013) and the development of a transformation system have been reported for the EPA-producing rhodophyte *Porphyridium purpureum* (Prasad et al. 2018; Li and Bock 2018), holding promise for modulation of its LC-PUFA biosynthesis. A nuclear transformation method was developed for *L. incisa*, a rare chlorophyte that is able to accumulate high amounts of ARA (Zorin et al. 2014). An *Agrobacterium*-mediated transformation method was reported for haptophytes *Isochrysis* sp. and *I. galbana*, facilitated by the development of the endogenous selectable marker, based on resistance to norflurazone (Prasad et al. 2014). The first stable genetic transformation of a coccolithophore, *Pleurochrysis carterae*, was established by means of polyethylene glycol (PEG)-mediated transfer of a bacterial hygromycin B-resistance gene (Endo et al. 2016). Another technology that has been successfully developed and demonstrated for different microalgae is based on episome delivery by biolistic bombardment, electroporation or bacterial conjugation (Karas et al. 2015; Diner et al. 2016; Slattery et al. 2018; Poliner et al. 2018a). Episomes can be a useful tool to develop marker-free algae strains.

Therefore, numerous works have demonstrated that species with high industrial potential can be engineered for desired traits. The current focus is on the stable and efficient integration of multiple genes for the expression of complex biosynthetic pathways.

Random transgene integration and position effect during nuclear transformation often render targeting of specific genes complicated. The rapid development of genome-editing techniques has greatly improved the potential for microalgal genetic engineering (Daboussi et al. 2014; Hopes et al. 2016; Nymark et al. 2016; Huang and Daboussi 2017). Studies in different laboratories have reported the successful application of CRISPR-Cas9-mediated gene editing in diatoms (*P. tricornutum* and *Thalassiosira pseudonana*) and eustigmatophytes (Hopes et al. 2016; Nymark et al. 2016; Poliner et al. 2018b; Russo et al. 2018). A model centric diatom *T. pseudonana* with a sequenced genome, established transformation systems and multiple biotechnology applications (Shrestha and Hildebrand 2017) has been shown amendable for CRISPR-Cas gene editing (Hopes et al. 2016). Genome editing has shown great promise in manipulating TAG storage in diatoms (see Sayanova et al. 2017) and *Nannochloropsis* (Ajjawi et al. 2017). Future research will show the potential of this tool regarding the manipulation of LC-PUFA biosynthesis.

7.4.1 Diatoms

7.4.1.1 Overexpressing Enzymes of Fatty Acid and TAG Biosynthetic Pathways

Recent studies have demonstrated the advantages of the genetic engineering approach to manipulate fatty acid composition and enhance the lipid accumulation in different species (Table 7.2). LC-PUFA biosynthesis could be improved through the overexpression of the biosynthesis enzymes and the regulation of genes involved in lipid metabolism. One of the first examples of genetic modification by direct targeting of lipid biosynthesis was the overexpression of an acetyl-CoA carboxylase (ACCase) isolated from the diatom *Cyclotella cryptica* in *C. cryptica* and *Navicula saprophila* (Dunahay et al. 1995). Although the ACCase activity enhanced to two- to threefolds, no significant increase in lipid accumulation was observed in transgenic diatoms. Therefore, it was considered that overexpression of ACC enzyme alone might not be sufficient to enhance the whole lipid biosynthesis pathway.

Overexpression of biosynthetic enzymes of LC-PUFA biosynthesis in *P. tricornutum* appears to be an effective way to improve the EPA and DHA content. *P. tricornutum* can accumulate EPA to levels up to 35% of TFA, representing a good source for the industrial production of this valuable omega-3 LC-PUFA. However, *P. tricornutum* does not accumulate significant levels of DHA. Heterologous expression of the *O. tauri* C20 Δ5-elongase in *P. tricornutum* to improve elongation of C20 EPA to C22 DPA resulted in an up to eightfold increase in DHA, and this level has been further increased with the co-expression of the acyl-CoA-dependent Δ6-desaturase (Hamilton et al. 2014). Importantly, no detrimental effects on *P. tricornutum* growth rates were observed in transgenic cells. This is the first report of enhancing the levels of omega-3 LC-PUFA in a diatom by iterative metabolic engineering. This successful example indicates that metabolic engineering of diatoms represents a promising strategy for optimizing accumulation of beneficial omega-3 LC-PUFA. The ability of an engineered strain (Pt_Elo5) to produce EPA and DHA was further tested during the scalable cultivation from the laboratory scale to a 550-L closed tubular photobioreactor (PBR) and a 1250-L raceway pond with artificial illumination (Hamilton et al. 2015). The transgenic line cultivated in PBR demonstrated a significant production of DHA over EPA, generating an EPA/DPA/DHA profile of 8.7%/4.5%/12.3% of TFA. A significant production of DHA has been observed in PBR, equivalent to ~2.1 mg L^{-1} DHA in a mid-exponentially growing algal culture. Cultivation of Pt_Elo5 in a raceway pond resulted in 36% of omega-3 LC-PUFA of TFA, with EPA and DHA constituting 23% and 10%, respectively. Interestingly, the transgenic strain grown in a raceway pond incorporated up to 12.8% EPA into neutral lipids. These results clearly demonstrate the efficacy of large-scale cultivation of the transgenic Pt_Elo5 and developing it as a platform for the industrial production of EPA and DHA.

A Δ5-desaturase gene, *PtD5b*, was cloned and overexpressed in *P. tricornutum* (Peng 2014). Fatty acid composition exhibited a significant alteration in the content of unsaturated fatty acids. Monounsaturated fatty acids and PUFA showed an increase of 75 and 64%, respectively. In transgenic cells, the levels of EPA showed an increase of 58%, and neutral lipid content was elevated up to 65%. Importantly, engineered cells showed a similar growth rate with the wild type; thus, biomass productivity was not affected. Another study from the same group reported that simultaneous expression of *PtD5b* and malonyl CoA-acyl carrier protein transacylase, catalysing the first committed step in type II fatty acid synthesis, further improved LC-PUFA levels; in particular, ca. 85 μg EPA/mg (dry weight) was achieved (Wang et al. 2017).

Recently, it was demonstrated that an overexpression of endogenous fatty acid elongase genes in *T. pseudonana* increased the levels of EPA and DHA up to 1.4- and 4.5-fold relatively (Cook and Hildebrand 2016).

Another potent approach to modifying the FA composition is through the expression *of* acyl-ACP thioesterases to control the chain length. Heterologous expression of plant acyl-ACP thioesterases specific towards the generation of short-chain lauric acid (C12:0) and myristic acid (C14:0) in *P. tricornutum* resulted in an increased accumulation of shorter chain length fatty acids (Radakovits et al. 2011). It is

Table 7.2 Recent advances in metabolic engineering of LC-PUFA-producing microalgae to modify fatty acid composition or enhance production of LC-PUFA

Approach	Target gene/pathway	Species	Outcome	References
Overexpression	ACCase	*C. cryptica*, *N. saprophila*	Two- to threefold increase in ACC activity, no change in lipid content	Dunahay et al. (1996)
	Δ5-elongase	*P. tricornutum*	Eightfold increase in DHA	Hamilton et al. (2014)
	Δ5-desaturase	*P. tricornutum*	Increase in neutral lipid content and EPA	Peng (2014)
	FA elongase Elo2	*T. pseudonana*	1.4-fold increase in EPA, 4.5-fold increase in DHA	Cook and Hildebrand (2016)
	Acyl-ACP thioesterase	*P. tricornutum*	Increased saturated fatty acids content	Radakovits et al. (2011)
	Thioesterase	*P. tricornutum*	Increased FA content up to 72%	Gong et al. (2011)
	GPAT	*P. tricornutum*	40% increase in EPA	Niu et al. (2016)
	DGAT2A	*P. tricornutum*	76% increase in LC-PUFA	Niu et al. (2013)
	Omega-3 desaturase	*Nannochloropsis*	Increase of EPA/ARA ratio	Kilian et al. (2013)
	Δ12-desaturase	*N. oceanica*	Increased production of omega-6 LC-PUFA under N starvation	Kaye et al. (2015)
	Expression of the desaturase genes in the EPA pathway using gene stacking	*N. oceanica*	A 25% increase in EPA (as %TFA)	Poliner et al. (2017)
	Δ5-desaturase	*L. incisa*	Complementation of the Δ5-desaturase mutant	Zorin et al. (2014)
Silencing	UGPase	*P. tricornutum*	45-fold increase in TAG	Daboussi et al. (2014)
	PDR	*P. tricornutum*	80% increase in lipid content	Ma et al. (2014)
	Nitrate reductase	*P. tricornutum*	43% increase in lipid content	Levitan et al. (2015a)
	Multi-functional lipase/phospholipase/acyltransferase	*T. pseudonana*	Increase in lipid yield without affecting growth	Trentacoste et al. (2013)
Knockout (Homologous recombination)	16:0 elongase	*N. oceanica*	Role of 16:0 to 18:0 elongation in regulating the flux towards EPA	Dolch et al. (2017)
	CTP:phosphocholine cytidylyltransferase 1 (*CCT1*) CTP: phosphoethanolamine cytidylyltransferase 1 (*PECT*1)	*N. oceanica*	Importance of PC and PE in EPA biosynthesis	Murakami et al. (2018)

plausible that overexpression of the acyl-ACP thioesterases with specificity for C18 FA may improve the LC-PUFA synthesis.

Overexpression of acyltransferases, mediating the assembly and exchange of acyl groups of glycerolipids and TAG, represents a valuable strategy not only for manipulating lipid content but also the fatty acid composition in microalgae. Engineering of glycerolipid biosynthesis by overexpressing the key enzymes involved in the TAG assembly may result in significant alterations in the LC-PUFA profile. Below some examples are given of when expression of acyltransferases coding genes resulted in the enhancement of LC-PUFA's relative proportions or content. Glycerol-3-phosphate acyltransferase (GPAT) is the critical enzyme that catalyses the first step of TAG biosynthesis: formation of lyso-phosphatidic acid by acylation of glycerol-3-phosphate. In *Arabidopsis thaliana,* out of nine GPATs, only the ER-localized GPAT9 is involved in TAG formation (Shockey et al. 2016; Singer et al. 2016). GPAT1 has been identified and overexpressed in *P. tricornutum* (Niu et al. 2016). Significant differences in lipid content and fatty acid composition were detected between transgenic and wild-type cells. Fatty acid composition showed higher proportions of unsaturated fatty acids compared to wild type. EPA levels increased by 40%, while those of C16:0 and monounsaturated fatty acids (MUFA) decreased. These results suggested that the GPAT1 could modulate TAG biosynthesis and its FA composition by mediating TAG formation from unsaturated acyl groups of chloroplast lipids (both 16:3 and 20:5 were increased in TAG). In the follow-up study, the authors co-expressed GPAT1 with lyso-phosphatidic acid acyltransferase LPAAT1 (Wang et al. 2018). The latter was shown to enhance TAG formation with enhanced deposition of LC-PUFA during nutrient-replete growth (Balamurugan et al. 2017). Both acyltransferases were localized to the chloroplast when expressed as GFP fusions (Wang et al. 2018). Two expression cassettes, harbouring either *GPAT1* or *LPAT1* and different selectable markers, were delivered into *P. tricornutum* by electroporation. When determined gravimetrically, a higher total lipid content (ca. 50% of dry biomass) was determined in the dual overexpression lines compared with the single overexpression GPAT1 or LPAT1 lines in previous studies (ca. 40% of dry biomass, respectively). However, no effect on LC-PUFA proportions or content was reported, which is difficult to explain by a cumulative effect.

The overexpression of type-2 diacylglycerol acyltransferase (DGAT2), which plays an important role in TAG assembly, could represent a valuable strategy for the manipulation of lipid content and fatty acid composition in microalgae. In higher plants, DGAT2 is considered to be key enzymes for incorporation of unusual fatty acids into TAG. The *DGAT2A* has been isolated from the marine diatom *P. tricornutum* (Niu et al. 2013). *DGAT2A* overexpression in *P. tricornutum* resulted in a 76% increase in LC-PUFA content of biomass and led to the production of more lipid droplets, while the neutral lipid content increased by 35%. The fatty acid composition showed a significant increase in the proportion of polyunsaturated fatty acids; in particular, EPA was increased by 76%. Moreover, the growth rate of transgenic microalgae remained similar, thereby maintaining a high biomass.

7.4.1.2 Blocking Competing Pathways

An alternative strategy to increase lipid accumulation is to block competing pathways to enhance the metabolic flux channelling to TAG biosynthesis.

Starch is a major carbon and energy storage compound in many microalgae. In diatoms, the fixed carbon is stored as lipids and the β-1,3-glucan, known as chrysolaminarin. Chrysolaminarin synthesis shares common carbon precursors with lipid synthesis, and UDP-glucose serves as the substrate for chrysolaminarin biosynthesis (Roessler 1987). The effects of the suppression of UDP-glucose pyrophosphorylase (UGPase) on chrysolaminarin biosynthesis and carbon allocation in *P. tricornutum* were investigated in two studies. The disruption of the UDP-glucose pyrophosphorylase using TALEN resulted in 45-fold increase in triacylglycerol accumulation in the transgenic strains (Daboussi et al. 2014). A complementary study based on antisense approach demonstrated that silencing of UDP-glucose pyrophosphorylase resulted in significant decrease in chrysolaminarin content and increase in lipid synthesis (Zhu et al. 2016). These results confirmed the feasibility of enhancing lipid production through redirecting photosynthetically assimilated carbon away from starch synthesis to neutral lipid synthesis.

The pyruvate dehydrogenase complex (PDC) catalyses the conversion of pyruvate into acetyl-CoA, NADH and CO_2, thus directing the carbon flow into TCA cycle and subsequently to oxidative phosphorylation or fatty acid synthesis. The PDC activity is primarily regulated by pyruvate dehydrogenase kinase (PDK), which phosphorylates and deactivates PDC. Ma et al. (2014) have identified a putative PDK gene (*PtPDK*) from *P. tricornutum* and generated a *PtPDK* antisense knockdown transgenic diatom (Ma et al. 2014). Neutral lipid content of transgenic cells increased up to 82% although the fatty acid composition was not altered. Transgenic cells showed slightly lower growth rate but similar to the wild-type cell size.

Another strategy to develop an efficient pathway to enhance lipid production is by suppressing a diatom's ability to incorporate nitrogen. Levitan et al. (2015a, b) have studied the effect of nitrogen stress on the remodelling process in *P. tricornutum* by knocking down the gene encoding for nitrate reductase, a key enzyme required for the assimilation of nitrate (Levitan et al. 2015a). A nitrate reductase knocked down resulted in a 43% increase in cellular lipid content, demonstrating that transgenic strains shunted ~40% more carbon towards TAGs than the wild type without losing photosynthetic capacity. This suggests that redirecting carbon towards lipid biosynthesis can be achieved by altering the expression of genes different from those directly involved with lipid biosynthesis.

A complimentary strategy to increase lipid accumulation is to decrease lipid catabolism. The targeted knockdown of lipid catabolism, and specifically lipases, could potentially improve the accumulation of free fatty acids and enhance lipid content with less impact on the primary carbon pathways associated with growth. Trentacoste et al. (2013) used antisense and RNAi constructs to disrupt lipid catabolism in *T. pseudonana* by targeting an enzyme with lipase activity. This targeted knockdown resulted in an increase in the accumulation of TAG and total lipid yield without impacting growth rate, under both continuous light and alternating light/dark and continuous growth and nutrient-replete condition vs. nutrient-deficient condition. Fatty acid analysis revealed that knockdown strains contained more 16:0 and 18:2 than wild-type species. These results demonstrate that disrupting lipid catabolism is a feasible approach to improve TAG accumulation in microalgae without affecting growth rates or biomass production.

These and other studies of the modulation of lipid biosynthesis in diatoms (Sayanova et al. 2017) support the feasibility of molecular engineering and pave the way for further improvements of LC-PUFA production in *P. tricornutum*, *T. pseudonana* and additional biotechnologically relevant diatoms.

7.4.2 Nannochloropsis

As noted in above sections, the eustigmatophytes of the *Nannochloropsis* genus are of high biotechnological and industrial interest given their inherent ability to attain high EPA contents in membrane lipids under favourable conditions and to accumulate high quantities of lipid reserves in the form of TAG under stressful conditions. *Nannochloropsis* microalgae were isolated from marine, brackish and fresh water environments, underlining the resilience of these microalgae to osmotic fluctuations. Regardless of the habitats, marine and freshwater *Nannochloropsis* species contain high concentrations of EPA, predominantly in the chloroplast membrane galactolipids, such as MGDG and DGDG, and in the extraplastidial lipids, in particular in DGTS (Schneider et al. 1995; Pal et al. 2013; Dolch et al. 2017). Different species of *Nannochloropsis* are cultivated worldwide in open ponds and photobioreactors. *Nannochloropsis gaditana* is regarded as an industrial species and has been considered for mass cultivation for biofuel production. Yet, the utility of *Nannochloropsis* will benefit from strain improvements, to enable better and more efficient exploitation in applied industrial settings.

Several studies used chemical mutagenesis to improve the EPA content or to study how the variations in LC-PUFA composition in *Nannochloropsis* impact the physiological performance. Random chemical and UV mutagenesis was shown to be a valuable tool for manipulating EPA levels in *Nannochloropsis* (Schneider et al. 1995). Mutants with enhancements in EPA content were reported in *Nannochloropsis oculata* ST-6 (Chaturvedi et al. 2004; Chaturvedi and Fujita 2006). The strains with a drastically reduced or a complete loss of linolenic acid 18:2 *n*-6 had a concomitant increase in EPA. Another study used *N*-methyl-*N*-nitrosourea-induced mutagenesis to isolate mutants resistant to quizalofop, an inhibitor of acetyl-CoA carboxylase (ACCase), which is the rate-limiting and committed enzyme in fatty acid synthesis. The tolerance to the herbicide and ACCase activity assays indicated that mutagenesis resulted in a mutation, conferring the resistance to quizalofop; however, the exact mutation was not reported. Herbicide-resistant mutants had elevated total fatty acid contents and also contained more omega-3 PUFA; this was accompanied by a concomitant increase in TAG. Altered lipid accumulation in EMS and UV-induced mutants of *N. salina* CCAP849/3 were obtained in a more recent study (Beacham et al. 2015). These works demonstrated that UV and chemical mutagenesis can be used for strain improvement and manipulating the fatty acid profile, implying that this method represents a potent tool for manipulating lipid and EPA contents in *Nannochloropsis*, avoiding or reducing the barrier of legislation for utilizing GMO organisms. Moreover, the potential of genetic modifications in *Nannochloropsis* via random mutagenesis was demonstrated by the selection of mutant strains that exhibited improved photosynthetic activity under the specific artificial conditions of a photobioreactor (Perin et al. 2015).

Nannochloropsis microalgae are currently at the forefront of microalgal research owing to several important advantages. Numerous biotechnological applications that target *Nannochloropsis* species as a robust producer of EPA and oils have benefited from next-generation sequencing technologies and omics studies. Multiple genomic sequences have recently become available for different *Nannochloropsis* species (Vieler et al. 2012; Wang et al. 2014; Starkenburg et al. 2014; Perin et al. 2015). The near-complete genome assembly in the industrial strain *N. gaditana* has been recently announced, which is expected to enable studies to better understand lipid metabolism, carbon partitioning, and photosynthetic efficiency and the ability to manipulate chromosomes (Schwartz et al. 2018). The wealth of genomic information paves the way for targeted genome modification for the enhancement of EPA content. Furthermore, *Nannochloropsis* microalgae are amendable for genetic engineering using homologous recombination (Jinkerson et al. 2013). Efficient bidirectional promoters that allow the compact assembly of co-regulated gene pairs have been developed (Kilian et al. 2011; Moog et al. 2015). Furthermore, the state-of-the-art techniques of genetic engineering for gene stacking and targeted gene disruption and repression in the *Nannochloropsis* genus are now available, opening a new avenue for synthetic biology in this microalga (Verruto et al. 2018; Poliner et al. 2018a). CRISP-CAS9-based tools were applied to enable the creation of loss-of-function mutants for strain development and functional studies (Poliner et al. 2018b). A Cas9-mediated genome-targeting platform (NgCas9+) was developed in *Nannochloropsis gaditana* (CCMP1894) and enabled the targeted disruption of the transcriptional factor that acts as a negative regulator of storage lipid formation (Ajjawi et al. 2017). A rational attenuation of expression of the transcriptional factor, Zn(ii)2Cys6 (ZnCys) encoded by Naga_100104g18, not only led to the enhancement of lipid productivity, but allowed for the significant storage lipid accumulation during growth in nutrient-replete cells. Hence, this study established metabolic engineering as a valuable tool to control and substantially improve phototrophic conversion of CO_2 to lipids with minor effect on growth.

A genetic engineering toolkit that includes an endogenous bidirectional promoter and optimized peptide bond skipping 2A peptides has been developed, allowing for multiple gene expression (Poliner et al. 2017). It was applied to enhance EPA production in *N. oceanica* CCMP1779 by co-expression of endogenous desaturases, mediating EPA biosynthesis. A set of genes encoding membrane-bound endoplasmic reticulum (ER)-localized enzymes involved in LC-PUFA biosynthesis in

N. oceanica were identified based on the genome information (Vieler et al. 2012), namely putative Δ12-, Δ6-, Δ5- and ω3-desaturases and Δ6-PUFA elongase. These were expressed as single or up to three desaturase-encoding cDNAs to demonstrate the feasibility of multiple gene stacking in *Nannochloropsis* (Poliner et al. 2017). In the earlier study by Kaye et al. (2015), overexpression of a single CDS for the native Δ12FAD under the control of the endogenous stress-inducible promoter resulted in an increased conversion of 18:1*n*-9 to 18:2*n*-6 and some shift towards the production of LC-PUFA under *N* starvation. The rationale for using the promoter of an N-starvation-inducible gene was to uncouple the expression of the Δ12FAD gene from its downregulation under nitrogen starvation, using only homologous genes and promoters (Kaye et al. 2015). Poliner et al. (2017) demonstrated that overproduction of the native Δ12 FAD, along with Δ5 FAD (converting DGLA to ARA), coding sequences under the control of the *EF* promoter led to an increase in EPA in *N. oceanica* CCMP1779 under nutrient-replete conditions (Poliner et al. 2017).

7.4.3 Systems Biology Approaches in Studying Algal Lipid LC-PUFA Production

During the last few years, nuclear genomes of various industrially important strains of microalgae have been sequenced, and information on mitochondrion and chloroplast genomes has become available (see Tirichine and Bowler 2011; Brodie et al. 2017). The genome sequences of *P. tricornutum*, *T. pseudonana*, *Emiliania huxleyi* and different ecotypes of *Ostreococcus tauri* have allowed the functional analysis of several novel activities involved in omega-3 LC-PUFA biosynthesis (Domergue et al. 2002, 2003, 2005; Tonon et al. 2005b; Sayanova et al. 2011; Vaezi et al. 2013; Shi et al. 2018). Based on genomic and transcriptomic data, several metabolic networks for lipid biosynthesis and PUFA biosynthesis have been constructed and pathways have been elucidated in *Nannochloropsis* (Vieler et al. 2012) and the oleaginous diatoms *P. tricornutum*, *F. solaris* and *Nitzschia* sp. (Cheng et al. 2014; Tanaka et al. 2015; Dolch and Maréchal 2015). Omics research represents a powerful tool for investigating changes occurring in algal species during LC-PUFA biosynthesis and TAG accumulation, and several transcriptomics, metabolomics and proteomics studies are now available (Jamers et al. 2009; Guarnieri and Pienkos 2015; Popko et al. 2016; Smith et al. 2016; León-Saiki et al. 2017; Remmers et al. 2018). An increasing number of multiomics studies have examined metabolic networks linking central carbon, nitrogen and lipid metabolic pathways (Abida et al. 2015; Levitan et al. 2015b; Levering et al. 2016; Longworth et al. 2016).

Advances in genetic transformation techniques and next-generation sequencing technology, together with transcriptomics, proteomics and metabolomics research, offer the possibility of enhancing the yields of omega-3 LC-PUFA in different microalgal species through metabolic engineering and synthetic biology approaches.

Genetic engineering requires procedures for introducing a set of genes into host genome, and the genetic techniques available for such integration represent a major bottleneck in the development of industrial sources for high-value lipids. Synthetic biology approach will allow to optimize genetic systems currently employed by designing robust and efficient DNA assembly methods based on a modular cloning toolbox. This will facilitate the optimization of LC-PUFA metabolic pathways in transgenic microalgae by improving metabolic flux and increasing stability of transgenes, resulting in significant accumulation of the desired product.

Synthetic biology may be described as the design of biological systems based on engineering principles. The successful engineering of complex biological systems involves an iterative process based on rational design, establishing biological parts and assembling them into genetic circuits and in vitro and in vivo testing (Fig. 7.2).

A basic requirement for synthetic biology is the availability of standardized and well-characterized building blocks, the bio-bricks, and efficient DNA assembly methods. In recent years, significant progress has been made in establishing standardized biological parts and assembling them into genetic circuits in microalgae (Scaife et al. 2015; Fu et al. 2017; Poliner et al. 2018a; Crozet et al. 2018). However, a Modular Cloning kit to use in diatoms similar to that developed for the green microalga *Chlamydomonas reinhardtii* is still under construction. DNA parts, including regulatory elements (promoters and terminators), reporter systems, selectable markers and delivery methods have been developed for diatoms and are available for genome engineering (Huang and Daboussi 2017). Approaches based on one-step assembly cloning technology, including BioBrick, Golden Gate and Gibson (Patron et al. 2015; Vazquez-Vilar et al. 2018), can be optimized and used in microalgae. Thus, there is an urgent need to build up a repository of standard genetic parts for use in diatom hosts for the sustainable production of valuable lipids.

7.4.4 Alternative Approaches: Adaptive Laboratory Evolution and Chemical Genetics for LC-PUFA-Producing Strains Improvement

Adaptive laboratory evolution (ALE) has become a frequent method in biotechnology to induce and accumulate adaptive changes in microbial cell populations during long-term selection under specific growth conditions or selection pressure (Dragosits and Mattanovich 2013). In regard to LC-PUFA-producers, ALE-guided approaches were instrumental to increase biomass and DHA production in the heterotrophic marine protist *Schizochytrium* sp. (Sun et al. 2016, 2018) and the heterotrophic microalga *Crypthecodinium cohnii* (Li et al. 2017b). In recent years, ALE showed promises in the development of biotechnologically valuable traits in photosynthetic microalgae (Perrineau et al. 2014; Su et al. 2018). Similarly, ALE approaches can be used to modify and improve LC-PUFA production. The use of chemical genetics with libraries of small biologically active molecules and chemically annotated drugs also has great potential in identifying and manipulating critical steps in microalgal lipid metabolism (Conte et al. 2018; Wase et al. 2018).

7.4.5 The Initial Risk Assessment of Genetically Modified Microalgae Cultivation for Large-Scale LC-PUFA Production

Dietary LC-PUFA supplements are becoming increasingly popular. The new sources of LC-PUFA in genetically modified organisms, such as oil-seed plants, heterotrophic microorganisms and photosynthetic microalgae, have great potential to play an important role in increasing the supply of *n*-3 LC-PUFA (Sayanova and Napier 2016; Sprague et al. 2017; Tocher et al. 2019). Whether or not microalgae, genetically modified for improved LC-PUFA production, would get to the food and feed markets largely depends on environmental, economic and human/animal health impacts, and public acceptance. Consumer acceptance is the major challenge both in EU countries and in the USA. The issues pertaining to large-scale cultivation of microalgae, their possible escapes to the environment and means of protection are discussed in Henley et al. (2013) and Szyjka et al. (2017). Current practices and the GMO directives within the EU define GM microalgae as genetically modified microorganisms (Spicer and Molnar 2018). As stated in Spicer and Molnar (2018), modified microalgae fall under umbrella of GMO but exemptions are provided for those produced through chemical or UV/radiation mutagenesis. Large-scale testing of GM strains of microalgae has begun in the USA, mainly for biofuel purposes, and few examples of joint ventures between industry and academic institutions have been established. To address the inevitable challenges posed by phototrophic GM microalgae cultivation as well as stability of engineered traits, Szyjka et al. (2017) examined the potential of genetically engineered algae to disperse, establish themselves in wild communities and persist in the presence of native species. The authors concluded that extensive testing of the stability of algal phenotypes in an open outdoor environment will be required since those phenotypes could vary from

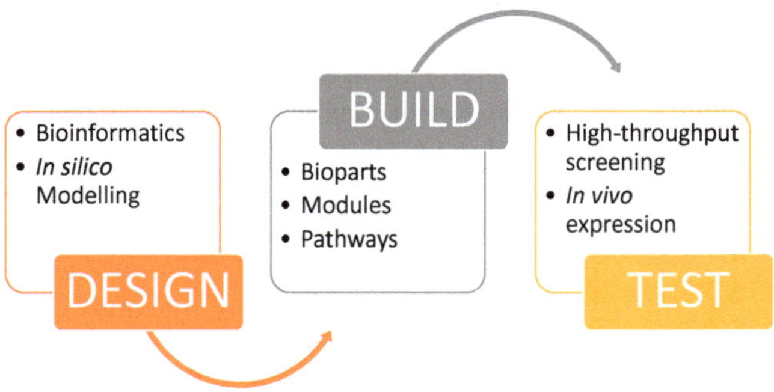

Fig. 7.2 Schematic representation of synthetic biology workflow

laboratory to outdoor conditions. Furthermore, neither of species tested in this work were able to outcompete or disrupt native populations of algae when inoculated in waters taken from five local lakes (Szyjka et al. 2017). It is apparent that further investigations are required to warrant safe and responsible cultivation of GM microalgae at industrial scales.

Perspectives and possible applications of GM organisms, including microalgae, in aquaculture are discussed in Sprague et al. (2017) and Tocher et al. (2019). In a series of trials performed with the EPA-enriched oil from transgenic Camelina as ingredients in fish (salmon and gilthead sea bream) feeds, no differences in growth performance, feed efficiency or survival between dietary treatments have been observed (Betancor et al. 2015; Tejera et al. 2016).

7.5 Perspectives and Conclusions

During the last decade, intense research has been performed to discover and maximize the mechanisms underlying the microalgal production of ω-3 LC-PUFA. Based on the recent advances described above, it should be clear that rational metabolic engineering of microalgae has a promising future for designing strains with improved desired fatty acid composition and content. These approaches will be facilitated by the rapid development of robust and efficient synthetic biology tools in microalgae (Georgianna and Mayfield 2012; Scaife et al. 2015; Fu et al. 2017; Crozet et al. 2018; Poliner et al. 2018a). We envisage that recent advances made in the field of synthetic biology will be successfully applied to the field of LC-PUFA production and metabolism. In the future, it is likely that a combination of strategies will provide the most feasible approach to improving LC-PUFA production and accumulation in TAG without inhibiting growth.

Acknowledgements I.K.G. acknowledges support by a grant from the Ministry of Science, Technology and Space, Israel. Rothamsted Research receives grant-aided support from the Biotechnology and Biological Sciences Research Council (BBSRC), UK. This project was supported by BBSRC grant BB/L002957/1.

References

Aarsetoey H, Aarsetoey R, Lindner T et al (2011) Low levels of the omega-3 index are associated with sudden cardiac arrest and remain stable in survivors in the subacute phase. Lipids 46:151–161. https://doi.org/10.1007/s11745-010-3511-3

Abida H, Dolch L-J, Meï C et al (2015) Membrane glycerolipid remodeling triggered by nitrogen and phosphorus starvation in *Phaeodactylum tricornutum*. Plant Physiol 167:118–136. https://doi.org/10.1104/pp.114.252395

Agostoni C (2008) Role of long-chain polyunsaturated fatty acids in the first year of life. J Pediatr Gastroenterol Nutr 47:S41–S44. https://doi.org/10.1097/01.mpg.0000338811.52062.b2

Ajjawi I, Verruto J, Aqui M et al (2017) Lipid production in *Nannochloropsis gaditana* is doubled by decreasing expression of a single transcriptional regulator. Nat Biotechnol 35:647–652. https://doi.org/10.1038/nbt.3865

Alves SP, Mendonça SH, Silva JL, Bessa RJB (2018) *Nannochloropsis oceanica*, a novel natural source of rumen-protected eicosapentaenoic acid (EPA) for ruminants. Sci Rep 8:10269. https://doi.org/10.1038/s41598-018-28576-7

Andrianasolo EH, Haramaty L, Vardi A et al (2008) Apoptosis-inducing galactolipids from a cultured marine diatom, *Phaeodactylum tricornutum*. J Nat Prod 71:1197–1201. https://doi.org/10.1021/np800124k.Apoptosis-Inducing

Apt KE, Grossman AR, Kroth-Pancic PG (1996) Stable nuclear transformation of the diatom *Phaeodactylum tricornutum*. Mol Gen Genet 252:572–579. https://doi.org/10.1007/BF02172403

Arao T, Yamada M (1994) Biosynthesis of polyunsaturated fatty acids in the marine diatom, *Phaeodactylum tricornutum*. Phytochemistry 35:1177–1181. https://doi.org/10.1016/S0031-9422(00)94817-9

Aussant J, Guihéneuf F, Stengel DB (2018) Impact of temperature on fatty acid composition and nutritional value in eight species of microalgae. Appl Microbiol Biotechnol 102:5279–5297. https://doi.org/10.1007/s00253-018-9001-x

Balamurugan S, Wang X, Wang H-L et al (2017) Occurrence of plastidial triacylglycerol synthesis and the potential regulatory role of AGPAT in the model diatom *Phaeodactylum tricornutum*. Biotechnol Biofuels 10:97. https://doi.org/10.1186/s13068-017-0786-0

Barclay WR, Meager KM, Abril JR (1994) Heterotrophic production of long chain omega-3 fatty acids utilizing algae and algae-like microorganisms. J Appl Phycol 6:123–129. https://doi.org/10.1007/BF02186066

Beacham TA, Macia VM, Rooks P et al (2015) Altered lipid accumulation in *Nannochloropsis salina* CCAP849/3 following EMS and UV induced mutagenesis. Biotechnol Rep (Amst) 7:87–94. https://doi.org/10.1016/j.btre.2015.05.007

Bell MV, Pond D (1996) Lipid composition during growth of motile and coccolith forms of *Emiliania huxleyi*. Phytochemistry 41:465–471. https://doi.org/10.1016/0031-9422(95)00663-X

Bell JG, Sargent JR (2003) Arachidonic acid in aquaculture feeds: current status and future opportunities. Aquaculture 218:491–499. https://doi.org/10.1016/S0044-8486(02)00370-8

Bendif EM, Probert I, Schroeder DC, de Vargas C (2013) On the description of *Tisochrysis lutea* gen. nov. sp. nov. and *Isochrysis nuda* sp. nov. in the Isochrysidales, and the transfer of Dicrateria to the Prymnesiales (Haptophyta). J Appl Phycol 25:1763–1776. https://doi.org/10.1007/s10811-013-0037-0

Betancor MB, Sprague M, Sayanova O et al (2015) Evaluation of a high-EPA oil from transgenic *Camelina sativa* in feeds for Atlantic salmon (*Salmo salar* L.): effects on tissue fatty acid composition, histology and gene expression. Aquaculture 444:1–12. https://doi.org/10.1016/j.aquaculture.2015.03.020

Betancor MB, Sprague M, Montero D et al (2016) Replacement of marine fish oil with de novo omega-3 oils from transgenic *Camelina sativa* in feeds for gilthead sea bream (*Sparus aurata* L.). Lipids 51:1171–1191. https://doi.org/10.1007/s11745-016-4191-4

Bhattacharya D, Price DC, Chan CX et al (2013) Genome of the red alga *Porphyridium purpureum*. Nat Commun 4:1941. https://doi.org/10.1038/ncomms2931

Bigogno C, Khozin-Goldberg I, Adlerstein D, Cohen Z (2002a) Biosynthesis of arachidonic acid in the oleaginous microalga *Parietochloris incisa* (Chlorophyceae): radiolabeling studies. Lipids 37:209–216. https://doi.org/10.1007/s11745-002-0882-6

Bigogno C, Khozin-Goldberg I, Boussiba S et al (2002b) Lipid and fatty acid composition of the green oleaginous alga *Parietochloris incisa*, the richest plant source of arachidonic acid. Phytochemistry 60:497–503. https://doi.org/10.1016/S0031-9422(02)00100-0

Bigogno C, Khozin-Goldberg I, Cohen Z (2002c) Accumulation of arachidonic acid-rich triacylglycerols in the microalga *Parietochloris incisa* (Trebouxiophyceae, Chlorophyta). Phytochemistry 60:135–143. https://doi.org/10.1016/S0031-9422(02)00037-7

Boller T, Felix G (2009) A renaissance of elicitors: perception of microbe-associated molecular patterns and danger signals by pattern-recognition receptors. Annu Rev Plant Biol 60:379–406. https://doi.org/10.1146/annurev.arplant.57.032905.105346

Boussiba S, Vonshak A, Cohen Z et al (1987) Lipid and biomass production by the halotolerant microalga *Nannochloropsis salina*. Biomass 12:37–47. https://doi.org/10.1016/0144-4565(87)90006-0

Brodie J, Chan CX, De Clerck O et al (2017) The algal revolution. Trends Plant Sci 22:726–738

Burja AM, Radianingtyas H, Windust A, Barrow CJ (2006) Isolation and characterization of polyunsaturated fatty acid producing Thraustochytrium species: screening of strains and optimization of omega-3 production. Appl Microbiol Biotechnol 72:1161–1169. https://doi.org/10.1007/s00253-006-0419-1

Calder PC (2015) Marine omega-3 fatty acids and inflammatory processes: effects, mechanisms and clinical relevance. BBA – Mol Cell Biol Lipids 1851:469–484. https://doi.org/10.1016/j.bbalip.2014.08.010

Camacho-Rodríguez J, Cerón-García MC, Fernández-Sevilla JM, Molina-Grima E (2015) Genetic algorithm for the medium optimization of the microalga *Nannochloropsis gaditana* cultured to aquaculture. Bioresour Technol 177:102–109. https://doi.org/10.1016/j.biortech.2014.11.057

Carrier G, Garnier M, Le Cunff L et al (2014) Comparative transcriptome of wild type and selected strains of the microalgae *Tisochrysis lutea* provides insights into the genetic basis, lipid metabolism and the life cycle. PLoS One 9:e86889. https://doi.org/10.1371/journal.pone.0086889

Chaturvedi R, Fujita Y (2006) Isolation of enhanced eicosapentaenoic acid producing mutants of *Nannochloropsis oculata* ST-6 using ethyl methane sulfonate induced mutagenesis techniques and their characterization at mRNA transcript level. Phycol Res 54:208–219. https://doi.org/10.1111/j.1440-1835.2006.00428.x

Chaturvedi R, Uppalapati SR, Alamsjah MA, Fujita Y (2004) Isolation of quizalofop-resistant mutants of *Nannochloropsis oculata* (Eustigmatophyceae) with high eicosapentaenoic acid following N-methyl-N-nitrosourea-induced random mutagenesis. J Appl Phycol 16:135–144. https://doi.org/10.1023/B:JAPH.0000044826.70360.8e

Chauton MS, Reitan KI, Norsker NH et al (2015) A techno-economic analysis of industrial production of marine microalgae as a source of EPA and DHA-rich raw material for aquafeed: research challenges and possibilities. Aquaculture 436:95–103. https://doi.org/10.1016/j.aquaculture.2014.10.038

Cheng J, Feng J, Cheng R et al (2014) Gene expression and metabolic pathways related to cell growth and lipid synthesis in diatom *Nitzschia* ZJU2 after two rounds of mutagenesis by γ-rays. RSC Adv 4:28463–28470. https://doi.org/10.1039/C4RA03359K

Conte M, Lupette J, Seddiki K et al (2018) Screening for biologically annotated drugs that trigger triacylglycerol accumulation in the diatom phaeodactylum. Plant Physiol 177:532–552. https://doi.org/10.1104/pp.17.01804

Cook O, Hildebrand M (2016) Enhancing LC-PUFA production in *Thalassiosira pseudonana* by overexpressing the endogenous fatty acid elongase genes. J Appl Phycol 28:897–905. https://doi.org/10.1007/s10811-015-0617-2

Crozet P, Navarro FJ, Willmund F et al (2018) Birth of a photosynthetic chassis: a MoClo toolkit enabling synthetic biology in the microalga *Chlamydomonas reinhardtii*. ACS Synth Biol 7:2074–2086. https://doi.org/10.1021/acssynbio.8b00251

Daboussi F, Leduc S, Maréchal A et al (2014) Genome engineering empowers the diatom *Phaeodactylum tricornutum* for biotechnology. Nat Commun 5:3831. https://doi.org/10.1038/ncomms4831

Dagar A, Zilberg D, Cohen Z et al (2010) Short-term dietary supplementation with the microalga *Parietochloris incisa* enhances stress resistance in guppies *Poecilia reticulata*. Aquac Res 41:267–277. https://doi.org/10.1111/j.1365-2109.2009.02329.x

Das UN (2006) Essential fatty acids: biochemistry, physiology and pathology. Biotechnol J 1:420–439. https://doi.org/10.1002/biot.200600012

Desbois AP, Mearns-Spragg A, Smith VJ (2009) A fatty acid from the diatom *Phaeodactylum tricornutum* is antibacterial against diverse bacteria including multi-resistant *Staphylococcus aureus* (MRSA). Mar Biotechnol 11:45–52. https://doi.org/10.1007/s10126-008-9118-5

Diner RE, Bielinski VA, Dupont CL et al (2016) Refinement of the diatom episome maintenance sequence and improvement of conjugation-based DNA delivery methods. Front Bioeng Biotechnol 4:65. https://doi.org/10.3389/fbioe.2016.00065

Dolch L-J, Maréchal E (2015) Inventory of fatty acid desaturases in the pennate diatom *Phaeodactylum tricornutum*. Mar Drugs 13:1317–1339. https://doi.org/10.3390/md13031317

Dolch L-J, Rak C, Perin G et al (2017) A palmitic acid elongase affects eicosapentaenoic acid and plastidial monogalactosyldiacylglycerol levels in Nannochloropsis. Plant Physiol 173:742–759. https://doi.org/10.1104/pp.16.01420

Domergue F, Lerchl J, Zähringer U, Heinz E (2002) Cloning and functional characterization of *Phaeodactylum tricornutum* front-end desaturases involved in eicosapentaenoic acid biosynthesis. Eur J Biochem 269:4105–4113

Domergue F, Abbadi A, Ott C et al (2003) Acyl carriers used as substrates by the desaturases and elongases involved in very long-chain polyunsaturated fatty acids biosynthesis reconstituted in yeast. J Biol Chem 278:35115–35126. https://doi.org/10.1074/jbc.M305990200

Domergue F, Abbadi A, Zähringer U et al (2005) In vivo characterization of the first acyl-CoA Delta6-desaturase from a member of the plant kingdom, the microalga *Ostreococcus tauri*. Biochem J 389:483–490. https://doi.org/10.1042/BJ20050111

Draaisma RB, Wijffels RH, (Ellen) Slegers P et al (2013) Food commodities from microalgae. Curr Opin Biotechnol 24:169–177. https://doi.org/10.1016/J.COPBIO.2012.09.012

Dragosits M, Mattanovich D (2013) Adaptive laboratory evolution – principles and applications for biotechnology. Microb Cell Fact 12:64. https://doi.org/10.1186/1475-2859-12-64

Dunahay TG, Jarvis EE, Roessler PG (1995) Genetic transformation of the diatoms *Cyclotella cryptica* and *Navicula saprophila*. J Phycol 31:1004–1012. https://doi.org/10.1111/j.0022-3646.1995.01004.x

Dunahay TG, Jarvis EE, Dais SS et al (1996) Manipulation of microalgal lipid production using genetic engineering. Appl Biochem Biotechnol 57:223. https://doi.org/10.1007/BF02941703

Endo H, Yoshida M, Uji T et al (2016) Stable nuclear transformation system for the coccolithophorid alga *Pleurochrysis carterae*. Sci Rep 6:1–10. https://doi.org/10.1038/srep22252

FAO (2016) The state of world fisheries and aquaculture 2016

Fischer H, Robl I, Sumper M, Kroger N (1999) Targeting and covalent modification of cell wall and membrane proteins heterologously expressed in the diatom *Cylindrotheca fusiformis* (Bacillariophyceae). J Phycol 35:113–120. https://doi.org/10.1046/j.1529-8817.1999.3510113.x

Freire I, Cortina-Burgueño A, Grille P et al (2016) *Nannochloropsis limnetica*: a freshwater microalga for marine aquaculture. Aquaculture 459:124–130. https://doi.org/10.1016/J.AQUACULTURE.2016.03.015

Frigolet ME, Gutiérrez-Aguilar R (2017) The role of the novel lipokine palmitoleic acid in health and disease. Adv Nutr An Int Rev J 8:173S–181S. https://doi.org/10.3945/an.115.011130

Fu HY, Picot D, Choquet Y et al (2017) Redesigning the QA binding site of Photosystem II allows reduction of exogenous quinones. Nat Commun 8:15274. https://doi.org/10.1038/ncomms15274

Fukuchi-Mizutani M, Tasaka Y, Tanaka Y et al (1998) Characterization of delta 9 acyl-lipid desaturase homologues from *Arabidopsis thaliana*. Plant Cell Physiol 39:247–253

Funk CD (2001) Prostaglandins and leukotrienes: advances in eicosanoid biology. Science 294:1871–1875

García JL, de Vicente M, Galán B (2017) Microalgae, old sustainable food and fashion nutraceuticals. Microb Biotechnol 10:1017–1024. https://doi.org/10.1111/1751-7915.12800
Georgianna DR, Mayfield SP (2012) Exploiting diversity and synthetic biology for the production of algal biofuels. Nature 488:329–335. https://doi.org/10.1038/nature11479
Gerber M (2012) Omega-3 fatty acids and cancers: a systematic update review of epidemiological studies. Br J Nutr 107:S228–S239. https://doi.org/10.1017/S0007114512001614
Gill S, Willette S, Dungan B et al (2018) Suboptimal temperature acclimation affects Kennedy pathway gene expression, lipidome and metabolite profile of *Nannochloropsis salina* during PUFA enriched TAG synthesis. Mar Drugs 16:425. https://doi.org/10.3390/md16110425
Gong Y, Guo X, Wan X et al (2011) Characterization of a novel thioesterase (PtTE) from *Phaeodactylum tricornutum*. J Basic Microbiol 51:666–672. https://doi.org/10.1002/jobm.201000520
Guarnieri MT, Pienkos PT (2015) Algal omics: unlocking bioproduct diversity in algae cell factories. Photosynth Res 123:255–263
Guihéneuf F, Stengel D (2013) LC-PUFA-enriched oil production by microalgae: accumulation of lipid and triacylglycerols containing n-3 LC-PUFA is triggered by nitrogen limitation and inorganic carbon availability in the marine haptophyte *Pavlova lutheri*. Mar Drugs 11:4246–4266. https://doi.org/10.3390/md11114246
Guihéneuf F, Mimouni V, Ulmann L, Tremblin G (2009) Combined effects of irradiance level and carbon source on fatty acid and lipid class composition in the microalga *Pavlova lutheri* commonly used in mariculture. J Exp Mar Bio Ecol 369:136–143. https://doi.org/10.1016/J.JEMBE.2008.11.009
Guihéneuf F, Fouqueray M, Mimouni V et al (2010) Effect of UV stress on the fatty acid and lipid class composition in two marine microalgae *Pavlova lutheri* (Pavlovophyceae) and *Odontella aurita* (Bacillariophyceae). J Appl Phycol 22:629–638. https://doi.org/10.1007/s10811-010-9503-0
Guschina IA, Harwood JL (2006) Lipids and lipid metabolism in eukaryotic algae. Prog Lipid Res 45:160–186. https://doi.org/10.1016/j.plipres.2006.01.001
Haimeur A, Ulmann L, Mimouni V et al (2012) The role of *Odontella aurita*, a marine diatom rich in EPA, as a dietary supplement in dyslipidemia, platelet function and oxidative stress in high-fat fed rats. Lipids Health Dis 11:147. https://doi.org/10.1186/1476-511X-11-147
Hamilton ML, Haslam RP, Napier JA, Sayanova O (2014) Metabolic engineering of *Phaeodactylum tricornutum* for the enhanced accumulation of omega-3 long chain polyunsaturated fatty acids. Metab Eng 22:3–9. https://doi.org/10.1016/j.ymben.2013.12.003
Hamilton ML, Warwick J, Terry A et al (2015) Towards the industrial production of omega-3 long chain polyunsaturated fatty acids from a genetically modified diatom *Phaeodactylum tricornutum*. PLoS One 10:e0144054. https://doi.org/10.1371/journal.pone.0144054
Harwood JL, Jones AL (1989) Lipid metabolism in algae. In: Callow JA (ed) Advances in botanical research. Academic, London, pp 1–53
Haslam RP, Sayanova O, Kim HJ et al (2016) Synthetic redesign of plant lipid metabolism. Plant J 87:76–86. https://doi.org/10.1111/tpj.13172
Henley WJ, Litaker RW, Novoveská L et al (2013) Initial risk assessment of genetically modified (GM) microalgae for commodity-scale biofuel cultivation. Algal Res 2:66–77. https://doi.org/10.1016/J.ALGAL.2012.11.001
Hodgson PA, Henderson RJ, Sargent JR, Leftley JW (1991) Patterns of variation in the lipid class and fatty acid composition of *Nannochloropsis oculata* (Eustigmatophyceae) during batch culture. J Appl Phycol 3:169–181. https://doi.org/10.1007/BF00003699
Hoffmann M, Wagner M, Abbadi A et al (2008) Metabolic engineering of ω3-very long chain polyunsaturated fatty acid production by an exclusively acyl-CoA-dependent pathway. J Biol Chem 283:22352–22362. https://doi.org/10.1074/jbc.M802377200
Hopes A, Nekrasov V, Kamoun S, Mock T (2016) Editing of the urease gene by CRISPR-Cas in the diatom *Thalassiosira pseudonana*. Plant Methods 12. https://doi.org/10.1186/s13007-016-0148-0

Horrocks LA, Yeo YK (1999) Health benefits of docosahexaenoic acid (DHA). Pharmacol Res 40:211–225

Hsiao TY, Blanch HW (2005) Physiological studies of eicosapentaenoic acid production in the marine microalga *Glossomastix chrysoplasta*. Biotechnol Bioeng 93:465–475. https://doi.org/10.1002/bit.20761

Huang W, Daboussi F (2017) Genetic and metabolic engineering in diatoms. Philos Trans R Soc B Biol Sci 372:20160411. https://doi.org/10.1098/rstb.2016.0411

Hubert F, Poisson L, Loiseau C et al (2017) Lipids and lipolytic enzymes of the microalga *Isochrysis galbana*. OCL 24:D407. https://doi.org/10.1051/ocl/2017023

Huerlimann R, de Nys R, Heimann K (2010) Growth, lipid content, productivity, and fatty acid composition of tropical microalgae for scale-up production. Biotechnol Bioeng 107:245–257. https://doi.org/10.1002/bit.22809

Hulatt CJ, Wijffels RH, Bolla S, Kiron V (2017) Production of fatty acids and protein by nannochloropsis in flat-plate photobioreactors. PLoS One 12:e0170440. https://doi.org/10.1371/journal.pone.0170440

Ifuku K, Yan D, Miyahara M et al (2015) A stable and efficient nuclear transformation system for the diatom *Chaetoceros gracilis*. Photosynth Res 123:203–211. https://doi.org/10.1007/s11120-014-0048-y

Innis SM (2000) The role of dietary n–6 and n–3 fatty acids in the developing brain. Dev Neurosci 22:474–480. https://doi.org/10.1159/000017478

Iskandarov U, Khozin-Goldberg I, Ofir R, Cohen Z (2009) Cloning and characterization of the Δ6 polyunsaturated fatty acid elongase from the green microalga *Parietochloris incisa*. Lipids 44:545–554. https://doi.org/10.1007/s11745-009-3301-y

Iskandarov U, Khozin-Goldberg I, Cohen Z (2010) Identification and characterization of Δ12, Δ6, and Δ5 desaturases from the green microalga *Parietochloris incisa*. Lipids 45:519–530. https://doi.org/10.1007/s11745-010-3421-4

Iskandarov U, Khozin-Goldberg I, Cohen Z (2011) Selection of a DGLA-producing mutant of the microalga *Parietochloris incisa*: I. Identification of mutation site and expression of VLC-PUFA biosynthesis genes. Appl Microbiol Biotechnol 90:249–256. https://doi.org/10.1007/s00253-010-3005-5

Jamers A, Blust R, De Coen W (2009) Omics in algae: paving the way for a systems biological understanding of algal stress phenomena? Aquat Toxicol 92:114–121. https://doi.org/10.1016/j.aquatox.2009.02.012

Janssen JH, Lamers PP, de Vos RCH et al (2019) Translocation and de novo synthesis of eicosapentaenoic acid (EPA) during nitrogen starvation in *Nannochloropsis gaditana*. Algal Res 37:138–144. https://doi.org/10.1016/j.algal.2018.11.025

Jiang M, Guo B, Wan X et al (2014) Isolation and characterization of the diatom phaeodactylum Δ5-elongase gene for transgenic LC-PUFA production in *Pichia pastoris*. Mar Drugs 12:1317–1334. https://doi.org/10.3390/md12031317

Jinkerson RE, Radakovits R, Posewitz MC (2013) Genomic insights from the oleaginous model alga *Nannochloropsis gaditana*. Bioengineered 4:37–43. https://doi.org/10.4161/bioe.21880

Karas BJ, Diner RE, Lefebvre SC et al (2015) Designer diatom episomes delivered by bacterial conjugation. Nat Commun 6:6925. https://doi.org/10.1038/ncomms7925

Kaulmann U, Hertweck C (2002) Biosynthesis of polyunsaturated fatty acids by polyketide synthases. Angew Chem Int Ed 41:1866. https://doi.org/10.1002/1521-3773

Kaye Y, Grundman O, Leu S et al (2015) Metabolic engineering toward enhanced LC-PUFA biosynthesis in *Nannochloropsis oceanica*: overexpression of endogenous Δ12 desaturase driven by stress-inducible promoter leads to enhanced deposition of polyunsaturated fatty acids in TAG. Algal Res 11:387–398. https://doi.org/10.1016/j.algal.2015.05.003

Khozin I, Adlerstein D, Bigogno C et al (1997) Elucidation of the biosynthesis of eicosapentaenoic acid in the microalga *Porphyridium cruentum*. Plant Physiol 114:223–230

Khozin-Goldberg I, Yu HZ, Adlerstein D et al (2000) Triacylglycerols of the red microalga *Porphyridium cruentum* can contribute to the biosynthesis of eukaryotic galactolipids. Lipids 35:881–889

Khozin-Goldberg I, Bigogno C, Shrestha P, Cohen Z (2002a) Nitrogen starvation induces the accumulation of arachidonic acid in the freshwater green alga *Parietochloris incisa* (Trebouxiophyceae). J Phycol 38:1–5. https://doi.org/10.1046/j.1529-8817.2002.01160.x

Khozin-Goldberg I, Didi-Cohen S, Shayakhmetova I, Cohen Z (2002b) Biosynthesis of eicosapentaenoic acid (EPA) in the freshwater eustigmatophyte *Monodus subterraneus* (Eustigmatophyceae). J Phycol 38:745–756. https://doi.org/10.1046/j.1529-8817.2002.02006.x

Khozin-Goldberg I, Iskandarov U, Cohen Z (2011) LC-PUFA from photosynthetic microalgae: occurrence, biosynthesis, and prospects in biotechnology. Appl Microbiol Biotechnol 91:905–915

Khozin-Goldberg I, Leu S, Boussiba S (2016) Microalgae as a source for VLC-PUFA production. In: Nakamura Y, Li-Beisson Y (eds) Lipids in plant and algae development, subcellular biochemistry. Springer, Cham, pp 471–510

Kilian O, Vick B (2013) United States Patent US 8.440,805 B2

Kilian O, Benemann CSE, Niyogi KK, Vick B (2011) High-efficiency homologous recombination in the oil-producing alga *Nannochloropsis* sp. Proc Natl Acad Sci 108:21265–21269. https://doi.org/10.1073/pnas.1105861108

Kiron V, Phromkunthong W, Huntley M et al (2012) Marine microalgae from biorefinery as a potential feed protein source for Atlantic salmon, common carp and whiteleg shrimp. Aquac Nutr 18:521–531. https://doi.org/10.1111/j.1365-2095.2011.00923.x

Koh HG, Kang NK, Kim EK et al (2018) Advanced multigene expression system for *Nannochloropsis salina* using 2A self-cleaving peptides. J Biotechnol 278:39–47. https://doi.org/10.1016/J.JBIOTEC.2018.04.017

Koven W, Barr Y, Lutzky S et al (2001) The effect of dietary arachidonic acid (20:4$n-6$) on growth, survival and resistance to handling stress in gilthead seabream (*Sparus aurata*) larvae. Aquaculture 193:107–122. https://doi.org/10.1016/S0044-8486(00)00479-8

Krienitz L, Wirth M (2006) The high content of polyunsaturated fatty acids in *Nannochloropsis limnetica* (Eustigmatophyceae) and its implication for food web interactions, freshwater aquaculture and biotechnology. Limnologica 36:204–210. https://doi.org/10.1016/j.limno.2006.05.002

Kyle DJ (2001) The large-scale production and use of a single-cell oil highly enriched in docosahexaenoic acid. In: Omega-3 fat acids, pp 92–107. https://doi.org/10.1021/bk-2001-0788.ch008

Lee Chang KJ, Nichols CM, Blackburn SI et al (2014) Comparison of thraustochytrids *Aurantiochytrium* sp., *Schizochytrium* sp., *Thraustochytrium* sp., and *Ulkenia* sp. for production of biodiesel, long-chain omega-3 oils, and exopolysaccharide. Mar Biotechnol 16:396–411. https://doi.org/10.1007/s10126-014-9560-5

León-Saiki GM, Remmers IM, Martens DE et al (2017) The role of starch as transient energy buffer in synchronized microalgal growth in *Acutodesmus obliquus*. Algal Res 25:160–167. https://doi.org/10.1016/j.algal.2017.05.018

Leu S, Boussiba S (2014) Advances in the production of high-value products by microalgae. Ind Biotechnol 10:169–183. https://doi.org/10.1089/ind.2013.0039

Levering J, Broddrick J, Dupont CL et al (2016) Genome-scale model reveals metabolic basis of biomass partitioning in a model diatom. PLoS One 11:e0155038. https://doi.org/10.1371/journal.pone.0155038

Levitan O, Dinamarca J, Zelzion E et al (2015a) An RNA interference knock-down of nitrate reductase enhances lipid biosynthesis in the diatom *Phaeodactylum tricornutum*. Plant J 84:963–973. https://doi.org/10.1111/tpj.13052

Levitan O, Dinamarca J, Zelzion E et al (2015b) Remodeling of intermediate metabolism in the diatom *Phaeodactylum tricornutum* under nitrogen stress. Proc Natl Acad Sci 112:412–417. https://doi.org/10.1073/pnas.1419818112

Leyland B, Leu S, Boussiba S, Voglmayr H (2017) Are Thraustochytrids algae? Fungal Biol 121:835–840. https://doi.org/10.1016/j.funbio.2017.07.006

Li Z, Bock R (2018) Replication of bacterial plasmids in the nucleus of the red alga *Porphyridium purpureum*. Nat Commun 9:3451. https://doi.org/10.1038/s41467-018-05651-1

Li X, Marella TK, Tao L et al (2017a) Optimization of growth conditions and fatty acid analysis for three freshwater diatom isolates. Phycol Res 65:177–187. https://doi.org/10.1111/pre.12174

Li X, Pei G, Liu L et al (2017b) Metabolomic analysis and lipid accumulation in a glucose tolerant *Crypthecodinium cohnii* strain obtained by adaptive laboratory evolution. Bioresour Technol 235:87–95. https://doi.org/10.1016/J.BIORTECH.2017.03.049

Liang Y, Maeda Y, Sunaga Y et al (2013) Biosynthesis of polyunsaturated fatty acids in the oleaginous marine diatom *Fistulifera* sp. strain JPCC DA0580. Mar Drugs 11:5008–5023. https://doi.org/10.3390/md11125008

Liang Y, Maeda Y, Yoshino T et al (2014) Profiling of polar lipids in marine oleaginous diatom *Fistulifera solaris* JPCC DA0580: prediction of the potential mechanism for eicosapentaenoic acid-incorporation into triacylglycerol. Mar Drugs 12:3218–3230. https://doi.org/10.3390/md12063218

Li-Beisson Y, Shorrosh B, Beisson F et al (2013) Acyl-lipid metabolism. Arab B 11:e0161. https://doi.org/10.1199/tab.0161

Liu C-P, Lin L-P (2005) Morphology and eicosapentaenoic acid production by *Monodus subterraneus* UTEX 151. Micron 36:545–550. https://doi.org/10.1016/j.micron.2005.05.001

Longworth J, Wu D, Huete-Ortega M et al (2016) Proteome response of *Phaeodactylum tricornutum*, during lipid accumulation induced by nitrogen depletion. Algal Res 18:213–224. https://doi.org/10.1016/j.algal.2016.06.015

Lopes-Marques M, Kabeya N, Qian Y et al (2018) Retention of fatty acyl desaturase 1 (fads1) in Elopomorpha and Cyclostomata provides novel insights into the evolution of long-chain polyunsaturated fatty acid biosynthesis in vertebrates. BMC Evol Biol 18:157. https://doi.org/10.1186/s12862-018-1271-5

Lupette J, Jaussaud A, Vigor C et al (2018) Non-enzymatic synthesis of bioactive isoprostanoids in the diatom phaeodactylum following oxidative stress. Plant Physiol 178:1344–1357. https://doi.org/10.1104/PP.18.00925

Ma Y-H, Wang X, Niu Y-F et al (2014) Antisense knockdown of pyruvate dehydrogenase kinase promotes the neutral lipid accumulation in the diatom *Phaeodactylum tricornutum*. Microb Cell Fact 13:100. https://doi.org/10.1186/s12934-014-0100-9

Ma XN, Chen TP, Yang B et al (2016) Lipid production from Nannochloropsis. Mar Drugs 14. https://doi.org/10.3390/md14040061

Ma H, Jin M, Zhu T et al (2018) Effect of dietary arachidonic acid levels on growth performance, fatty acid profiles and lipid metabolism of juvenile yellow catfish (*Pelteobagrus fulvidraco*). Aquaculture 486:31–41. https://doi.org/10.1016/J.AQUACULTURE.2017.11.055

Martins DA, Custódio L, Barreira L et al (2013) Alternative sources of *n*-3 long-chain polyunsaturated fatty acids in marine microalgae. Mar Drugs 11(7):2259–2281. https://doi.org/10.3390/md11072259

Matsuda T, Sakaguchi K, Hamaguchi R et al (2012) Analysis of Δ12-fatty acid desaturase function revealed that two distinct pathways are active for the synthesis of PUFAs in *T. aureum* ATCC 34304. J Lipid Res 53:1210–1222. https://doi.org/10.1194/jlr.M024935

Mazzocchi A, D'Oria V, Cosmi V et al (2018) The role of lipids in human milk and infant formulae. Nutrients 10:567. https://doi.org/10.3390/nu10050567

Meesapyodsuk D, Qiu X (2012) The front-end desaturase: structure, function, evolution and biotechnological use. Lipids 47:227–237. https://doi.org/10.1007/s11745-011-3617-2

Meng Y, Jiang J, Wang H et al (2015) The characteristics of TAG and EPA accumulation in *Nannochloropsis oceanica* IMET1 under different nitrogen supply regimes. Bioresour Technol 179:483–489. https://doi.org/10.1016/j.biortech.2014.12.012

Metz JG, Roessler P, Facciotti D et al (2001) Production of polyunsaturated fatty acids by polyketide synthases in both prokaryotes and eukaryotes. Science 293:290–293. https://doi.org/10.1126/science.1059593

Metz JG, Kuner J, Rosenzweig B et al (2009) Biochemical characterization of polyunsaturated fatty acid synthesis in Schizochytrium: release of the products as free fatty acids. Plant Physiol Biochem 47:472–478. https://doi.org/10.1016/j.plaphy.2009.02.002

Miyagawa-Yamaguchi A, Okami T, Kira N et al (2011) Stable nuclear transformation of the diatom Chaetoceros sp. Phycol Res 59:113–119. https://doi.org/10.1111/j.1440-1835.2011.00607.x

Miyahara M, Aoi M, Inoue-Kashino N et al (2013) Highly efficient transformation of the diatom *Phaeodactylum tricornutum* by multi-pulse electroporation. Biosci Biotechnol Biochem 77:874–876. https://doi.org/10.1271/bbb.120936

Moog D, Stork S, Reislöhner S et al (2015) In vivo localization studies in the stramenopile alga *Nannochloropsis oceanica*. Protist 166:161–171. https://doi.org/10.1016/j.protis.2015.01.003

Murakami H, Nobusawa T, Hori K et al (2018) Betaine lipid is crucial for adapting to low temperature and phosphate deficiency in Nannochloropsis. Plant Physiol 177:181–193. https://doi.org/10.1104/pp.17.01573

Muto M, Fukuda Y, Nemoto M et al (2013) Establishment of a genetic transformation system for the marine pennate diatom *Fistulifera* sp. strain JPCC DA0580-a high triglyceride producer. Mar Biotechnol 15:48–55. https://doi.org/10.1007/s10126-012-9457-0

Nath PRR, Khozin-Goldberg I, Cohen Z et al (2012) Dietary supplementation with the microalgae *Parietochloris incisa* increases survival and stress resistance in guppy (*Poecilia reticulata*) fry. Aquac Nutr 18:167–180. https://doi.org/10.1111/j.1365-2095.2011.00885.x

Nayak S, Koven W, Meiri I et al (2017) Dietary arachidonic acid affects immune function and fatty acid composition in cultured rabbitfish *Siganus rivulatus*. Fish Shellfish Immunol 68:46–53. https://doi.org/10.1016/j.fsi.2017.07.003

Nayak S, Khozin-Goldberg I, Cohen G, Zilberg D (2018) Dietary supplementation with ω6 LC-PUFA-rich algae modulates zebrafish immune function and improves resistance to streptococcal infection. Front Immunol 9:1960. https://doi.org/10.3389/fimmu.2018.01960

Neumann U, Derwenskus F, Gille A et al (2018) Bioavailability and safety of nutrients from the microalgae *Chlorella vulgaris*, *Nannochloropsis oceanica* and *Phaeodactylum tricornutum* in C57BL/6 Mice. Nutrients 10:965. https://doi.org/10.3390/nu10080965

Nguyen HM, Cuiné S, Beyly-Adriano A et al (2013) The green microalga *Chlamydomonas reinhardtii* has a single ω-3 fatty acid desaturase that localizes to the chloroplast and impacts both plastidic and extraplastidic membrane lipids. Plant Physiol 163:914–928. https://doi.org/10.1104/pp.113.223941

Niu YF, Zhang MH, Li DW et al (2013) Improvement of neutral lipid and polyunsaturated fatty acid biosynthesis by overexpressing a type 2 diacylglycerol acyltransferase in marine diatom *Phaeodactylum tricornutum*. Mar Drugs 11:4558–4569. https://doi.org/10.3390/md11114558

Niu Y-F, Wang X, Hu D-X et al (2016) Molecular characterization of a glycerol-3-phosphate acyltransferase reveals key features essential for triacylglycerol production in *Phaeodactylum tricornutum*. Biotechnol Biofuels 9:60. https://doi.org/10.1186/s13068-016-0478-1

Nugent AP (2004) The metabolic syndrome. Nutr Bull 29:36–43

Nymark M, Sharma AK, Sparstad T et al (2016) A CRISPR/Cas9 system adapted for gene editing in marine algae. Sci Rep 6:24951. https://doi.org/10.1038/srep24951

Ohlrogge J, Browse J (1995) Lipid biosynthesis. Plant Cell 7:957–970. https://doi.org/10.1105/tpc.7.7.957

Pal D, Khozin-Goldberg I, Cohen Z, Boussiba S (2011) The effect of light, salinity, and nitrogen availability on lipid production by *Nannochloropsis* sp. Appl Microbiol Biotechnol 90:1429–1441. https://doi.org/10.1007/s00253-011-3170-1

Pal D, Khozin-Goldberg I, Didi-Cohen S et al (2013) Growth, lipid production and metabolic adjustments in the euryhaline eustigmatophyte *Nannochloropsis oceanica* CCALA 804 in response to osmotic downshift. Appl Microbiol Biotechnol 97:8291–8306. https://doi.org/10.1007/s00253-013-5092-6

Pal-Nath D, Didi-Cohen S, Shtaida N et al (2017) Improved productivity and oxidative stress tolerance under nitrogen starvation is associated with the ablated Δ5 desaturation in the green microalga *Lobosphaera incisa*. Algal Res 26:25–38. https://doi.org/10.1016/J.ALGAL.2017.06.026

Pasquet V, Ulmann L, Mimouni V et al (2014) Fatty acids profile and temperature in the cultured marine diatom *Odontella aurita*. J Appl Phycol 26:2265–2271. https://doi.org/10.1007/s10811-014-0252-3

Patron NJ, Orzaez D, Marillonnet S et al (2015) Standards for plant synthetic biology: a common syntax for exchange of DNA parts. New Phytol 208:13–19. https://doi.org/10.1111/nph.13532

Peng K-T (2014) Delta 5 fatty acid desaturase upregulates the synthesis of polyunsaturated fatty acids in the marine diatom *Phaeodactylum tricornutum*. J Agric Food Chem 62:8773–8776

Perez-Garcia O, Escalante FME, de Bashan LE, Bashan Y (2011) Heterotrophic cultures of microalgae: metabolism and potential products. Water Res 45:11–36. https://doi.org/10.1016/j.watres.2010.08.037

Perin G, Bellan A, Segalla A et al (2015) Generation of random mutants to improve light-use efficiency of *Nannochloropsis gaditana* cultures for biofuel production. Biotechnol Biofuels 8:161. https://doi.org/10.1186/s13068-015-0337-5

Perrineau M-M, Zelzion E, Gross J et al (2014) Evolution of salt tolerance in a laboratory reared population of *Chlamydomonas reinhardtii*. Environ Microbiol 16:1755–1766. https://doi.org/10.1111/1462-2920.12372

Petrie JR, Liu Q, Mackenzie AM et al (2010a) Isolation and characterisation of a high-efficiency desaturase and elongases from microalgae for transgenic LC-PUFA production. Mar Biotechnol 12:430–438. https://doi.org/10.1007/s10126-009-9230-1

Petrie JR, Shrestha P, Mansour MP et al (2010b) Metabolic engineering of omega-3 long-chain polyunsaturated fatty acids in plants using an acyl-CoA Δ6-desaturase with ω3-preference from the marine microalga *Micromonas pusilla*. Metab Eng 12:233–240. https://doi.org/10.1016/j.ymben.2009.12.001

Poliner E, Pulman JA, Zienkiewicz K et al (2017) A toolkit for *Nannochloropsis oceanica* CCMP1779 enables gene stacking and genetic engineering of the eicosapentaenoic acid pathway for enhanced long-chain polyunsaturated fatty acid production. Plant Biotechnol J 16:298–309. https://doi.org/10.1111/pbi.12772

Poliner E, Farré EM, Benning C (2018a) Advanced genetic tools enable synthetic biology in the oleaginous microalgae *Nannochloropsis* sp. Plant Cell Rep 37:1383–1399. https://doi.org/10.1007/s00299-018-2270-0

Poliner E, Takeuchi T, Du ZY et al (2018b) Nontransgenic marker-free gene disruption by an episomal CRISPR system in the oleaginous microalga, *Nannochloropsis oceanica* CCMP1779. ACS Synth Biol 7:962–968. https://doi.org/10.1021/acssynbio.7b00362

Popko J, Herrfurth C, Feussner K et al (2016) Metabolome analysis reveals betaine lipids as major source for triglyceride formation, and the accumulation of sedoheptulose during nitrogen-starvation of *Phaeodactylum tricornutum*. PLoS One 11(10):e0164673. https://doi.org/10.1371/journal.pone.0164673

Poulsen N, Kröger N (2005) A new molecular tool for transgenic diatoms: control of mRNA and protein biosynthesis by an inducible promoter-terminator cassette. FEBS J 272:3413–3423. https://doi.org/10.1111/j.1742-4658.2005.04760.x

Prasad B, Vadakedath N, Jeong HJ et al (2014) *Agrobacterium tumefaciens*-mediated genetic transformation of haptophytes (Isochrysis species). Appl Microbiol Biotechnol 98:8629–8639. https://doi.org/10.1007/s00253-014-5900-7

Prasad B, Lein W, Thiyam G et al (2018) Stable nuclear transformation of rhodophyte species *Porphyridium purpureum*: advanced molecular tools and an optimized method. Photosynth Res 140:173–188. https://doi.org/10.1007/s11120-018-0587-8

Qi B, Fraser TCM, Bleakley CL et al (2003) The variant "his-box" of the C18-Delta9-PUFA-specific elongase IgASE1 from *Isochrysis galbana* is essential for optimum enzyme activity. FEBS Lett 547:137–139. https://doi.org/10.1016/S0014-5793(03)00676-8

Qi BX, Fraser T, Mugford S et al (2004) Production of very long chain polyunsaturated omega-3 and omega-6 fatty acids in plants. Nat Biotechnol 22:739–745. https://doi.org/10.1038/Nbt972

Qiu X (2003) Biosynthesis of docosahexaenoic acid (DHA, 22:6-4, 7,10,13,16,19): two distinct pathways. Prostaglandins Leukot Essent Fat Acids 68:181–186. https://doi.org/10.1016/S0952-3278(02)00268-5

Radakovits R, Eduafo PM, Posewitz MC (2011) Genetic engineering of fatty acid chain length in *Phaeodactylum tricornutum*. Metab Eng 13:89–95. https://doi.org/10.1016/j.ymben.2010.10.003

Ratledge C, Kanagachandran K, Anderson AJ et al (2001) Production of docosahexaenoic acid by *Crypthecodinium cohnii* grown in a pH-auxostat culture with acetic acid as principal carbon source. Lipids 36:1241–1246. https://doi.org/10.1007/s11745-001-0838-x

Recchiuti A, Serhan CN (2012) Pro-resolving lipid mediators (SPMs) and their actions in regulating miRNA in novel resolution circuits in inflammation. Front Immunol 3:298. https://doi.org/10.3389/fimmu.2012.00298

Remmers IM, D'Adamo S, Martens DE et al (2018) Orchestration of transcriptome, proteome and metabolome in the diatom *Phaeodactylum tricornutum* during nitrogen limitation. Algal Res 35:33–49. https://doi.org/10.1016/j.algal.2018.08.012

Řezanka T, Kolouchová I, Nedbalová L, Sigler K (2018) Enantiomeric separation of triacylglycerols containing very long chain fatty acids. J Chromatogr A 1557:9–19. https://doi.org/10.1016/j.chroma.2018.04.064

Rodolfi L, Biondi N, Guccione A et al (2017) Oil and eicosapentaenoic acid production by the diatom *Phaeodactylum tricornutum* cultivated outdoors in Green Wall Panel (GWP®) reactors. Biotechnol Bioeng 114:2204–2210. https://doi.org/10.1002/bit.26353

Roessler PG (1987) UDP-glucose pyrophosphorylase activity in the diatom *Cyclotella cryptica* pathway of chrysolaminarin synthesis. J Phycol 23:494–498. https://doi.org/10.1111/j.1529-8817.1987.tb02536.x

Ruiz-Lopez N, Haslam RP, Usher S et al (2015) An alternative pathway for the effective production of the omega-3 long-chain polyunsaturates EPA and ETA in transgenic oilseeds. Plant Biotechnol J 13:1–12. https://doi.org/10.1111/pbi.12328

Russo MT, Aiese Cigliano R, Sanseverino W, Ferrante MI (2018) Assessment of genomic changes in a CRISPR/Cas9 *Phaeodactylum tricornutum* mutant through whole genome resequencing. Peer J 6:e5507. https://doi.org/10.7717/peerj.5507

Ryckebosch E, Bruneel C, Muylaert K, Foubert I (2012) Microalgae as an alternative source of omega-3 long chain polyunsaturated fatty acids. Lipid Technol 24:128–130. https://doi.org/10.1002/lite.201200197

Safafar H, Hass M, Møller P et al (2016) High-EPA biomass from *Nannochloropsis salina* cultivated in a flat-panel photo-bioreactor on a process water-enriched growth medium. Mar Drugs 14:144. https://doi.org/10.3390/md14080144

Sarker PK, Kapuscinski AR, Bae AY et al (2018) Towards sustainable aquafeeds: evaluating substitution of fishmeal with lipid-extracted microalgal co-product (*Nannochloropsis oculata*) in diets of juvenile Nile tilapia (*Oreochromis niloticus*). PLoS One 13:e0201315. https://doi.org/10.1371/journal.pone.0201315

Sayanova O, Napier JA (2016) Metabolic engineering of microalgae for sustainable production of omega-3 long chain polyunsaturated fatty acids. Curr Biotechnol 5:198–212. https://doi.org/10.2174/2211550105666160223214550

Sayanova O, Smith MA, Lapinskas P et al (1997) Expression of a borage desaturase cDNA containing an N-terminal cytochrome b5 domain results in the accumulation of high levels of delta6-desaturated fatty acids in transgenic tobacco. Proc Natl Acad Sci USA 94:4211–4216. https://doi.org/10.1073/PNAS.94.8.4211

Sayanova O, Haslam RP, Calerón MV et al (2011) Identification and functional characterisation of genes encoding the omega-3 polyunsaturated fatty acid biosynthetic pathway from the coccolithophore *Emiliania huxleyi*. Phytochemistry 72:594–600. https://doi.org/10.1016/j.phytochem.2011.01.022

Sayanova O, Ruiz-Lopez N, Haslam RP, Napier JA (2012) The role of Δ6-desaturase acyl-carrier specificity in the efficient synthesis of long-chain polyunsaturated fatty acids in transgenic plants. Plant Biotechnol J 10:195–206. https://doi.org/10.1111/j.1467-7652.2011.00653.x

Sayanova O, Mimouni V, Ulmann L et al (2017) Modulation of lipid biosynthesis by stress in diatoms. Philos Trans R Soc B Biol Sci 372:20160407. https://doi.org/10.1098/rstb.2016.0407

Scaife MA, Nguyen GTDT, Rico J et al (2015) Establishing *Chlamydomonas reinhardtii* as an industrial biotechnology host. Plant J 82:532–546. https://doi.org/10.1111/tpj.12781

Schneider JC, Roessler P (1994) Radiolabeling studies of lipids and fatty acids in Nannochloropsis (Eustigmatophyceae), an oleaginous marine alga. J Phycol 30:594–598

Schneider JC, Livne A, Sukenik A, Roessler PG (1995) A mutant of Nannochloropsis deficient in eicosapentaenoic acid production. Phytochemistry 40:807–814. https://doi.org/10.1016/0031-9422(95)00365-E

Schwartz AS, Brown R, Ajjawi I et al (2018) Complete genome sequence of the model oleaginous alga *Nannochloropsis gaditana* CCMP1894. Genome Announc 6. https://doi.org/10.1128/genomeA.01448-17

Serhan CN (2014) Pro-resolving lipid mediators are leads for resolution physiology. Nature 510:92–101. https://doi.org/10.1038/nature13479

Shanklin J, Guy JE, Mishra G, Lindqvist Y (2009) Desaturases: emerging models for understanding functional diversification of diiron-containing enzymes. J Biol Chem 284:18559–18563. https://doi.org/10.1074/jbc.R900009200

Shepherd CJ, Jackson AJ (2013) Global fishmeal and fish-oil supply: inputs, outputs and markets. J Fish Biol 83:1046–1066. https://doi.org/10.1111/jfb.12224

Shi T, Yu A, Li M et al (2012) Identification of a novel C22-Δ4-producing docosahexaenoic acid (DHA) specific polyunsaturated fatty acid desaturase gene from *Isochrysis galbana* and its expression in *Saccharomyces cerevisiae*. Biotechnol Lett 34:2265–2274. https://doi.org/10.1007/s10529-012-1028-y

Shi H, Wu R, Zheng Y, Yue X (2018) Molecular mechanisms underlying catalytic activity of delta 6 desaturase from *Glossomastix chrysoplasta* and *Thalassiosira pseudonana*. J Lipid Res 59:79–88. https://doi.org/10.1194/jlr.M079806

Shields RJ, Lupatsch I (2012) Algae for aquaculture and animal feeds. Tech – Theor Prax 21:23–37

Shockey J, Regmi A, Cotton K et al (2016) Identification of Arabidopsis GPAT9 (At5g60620) as an essential gene involved in triacylglycerol biosynthesis. Plant Physiol 170:163–179. https://doi.org/10.1104/pp.15.01563

Shrestha RP, Hildebrand M (2017) Development of a silicon limitation inducible expression system for recombinant protein production in the centric diatoms *Thalassiosira pseudonana* and *Cyclotella cryptica*. Microb Cell Factories 16:145. https://doi.org/10.1186/s12934-017-0760-3

Simopoulos AP (2002) The importance of the ratio of omega-6/omega-3 essential fatty acids. Biomed Pharmacother 56:365–379

Simopoulos AP (2008) The importance of the omega-6/omega-3 fatty acid ratio in cardiovascular disease and other chronic diseases. Exp Biol Med (Maywood) 233:674–688. https://doi.org/10.3181/0711-MR-311

Singer SD, Chen G, Mietkiewska E et al (2016) Arabidopsis GPAT9 contributes to synthesis of intracellular glycerolipids but not surface lipids. J Exp Bot 67:4627–4638. https://doi.org/10.1093/jxb/erw242

Skrede A, Mydland L, Ahlstrøm Ø et al (2011) Evaluation of microalgae as sources of digestible nutrients for monogastric animals. J Anim Feed Sci 20:131–142. https://doi.org/10.22358/jafs/66164/2011

Slattery SS, Diamond A, Wang H et al (2018) An expanded plasmid-based genetic toolbox enables Cas9 genome editing and stable maintenance of synthetic pathways in *Phaeodactylum tricornutum*. ACS Synth Biol 7:328–338. https://doi.org/10.1021/acssynbio.7b00191

Smith SR, Gillard JTF, Kustka AB et al (2016) Transcriptional orchestration of the global cellular response of a model pennate diatom to diel light cycling under iron limitation. PLoS Genet 12:e1006490. https://doi.org/10.1371/journal.pgen.1006490

Solovchenko AE, Khozin-Goldberg I, Didi-Cohen S et al (2008) Effects of light intensity and nitrogen starvation on growth, total fatty acids and arachidonic acid in the green microalga *Parietochloris incisa*. J Appl Phycol 20:245–251. https://doi.org/10.1007/s10811-007-9233-0
Spicer A, Molnar A (2018) Gene editing of microalgae: scientific progress and regulatory challenges in Europe. Biology (Basel) 7:21. https://doi.org/10.3390/biology7010021
Sprague M, Betancor MB, Tocher DR (2017) Microbial and genetically engineered oils as replacements for fish oil in aquaculture feeds. Biotechnol Lett 39:1599–1609. https://doi.org/10.1007/s10529-017-2402-6
Starkenburg SR, Kwon KJ, Jha RK et al (2014) A pangenomic analysis of the Nannochloropsis organellar genomes reveals novel genetic variations in key metabolic genes. BMC Genomics 15:212. https://doi.org/10.1186/1471-2164-15-212
Steinrücken P, Prestegard SK, de Vree JH et al (2018) Comparing EPA production and fatty acid profiles of three *Phaeodactylum tricornutum* strains under western Norwegian climate conditions. Algal Res 30:11–22. https://doi.org/10.1016/j.algal.2017.12.001
Su Y, Yi Z, Björnsdóttir SH et al (2018) Adaptive laboratory evolution for enhanced carotenoid production in microalgae. Methods Mol Biol 1852:117–126
Sukenik A (1991) Ecophysiological considerations in the optimization of eicosapentaenoic acid production by *Nannochloropsis* sp. (Eustigmatophyceae). Bioresourc Technol 35:263–269
Sukenik A, Beardall J, Kromkamp JC et al (2009) Photosynthetic performance of outdoor Nannochloropsis mass cultures under a wide range of environmental conditions. Aquat Microb Ecol 56:297–308. https://doi.org/10.3354/ame01309
Sun X-M, Ren L-J, Ji X-J et al (2016) Adaptive evolution of *Schizochytrium* sp. by continuous high oxygen stimulations to enhance docosahexaenoic acid synthesis. Bioresour Technol 211:374–381. https://doi.org/10.1016/J.BIORTECH.2016.03.093
Sun X-M, Ren L-J, Bi Z-Q et al (2018) Adaptive evolution of microalgae *Schizochytrium* sp. under high salinity stress to alleviate oxidative damage and improve lipid biosynthesis. Bioresour Technol 267:438–444. https://doi.org/10.1016/J.BIORTECH.2018.07.079
Szyjka SJ, Mandal S, Schoepp NG et al (2017) Evaluation of phenotype stability and ecological risk of a genetically engineered alga in open pond production. Algal Res 24:378–386. https://doi.org/10.1016/J.ALGAL.2017.04.006
Tanaka T, Maeda Y, Veluchamy A et al (2015) Oil accumulation by the oleaginous diatom *Fistulifera solaris* as revealed by the genome and transcriptome. Plant Cell 27:162–176. https://doi.org/10.1105/tpc.114.135194
Taoka Y, Nagano N, Okita Y et al (2009) Influences of culture temperature on the growth, lipid content and fatty acid composition of *Aurantiochytrium* sp. Strain mh0186. Mar Biotechnol (NY) 11:368–374. https://doi.org/10.1007/s10126-008-9151-4
Tejera N, Vauzour D, Betancor MB et al (2016) A transgenic *Camelina sativa* seed oil effectively replaces fish oil as a dietary source of eicosapentaenoic acid in mice. J Nutr 146:227–235. https://doi.org/10.3945/jn.115.223941.227
Tirichine L, Bowler C (2011) Decoding algal genomes: tracing back the history of photosynthetic life on Earth. Plant J 66:45–57. https://doi.org/10.1111/j.1365-313X.2011.04540.x
Tocher DR (2015) Omega-3 long-chain polyunsaturated fatty acids and aquaculture in perspective. Aquaculture 449:94–107. https://doi.org/10.1016/j.aquaculture.2015.01.010
Tocher DR, Glencross BD (2015) Lipids and fatty acids. In: Lee C-S, Lim C, Webster C, Gatlin DM (eds) Dietary nutrients, additives, and fish health. Wiley, Hoboken, pp 47–94
Tocher D, Betancor M, Sprague M et al (2019) Omega-3 long-chain polyunsaturated fatty acids, EPA and DHA: bridging the gap between supply and demand. Nutrients 11:89. https://doi.org/10.3390/nu11010089
Tonon T, Harvey D, Larson TR, Graham IA (2002) Long chain polyunsaturated fatty acid production and partitioning to triacylglycerols in four microalgae. Phytochemistry 61:15–24
Tonon T, Qing R, Harvey D et al (2005a) Identification of a long-chain polyunsaturated fatty acid acyl-coenzyme A synthetase from the diatom *Thalassiosira pseudonana*. Plant Physiol 138:402–408. https://doi.org/10.1104/pp.104.054528.activity

Tonon T, Sayanova O, Michaelson LV et al (2005b) Fatty acid desaturases from the microalga *Thalassiosira pseudonana*. FEBS J 272:3401–3412. https://doi.org/10.1111/j.1742-4658.2005.04755.x

Trentacoste EM, Shrestha RP, Smith SR et al (2013) Metabolic engineering of lipid catabolism increases microalgal lipid accumulation without compromising growth. Proc Natl Acad Sci USA 110:1–6. https://doi.org/10.1073/pnas.1309299110

Usher S, Han L, Haslam RP et al (2017) Tailoring seed oil composition in the real world: optimising omega-3 long chain polyunsaturated fatty acid accumulation in transgenic *Camelina sativa*. Sci Rep 7:6570. https://doi.org/10.1038/s41598-017-06838-0

Vaezi R, Napier J, Sayanova O et al (2013) Identification and functional characterization of genes encoding omega-3 polyunsaturated fatty acid biosynthetic activities from unicellular microalgae. Mar Drugs 11:5116–5129. https://doi.org/10.3390/md11125116

Vazquez-Vilar M, Orzaez D, Patron N (2018) DNA assembly standards: setting the low-level programming code for plant biotechnology. Plant Sci 273:33–41. https://doi.org/10.1016/J.PLANTSCI.2018.02.024

Verruto J, Francis K, Wang Y et al (2018) Unrestrained markerless trait stacking in *Nannochloropsis gaditana* through combined genome editing and marker recycling technologies. Proc Natl Acad Sci 115:E7015–E7022. https://doi.org/10.1073/pnas.1718193115

Vieler A, Wu G, Tsai C-H et al (2012) Genome, functional gene annotation, and nuclear transformation of the heterokont oleaginous alga *Nannochloropsis oceanica* CCMP1779. PLoS Genet 8:e1003064. https://doi.org/10.1371/journal.pgen.1003064

Vigani M, Parisi C, Rodr Iguez-Cerezo E et al (2015) Food and feed products from micro-algae: market opportunities and challenges for the EU. Trends Food Sci Technol 42:81–92. https://doi.org/10.1016/j.tifs.2014.12.004

Wallis JG, Browse J (1999) The Delta8-desaturase of *Euglena gracilis*: an alternate pathway for synthesis of 20-carbon polyunsaturated fatty acids. Arch Biochem Biophys 365:307–316. https://doi.org/10.1006/abbi.1999.1167

Wang D, Ning K, Li J et al (2014) Nannochloropsis genomes reveal evolution of microalgal oleaginous traits. PLoS Genet 10:e1004094. https://doi.org/10.1371/journal.pgen.1004094

Wang X, Liu Y-H, Wei W et al (2017) Enrichment of long-chain polyunsaturated fatty acids by coordinated expression of multiple metabolic nodes in the oleaginous microalga *Phaeodactylum tricornutum*. J Agric Food Chem 65(35):7713–7720. https://doi.org/10.1021/acs.jafc.7b02397

Wang X, Dong H-P, Wei W et al (2018) Dual expression of plastidial GPAT1 and LPAT1 regulates triacylglycerol production and the fatty acid profile in *Phaeodactylum tricornutum*. Biotechnol Biofuels 11:318. https://doi.org/10.1186/s13068-018-1317-3

Wase N, Black P, DiRusso C (2018) Innovations in improving lipid production: algal chemical genetics. Prog Lipid Res 71:101–123. https://doi.org/10.1016/J.PLIPRES.2018.07.001

Wen Z-Y, Chen F (2003) Heterotrophic production of eicosapentaenoic acid by microalgae. Biotechnol Adv 21:273–294. https://doi.org/10.1016/S0734-9750(03)00051-X

Wu G, Truksa M, Datla N et al (2005) Stepwise engineering to produce high yields of very long-chain polyunsaturated fatty acids in plants. Nat Biotechnol 23:1013–1017. https://doi.org/10.1038/nbt1107

Xie D, Jackson EN, Zhu Q (2015) Sustainable source of omega-3 eicosapentaenoic acid from metabolically engineered *Yarrowia lipolytica*: from fundamental research to commercial production. Appl Microbiol Biotechnol 99:1599–1610. https://doi.org/10.1007/s00253-014-6318-y

Xue Z, Sharpe PL, Hong S-P et al (2013) Production of omega-3 eicosapentaenoic acid by metabolic engineering of *Yarrowia lipolytica*. Nat Biotechnol 31:734–740. https://doi.org/10.1038/nbt.2622

Yokochi T, Honda D, Higashihara T, Nakahara T (1998) Optimization of docosahexaenoic acid production by *Schizochytrium limacinum* SR21. Appl Microbiol Biotechnol 49:72–76. https://doi.org/10.1007/s002530051139

Yongmanitchai W, Ward OP (1991) Growth of and omega-3 fatty acid production by *Phaeodactylum tricornutum* under different culture conditions. Appl Environ Microbiol 57:419–425

Yu R, Yamada A, Watanabe K et al (2000) Production of eicosapentaenoic acid by a recombinant marine cyanobacterium, *Synechococcus* sp. Lipids 35:1061–1064. https://doi.org/10.1007/s11745-000-0619-6

Zaslavskaia LA, Lippmeier JC, Kroth PG et al (2000) Transformation of the diatom *Phaeodactylum tricornutum* with a variety of selectable marker and reporter genes. J Phycol 386:379–386. https://doi.org/10.1046/j.1529-8817.2000.99164.x

Zhou X-R, Robert SS, Petrie JR et al (2007) Isolation and characterization of genes from the marine microalga *Pavlova salina* encoding three front-end desaturases involved in docosahexaenoic acid biosynthesis. Phytochemistry 68:785–796. https://doi.org/10.1016/j.phytochem.2006.12.016

Zhu BH, Shi HP, Yang GP et al (2016) Silencing UDP-glucose pyrophosphorylase gene in *Phaeodactylum tricornutum* affects carbon allocation. New Biotechnol 33:237–244. https://doi.org/10.1016/j.nbt.2015.06.003

Zhukova NV, Aizdaicher NA (1995) Fatty acid composition of 15 species of marine microalgae. Phytochemistry 39:351–356. https://doi.org/10.1016/0031-9422(94)00913-E

Zittelli GC, Lavista F, Bastianini A et al (1999) Production of eicosapentaenoic acid by *Nannochloropsis* sp. cultures in outdoor tubular photobioreactors. J Biotechnol 70:299–312

Zorin B, Grundman O, Khozin-Goldberg I et al (2014) Development of a nuclear transformation system for oleaginous green alga *Lobosphaera* (*Parietochloris*) *incisa* and genetic complementation of a mutant strain, deficient in arachidonic acid biosynthesis. PLoS One 9:e105223. https://doi.org/10.1371/journal.pone.0105223

Zorin B, Pal-Nath D, Lukyanov A et al (2017) Arachidonic acid is important for efficient use of light by the microalga *Lobosphaera incisa* under chilling stress. Biochim Biophys Acta Mol Cell Biol Lipids 1862:853–868. https://doi.org/10.1016/j.bbalip.2017.04.008

Part III
Integrated Approaches

Chapter 8
Integrated Biorefineries for Algal Biomolecules

Edgar Suarez Garcia, Giuseppe Olivieri, Lolke Sijtsma, Marian H. Vermuë, Maria Barbosa, J. Hans Reith, Corjan van den Berg, Michel H. M. Eppink, and René H. Wijffels

Abstract Algae are a renewable source of biomolecules for multiple applications ranging from fuels to specialties. However, their implementation as feedstock in industrial processes has only been achieved for few high-value products. This is due to the elevated costs in cultivation and downstream processing. In order to decrease the biorefinery costs and to enhance the overall process profitability, new separation processes need to be developed. Such processes must start from the understanding of the cell architecture, as a basis to develop an optimal fractionation strategy, and must include selective and mild disentanglement processes, in order to preserve the functionality of the target molecules. In this regard, we propose novel integration concepts such as self-disintegration, simultaneous disintegration and disentanglement, and self-separating systems within the framework of process intensification, in such a way that auxiliary chemicals, solvents and numerous unit operations become redundant.

E. S. Garcia (✉) · G. Olivieri · M. H. Vermuë · M. Barbosa · J. H. Reith · C. van den Berg · M. H. M. Eppink
Bioprocess Engineering, AlgaePARC, Wageningen University and Research, Wageningen, The Netherlands
e-mail: edgar.suarezgarcia@wur.nl

L. Sijtsma
Wageningen Food and Biobased Research, Wageningen University and Research, Wageningen, The Netherlands

R. H. Wijffels
Bioprocess Engineering, AlgaePARC, Wageningen University and Research, Wageningen, The Netherlands

Faculty of Biosciences and Aquaculture, Nord University, Bodø, Norway

A. Hallmann, P. H. Rampelotto (eds.), *Grand Challenges in Algae Biotechnology*, Grand Challenges in Biology and Biotechnology,
https://doi.org/10.1007/978-3-030-25233-5_8

8.1 Introduction

Algae are photosynthetic organisms which produce large amounts of biomolecules (Bharathiraja et al. 2015), making them promising feedstock for the sustainable production of food, feed and non-food products (Draaisma et al. 2013; Milledge 2011; Wijffels and Barbosa 2010).

Algae can be grown on land unsuitable for agriculture using residual water or seawater and CO_2 from flue gas. Algal cultures have a high areal productivity, and the produced biomass is a rich source of proteins, oils (including tri-acyl glycerides, phospholipids and glycolipids, and poly unsaturated fatty acids), polysaccharides (e.g. starch, xylans, pectins and β-glucans) and other high-added value compounds, such as colorants, anti-oxidants and bioactive ingredients (Behera et al. 2015; Cuellar-Bermudez et al. 2015; Vigani et al. 2015). Algal biorefineries may become advantageous for regions with limited biomass availability, extreme climate conditions and land unsuitable for agriculture including desert areas (Subhadra and Grinson-George 2011).

The range of algal products covers bulk commodities, chemical building blocks, fuels, feed, pigments, food, cosmetics and specialties (Behera et al. 2015; Cuellar-Bermudez et al. 2015; Draaisma et al. 2013; Jung et al. 2013; Ruiz Gonzalez et al. 2016). Despite the potential of algae as a production platform, at present, applications are limited to various niche markets such as food supplements, in aquaculture, as a source of PUFA and in personal care products ('t Lam et al. 2018). The implementation of algae in other sectors is limited mainly due to unfavourable economics (Ruiz Gonzalez et al. 2016).

Either the costs of algal biomass production need to be reduced and the scale of production needs to be increased significantly or the overall revenue must increase. The separation of all the biomass components and their consequent exploitation in the most remunerative market can compensate the high cultivation costs. Norsker et al. (2011) estimated future projected production costs at ca. 650 €/tonne biomass dry weight. Via biorefinery, the value of the biomass was projected at 1650 €/tonne dry weight (Norsker et al. 2011; Ruiz Gonzalez et al. 2016; Wijffels et al. 2010). Indeed, the fractionation of algae biomass into multiple products via biorefinery enhances economic revenues per ton of biomass. However, the biorefinery process comes at additional costs which can still turn the whole process back to be economically inconvenient. Indeed, also the biorefinery costs, at current 50–60% of the total costs (Ruiz Gonzalez et al. 2016), must be reduced to a typical 30% for a biotechnological process ('t Lam et al. 2018). Thus, cost reductions in biomass production and development of effective biorefinery technologies and novel products are all essential for the successful further development and scale-up of the algal sector. Since algae are mostly used in the form of whole biomass, technologies for fractionation are largely unutilized and under-developed. This implies that appropriate technologies for multi-product biorefinery of algae are lacking.

To achieve economic feasibility, a true mild and multi-product biorefinery is needed. This requires an approach that does not focus on one main product plus one or more residues, which may be sold for low-value applications. Instead, all fractions should be considered as functional co-products in order to optimize product value and revenues. Already several recent publications have addressed these issues and have proposed the development of novel cell disintegration and extraction techniques ('t Lam et al. 2018), multi-product valorization (Gifuni et al. 2018), cascade of extractions (Soto-Sierra et al. 2018) and mild processing (Dixon and Wilken 2018).

In addition to these novel trends, we propose that the design of a biorefinery chain must be based on the knowledge of the cell structure as a first step to develop the right disintegration–extraction–fractionation methods. Furthermore, we discuss other fractionation concepts such as self-disintegration, simultaneous disintegration-disentanglement and self-separating systems as future directions in algae biorefinery in a framework of process intensification. The term disentanglement is used to refer to the process of gentle separation of molecules (e.g. proteins) within larger molecular complex (e.g. thylakoid membrane).

8.2 The Challenge of Biomolecule Extraction from Algae

The extraction of biomolecules from algal biomass can be evaluated from two main dimensions (Fig. 8.1): unit operations and cell structure. The unit operations dimension (Fig. 8.1a) includes technologies for cell disintegration, extraction, fractionation and purification. Typically, they are applied in a sequential way, leading to multiple side streams. This corresponds to the traditional 'empirical approach' (Fig. 8.1c), where the available unit operations are implemented based on common process engineering knowledge and where the cell is seen as a black box.

The cell-structure dimension (Fig. 8.1b) comprises morphology and cell physiology, aspects which are often overlooked (Dixon and Wilken 2018; Gifuni et al. 2018). Knowledge on the cell composition, cell-wall strength, localization of cellular compartments and organelles, hydrophobicity and charge can be used to devise an optimal biorefinery. New biorefineries should be designed following a 'mechanistic approach' (Fig. 8.1d), where the cell structure provides the foundations to define the unit operations and the operational settings such as the energy input and process times.

8.2.1 Unit Operations Dimension

A broad range of unit operations has been reported for biorefinery of algae (Fig. 8.1a) (Günerken et al. 2015; Hu et al. 2019). For cell disintegration, the most common methods include mechanical methods such as bead milling (Postma et al. 2017; Suarez Garcia et al. 2018a) and homogenization (Safi et al. 2014a, 2017), which, due to their controllability, technological readiness level and efficiency, make

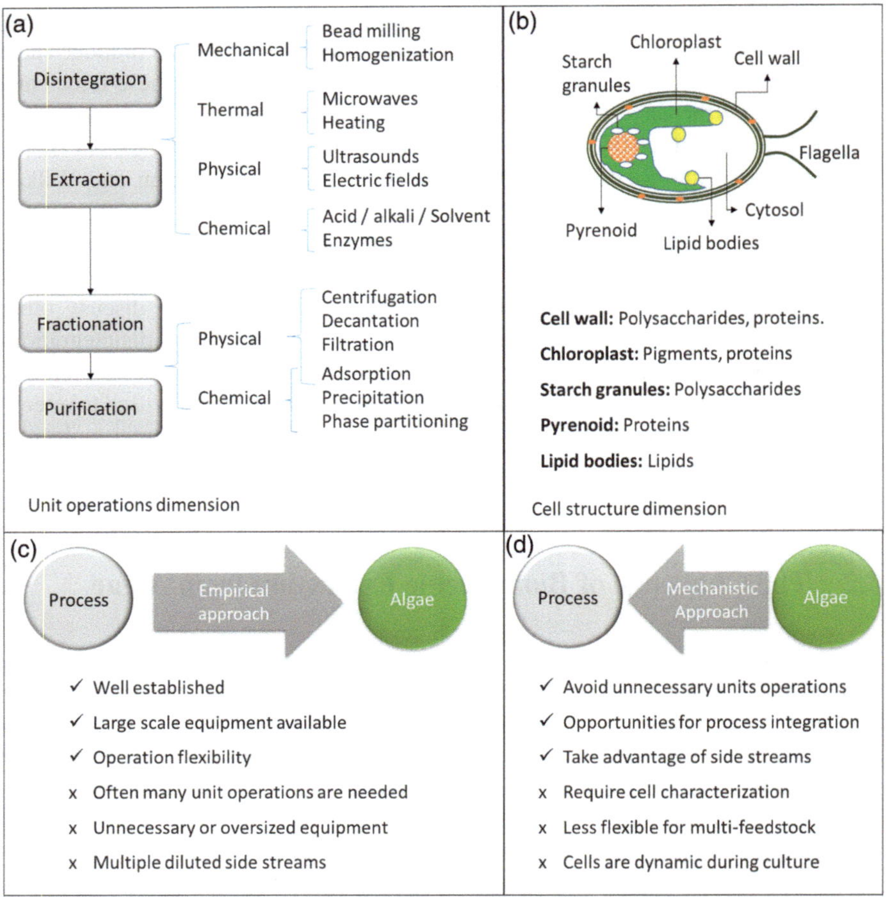

Fig. 8.1 (**a**) Unit operations dimension, showing an overview of typical process units used in algae biorefinery. (**b**) Cell-structure dimension, with schematic representation of an algal cell and its composition. (**c**) Empirical biorefinery approach: application of technologies. (**d**) Proposed mechanistic approach based on the cell structure. Checkmarks and crosses refer to advantages and disadvantages of each approach

them suitable for large-scale cell disruption (Balasundaram et al. 2009). Other methods include high-voltage electric discharges (Zhang et al. 2018), pulsed electric fields and ultrasounds ('t Lam et al. 2017a, b; Carullo et al. 2018; Grimi et al. 2014; Parniakov et al. 2015), thermal processing and microwaves (Jazrawi et al. 2015; Passos et al. 2015), chemical and enzymatic hydrolysis (Safi et al. 2017; Sari et al. 2015) and dissolution using ionic liquids (Desai et al. 2016; Fujita et al. 2013; Teixeira 2012). Other innovative technologies have been proposed, among which hydrodynamic cavitation (Lee et al. 2015), microfluidization (Cha et al. 2012) and explosive decompression (Lorente et al. 2018; Phong et al. 2018b; Günerken et al. 2019).

In general, mechanical methods lead to nearly complete cell disintegration and moderate-to-high extraction yields (Safi et al. 2014a, 2017; Suarez Garcia et al. 2018a) at a high-energy input. It has been reported that by optimizing the operation parameters and by controlling the disintegration time, the energy consumption can be reduced from ~10 to 0.45 kWh kg_{DW}^{-1} (DW, dry weight) (Postma et al. 2017; Safi et al. 2017; Suarez Garcia et al. 2018a). However, these are still orders of magnitudes higher than the energy required to break the cells ($1.9 \times 10^{-4} - 2.5 \times 10^{-3}$ kWh kg_{DW}^{-1}) (Günther et al. 2016; Lee et al. 2013).

Ultrasound and electric fields are external fields which generally result in low disintegration efficiencies and low extraction yields ('t Lam et al. 2017a, b), which, as demonstrated for *C. reinhardtii*, is due to the presence of a rigid polysaccharide-rich cell wall in algae. Alkaline extraction and enzymatic methods usually lead to high disintegration and extraction yields and lower energy consumptions [0.3 kWh kg_{DW}^{-1} (Safi et al. 2017; Sari et al. 2013)]. In spite of some recent publications (Rashidi and Trindade 2018), information on cell-wall composition and structure of microalgae is still limited, which hinders a tailor-made approach for cell disruption. Furthermore, the costs associated with the enzymes are still prohibitive (Sari et al. 2015). Similarly, the use of ionic liquids as green solvents has shown promising results (Suarez Garcia et al. 2019), but their high costs and toxicity (Brandt et al. 2013) prevent large-scale applications.

In general, the high-energy consumption observed with existing unit operations is due to the high water content of the algal suspensions and the long processing times that are used. Most of the energy input is dissipated in the water medium. Moreover, most of the disintegration methods investigated so far are unspecific, don't take into consideration the cell-wall properties (Suarez Garcia et al. 2019), and are based on the introduction of energy from an external source into the diluted cell suspension (empirical approach).

A second problem commonly observed with traditional mechanical disintegration unit operations is the excessive comminution, which leads to stable oil-in-water emulsions (Huimin et al. 2014; Law et al. 2018) or Pickering emulsions (Suarez Garcia et al. 2018b; Tenorio et al. 2017). As a consequence, the biomolecules are more difficult to separate and purify (Van Hee et al. 2004; Zinkoné et al. 2018). Comminution indicates the progressive reduction in size of the cell debris, which is caused mainly by shear forces. Pickering emulsions are emulsions which are stabilized by solid particles, in this particular case, by cell debris in the colloidal range (e.g. fragments from thylakoid membranes).

Besides aqueous (Safi et al. 2014c), alkaline (Sari et al. 2015) and enzymatic extraction (Sari et al. 2013), ionic liquids (ILs) have been investigated for the direct dissolution of algal biomass (Fujita et al. 2013) and for the extraction of carbohydrates (Pezoa-Conte et al. 2015; Teixeira 2012), lipids (Kim et al. 2012; Orr et al. 2016; Orr and Rehmann 2016), pigments (Desai et al. 2016) and proteins (Lee et al. 2017). The main advantage of ILs is their vast chemical diversity, for which they are considered designer solvents, and the high extraction yields that can be achieved.

There are several drawbacks associated with ILs, including high costs (Brandt et al. 2013), potential toxicity (Ventura et al. 2017), high temperatures required for effective extraction (Teixeira 2012) and difficult back-extraction (Mai et al. 2014). In this regard, ILs from algal extracts are typically recovered using extraction followed by centrifugation to remove the insoluble biomass and distillation to remove the excess solvent (Desai et al. 2016; Orr et al. 2016), unit operations which are expensive and energy intensive. Suarez Garcia (2019) demonstrated that ILs can be used for selective extraction under mild condition, and that ultrafiltration is a promising method for their recovery. This approach, however, is still undesirable, as multiple filtration steps will be needed to guarantee that the ILs are completely recovered from the extract.

The unit operations commonly applied for the separation and purification of algae biomolecules are isoelectric precipitation (Ba et al. 2016; Benelhadj et al. 2016; Cavonius et al. 2015; Chronakis 2001; Gerde et al. 2013; Ursu et al. 2014), (ultra) filtration (Suarez Garcia et al. 2018b; Lorente et al. 2018; Safi et al. 2014b; Ursu et al. 2014), aqueous two-phase partitioning (Suarez Garcia 2019; Suarez Ruiz et al. 2018) and sequences of unit operations (Chronakis 2001; Schwenzfeier et al. 2011; Waghmare et al. 2016). Often, these methods are not specific for the cell structure of algae (Knoetzel et al. 1988; Schwenzfeier et al. 2014; Tamayo et al. 2018), target a single product ('t Lam et al. 2018) and follow the traditional sequence of unit operations (Fig. 8.1a), resulting in low yields/purities, undervalued side streams and high costs ('t Lam et al. 2018).

8.2.2 *Cell-Structure Dimension*

In Fig. 8.2a, a representation of the main structures of two model organisms is shown, the microalga *Tetraselmis suecica* and the macroalga *Ulva lactuca*.

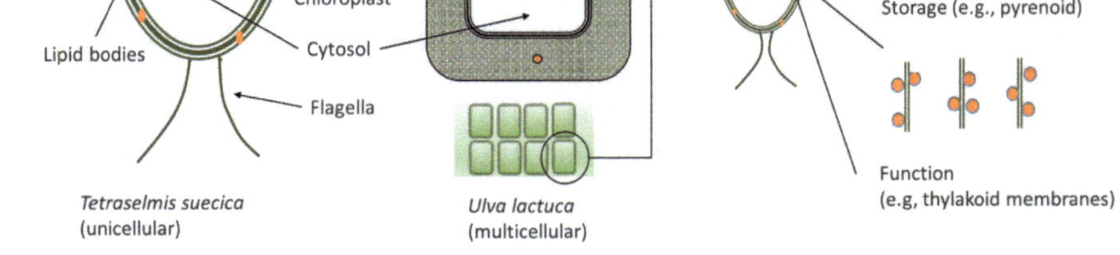

Fig. 8.2 (**a**) Representation of the main structures in micro- and macroalgae. Two model strains are represented: *Tetraselmis suecica* and *Ulva lactuca*. (**b**) Schematic view of three categories of algal biomolecules according to cell morphology and function

For microalgae, three main structures are highlighted: cell wall, chloroplast and pyrenoid. The cell wall is mainly composed of polysaccharides, but it also contains a small percentage of proteins and ash (Becker et al. 1998; Kermanshahi-pour et al. 2014). In the chloroplast, the main metabolic functions take place. It is a complex cellular compartment rich in pigments, glyco- and phospholipids and proteins (Gibbs 1962; Yang et al. 2017). Neutral lipids are usually accumulated in lipid bodies, which also contain specific proteins (Goold et al. 2015). The pyrenoid is embedded in the chloroplast and surrounded, in green algae, by a starch sheath. The pyrenoid is densely packed with the enzyme RuBisCO (Meyer et al. 2017), which is important for carbon fixation.

Green macroalgae are, in general, poor in lipids, contain low amounts of proteins and are rich in carbohydrates. The cell wall contains an intricate arrangement of polysaccharides (cellulose, ulvan and β-glucans) in which proteins are embedded (Lahaye and Robic 2007). The morphological differences among algal strains and their connection to fractionation methods have not been thoroughly considered as a basis for the biorefinery.

Several studies have addressed the influence of cell-wall composition on the disintegration and extraction performance (Alhattab et al. 2018; Baudelet et al. 2017; Safi et al. 2014c; Wang and Lan 2018). However, the fractionation and purification steps are rarely covered. Moreover, most investigations do not take into consideration the purity of the extracted molecules and mostly focus on yields and efficiencies. Purity is a key indicator as it provides a precise mass balance of the extracted biomolecules and allows a more comprehensive evaluation of a biorefinery process (Schwenzfeier et al. 2011; Soto-Sierra et al. 2018; Suarez Garcia et al. 2018b).

Suarez Garcia et al. (2018a) proposed that the low yields and purities of proteins extracted during aqueous bead milling can be explained by their function within the cells (Fig. 8.2b). Proteins present in the cytoplasm ('storage') are easily extracted upon cell disintegration, as they are water soluble and readily available. Other proteins, associated with membranes and organelles, have a more metabolic 'function' and are often bound to lipids and pigments (e.g. thylakoid membranes). A third group of proteins conduct a more 'structural' role, for example, as part of the outer cell wall and membranes. Proteins associated with membranes and pigments display a more hydrophilic nature or are embedded in the cell-wall polysaccharides and, therefore, they cannot be extracted in an aqueous medium by simple mechanical shear.

The categories storage, function and structure (Fig. 8.2b) also apply to other biomolecules. Lipids, for example, are mainly found in lipid bodies (storage) or membrane lipids (structure), depending on the algal strain and cultivation conditions. Similarly, carbohydrates usually play a structural (cell wall) or storage role (starch granules). This categorization is important for the design of an optimal extraction–fractionation process.

Because the biomolecules in algae are usually located in several structures within the cell and are often interconnected or forming macromolecular complexes, the application of traditional unit operations leads to low yield and low purities. Tamayo

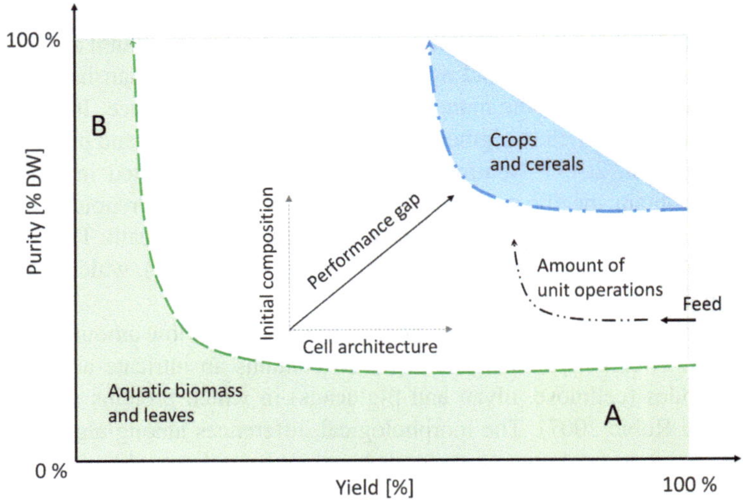

Fig. 8.3 Yield-purity diagram for protein extraction from aquatic biomass and crops. Area in green (bottom-left) and blue (top-right) corresponds to data reported by Tamayo et al. (2018). A and B are opposing regions common in algae processing. The shape of the curve is dictated by the number of unit operations. The performance gap (difference in efficiency between the biorefinery of crops and that of aquatic biomass) is determined by two main factors: the initial protein concentration and cell architecture

et al. (2018) presented a complete overview of the yield-purity (Y-P) diagram for proteins extracted using a wide range of technologies. In their work, aquatic biomass (including micro- and macroalgae) and leaves were compared to crops and cereals (Fig. 8.3). The authors found that the Y-P curve for algae falls in the bottom-left corner while that for crops is located in the upper-right corner.

The performance gap (distance between Y-P curve for crops and algae, Fig. 8.3) can be explained by three main factors: initial composition, cell structure and unit operations. The initial composition is crucial, as it sets the start of the Y-P curve (feed in Fig. 8.3). The amount of unit operations is defined by the target product application, which also sets the window of operation, and by the cell structure, which dictates how the biomolecules can be extracted.

Crops are usually located in the top-right region of the Y-P curve because of the high initial composition and simple structure: starch granules, lipid and/or protein bodies enclosed by a rigid shell. The fractionation can, therefore, be conducted in a few steps only by dry or wet processing (Berghout et al. 2014; Wang et al. 2004) and results in high yields and purities.

Methods reporting high yields of algal molecules (Ba et al. 2016; Cavonius et al. 2015; Safi et al. 2014c) are often unspecific and lead to the co-extraction of other biomolecules (high yield, low purity, few unit operations: Region A in Fig. 8.3). On the opposite, high purities and low yields (Region B, Fig. 8.3) can be obtained after multiple unit operations and intensive processing (Chronakis 2001; Schwenzfeier et al. 2011; Waghmare et al. 2016).

8.3 Towards a Mechanistic Approach in Algae Biorefinery

Several examples from literature are presented in this next section, which illustrate the transition from empirical to a mechanistic biorefinery approach.

8.3.1 Design the Right Cell Architecture

Cells can be genetically engineered to simplify the biorefinery process, reduce energy consumption or enhance the fractionation performance (Spolaore et al. 2006). For example, mutant strains which overproduce and accumulate lipids have long been used to improve the economic feasibility of biofuel production (Breuer et al. 2014). Similarly, cyanobacteria strains have been transformed to overproduce proteins (Trautmann et al. 2016), leading to superior protein extraction yields and purities (Fig. 8.4a). Recently, ’t Lam et al. (2017a, b) reported that the extraction yield of proteins using pulsed electric fields (PEF) increased fourfold at 0.05 kWh kg_{DW}^{-1} for a cell-wall-deficient mutant of *C. reinhardtii* in comparison to the wild type (Fig. 8.4a). Disadvantages of this approach include time and effort required to produce the mutants and the poor performance that the transformed strains could display during culture.

8.3.2 Cell Disintegration by Lytic Organisms, Enzymes and Selective Chemicals

Traditional disintegration methods depend on the supply of energy from an external source, on expensive enzyme cocktails or on chemicals which need to be back-extracted. Alternatively, several microorganisms (bacteria, algae, cyanobacteria and viruses), when co-cultured with algal cells, can secrete lytic enzymes which disrupt

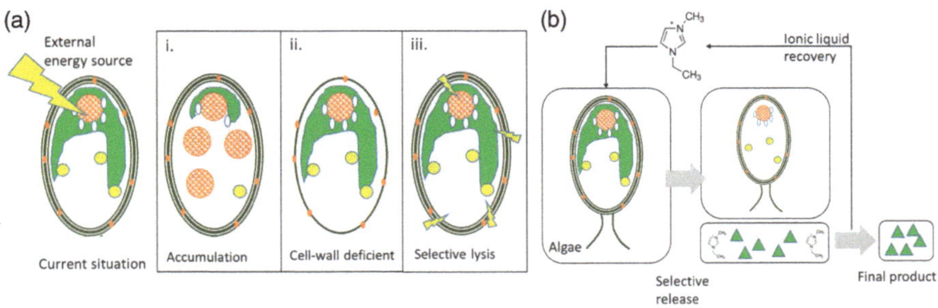

Fig. 8.4 (**a**) Novel approaches based on genetic transformation (i and ii) and selective lysis with microorganisms, enzymes and chemicals (iii). (**b**) Conceptual representation of in situ extraction using ILs

the cell walls (Lenneman et al. 2014). This method is highly selective and mild and can be performed in a single step. The main limitations are the effort required to select the right lytic organism, difficulties with the accurate control of the co-culture and the time needed for complete cell disintegration.

The use of lytic enzymes (Sari et al. 2013; Soto Sierra et al. 2017), chemical hydrolysis (Safi et al. 2017) and cell-wall permeabilization with ILs (Fig. 8.4a-iii) (Suarez Garcia 2019; Desai et al. 2016) are highly specific and energy efficient. However, chemical hydrolysis could lead to denaturation of sensitive molecules, enzymes and ILs can be costly (George et al. 2015), and additional separation steps are required to recover the ILs (Suarez Garcia 2019).

8.3.3 Demulsification

Excessive comminution during mechanical disintegration causes the formation of small cell debris, which are highly stable in suspension and, therefore, difficult to fractionate (Zinkoné et al. 2018). Such stability can be broken (demulsification) by thermal processing and pH adjustment and by the addition of medium-chain alcohols, electrolytes, free oils, enzymes, organic coagulants and air bubbles (Chabrand and Glatz 2009; Luthy et al. 1978). The addition of external agents is disadvantageous as it implies additional operational costs.

Demulsification by means of external fields is another promising approach, as it avoids the use of auxiliary chemicals. Besides centrifugation (Law et al. 2018), 't Lam et al. (2018) summarized several technologies including electric fields, ultrasounds and lasers. These unit operations have been demonstrated for the separation of oil fractions from algae, but they could be potentially implemented in order to prevent emulsion formation during mechanical disintegration. The main drawback of external fields is the additional capital expenditures required to generate the field through the extraction unit. Furthermore, because algae suspensions are normally diluted systems, high-energy inputs could be necessary.

8.3.4 Membraneless Osmosis

The main disadvantage of traditional aqueous two-phase systems (ATPS) is the utilization of external chemicals (organic solvents, polymers, salts, ionic liquids) which need to be back-extracted (Suarez Garcia 2019; Ventura et al. 2017). Alternatively, phase separation can be achieved without the need of external chemicals, a fractionation concept called membraneless osmosis (Antonov 2000).

The phenomena behind phase separation in membraneless osmosis are based on the differences in the chemical potential of molecules present in the mixture (Antonov et al. 1982). Proteins are separated from sugars in a similar way as the salting-out phenomena in salt-rich solutions and, therefore, the term sugaring-out has been used (Sankaran et al. 2018a). Once in solution, proteins and polysaccharides are separated into two distinct phases and the water molecules migrate, across the interphase, towards the more hydrophilic solution. In overall, the more hydrophobic phase (protein-rich) is concentrated, while the polysaccharide-rich phase is diluted (Antonov et al. 1982).

This principle has been proven for colloidal-dispersed systems, such as skimmed milk, blood plasma, microbial broths, plant-based extracts and chloroplasts (Antonov 2000). Membraneless osmosis can be conducted at room temperature, and phase formation takes place rapidly (within seconds) regardless of the system volume. Besides, the separation-concentration performance and the energy consumption are greatly improved compared to other methods, without adding chemically based polymers or salts (Antonov 2000).

8.3.5 Ionic Liquid Recovery

IL recovery can be accomplished by making use of their physicochemical and thermophysical properties. Du et al. (2017) reported a process to extract lipids from microalgae using switchable ILs. These ILs can be recovered by switching the solvent hydrophilicity using CO_2. A second interesting approach is to induce phase changes by means of temperature variations. Several thermo-morphic ILs have been reported with critical temperatures ranging from −40 to 157 °C (Qiao et al. 2017). For example, distillable ILs can be removed by evaporation at 45 °C (Song et al. 2018), a temperature which could still be considered mild. In recent years, a new type of ILs was proposed as new platform in green chemistry: sponge-like ILs (Lozano et al. 2015). The solid–liquid character of these ILs can be manipulated by cooling and centrifugation, which provides a unique tool for integrated and selective extraction processes.

Natural deep eutectic solvents (NADESs) have also been proposed in recent years as an alternative to traditional deep eutectic solvents (DESs) and ILs, since they can be obtained from renewable, low-cost sources (Jablonský et al. 2018). NADESs have been successfully applied for the recovery of lipids from algae (Lu et al. 2016; Pan et al. 2017; Tommasi et al. 2017), but the unique properties of the DES can be used for the extraction of proteins, carbohydrates and high-value compounds (Jablonský et al. 2018).

8.3.6 Multi-product Biorefinery and Functionality

The market potential of algal products can become an important driver in algae biorefinery. The works by ’t Lam et al. (2018), Gifuni et al. (2018), Khanra et al. (2018), Koller et al. (2014), Soto-Sierra et al. (2018) and Law et al. (2018) already show a general trend towards multi-products, functionality and economic revenues.

Suarez Garcia et al. (2018b) have shown that crude fractions from algae can display similar functional properties in comparison to commercial protein isolates. In this regard, new biorefineries should be developed from the functional properties and applications, rather than on the purity of the target products. This, in turn, means that mild technologies and high-value products are to be targeted first.

8.3.7 Process Integration

Many authors have investigated sequences of unit operations to purify biomolecules from algae (Chronakis 2001; Kadam et al. 2016; Schwenzfeier et al. 2011; Waghmare et al. 2016; Wong and Cheung 2001), often resulting in low yields and purities (regions A and B of Fig. 8.3). Recent reviews have also proposed sequences of unit operations, but with a more clear mechanistic approach. Dixon and Wilken (2018), Gifuni et al. (2018), Soto-Sierra et al. (2018) and Khanra et al. (2018) proposed cascades of extractions based on the composition of the biomass and on the value of the target molecules and their potential market application.

Mild and highly specific methods are preferred, so the extraction steps do not compromise the functionality of the molecules in later stages of the refining process (Dixon and Wilken 2018; Gifuni et al. 2018). The overall extraction yields can be optimized by selecting the order of biomolecule extraction (Ansari et al. 2017) and by fully utilizing the side products (Gifuni et al. 2018; Tamayo et al. 2018).

Besides the traditional cascade of extractions, process integration has not been explored sufficiently. Only few reports have presented integration alternatives for cultivation harvesting and harvesting disintegration ('t Lam et al. 2018). In Table 8.1, an overview of reported integration concepts for the steps of extraction and fractionations is presented.

Table 8.1 Overview of reported process integration for extraction–fractionation of algal biomolecules

Technology	Target	References
IL-assisted bead milling and two-phase partitioning	Proteins	Suarez Garcia (2019)
In situ selective extraction with ILs	Proteins	Suarez Garcia (2019)
Sugaring-out biphasic flotation + ultrasounds	Proteins	Sankaran et al. (2018b)
Alkali-assisted ultrasound	Proteins	Phong et al. (2018a)
IL-assisted ultrasound	Proteins	Lee et al. (2017)
Ultrasound and two-phase partitioning	Proteins	Phong et al. (2017)
Extraction and transesterification	Lipids	Salam et al. (2016)
IL-assisted microwave transesterification	Lipids	Wahidin et al. (2016)
Enzyme-assisted alkaline solubilization	Proteins	Sari et al. (2013)
Bead milling and two-phase partitioning	Proteins	Cisneros and Rito-Palomares (2004)

As can be seen, most of the reports combine a cell disintegration method under the presence of an agent (enzymes, ILs, salts, polymers), which favours extraction and/or allows two-phase fractionation. This is the same approach investigated by Suarez Garcia (2019) [IL-assisted bead milling and in situ extraction (Fig. 8.4b)].

Although these processes result in high performances and low-energy consumption, they still rely on external chemicals or additional unit operations, ultimately resembling the typical biorefinery chain. Future research must be directed towards auxiliary-free simplified systems, which make use of the cells' biochemical properties to self-separate.

8.4 Future Directions: Biorefinery Intensification

We have highlighted several challenges in algae biorefinery (Sect. 8.2) and various potential solutions through innovative concepts reported in literature (Sect. 8.3). In this section, we expand these ideas and present three approaches which may contribute towards intensification of algae biorefineries: self-disintegration, simultaneous disintegration and disentanglement, and self-separation.

8.4.1 Self-Disintegration

Genetic manipulations can be conducted to engineer strains which self-disintegrate at the end of the cultivation period. This would allow cultivation, disintegration and potentially harvesting to be conducted in singe step. In addition, it will guarantee minimum energy consumption and high specificity, without the need of auxiliary chemical or unit operations.

8.4.2 Simultaneous Disruption and Disentanglement

At the end of the cultivation period, algae can be cultured with microorganisms which can produce the specific enzymes required for disintegration and disentanglement, in such a way that several products can be solubilized sequentially from the biomass in a single-unit operation (Fig. 8.5). Such process would be mild, highly selective and energy efficient, requiring simple intermediate washing steps to the recover the soluble products. This approach, however, requires extensive screening of microorganisms which can disentangle different cell structures (e.g. cell walls, chloroplast, pyrenoid, membranes), and can be disadvantageous in terms of process flexibility, as the hydrolytic enzymes will be selective only towards certain algal strains.

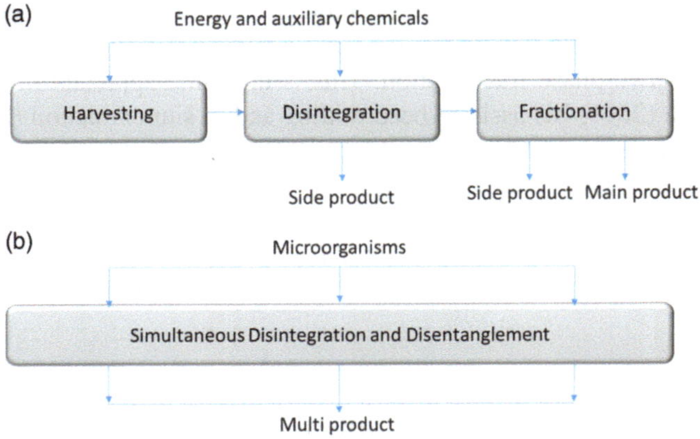

Fig. 8.5 Typical biorefinery chain (**a**) in comparison to simultaneous disintegration and disentanglement (**b**)

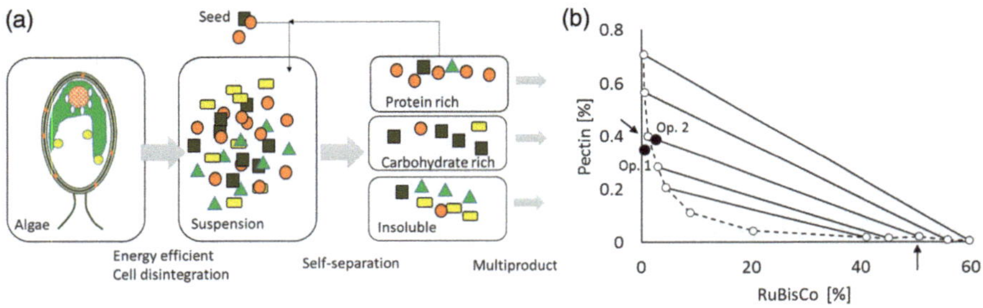

Fig. 8.6 (**a**) Schematic representation of self-separating systems using membraneless osmosis. (**b**) Binodal curve (empty markers) and tie lines (solid lines) for the system RuBisCO-Pectin (Antonov 2000). Filled markers are hypothetical operation points (Op1: After cell disintegration; Op2: after addition of 'seed' biomolecules). Arrows point out the composition of top and bottom phases after phase splitting

8.4.3 Self-Separating Systems

In this approach, the polysaccharides and proteins present in algae could be used to induce their thermodynamic incompatibility, leading to the formation of two phases in which proteins and carbohydrates are fractionated. In order to achieve this, after harvesting and cell disruption, the mixture of suspended molecules is mixed with 'seed' molecules in order to trigger phase splitting. After phase separation, a portion of the separated molecules are recycled to the extraction in order to keep system under continuous operation (Fig. 8.6a).

In order to achieve self-separation, the algae suspension after cell disintegration must contain the right type of molecules, above a critical concentration necessary for two-phase formation. Antonov (2000) presented an overview of phase diagrams for protein-polysaccharide pairs, among which pectin and RuBisCO (ribulose-1,5-

bisphosphate carboxylase/oxygenase). Certain algal strains are rich in pectin-like polysaccharides (Domozych et al. 2012; Kermanshahi-pour et al. 2014), while RuBisCO is one of the most abundant proteins in algae. Pectin-like polysaccharides from the cell wall of *T. suecica* can reach approximately 2–5 wt% (DW) of the total biomass (Kermanshahi-pour et al. 2014; Scholz et al. 2014). Regarding RuBisCO, the content for green algae is ~3 mg $mg_{Chlorophyl}^{-1}$ (Yokota and Canvin 1985). Since green algae contain around 2 wt% chlorophyll, the expected total content of RuBisCO is ~5 wt% (DW).

For an algal suspension containing 10 wt% biomass (Postma et al. 2017), the concentration of pectin polysaccharides and RuBisCo in suspension after bead milling will be 0.35 and 0.5%, respectively. Under these conditions, the system theoretically will remain in the single-phase region (Op1, Fig. 8.6a). It is necessary to add an extra amount of biomolecules ('seed', Fig. 8.6a) to reach a concentration of 0.4% polysaccharides and 2–2.5% proteins (Op2, Fig. 8.6b), where the system will form two phases. In this point, the bottom phase will contain nearly 50% of proteins and the upper phase will contain close to 0.4% polysaccharides (Fig. 8.6b). A fraction of the protein-rich phase can then be recycled back in the system for the next fractionation cycle (Fig. 8.6), which can be conducted by common operation such as acidification and ultrafiltration.

Despite the evident advantages of this method, significant research is needed in order to understand the complex interaction of biomolecules in suspension and their thermodynamic incompatibility and the effect of other proteins and sugars present in the system. In addition, since the phase diagram is a fingerprint for every polysaccharide-protein pair, a new binodal curve will be required for every algal strain under consideration.

8.5 Techno-economic Analysis

The basic concept of the biorefinery is to turn the process loss into profit by increasing the overall revenue coming from the algal biomass, in order to compensate the cultivation cost. The necessary condition to achieve this target is that the additional costs of the biorefinery process should not negatively compromise the whole economic feasibility. Therefore, some techno-economic analysis appeared in literature to estimate the additional cost of the downstream (Asiedu et al. 2018; Barlow et al. 2016; Chauton et al. 2015; Cheali et al. 2014, 2016; Delrue et al. 2013; Dibenedetto et al. 2016; Dong et al. 2016; Fasaei et al. 2018; Ferreira et al. 2013; Fozer et al. 2017; García-Prieto et al. 2017; Gorry et al. 2018; Laurens et al. 2015; Panis and Carreon 2016; Penloglou et al. 2016; Posada et al. 2016; Psycha and Kokossis 2017; Ríos et al. 2013; Rizwan et al. 2015; Ruiz Gonzalez et al. 2016; Thomassen et al. 2016; Thomassen et al. 2018; Vlysidis et al. 2011; Walsh et al. 2018). The framework is very broad and significantly depends on chosen products and related extraction techniques.

However, many of them focus only on one main/only products and, in few cases, additional co-products are considered from the further economic exploitation of the otherwise considered 'by-products' (Barlow et al. 2016; Dong et al. 2016; Gorry et al. 2018; Posada et al. 2016). The case of biodiesel is emblematic from this point of view: attempting to turn profitable the process, having still fuel as first product as always, resulted failing: either the production scale of the co-products was exceeding the current market demand or the additional costs due to the biorefinery were too high (Davis et al. 2014; Delrue et al. 2013; Ferreira et al. 2013; Ríos et al. 2013; Rizwan et al. 2015). Other analysis focused on pigments (Thomassen et al. 2016) or PUFA (Chauton et al. 2015) as main products.

Very few contributions in literature addressed the techno-economic analysis of a more integrated biorefinery, where the selection of the products to be extracted was priority (Laurens et al. 2015; Posada et al. 2016; Ruiz Gonzalez et al. 2016). Very little attention is given in literature to the market analysis of the biorefinery. All the techno-economic analyses end in a performance indicator of profitability such as net present value or return of investment, or just the whole process cost, but still all of them assume the product price as a fixed number, while market demand is in some cases vary sensitive to internal (offer rate) and external factors (price of competitive products). One of the few studies about the economic risk of algal biorefinery is given by Cheali et al. (2016).

For this reason, a top-bottom approach has been followed in the last years in some EU projects and related publications (MIRACLES EU, D-FACTORY EU, SABANA EU, TASAB EU, BISIGODOS EU, PUFA CHAIN EU, MAGNIFICENT EU). The new proposed criteria for the optimal biorefinery process are given by the following steps:

1. Start from the selection of product portfolio: the simultaneous production of proteins, lipids, pigments and polysaccharides from an algal strain is unfeasible. The choice of the product portfolio must come from a potentially high co-accumulation of the some selected products in the target strain and with the same cultivation conditions, as well as efficient (high yield) extraction of the products themselves
2. Design the process according to a cascade approach: choose the sequence of extraction operation based on the properties of the products. Optimize the extraction yield of the target product in a selected unit operation without significantly affecting the properties/quality of the residual components in the biomass. For instance, thermal operation or the use of organic solvent for lipid extraction should be postponed when soluble proteins also belong to the chosen product portfolio (Vanthoor-koopmans et al. 2012).
3. Choose extraction operations that can be performed in the same equipment: the case of the use of supercritical extraction with a sequence of solvent characterized by different polarity is one of the most interesting (Gilbert-López et al. 2015, 2017; Herrero and Ibañez 2018; Sánchez-Camargo et al. 2018).

8.6 Concluding Remarks

Recent publications have addressed key elements in the field of algae biorefinery which need to be taken into consideration in future research and commercial developments. However, other crucial aspects are often overlooked. Algae biorefinery must start from the understanding of the cell structure and composition and must include the complete spectrum of potential market applications, as a basis for process design. The cascade of separation steps must prioritize high-value products and mild processing. Beyond a mere sequence of unit operations, process intensification should receive more attention. Emphasis should be given to self-disintegration and self-separating systems, mild and in situ selective disentanglement, and process intensification, as they guarantee minimum energy consumption, chemicals-free separations and minimum processing.

Acknowledgements This project was financed by the Dutch Technology Foundation STW under the project AlgaePro4You, no. 12635. From January 2017, STW continued its activities as NWO (Dutch national science foundation) Applied and Engineering Sciences (TTW).

References

Alhattab M, Kermanshahi-Pour A, Brooks MSL (2018) Microalgae disruption techniques for product recovery: influence of cell wall composition. J Appl Phycol 1:1–28. https://doi.org/10.1007/s10811-018-1560-9

Ansari FA, Shriwastav A, Gupta SK, Rawat I, Bux F (2017) Exploration of microalgae biorefinery by optimizing sequential extraction of major metabolites from *Scenedesmus obliquus*. Ind Eng Chem Res 56:3407–3412. https://doi.org/10.1021/acs.iecr.6b04814

Antonov YA (2000) Use of membraneless osmosis for concentration of proteins from molecular-dispersed and colloidal-dispersed solutions (review). Appl Biochem Microbiol 36:325–337

Antonov Y, Grinberg Y, Zhuravskaya N, Tolstoguzov V (1982) The concentration of the proteins from skimmed milk by membraneless isobaric osmosis. Carbohydrates Polym 2:81–90

Asiedu A, Ben S, Resurreccion E, Kumar S (2018) Techno-economic analysis of protein concentrate produced by flash hydrolysis of microalgae. Environ Prog Sustain Energy 37:881–890. https://doi.org/10.1002/ep.12722

Ba F, Violeta A, Laroche C, Djelveh G (2016) *Haematococcus pluvialis* soluble proteins: extraction, characterization, concentration / fractionation and emulsifying properties. Bioresour Technol 200:147–152

Balasundaram B, Harrison S, Bracewell DG (2009) Advances in product release strategies and impact on bioprocess design. Trends Biotechnol 27:477–485. https://doi.org/10.1016/j.tibtech.2009.04.004

Barlow J, Sims RC, Quinn JC (2016) Techno-economic and life-cycle assessment of an attached growth algal biorefinery. Bioresour Technol 220:360–368. https://doi.org/10.1016/j.biortech.2016.08.091

Baudelet PH, Ricochon G, Linder M, Muniglia L (2017) A new insight into cell walls of Chlorophyta. Algal Res 25:333–371. https://doi.org/10.1016/j.algal.2017.04.008

Becker B, Melkonian M, Kamerling JP (1998) The cell wall (theca) of *Tetraselmis striata* (Chlorophyta): macromolecular composition and structural elements of the complex polysaccharides. J Phycol 34:779–787

Behera S, Singh R, Arora R, Sharma NK, Shukla M, Kumar S (2015) Scope of algae as third generation biofuels. Front Bioeng Biotechnol 2:1–13. https://doi.org/10.3389/fbioe.2014.00090

Benelhadj S, Gharsallaoui A, Degraeve P, Attia H, Ghorbel D (2016) Effect of pH on the functional properties of Arthrospira (Spirulina) platensis protein isolate. Food Chem 194:1056–1063. https://doi.org/10.1016/j.foodchem.2015.08.133

Berghout JAM, Boom RM, Van Der Goot AJ (2014) The potential of aqueous fractionation of lupin seeds for high-protein foods. Food Chem 159:64–70. https://doi.org/10.1016/j.foodchem.2014.02.166

Bharathiraja B, Chakravarthy M, Ranjith Kumar R, Yogendran D, Yuvaraj D, Jayamuthunagai J, Praveen Kumar R, Palani S (2015) Aquatic biomass (algae) as a future feed stock for bio-refineries: a review on cultivation, processing and products. Renew Sust Energ Rev 47:635–653. https://doi.org/10.1016/j.rser.2015.03.047

Brandt A, Gräsvik J, Hallett JP, Welton T (2013) Deconstruction of lignocellulosic biomass with ionic liquids. Green Chem 15:550–583. https://doi.org/10.1039/c2gc36364j

Breuer G, De Jaeger L, Artus VPG, Martens DE, Springer J, Draaisma RB, Eggink G, Wijffels RH, Lamers PP (2014) Superior triacylglycerol (TAG) accumulation in starchless mutants of *Scenedesmus obliquus*: (II) evaluation of TAG yield and productivity in controlled photobioreactors. Biotechnol Biofuels 7:1–11

Carullo D, Abera BD, Casazza AA, Donsì F, Perego P, Ferrari G, Pataro G (2018) Effect of pulsed electric fields and high pressure homogenization on the aqueous extraction of intracellular compounds from the microalgae *Chlorella vulgaris*. Algal Res 31:60–69. https://doi.org/10.1016/j.algal.2018.01.017

Cavonius LR, Albers E, Undeland I (2015) pH-shift processing of *Nannochloropsis oculata* microalgal biomass to obtain a protein-enriched food or feed ingredient. Algal Res 11:95–102

Cha KH, Koo SY, Song DG, Pan CH (2012) Effect of microfluidization on bioaccessibility of carotenoids from *Chlorella ellipsoidea* during simulated digestion. J Agric Food Chem 60:9437–9442. https://doi.org/10.1021/jf303207x

Chabrand RM, Glatz CE (2009) Destabilization of the emulsion formed during the enzyme-assisted aqueous extraction of oil from soybean flour. Enzym Microb Technol 45:28–35. https://doi.org/10.1016/j.enzmictec.2009.03.008

Chauton MS, Reitan KI, Norsker NH, Tveterås R, Kleivdal HT (2015) A techno-economic analysis of industrial production of marine microalgae as a source of EPA and DHA-rich raw material for aquafeed: research challenges and possibilities. Aquaculture 436:95–103. https://doi.org/10.1016/j.aquaculture.2014.10.038

Cheali P, Quaglia A, Gernaey KV, Sin G (2014) Effect of market price uncertainties on the design of optimal biorefinery systems—a systematic approach. Ind Eng Chem Res 53:6021–6032. https://doi.org/10.1021/ie4042164

Cheali P, Posada JA, Gernaey KV, Sin G (2016) Economic risk analysis and critical comparison of optimal biorefinery concepts. Biofuels Bioprod Biorefin 10:435–445. https://doi.org/10.1002/bbb.1654

Chronakis IS (2001) Gelation of edible blue-green algae protein isolate (*Spirulina platensis* strain Pacifica): thermal transitions, rheological properties, and molecular forces involved. J Agric Food Chem 49:888–898. https://doi.org/10.1021/jf0005059

Cisneros M, Rito-Palomares M (2004) A simplified strategy for the release and primary recovery of c-phycocyanin produced by Spirulina maxima. Chem Biochem Eng Q 18:385–390

Cuellar-Bermudez SP, Aguilar-Hernandez I, Cardenas-Chavez DL, Ornelas-Soto N, Romero-Ogawa MA, Parra-Saldivar R (2015) Extraction and purification of high-value metabolites from microalgae: essential lipids, astaxanthin and phycobiliproteins. Microb Biotechnol 8:190–209. https://doi.org/10.1111/1751-7915.12167

Davis RE, Fishman DB, Frank ED, Johnson MC, Jones SB, Kinchin CM, Skaggs RL, Venteris ER, Wigmosta MS (2014) Integrated evaluation of cost, emissions, and resource potential for algal biofuels at the national scale. Environ Sci Technol 48:6035–6042. https://doi.org/10.1021/es4055719

Delrue F, Li-Beisson Y, Setier P-A, Sahut C, Roubaud A, Froment A-K, Peltier G (2013) Comparison of various microalgae liquid biofuel production pathways based on energetic, economic and environmental criteria. Bioresour Technol 136:205–212. https://doi.org/10.1016/j.biortech.2013.02.091

Desai RK, Streefland M, Wijffels RH, Eppink MHM (2016) Novel astaxanthin extraction from *Haematococcus pluvialis* using cell permeabilising ionic liquids. Green Chem 18:1261–1267. https://doi.org/10.1039/c5gc01301a

Dibenedetto A, Colucci A, Aresta M (2016) The need to implement an efficient biomass fractionation and full utilization based on the concept of "biorefinery" for a viable economic utilization of microalgae. Environ Sci Pollut Res Int 23:22274–22283. https://doi.org/10.1007/s11356-016-6123-5

Dixon C, Wilken LR (2018) Green microalgae biomolecule separations and recovery. Bioresour Bioprocess 5:14. https://doi.org/10.1186/s40643-018-0199-3

Domozych DS, Ciancia M, Fangel JU, Mikkelsen MD, Ulvskov P, Willats WGT (2012) The cell walls of green algae: a journey through evolution and diversity. Front Plant Sci 3:1–7. https://doi.org/10.3389/fpls.2012.00082

Dong T, Knoshaug EP, Davis R, Laurens LML, Van Wychen S, Pienkos PT, Nagle N (2016) Combined algal processing: a novel integrated biorefinery process to produce algal biofuels and bioproducts. Algal Res 19:316–323. https://doi.org/10.1016/j.algal.2015.12.021

Draaisma RB, Wijffels RH, Slegers PM, Brentner LB, Roy A, Barbosa MJ (2013) Food commodities from microalgae. Curr Opin Biotechnol 24:169–177. https://doi.org/10.1016/j.copbio.2012.09.012

Du Y, Schuur B, Brilman DWF (2017) Maximizing lipid yield in *Neochloris oleoabundans* algae extraction by stressing and using multiple extraction stages with N-ethylbutylamine as switchable solvent. Ind Eng Chem Res 56:8073–8080. https://doi.org/10.1021/acs.iecr.7b01032

Fasaei F, Bitter JH, Slegers PM, van Boxtel AJB (2018) Techno-economic evaluation of microalgae harvesting and dewatering systems. Algal Res 31:347–362. https://doi.org/10.1016/j.algal.2017.11.038

Ferreira AF, Ribeiro LA, Batista AP, Marques PASS, Nobre BP, Palavra AMF, da Silva PP, Gouveia L, Silva C (2013) A biorefinery from *Nannochloropsis* sp. microalga – energy and CO_2 emission and economic analyses. Bioresour Technol 138:235–244. https://doi.org/10.1016/j.biortech.2013.03.168

Fozer D, Valentinyi N, Racz L, Mizsey P (2017) Evaluation of microalgae-based biorefinery alternatives. Clean Techn Environ Policy 19:501–515. https://doi.org/10.1007/s10098-016-1242-8

Fujita K, Kobayashi D, Nakamura N, Ohno H (2013) Direct dissolution of wet and saliferous marine microalgae by polar ionic liquids without heating. Enzym Microb Technol 52:199–202. https://doi.org/10.1016/j.enzmictec.2012.12.004

García-Prieto CV, Ramos FD, Estrada V, Villar MA, Diaz MS (2017) Optimization of an integrated algae-based biorefinery for the production of biodiesel, astaxanthin and PHB. Energy 139:1159–1172. https://doi.org/10.1016/j.energy.2017.08.036

George A, Brandt A, Tran K, Zahari SMSNS, Klein-Marcuschamer D, Sun N, Sathitsuksanoh N, Shi J, Stavila V, Parthasarathi R, Singh S, Holmes BM, Welton T, Simmons BA, Hallett JP (2015) Design of low-cost ionic liquids for lignocellulosic biomass pretreatment. Green Chem 17:1728–1734. https://doi.org/10.1039/C4GC01208A

Gerde JA, Wang T, Yao L, Jung S, Johnson LA, Lamsal B (2013) Optimizing protein isolation from defatted and non-defatted Nannochloropsis microalgae biomass. Algal Res 2:145–153

Gibbs SP (1962) The ultrastructure of the chloroplasts of algae. J Ultrastruct Res 7:418–435. https://doi.org/10.1016/S0022-5320(62)90038-2

Gifuni I, Pollio A, Safi C, Marzocchella A, Olivieri G (2018) Current bottlenecks and challenges of the microalgal biorefinery. Trends Biotechnol 37:242–252. https://doi.org/10.1016/j.tibtech.2018.09.006

Gilbert-López B, Mendiola JA, Fontecha J, Van Den Broek LAM, Sijtsma L, Cifuentes A, Herrero M, Ibáñez E (2015) Downstream processing of *Isochrysis galbana*: a step towards microalgal biorefinery. Green Chem 17:4599–4609. https://doi.org/10.1039/c5gc01256b

Gilbert-López B, Mendiola JA, van den Broek LAM, Houweling-Tan B, Sijtsma L, Cifuentes A, Herrero M, Ibáñez E (2017) Green compressed fluid technologies for downstream processing of *Scenedesmus obliquus* in a biorefinery approach. Algal Res 24:111–121. https://doi.org/10.1016/j.algal.2017.03.011

Goold H, Beisson F, Peltier G, Li-Beisson Y (2015) Microalgal lipid droplets: composition, diversity, biogenesis and functions. Plant Cell Rep 34:545–555. https://doi.org/10.1007/s00299-014-1711-7

Gorry P, Sanchez L, Morales M (2018) Microalgae biorefineries for energy and coproduct production. In: Jacob-Lopes E, Queiroz ZL, Queiroz M (eds) Energy from microalgae. Springer, Cham, pp 89–140

Grimi N, Dubois A, Marchal L, Jubeau S, Lebovka NI, Vorobiev E (2014) Selective extraction from microalgae *Nannochloropsis* sp. using different methods of cell disruption. Bioresour Technol 153:254–259. https://doi.org/10.1016/j.biortech.2013.12.011

Günerken E, D'Hondt E, Eppink MHM, Garcia-Gonzalez L, Elst K, Wijffels RH (2015) Cell disruption for microalgae biorefineries. Biotechnol Adv 33:243–260. https://doi.org/10.1016/j.biotechadv.2015.01.008

Günther S, Gernat D, Overbeck A, Kampen I, Kwade A (2016) Micromechanical properties and energy requirements of microalgae *Chlorella vulgaris* for cell disruption. Chem Eng Technol 39:1693–1699. https://doi.org/10.1002/ceat.201400632

Günerken E, D'Hondt E, Eppink MHM, Wijffels RH, Elst K (2019) Disruption of microalgae with a novel continuous explosive decompression device. Algal Res 39:101376. https://doi.org/10.1016/j.algal.2018.12.001

Herrero M, Ibañez E (2018) Green extraction processes, biorefineries and sustainability: recovery of high added-value products from natural sources. J Supercrit Fluids 134:252–259. https://doi.org/10.1016/j.supflu.2017.12.002

Ḥu Y, Gong M, Feng S, Xu C (Charles), Bassi A (2019) A review of recent developments of pre-treatment technologies and hydrothermal liquefaction of microalgae for bio-crude oil production. Renew Sustain Energy Rev 101:476–492. https://doi.org/10.1016/j.rser.2018.11.037

Huimin X, Lin L, Shilin G, Elfalleh W, Shenghua H, Qinghai S, Ying M (2014) Formation, stability, and properties of an algae oil emulsion for application in UHT milk. Food Bioprocess Technol 7:567–574. https://doi.org/10.1007/s11947-013-1054-3

Jablonský M, Škulcová A, Malvis A, Šima J (2018) Extraction of value-added components from food industry based and agro-forest biowastes by deep eutectic solvents. J Biotechnol 282:46–66. https://doi.org/10.1016/j.jbiotec.2018.06.349

Jazrawi C, Biller P, He Y, Montoya A, Ross AB, Maschmeyer T, Haynes BS (2015) Two-stage hydrothermal liquefaction of a high-protein microalga. Algal Res 8:15–22. https://doi.org/10.1016/j.algal.2014.12.010

Jung KA, Lim SR, Kim Y, Park JM (2013) Potentials of macroalgae as feedstocks for biorefinery. Bioresour Technol 135:182–190. https://doi.org/10.1016/j.biortech.2012.10.025

Kadam SU, Álvarez C, Tiwari BK, O'Donnell CP (2016) Extraction and characterization of protein from Irish brown seaweed *Ascophyllum nodosum*. Food Res Int 99:1021–1027. https://doi.org/10.1016/j.foodres.2016.07.018

Kermanshahi-pour A, Sommer TJ, Anastas PT, Zimmerman JB (2014) Enzymatic and acid hydrolysis of *Tetraselmis suecica* for polysaccharide characterization. Bioresour Technol 173:415–421. https://doi.org/10.1016/j.biortech.2014.09.048

Khanra S, Mondal M, Halder G, Tiwari ON, Gayen K, Bhowmick TK (2018) Downstream processing of microalgae for pigments, protein and carbohydrate in industrial application: a review. Food Bioprod Process 110:60–84. https://doi.org/10.1016/j.fbp.2018.02.002

Kim YH, Choi YK, Park J, Lee S, Yang YH, Kim HJ, Park TJ, Hwan Kim Y, Lee SH (2012) Ionic liquid-mediated extraction of lipids from algal biomass. Bioresour Technol 109:312–315. https://doi.org/10.1016/j.biortech.2011.04.064
Knoetzel J, Braumann T, Grimme LH (1988) Pigment-protein complexes of green algae: improved methodological steps for the quantification of pigments in pigment-protein complexes derived from the green algae Chlorella and Chlamydomonas. J Photochem Photobiol B Biol 1:475–491. https://doi.org/10.1016/1011-1344(88)85009-7
Koller M, Muhr A, Braunegg G (2014) Microalgae as versatile cellular factories for valued products. Algal Res 6:52–63. https://doi.org/10.1016/j.algal.2014.09.002
Lahaye M, Robic A (2007) Structure and function properties of Ulvan, a polysaccharide from green seaweeds. Biomacromolecules 8:1765–1774. https://doi.org/10.1021/bm061185q
Laurens LML, Slaby EF, Clapper GM, Howell S, Scott D (2015) Algal biomass for biofuels and bioproducts: overview of boundary conditions and regulatory landscape to define future algal biorefineries. Ind Biotechnol 11:221–228. https://doi.org/10.1089/ind.2015.0007
Law SQK, Mettu S, Ashokkumar M, Scales PJ, Martin GJO (2018) Emulsifying properties of ruptured microalgae cells: barriers to lipid extraction or promising biosurfactants? Colloids Surf B Biointerfaces 170:438–446. https://doi.org/10.1016/j.colsurfb.2018.06.047
Lee AK, Lewis DM, Ashman PJ (2013) Force and energy requirement for microalgal cell disruption: an atomic force microscope evaluation. Bioresour Technol 128:199–206. https://doi.org/10.1016/j.biortech.2012.10.032
Lee AK, Lewis DM, Ashman PJ (2015) Microalgal cell disruption by hydrodynamic cavitation for the production of biofuels. J Appl Phycol 27:1881–1889. https://doi.org/10.1007/s10811-014-0483-3
Lee SY, Show PL, Ling TC, Chang JS (2017) Single-step disruption and protein recovery from *Chlorella vulgaris* using ultrasonication and ionic liquid buffer aqueous solutions as extractive solvents. Biochem Eng J 124:26–35. https://doi.org/10.1016/j.bej.2017.04.009
Lenneman EM, Wang P, Barney BM (2014) Potential application of algicidal bacteria for improved lipid recovery with specific algae. FEMS Microbiol Lett 354:102–110. https://doi.org/10.1111/1574-6968.12436
Lorente E, Hapońska M, Clavero E, Torras C, Salvadó J (2018) Steam explosion and vibrating membrane filtration to improve the processing cost of microalgae cell disruption and fractionation. Processes 6:28. https://doi.org/10.3390/pr6040028
Lozano P, Bernal JM, Garcia-verdugo E, Sanchez-gomez G, Vaultier M, Isabel M, Luis SV (2015) Sponge-like ionic liquids: a new platform for green biocatalytic chemical processes. Green Chem 17:3706–3717. https://doi.org/10.1039/c5gc00894h
Lu W, Alam MA, Pan Y, Wu J, Wang Z, Yuan Z (2016) A new approach of microalgal biomass pretreatment using deep eutectic solvents for enhanced lipid recovery for biodiesel production. Bioresour Technol 218:123–128. https://doi.org/10.1016/j.biortech.2016.05.120
Luthy RG, Selleck RE, Galloway TR, Selleck RE, Galloway TR (1978) Removal of emulsified oil with dissolved air flotation organic coagulants and. J Water Pollut Control Fed 50:331–346
Mai NL, Ahn K, Koo Y (2014) Methods for recovery of ionic liquids – a review. Process Biochem 49:872–881. https://doi.org/10.1016/j.procbio.2014.01.016
Meyer MT, Whittaker C, Griffiths H (2017) The algal pyrenoid: key unanswered questions. J Exp Bot 68:3739–3749. https://doi.org/10.1093/jxb/erx178
Milledge JJ (2011) Commercial application of microalgae other than as biofuels: a brief review. Rev Environ Sci Biotechnol 10:31–41. https://doi.org/10.1007/s11157-010-9214-7
Norsker NH, Barbosa MJ, Vermuë MH, Wijffels RH (2011) Microalgal production – a close look at the economics. Biotechnol Adv 29:24–27. https://doi.org/10.1016/j.biotechadv.2010.08.005
Orr VCA, Rehmann L (2016) Ionic liquids for the fractionation of microalgae biomass. Curr Opin Green Sustain Chem 2:22–27. https://doi.org/10.1016/j.cogsc.2016.09.006
Orr VCA, Plechkova NV, Seddon KR, Rehmann L (2016) Disruption and wet extraction of the microalgae *Chlorella vulgaris* using room-temperature ionic liquids. ACS Sustain Chem Eng 4:591–600. https://doi.org/10.1021/acssuschemeng.5b00967

Pan Y, Alam MA, Wang Z, Huang D, Hu K, Chen H, Yuan Z (2017) One-step production of biodiesel from wet and unbroken microalgae biomass using deep eutectic solvent. Bioresour Technol 238:157–163. https://doi.org/10.1016/j.biortech.2017.04.038

Panis G, Carreon JR (2016) Commercial astaxanthin production derived by green alga *Haematococcus pluvialis*: a microalgae process model and a techno-economic assessment all through production line. Algal Res 18:175–190. https://doi.org/10.1016/j.algal.2016.06.007

Parniakov O, Barba FJ, Grimi N, Marchal L, Jubeau S, Lebovka N, Vorobiev E (2015) Pulsed electric field and pH assisted selective extraction of intracellular components from microalgae nannochloropsis. Algal Res 8:128–134. https://doi.org/10.1016/j.algal.2015.01.014

Passos F, Carretero J, Ferrer I (2015) Comparing pretreatment methods for improving microalgae anaerobic digestion: thermal, hydrothermal, microwave and ultrasound. Chem Eng J 279:667–672. https://doi.org/10.1016/j.cej.2015.05.065

Penloglou G, Chatzidoukas C, Kiparissides C (2016) A microalgae-based biorefinery plant for the production of valuable biochemicals: design and economics. Comput Aided Chem Eng 38:1731–1736. https://doi.org/10.1016/B978-0-444-63428-3.50293-9

Pezoa-Conte R, Leyton A, Anugwom I, von Schoultz S, Paranko J, Mäki-Arvela P, Willför S, Muszyński M, Nowicki J, Lienqueo ME, Mikkola JP (2015) Deconstruction of the green alga *Ulva rigida* in ionic liquids: closing the mass balance. Algal Res 12:262–273. https://doi.org/10.1016/j.algal.2015.09.011

Phong WN, Le CF, Show PL, Chang J-S, Ling TC (2017) Extractive disruption process integration using ultrasonication and an aqueous two-phase system for protein recovery from *Chlorella sorokiniana*. Eng Life Sci 17:357–369. https://doi.org/10.1002/elsc.201600133

Phong WN, Show PL, Le CF, Tao Y, Chang JS, Ling TC (2018a) Improving cell disruption efficiency to facilitate protein release from microalgae using chemical and mechanical integrated method. Biochem Eng J 135:83–90. https://doi.org/10.1016/j.bej.2018.04.002

Phong WN, Show PL, Ling TC, Juan JC, Ng EP, Chang JS (2018b) Mild cell disruption methods for bio-functional proteins recovery from microalgae—recent developments and future perspectives. Algal Res 31:506–516. https://doi.org/10.1016/j.algal.2017.04.005

Posada JA, Brentner LB, Ramirez A, Patel MK (2016) Conceptual design of sustainable integrated microalgae biorefineries: parametric analysis of energy use, greenhouse gas emissions and techno-economics. Algal Res 17:113–131. https://doi.org/10.1016/j.algal.2016.04.022

Postma PR, Suarez Garcia E, Safi C, Yonathan K, Olivieri G, Barbosa MJ, Wijffels RH, Eppink MHM (2017) Energy efficient bead milling of microalgae: effect of bead size on disintegration and release of proteins and carbohydrates. Bioresour Technol 224:670–679. https://doi.org/10.1016/j.biortech.2016.11.071

Psycha M, Kokossis A (2017) Techno-economic evaluation of an integrated microalga biorefinery targeting the co-production of specialty chemicals. In: Espuña A, Graells M, Puigjaner L (eds) 27th European symposium on computer aided process engineering, computer aided chemical engineering. Elsevier, pp 1981–1986. https://doi.org/10.1016/B978-0-444-63965-3.50332-9

Qiao Y, Ma W, Theyssen N, Chen C, Hou Z (2017) Temperature-responsive ionic liquids: fundamental behaviors and catalytic applications. Chem Rev 117:6881–6928. https://doi.org/10.1021/acs.chemrev.6b00652

Rashidi B, Trindade LM (2018) Detailed biochemical and morphologic characteristics of the green microalga *Neochloris oleoabundans* cell wall. Algal Res 35:152–159. https://doi.org/10.1016/j.algal.2018.08.033

Ríos SD, Torres CM, Torras C, Salvadó J, Mateo-Sanz JM, Jiménez L (2013) Microalgae-based biodiesel: economic analysis of downstream process realistic scenarios. Bioresour Technol 136:617–625. https://doi.org/10.1016/j.biortech.2013.03.046

Rizwan M, Zaman M, Lee JH, Gani R (2015) Optimal processing pathway selection for microalgae-based biorefinery under uncertainty. Comput Chem Eng 82:362–373. https://doi.org/10.1016/j.compchemeng.2015.08.002

Ruiz Gonzalez J, Olivieri G, de Vree J, Bosma R, Willems P, Reith H, Eppink M, Kleinegris DMM, Wijffels RH, Barbosa M (2016) Towards industrial products from microalgae. Energy Environ Sci 9:3036–3043. https://doi.org/10.1039/C6EE01493C

Safi C, Charton M, Ursu AV, Laroche C, Zebib B, Pontalier PY, Vaca-Garcia C (2014a) Release of hydro-soluble microalgal proteins using mechanical and chemical treatments. Algal Res 3:55–60. https://doi.org/10.1016/j.algal.2013.11.017

Safi C, Liu DZ, Yap BHJ, Martin GJO, Vaca-Garcia C, Pontalier PY (2014b) A two-stage ultrafiltration process for separating multiple components of *Tetraselmis suecica* after cell disruption. J Appl Phycol 26:2379–2387. https://doi.org/10.1007/s10811-014-0271-0

Safi C, Ursu AV, Laroche C, Zebib B, Merah O, Pontalier PY, Vaca-Garcia C (2014c) Aqueous extraction of proteins from microalgae: effect of different cell disruption methods. Algal Res 3:61–65. https://doi.org/10.1016/j.algal.2013.12.004

Safi C, Cabas Rodriguez L, Mulder WJ, Engelen-Smit N, Spekking W, van den Broek LAM, Olivieri G, Sijtsma L (2017) Energy consumption and water-soluble protein release by cell wall disruption of *Nannochloropsis gaditana*. Bioresour Technol 239:204–210. https://doi.org/10.1016/j.biortech.2017.05.012

Salam KA, Velasquez-Orta SB, Harvey AP (2016) A sustainable integrated in situ transesterification of microalgae for biodiesel production and associated co-products – a review. Renew Sust Energ Rev 65:1179–1198. https://doi.org/10.1016/j.rser.2016.07.068

Sánchez-Camargo A d P, Pleite N, Mendiola JA, Cifuentes A, Herrero M, Gilbert-López B, Ibáñez E (2018) Development of green extraction processes for *Nannochloropsis gaditana* biomass valorization. Electrophoresis 39:1875–1883. https://doi.org/10.1002/elps.201800122

Sankaran R, Manickam S, Jiun Y, Chuan T, Chang J (2018a) Extraction of proteins from microalgae using integrated method of sugaring-out assisted liquid biphasic flotation (LBF) and ultrasound. Ultrason Sonochem 48:231–239. https://doi.org/10.1016/j.ultsonch.2018.06.002

Sankaran R, Manickam S, Yap YJ, Ling TC, Chang JS, Show PL (2018b) Extraction of proteins from microalgae using integrated method of sugaring-out assisted liquid biphasic flotation (LBF) and ultrasound. Ultrason Sonochem 48:231–239. https://doi.org/10.1016/j.ultsonch.2018.06.002

Sari YW, Bruins ME, Sanders JPM (2013) Enzyme assisted protein extraction from rapeseed, soybean, and microalgae meals. Ind Crop Prod 43:78–83. https://doi.org/10.1016/j.indcrop.2012.07.014

Sari YW, Mulder WJ, Sanders JPM, Bruins ME (2015) Towards plant protein refinery: review on protein extraction using alkali and potential enzymatic assistance. Biotechnol J 10:1138–1157. https://doi.org/10.1002/biot.201400569

Scholz MJ, Weiss TL, Jinkerson RE, Jing J, Roth R, Goodenough U, Posewitz MC, Gerken HG (2014) Ultrastructure and composition of the *Nannochloropsis gaditana* cell wall. Eukaryot Cell 13:1450–1464. https://doi.org/10.1128/EC.00183-14

Schwenzfeier A, Wierenga PA, Gruppen H (2011) Isolation and characterization of soluble protein from the green microalgae Tetraselmis sp. Bioresour Technol 102:9121–9127. https://doi.org/10.1016/j.biortech.2011.07.046

Schwenzfeier A, Wierenga PA, Eppink MHM, Gruppen H (2014) Effect of charged polysaccharides on the techno-functional properties of fractions obtained from algae soluble protein isolate. Food Hydrocoll 35:9–18. https://doi.org/10.1016/j.foodhyd.2013.07.019

Song CP, Yap QY, Yin M, Nagasundara RR (2018) Environmentally benign and recyclable aqueous two-phase system composed of distillable CO_2-based alkyl carbamate ionic liquids. ACS Sustain Chem Eng 6:10344–10354. https://doi.org/10.1021/acssuschemeng.8b01685

Soto Sierra L, Dixon CK, Wilken LR (2017) Enzymatic cell disruption of the microalgae *Chlamydomonas reinhardtii* for lipid and protein extraction. Algal Res 25:149–159. https://doi.org/10.1016/j.algal.2017.04.004

Soto-Sierra L, Stoykova P, Nikolov ZL (2018) Extraction and fractionation of microalgae-based protein products. Algal Res 36:175–192. https://doi.org/10.1016/j.algal.2018.10.023

Spolaore P, Joannis-Cassan C, Duran E, Isambert A (2006) Commercial applications of microalgae. Pauline J Biosci Bioeng 101:87–96. https://doi.org/10.1263/jbb.101.87

Suarez Garcia E (2019) Biorefinery of functional biomolecules from algae. PhD thesis Wageningen University and Research (submitted)

Suarez Garcia E, Leeuwen J v, Safi C, Sijtsma L, Eppink MHM, Wijffels RH, van den Berg C (2018a) Selective and energy efficient extraction of functional proteins from microalgae for food applications. Bioresour Technol 268:197–203. https://doi.org/10.1016/j.biortech.2018.07.131

Suarez Garcia E, Van Leeuwen JJA, Safi C, Sijtsma L, Van Den Broek LAM, Eppink MHM, Wijffels RH, Van Den Berg C (2018b) Techno-functional properties of crude extracts from the green microalga *Tetraselmis suecica*. J Agric Food Chem 66:7831–7838. https://doi.org/10.1021/acs.jafc.8b01884

Suarez Garcia E, Lo C, Eppink MHM, Wijffels RH, van den Berg C (2019) Understanding mild cell disintegration of microalgae in bead mills for the selective release of biomolecules. Submitted Chemical Engineering Science

Suarez Ruiz CA, Emmery DP, Wijffels RH, Eppink MHM, van den Berg C (2018) Selective and mild fractionation of microalgal proteins and pigments using aqueous two-phase systems. J Chem Technol Biotechnol 93:2774–2783. https://doi.org/10.1002/jctb.5711

Subhadra B, Grinson-George (2011) Algal biorefinery-based industry: an approach to address fuel and food insecurity for a carbon-smart world. J Sci Food Agric 91:2–13. https://doi.org/10.1002/jsfa.4207

't Lam GP, van der Kolk JA, Chordia A, Vermuë MH, Olivieri G, Eppink MHM, Wijffels RH (2017a) Mild and selective protein release of cell wall deficient microalgae with pulsed electric field. ACS Sustain Chem Eng 5:6046–6053

't Lam GP, Postma PR, Fernandes DA, Timmermans RAH, Vermuë MH, Barbosa MJ, Eppink MHM, Wijffels RH, Olivieri G (2017b) Pulsed electric field for protein release of the microalgae *Chlorella vulgaris* and *Neochloris oleoabundans*. Algal Res 24:181–187. https://doi.org/10.1016/j.algal.2017.03.024

't Lam GP, Vermuë MH, Eppink MHM, Wijffels RH, van den Berg C (2018) Multi-product microalgae biorefineries: from concept towards reality. Trends Biotechnol 36:216–227. https://doi.org/10.1016/j.tibtech.2017.10.011

Tamayo A, Kyriakopoulou KE, Suarez-garcia E, Berg C, Van Den Jan A, Der Goot V (2018) Understanding differences in protein fractionation from conventional crops, and herbaceous and aquatic biomass. Trends Food Sci Technol 71:235–245. https://doi.org/10.1016/j.tifs.2017.11.010

Teixeira RE (2012) Energy-efficient extraction of fuel and chemical feedstocks from algae. Green Chem 14:419–427. https://doi.org/10.1039/c2gc16225c

Tenorio AT, de Jong EWM, Nikiforidis CV, Boom RM, van der Goot AJ (2017) Interfacial properties and emulsification performance of thylakoid membrane fragments. Soft Matter 1:15161. https://doi.org/10.1039/C6SM02195F

Thomassen G, Egiguren Vila U, Van Dael M, Lemmens B, Van Passel S (2016) A techno-economic assessment of an algal-based biorefinery. Clean Techn Environ Policy 18:1849–1862. https://doi.org/10.1007/s10098-016-1159-2

Thomassen G, Van Dael M, Van Passel S (2018) The potential of microalgae biorefineries in Belgium and India: an environmental techno-economic assessment. Bioresour Technol 267:271–280. https://doi.org/10.1016/j.biortech.2018.07.037

Tommasi E, Cravotto G, Galletti P, Grillo G, Mazzotti M, Sacchetti G, Samorì C, Tabasso S, Tacchini M, Tagliavini E (2017) Enhanced and selective lipid extraction from the microalga *P. tricornutum* by dimethyl carbonate and supercritical CO_2 using deep eutectic solvents and microwaves as pretreatment. ACS Sustain Chem Eng 5:8316–8322. https://doi.org/10.1021/acssuschemeng.7b02074

Trautmann A, Watzer B, Wilde A, Forchhammer K, Posten C (2016) Effect of phosphate availability on cyanophycin accumulation in Synechocystis sp. PCC 6803 and the production strain BW86. Algal Res 20:189–196. https://doi.org/10.1016/j.algal.2016.10.009

Ursu A, Marcati A, Sayd T, Sante-lhoutellier V, Djelveh G, Michaud P (2014) Bioresource technology extraction, fractionation and functional properties of proteins from the microalgae *Chlorella vulgaris*. Bioresour Technol 157:134–139. https://doi.org/10.1016/j.biortech.2014.01.071

Van Hee P, Middelberg APJ, Van Der Lans RGJM, Van Der Wielen LAM (2004) Relation between cell disruption conditions, cell debris particle size, and inclusion body release. Biotechnol Bioeng 88:100–110. https://doi.org/10.1002/bit.20343

Vanthoor-koopmans M, Wijffels RH, Barbosa MJ, Eppink MHM (2012) Biorefinery of microalgae for food and fuel. Bioresour Technol 135:142–149. https://doi.org/10.1016/j.biortech.2012.10.135

Ventura SPM, e Silva FA, Quental MV, Mondal D, Freire MG, Coutinho JAP (2017) Ionic-liquid-mediated extraction and separation processes for bioactive compounds: past, present, and future trends. Chem Rev 117:6984–7052. https://doi.org/10.1021/acs.chemrev.6b00550

Vigani M, Parisi C, Rodr E, Barbosa MJ, Sijtsma L, Ploeg M (2015) Food and feed products from micro-algae: market opportunities and challenges for the EU. Food Sci Technol 42:81–92. https://doi.org/10.1016/j.tifs.2014.12.004

Vlysidis A, Binns M, Webb C, Theodoropoulos C (2011) A techno-economic analysis of biodiesel biorefineries: assessment of integrated designs for the co-production of fuels and chemicals. Energy 36:4671–4683. https://doi.org/10.1016/j.energy.2011.04.046

Waghmare AG, Salve MK, LeBlanc JG, Arya SS (2016) Concentration and characterization of microalgae proteins from *Chlorella pyrenoidosa*. Bioresour Bioprocess 3:16. https://doi.org/10.1186/s40643-016-0094-8

Wahidin S, Idris A, Shaleh SRM (2016) Ionic liquid as a promising biobased green solvent in combination with microwave irradiation for direct biodiesel production. Bioresour Technol 206:150–154. https://doi.org/10.1016/j.biortech.2016.01.084

Walsh MJ, Van Doren LG, Shete N, Prakash A, Salim U (2018) Financial tradeoffs of energy and food uses of algal biomass under stochastic conditions. Appl Energy 210:591–603. https://doi.org/10.1016/j.apenergy.2017.08.060

Wang C, Lan CQ (2018) Effects of shear stress on microalgae – a review. Biotechnol Adv 36:986–1002. https://doi.org/10.1016/j.biotechadv.2018.03.001

Wang H, Johnson LA, Wang T (2004) Preparation of soy protein concentrate and isolate from extruded-expelled soybean meals. J Am Oil Chem Soc 81:713–717. https://doi.org/10.1007/s11746-004-966-8

Wijffels R, Barbosa M (2010) An outlook on microalgal biofuels. Science 329:796–799

Wijffels RH, Barbosa MJ, Eppink MHM (2010) Microalgae for the production of bulk chemicals and biofuels. Biofuels Bioprod Biorefin 4:287–295. https://doi.org/10.1002/bbb

Wong K, Cheung PC (2001) Nutritional evaluation of some subtropical red and green seaweeds Part II. In vitro protein digestibility and amino acid profiles of protein concentrates. Food Chem 72:11–17. https://doi.org/10.1016/S0308-8146(00)00176-X

Yang M, Jiang J-P, Xie X, Chu Y-D, Fan Y, Cao X-P, Xue S, Chi Z-Y (2017) Chloroplasts isolation from *Chlamydomonas reinhardtii* under nitrogen stress. Front Plant Sci 8:1–12. https://doi.org/10.3389/fpls.2017.01503

Yokota A, Canvin DT (1985) Ribulose bisphosphate carboxylase/oxygenase content determined with [14C]Carboxypentitol bisphosphate in plants and algae. Plant Physiol 77:735–739. https://doi.org/10.1104/pp.77.3.735

Zhang R, Grimi N, Marchal L, Vorobiev E (2018) Application of high-voltage electrical discharges and high-pressure homogenization for recovery of intracellular compounds from microalgae *Parachlorella kessleri*. Bioprocess Biosyst Eng 42:29–36. https://doi.org/10.1007/s00449-018-2010-4

Zinkoné TR, Gifuni I, Lavenant L, Pruvost J, Marchal L (2018) Bead milling disruption kinetics of microalgae: process modeling, optimization and application to biomolecules recovery from *Chlorella sorokiniana*. Bioresour Technol 267:458–465. https://doi.org/10.1016/j.biortech.2018.07.080

Chapter 9
Combining Microalgae-Based Wastewater Treatment with Biofuel and Bio-Based Production in the Frame of a Biorefinery

Alice Ferreira, Alberto Reis, Senka Vidovic, Jelena Vladic, Spyros Gkelis, Lusine Melkonyan, Gayane Avetisova, Roberta Congestri, Gabriel Acién, Raul Muñoz, Pierre Collet, and Luisa Gouveia

Abstract Conventional wastewater treatment (WWT) is currently based on the action of heterotrophic and nitrifying bacteria supported by mechanical aeration in a sequence of interconnected aerobic, anoxic, and anaerobic processes. It provided satisfactory levels of pollutant removal, but at the expenses of high-energy consumption and environmental impacts (high CO_2 footprint, nutrient losses, and secondary pollution due to the use of chemicals).

Algal-bacterial systems constitute a sustainable and cost-effective alternative WWT based in solar-driven oxygenation (mediated by photosynthesis), enhanced

---

A. Ferreira · A. Reis · L. Gouveia (✉)
LNEG, National Laboratory of Energy and Geology I.P., Bioenergy Unit, Lisbon, Portugal
e-mail: luisa.gouveia@lneg.pt

S. Vidovic · J. Vladic
Faculty of Technology, Department of Biotechnology and Pharmaceutical Engineering, University of Novi Sad, Novi Sad, Serbia

S. Gkelis
Department of Botany, School of Biology, Aristotle University of Thessaloniki, Thessaloniki, Greece

L. Melkonyan · G. Avetisova
SPC "Armbiotechnology" NAS RA, Yerevan, Armenia

R. Congestri
Laboratory of Biology of Algae, Department of Biology, University of Rome 'Tor Vergata', Rome, Italy

G. Acién
Department of Chemical Engineering, University of Almería, Almería, Spain

R. Muñoz
Institute of Sustainable Processes, Valladolid University, Valladolid, Spain

P. Collet
IFP Energies nouvelles, Rueil-Malmaison, France

A. Hallmann, P. H. Rampelotto (eds.), *Grand Challenges in Algae Biotechnology*, Grand Challenges in Biology and Biotechnology,
https://doi.org/10.1007/978-3-030-25233-5_9

nutrient assimilation into the biomass (resulting from combined auto-, mixo-, and heterotrophic growth), and efficient pathogen removal (due to high pH and O_2 concentrations). Moreover, microalgae-based WWT can be combined with flue gas treatment since microalgae are able to bioconvert CO_2 into biomass that can be transformed efficiently into biofuels, biofertilizers, and bioplastics or allow the recovery of high-value products using subcritical water extraction.

Many industries generate huge amounts of wastewater in their processes (e.g., urban, aquaculture, poultry, cattle, swine, dairy, brewery, tobacco, etc.) with high loads of organics and nutrients that need to be removed before they can be discharged safely into the environment.

The whole process can be envisaged as a microalgae-based biorefinery, which is crucial for the full use and commercialization of microalgae in various areas, allowing the incorporation of a large range of products and associated industries, which brings enormous benefits for bioeconomy and emerging societies.

This chapter provides the view and results from the authors concerning the treatment of several WWs, the characterization of the produced microalgae biomass, the biochemical and thermochemical conversion of the biomass, as well as the extraction of compounds toward different biofuels and bio-based products. The importance of these coupling processes into a biorefinery and bioeconomy frames will be validated by a literature review of life cycle assessment to highlight the added value from an environmental perspective.

9.1 Introduction

9.1.1 Wastewater: Industries/Economic Sectors

Most human and industrial activities use a large percentage of the available freshwater, returning much of that to the environment with additional nutrients and pollutants, as a wastewater (WW). With the overall growing of global water demand and the constant increase of the amount of produced WW, it is urgent to develop a new generation of energy-efficient and resource-recovery processes, while complying with the ever-stricter environmental regulations.

At the beginning of the twenty-first century, the world faced a water crisis following the continuous population growth, urbanization, land use change, industrialization, intensification of food production, improved living standards, economic development, and poor water use practices and WW management strategies (UN-Water Secretariat 2015).

Wastewater can be contaminated with a vast number of different components such as pathogens, organic compounds, synthetic chemicals, nutrients such as phosphorus and nitrogen, organic matter, and heavy metals. They are either in solution or as particulate matter affecting water quality. These components can have bio-cumulative, persistent, and synergistic effects, contaminating freshwater, brackish, and marine ecosystems, threatening food security and access to safe drinking and bathing water, and generating major health and environmental problems (Corcoran et al. 2010).

Wastewater treatment can also have a large impact on climate change. Conventional WW treatment techniques generate methane (CH_4), nitrous oxide (N_2O), and carbon dioxide (CO_2). It is noteworthy that CH_4 has an impact 28 times greater than CO_2 and N_2O is 265 times more pollutant (IPCC 2013). In addition, it is expected that CH_4 and N_2O emissions from WW will increase almost 50 and 25%, respectively, until 2020 (IPCC 2013). Therefore, there is a demand to investigate and implement alternatives to current WW treatment techniques, in order to minimize the production/emission of greenhouse gases and power consumption (Corcoran et al. 2010).

According to the fourth World Water Development Report, only 20% of globally generated WW receives adequate treatment (UNESCO 2012). For highly developed countries, the treatment capacity is 70% of the generated WW, while for developing countries it drops to only 8%. Most of the WW generated in Europe (71%) undergoes treatment, mainly due to the high public awareness regarding health and environment protection, technological advancement, and investments by most governments in WWT systems (Sato et al. 2013).

Nutrient enrichment of water bodies (eutrophication) is often associated with nitrogen and phosphorus from agriculture practices, but it is also caused by human and industrial activities. This can lead to the overgrowth of vascular plants, algal blooms, and the depletion of dissolved oxygen in the water, which can stress or kill aquatic organisms. Some algae such as cyanobacteria can produce toxins that are hazardous for human, livestock, and wildlife. Nutrient enrichment can also cause acidification of freshwater ecosystems with negative impact on biodiversity (Palaniappan et al. 2010).

Urban WW consists of blackwater (excreta, urine, and associated sludge) and graywater (kitchen and bathroom WW) or the mixture of domestic, commercial, including hospitals, industrial, and runoff rainwater. These WWs are frequently discharged into sewers. Urban WW is mainly composed of water (99.9%) with suspended and dissolved organic solids (carbohydrates, lignin, fats, soaps, synthetic detergents, and synthetic organic chemicals from industries), inorganic solids including metals, and pathogens virus and bacteria (Bai et al. 2010) (Table 9.1).

The vast extent of agricultural activities around the world contributes significantly to water-pollutant loads (Table 9.1). Agriculture is the single largest user of freshwater resources, accounting for almost 70% of global water withdrawals, and it is also a significant source of water pollution (UN-Water Secretariat 2015).

Agriculture, intensive cultivation, and the growing "factory" of livestock and aquaculture release also a large nonpoint source contribution of pollutants to surface and groundwaters (Corcoran et al. 2010) (Fig. 9.1).

Nowadays, aquaculture is a fast-growing industry (Fig. 9.2). In 2016, fish harvested from aquaculture to the global human fish consumption amounted to 80 million tons (FAO 2018), and it is expected a consistent yearly growth around 4–5% (Fig. 9.2).

Freshwater aquaculture is an important economic activity in Europe (EU), with over 18,000 direct jobs and an annual turnover of ~1000 million euros. Freshwater aquaculture represents 23% of the EU aquaculture, which itself only satisfies ~9% of the EU 25.3 billion euros/year fish market (EUMOFA 2015). The high production costs, lack of species diversification, regulatory site restrictions, and environmental

Table 9.1 Wastewater produced in Europe by economic sector and associated nutrient (nitrogen and phosphorus) and chemical oxygen demand (COD) loads (Eurostat)

Economic sector	Wastewater flow rate (million m^3/year)	COD (ton/year)	Nitrogen (ton/year)	Phosphorus (ton/year)
Urban	20,400	8,200,000	816,000	200,000
Agriculture	1,257,000[a]	–	–	–
Dairy	538	8,100,000	322,000	–
Brewery	1040[b]	4,160,000[c]	54,500[c]	31,200[c]
Slaughterhouse				
Swine	35	268,000	22,000	–
Poultry	21	161,000	13,000	–
Cattle	12	92,000	7300	–

[a]Obtained from WWAP (United Nations World Water Assessment Programme) (WWAP 2017)
[b]Estimated considering a global beer production (in 2015) of 1.93 billion hL (Statista 2017) and a wastewater/beer ratio of 5.39 (Brewers Association 2013)
[c]Estimated considering the general composition of brewery wastewater reported by Simate et al. (2011)

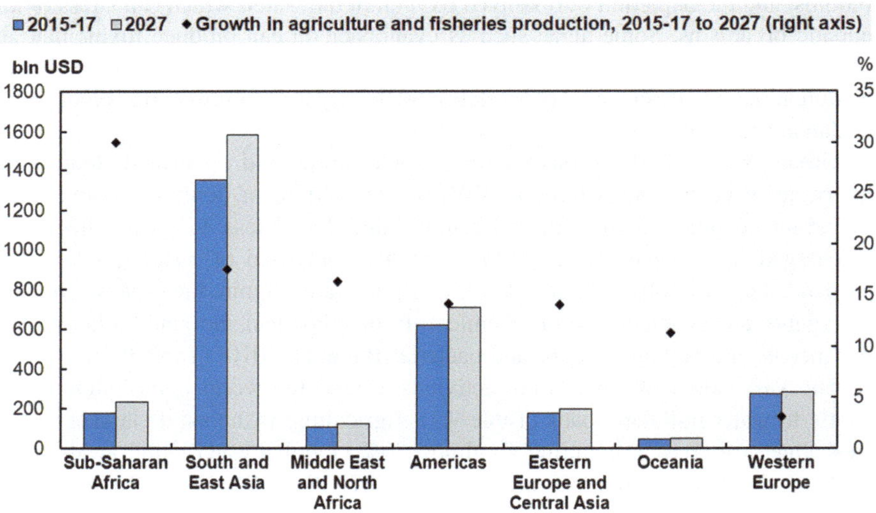

Fig. 9.1 Estimated net value of agricultural and fish production (in millions $US) (OECD/FAO 2018)

regulations have resulted in a lower competitiveness of EU aquaculture compared to their Asian or African counterparts. The high production costs in EU aquaculture derive from the use of costly fish feed, an intensive use of expensive (and scarce) water, and the ever-stricter environmental regulations for aquaculture WW discharge and recycling (which entails the need for costly AWW treatments) (FAO 2014). In this sense, microalgae biomass production from aquaculture WW will result in a more competitive and environmentally friendly EU fish farming industry and even higher standards of product quality and production methods.

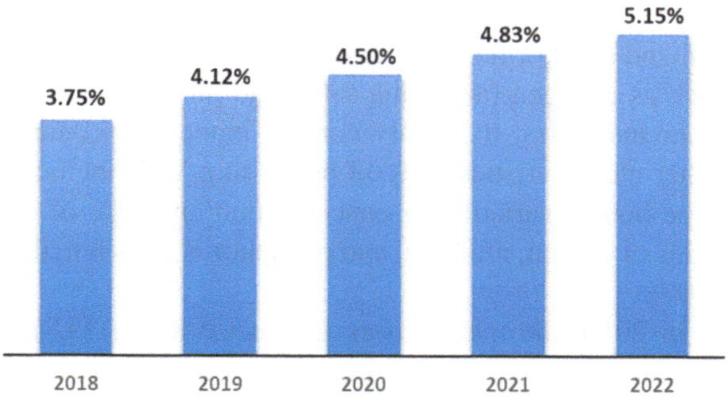

Fig. 9.2 Global aquaculture market growth (%) (White 2018)

The growth in aquaculture activity implies an increase of discharged effluent, as a result of the operations from hatcheries and farming systems. The major waste compounds are organic substances, nitrogen-containing compounds, and phosphates, mainly from uneaten food and fish feces. Different approaches have been applied from fish keeping in ponds for easier harvesting to high technological fish farms extensively using feed, hormones, and often antibiotics with a known impact on the environment. Regarding the pollution generated by aquaculture, nitrogen and phosphorus are considered as the major waste components. In addition, several fish excrete nitrogenous waste products by diffusion and ion exchange through the gills, urine, and feces. Various physical, chemical, and biological methods used in conventional WWT have been applied in aquaculture systems. Solid removal is accomplished by sedimentation and sand or mechanical filtration. Biological processes such as submerged biofilters, trickling filters, rotating biological contactors, and fluidized bed reactors have been employed for the oxidation of organic matter, nitrification, or denitrification (Turcios et al. 2014).

The food and beverage industry is one of the most important sectors in the European Union in terms of financial and social significance owing to its continuous growth. Therefore, it requires considerable resources such as water and energy and generates solid and liquid wastes, which inevitably has an impact on the environment. Compared to other industrial sectors, the food industry uses a much higher amount of water for each ton of product. In Europe, the food and drink sector is responsible for approximately 1.8% of total water use, as water is an essential input as an ingredient, as a key processing element, and as a cooling and heating agent in many production processes (Valta et al. 2015). Food WW is one of the most highly released throughout the world. In general, due to their high organic content, the WW in food processing factories is treated using biological processes (mainly via anaerobic digestion).

Food industries include meat processing, dairy products, seafood and fish processing, fruits and vegetable processing, starch and gluten products, confectionery, sugar processing, alcoholic and nonalcoholic beverages, and rice and bean products among others. The characteristics and volume of WW discharged from food processing factories vary with the products and production procedures. WWs released from these industries are typically turbid, with high concentrations of total suspended solids (TSS); chemical oxygen demands (COD); biological oxygen demand (BOD); fats, oils, and grease (FOG); minerals; sugars; starch; and nutrients such as nitrogen (ammonia salts) and phosphate. Nevertheless, when compared to other industries, they only contain innocuous contents of hazardous compounds. Other characteristics of food processing WW include the large seasonal variation, large hourly variation in concentration and type of contaminants, and unbalanced ratios of carbon, nitrogen, and phosphorus, which make the treatment difficult (Colic 2007).

In 2010, the meat processing sector in the EU-27 was the largest sub-sector, representing 20% of the total turnover of European food and drink industry. Beef, pork, and poultry are the main types processed in Europe (Valta et al. 2015). Increased livestock production and the whole chain of the meat processing industry consume large volumes of water and produce significant amounts of contaminated WW. This is mainly due to the slaughtering of animals, cleaning of the slaughterhouse facilities, as well as meat processing plants. The WW generated by slaughterhouses has a high organic strength, containing blood manure, hair, fat, feathers, and bones. The primary sources of BOD are blood, solubilized fat, urine, and feces. Blood and manure are also significant sources of nitrogen and phosphorus (EPA 2002). This WW is usually subjected to a primary treatment, including the use of screens, settlers, and fat separators. Piggery WW can be characterized by its high concentration of organic N and high COD and BOD values caused by its high organic matter content, which makes piggery WW one of the most polluted agro-industrial wastes (Salama et al. 2017).

The potato processing industry belongs to the sector of processing and preserving of fruits and vegetables. Various types of products can derive from potato including potato chips, French fries, potato starch, etc. As a consequence, the process lines and the water used can be quite different depending on the end product. According to Israilides et al. (2008) for a chip potato processing industry, around 5 tons of water is required per ton of potato. The WW generated contains high concentrations of biodegradable components like starch and proteins, in addition to significant concentrations of total suspended solids (TSS) and organic and inorganic nitrogenous compounds (Total Kjeldahl Nitrogen—TKN) and phosphates (Valta et al. 2015). Foreign components that may accompany the potato include dirt, caustic, fat, and cleaning and preserving chemicals (Hung et al. 2006).

Beer is the main alcoholic beverage consumed in the world, and its production releases about 140 Mt. of CO_2 every year, almost three times the amount of CO_2 released in the production of biofuels, and its purity reaches 99% (Xu et al. 2010). Beer production is a water-consuming process as water is the most important raw material used by the brewing sector, with almost 70% of used water ending as WW (Valta et al. 2015). It is widely estimated that 3–10 L of WW are generated per liter of produced beer (Simate et al. 2011). This WW typically has a high organic load from the organic components present such as sugars, soluble starch, ethanol, and fatty acids, which are easily biodegradable. Heavy metals are normally present in very low concentrations (Simate et al. 2011).

The dairy industry involves processing raw milk into products such as consumer milk, butter, cheese, yogurt, condensed milk, milk powder, and ice cream. Europe is the largest exporter of dairy products in the world. In 2010, the dairy industry was approximately 13% of the European food and drink sector turnover. The WWs generated from this industry are distinguished by their high concentration of nutrients such as N, P, and organic content (presence of whey) and a wide range of pH, which requires costly physicochemical-based treatments. In dairy plants, the effluents present great fluctuations in quality and quantity because each product needs a separate technological line (Slavov 2017). Depending on the product, 40–90% of the water ends up as WW together with 0.5–3% of the milk intake. Conventional treatment methods are not efficient when high amounts of proteins and lipids are present, which results in a set of problems, such as alkaline pH, dark coloration, and high levels of both COD and BOD (Drogui et al. 2008).

9.1.2 Wastewater Treatment

Most conventional WWT techniques currently rely on the action of heterotrophic and nitrifying bacteria supported by mechanical aeration in a series of interconnected aerobic, anoxic, and anaerobic processes (Tchobanoglous et al. 2003). Nitrogen and phosphorus in WWs are mainly removed through biological processes such as nitrification, denitrification, or biological enhanced P removal. However, although these technologies have constantly provided satisfactory results for pollutant removal, several treatment stages and recirculation are required to meet the stricter environmental regulations imposed by EU legislation. In addition, the overall increase of process costs, complexity and energy input, and the environmental impacts, such as GHG emissions, nutrient losses, and secondary pollution from the use of chemical reagents (Fig. 9.3), force the existing facilities to explore more sustainable alternatives (Cuellar-Bermudez et al. 2017; Gonçalves et al. 2017).

In this context, the future technologies to carry out WWT must be able to maximize both nutrient recovery and bioenergy/bio-product production from WW with minimum environmental impact, while providing an efficient nutrient removal (Alcántara et al. 2013).

9.1.3 Microalgae-Based Wastewater Treatment

Presently, the approach with algal-bacterial symbiosis has arisen as a promising platform to support a sustainable and low-cost WWT due to the ability of microalgae to grow in nutrient-rich environments together with the accumulation of nutrients from wastewaters (WWs) and the need of reducing microalgae production costs (Posadas et al. 2017; Ferreira et al. 2018).

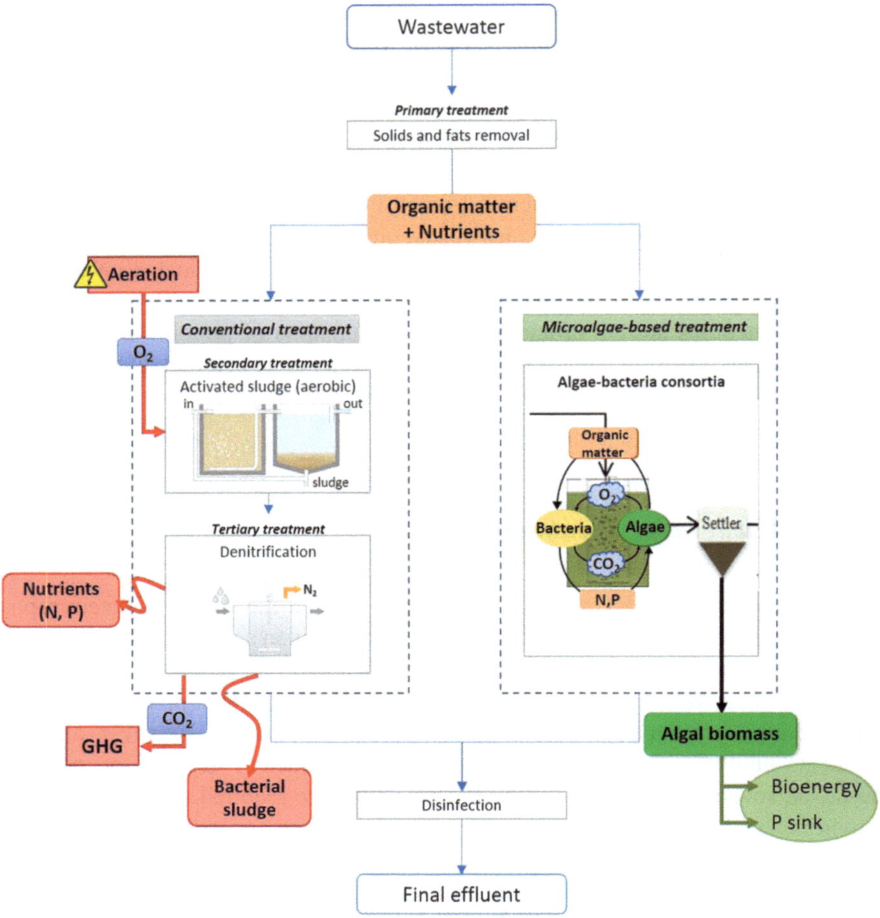

Fig. 9.3 Comparison between conventional wastewater treatment and wastewater treatment-based microalgae [Adapted from Muñoz and Guieysse (2006) and Cuellar-Bermudez et al. (2017)]

Microalgal-bacterial processes offer a cost-effective, more efficient, and sustainable alternative to conventional tertiary treatment since they can play a dual role of bioremediation of WW due to their potential for cost-free oxygenation and efficient nutrient removal as well as generating valuable biomass with concomitant CO_2 sequestration. In addition, WW remediation by microalgae is an eco-friendly process, allowing nutrient recycling and with no secondary pollution as long as the biomass could be reused for different applications (Fig. 9.3) (Rawat et al. 2011; Ferreira et al. 2018). This strategy is a win-win process that benefits both parts, as microalgae cultivation represents an option to clean the WW, and whereby WW can be used as a readily available and cost-effective source of water and nutrients for microalgae growth (Cuellar-Bermudez et al. 2017). The main advantage is that, while the microalgae will remove excess nutrients from the WW, there will be the simultaneous accumulation of biomass for downstream processing (Rawat et al. 2011). This biomass can, then, be commercialized, for example, as food and feed, biofuels, biofertilizers, and bioplastics and for the extraction of high-value products with cosmetic and pharmaceutical interest. It is worth noticing that the number of publications in microalgae-based WW treatment has increased from 60 in 2007 to >1000 in 2017.

The combination of wastewater treatment (WWT) with microalgal cultivation was first highlighted in the early 1950s by Oswald and Gotaas (1957). Nowadays, the use of a wide range of microalgae such as *Chlorella*, *Scenedesmus*, *Phormidium*, *Botryococcus*, *Chlamydomonas*, and *Spirulina* for treating different WWs has already been widely reported, and the achieved results seem promising due to their fast growth and good tolerance of different environmental conditions (Martínez et al. 2000; Kong et al. 2010; Wang et al. 2010; Mata et al. 2014; Ferreira et al. 2018; Gao et al. 2018). The studies show that microalgae cultivation using WWs promotes an effective treatment of these waters, while contributing to the production of microalgal biomass at reduced costs and with lower environmental impacts since no nutrient or freshwater supplies are required.

Microalgal-based WWT is achieved through photosynthesis, by which microalgae are able to provide O_2 to heterotrophic aerobic bacteria to mineralize organic pollutants, using in turn the released CO_2 from bacterial respiration. This synergistic relationship is advantageous for both microalgae and bacteria, with photosynthetic aeration being an interesting alternative to reduce operation costs, and limits the risks for pollutant volatilization under intensive mechanical aeration (Muñoz and Guieysse 2006). Moreover, most conventional approaches do not recycle phosphorus, which is retrieved along with various other waste products, some of which are toxic. However, in the case of microalgal-based WWT, these nutrients are incorporated into the microalgal-bacterial biomass, resulting from their combined auto- and heterotrophic growth. In addition, the application of microalgae eliminates the sludge treatment, which can be logistically challenging (Chen et al. 2015) (Fig. 9.3).

9.1.4 Algal-Bacterial Consortia in Wastewater Treatment

High microalgae diversity is often found in both naturally occurring and artificial WWT substrata, with cyanobacteria, diatoms, and green algae being the predominant groups in both cases (Fig. 9.4). More than 130 microalgal taxa have been reported in recent WWT studies, whereas only a few strains have been isolated directly from those environments (Table 9.2). Filamentous cyanobacteria such as *Arthrospira*, *Leptolyngbya*, *Oscillatoria*, *Phormidium*, and *Pseudanabaena*; various green algae such as *Chlorella*, *Chlorococcum*, and *Scenedesmus*; and chain-forming (*Achnanthes* and *Diadesmis* spp.) or solitary elongated diatoms such as *Nitzschia*, *Navicula*, and *Craticula* are found within the microalgae comprising the WWT consortia.

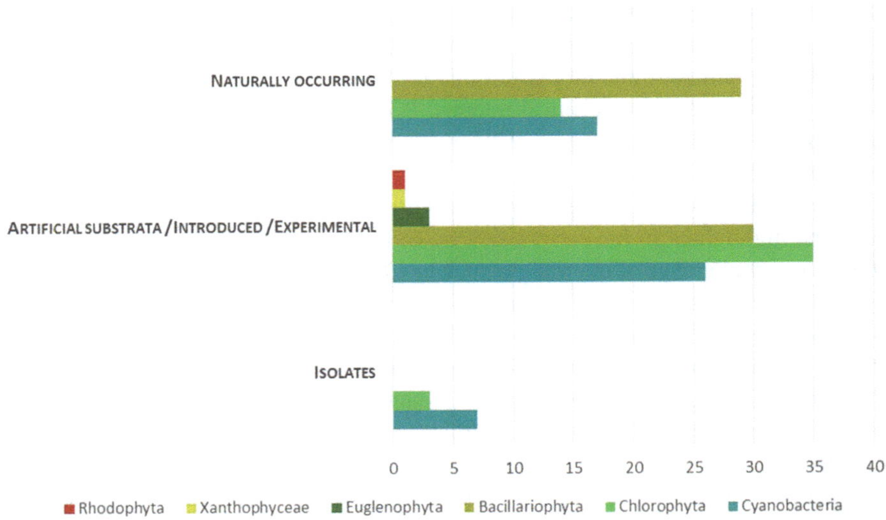

Fig. 9.4 Microalgal taxa naturally occurring or used in artificial and/or experimental substrata in WWT and strains isolated from WWT. The figure was generated based on data given in Table 9.2

These consortia grow spontaneously as biofilms, which also include heterotrophic microbes, on submerged surfaces (tank walls, overflow weirs, and troughs) of the tanks where a permanently high nutrient regime and solar irradiance favors the rapid development of taxa adapted to enriched/hypereutrophic environments (Davis et al. 1990a, b; Hoffmann 1998; Albertano et al. 1999; Congestri et al. 2003, 2005, 2006). The flow regime and turbidity of water in the secondary and tertiary tanks and the presence of sludge flocs prevent the growth of a stable phytoplankton assemblage (Albertano et al. 1999).

Wastewater treatment plant (WWTP) microalgal communities have been studied to assess the functioning of these facilities in east Europe (Sládečková 1994; Sládečková and Matulová 1998; Sládečková et al. 2001), while their potential application in the tertiary treatment of effluents was studied for cooler climate plants (Davis et al. 1990b; Tang et al. 1997).

Although the ability of microalgae and cyanobacteria to absorb/adsorb and store inorganic nutrients from WWs in stabilization ponds was anticipated in the 1950 by Oswald and Gotaas (1957), nevertheless investigations aimed at assessing the potential of native, WWTP adapted, communities in WW only started in late 1990 with studies focusing on algal turf scrubbers (Craggs et al. 1996) and on cultured WWTP biofilms in specially designed microcosms. The latter investigations evidenced that diversity decreased during biofilm development in culture with *Diadesmis confervacea*, *Phormidium* spp., *Scenedesmus* spp., and *Synechocystis* spp. acting as key taxa in the cultured biofilm (Congestri and Albertano 2011; Di Pippo et al. 2014; Guzzon et al. 2019). In addition, biofilm biomass and phosphorus accumulation are strictly dependent on the light regime with green algae as *Chlorococcum* sp., *Scenedesmus* sp., and *Desmodesmus* sp. showing maximal

Table 9.2 Microalgal taxa naturally occurring or used in artificial and/or experimental substrata in WWT and strains isolated from WWT

Taxa	Naturally occurring	Artificial substrata/introduced/ experimental	Isolates
Cyanobacteria			
Arthrospira jenneri		+	
Anabaena augstumalis			+
Anabaena sp.	+	+	
Aphanothece stagnina	+		
Aphanothece sp.	+		
Calothrix parietina	+		
Calothrix sp.		+	+
Chroococcus obliteratus	+	+	
Chroococcus vacuolatus	+		
Fischerella sp.			
Geitlerinema sp.		+	
Leptolyngbya benthonica		+	
Leptolyngbya foveolara		+	
Leptolyngbya sp.	+	+	
Limnothrix redekei	+	+	
Limnothrix planctonica			
Limnothrix sp.			
Lyngbya cf. *martensiana*	+		
Lyngbya spp.	+		
Microcystis aeruginosa		+	
Nostoc sp.		+	+
Oscillatoria limosa	+	+	
Oscillatoria cf. *subbrevis*	+		
Oscillatoria tenuis	+	+	
Phormidium autumnale		+	+
Phormidium cf. *laetevirens*			
Phormidium nigrum		+	
Phormidium cf. *pseudocutissimum*	+	+	
Phormidium spp.	+	+	
Planktolyngbya sp.		+	
Planktothrix isothrix		+	
Planktothrix cf. *prolifica*		+	
Pseudanabaena limnetica		+	
Pseudanabaena minima			
Pseudanabaena sp.		+	
Synechocystis aquatilis	+		+
Synechocystis salina		+	
Synechocystis spp.		+	+

(continued)

Table 9.2 (continued)

Taxa	Naturally occurring	Artificial substrata/introduced/ experimental	Isolates
Trichormus variabilis			+
Westiellopsis sp.			
Woronichinia sp.		+	
Chroococcal cyanobacteria	+	+	
Chlorophyta			
Actinastrum sp.	+	+	
Ankistrodesmus sp.	+	+	
Chlamydomonas reinhardtii		+	
Chlamydomonas sp.	+	+	
Chlorella sorokiniana	+	+	
Chlorella vulgaris	+	+	+
Chlorella zofingiensis		+	
Chlorella sp.	+	+	
Chlorococcum humicola	+	+	
Chlorococcum sp.		+	
Closterium sp.	+	+	+
Dictyosphaerium sp.		+	
Desmodesmus sp.		+	+
Golenkinia sp.	+		
Micractinium sp.	+	+	
Oocystis sp.		+	
Oedogonium sp.	+	+	
Pediastrum sp.			
Protoderma sp.		+	
Pseudococcomyxa sp.		+	
Raphidocelis subcapitata		+	
Scenedesmus acutus		+	
Scenedesmus armatus		+	
Scenedesmus ecornis		+	
Scenedesmus quadricauda	+	+	
Scenedesmus obliquus	+	+	
Scenedesmus obtusus		+	
Scenedesmus rubescens			
Scenedesmus spp.		+	
Selenastrum sp.		+	
Sphaerocystis sp.		+	
Spirogyra sp.	+		
Stigeoclonium tenue		+	
Stigeoclonium setigerum		+	

(continued)

Table 9.2 (continued)

Taxa	Naturally occurring	Artificial substrata/introduced/ experimental	Isolates
Stigeoclonium sp.		+	
Rhizoclonium hieroglyphicum		+	
Ulothrix aequalis		+	
Ulothrix zonata		+	
Bacillariophyta			
Achnanthes exigua		+	
Achnanthidium exiguum	+	+	
Adlafia minuscola var. *muralis*	+		
Amphora coffeaeformis	+	+	
Aulacoseira sp.			
Bacillaria paxillifera		+	
Cocconeis sp.			
Craticula accomoda	+	+	
Craticula cuspidata	+	+	
Cyclotella meneghiniana		+	
Cymbopleura amphicephala	+	+	
Cymbella minuta		+	
Denticula tenuis	+		
Diadesmis confervacea	+	+	
Encyonema minutum	+		
Eolimna minima			
Eolimna subminuscula	+	+	
Eunotia bilunaris			
Fallacia pygmaea	+		
Fragilaria ulna		+	
Gomphonema parvulum	+	+	
Gyrosigma sp.		+	
Lemnicola hungarica	+		
Luticola mutica	+		
Navicula gregaria	+	+	
Navicula radiosa	+		
Navicula salinarum	+		
Nitzschia amphibia	+	+	
Nitzschia aurariae			
Nitzschia palea		+	
Nitzschia palea var. *debilis*	+	+	
Nitzschia palea var. *minuta*		+	
Nitzschia sinuata	+		
Nitzschia sp.	+	+	

(continued)

Table 9.2 (continued)

Taxa	Naturally occurring	Artificial substrata/introduced/ experimental	Isolates
Nitzschia umbonata	+	+	
Phaeodactylum tricornutum		+	
Pinnularia gibba	+	+	
Sellaphora pupula	+	+	
Staurosirella pinnata	+	+	
Stephanocyclus meneghiniana	+	+	
Synedra acus	+		
Synedra ulna sensu lato			
Euglenophyta			
Euglena viridis		+	
Euglena sp.		+	
Trachelomonas sp.		+	
Xanthophyceae			
Tribonema sp.		+	
Rhodophyta			
Galdieria sulphuraria		+	

phosphorus removal (Guzzon et al. 2008). A different strain of *Desmodesmus* sp., isolated from the outflow of a secondary sedimentation tank of a municipal WWTP in the Rome (Italy) area showed promising capability of tolerating high heavy metal concentrations and removed Cu and Ni from aqueous media up to 95% within a 4-day incubation (Rugnini et al. 2017).

The fast growth ability of unicellular green algae and their nutrient stripping capacity make *Chlorella* and *Scenedesmus* spp. naturally dominate most continuous microalgal-based treatment systems (Abdel-Raouf et al. 2012). Selection of proper microalgae species is crucial to the success of an effective combination of WW treatment and algae-based biofuel production; native microalgae species perform better than most other species in commercial scale cultivation with WWs (Chen et al. 2015; references therein).

In wastewater, microalgae usually occur in complex consortia comprised by microalgae and prokaryotic microbial communities (including bacteria and archaea) naturally grown or specifically inoculated from previous cultures. Algal-prokaryotic consortia have been used to treat WW in different types of reactors, including waste stabilization ponds, high rate algal ponds, and closed photobioreactors (Wang et al. 2018). The use of these consortia in the remediation of WWs can be very advantageous because (1) cooperative interactions between the co-cultivated microorganisms can occur, enhancing the overall uptake of nutrients, and (2) these systems tend to be more resistant to environmental conditions oscillations (Gonçalves et al. 2017). The diversity of heterotrophic bacteria on WWT consortia is still not well-known. The most common phyla of bacteria identified in WWTs are *Proteobacteria*,

Firmicutes, and *Chlamydiae* (García et al. 2018), but few examples are known at species or genus level: The oil-degrading β-proteobacterium *Burkholderia cepacia* has been reported in a stable consortium with cyanobacteria species of *Phormidium*, *Oscillatoria*, and *Chroococcus* (Chavan and Mukherji 2008). Co-immobilization of *Chlorella vulgaris* in alginate beads with the microalgae growth-promoting α-proteobacterium *Azospirillum brasilense* under semicontinuous synthetic WW culture conditions has been shown to significantly increase the removal of ammonium and soluble phosphorus ions compared to immobilization of the microalgae alone (de-Bashan et al. 2002). The actinobacteria *Rhodococcus* sp. and *Kibdelosporangium aridum* have been isolated from WWs and then effectively used for biotreatment along with the algae *Chlorella*, *Scenedesmus*, *Stichococcus*, and *Phormidium* (Safonova et al. 2004). A *Chlorella vulgaris*-*Bacillus licheniformis* consortium has effectively eliminated nitrogen, phosphorus, and soluble COD from synthetic WW (Ji et al. 2018). Recently, Limayem et al. (2018) performed a prokaryotic community profiling and found that the most abundant classes of bacteria found in local swine and municipal WWs with and without algae were *Clostridia*, *Flavobacteria*, *Sphingobacteria*, γ-proteobacteria, δ-proteobacteria, α-proteobacteria, *Anaerolineae*, β-proteobacteria, and *Actinobacteria*.

The proportion of microalgae and bacteria and the dominant species are a function of WW composition and operational conditions. In terms of bacteria population, the similar metabolisms and microorganisms found in conventional activated sludge processes have been identified in algal-bacterial processes (García et al. 2018). The complex interactions between bacteria and microalgae in WWT are not yet fully understood; however, typically bacteria are responsible for COD degradation to mineral components, consuming photosynthetic oxygen and releasing CO_2, whereas microalgae consume the CO_2 and mineral nutrients to produce microalgae biomass and the O_2 requested by bacteria (Muñoz and Guieysse 2006; Gonçalves et al. 2017; Wang et al. 2018). The exchange of metabolites results in an overall increase in biomass productivities and, hence, nutrients removal efficiencies. On the other hand, negative effects of this interaction may occur: the heterotrophic prokaryotes may compete with algae for nutrients and produce algicidal metabolites, or shading and the photosynthesis of the algae may increase pH and O_2 to inhibiting levels for the prokaryotes (Wang et al. 2018).

9.1.5 Subcritical Water Extraction of Bioactive Compounds

Pressurized fluid extractions are technologies where pressure is applied to allow the use of liquids at temperatures higher than their normal boiling point (Mustafa and Turner 2011). Different solvents can be used in the process; however, water has ecological and economic advantages over other solvents. Therefore, by improving its solvation properties, subcritical water extraction (SWE) becomes the most promising alternative technique, which completely meets the requirements and principles of green technologies.

SWE is conducted at temperatures between 100 and 374 °C and pressure high enough to keep water in its liquid state. Under these conditions, water's physical and chemical properties are modified. Its permittivity, viscosity, and surface tension decrease, while its diffusion and dielectric characters increase. These modifications altogether improve the solvation properties of water. At room temperature, the solvation power of water is limited to polar components because of the high dielectric constant ($\varepsilon = 80$). However, by increasing the temperature, the dielectric constant of water approaches the values of organic solvents such as ethanol ($\varepsilon = 25$) and methanol ($\varepsilon = 33$) at 220 °C (Miller and Hawthorne 1998; Mustafa and Turner 2011; Khoddami et al. 2013; Xu et al. 2015). Therefore, by adjusting the parameters of extraction, it is possible to adjust the characteristics and affinities of subcritical water for components of different polarities. Factors that can impact the extraction efficiency and the quality of obtained products are temperature, pressure, extraction time, presence of additives, and characteristics of the material. Furthermore, temperature is the most dominant parameter affecting the product's properties and extraction process (Teo et al. 2010; Mustafa and Turner 2011). SWE is successfully used for the extraction of bioactive compounds from different materials such as food, medicinal plants, waste, by-products, and marine products (Ibañez et al. 2003; García-Marino et al. 2006; Sereewatthanawut et al. 2008; Rodríguez-Meizoso et al. 2010; Singh and Saldaña 2011; Pramote et al. 2012; Vladić et al. 2017).

Although it has been determined that SWE can be used successfully for the extraction of bioactive components, there is still insufficient investigation in the area of extraction of bioactive components from microalgae, and only a few studies have been conducted so far. Rodríguez-Meizoso et al. (2010) performed SWE at 50–200 °C followed by the characterization of compounds with antioxidant and antimicrobial activity from *Haematococcus pluvialis* microalga. No connection between the applied temperature and antimicrobial potential of extracts was recorded. However, they determined the positive impact of temperature on the extraction yield and the antioxidant potential of extracts as well. The presence of simple phenols with medium-high polarity, such as gallic acid, was found in the extracts, while carotenoids were not present, probably due to their nonpolar nature. At 200 °C, the obtained extracts presented the highest antioxidant activity. In addition, only at this temperature, there was the presence of vitamin E on the extracts. Hence, authors state that the antioxidant value was correlated with the presence of vitamin E, phenols, and products of caramelization and possible Maillard reaction products obtained during the extraction at high temperatures.

Several studies reported that the increase in temperature of subcritical water resulted in an increase of antioxidant activity of extracts, most likely due to the formation of new antioxidant components (Rodríguez-Meizoso et al. 2006, 2010; Plaza et al. 2010a; Vladić et al. 2017). The generation of new bioactive compounds/antioxidants during the extraction process via Maillard, caramelization, and thermos-oxidation reactions could be the result of high extraction temperatures in the process. Also, it was proven that certain products of the Maillard reaction possess an antioxidant activity (Plaza et al. 2010a). The neoformation of antioxidants during exposure of the material to high temperatures during the subcritical water extraction has been verified in microalgae *Chlorella vulgaris* and seaweeds (*Sargassum vulgare*, *Sargassum muticum*, *Porphyra* spp., *Cystoseira abies-marina*, *Undaria pinnatifida*, and *Halopitys incurvus*) (Plaza et al. 2010a). Plaza et al. (2010b) suggested that the antioxidant capacity of the newly formed compounds from those processes during the extraction depends on the nature of the sample. Nevertheless, the application of higher temperatures can cause degradation of certain components of interest; therefore, optimization in accordance with the desired components is important.

Zakaria et al. (2017) investigated the effect of extraction temperature (100–250 °C) and time (5–20 min) and microalgae concentration (5–20 wt.%) on content of phenols and antioxidant activity of *Chlorella* sp. extracts obtained by subcritical water. The authors also stated that increasing the temperature from 100 to 175 °C further increased the content of phenols in extracts, with 175 °C achieving the highest content of phenols. However, for temperatures higher than 175 °C, the yield of phenols decreased most likely due to degradation. Also, in terms of antioxidant activity of extracts, the same trend was noted. The extract with the highest phenolic content also showed the highest antioxidant potential, which further indicates that phenols are carriers of antioxidant activity in obtained extracts. Longer extraction time is not favorable for the yield of phenols due to the elevated extraction temperature that causes decomposition of the phenolic compounds during extended extraction time. Awaluddin et al. (2016) investigated optimal conditions of SWE for the extraction of biochemical compounds (carbohydrates and proteins) from *C. vulgaris*. Process parameters which were altered were temperature (180–374 °C), extraction time (1–20 min), particle size (38–250 μm), and biomass loading (5–40 wt.%). The authors determined that microalgal loading concentration is a significant parameter affecting protein yield, whereas extraction temperature, biomass loading, and extraction time affected carbohydrate yield. However, temperature represented the most critical and dominant parameter in the extraction. The optimal content of proteins and carbohydrates was achieved by applying the following conditions: 5 min, 277 °C, and 5% microalgal loading.

As a result, it was determined that when it comes to extraction of high-value compounds from microalgae biomass, SWE represents a very successful option. Although the biomass obtained from WW treatment possesses valuable bioactive components, the extraction of such compounds has not been studied extensively, mainly due to safety issues. On the other hand, considering that drastic conditions such as temperature between 100 and 374 °C and pressure are characteristics of the SWE process, this extraction method can be beneficial for microalgae biomass obtained from WW treatment as it potentially leads to elimination and/or destruction of various bacteria types and pathogens present in the biomass.

9.1.6 Microalgae-Based Bioenergy

The current increase in global energy demand and the negative long-term impact of fossil fuels-based energy sources on the environment and sustainable development have led to a reassured interest in renewable energy resources (Quinn and Davis

2015). Various feedstock for biofuel production are being investigated as viable alternatives for traditional energy sources.

Microalgae represent a promising alternative based on inherent advantages such as rapid growth rate, high lipid yields, high CO_2 uptake rate, lower land use, lower water consumption, and harvesting could be done in a daily basis, avoiding the storage requirement. However, microalgae production cost is not yet sustainable. The possibility of using nutrients from waste streams (e.g., WWs and/or CO_2 flue gas emissions) can help to decrease the production costs, in addition to the reduction of environmental impact and disposal problems (Mata et al. 2013). In fact, Pittman et al. (2011) and Lundquist et al. (2010) found that using microalgae cultivation for WW treatment coupled with biofuel production is mandatory in order to turn the process economically viable or provide a positive energy return. Efforts to advance the commercial feasibility of microalgae-derived biofuels have focused on improvements to the various processing steps, from the production of feedstock to fuel conversion processes (Quinn and Davis 2015).

Biomass conversion technologies can be classified into thermochemical, biochemical, and chemical conversion (Fig. 9.5). The choice of conversion process depends on different factors such as the type and quantity of biomass feedstock, the desired form of the energy, economic considerations, project-specific aspects, and the desired end form of the product (Brennan and Owende 2010).

Thermochemical conversion consists of the thermal decomposition of organic components in the biomass to yield fuel products (Brennan and Owende 2010). The main thermochemical processes include liquefaction, pyrolysis, and gasification.

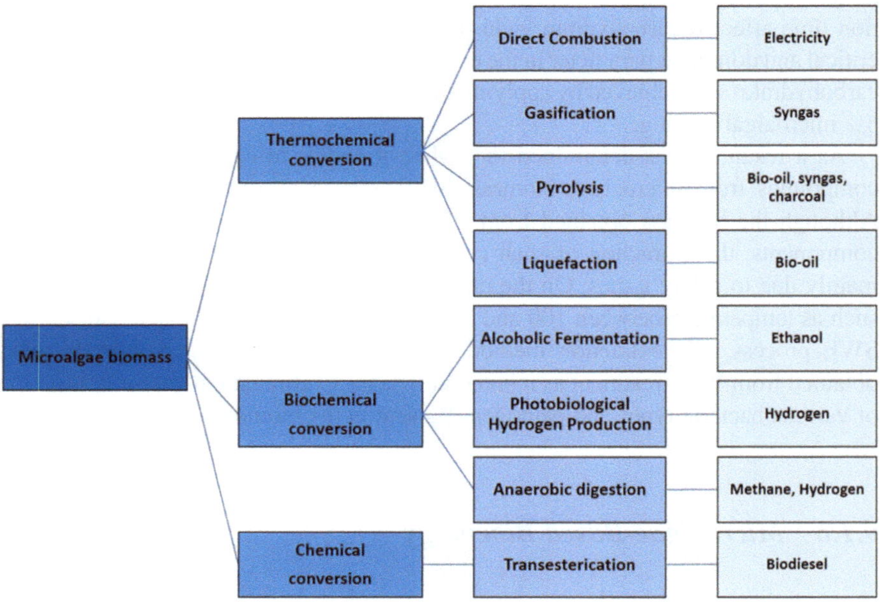

Fig. 9.5 Microalgal biomass conversion processes (Adapted from Wang et al. 2018)

The thermochemical conversion techniques are a promising pathway to process the microalgae and separate different compounds (Ferreira et al. 2015), since they present a small footprint, and efficient nutrient recovery, and no fugitive gas emissions. In addition, they have short processing times and can handle a variety of feedstock and blends, and the high temperatures eliminate possible pathogens and pharmaceutically active compounds. After conversion, only minor residues such as ash are left (Razzak et al. 2013).

In a direct combustion process, biomass is burned in the presence of air, usually in a furnace, boiler, or steam turbine at temperatures above 800 °C, to convert the stored chemical energy in the biomass into hot gases. Any type of biomass can be burned, if the moisture content is lower than 50% on a dry basis. The produced heat must be immediately used since storage is not possible (Brennan and Owende 2010).

Gasification involves the partial oxidation of biomass into a combustible gas mixture at high temperatures (800–1000 °C). In this process, the biomass reacts with O_2 and water (steam) to produce syngas. Syngas is a mixture of CO, H_2, CO_2, N, and traces of CH_4. It has a low calorific value (typically 4–6 MJ m^{-3}) that can be burned directly or used as a fuel for gas engines or turbines (Brennan and Owende 2010).

Thermochemical liquefaction consists in a low-temperature (300–500 °C) and high-pressure (5–20 MPa) process aided by a catalyst, in the presence or not of hydrogen, to generate bio-oil. Its major advantage is the possibility of converting wet algal biomass into liquid fuel through a complex sequence of physical structure and chemical changes, reducing the associated costs with drying the biomass (Brennan and Owende 2010).

Pyrolysis involves the thermal decomposition of raw feedstock in an oxygen-free atmosphere (usually nitrogen) under medium to high temperatures (350–700 °C). Under these conditions the different components of the biomass are converted into a variety of compounds that are collected as gas (biogas), liquid (bio-oil), and solid fractions (bio-char) (Razzak et al. 2013; Silva et al. 2016). The pyrolysis gases usually contain CO, CO_2, light hydrocarbons (C_1–C_4), and H_2. Regarding the bio-char, this presents a high content of C, some H, and a minimum of O. Bio-char can be used in various ways such as a soil amendment, energy carrier, adsorbents, and catalyst support. Finally, bio-oil is a complex mixture of oxygenated compounds, water (15–40 wt%), and some fine char particles (Fermoso et al. 2017). The production of bio-oil by biomass pyrolysis is an interesting option since it allows the transformation of solid biomass feedstock into liquid biofuels. Microalgae have been pointed out as a promising feedstock for pyrolysis processes. They produce a large variety of lipids that can be extracted before their conversion into biofuels or be used directly in pyrolysis processes (Ferreira et al. 2015). In fact, bio-oil similar to fossil oil can be obtained from microalgae, and it may be used directly as a liquid fuel, added to petroleum refinery feedstocks, or catalytically upgraded into transport grade fuels to be applied in the transportation sector, contributing to reduce the overall GHG emissions (Razzak et al. 2013).

Biochemical conversion uses biocatalysts like microorganisms and enzymes, in addition to heat and other chemicals to convert the biomass into biofuels (Aslanzadeh et al. 2014). Biochemical conversion of microalgal biomass can be carried out using several different approaches such as fermentation, anaerobic digestion, and photobiological production techniques.

Anaerobic digestion is the conversion of organic wastes into a biogas, which is mainly composed of CH_4 (55–75%) and CO_2 (25–45%), with traces of other gases such as hydrogen sulfide (Molinuevo-Salces et al. 2016). This process is many times preferred over other conversion processes because it is appropriate for high moisture content (80–90% moisture) organic wastes, which can be useful for wet algal biomass, avoiding the energy demanding drying step (Brennan and Owende 2010).

Alcoholic fermentation consists in the conversion of biomass materials that contain sugars, starch, or cellulose into ethanol, through the action of yeasts. In this process, the biomass is mashed, and the starch is converted to sugars that are then mixed with water and yeast (commonly *Saccharomyces cerevisiae*) in fermenters at a given temperature. Following this, a purification process (distillation) is used to remove the water and other impurities present in the diluted alcohol product (10–15% ethanol). The concentrated ethanol (95% volume for one distillation) is removed and condensed into liquid form, which can be used as a supplement or substitute for petrol in cars. The remaining solid residue can be used for cattle feed or for gasification (Brennan and Owende 2010).

Hydrogen (H_2) is a naturally occurring molecule that is a clean and efficient energy carrier, with the highest energy content per unit weight of any known fuel (142 kJ/kg). In addition, it can be transported for domestic/industrial consumption through conventional means, being safer to handle than domestic natural gas (Das and Veziroglu 2008). Biological hydrogen (bioH_2) can be produced mainly by two routes: photobiologically—biophotolysis of water using green algae and cyanobacteria and photo-decomposition of organic compounds by photosynthetic bacteria (Das and Veziroglu 2008)—and by bacterial fermentative processes such as dark fermentation. The first one is based on the uptake of CO_2 or other organic substrates and water by photosynthetic organisms, which means that it requires a constant light source supply that, in addition to low yields, is the major drawback of the process (Ortigueira et al. 2015). In recent years, bioH_2 production through dark fermentation has received increased attention due to its many advantages, such as the high H_2 production rates, the potential to convert biomass or bio-wastes into H_2, and the feasibility of the process design and control (Batista et al. 2014). Unlike the first two routes, dark fermentation is an indirect technology in which several genera of bacteria (namely, *Clostridium* and *Enterobacter*) are capable of using the carbohydrates, proteins, and lipids present in the microalgae as substrates to produce H_2, CO_2, and organic acids, through the acidogenic pathway.

Both bioethanol and bioH_2 can be produced from any sugars-containing biomass feedstock. The appeal of the feedstock increases if it is readily available and low-cost. In this context, microalgae are gaining wide attention since they are able to absorb solar energy and CO_2 and synthesize and accumulate large quantities of carbon compounds, such as starch (Ortigueira et al. 2015). Moreover, microalgal cells are buoyant, avoiding the need for structural biopolymers such as hemicellulose and lignin, which simplifies the pretreatment steps (Singh and Olsen 2011).

Other important conversion process is a chemical reaction called transesterification, which occurs between triglycerides and alcohol (usually methanol) in the presence of a catalyst to produce monoesters that are termed as biodiesel. Biodiesel is a derivative of oil crops and biomass that can be used directly in conventional diesel engines. It is a mixture of monoalkyl esters of long-chain fatty acids (FAME) derived from a renewable lipid feedstock such as algal oil (Brennan and Owende 2010). Some microalgae species can present a wide lipid content, which may vary between 1 and 70%. The produced lipids are chemically similar to common vegetable oils and are therefore a potential source of biodiesel (Farrelly et al. 2013). In fact, algal biodiesel was found to be in accordance with the standard values imposed by the International Biodiesel Standard for Vehicles (EN14214), with lower heating value, lower viscosity, and higher density. Moreover, it presents major advantages since it is renewable; biodegradable; nontoxic; containing only minimum levels of particulates, carbon monoxide, soot, hydrocarbons, and Sox; and with a potential reduction of 78% of CO_2 emissions, when compared to petroleum diesel (Brennan and Owende 2010).

In order to decrease energy demand, time, and costs from lipid extraction to the subsequent conversion into biodiesel, several works have been carried out on the development of an in situ transesterification directly from the biomass (Jazzar et al. 2015; Gouveia et al. 2016b; Bharathiraja et al. 2016). This emerging technique consists in a single-step method, where the extraction of the lipids and their transesterification occurs simultaneously in one reactor. The alcohol acts as a solvent as well as a transesterification agent, which allows the reduction in size of biodiesel production units and thus the production costs, in addition to better yields (Gouveia et al. 2016b). Nevertheless, the microalgae grown in WW generally have a low content of lipids (<20%), being the biodiesel not an interesting fuel-based microalga derived from WW (Batista et al. 2015; Gouveia et al. 2016a; Ferreira et al. 2018; Oliveira et al. 2018).

9.1.7 Microalgae-Based Biofertilizers

Modern agricultural practices rely on an intensive use of chemical fertilizers and pesticides in order to achieve maximum yields. Thus, this overdependency on synthetic agrochemicals, which are not only finite resources and toxic, but their price is rising, has resulted in massive ecological degradation, including ocean dead zones, eutrophication, soil infertility, biodiversity losses, and human diseases from food and water contamination (Chakhalyan et al. 2008).

The analysis of soil has shown that the widespread and permanent use of chemical fertilizers has led to a catastrophic situation, particularly to full pauperization of soil and multiple excess of critical concentration of nitrates and nitrites. To avoid these problems, there is an actual concern for replacing the harmful chemicals for ecologically safer fertilizers to remediate soil and enhance its fertility (Avetisova 2013; Mishra et al. 2013; Uysal et al. 2015).

One possible way to solve these problems is the development and manufacturing of biofertilizers. Biofertilizers play a major role and are essential for the controlled mineralization and fertilization processes. They are cost-effective, eco-friendly, and a renewable resource, making then a sustainable alternative for the presently used cost- and energy-intensive chemical fertilizers, not only to enhance agriculture production but also to diminish environmental pollution (Kawalekar 2013). Biological fertilizers contain living or latent microorganisms or natural compounds derived from organisms such as algae, bacteria, and fungi. These organisms have the ability to transform numerous organic and inorganic compounds containing macro- and microelements into other compounds easily digestible by plants (Avetisova et al. 2014; Kumar et al. 2017; Melkonyan et al. 2017).

The algal biomass can itself act as both a nutrient and carrier material for biofertilizers, as it presents a better shelf life (post drying), being less thermo-sensitive and more robust for storage and transportation, which make it a better choice as a biofertilizer (Mahapatra et al. 2018).

The majority of cyanobacteria is capable of fixing atmospheric nitrogen and can be effectively used as biofertilizers. They can help build up soil fertility, thereby enhancing crop yield as demonstrated with rice (Song et al. 2005). A study showed that the dry biomass obtained from the green microalga *Acutodesmus dimorphus* applied as a biofertilizer was able to increase the growth of plant and production of flowers in Roma tomato plants (Garcia-Gonzalez and Sommerfeld 2016). Faheed and Fattah (2008) also studied the growth parameters in lettuce plant using *Chlorella vulgaris* as biofertilizer. This study revealed that the dry powder of *C. vulgaris* was a suitable substrate for germinating the seeds of *Lactuca sativa* as it significantly enhanced the process of germination when compared to unfertilized control. It also increased significantly the amount of pigments (chlorophyll a, chlorophyll b, and carotenoids) in the lettuce seedlings (Faheed and Fattah 2008). Agwa et al. (2017) also investigated the potential of *C. vulgaris* as a biofertilizer for *Hibiscus esculentus* and its role in enhancing soil fertility. This study suggested that this microalga is an efficient, economical, and safe biofertilizer to substitute chemical fertilizers in enhancing plant growth and having no detrimental effect on the plant.

Some studies were also done using WWs to grow microalgae at a lower cost and make the bioavailability of nutrients later for enhancing growth of crops through controlled release of nutrients. Renuka et al. (2016) compared unicellular and filamentous microalgae as a biofertilizer for wheat crop, with the latter achieving the better results. Dineshkumar et al. (2017) grew *Chlorella vulgaris* and *Spirulina platensis* in cow manure and used the biomass on maize resulting in an increase on its yield and growth up to 51%.

9.2 Wastewater Conversions Toward Biofuels and Bio-Based Products

9.2.1 Urban

Sewage is produced by human activities at ratios around 100 L/day per equivalent person in developed countries, this figure largely modifying as a function of social and economic development, but also according to location and environmental factors

(Wheeler and Beatley 2004). Sewage is a contaminant if it is not adequately treated; hence different conventional technologies are largely used. Activated sludge process is one of the most extended ones, and it is divided into three major steps: (1) primary sedimentation, (2) biological treatment plus secondary sedimentation, and (3) tertiary treatment. The composition of water stream after each one of these steps modifies, mainly removing from the WW the organic matter or COD, in addition to nitrogen (both as ammonium and nitrate) and phosphorus. Comparing the composition of different WW types with that of a standard microalgae culture medium as Arnon medium, it can be observed that the differences are minor (Acién et al. 2016) (Table 9.3). Thus, sewage contains carbon, nitrogen, and phosphorus in the same ranges that of Arnon culture medium. On the basis of this composition, the achievable potential microalgae biomass concentration if completely consuming the nutrients contained into the WW can be easily calculated. Figure 9.6 shows that, in spite of the high concentration of organic matter (COD), in all the WW types, the cultures become carbon limited if not additional carbon, such as CO_2, is provided.

The next nutrient limiting the production of biomass is nitrogen, especially for WW after secondary treatment. When considering WW from primary treatment, the ratio N/P is close to the optimal one. Only when using centrate from anaerobic digestion of activated sludge the cultures become phosphorus limited. This simple analysis is critical when designing whatever microalgae-based WW treatment process because unbalanced nutrients at the inlet make the complete removal of these compounds at the outlet difficult. In some cases it is necessary to provide additional nutrients to optimize the performance of the system (Sepúlveda et al. 2015).

Microalgae-based processes can be used to treat all these effluents, so the incorporation of microalgae production after primary treatment completely would replace conventional WWTP, whereas the utilization of microalgae for the treatment of WW after secondary treatment or including to treat centrate from anaerobic digestion implies the combination of both conventional and microalgae-based technologies (Acién et al. 2016). Whatever the effluent type, the utilization of open reactors, mainly raceways, is mandatory. These reactors are the cheapest ones, and more importantly, they can be scaled up at large scale, up to several hectares. It has been reported that sewage can be adequately treated, using raceway reactors. For the effluent to accomplish regulation limits, large hydraulic retention times up to 10 days are necessary (Martínez et al. 2000; Park et al. 2011; Olguín 2012).

Table 9.3 Mean composition of sewage produced by human activities (Acién et al. 2016)

Conc. (mg/L)	Arnon	Primary treatment	Secondary treatment	Anaerobic digestion
COD	–	500.0	110.0	300.0
$N-NO_3$	114.0	2.4	0.0	5.3
$N-NH_4$	0.0	62.6	20.8	506.5
$P-PO_4$	41.0	11.3	10.0	12.0
TKN	114	65	20	511
TP	41	11	10	12
TC	47	296	82	247

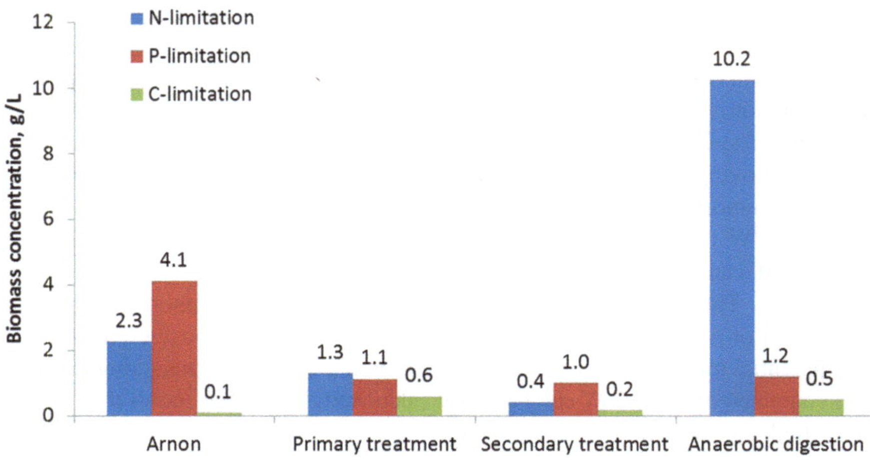

Fig. 9.6 Variation of the achievable microalgae biomass concentration as function of nutrient content of different considered WW types. Obtained data considering an elemental composition of the biomass such as C = 50% d.wt., N = 5% d.wt., P = 1% d.wt.

This large time is due to the high water depth commonly used on these reactors, which reduces the penetration of the light, thus reducing the performance of microalgae cells. It must be taken into account that in an ideal system, the microalgae must uptake all the nutrients contained into the WW. For that, maximal light availability is required. If providing WW after secondary treatment, the problem is the opposite, because the nutrient content is too low, and to maintain a high microalgae biomass productivity, the hydraulic retention time must be low, in the range of 1 day. This time is much shorter than the growth rate of microalgae; then to avoid the washout of the system, it is mandatory to use membrane bioreactors or to recirculate the microalgae biomass from the outlet (Gómez-Serrano et al. 2015). The third option is to use centrate from anaerobic digestion. In this case the nutrient content is much higher, and dilution might be required to avoid inhibition of the microalgae growth (Morales-Amaral et al. 2015).

If adequately managed, microalgae can be used to treat all these types of WW, achieving high biomass productivity, while, at the same time, accomplishing regulation in terms of quality of effluents, and recovering the nutrients from the WW. COD removal is always higher than 90% because it is mainly performed by bacteria naturally appearing on these systems; however, to optimize the nitrogen and phosphorus removal, the reactors must be operated at the optimal dilution rate as a function of environmental conditions (Morales-Amaral et al. 2015). Thus, in winter time the nutrient load must be lower than in summer time. To maximize the performance of microalgae-based WWT process, it is recommendable to optimize the performance of microalgae cells, by reducing the water depth and providing optimal pH for the microalgae cells (Posadas et al. 2015a). In the case of pH, it must be kept at 8.0–8.5; the required amount of CO_2 is low because the organic carbon contained into the WW is efficiently transformed into inorganic carbon by the bacteria. However, if in whatever circumstances the pH increases above 9, the biological system can be damaged conducing to the washout of the system. Other relevant factors are the temperature and dissolved oxygen, but these parameters are scarcely controlled in large raceway reactors (Mendoza et al. 2013).

Considering the production of biomass, WWT processes are potentially a source of large amounts of low-cost microalgae biomass. Thus, when combining microalgae production with WW and flue gas treatment, the biomass production cost can be reduced below 0.5 €/kg, and up to 15,000 t/year can be produced in a town of 150,000 inhabitants (Acién et al. 2016). Moreover, the biochemical composition of this biomass is similar compared to produced biomass using clean water and fertilizers. It must be noted that when using WW no specific microalgae species and strains are produced, but instead a consortium of microalgae and bacteria are produced. However, usually there is always a microalga species prevailing into the culture, such as *Scenedesmus* or *Chlorella*, with the proportion of bacteria being really minor, below 10% of dry matter (Gómez-Serrano et al. 2015). The biochemical composition of the produced biomass mainly contains proteins (30–50%), carbohydrates (20–40%), and lipids (10–20%). On the basis of this composition, a large portfolio of potential applications can be considered. However, EU regulation forbides to use of whatever material obtained from WW treatment for human-related applications. Moreover, severe restrictions exist about the utilization of this microalgae biomass in feed-related applications, which leaves only the agriculture- and biofuels-related applications.

The production of biofuels is today an unsolved challenge both for economic and technological reasons. The most promising strategy could be the production of biogas and upgrading the biogas to biomethane for transport, because it is a simple technology which is able to transform up to 30% of the total produced biomass into final product (Cabanelas et al. 2013; Milano et al. 2016). The centrate from the anaerobic digestion of microalgae biomass must be valorized as biofertilizers; otherwise it is a residue that should be adequately managed/disposed. Alternatively, the whole microalgae biomass can be transformed into biofertilizers by enzymatic hydrolysis (Romero García et al. 2012). This type of products has been demonstrated to be really useful and interesting to improve the yield of food production by agriculture (Ördög et al. 2004; Plaza et al. 2018).

Batista et al. (2015) and Ferreira et al. (2018) obtained bioH_2 by dark fermentation from the biomass that treated urban WW. The same authors also obtained bio-char, biogas, and bio-oil from the same biomass through pyrolysis (data not published yet).

9.2.2 *Food*

9.2.2.1 Dairy

Dairy effluents (from milk, cheese, yoghurt, and other products) are known to contain high organic load, high BOD, COD, and nutrients such as phosphates, ammonia, and/or nitrate, which can be detrimental to the water quality and aquatic life. For instance, cheese whey, the liquid leftover remaining from the cheese production process, constitutes a serious environmental threat (Dragone et al. 2009). A list of the major components of whey includes disaccharide lactose, which can eventually yield glucose and galactose, conducting to BOD and COD values in the range of 30,000–50,000 mg/L and 60,000–80,000 mg/L, respectively. In addition, this effluent also contains proteins and lipids together with water-soluble vitamins and minerals (Siso 1996). The typical composition of dairy WW is depicted in Table 9.4.

A dairy effluent was successfully treated by immobilized *Chlorella pyrenoidosa* and *Euglena gracilis* in either open or closed systems (Yadavalli et al. 2014). *Euglena gracilis* presented a lipid content of 11% after 7 days of cultivation and *C. pyrenoidosa* 7% for the same parameter after 10 days. Simultaneously *Chlorella* and *Euglena* protein contents were 44% and 29%, respectively. On the other hand, a significant rise in carbohydrates in *C. pyrenoidosa* was reported. NH_4^+-N was completely removed by immobilized *Chlorella* sp., while 96% reduction was observed for *Euglena*. Moreover, 98% PO_4^{3-}-P removal was achieved with both species regardless of the modes of operation. BOD and COD also decreased significantly (80–96%).

Abreu et al. (2012) have studied the mixotrophic cultivation of *Chlorella vulgaris* using industrial dairy waste as organic carbon source. Under such strategy microalgae exhibited higher specific growth rate, final biomass concentration, as well as higher productivities of lipids, starch, and proteins compared to microalgae cultivated under photoautotrophic conditions. Moreover, the supplementation of the inorganic culture medium with hydrolyzed cheese whey powder yielded a clear improvement in biomass production and carbohydrate utilization when compared with the culture enriched with a mixture of pure glucose and galactose, highlighting the presence of beneficial growth-promoting nutrients in cheese whey. This main dairy industry by-product could be considered a feasible alternative to decrease substantially microalgal biomass production costs.

Ferreira et al. (2018) used microalga *S. obliquus* to treat dairy wastewater, obtaining removal efficiencies of 72 and 66% for phosphorus and COD, respectively, and achieving a complete removal of N-NH_4^+. The microalgal biomass

Table 9.4 Typical composition of dairy wastewater (Slavov 2017)

Parameters	Concentrations Dairy WW
COD (mg O_2 L^{-1})	500–10,400
TOC (mg C L^{-1})	1400–1500
TN (mg N L^{-1})	10–660
TP (mg P L^{-1})	0–600
PO_4^{3-} (mg P-PO_4^{3-} L^{-1})	10–326
TSS (mg L^{-1})	60–5400
pH	4–11
Color	White; yellowish-green (whey)

obtained contained 53% protein, 18% lipids, and 14% sugars in its composition. The biomass was then used as feedstock for dark fermentation to produce bioH_2 (Ferreira et al. 2018) and for pyrolysis to obtain bio-oil, bio-char, and biogas (Ferreira et al. 2020).

9.2.2.2 Brewery

The brewing process generates alcohol, sugars, soluble starch, proteins, and volatile fatty acids that all end up in WW (Raposo et al. 2010). This results in a WW that can be classified as a high-strength WW according to the values shown in Table 9.5.

Studies that have coupled microalgae cultivation with the treatment of brewery WW have been done using *Chlorella* species (Raposo et al. 2010; Darpito et al. 2015) and *Scenedesmus* (Ferreira et al. 2017; Marchão et al. 2018). *Chlorella*-based treatment of brewery WW (Raposo et al. 2010; Darpito et al. 2015) achieved removal efficiencies above 85% for total nitrogen and 90 and 66% for phosphorus, respectively. In addition, both showed that the use of brewery WW increased lipid accumulation and improved fatty acid profiles, which is advantageous for the production of biodiesel.

Scenedesmus obliquus-based treatment has achieved removal efficiencies of 80–90% for total nitrogen, 18–40% for total phosphorus, and 50–61% for COD (Ferreira et al. 2017). The treatment has succeeded in removing pollutants to levels in accordance to the Portuguese legislation (Decree-Law No 236/98 1998). The microalgal biomass generated contained an average of 36% protein, 23% sugars, and 10% lipids.

Subcritical Water Extraction of the *Scenedesmus* Biomass Grown in the Brewery Effluent

In a recent study, subcritical water extraction has been employed to extract bioactive constituents from *Scenedesmus obliquus* microalgae biomass obtained from brewery WW treatment (Ferreira et al. 2019). The SWE process was performed at three

Table 9.5 Typical composition of brewery wastewater (Simate et al. 2011; AL-Rajhia et al. 2012)

Parameters	Concentrations Brewery WW
COD (mg O_2 L^{-1})	565–7837
TOC (mg C L^{-1})	1400–1500
TN (mg N L^{-1})	30–75
TKN (mg N L^{-1})	25–80
NH_4^+ (mg N-NH_4^+ L^{-1})	3–106
TP (mg P L^{-1})	12–20
PO_4^{3-} (mg P-PO_4^{3-} L^{-1})	10–326
TSS (mg L^{-1})	61–3000
pH	3–12
Color	Yellow

different temperatures: 120, 160, and 200 °C. The parameters which were constant in these three experiments were time, pressure, and microalgae load. The yield of extracted phenols was in the range from 0.249 to 1.016 mg GAE (gallic acetic equivalent) per mL of obtained extract. In addition, the yield of extracted flavonoids was much lower, from 0.05 to 0.167 mg CE (catechin equivalent) per mL. Temperature demonstrated to have a positive impact on the extraction efficacy, and the result was in line with previously published studies focused on the extraction of bioactive compounds from microalgae by applying SWE. Higher yield of phenols correlated with higher temperatures. Moreover, the yield of phenols was around fivefold higher with an increase in temperature from 120 to 200 °C, whereas the temperature jump from 120 to 160 °C resulted in an approximately double yield. In accordance with these findings, subcritical water at a temperature of 200 °C was the optimal condition for obtaining the highest quantity of polyphenols from microalgae extracts (Ferreira et al. 2019).

Furthermore, in order to determine whether SWE, apart from extracting bioactive components, can be utilized as a "purifying" technology and provide microbiologically safe extracts from the microalgae biomass, the microbiological profile of the initial lyophilized microalgae biomass (obtained in WWT) and liquid extracts obtained by subcritical water was conducted. Microbiological evaluation focused on total number of microorganisms, presence of molds and yeasts, spores of anaerobic bacteria, and indicators of fecal contamination—*Enterobacteriaceae* and *Escherichia coli*. The results showed that the total number of microorganisms in the freeze-dried biomass was 64,000 cfu/g, the presence of molds and yeasts was 100–200 cfu/g, and spores of anaerobic bacteria were also present. However, preliminary results of liquid extracts from *Scenedesmus obliquus* cultivated in brewery effluent showed that there was a complete elimination of microorganisms at all investigated temperatures. No microorganisms were detected in liquid extracts obtained by subcritical water unlike in the biomass which was used as the extraction material. It can be concluded that subcritical water at all temperatures can be employed for obtaining "clean" and microbiologically safe products. These results are highly significant and indicate that subcritical water as an extraction and purification process can be used in order to valorize microalgae biomass obtained by WW treatment (Ferreira et al. 2019).

Biofuels from *Scenedesmus* Biomass Grown in the Brewery Effluent

This biomass obtained from the brewery effluents treatment was also upgraded into biofuels. The biomass was used as substrate for a dark fermentation process to produce bioH_2, using *Enterobacter aerogenes*. The fermentation conditions were $T = 30$ °C, 150 rpm, and 6 h, and the yield obtained was 67.1 mL H_2 g^{-1} (VS) and purity (H_2/CO_2) = 1.02 (Ferreira et al. 2017).

The biomass was also submitted to pyrolysis in a quartz fixed bed reactor (internal diameter of 16 mm and a length of 150 mm), in a nitrogen atmosphere for15 min. The amount of raw microalga was about 2.5 g, and the reactor was placed in the circular furnace preheated to 475 °C. Nitrogen with purity of 99.99% was used as the carrier gas, and its flow (200 mL/min) was monitored using a mass flow controller.

The pyrolysis liquid amount was determined by the mass difference of the condensers before and after a pyrolysis test. The bio-oil was recovered by washing the reactor with acetone followed by rotating evaporation of acetone under a reduced pressure. It was obtained bio-oil (64%), bio-char (30%), and biogas (6%). Both conversion routes seem to have potential (Ferreira et al. 2017, 2020).

Biofertilizer Potential of *Scenedesmus* Biomass Grown in the Brewery Effluent *Scenedesmus obliquus* biomass from brewery treatment was also evaluated for its biofertilizer/germination activity potential in barley and wheat seeds.

To determine the growth-stimulating action of *S. obliquus* microalga under laboratory conditions, the effect of the culture in synthetic medium and in brewery wastewater and the pellet of microalga biomass (after centrifugation) on the seeds of barley and wheat was studied (Ferreira et al. 2020).

The study was carried out on filter paper and soil in Petri dishes. Seventeen seeds of each plant were placed on the filter paper and on 50 g of soil. Each sample was wetted with 8 mL of tap water-diluted microalga culture (dilution rate = 1:8) or resuspended microalgal pellet. Microalgal culture, such as synthetic medium (SM) and brewery wastewater (BWW), was pelleted by centrifugation (4000 rpm, 10 min) and resuspended in the tap water (12.5%, v/v). The tap water (without any culture addition) was used as control. Dishes with sown seeds were placed under room conditions (a daily cycle of 12/12 h darkness/light) at 25 °C (Fig. 9.7). Samples were daily watered with the same required amount of tap water. As depicted from Fig. 9.7a when seeds of barley were treated with biomass pellet obtained in the WW, all sprouts on sixth days were emerged (against ten of biomass pellet from synthetic medium and five from control—only water). The average length of sprouts was 13.5 cm (Ferreira et al. 2019).

The treatment more efficient on germination/growth appeared to be the pellet biomass from the brewery wastewater treatment, which provided 100% seeds germination of barley, higher length of sprout, and better color saturation.

On the other hand, seeds of wheat were treated equally; however only in the case of pellet water solutions 6 sprouts from 17 seeds were emerged, while, in the control, only 3 sprouts (Fig. 9.7b).

The same tests were done with 17 seeds (barley and wheat) placed on wet soil in Petri dish. The same behavior was observed for both seeds, with biomass pellet from BWW giving the better results. This observation was especially evident for barley seeds (Fig. 9.8a, b).

Concluding, the results showed that the pellet of *S. obliquus* has a positive effect on seeds germination/growth of barley and wheat, with special emphasis on barley.

9.2.2.3 Potato

Potato processing generates a medium-strength WW according to Tchobanoglous et al. (2003) (Table 9.6). The soluble fraction of this WW matches the optimum C/N/P ratio of 100/18/2 proposed by Oswald (1988) for microalgae growth, which

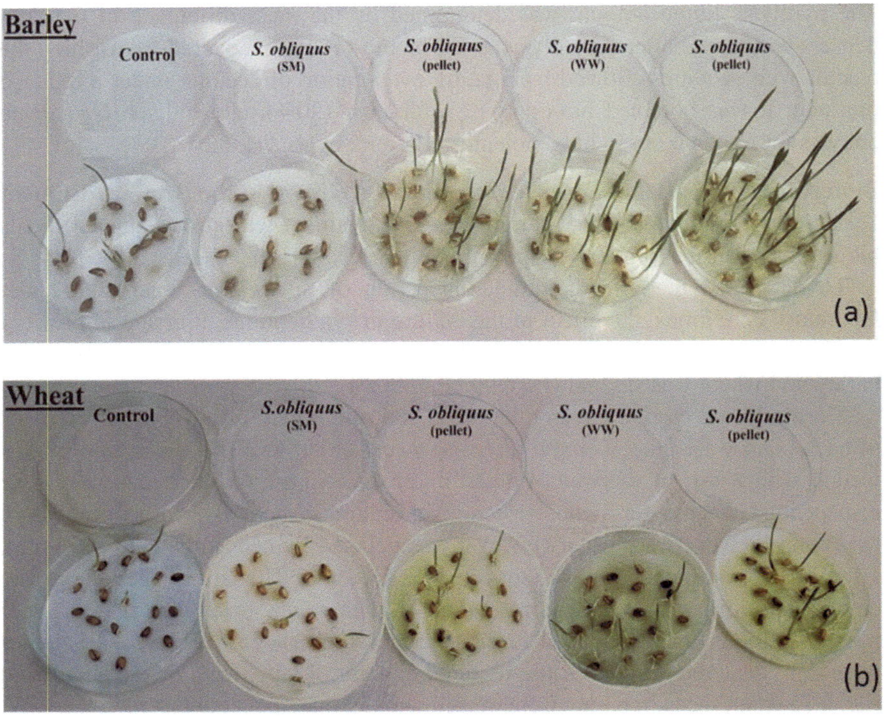

Fig. 9.7 The effect of biomass pellet of *Scenedesmus obliquus* and its culture medium (synthetic, SM, and brewery wastewater, BWW) on barley (**a**) and wheat (**b**) seeds germination in wet filter paper, after 6 days

highlights the potential of this WW to support a low-cost mass cultivation of microalgae (Posadas et al. 2014b). Despite its carbohydrate-based origin, the organic carbon in this type of WW presents a low-medium biodegradability (30–50%) by algal-bacterial consortia, which correlates with total nitrogen (TN) biodegradability (20–60%). However, the low total phosphorus (TP) concentration in this WW can be easily removed with removal efficiencies of ≈90%. The number of studies devoted to evaluating the potential of this type of WW to produce microalgae biomass is scarce and limited to small-batch cultivations focused on WW treatment rather than biofuel or bio-product synthesis. The carbon and nitrogen content of this biomass (43–51% and 6.5–8.8%, respectively) suggests a balanced biomass composition, suitable for conventional microalgae conversion strategies.

9.2.2.4 Coffee

The high total organic carbon (TOC), TN, and TP concentrations in the WW originated from coffee manufacture render it a high-strength effluent. Interestingly, most of the dissolved nitrogen was present as organic nitrogen (Table 9.6). The dark

Fig. 9.8 The effect of pellet of *Scenedesmus obliquus* and its culture medium (synthetic and brewery wastewater) on barley (**a**) and wheat (**b**) seeds germination in soil, after 7 days

coloration of this type of WW hinders light penetration, entailing the need for either a pretreatment to reduce the color or a dilution of the WW with freshwater. Algal-bacterial systems can support TOC and TN removals of 18–56% and 8–80%, respectively, WW dilution increasing its biodegradability as a result of a reduction in toxicity and the enhancement in light penetration. The unbalanced C/N/P ratio of this wastewater (100/9/1) does not significantly modify the C and N of the biomass (42–57% and 5.7–8.5%, respectively). The production of microalgae biomass from coffee WW has been only assessed batch-wise (Posadas et al. 2014b), with the sole purpose of WW treatment.

9.2.2.5 Yeast

Wastewater from a yeast production factory can be also classified as a high-strength WW with an unbalanced C/N/P ratio (100/28/0.3). Interestingly, most of the nitrogen is typically present as ammonium, while carbon is present mainly as inorganic carbon (which correlates with the relatively high pH of this WW) (Table 9.6). This

Table 9.6 Typical composition of wastewaters originated from potato and coffee processing and yeast production

	Concentrations		
Parameters	Potato WW	Coffee WW	Yeast WW
TOC (mg L^{-1})	327	8532	1186
IC (mg L^{-1})	54	171	1353
TN (mgN L^{-1})	69	766	703
N-NH_4^+ (mg L^{-1})	13	101	565
N-NO_2^- (mg L^{-1})	<D.L.	3	<D.L.
N-NO_3^- (mg L^{-1})	<D.L.	<D.L.	<D.L.
$P_{soluble}$ (mg P L^{-1})	6	59	7
pH	7.1	7.0	8.1
C/N/P (g/g/g)	100/18/2	100/9/1	100/28/0.3
Color	Yellow	Black	Brown

Data refer to the soluble fraction (Posadas et al. 2014a)

composition results from a previous anaerobic digestion step, which reduces the content of biodegradable carbon, ammonifies most of the nitrogen, and mediates a high alkalinity. The biodegradability of carbon and nitrogen in this type of WW by algal-bacterial consortia is low (27–33% and 12–50%) and was only achieved after dilution at ratio >4. This low carbon and nitrogen biodegradability can be explained by the low phosphorus concentration in this WW and the high recalcitrance of the organic matter following the anaerobic digestion stage (Posadas et al. 2014b). Similarly, to the extent of the investigations conducted with potato and coffee WWs, the only studies addressing microalgae production from yeast WW have been conducted batch-wise.

9.2.3 Livestock Production

9.2.3.1 Swine

Livestock agriculture in Europe has developed into an efficient industry during the last decades, but its future is tarnished by several environmental problems (e.g., water, air, and soil pollution) that stem from the large quantities of wastes produced within several intensively farming regions. The entire manure production in the EU that potentially is available for manure processing is estimated to 1.4 billion tons. Germany is the first producer with up to 202 million tons, whereas Spain produces 117 million tons, and Italy up to 88 million tons per year. Most of the manure is stored and land spread contributing to contamination of water reservoirs and emissions to the atmosphere. Alternatively, manure can be processed to change its physical and/or chemical properties as an objective itself or in order to recover energy from the livestock manure, make the livestock manure more stable, or remove nutrients from the mainstream (Liao et al. 1995; Burns et al. 2003; Cantrell et al. 2008).

Different technologies can be applied to manage manure, but no universal solutions exist. In large farms or concentrated areas, manures are processed mainly by anaerobic digestion to produce biogas, whereas in small farms or not concentrated areas, they are stored and then released in the environment. Processed and unprocessed manures are usually land spread, but production of these residues overpasses the capacity of arable land available. Also, only a small fraction of the nutrients contained in the slurry (particularly N and P) are being consumed by plants, causing enormous environmental problems related to leaching, pollution of surface waters, emissions of noxious gases into the atmosphere, and surface water eutrophication. Moreover, the Nitrates Directive (1991) establishes action programs to cut down nitrogen pollution and states the limit of 0.170 tonN/ha year from livestock manure to control pollution, improve water quality, and protect human health.

Alternative technologies can be used to produce valuable products from manure, thus avoiding the necessity of land spreading, one of them being the production of microalgae that was proposed in the 1970s (Wilson and Houghton 1974). The application of microalgae-bacteria consortia allows using solar energy to transform manure nutrients (C, N, P) into valuable biomass that can be used to produce commodities as feed and biofertilizers. Microalgae hold a great potential because of their high growth rate, easy production, better utilization of sunlight compared to conventional plants, shorter life cycles, and independence from fertile agricultural land. Production of microalgae using pig manure is not easy. Pig farms produce manure containing up to 9000 mg N/L and 500 mg P/L, in addition to high COD up to 16,000 mg/L (Table 9.7). Anaerobic digestion reduced the COD content to 9000 mg/L, but N and P remained almost unchanged, at 8000 mg N/L and 400 mg P/L, respectively.

The major problem of pig manure is the content of solids that can reach values up to 20 g/L. This results in a really dark sludge that makes light penetration difficult and hinders the use of raw manure to produce microalgae. Farmers usually store the pig manure in large reservoirs, the solids content decreasing by natural settling of large particles; thus the solids content can reduce ten times up to 2 g/L, which is still too much solids for microalgae production. However, because pig manure still contains too much N, it must be diluted with water as previous step to be used as microalgae culture medium. Ammonium has been reported to be toxic for microalgae at concentrations upper to 100 mg/L (Collos and Harrison 2014). Thus, pig manure must be diluted more than ten times to avoid ammonium inhibition, reducing at the same time the phosphorus and COD concentration, in addition to solids content decreasing.

Once the problems related with inadequate composition were solved, different microalgae have been demonstrated to growth adequately on this effluent. The most typical genus prevailing on these systems is *Scenedesmus* (Sevrin-Reyssac 1998), although also *Chlorella* and *Phormidium* genera have been demonstrated to be able to be produced using this effluent (de la Noüe and Bassères 1989). Whatever the microalgae and the specific concentration of the manure on these processes, always a

Table 9.7 Typical composition of swine wastewater (Ledda et al. 2016)

Parameters	Concentrations
	Swine WW
COD (mg O_2 L^{-1})	16,000
TOC (mg C L^{-1})	5000
IC (mg C L^{-1})	400
TN (mg N L^{-1})	6000
TKN (mg N L^{-1})	9000
NH_4^+ (mg N-NH_4^+ L^{-1})	6800
TP (mg P L^{-1})	500
PO_4^{3-} (mg P-PO_4^{3-} L^{-1})	1550
C:N:P[a]	1:1.7:0.1
TSS (g L^{-1})	30
pH	7.8
Color	Dark

consortium of microalgae and bacteria prevails; in the case of manure, the presence of nitrifying bacteria is widely reported (Mezzari et al. 2013). Produced biomass is suitable for whatever nonhuman-related application such as biofertilizers, energy, etc. (Acién Fernández et al. 2018). Ferreira et al. (2018) obtained bioH_2 by dark fermentation, achieving higher yields (390 mL H_2/g volatile solids).

9.2.3.2 Poultry

Poultry livestock production generates significant quantities of organic waste as it includes manure, wasted feed, bedding material, feathers, fats, blood, skin, and broken eggs; it also contains high levels of nitrogen, phosphorus, and chlorine and pathogens such as *Salmonella* and *Campylobacter* (AL-Rajhia et al. 2012). Poultry production effluent could contain also used residual compounds, such as antibiotics, antiprotozoal drugs, antioxidants, mold inhibitors, probiotics, polychlorinated phenols, or tetrachlorodibenzo-p-dioxin (Markou et al. 2016).

Poultry industry effluent can be considered as a high-strength WW (Table 9.8) with potential to be treated by microalgae, as reported by some authors (Markou et al. 2016; Oliveira et al. 2018).

The recovery of nutrients is normally efficient (P-PO_4^{3-} almost complete and TP of 80–99%) (Oliveira et al. (2018) using *Scenedesmus obliquus*; Markou et al. (2016) using *Chlorella vulgaris*), 97% for N-NH_4^+ (Oliveira et al. 2018), and COD removal was almost total, for around 10 days, in both studies.

Poultry litter extract (PLE) [prepared from shade-dried poultry litter with deionized water at 1.25% (w/v)] as growth medium recorded up to 180% more biomass growth compared to standard growth medium BG11 (Bhatnagar et al. 2011).

The microalgae biomass composition was around 35% of protein, 20–35% of carbohydrates, and 15–25% of lipids (Markou et al. 2016; Oliveira et al. 2018).

Table 9.8 Typical composition of poultry industry wastewater (AL-Rajhia et al. 2012)

Parameters	Concentrations
	Poultry WW
COD (mg O_2 L^{-1})	1030–3000
TKN (mg N L^{-1})	110–325
NH_4^+ (mg N-NH_4^+ L^{-1})	16–280
TP (mg P L^{-1})	16–32
TSS (mg L^{-1})	300–950
pH	7 ± 1.0
Color	Brownish gray

The produced microalgae biomass from the poultry WW treatment has a potential interest to poultry feed supplement, improving poultry health and quality of the related products (Lemahieu et al. 2014). Other applications could be biofuels, such as biodiesel, bioethanol, and biogas, improving the economics and environmental footprint of the whole production process (Muylaert et al. 2015). Ferreira et al. (2018) used the microalgae-based poultry WWT as a feedstock for bioH_2 production through dark fermentation. The same authors also performed pyrolysis to obtain bio-oil, bio-char, and biogas (Ferreira et al. 2020).

9.2.3.3 Cattle

The recent expansion of cattle raising activity has generated a necessity for intensive feeding in limited areas as opposed to free-range grazing, raising the production of cattle WW (Mouri and Aisaki 2015). A typical composition of this kind of effluent is presented on Table 9.9.

Mendonça et al. (2018) studied the microalgae-mediated bioremediation and valorization of cattle WW previously digested in a hybrid anaerobic reactor using a vertical alveolar flat panel photobioreactor. A comparison between batch and continuous operation modes for the cultivation of the microalga *Scenedesmus obliquus* with anaerobic cattle WW was investigated. Maximum COD removal in batch and continuous operation was 70 and 61%, respectively. Moreover, removal efficiencies ranges were 98–99% of NH_4^+ and 69–77.5% of PO_4^{3-} after 12 days of operation in batch mode. The corresponding recorded ranges for continuous operation were 94–96% of NH_4^+ and 65–70% of PO_4^{3-}. The best biomass productivity (358 mg $L^{-1}$$day^{-1}$) was found in batch. Concerning the biochemical macro composition of the produced biomass, the highest range in protein content (53–55%) was achieved in continuous mode compared to batch mode (32–42%). The opposite trend was observed for lipid content: 18–29% for batch mode vs. 13–15% for continuous operation. This microalgae-mediated process can be considered promising for the bioremediation and valorization of cattle-based effluents yielding a protein-rich microalgal biomass that can be eventually used as cattle feed, in the frame of a circular economy.

Table 9.9 Typical composition of cattle industry wastewater (Banks and Wang 2005)

Parameters	Concentrations
	Cattle WW
COD (mg O_2 L^{-1})	2100–12,873
TN (mg N L^{-1})	93–378
TP (mg P L^{-1})	26–79
TSS (mg L^{-1})	467–6090
pH	6.0–9.3
Color	Brownish

Scenedesmus obliquus was able to remove nutrients from cattle breeding WWs among others with removal ranges of 95–100% for nitrogen, 63–99% for phosphorus, and 48–70% for COD (Ferreira et al. 2018). The biomass productivity was higher than in synthetic medium (Bristol), highlighting the mixotrophic nutritional mechanism involved (Ferreira et al. 2018).

The potential of the produced *Scenedesmus obliquus* biomass for the generation of bioH_2 through batch dark fermentation processes with *Enterobacter aerogenes* was evaluated. The obtained yield was 50.1 mL H_2/g volatile solids (VS), the lowest for all biomasses tested (from poultry, swine, brewery, dairy, and urban WWs as well as from Bristol medium). The biogas volume purity ratio between H_2 and CO_2 was around 1, far below the higher purity of the produced biogas from swine and poultry (7.0 and 6.7, respectively) (Ferreira et al. 2018). *S. obliquus* biomass was also subjected to pyrolysis to obtain bio-char, bio-oil, and biogas (Ferreira et al. 2020).

9.2.4 Agriculture

Although agriculture and food and fuel production have frequently been perceived as conflicting, there is a substantial potential to deal with food and fuel production all together.

As food production grows to meet the nutritional necessities of a growing population, there is also a simultaneous increase in the production of agricultural residues (IRENA 2016).

For every ton of crop produced, a significant amount of residues is obtainable in the field after harvest. A fraction between a quarter and a half can be practically and sustainably collected. The remaining fraction makes part of wastes in which a fraction is in the form of WW (FAOSTAT 2015).

Agriculture, one of the major activities worldwide, generates a range of WWs usually rich in nutrients such as N and P. According to Guldhe et al. (2017), several studies have shown that microalgae were able to grow and remove nutrients efficiently from different agricultural WWs. The list of major constituents of agricultural WW may include plant stalks, hulls, leaves, etc. (Chiu et al. 2015). The nutrient concentrations of agricultural WW however are purely source dependent and quite variable.

The challenges concerning the cultivation of microalgae in agricultural WWs include (1) significantly higher or lower nutrient concentration (Cheng et al. 2013; Zhou et al. 2014; Chen et al. 2015) that could limit or inhibit microalgal growth; (2) unreachability of carbon for microalgal absorption as a huge portion of C is locked in insoluble organic compounds such as agricultural wastes (Zhou et al. 2014); (3) high turbidity that impedes light penetration during phototrophic microalgal cultivation (Zhou et al. 2014; Chen et al. 2015); and (4) the ever accumulative use of pesticides, insecticides, antibiotics, and other toxic chemical compounds in agriculture, and agriculture-based industries are expected to impact microalgal growth and the quality of biomass produced.

The microalgae-mediated treatment of several WW streams from agro-industrial sources has been reported in the last years and reviewed by Markou et al. (2018). Jayakumar et al. (2017) reviewed the WW availability in Malaysia with special emphasis on two feedstocks (palm oil mill effluent and rubber mill effluent) for further production of microalgae directed to biodiesel, boosting future biodiesel industries including opportunities and challenges. Malaysia currently generates 12.0 million tons of crude palm that yield 8.9 million tons of palm oil. From 5 to 7.5 tons of clean water is needed to process 1 ton of palm oil, in which half will become palm oil mill effluent (Abdul et al. 2003), offering huge opportunities of treatment and valorization. *Spirulina platensis* was reported to reduce the COD, nitrogen, and phosphorus up to 50.8, 96.5, and 85.9%, respectively, achieving a specific growth rate of 0.41 day^{-1}. *Chlamydomonas incerta*, *Chlorella sorokiniana*, *Chlorella vulgaris*, and *Botryococcus sudeticus* have been studied for WW treatment with biomass productivities from 8.0 to 5.3 mg L^{-1} day^{-1}. Lipid content in the biomass ranged from 21 to 31% (Hadiyanto et al. 2013). On the other hand, *Micractinium* sp. and *Mychonastes rotundu* achieved higher biomass productivities (57.2 and 60.0 mg L^{-1} day^{-1}, respectively) with lipid content in between 26.7 and 27.7%.

Malaysian water bodies are polluted from rubber industries as massive quantities of skim latex, serum, uncoagulated latex, and washing effluents are released daily. It is estimated that about 80 million liters of untreated rubber waste effluents are released per day into rivers in Malaysia (Massoudinejad et al. 2015). The latex washing and serum contain proteins, lipids, and sugars which are the main constituents in the rubber mill effluent together with ammonium and other plant nutrients adequate for microalgae growth (Nur and Hadiyanto 2013). *Ankistrodesmus convolutes*, *Oscillatoria salina*, and *Microcystis aeruginosa* performed better in terms of biomass productivity as well as lipid productivity with this kind of effluent compared to standard known medium composition (Habib et al. 2004). Moreover, experiments in high-rate oxidation ponds (HROP) conducted with *Chlorella* sp. yielded a biomass rich in protein (up to 55% of dry weight) and carotenoids (up to 2.4 mg g^{-1}). In another study with *Chlorella vulgaris*, 93.4% COD and 79.3% TKN have been removed for a microalgal biomass productivity up to 12.0 g^{-2} day^{-1} (Bich et al. 1999).

Table 9.10 Typical composition of aquaculture wastewater (Posadas et al. 2015b)

Parameters	Concentrations Aquaculture WW
COD (mg O_2 L^{-1})	678 ± 249
TOC (mg C L^{-1})	161 ± 67
IC (mg C L^{-1})	65 ± 37
TN (mg N L^{-1})	31 ± 10
TKN (mg N L^{-1})	33 ± 17
NH_4^+ (mg N-NH_4^+ L^{-1})	10 ± 8
TP (mg P L^{-1})	19 ± 5
PO_4^{3-} (mg P-PO_4^{3-} L^{-1})	14 ± 7
C:N:P[a]	100:14:6
TSS (g L^{-1})	0.2 ± 0.2
pH	6.7 ± 0.4
Color	Gray

9.2.5 *Aquaculture*

Aquaculture WW can be assimilated to a medium-strength WW similar to domestic WW and suitable for mass cultivation of microalgae (Table 9.10). A moderate-high biodegradability of the organic matter and nitrogen present in this type of WW can be expected from preliminary batch biodegradation tests (40–60% and 64–85%, respectively) (Posadas et al. 2014a). The continuous treatment of aquaculture WW at 10 days of hydraulic retention time has resulted in removals of organic matter, nitrogen, and phosphorus of 77%, 83%, and 94%, respectively (Posadas et al. 2015b), with biomass productivities limited to 5 g m^{-2} day^{-1} as a result of the high extent of carbon and NH_3 stripping from the open HRAP. The biomass produced during microalgae-based aquaculture WW treatment can be used as a feedstock for fish production based on its high protein (50–60% on a dry weight basis) and polyunsaturated fatty acids (PUFA) content (Safafar et al. 2016).

9.3 Environmental Benefits of Coupling Algae-Based Wastewater Treatment with Production of Biofuels and/or Bio-Products Through LCA Study

Algae for biofuels have many advantages in comparison to first- and second-generation biofuels, notably the fact that they can grow in unproductive land with low economic values, which promises the possibility of biofuel with a low competition with food crop for arable land (Chisti 2007). They also can use non-potable water such as brackish, saline water or even WW, avoiding freshwater and fertilizers requirements. To assess environmental impacts linked to the production of bioenergy and/or bio-products from microalgae, the most commonly used methodology is LCA.

It is a well-established approach to quantify all relevant emissions and consumed resources, as well as the related environmental impacts associated with a product's life cycle. The method takes into account the entire life cycle from the extraction of resources, through production, use, and recycling, to disposal of the remaining waste.

Most LCA studies on production of bioenergy/bio-products from microalgae consider that nutrients requirement is fulfilled by mineral fertilizers as in intensive agriculture (Collet et al. 2014), but some authors such as Clarens et al. (2010), Fortier et al. (2014), or Beal et al. (2015) also consider that the needs in nitrogen and phosphorus can be totally or partially covered by the addition of WW to the growth medium. The use of WW offers the double advantage of a non-reclaimed source of water and nutrients, as water and nutrients consumptions have been identified as some of the main environmental concerns of bioenergy production from microalgae (Collet et al. 2014). However, some authors point out that nutrients availability could be limiting (Orfield et al. 2014; Beal et al. 2015), as mineral content in WW can strongly vary depending on the place and the period of the year. Also, having a WW treatment plant or a sewage collecting network next to the microalgae plant is very restrictive. Furthermore, it should be acknowledged that microalgae grown on WW could not be used afterward as feedstock for fish or cattle.

Wastewaters considered in LCA studies are most of the time urban WW. Aquaculture WW has also been considered (Sfez et al. 2015), as well as livestock WW (Jez et al. 2017). To our knowledge, no other WW sources have been integrated in LCA models.

Microalgae cultivated on WWs are mainly dedicated to bioenergy production. It could be bioelectricity produced by direct combustion (Clarens et al. 2010); biodiesel by transesterification (Clarens et al. 2010; Beal et al. 2015; Jez et al. 2017); jet fuel, naphtha, or biodiesel by hydrothermal liquefaction (Fortier et al. 2014; Beal et al. 2015; Mu et al. 2017; Cruce and Quinn 2019); bio-oil by pyrolysis (Roostaei and Zhang 2017); or biogas (Arashiro et al. 2018; Lopes et al. 2018). In some publications, oilcakes obtained after oil extraction are also converted into bioenergy via anaerobic digestion (Mu et al. 2017; Jez et al. 2017). Some authors also suggest the production of biofertilizers from microalgae cultivated on WW (Arashiro et al. 2018).

From an environmental point of view, the use of WW strongly reduces negative effects on several impact categories. In Clarens et al. (2011), supplying nutrients by urban WW instead of chemical fertilizers multiplies EROI by two to four depending on the bioenergy produced (biodiesel or bioelectricity). In Fortier et al. (2014), the GHG emissions related to the production of biojet fuel are more than halved by using WW. However, smaller GHG reductions (10–40%) are reported in Collotta et al. (2018). Furthermore, when the credits of WW treatment are taken into account, the production of bioenergy from microalgae could provide absolute negative values for climate change (Mu et al. 2014). In this paper, use of WW as input for nutrients has a double effect: it avoids the use of nutrients fertilizers (which is most of the time taken into account in the literature), but it also displaces the need of WW treatments (which is rarely considered). This double benefit leads to negative values for gasoline and/or biodiesel production from microalgae. On other impacts directly linked to nutrients production and use, such as eutrophication, the use of WW, whether only benefits from nutrient substitution or both nutrient substitution and WW treatments are considered, allows a net positive impact on the environment (Mu et al. 2014; Collotta et al. 2018).

Use of WW always decreased the environmental cost of the analyzed system, whether it is bio-products (Arashiro et al. 2018) or biofuels (Mu et al. 2014) production. However, it could be a limiting factor for biomass production and exclude biomass valorization from animal feed in the actual context. Furthermore, needed combination of a sufficient source of industrial CO_2 and WW near microalgae cultivation location limits the production of high volumes of bioenergy/bio-products. Inclusion of biogenic carbon dynamics could be further investigated to broaden the scope of climate change assessment.

9.4 Conclusions

The utilization of wastes and wastewaters to cultivate algae could simultaneously solve the problems of (1) freshwater demand, (2) high cost of nutrients, (3) CO_2 emissions, and (4) the need to remediate wastes. The versatile metabolism of microalgae along with the possibility of using multiple process configurations allows this technology to be implemented for the treatment of most types of wastewater.

Efficient nutrient recovery from WW could be performed by using microalgae concomitantly with the production of a useful biomass for biofuels, biofertilizers, bioplastics, and high-value compounds with environmental and economic advantages (e.g., lower energy demand for WWT, simultaneous CO_2 mitigation, etc.).

In this sense, WW can be a cost-efficient and sustainable source of energy, nutrients, organic matter, and other useful by-products through the application of microalgae into the WW treatment process. In the context of a circular economy, whereby economic development is balanced with the protection of natural resources and environmental sustainability, WW represents a widely available and valuable resource.

An overview of LCA studies using wastewater to grow microalgae has been done. Whether it is for the production of bioenergy or bio-products with high added value, the use of WW leads to a reduction of several environmental burdens, including climate change and eutrophication of the use of freshwater. At last, environmental impacts are even lower when considering the double benefit of avoiding the use of chemical fertilizers and displacing the need of WW treatments.

Acknowledgment The authors would like to acknowledge the following European projects: COST Action 1408-EUALGAE-European network for algal-bio-products, ERANETLAC/0001/2014 (ELAC2014/BEE0357) GREENBIOREFINERY-Processing of brewery wastes with microalgae for producing valuable compounds. The GREENBIOREFINERY project is implemented in the framework of ERANet-LAC, a Network of the European Union (EU), Latin America and the Caribbean Countries (CELAC) co-funded by the European Commission within the 7th Framework Programme for Research and Technology Development (FP7). Support is provided by the following national funding organizations: MINCYT, Argentina; COLCIENCIAS, Colombia; FCT, Portugal; and ISCIII, Spain; SABANA—Sustainable integrated Algae Biorefinery for the production of bioactive compounds for Agriculture and Aquaculture Grant Agreement No. 727874. PHOENIX—People for the European Bio-Energy Mix project (H2020-MSCA-RISE-2015) and special thanks to his Portuguese Coordinator Dr. Luis Duarte (LNEG); ALGAVALOR - Lisboa-01-0247-FEDER-035234, supported by Operational Programme for Competitiveness and Internationalization (COMPETE2020), by Lisbon Portugal Regional Operational Programme (Lisboa 2020) and by Algarve Regional Operational Programme (Algarve 2020) under the Portugal 2020 Partnership Agreement, through the European Regional Development Fund (ERDF). The financial support from the Regional Government of Castilla y León and the EU-FEDER programme (CLU 2017-09 and UIC 71) are gratefully acknowledged.

References

Abdel-Raouf N, Al-Homaidan AA, Ibraheem IBM (2012) Microalgae and wastewater treatment. Saudi J Biol Sci 19:257–275. https://doi.org/10.1016/J.SJBS.2012.04.005

Abdul LA, Suzylawati I, Subhash B (2003) Water recycling from palm oil mill effluent (POME) using membrane technology. Desalination 157:87–95. https://doi.org/10.1016/S0011-9164(03)00387-4

Abreu AP, Fernandes B, Vicente AA et al (2012) Mixotrophic cultivation of *Chlorella vulgaris* using industrial dairy waste as organic carbon source. Bioresour Technol 118:61–66. https://doi.org/10.1016/J.BIORTECH.2012.05.055

Acién Fernández FG, Gómez-Serrano C, Fernández-Sevilla JM (2018) Recovery of nutrients from wastewaters using microalgae. Front Sustain Food Syst 2:59. https://doi.org/10.3389/fsufs.2018.00059

Acién FG, Gómez-Serrano C, Morales-Amaral MM et al (2016) Wastewater treatment using microalgae: how realistic a contribution might it be to significant urban wastewater treatment? Appl Microbiol Biotechnol 100:9013–9022. https://doi.org/10.1007/s00253-016-7835-7

Agwa OK, Ogugbue CJ, Williams EE (2017) Field evidence of *Chlorella vulgaris* potentials as a biofertilizer for *Hibiscus esculentus*. Int J Agric Res 12:181–189. https://doi.org/10.3923/ijar.2017.181.189

Albertano P, Congestri R, Elliot Shubert LE (1999) Cyanobacterial biofilms in sewage treatment plants along the Thyrrenian coast (Mediterranean Sea), Italy. Arch Hydrobiol Suppl Algol Stud 94:13–24. https://doi.org/10.1127/algol_stud/94/1999/13

Alcántara C, García-Encina PA, Muñoz R (2013) Evaluation of mass and energy balances in the integrated microalgae growth-anaerobic digestion process. Chem Eng J 221:238–246. https://doi.org/10.1016/J.CEJ.2013.01.100

AL-Rajhia S, Raut N, AL-Qasmi F et al (2012) Treatments of industrials wastewater by using microalgae. In: 2012 International conference on environmental, biomedical and biotechnology IPCBEE. IACSIT Press, Singapore, pp 217–221

Arashiro LT, Montero N, Ferrer I et al (2018) Life cycle assessment of high rate algal ponds for wastewater treatment and resource recovery. Sci Total Environ 622–623:1118–1130. https://doi.org/10.1016/J.SCITOTENV.2017.12.051

Aslanzadeh S, Ishola MM, Richards T, Taherzadeh MJ (2014) An overview of existing individual unit operations. Biorefineries 3–36. https://doi.org/10.1016/B978-0-444-59498-3.00001-4

Avetisova G (2013) Method of producing biological fertilizer based on nitrogen-fixing bacteria (versions). Patent RA №2 713A:1–5. http://www.aipa.am/search_mods/patents/view_item.php?id=2713AAM20120133&language=en

Avetisova GY, Melkonyan LH, Chakhalyan AK et al (2014) New complex nitrogen biofertilizer for agriculture. In: International conference "translational biocatalysts." London, December 16, p 12

Awaluddin SA, Thiruvenkadam S, Izhar S et al (2016) Subcritical water technology for enhanced extraction of biochemical compounds from *Chlorella vulgaris*. Biomed Res Int 2016:1–10. https://doi.org/10.1155/2016/5816974

Bai S, Srikantaswamy S, Shivakumar D (2010) Urban wastewater characteristic and its management in urban areas—a case study of Mysore City, Karnataka, India. J Water Resour Prot 02:717–726. https://doi.org/10.4236/jwarp.2010.28082

Banks CJ, Wang Z (2005) Treatment of meat wastes. In: Wang LK, Hung Y-T, Lo HH, Yapijakis C (eds) Waste treatment in the food processing industry. CRC, Boca Raton, pp 67–100

Batista AP, Moura P, Marques PASS et al (2014) *Scenedesmus obliquus* as feedstock for biohydrogen production by *Enterobacter aerogenes* and *Clostridium butyricum*. Fuel 117:537–543. https://doi.org/10.1016/j.fuel.2013.09.077

Batista AP, Ambrosano L, Graça S et al (2015) Combining urban wastewater treatment with biohydrogen production – an integrated microalgae-based approach. Bioresour Technol 184:230–235. https://doi.org/10.1016/J.BIORTECH.2014.10.064

Beal CM, Gerber LN, Sills DL et al (2015) Algal biofuel production for fuels and feed in a 100-ha facility: a comprehensive techno-economic analysis and life cycle assessment. Algal Res 10:266–279. https://doi.org/10.1016/J.ALGAL.2015.04.017

Bharathiraja B, Ranjith Kumar R, PraveenKumar R et al (2016) Biodiesel production from different algal oil using immobilized pure lipase and tailor made *Pichia pastoris* with Cal A and Cal B genes. Bioresour Technol 213:69–78. https://doi.org/10.1016/J.BIORTECH.2016.02.041

Bhatnagar A, Chinnasamy S, Singh M, Das KC (2011) Renewable biomass production by mixotrophic algae in the presence of various carbon sources and wastewaters. Appl Energy 88:3425–3431. https://doi.org/10.1016/J.APENERGY.2010.12.064

Bich NN, Yaziz MI, Bakti NAK (1999) Combination of *Chlorella vulgaris* and *Eichhornia crassipes* for wastewater nitrogen removal. Water Res 33:2357–2362. https://doi.org/10.1016/S0043-1354(98)00439-4

Brennan L, Owende P (2010) Biofuels from microalgae—a review of technologies for production, processing, and extractions of biofuels and co-products. Renew Sustain Energy Rev 14:557–577. https://doi.org/10.1016/J.RSER.2009.10.009

Brewers Association (2013) Wastewater management guidance manual. https://www.brewersassociation.org/. Accessed 18 Dec 2017

Burns RT, Moody LB, Celen I, Buchanan JR (2003) Optimization of phosphorus precipitation from swine manure slurries to enhance recovery. Water Sci Technol 48:139–146

Cabanelas ITD, Arbib Z, Chinalia FA et al (2013) From waste to energy: microalgae production in wastewater and glycerol. Appl Energy 109:283–290. https://doi.org/10.1016/J.APENERGY.2013.04.023

Cantrell KB, Ducey T, Ro KS, Hunt PG (2008) Livestock waste-to-bioenergy generation opportunities. Bioresour Technol 99:7941–7953. https://doi.org/10.1016/J.BIORTECH.2008.02.061

Chakhalyan A, Avetisova G, Saghiyan A et al (2008) Zeolites in complex nitrogen biofertilizers. Stud Surf Sci Catal 174:521–524. https://doi.org/10.1016/S0167-2991(08)80254-6

Chavan A, Mukherji S (2008) Treatment of hydrocarbon-rich wastewater using oil degrading bacteria and phototrophic microorganisms in rotating biological contactor: effect of N:P ratio. J Hazard Mater 154:63–72. https://doi.org/10.1016/J.JHAZMAT.2007.09.106

Chen G, Zhao L, Qi Y (2015) Enhancing the productivity of microalgae cultivated in wastewater toward biofuel production: a critical review. Appl Energy 137:282–291. https://doi.org/10.1016/J.APENERGY.2014.10.032

Cheng W, Chen H, Su C, Yan S (2013) Abundance and persistence of antibiotic resistance genes in livestock farms: a comprehensive investigation in eastern China. Environ Int 61:1–7. https://doi.org/10.1016/J.ENVINT.2013.08.023

Chisti Y (2007) Biodiesel from microalgae. Biotechnol Adv 25:294–306. https://doi.org/10.1016/j.biotechadv.2007.02.001

Chiu S-Y, Kao C-Y, Chen T-Y et al (2015) Cultivation of microalgal *Chlorella* for biomass and lipid production using wastewater as nutrient resource. Bioresour Technol 184:179–189. https://doi.org/10.1016/J.BIORTECH.2014.11.080

Clarens AF, Resurreccion EP, White MA, Colosi LM (2010) Environmental life cycle comparison of algae to other bioenergy feedstocks. Environ Sci Technol 44:1813–1819. https://doi.org/10.1021/es902838n
Clarens AF, Nassau H, Resurreccion EP et al (2011) Environmental impacts of algae-derived biodiesel and bioelectricity for transportation. Environ Sci Technol 45:7554–7560. https://doi.org/10.1021/es200760n
Colic M (2007) Special treatment: special treatment: the food manufacturing wastewater challenge. Clean Water Technology. https://www.manufacturing.net/article/2007/02/special-treatment-food-manufacturing-wastewater-challenge. Accessed 3 Oct 2018
Collet P, Lardon L, Hélias A et al (2014) Biodiesel from microalgae – life cycle assessment and recommendations for potential improvements. Renew Energy 71:525–533. https://doi.org/10.1016/J.RENENE.2014.06.009
Collos Y, Harrison PJ (2014) Acclimation and toxicity of high ammonium concentrations to unicellular algae. Mar Pollut Bull 80:8–23. https://doi.org/10.1016/J.MARPOLBUL.2014.01.006
Collotta M, Champagne P, Mabee W, Tomasoni G (2018) Wastewater and waste CO_2 for sustainable biofuels from microalgae. Algal Res 29:12–21. https://doi.org/10.1016/J.ALGAL.2017.11.013
Congestri R, Albertano P (2011) Benthic diatoms in biofilm culture. In: Seckbach J, Kociolek P (eds) The diatom world. Cellular origin, life in extreme habitats and astrobiology, vol 19. Springer, Dordrecht, pp 227–243
Congestri R, Sangiorgi VC, Albertano P (2003) Cytomorphology and distribution of periphytic cyanobacteria in one Italian waste water treatment plant. Arch Hydrobiol Suppl Algol Stud 109:185–195. https://doi.org/10.1127/1864-1318/2003/0109-0185
Congestri R, Cox EJ, Cavacini P, Albertano P (2005) Diatoms (Bacillariophyta) in phototrophic biofilms colonising an Italian wastewater treatment plant. Diatom Res 20:241–255. https://doi.org/10.1080/0269249X.2005.9705634
Congestri R, Di Pippo F, De Philippis R et al (2006) Seasonal succession of phototrophic biofilms in an Italian wastewater treatment plant: biovolume, spatial structure and exopolysaccharides. Aquat Microb Ecol 45:301–312. https://doi.org/10.3354/ame045301
Corcoran E, Nelleman C, Baker E et al (eds) (2010) Sick water?: the central role of wastewater management in sustainable development. A rapid response assessment. UNEP/UNHABITAT
Craggs RJ, Adey WH, Jenson KR et al (1996) Phosphorus removal from wastewater using an algal turf scrubber. Water Sci Technol 33:191–198. https://doi.org/10.1016/0273-1223(96)00354-X
Cruce JR, Quinn JC (2019) Economic viability of multiple algal biorefining pathways and the impact of public policies. Appl Energy 233–234:735–746. https://doi.org/10.1016/J.APENERGY.2018.10.046
Cuellar-Bermudez SP, Aleman-Nava GS, Chandra R et al (2017) Nutrients utilization and contaminants removal. A review of two approaches of algae and cyanobacteria in wastewater. Algal Res 24:438–449. https://doi.org/10.1016/J.ALGAL.2016.08.018
Darpito C, Shin WS, Jeon S et al (2015) Cultivation of *Chlorella protothecoides* in anaerobically treated brewery wastewater for cost-effective biodiesel production. Bioprocess Biosyst Eng 38:523–530. https://doi.org/10.1007/s00449-014-1292-4
Das D, Veziroglu TN (2008) Advances in biological hydrogen production processes. Int J Hydrog Energy 33:6046–6057. https://doi.org/10.1016/J.IJHYDENE.2008.07.098
Davis LS, Hoffmann JP, Cook PW (1990a) Seasonal succession of algal periphyton from a wastewater treatment facility. J Phycol 26:611–617. https://doi.org/10.1111/j.0022-3646.1990.00611.x
Davis LS, Hoffmann JP, Cook PW (1990b) Production and nutrient accumulation by periphyton in a wastewater treatment facility. J Phycol 26:617–623. https://doi.org/10.1111/j.0022-3646.1990.00617.x
de la Noüe J, Bassères A (1989) Biotreatment of anaerobically digested swine manure with microalgae. Biol Wastes 29:17–31. https://doi.org/10.1016/0269-7483(89)90100-6

de-Bashan LE, Moreno M, Hernandez J-P, Bashan Y (2002) Removal of ammonium and phosphorus ions from synthetic wastewater by the microalgae *Chlorella vulgaris* coimmobilized in alginate beads with the microalgae growth-promoting bacterium *Azospirillum brasilense*. Water Res 36:2941–2948. https://doi.org/10.1016/S0043-1354(01)00522-X

Decree-Law No 236/98 (1998) Decree-Law 236/98 of the Portuguese Ministry of the Environment of 1 August establishing water quality standards. Diário da República I Série-A 3676–3722

Di Pippo F, Ellwood NTW, Guzzon A et al (2014) Diversity and biomass accumulation in cultured phototrophic biofilms. Eur J Phycol 49:384–394. https://doi.org/10.1080/09670262.2014.948075

Dineshkumar R, Kumaravel R, Gopalsamy J et al (2017) Microalgae as bio-fertilizers for rice growth and seed yield productivity. Waste Biomass Valoriz 9:793–800. https://doi.org/10.1007/s12649-017-9873-5

Dragone G, Mussatto SI, Oliveira JM, Teixeira JA (2009) Characterisation of volatile compounds in an alcoholic beverage produced by whey fermentation. Food Chem 112:929–935. https://doi.org/10.1016/J.FOODCHEM.2008.07.005

Drogui P, Asselin M, Brar SK et al (2008) Electrochemical removal of pollutants from agro-industry wastewaters. Sep Purif Technol 61:301–310. https://doi.org/10.1016/J.SEPPUR.2007.10.013

EPA (2002) Development document for the proposed effluent limitations guidelines and standards for the meat and poultry products industry point source category (40 CFR 432). Environ Prot Agency

EUMOFA (2015) No title. http://www.eumofa.eu/eumofa-publications. Accessed 27 Aug 2016

Eurostat. Eurostat 2013–2015. http://ec.europa.eu/eurostat

Faheed FA, Fattah ZA-E (2008) Effect of *Chlorella vulgaris* as bio-fertilizer on growth parameters and metabolic aspects of lettuce plant. J Agric Soc Sci 4:165–169

FAO (2014) No title. http://www.fao.org/fishery/sofia/en. Accessed 27 Aug 2016

FAO (2018) The state of world fisheries and aquaculture 2018 – meeting the sustainable development goals. Rome

FAOSTAT (2015) No title. http://www.fao.org/faostat/en/

Farrelly DJ, Everard CD, Fagan CC, McDonnell KP (2013) Carbon sequestration and the role of biological carbon mitigation: a review. Renew Sustain Energy Rev 21:712–727. https://doi.org/10.1016/J.RSER.2012.12.038

Fermoso J, Coronado JM, Serrano DP, Pizarro P (2017) Pyrolysis of microalgae for fuel production. In: Microalgae-based biofuels and bioproducts. Woodhead, Cambridge, pp 259–281

Ferreira AF, Soares Dias AP, Silva CM, Costa M (2015) Evaluation of thermochemical properties of raw and extracted microalgae. Energy 92:365–372. https://doi.org/10.1016/J.ENERGY.2015.04.078

Ferreira A, Ribeiro B, Marques PASS et al (2017) *Scenedesmus obliquus* mediated brewery wastewater remediation and CO_2 biofixation for green energy purposes. J Clean Prod 165:1316–1327. https://doi.org/10.1016/j.jclepro.2017.07.232

Ferreira A, Marques P, Ribeiro B et al (2018) Combining biotechnology with circular bioeconomy: from poultry, swine, cattle, brewery, dairy and urban wastewaters to biohydrogen. Environ Res 164:32–38. https://doi.org/10.1016/j.envres.2018.02.007

Ferreira A, Ribeiro B, Tavares LAM, Vladic J, Vidovic S, Cvetkovic D, Melkonyan L, Avetisova G, Goginyan V, Gouveia L (2019) *Scenedesmus obliquus* microalga-based biorefinery – from brewery effluent to bioactive compounds, biofuels and biofertilizers – aiming a circular bioeconomy. Biofuel Bioprod Biorefin 13:1169–1186. https://doi.org/10.1002/bbb.2032

Ferreira AF, Ferreira A, Dias APS, Gouveia L (2020) Pyrolysis of *Scenedesmus obliquus* biomass following the treatment of different wastewaters

Fortier M-OP, Roberts GW, Stagg-Williams SM, Sturm BSM (2014) Life cycle assessment of bio-jet fuel from hydrothermal liquefaction of microalgae. Appl Energy 122:73–82. https://doi.org/10.1016/J.APENERGY.2014.01.077

Gao S, Hu C, Sun S et al (2018) Performance of piggery wastewater treatment and biogas upgrading by three microalgal cultivation technologies under different initial COD concentration. Energy 165:360–369. https://doi.org/10.1016/J.ENERGY.2018.09.190

García D, Posadas E, Blanco S et al (2018) Evaluation of the dynamics of microalgae population structure and process performance during piggery wastewater treatment in algal-bacterial photobioreactors. Bioresour Technol 248:120–126. https://doi.org/10.1016/J.BIORTECH.2017.06.079

Garcia-Gonzalez J, Sommerfeld M (2016) Biofertilizer and biostimulant properties of the microalga *Acutodesmus dimorphus*. J Appl Phycol 28:1051–1061. https://doi.org/10.1007/s10811-015-0625-2

García-Marino M, Rivas-Gonzalo JC, Ibáñez E, García-Moreno C (2006) Recovery of catechins and proanthocyanidins from winery by-products using subcritical water extraction. Anal Chim Acta 563:44–50. https://doi.org/10.1016/J.ACA.2005.10.054

Gómez-Serrano C, Morales-Amaral MM, Acién FG et al (2015) Utilization of secondary-treated wastewater for the production of freshwater microalgae. Appl Microbiol Biotechnol 99:6931–6944. https://doi.org/10.1007/s00253-015-6694-y

Gonçalves AL, Pires JCM, Simões M (2017) A review on the use of microalgal consortia for wastewater treatment. Algal Res 24:403–415. https://doi.org/10.1016/J.ALGAL.2016.11.008

Gouveia L, Graça S, Sousa C et al (2016a) Microalgae biomass production using wastewater: treatment and costs: scale-up considerations. Algal Res 16:167–176. https://doi.org/10.1016/J.ALGAL.2016.03.010

Gouveia L, Janelas J, Tropecêlo AI, Oliveira AC (2016b) Microalga *Nannochloropsis* sp. biomass for biodiesel production: conventional (cell disruption) and in situ transesterification. J Mar Biol Oceanogr 5:1–7. https://doi.org/10.4172/2324-8661.1000152

Guldhe A, Kumari S, Ramanna L et al (2017) Prospects, recent advancements and challenges of different wastewater streams for microalgal cultivation. J Environ Manag 203:299–315. https://doi.org/10.1016/J.JENVMAN.2017.08.012

Guzzon A, Bohn A, Diociaiuti M, Albertano P (2008) Cultured phototrophic biofilms for phosphorus removal in wastewater treatment. Water Res 42:4357–4367. https://doi.org/10.1016/J.WATRES.2008.07.029

Guzzon A, Di Pippo F, Congestri R (2019) Wastewater biofilm photosynthesis in photobioreactors. Microorganisms 7(8):e252. https://doi.org/10.3390/microorganisms7080252

Habib MAB, Yusoff FM, Phang SM et al (2004) Growth and nutritional value of a tropical green alga, *Ankistrodesmus convolutus* Corda, in agro-industrial effluents. Pertanika J Trop Agric Sci 27:79–89

Hadiyanto H, Christwardana M, Soetrisnanto D (2013) Phytoremediations of palm oil mill effluent (POME) by using aquatic plants and microalgae for biomass production. J Environ Sci Technol 6:79–90. https://doi.org/10.3923/jest.2013.79.90

Hoffmann JP (1998) Wastewater treatment with suspended and nonsuspended algae. J Phycol 34:757–763. https://doi.org/10.1046/j.1529-8817.1998.340757.x

Hung Y-T, Lo HH, Awad A, Salman H (2006) Potato wastewater treatment. In: Wang LK, Hung YT, Lo HH, Yapijakis C (eds) Waste treatment in the food processing industry. CRC, London, pp 193–254

Ibañez E, Kubátová A, Señoráns FJ et al (2003) Subcritical water extraction of antioxidant compounds from rosemary plants. J Agric Food Chem 51:375–382. https://doi.org/10.1021/JF025878J

IPCC (2013) Climate change 2013. The physical science basis. Working Group I contribution on the fifth assessment report of the intergovernmental panel on climate change. Cambridge University Press, Cambridge

IRENA (2016) Boosting biofuels: sustainable paths to greater energy security. International Renewable Energy Agency

Israilides C, Vlyssides AG, Arapoglou D et al (2008) Integrated management of potato starch wastes. In: Proceeding of the waste and resource management – a shared responsibility (Waste 2008). Sytatford-upon-Avon, UK, No. 36, September 16–17

Jayakumar S, Yusoff MM, Rahim MHA et al (2017) The prospect of microalgal biodiesel using agro-industrial and industrial wastes in Malaysia. Renew Sustain Energy Rev 72:33–47. https://doi.org/10.1016/J.RSER.2017.01.002

Jazzar S, Olivares-Carrillo P, Pérez de los Ríos A et al (2015) Direct supercritical methanolysis of wet and dry unwashed marine microalgae (*Nannochloropsis gaditana*) to biodiesel. Appl Energy 148:210–219. https://doi.org/10.1016/J.APENERGY.2015.03.069

Jez S, Spinelli D, Fierro A et al (2017) Comparative life cycle assessment study on environmental impact of oil production from micro-algae and terrestrial oilseed crops. Bioresour Technol 239:266–275. https://doi.org/10.1016/J.BIORTECH.2017.05.027

Ji X, Jiang M, Zhang J et al (2018) The interactions of algae-bacteria symbiotic system and its effects on nutrients removal from synthetic wastewater. Bioresour Technol 247:44–50. https://doi.org/10.1016/J.BIORTECH.2017.09.074

Kawalekar JS (2013) Role of biofertilizers and biopesticides for sustainable agriculture. J Bio Innov 2:73–78

Khoddami A, Wilkes M, Roberts T et al (2013) Techniques for analysis of plant phenolic compounds. Molecules 18:2328–2375. https://doi.org/10.3390/molecules18022328

Kong QX, Li L, Martinez B et al (2010) Culture of microalgae *Chlamydomonas reinhardtii* in wastewater for biomass feedstock production. Appl Biochem Biotechnol 160:9–18. https://doi.org/10.1007/s12010-009-8670-4

Kumar R, Kumawat N, Sahu YK (2017) Role of biofertilizers in agriculture. Pop Kheti 5:63–66

Ledda C, Schievano A, Scaglia B et al (2016) Integration of microalgae production with anaerobic digestion of dairy cattle manure: an overall mass and energy balance of the process. J Clean Prod 112:103–112. https://doi.org/10.1016/J.JCLEPRO.2015.07.151

Lemahieu C, Bruneel C, Termote-Verhalle R et al (2014) Effect of different microalgal n−3 PUFA supplementation doses on yolk color and n−3 LC-PUFA enrichment in the egg. Algal Res 6:119–123. https://doi.org/10.1016/J.ALGAL.2014.10.007

Liao PH, Chen A, Lo KV (1995) Removal of nitrogen from swine manure wastewaters by ammonia stripping. Bioresour Technol 54:17–20. https://doi.org/10.1016/0960-8524(95)00105-0

Limayem A, Micciche A, Nayak B, Mohapatra S (2018) Prokaryotic community profiling of local algae wastewaters using advanced 16S rRNA gene sequencing. Environ Sci Pollut Res 25:704–711. https://doi.org/10.1007/s11356-017-0078-z

Lopes AC, Valente A, Iribarren D, González-Fernández C (2018) Energy balance and life cycle assessment of a microalgae-based wastewater treatment plant: a focus on alternative biogas uses. Bioresour Technol 270:138–146. https://doi.org/10.1016/J.BIORTECH.2018.09.005

Lundquist T, Woertz I, Quinn N, Benemann J (2010) A realistic technology and engineering assessment of algae biofuel production. Energy Biosciences Institute, University California, pp 1–178

Mahapatra DM, Chanakya HN, Joshi NV et al (2018) Algae-based biofertilizers: a biorefinery approach. In: Panpatte DG, Jhala YK, Shelat HN, Vyas RV (eds) Microorganisms for green revolution, vol 2. Microbes for sustainable agro-ecosystem. Springer, Singapore, pp 177–196

Marchão L, da Silva TL, Gouveia L, Reis A (2018) Microalgae-mediated brewery wastewater treatment: effect of dilution rate on nutrient removal rates, biomass biochemical composition, and cell physiology. J Appl Phycol 30:1583–1595. https://doi.org/10.1007/s10811-017-1374-1

Markou G, Iconomou D, Muylaert K (2016) Applying raw poultry litter leachate for the cultivation of *Arthrospira platensis* and *Chlorella vulgaris*. Algal Res 13:79–84. https://doi.org/10.1016/J.ALGAL.2015.11.018

Markou G, Wang L, Ye J, Unc A (2018) Using agro-industrial wastes for the cultivation of microalgae and duckweeds: contamination risks and biomass safety concerns. Biotechnol Adv 36:1238–1254. https://doi.org/10.1016/J.BIOTECHADV.2018.04.003

Martínez M, Sánchez S, Jiménez J et al (2000) Nitrogen and phosphorus removal from urban wastewater by the microalga *Scenedesmus obliquus*. Bioresour Technol 73:263–272. https://doi.org/10.1016/S0960-8524(99)00121-2

Massoudinejad M, Mehdipour-Rabori M, Dehghani MH (2015) Treatment of natural rubber industry wastewater through a combination of physicochemical and ozonation processes. J Adv Environ Heal Res 3:242–249

Mata M, Melo A, Meireles S (2013) Potential of microalgae *Scenedesmus obliquus* grown in brewery wastewater for biodiesel production. Chem Eng Trans 32:901–906. https://doi.org/10.3303/CET1332151

Mata T, Santos J, Mendes A, Caetano N (2014) Sustainability evaluation of biodiesel produced from microalgae *Chlamydomonas* sp grown in brewery wastewater. Chem Eng Trans 37:823–828. https://doi.org/10.3303/CET1437138

Melkonyan L, Avetisova G, Chakhalyan A et al (2017) Biotechnological approaches for obtaining of new complex biopreparation for organic agriculture. In: 7th Congress of European Microbiologists (FEMS). Valencia, Spain, July 9–13, pp FEMS7-1142

Mendonça HV, Ometto JPHB, Otenio MH et al (2018) Microalgae-mediated bioremediation and valorization of cattle wastewater previously digested in a hybrid anaerobic reactor using a photobioreactor: comparison between batch and continuous operation. Sci Total Environ 633:1–11. https://doi.org/10.1016/J.SCITOTENV.2018.03.157

Mendoza JL, Granados MR, de Godos I et al (2013) Oxygen transfer and evolution in microalgal culture in open raceways. Bioresour Technol 137:188–195. https://doi.org/10.1016/J.BIORTECH.2013.03.127

Mezzari MP, da Silva MLB, Nicoloso RS et al (2013) Assessment of N_2O emission from a photobioreactor treating ammonia-rich swine wastewater digestate. Bioresour Technol 149:327–332. https://doi.org/10.1016/J.BIORTECH.2013.09.065

Milano J, Ong HC, Masjuki HH et al (2016) Microalgae biofuels as an alternative to fossil fuel for power generation. Renew Sustain Energy Rev 58:180–197. https://doi.org/10.1016/J.RSER.2015.12.150

Miller DJ, Hawthorne SB (1998) Method for determining the solubilities of hydrophobic organics in subcritical water. Anal Chem 70:1618–1621. https://doi.org/10.1021/ac971161x

Mishra D, Rajvir S, Mishra U, Kumar S (2013) Role of bio-fertilizer in organic agriculture: a review. Res J Recent Sci 2:39–41. https://doi.org/10.1039/C3PP50325A

Molinuevo-Salces B, Mahdy A, Ballesteros M, González-Fernández C (2016) From piggery wastewater nutrients to biogas: microalgae biomass revalorization through anaerobic digestion. Renew Energy 96:1103–1110. https://doi.org/10.1016/j.renene.2016.01.090

Morales-Amaral M del M, Gómez-Serrano C, Acién FG et al (2015) Outdoor production of *Scenedesmus* sp. in thin-layer and raceway reactors using centrate from anaerobic digestion as the sole nutrient source. Algal Res 12:99–108. https://doi.org/10.1016/J.ALGAL.2015.08.020

Mouri G, Aisaki N (2015) Using land-use management policies to reduce the environmental impacts of livestock farming. Ecol Complex 22:169–177. https://doi.org/10.1016/J.ECOCOM.2015.03.003

Mu D, Min M, Krohn B et al (2014) Life cycle environmental impacts of wastewater-based algal biofuels. Environ Sci Technol 48:11696–11704. https://doi.org/10.1021/es5027689

Mu D, Ruan R, Addy M et al (2017) Life cycle assessment and nutrient analysis of various processing pathways in algal biofuel production. Bioresour Technol 230:33–42. https://doi.org/10.1016/J.BIORTECH.2016.12.108

Muñoz R, Guieysse B (2006) Algal–bacterial processes for the treatment of hazardous contaminants: a review. Water Res 40:2799–2815. https://doi.org/10.1016/J.WATRES.2006.06.011

Mustafa A, Turner C (2011) Pressurized liquid extraction as a green approach in food and herbal plants extraction: a review. Anal Chim Acta 703:8–18. https://doi.org/10.1016/J.ACA.2011.07.018

Muylaert K, Beuckels A, Depraetere O et al (2015) Wastewater as a source of nutrients for microalgae biomass production. In: Moheimani NR, McHenry MP, de Boer K, Bahri PA (eds) Biomass and biofuels from microalgae. Springer, Cham, pp 75–94

Nur MMA, Hadiyanto H (2013) Agroindustry as growth medium for microalgae based bioenergy feedstock in Indonesia (an overview). Int J Sustain Futur Hum Secur 1:3–7

OECD/FAO (2018) OECD-FAO agricultural outlook 2018–2027. OECD, Paris/Food and Agriculture Organization of the United Nations, Rome

Olguín EJ (2012) Dual purpose microalgae–bacteria-based systems that treat wastewater and produce biodiesel and chemical products within a biorefinery. Biotechnol Adv 30:1031–1046. https://doi.org/10.1016/J.BIOTECHADV.2012.05.001

Oliveira AC, Barata A, Batista AP, Gouveia L (2018) *Scenedesmus obliquus* in poultry wastewater bioremediation. Environ Technol. https://doi.org/10.1080/09593330.2018.1488003

Ördög V, Stirk WA, Lenobel R et al (2004) Screening microalgae for some potentially useful agricultural and pharmaceutical secondary metabolites. J Appl Phycol 16:309–314. https://doi.org/10.1023/B:JAPH.0000047789.34883.aa

Orfield ND, Keoleian GA, Love NG (2014) A GIS based national assessment of algal bio-oil production potential through flue gas and wastewater co-utilization. Biomass Bioenergy 63:76–85. https://doi.org/10.1016/J.BIOMBIOE.2014.01.047

Ortigueira J, Alves L, Gouveia L, Moura P (2015) Third generation biohydrogen production by *Clostridium butyricum* and adapted mixed cultures from *Scenedesmus obliquus* microalga biomass. Fuel 153:128–134. https://doi.org/10.1016/j.fuel.2015.02.093

Oswald WJ (1988) Microalgae and wastewater treatment. In: Borowitzka MA, Borowitska LJ (eds) Microalgal biotechnology. Cambridge University Press, New York, pp 357–394

Oswald WJ, Gotaas HB (1957) Photosynthesis in sewage treatment. Trans Am Soc Civ Eng 122:73–97

Palaniappan M, Gleick PH, Allen L et al (2010) Clearing the waters: a focus on water quality solutions. United Nations Environment Programme (UNEP), Nairobi

Park JBK, Craggs RJ, Shilton AN (2011) Wastewater treatment high rate algal ponds for biofuel production. Bioresour Technol 102:35–42. https://doi.org/10.1016/J.BIORTECH.2010.06.158

Pittman JK, Dean AP, Osundeko O (2011) The potential of sustainable algal biofuel production using wastewater resources. Bioresour Technol 102:17–25. https://doi.org/10.1016/j.biortech.2010.06.035

Plaza M, Amigo-Benavent M, del Castillo MD et al (2010a) Facts about the formation of new antioxidants in natural samples after subcritical water extraction. Food Res Int 43:2341–2348. https://doi.org/10.1016/J.FOODRES.2010.07.036

Plaza M, Amigo-Benavent M, del Castillo MD et al (2010b) Neoformation of antioxidants in glycation model systems treated under subcritical water extraction conditions. Food Res Int 43:1123–1129. https://doi.org/10.1016/J.FOODRES.2010.02.005

Plaza BM, Gómez-Serrano C, Acién-Fernández FG, Jimenez-Becker S (2018) Effect of microalgae hydrolysate foliar application (*Arthrospira platensis* and *Scenedesmus* sp.) on *Petunia x hybrida* growth. J Appl Phycol 30:2359–2365. https://doi.org/10.1007/s10811-018-1427-0

Posadas E, Bochon S, Coca M et al (2014a) Microalgae-based agro-industrial wastewater treatment: a preliminary screening of biodegradability. J Appl Phycol 26:2335–2345. https://doi.org/10.1007/s10811-014-0263-0

Posadas E, García-Encina PA, Domínguez A et al (2014b) Enclosed tubular and open algal–bacterial biofilm photobioreactors for carbon and nutrient removal from domestic wastewater. Ecol Eng 67:156–164. https://doi.org/10.1016/J.ECOLENG.2014.03.007

Posadas E, Morales M del M, Gomez C et al (2015a) Influence of pH and CO_2 source on the performance of microalgae-based secondary domestic wastewater treatment in outdoors pilot raceways. Chem Eng J 265:239–248. https://doi.org/10.1016/J.CEJ.2014.12.059

Posadas E, Muñoz A, García-González MC et al (2015b) A case study of a pilot high rate algal pond for the treatment of fish farm and domestic wastewaters. J Chem Technol Biotechnol 90:1094–1101. https://doi.org/10.1002/jctb.4417

Posadas E, Alcántara C, García-Encina PA et al (2017) Microalgae cultivation in wastewater. Microalgae-Based Biofuels Bioprod 67–91. https://doi.org/10.1016/B978-0-08-101023-5.00003-0

Pramote K, Nucha S, Suched S et al (2012) Subcritical water extraction of flavoring and phenolic compounds from cinnamon bark *Cinnamomum zeylanicum*. J Oleo Sci 61:349–355. https://doi.org/10.5650/jos.61.349

Quinn JC, Davis R (2015) The potentials and challenges of algae based biofuels: a review of the techno-economic, life cycle, and resource assessment modeling. Bioresour Technol 184:444–452. https://doi.org/10.1016/J.BIORTECH.2014.10.075

Raposo MF d J, Oliveira SE, Castro PM et al (2010) On the utilization of microalgae for brewery effluent treatment and possible applications of the produced biomass. J Inst Brew 116:285–292. https://doi.org/10.1002/j.2050-0416.2010.tb00433.x

Rawat I, Ranjith Kumar R, Mutanda T, Bux F (2011) Dual role of microalgae: phycoremediation of domestic wastewater and biomass production for sustainable biofuels production. Appl Energy 88:3411–3424. https://doi.org/10.1016/J.APENERGY.2010.11.025

Razzak SA, Hossain MM, Lucky RA et al (2013) Integrated CO_2 capture, wastewater treatment and biofuel production by microalgae culturing—a review. Renew Sustain Energy Rev 27:622–653. https://doi.org/10.1016/J.RSER.2013.05.063

Renuka N, Prasanna R, Sood A et al (2016) Exploring the efficacy of wastewater-grown microalgal biomass as a biofertilizer for wheat. Environ Sci Pollut Res 23:6608–6620. https://doi.org/10.1007/s11356-015-5884-6

Rodríguez-Meizoso I, Marin FR, Herrero M et al (2006) Subcritical water extraction of nutraceuticals with antioxidant activity from oregano. Chemical and functional characterization. J Pharm Biomed Anal 41:1560–1565. https://doi.org/10.1016/J.JPBA.2006.01.018

Rodríguez-Meizoso I, Jaime L, Santoyo S et al (2010) Subcritical water extraction and characterization of bioactive compounds from *Haematococcus pluvialis* microalga. J Pharm Biomed Anal 51:456–463. https://doi.org/10.1016/J.JPBA.2009.03.014

Romero García JM, Acién Fernández FG, Fernández Sevilla JM (2012) Development of a process for the production of l-amino-acids concentrates from microalgae by enzymatic hydrolysis. Bioresour Technol 112:164–170. https://doi.org/10.1016/J.BIORTECH.2012.02.094

Roostaei J, Zhang Y (2017) Spatially explicit life cycle assessment: opportunities and challenges of wastewater-based algal biofuels in the United States. Algal Res 24:395–402. https://doi.org/10.1016/J.ALGAL.2016.08.008

Rugnini L, Costa G, Congestri R, Bruno L (2017) Testing of two different strains of green microalgae for Cu and Ni removal from aqueous media. Sci Total Environ 601–602:959–967. https://doi.org/10.1016/J.SCITOTENV.2017.05.222

Safafar H, Hass M, Møller P et al (2016) High-EPA biomass from *Nannochloropsis salina* cultivated in a flat-panel photo-bioreactor on a process water-enriched growth medium. Mar Drugs 14:144. https://doi.org/10.3390/md14080144

Safonova E, Kvitko KV, Iankevitch MI et al (2004) Biotreatment of industrial wastewater by selected algal-bacterial consortia. Eng Life Sci 4:347–353. https://doi.org/10.1002/elsc.200420039

Salama E-S, Kurade MB, Abou-Shanab RAI et al (2017) Recent progress in microalgal biomass production coupled with wastewater treatment for biofuel generation. Renew Sustain Energy Rev 79:1189–1211. https://doi.org/10.1016/J.RSER.2017.05.091

Sato T, Qadir M, Yamamoto S et al (2013) Global, regional, and country level need for data on wastewater generation, treatment, and use. Agric Water Manag 130:1–13. https://doi.org/10.1016/J.AGWAT.2013.08.007

Sepúlveda C, Acién FG, Gómez C et al (2015) Utilization of centrate for the production of the marine microalgae *Nannochloropsis gaditana*. Algal Res 9:107–116. https://doi.org/10.1016/J.ALGAL.2015.03.004

Sereewatthanawut I, Prapintip S, Watchiraruji K et al (2008) Extraction of protein and amino acids from deoiled rice bran by subcritical water hydrolysis. Bioresour Technol 99:555–561. https://doi.org/10.1016/J.BIORTECH.2006.12.030

Sevrin-Reyssac J (1998) Biotreatment of swine manure by production of aquatic valuable biomasses. Agric Ecosyst Environ 68:177–186. https://doi.org/10.1016/S0167-8809(97)00070-4

Sfez S, Van Den Hende S, Taelman SE et al (2015) Environmental sustainability assessment of a microalgae raceway pond treating aquaculture wastewater: from up-scaling to system integration. Bioresour Technol 190:321–331. https://doi.org/10.1016/J.BIORTECH.2015.04.088

Silva CM, Ferreira AF, Dias AP, Costa M (2016) A comparison between microalgae virtual biorefinery arrangements for bio-oil production based on lab-scale results. J Clean Prod 130:58–67. https://doi.org/10.1016/J.JCLEPRO.2015.09.053
Simate GS, Cluett J, Iyuke SE et al (2011) The treatment of brewery wastewater for reuse: state of the art. Desalination 273:235–247. https://doi.org/10.1016/j.desal.2011.02.035
Singh A, Olsen SI (2011) A critical review of biochemical conversion, sustainability and life cycle assessment of algal biofuels. Appl Energy 88:3548–3555. https://doi.org/10.1016/J.APENERGY.2010.12.012
Singh PP, Saldaña MDA (2011) Subcritical water extraction of phenolic compounds from potato peel. Food Res Int 44:2452–2458. https://doi.org/10.1016/J.FOODRES.2011.02.006
Siso MIG (1996) The biotechnological utilization of cheese whey: a review. Bioresour Technol 57:1–11. https://doi.org/10.1016/0960-8524(96)00036-3
Sládečková A (1994) The role of periphyton in waste treatment technology. SIL Proc 1922–2010 25:1929–1932. https://doi.org/10.1080/03680770.1992.11900527
Sládečková A, Matulová D (1998) Periphyton as bioeliminator. SIL Proc 1922–2010 26:1777–1780. https://doi.org/10.1080/03680770.1995.11901043
Sládečková A, Sládečk V, Koči V, Ambrožová J (2001) Bio-monitoring of impacts of wastewater treatment plant effluents on receiving streams. In: IWA, 2nd World Water Congress. Berlin, p 1364
Slavov AK (2017) Dairy wastewaters – general characteristics and treatment possibilities – a review. Food Technol Biotechnol 55:14–28. https://doi.org/10.17113/ftb.55.01.17.4520
Song T, Mårtensson L, Eriksson T et al (2005) Biodiversity and seasonal variation of the cyanobacterial assemblage in a rice paddy field in Fujian, China. FEMS Microbiol Ecol 54:131–140. https://doi.org/10.1016/j.femsec.2005.03.008
Statista (2017) Beer production worldwide from 1998 to 2015 (in billion hectoliters). https://www.statista.com/statistics/270275/worldwide-beer-production/. Accessed 30 May 2017
Tang EPY, Vincent WF, Proulx D et al (1997) Polar cyanobacteria versus green algae for tertiary waste-water treatment in cool climates. J Appl Phycol 9:371–381. https://doi.org/10.1023/A:1007987127526
Tchobanoglous G, Burton FL, Stensel HD (2003) Wastewater engineering: treatment and reuse, 4th edn. McGraw Hill, New York
Teo CC, Tan SN, Yong JWH et al (2010) Pressurized hot water extraction (PHWE). J Chromatogr A 1217:2484–2494. https://doi.org/10.1016/J.CHROMA.2009.12.050
Turcios A, Papenbrock J, Turcios AE, Papenbrock J (2014) Sustainable treatment of aquaculture effluents—what can we learn from the past for the future? Sustainability 6:836–856. https://doi.org/10.3390/su6020836
UNESCO (2012) Managing water under uncertainty and risk. UNESCO, Paris
UN-Water Secretariat (2015) Wastewater management. New York
Uysal O, Uysal FO, Ekinci K (2015) Evaluation of microalgae as microbial fertilizer. Eur J Sustain Dev 4:77–82. https://doi.org/10.14207/ejsd.2015.v4n2p77
Valta K, Kosanovic T, Malamis D et al (2015) Overview of water usage and wastewater management in the food and beverage industry. Desalin Water Treat 53:3335–3347. https://doi.org/10.1080/19443994.2014.934100
Vladić J, Canli O, Pavlić B et al (2017) Optimization of *Satureja montana* subcritical water extraction process and chemical characterization of volatile fraction of extracts. J Supercrit Fluids 120:86–94. https://doi.org/10.1016/J.SUPFLU.2016.10.016
Wang L, Min M, Li Y et al (2010) Cultivation of green algae *Chlorella* sp. in different wastewaters from municipal wastewater treatment plant. Appl Biochem Biotechnol 162:1174–1186. https://doi.org/10.1007/s12010-009-8866-7
Wang M, Keeley R, Zalivina N et al (2018) Advances in algal-prokaryotic wastewater treatment: a review of nitrogen transformations, reactor configurations and molecular tools. J Environ Manag 217:845–857. https://doi.org/10.1016/J.JENVMAN.2018.04.021

Wheeler SM, Beatley T (eds) (2004) Sustainable urban development reader, 3rd edn. New York, Routledge
White C (2018) Technavio report: global aquaculture market's growth accelerating through 2022. Seafood. https://www.seafoodsource.com/features/technavio-report-global-aquaculture-markets-growth-accelerating-through-2022. Accessed 18 Dec 2018
Wilson M, Houghton JA (1974) Growth of algae on pig manure. Irish J Agric Res 13:49–60
WWAP (2017) Wastewater: the untapped resource. United Nations world water development report Wastewater untapped resource. https://doi.org/10.1017/CBO9781107415324.004
Xu Y, Isom L, Hanna MA (2010) Adding value to carbon dioxide from ethanol fermentations. Bioresour Technol 101:3311–3319. https://doi.org/10.1016/J.BIORTECH.2010.01.006
Xu H, Wang W, Jiang J et al (2015) Subcritical water extraction and antioxidant activity evaluation with on-line HPLC-ABTS·+ assay of phenolic compounds from marigold (*Tagetes erecta* L.) flower residues. J Food Sci Technol 52:3803–3811. https://doi.org/10.1007/s13197-014-1449-9
Yadavalli R, Rao CS, Rao RS, Potumarthi R (2014) Dairy effluent treatment and lipids production by *Chlorella pyrenoidosa* and *Euglena gracilis*: study on open and closed systems. Asia-Pac J Chem Eng 9:368–373. https://doi.org/10.1002/apj.1805
Zakaria SM, Kamal SMM, Harun MR et al (2017) Subcritical water technology for extraction of phenolic compounds from *Chlorella* sp. microalgae and assessment on its antioxidant activity. Molecules 22:1105. https://doi.org/10.3390/molecules22071105
Zhou W, Chen P, Min M et al (2014) Environment-enhancing algal biofuel production using wastewaters. Renew Sustain Energy Rev 36:256–269. https://doi.org/10.1016/J.RSER.2014.04.073

Chapter 10
Microalgal Consortia: From Wastewater Treatment to Bioenergy Production

Ana L. Gonçalves, Francisca M. Santos, and José C. M. Pires

Abstract Cultivation of microalgae has been the focus of several research studies worldwide, due to the huge potential of these photosynthetic microorganisms in a wide range of applications, namely environmental and biotechnological ones. Regarding environmental applications, these microorganisms can play an important role in CO_2 uptake and wastewater treatment processes and can be used as raw materials for bioenergy production. However, cultivation of these microorganisms for these applications still faces some problems: (1) it is very difficult to maintain pure cultures of these microorganisms in wastewater treatment processes and (2) bioenergy production process using these microorganisms is still not economically viable. To face these challenges, several studies have reported the use of microalgal consortia. When using microalgal consortia, cooperative interactions can occur, enhancing biomass productivities and therefore nutrients uptake and lipids content. Additionally, these systems tend to be more resistant to environmental conditions' oscillations, facilitating the overall production process. In this study, an overview on the use of microalgal consortia for CO_2 capture, wastewater treatment and bioenergy production is provided, focusing on the interactions that can occur between these microorganisms and how they can improve these environmental applications.

10.1 Introduction

Anthropogenic activities, such as agricultural practices, urbanization and industrialization, as well as fossil fuel-based economies, have contributed to the degradation of air and water quality and to the depletion of fossil fuel resources (O'Neill and Oppenheimer 2002; Aslan and Kapdan 2006; Demirbas 2011). These activities have been responsible for the increase of atmospheric CO_2 concentration in the last

A. L. Gonçalves · F. M. Santos · J. C. M. Pires (✉)
Faculty of Engineering, LEPABE—Laboratory for Process Engineering, Environment, Biotechnology and Energy, University of Porto, Porto, Portugal
e-mail: jcpires@fe.up.pt

A. Hallmann, P. H. Rampelotto (eds.), *Grand Challenges in Algae Biotechnology*, Grand Challenges in Biology and Biotechnology,
https://doi.org/10.1007/978-3-030-25233-5_10

decades, which has resulted in negative environmental impacts, such as the greenhouse effect and oceans' acidification (O'Neill and Oppenheimer 2002). Additionally, population increase has contributed to an augmented disposal of wastes into natural water resources, resulting in their contamination with nitrogen and phosphorus and in the scarcity of freshwater (Aslan and Kapdan 2006; Rawat et al. 2011; Renuka et al. 2013). Finally, the continuous increase in the world energy demand and the use of non-renewable energy resources, such as fossil fuels, has resulted in the depletion of this energy resource (Ranjan et al. 2010; Demirbas 2011).

In opposition to the negative impacts mainly caused by anthropogenic activities, it becomes urgent for world economies to: (1) reduce CO_2 emissions and provide new strategies to reduce CO_2 concentration in the atmosphere and in flue gas emissions (Pielke 2009); (2) provide new methods to reduce nitrogen and phosphorus concentrations present in discharged effluents and avoid eutrophication (Renuka et al. 2013; Ruiz et al. 2013); and (3) search for different energy supplies that are renewable and environmentally friendly (carbon neutral) (Chisti 2007; Demirbas 2011).

Due to the huge potential of microalgae and cyanobacteria in several applications, especially in environmental ones, such as nutrients uptake and bioenergy production, cultivation of these photosynthetic microorganisms for these purposes has attracted researchers worldwide. Through autotrophy, microalgae also fix CO_2 from the atmosphere or from flue gas emissions, reducing the concentrations of this greenhouse gas in the atmosphere (Allen et al. 2009; Ho et al. 2011; Tang et al. 2011). Since microalgal growth depends on the presence of inorganic forms of nitrogen and phosphorus (microalgal biomass is mainly composed of carbon, nitrogen and phosphorus—macronutrients), these microorganisms can be grown in nitrogen- and phosphorus-rich wastewaters, assimilating these nutrients and reducing their concentration in these effluents (Rawat et al. 2011; Silva-Benavides and Torzillo 2012). In addition to the remediation potential described for these microorganisms, depending on the culturing conditions and on the culture medium used, microalgal biomass can be further applied to human food and animal feed and in the production of drugs, cosmetics, functional food, biofuels and fertilizers (Allen et al. 2009; Brennan and Owende 2010; Parmar et al. 2011; Odjadjare et al. 2015). Besides the wide variety of applications described for microalgae, cultivation of these microorganisms presents other advantages: (1) they present higher growth rates, higher biomass and lipid productivities and higher nutritional values (on a per unit area basis) than other photosynthetic organisms, such as terrestrial crops; (2) they can be grown in non-arable land and require far less land than terrestrial crops, thus not competing with land required for agricultural practices and food production; and (3) they can also grow in a wide variety of environmental conditions and in low-quality waters, reducing the requirements for freshwater and nutrients (Pulz and Gross 2004; Chisti 2007). However, cultivation of these microorganisms still presents some challenges regarding the achievement of high biomass and lipid productivities and high nutrients removal efficiencies at reduced costs.

One possibility to face these challenges includes the use of microalgal consortia (microalgal and microalgal–bacterial), in order to establish an effective system in terms of CO_2 capture, nutrients removal and bioenergy production. Recently, several studies have reported the potential of these consortia in different applications, including biomass and lipids production, CO_2 capture and nutrients removal (Muñoz and Guieysse 2006; Rawat et al. 2011; Subashchandrabose et al. 2011; Olguín 2012; Unnithan et al. 2014; Ramanan et al. 2016). This study presents an updated review on the use of microalgal consortia for biotechnological applications, such as CO_2 capture, nutrients removal from wastewaters and lipids production for biofuels.

10.2 Applications of Microalgae

Due to the huge taxonomic diversity of microalgae and to their extensive environmental distribution, these photosynthetic microorganisms have numerous applications in diversified areas, such as environment (CO_2 removal and wastewater treatment), energy (biofuels production), pharmaceutical and cosmetics industries, aquaculture, animal feed and human food (Spolaore et al. 2006; Allen et al. 2009; Brennan and Owende 2010; Show et al. 2017; Khan et al. 2018). Through photosynthesis, microalgae are able to assimilate CO_2 from the atmosphere, as well as from flue gas emissions. Thus, microalgae can be applied to mitigate the increasing tendency of atmospheric CO_2 concentration that has been observed since Industrial Revolution (Allen et al. 2009; Ho et al. 2011; Tang et al. 2011; Show et al. 2017; Khan et al. 2018). Additionally, these photosynthetic microorganisms assimilate other compounds, such as nitrogen and phosphorus. These nutrients are frequently found in wastewaters, meaning that microalgae can be a promising alternative in wastewater treatment processes (Rawat et al. 2011; Silva-Benavides and Torzillo 2012; Show et al. 2017; Khan et al. 2018). For human food and animal feed, algal biomass is suitable because they are an important source of natural vitamins, minerals and fatty acids. They can be used to feed different animals, such as cats, dogs, aquarium fish, birds, horses and cows (Hu 2004; Spolaore et al. 2006). Several compounds, such as pigments, antioxidants, β-carotenes, proteins, polysaccharides, triglycerides, fatty acids and vitamins, can be extracted and used as raw materials for the production of cosmetics, drugs and functional food (Hu 2004; Singh et al. 2005; Bhalamurugan et al. 2018). Finally, the fatty acids produced by microalgae can be extracted and used for biodiesel production and residual biomass can be fermented to produce ethanol or methane and fertilizers (Brennan and Owende 2010; John et al. 2011; Parmar et al. 2011). Figure 10.1 presents a schematic representation of the main applications described for microalgae. Although microalgae can be used in all these applications, only a few of them are currently applied at a commercial scale (Table 10.1). This is a result of the high costs associated with microalgal biomass production, which limits microalgal biomass applications to the commercialization of high-valued products.

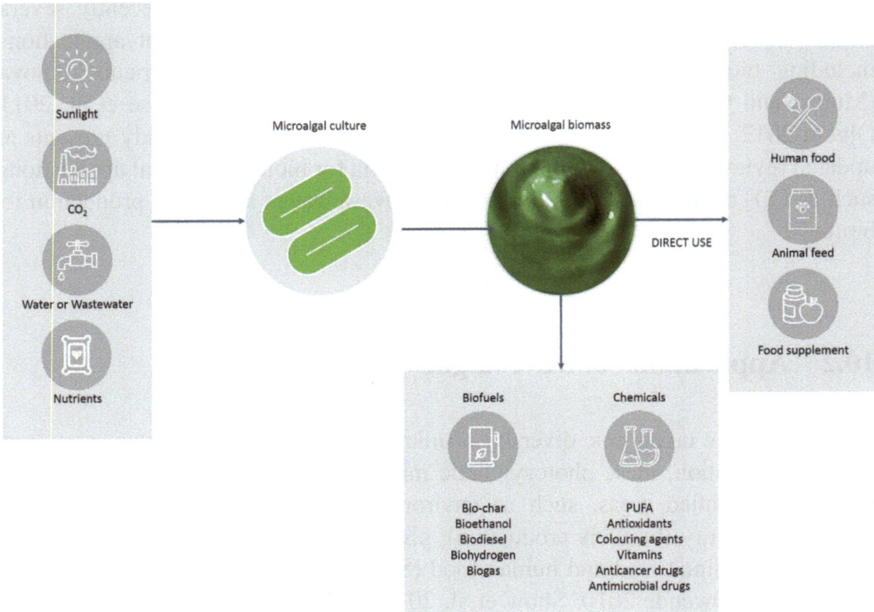

Fig. 10.1 Main applications described for microalgae [adapted from Khan et al. (2018)]

Among the applications described for microalgae, this study focuses on the use of microalgae and microalgal consortia for CO_2 capture, nutrients removal from wastewaters and bioenergy production.

10.2.1 CO_2 Capture

Carbon is the most important element for microalgal growth, followed by nitrogen and phosphorus (microalgal biomass contains approximately 50% w/w of carbon, which is all derived from CO_2). Accordingly, the production of 1 g of microalgal biomass corresponds to a CO_2 fixation of approximately 1.83 g, which means that these microorganisms can be effectively applied in CO_2 capture (Cheah et al. 2015).

Microalgae can fix CO_2 from both the atmosphere or flue gas emissions. Table 10.2 presents CO_2 uptake rates determined for different microalgae using both CO_2 sources. The use of atmospheric CO_2 allows higher flexibility when selecting the location of the microalgal facility, since it does not need to be located close to a CO_2 emission source and it does not require CO_2 transporting systems (Cheah et al. 2015; Moreira and Pires 2016). Several studies have reported the use of microalgae in CO_2 capture from the atmosphere (Cheng et al. 2006; López et al. 2009; Hulatt and Thomas 2011; Arbib et al. 2014). In the study performed by López et al. (2009), the growth of the cyanobacterium *Anabaena* sp. in bubble-column

Table 10.1 Current status of large-scale applications of microalgae

Microalgae	Applications	Country	Status
Aphanizomenon flos-aquae	Human nutrition	USA	Commercial
Chaetoceros muelleri	Aquaculture feed	Global	Commercial
Chlorella spp.	Human nutrition, aquaculture feed, cosmetics	Japan, Taiwan, Czech Republic, Germany	Commercial
Chlorella vulgaris	Human nutrition, aquaculture feed, animal feed, cosmetics	Portugal, USA, Japan, Taiwan, Czech Republic, Germany	Commercial
Crypthecodinium cohnii	Docosahexaenoic acid	USA	Commercial
Dunaliella salina	Human nutrition, aquaculture feed, β-carotene	Australia, Israel, China, India, USA	Commercial
Dunaliella tertiolecta	Aquaculture feed	Global	Commercial
Haematococcus pluvialis	Aquaculture feed, astaxanthin	USA, Sweden, Israel	Commercial
Isochrysis spp.	Aquaculture feed	Global	Commercial
Monochrysis lutheri	Aquaculture feed	Global	Commercial
Nannochloropsis oculata	Lipids, PUFA	Portugal, USA, Spain, Germany, Belgium, Italy	Commercial
Nannochloropsis spp.	Aquaculture feed	Global	Commercial
Pavlova spp.	Aquaculture feed	Global	Commercial
Shizochytrium sp.	Docosahexaenoic acid	USA	Commercial
Skeletonema spp.	Aquaculture feed	Global	Commercial
Spirulina platensis	Human and animal nutrition, cosmetics, phycobiliproteins	Thailand, USA, China, India, Vietnam, Japan	Commercial
Tetraselmis suecica	Aquaculture feed	Global	Commercial
Thalassiosira pseudonana	Aquaculture feed	Global	Commercial

Adapted from Spolaore et al. (2006), Paul et al. (2013), Zhu (2015)

PBRs (working volume of 1.8 L), using atmospheric air, has resulted in a CO_2 fixation rate of 1.45×10^3 mg CO_2 L^{-1} day^{-1}. Arbib et al. (2014) have evaluated the potential of three microalgal species (*Chlorella kessleri*, *Chlorella vulgaris* and *Scenedesmus obliquus*) in CO_2 capture from the atmosphere using 2-L flasks as PBRs. After 10 days of culturing, CO_2 fixation rates determined for the studied microorganisms were 320, 297 and 418 mg CO_2 L^{-1} day^{-1}, respectively. However, atmospheric CO_2 concentration (approximately 0.04% v/v) can be limiting to microalgal growth, due to low mass transfer rate of CO_2 from the gaseous stream to the liquid medium. According to McGinn et al. (2011), CO_2 diffuses into the liquid medium 10^4 times slower than through the gaseous medium. Accordingly,

Table 10.2 Application of microalgae in CO_2 capture from the atmosphere and flue gases and respective fixation rates

Microorganisms	CO_2 source	% CO_2 (v/v)	System and operation mode	Time (days)	CO_2 fixation rate (mg CO_2 L^{-1} day^{-1})	References
Anabaena sp.	Atmospheric air	≈0.04	Closed suspended system, continuous mode; V = 1.8 L	–	1.45×10^3	López et al. (2009)
Anabaena sp.	Atmospheric air + pure CO_2	10	Closed suspended system, batch mode; V = 5 L	4	1.01×10^3	Chiang et al. (2011)
Botryococcus braunii	Atmospheric air + pure CO_2	5	Closed suspended system, batch mode; V = 8 L	15	497	Sydney et al. (2010)
Chlorella sp.	Atmospheric air + pure CO_2	5	Closed suspended system, batch mode; V = 0.6 L	7	583	Ryu et al. (2009)
Chlorella kessleri	Atmospheric air	≈0.04	Closed suspended system, batch mode, V = 2 L	10	320	Arbib et al. (2014)
Chlorella sorokiniana	Flue gas	15.6	Closed suspended system, batch mode, V = 1.4 L	8	124	Kumar et al. (2014)
Chlorella vulgaris	Atmospheric air	≈0.04	Closed suspended system, batch mode, V = 2 L	10	297	Arbib et al. (2014)
Chlorella vulgaris	Atmospheric air + pure CO_2	2	Closed suspended system, batch mode; V = 1 L	10	430	Yeh and Chang (2011)
Chlorella vulgaris	Atmospheric air + pure CO_2	5	Closed suspended system, batch mode; V = 8 L	15	252	Sydney et al. (2010)
Chlorella vulgaris	Atmospheric air + pure CO_2	6.5	Closed suspended system, batch mode; V = 0.09 L	7	2.22×10^3	Anjos et al. (2013)
Dunaliella tertiolecta	Atmospheric air + pure CO_2	5	Closed suspended system, batch mode; V = 8 L	15	272	Sydney et al. (2010)
Scenedesmus obliquus	Atmospheric air	≈0.04	Closed suspended system, batch mode, V = 2 L	10	437	Arbib et al. (2014)

Scenedesmus obliquus	Atmospheric air + pure CO_2	10	Closed suspended system, batch mode, V = 1 L	12	550	Ho et al. (2010)
Spirulina platensis	Atmospheric air + pure CO_2	5	Closed suspended system, batch mode; V = 8 L	15	318	Sydney et al. (2010)
Microalgal consortium obtained from a river	Atmospheric air	≈0.04	Closed suspended system, batch mode, V = 2 L	10	418	Arbib et al. (2014)

V Working volume

costly CO_2 sparging might be required to increase the retention time of CO_2 in the culture medium. As an alternative to the use of pure CO_2 to feed microalgal cultures, several authors have reported the use of flue gases. CO_2 concentration in flue gases typically ranges between 6 and 15% (v/v). For this reason, several authors have evaluated the effect of different CO_2 concentrations on microalgal growth and CO_2 uptake. For example, in the study performed by Morais and Costa (2007a), the effect of CO_2 concentrations of 0, 6 and 12% (v/v) on biomass productivities and CO_2 fixation rates of *Spirulina* sp. and *S. obliquus* was evaluated. In this study, the authors have reported higher biomass productivities, specific growth rates and CO_2 fixation rates in cultures performed at 6 and 12% (v/v) of CO_2, with maximum values obtained at 6% (v/v). As a result, current microalgal production plants are being projected near large CO_2 emission sources, such as power plants and refineries (Moreira and Pires 2016). However, flue gases usually contain large amounts of nitrogen oxides (NO_x) and sulphur oxides (SO_x), which can significantly reduce the pH of the culture medium, thus inhibiting microalgal growth (Pires et al. 2011; Cheah et al. 2015; Moreira and Pires 2016). In the case of NO_x, these compounds can be used by microalgae as nitrogen source, thus not presenting such a negative impact for microalgal growth. On the other hand, high SO_2 concentrations (between 100 and 250 mg L^{-1}) can be harmful to microalgae, due to the formation of bisulphite (HSO_3^-), sulphite (SO_3^{2-}) and sulphate (SO_4^{2-}), which drastically decrease the pH of the culture medium to values between 2.5 and 3.5 (Lam et al. 2012; Cheah et al. 2015). Additionally, high CO_2 concentrations (between 10 and 20% v/v) also contribute to a decrease in pH to approximately 5.5. Although some microalgal species are able to perform photosynthesis in these conditions, thus counterbalancing the pH decrease due to high CO_2 levels, other species are unable to withstand this acidic environment (Cheah et al. 2015). The presence of particulate matter and high temperatures are two additional characteristics of flue gases that should be taken into account when using these gases for microalgal growth. Particulate material resulting from combustion processes presents a quite variable composition, depending on the emission sources. These materials can be divided in aerosols, fly ash and soot, and their effect on microalgal growth was already assessed by few authors (Costa et al. 2017). For example, in the study performed by Vaz et al. (2016), the effect of flue gas ashes on the growth of *Spirulina* sp. LEB 18 and *Chlorella fusca* LEB 111 was evaluated. In this study, the authors have concluded that addition of 40 ppm ashes from coal burning has not inhibited the growth of these microorganisms and that this particulate matter can act as a source of trace elements for microalgal metabolism. Regarding temperature, flue gases from power plants can reach temperatures up to 120 °C, which limits its application for microalgal growth to thermophilic microalgal strains or requires a previous cooling step, which is energy-demanding and costly (Costa et al. 2017).

Due to these constraints, only a few studies have reported the use of real flue gases for microalgal growth. When growing *Chlorella* sp. in an outdoor open thin-layer PBR using a flue gas containing 6–8% (v/v) of CO_2, Doucha et al. (2005) have reported CO_2 removal efficiencies between 10 and 50%. A CO_2 removal efficiency of 40.2% was obtained by Li et al. (2011) when growing *S. obliquus* in a flue gas containing 6–18% (v/v) of CO_2. Furthermore, some authors have referred that native species isolated near thermal power plants are the best candidates for CO_2 capture from these gaseous streams (Morais and Costa 2007b; Radmann et al. 2011).

10.2.2 Nutrients Removal from Wastewaters

As mentioned above, the main nutrients required for microalgal growth are carbon, nitrogen and phosphorus. Regarding carbon uptake, although microalgae are mainly autotrophic, some microalgae are heterotrophic, using only organic carbon (e.g. acetate, glucose, glycerol and ethanol) as carbon source, whereas others are mixotrophic, using facultatively an organic carbon source in addition to CO_2. In the case of nitrogen, eukaryotic microalgae are able to assimilate fixed nitrogen, such as ammonium-nitrogen (NH_4-N), nitrate-nitrogen (NO_3-N) and nitrite-nitrogen (NO_2-N). Finally, phosphorus is assimilated by microalgae in the forms of $H_2PO_4^-$ and HPO_4^{2-} (Gonçalves et al. 2017).

Due to the high concentrations of these nutrients in wastewaters from different sources and to the need of reducing microalgal production costs and environmental impact (reduce the requirements for freshwater), several studies have reported the use of wastewaters for microalgal growth. In addition, the use of microalgae to remove nitrogen and phosphorus from wastewaters is seen as a viable alternative to overcome the drawbacks associated with currently used methods. Nitrogen and phosphorus present in wastewaters are mainly removed in the tertiary treatment phase and the most commonly used methods include biological processes, such as anaerobic digestion followed by nitrification and denitrification (Queiroz et al. 2007; Renuka et al. 2013). However, these methods require several tanks and internal recycles of activated sludge, resulting in an overall increase of process costs, complexity and energy input (Foess et al. 1998; Jeyanayagam 2005; Larsdotter 2006; Singh and Thomas 2012). Alternatively, nitrogen and phosphorus removal may be achieved by chemical methods, such as precipitation using aluminium and iron salts. However, these methods are costly and produce large amounts of sludge contaminated with chemical compounds that require further treatment (Wang et al. 2006). The use of microalgae for nutrients removal presents several advantages over those commonly applied in the tertiary treatment step because: (1) nitrogen and phosphorus assimilated by microalgae can be recycled by the production of fertilizers from microalgal biomass; (2) the resulting biomass can be used for the production of bioenergy, food, animal feed and pharmaceuticals; and (3) an oxygenated effluent is discharged into the water bodies (Aslan and Kapdan 2006; Rawat et al. 2011; Renuka et al. 2013).

Domestic, leachate, agricultural, refinery and industrial wastewaters are examples of wastewater sources that have already been used for microalgal growth (Chojnacka et al. 2004; Safonova et al. 2004; Mustafa et al. 2012; Hernández et al. 2013; Posadas et al. 2013). Table 10.3 presents nitrogen and phosphorus removal rates determined when culturing microalgae in different wastewaters.

Although the majority of the studies refer to the use of suspended-growth systems, Shi et al. (2007) have assessed nitrogen and phosphorus removal from a municipal wastewater collected in Cologne (Germany) using an immobilization method—the twin-layer system. In this method, the microalgae *C. vulgaris* and *Scenedesmus rubescens* were immobilized through self-adhesion on a substrate layer and another layer provided the growth medium required for microalgal growth. Using this system, microalgae remained 100% immobilized, being able to completely remove NO_3-N (initial concentration between 3.7 and 6.2 mg N L^{-1}) after an exposure period of 4 days. More recently, Gouveia et al. (2016) have cultured *C. vulgaris*, *S. obliquus* and a native consortium in a municipal wastewater collected from Figueira da Foz (Portugal), aiming to determine the best candidate in terms of wastewater remediation, biomass productivity and quality for further uses, such as biofuels, biofertilizers and bioplastics production. The studied cultures have effectively removed nitrogen and phosphorus from the wastewater, reaching nitrogen removal efficiencies of 84–98% and phosphorus removal efficiencies of 95–100%. Taking into account these results, the authors have proposed the native consortium as the best option for nutrients removal and biomass production.

Valderrama et al. (2002) have cultured *C. vulgaris* in an industrial effluent resulting from ethanol and citric acid production, achieving NH_4-N and phosphate-phosphorus (PO_4-P) removal efficiencies of 71.6 and 28%, respectively (initial NH_4-N concentration in this effluent ranged between 3 and 8 mg N L^{-1}, whereas initial PO_4-P concentration ranged between 0 and 0.36 mg P L^{-1}). Similarly, Lim et al. (2010) have grown *C. vulgaris* in high rate algal ponds (HRAPs) fed with a textile industry wastewater to evaluate the potential of this microalga in nitrogen and phosphorus removal. Although *C. vulgaris* was able to grow in the textile wastewater (NH_4-N and PO_4-P initial concentrations of 6.50 mg L^{-1} and 7.14 mg L^{-1}, respectively), nitrogen and phosphorus removal efficiencies achieved were not very high: 44.4–45.1% and 33.1–33.3%, respectively.

In the study performed by Hernández et al. (2013), *Chlorella sorokiniana* was grown in a potato-processing wastewater presenting an initial NH_4-N concentration of 12.1 mg N L^{-1} and PO_4-P concentration of 3.4 mg P L^{-1}. After a cultivation period of 10 days, nitrogen and phosphorus removal efficiencies achieved were 95 and 80.7%, respectively. Liu et al. (2016) have determined the remediation potential of the filamentous microalgae *Klebsormidium* sp. and *Stigeoclonium* spp. grown in an outdoor Algal Turf Scrubber using horticultural wastewater as culture medium (NO_3-N and PO_4-P initial concentrations of 47.2 mg L^{-1} and 11.6 mg L^{-1}, respectively). With this study, the authors have demonstrated that these microalgae can effectively remove nitrogen and phosphorus from this wastewater, since nitrogen removal efficiencies achieved oscillated between 88 and 99% and phosphorus removal efficiencies were higher than 99%.

The use of microalgae for the remediation of different anaerobically digested effluents has also been reported in the literature. For example, Wilkie and Mulbry

Table 10.3 Application of microalgae in nitrogen and phosphorus removal from different wastewaters and respective removal efficiencies

Microorganisms	Waste stream	System and operation mode	Removal time (days)	Nitrogen		Phosphorus		References
				S_i (mg N L^{-1})	*R* (%)	S_i (mg P L^{-1})	*R* (%)	
Chlorella vulgaris/ *Scenedesmus rubescens*	Municipal wastewater	Immobilized system, continuous mode, V = 2 L	4	3.7–6.2 NO_3-N	≈100	0.4–0.7 TP	n.a.	Shi et al. (2007)
Chlamydomonas reinhardtii	Municipal wastewater	Closed suspended system, batch mode, V = 2 L	10	128.6 TN	33–43	120.6 TP	10–13	Kong et al. (2009)
Chlorella vulgaris/ *Scenedesmus obliquus*	Municipal wastewater	Closed suspended system, semi-continuous mode, V = 150 L	13	119.3–346.6 NH_4-N	84–95	4.6–8.3 TP	92–95	Gouveia et al. (2016)
Chlorella vulgaris	Ethanol and citric acid industry wastewater	Closed suspended system, batch mode, V = 16.5 L	6	3–8 NH_4-N	71.6	1.5–3.5 PO_4-P	28	Valderrama et al. (2002)
Chlorella vulgaris	Textile industry wastewater	Open suspended system, batch mode, V = 36 L	10	6.50 NH_4-N	44.4–45.1	7.14 PO_4-P	33.1–33.3	Lim et al. (2010)
Chlorella sorokiniana	Potato-processing wastewater	Closed suspended system, batch mode, V = 5 L	30	12.1 NH_4-N	>95	3.4 PO_4-P	80.7	Hernández et al. (2013)
Klebsormidium sp./ *Stigeoclonium* spp.	Horticultural wastewater	Immobilized system, semi-continuous mode, V = 65 L	9	47.2 NO_3-N	≈100	11.6 PO_4-P	70–84	Liu et al. (2016)
Chlorella zofingiensis	Piggery wastewater	Closed suspended system, batch 23 mode, V = 1.37 L	10	148.0 TN	78.72	156.0 TP	85.00	Zhu et al. (2013)
Native microalgae	Anaerobically digested dairy industry wastewater	Immobilized system, semi-continuous mode, V = 75 L	23	78 TN	62	7 TP	70	Wilkie and Mulbry (2002)
Spirulina sp.	Anaerobically digested piggery wastewater	Open suspended system, semi-continuous mode	6–7	1209–1481 NH_4-N	84–96	164–620 PO_4-P	72–87	Olguín et al. (2003)

Note: *n.a.* not applicable, *NH_4-N* ammonium-nitrogen, *NO_3-N* nitrate-nitrogen, *PO_4-P* phosphate-phosphorus, *R* removal efficiency (in %), S_i initial concentration (in mg L^{-1}), *TN* total nitrogen, *TP* total phosphorus, *V* working volume

(2002) have evaluated nutrients recovery in an anaerobically digested dairy industry effluent using native microalgae, reporting total nitrogen (TN) and total phosphorus (TP) removal efficiencies of 62 and 70%, respectively (initial TN and TP concentrations were 78 mg L^{-1} and 7 mg L^{-1}, respectively). On the other hand, Olguín et al. (2003) and Ledda et al. (2015) have focused on nutrients removal and biomass production in anaerobically digested piggery wastewaters. In the study performed by Olguín et al. (2003), cultivation of *Spirulina* sp. in outdoor conditions and semi-continuous mode has resulted in biomass productivities between 11.8 and 15.1 g m^{-2} day^{-1}. In the same conditions, NH_4-N removal efficiencies ranged between 84 and 96% and PO_4-P removal efficiencies ranged between 72 and 87% (initial NH_4-N and PO_4-P concentrations were about 1209–1481 mg L^{-1} and 164–620 mg L^{-1}, respectively). Similarly, Ledda et al. (2015) have grown *Chlorella* sp. in an anaerobically digested piggery wastewater (with NH_4-N and PO_4-P initial concentrations of 60 mg L^{-1} and 18 mg L^{-1}, respectively), reporting biomass productivities of 0.10 g L^{-1} day^{-1} and nitrogen and phosphorus removal efficiencies of 95 and 85%, respectively.

Although microalgae have been successfully applied in nutrients removal from different wastewaters, its application on an industrial scale presents some challenges, especially regarding contaminations control. For this, the following options must be taken into account: (1) selection of fast-growing and highly resistant microalgae, such as *Chlorella* or *Scenedesmus* and (2) manipulation of the operational conditions, such as hydraulic residence times and recirculation of biomass to sustain specific microalgal populations (Benemann et al. 1980; Wood 1987; Muñoz and Guieysse 2006).

10.2.3 Bioenergy Production

Microalgae have been proposed as a potential renewable source of fuel, replacing the first- and second-generation feedstocks (e.g. food, oil crops and lignocellulosic residues). In an industrial scale, the production of biofuels using microalgae is particularly interesting because: (1) it does not compete for land with agricultural crops (thus not interfering with food production); (2) it does not require arable or fertile land for cultivation; (3) microalgae present high biomass productivities and high growth rates; and (4) microalgae have higher lipid contents and productivities compared to terrestrial crops (Rawat et al. 2013; Odjadjare et al. 2015; Khan et al. 2018; Mathimani et al. 2019). In fact, microalgae can provide feedstock for several different types of biofuels including biodiesel, bio-oil, biosyngas and biohydrogen. Nowadays, the most attractive application and the one with higher research efforts is biodiesel, since microalgae have the potential to completely displace fossil diesel (Chisti 2007). The average oil content in microalgae varies from 20 to 50% (Table 10.4), but 85% can be achieved with some species (Chisti 2007; Rawat et al. 2013). Different microalgal species produce many different types of lipids and, although the majority are suitable for biodiesel production, others are not

Table 10.4 Oil content of different microalgae

Microalgae	Oil content (% dry wt.)	References
Botryococcus braunii	25–86	Chisti (2007), Dayananda et al. (2007)
Chlorella protothecoides	55	Xu et al. (2006)
Chlorella sp.	28–35	Chisti (2007), Andruleviciute et al. (2014)
Crypthecodinium cohnii	20	Chisti (2007)
Cylindrotheca sp.	16–37	Chisti (2007)
Dunaliella primolecta	23	Chisti (2007)
Haematococcus sp.	23	Andruleviciute et al. (2014)
Isochrysis sp.	25–33	Chisti (2007)
Nannochloris sp.	20–35	Chisti (2007)
Nannochloropsis sp.	31–68	Chisti (2007)
Neochloris oleoabundans	29–65	Maity et al. (2014)
Nitzschia sp.	45–47	Chisti (2007)
Phaeodactylum tricornutum	20–30	Chisti (2007)
Scenedesmus sp.	67	Andruleviciute et al. (2014)
Schizochytrium sp.	50–77	Chisti (2007)
Tetraselmis suecica	15–23	Chisti (2007)

satisfactory. *Chlorella* sp., *Chlorococcum* sp., *Haematococcus pluvialis* and *Neochloris oleoabundans* were considered a good option for biodiesel production (Maity et al. 2014).

The production of biodiesel requires the extraction of the lipids and fatty acids from the microalgal biomass. This step is usually done in lyophilized biomass using a solvent, such as hexane, ethanol (96%), or a mixture of the previous ones. Then, the algal oil is submitted to a process known as transesterification (Odjadjare et al. 2015; Show et al. 2017; Bhalamurugan et al. 2018). This process is a multiple step reaction where the triglycerides react with alcohol (usually methanol) in the presence of a catalyst, producing fatty acid methyl esters (FAME) or biodiesel and glycerol (Chisti 2007). The resulting biodiesel has very similar characteristics to the conventional diesel fuel (e.g. flash point, kinematic viscosity and higher heating value—HHV) (Meher et al. 2006; Raheem et al. 2018). Miao and Wu (2006) evaluated the effect of the catalyst concentration, the molar ratio methanol:oil and temperature on the quality of the biodiesel produced from the microalga *Chlorella protothecoides*. Microalgal cultivation was performed at 26 °C under a continuous illumination intensity of 40 $\mu mol\ m^{-2}\ s^{-1}$. The aeration was provided through the injection of air bubbles. *C. protothecoides* was cultivated under autotrophic and heterotrophic conditions, with the latter using glucose as a substrate. The experiments used sulphuric acid (H_2SO_4) as a catalyst and different concentrations were tested (25, 50, 60 and 100%). Additionally, different methanol:oil molar ratios (25:1, 30:1, 45:1, 56:1, 70:1 and 84:1) and different temperatures (30, 50 and 90 °C) were tested. The authors concluded that heterotrophic growth of *C. protothecoides* resulted in higher lipid content (55.2%) and that higher biodiesel quality can be obtained in the presence of

100% catalyst at the highest evaluated temperature (90 °C). On the economic point of view, the authors concluded that the best factor combination was a catalyst concentration of 100% with molar ratio of 56:1 at 30 °C. Xu et al. (2006) evaluated the potential to produce high-quality biodiesel from the microalga *C. protothecoides*. The microalga was heterotrophically cultivated in a 5-L fermenter tank where corn powder hydrolysate was used as substrate. The cultivation temperature was controlled at 28 °C and the aeration rate and the agitation speed were set at 0.5 volume per volume liquid per minute (vvm) and 300 rotations per minute (rpm), respectively. The biodiesel was obtained through acidic transesterification with 56:1 molar ratio of methanol:oil at 30 °C. The authors obtained a lipid content of 46.1% and concluded that the proprieties of the biodiesel from microalgal oil were comparable to the ones of diesel fuel, with a HHV of 41 MJ kg^{-1}, a density of 0.864 kg L^{-1} and a viscosity of 5.2×10^{-4} Pa s (at 40 °C). Johnson and Wen (2009) evaluated the potential of producing a high-quality biodiesel from the microalga *Schizochytrium limacinum*. The biodiesel was produced through two methods: (1) oil solvent extraction followed by transesterification; and (2) direct transesterification. For the transesterification, a mixture of methanol (3.4 mL), H_2SO_4 (0.6 mL) and a solvent (chloroform, hexane or petroleum ether—4.0 mL) was heated at 90 °C for 40 min. For each process, wet and dry biomass was used as feedstock. The authors concluded that when the dry biomass was used, the two-stage process led to a biodiesel yield of 57% and a FAME content of 66.4%. The one-stage process resulted in higher biodiesel yield; however, higher FAME content (63.5%) was only obtained when chloroform was used as a solvent. The authors concluded that the direct transesterification using dry biomass of the microalga *S. limanicum* is suitable to produce good-quality biodiesel.

10.3 Interactions and Benefits of Using Microalgal Consortia

Although microalgae have been successfully applied in several environmental applications, maintaining microalgal monocultures can be a hard task to achieve (Padmaperuma et al. 2018). To overcome this problem, several studies have exploited the potential of microalgal consortia in these applications, reporting several advantages over single-species cultures (Wilkie and Mulbry 2002; Muñoz and Guieysse 2006; González-Fernández and Ballesteros 2012; Subashchandrabose et al. 2011; He et al. 2013). These cultures can result in the development of robust systems able to resist to stress conditions, thus promoting effective degradation processes and improved biomass and bioenergy productivities (Paerl and Pinckney 1996; Subashchandrabose et al. 2011; Gonçalves et al. 2017; Nath et al. 2017a).

Microalgal consortia can naturally occur in the environment or can be artificially engineered/designed for a specific application (Jagmann and Philipp 2014; Padmaperuma et al. 2018). These consortia can be constituted exclusively by photosynthetic microorganisms (microalgal consortia) or by photosynthetic microorganisms and heterotrophic bacteria (microalgal–bacterial consortia) (Gonçalves et al.

2017). The following sections describe the main interactions established in both types of consortia (microalgal and microalgal–bacterial), with emphasis on how these interactions can improve CO_2 capture, nutrients removal and bioenergy production.

10.3.1 Microalgal Consortia

Interactions between photosynthetic microorganisms are not well documented in the literature (Qin et al. 2016). However, it is thought that both cooperative and competitive interactions can occur in these consortia (Nath et al. 2017a). Regarding cooperative interactions, metabolites' exchange between the microorganisms integrating the consortium is the most common and can be very advantageous in biomass production and hence, nutrients uptake and bioenergy production (Mendes and Vermelho 2013). As competitive interactions, several studies have referred the excretion of metabolites, also known as allelochemicals, that exhibit a negative effect towards the co-cultivated microorganisms (Cembella 2003; Gross 2003; Mendes and Vermelho 2013). For example, when growing a microalgal consortium composed of *C. vulgaris* and *Pseudokirchneriella subcapitata*, Fergola et al. (2007) have reported that *C. vulgaris* excreted a fatty acids mixture (also known as chlorellin), which was responsible for the inhibition of *P. subcapitata* growth. Allelochemicals production can be influenced by both abiotic and biotic factors. Regarding abiotic factors, nutrients starvation, low light intensities and temperatures and high pH values promote the excretion of these secondary metabolites. The influence of biotic factors is related to the concentration/predominance of microalgal species that produce these toxic compounds in the consortium (Mendes and Vermelho 2013).

In CO_2 capture, wastewater treatment and bioenergy production processes, interactions between photosynthetic microorganisms can have the following advantages: (1) enhancement of the overall biomass productivities and lipids production; (2) enhancement of the overall nutrients uptake, providing that sufficient nutrients are supplied; (3) resistance to contaminants and predators through the induction of allelochemicals production; and (4) the development of a settleable system (by combining single-cell microorganisms with flocculating ones), thus avoiding the requirements for a harvesting method and reducing biomass recovery costs (Gonçalves et al. 2017; Nath et al. 2017a, b). Additionally, the use of microalgal consortia for these purposes ensures the viability of a remediation process because the loss of one microorganism can be compensated by the other microorganisms integrating the consortia (Renuka et al. 2013).

10.3.2 Microalgal–Bacterial Consortia

Regarding microalgal–bacterial consortia, several authors have reported that both cooperative and competitive interactions can occur (Muñoz and Guieysse 2006;

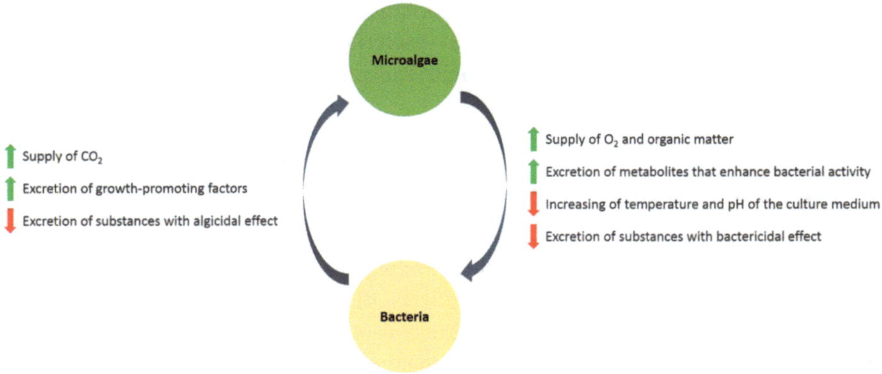

Fig. 10.2 Schematic representation of the cooperative end competitive interactions that can occur in microalgal–bacterial consortia [adapted from Gonçalves et al. (2017)]

Natrah et al. 2014; Unnithan et al. 2014; Solimeno and García 2017). Figure 10.2 presents the main cooperative and competitive interactions already described for microalgal–bacterial consortia.

In competitive interactions, both microalgae and bacteria can have adverse effects on each other: microalgae can excrete secondary metabolites presenting bactericidal effect (Pratt et al. 1944; Kellam and Walker 1989; Najdenski et al. 2013; Natrah et al. 2014) and bacteria, in turn, can excrete metabolites with algicidal activity (Natrah et al. 2014). For example, chlorellin produced by *Chlorella* presents bactericidal activity against Gram-positive and Gram-negative bacteria, such as *Staphylococcus aureus*, *Bacillus subtilis*, *Escherichia coli* and *Pseudomonas aeruginosa* (Pratt et al. 1944). Besides metabolites' excretion with antibacterial activity, the increase in pH due to photosynthetic growth inhibits bacterial growth (Muñoz and Guieysse 2006; Unnithan et al. 2014; Gonçalves et al. 2017).

As cooperative interactions between microalgae and bacteria, it is possible to refer nutrients' exchange between these microorganisms: (1) microalgae supply bacteria with organic compounds that can be used as carbon and energy source and with O_2, which is required for the oxidation of organic matter; and (2) bacteria provide microalgae with the CO_2 required for photosynthetic activity (Paerl and Pinckney 1996; Bordel et al. 2009; Godos et al. 2009; Fouilland 2012; Nath et al. 2017a). Besides cooperative interactions through nutrients' exchange, other symbiotic relationships can occur between these microorganisms. Microalgae can enhance bacterial growth through the release of extracellular metabolites, such as extracellular polymeric substances, and can serve as a habitat for bacteria, protecting them from adverse environmental conditions (Unnithan et al. 2014). Mandal et al. (2011) have demonstrated that extracellular polymeric substances produced by the microalga *Amphidinium carterae* have stimulated the growth of the bacterium *Bacillus pumilus*. On the other hand, bacteria can excrete growth-promoting factors, such as vitamins and siderophores (chelating agents for microalgal growth under iron-limiting conditions), which can promote microalgal growth (Subashchandrabose et al. 2011;

Gonçalves et al. 2017). For example, in the study performed by De-Bashan et al. (2004), co-cultivation of *Azospirillum brasilense* with *C. vulgaris* and *C. sorokiniana* in alginate beads enhanced microalgal growth and improved nitrogen and phosphorus removal from a municipal wastewater used as culture medium.

Apart from being effective in nutrients removal, these systems can further improve current wastewater treatment processes because (1) the costs associated with the oxygenation of activated sludge tanks can be significantly reduced and (2) the greenhouse effects associated with wastewater treatment plants can be considered negligible, since the CO_2 released by bacteria is converted into organic matter by microalgae (Godos et al. 2009; Quijano et al. 2017). Due to the symbiotic interactions that can occur and consequent increase in biomass productivities, the use of these consortia can also improve fatty acids productivities and, hence, bioenergy production.

10.4 Applications of Microalgal Consortia

10.4.1 CO_2 Capture

Although microalgal consortia have been mostly applied in nutrients removal processes, some authors have also reported their application in CO_2 capture. Boonma et al. (2014) have cultivated a microalgal consortium (composed of 65.7% *Scenedesmus* spp., 25.4% *Micractinium* sp., 3.6% *Dictyosphaerium* sp., 2.7% *Pseudanabaena* sp., 0.8% *Monoraphidium* sp., 1% *Chlamydomonas* sp., 0.4% *Chlorella* sp. and 0.4% *Euglena* sp.) using different CO_2 concentrations (0.04, 10 and 30% v/v), achieving the highest CO_2 fixation rates in the cultures supplemented with 30% (v/v) CO_2: 0.0271 g CO_2 L^{-1} day^{-1}. When growing a *Chlorella* sp./*Scenedesmus* sp. consortium in a primary-treated municipal wastewater, Koreivienė et al. (2014) have demonstrated an accumulation of CO_2 in microalgal biomass ranging between 0.65 and 1.37 g CO_2 L^{-1} day^{-1}. Bhakta et al. (2015) aimed to isolate a highly CO_2-tolerant microalgal consortium and evaluate its potential on CO_2 fixation. The isolated consortium was mainly composed of *Chlorella* sp., *Scenedesmus* sp., *Sphaerocystis* sp. and *Spirulina* sp. and it was able to grow in CO_2 concentrations up to 50% (v/v). CO_2 removal efficiencies of the isolated consortium oscillated between 53 and 100%, which corresponds to 150–291 mg g^{-1}. Besides being effective in CO_2 uptake, this consortium was also effective in nutrients (nitrogen and phosphorus) removal and lipids production. Nath et al. (2017b) evaluated biomass productivities and carbonic anhydrase activity (the enzyme responsible for the active transport of CO_2 into microalgal cells) of two microalgal consortia (A and B) grown in different culture media: BG-11 medium and Zarrouk medium. Consortium A was composed of *Synechococcus* PCC 7942, *Chlorella* sp., *Nostoc muscorum*, *Oscillatoria* sp. and *Spirulina platensis*, whereas consortium B was composed of *Synechocystis* PCC6803, *Scenedesmus dimorphus*, *Anabaena cylindrica*, *Lyngbya* sp. and *S. platensis*. When comparing microalgal biomass productivities obtained in these consortia with biomass productivities obtained for monocultures of each microorganism present in the consortia, the authors have concluded that microalgal growth in the consortia was higher. Additionally, the authors have determined higher carbonic anhydrase activity in microalgal consortia, which results in a better CO_2 fixation in these cultures.

10.4.2 Nutrients Removal (Wastewater Polishing)

The use of microalgal consortia for wastewater polishing (nitrogen and phosphorus removal) using wastewaters from different sources is well documented in the literature. When using a native microalgal consortium from a carpet mill industry effluent, Chinnasamy et al. (2010) have reported an almost complete removal of NO_3-N and PO_4-P, with removal efficiencies ranging between 96.6 and 99.8%. In the study performed by Koreivienė et al. (2014), a non-native consortium composed of *Chlorella* sp. and *Scenedesmus* sp. was effectively applied in nitrogen and phosphorus removal from a primary-treated municipal wastewater. In this study, the authors have reported TN and TP removal efficiencies ranging between 88.6–96.4% and 99.7–99.9%, respectively. Gonçalves et al. (2016a) have evaluated nitrogen and phosphorus removal efficiencies of three artificially designed microalgal consortia: (1) *Synechocystis salina* + *C. vulgaris*; (2) *S. salina* + *P. subcapitata*; and (3) *S. salina* + *Microcystis aeruginosa*. In this study, the authors have grown single and dual-species cultures using a synthetic medium containing initial NO_3-N and PO_4-P concentrations of approximately 45 mg N L^{-1} and 10 mg P L^{-1}, respectively. Results from this study have revealed that microalgal growth in the consortia have improved both nitrogen and phosphorus removal efficiencies. NO_3-N and PO_4-P removal efficiencies obtained for single cultures were 52–71% and 50–81%, respectively, whereas NO_3-N and PO_4-P removal efficiencies determined for the consortia ranged between 85–78% and 86–97%, respectively.

Microalgal–bacterial consortia have also been successfully applied in nitrogen, phosphorus and organic matter removal from different wastewater sources (Solimeno and García 2017; Quijano et al. 2017). When growing a microalgal–bacterial consortium composed of *C. vulgaris* and primary-treated municipal wastewater native bacteria in tubular PBRs processing a primary-treated municipal wastewater (NH_4-N and TP initial concentrations of 17–207 mg L^{-1} and 1.4–19.5 mg L^{-1}, respectively), He et al. (2013) have demonstrated removal efficiencies ranging between 30.9 and 100% for nitrogen and between 65 and 98% for phosphorus. Alcántara et al. (2015) have used a closed tank to evaluate the performance of a microalgal consortium from a HRAP treating diluted vinasse and activated sludge native bacteria in nitrogen and organic matter removal from a synthetic wastewater containing 120 mg L^{-1} of NH_4-N and 200 mg L^{-1} of total organic carbon (TOC). These authors have reported NH_4-N removal efficiencies ranging between 75 and 96% and TOC removal efficiencies ranging between 86 and 90%. Gonçalves et al. (2016b) have evaluated the potential of dual-species cultures of the microalga *C. vulgaris* and a bacterium isolated from a municipal wastewater treatment plant (*Enterobacter asburiae*, *Klebsiella* sp. or *Raoultella ornithinolytica*) in biomass production and wastewater polishing from a synthetic medium that mimics a secondary-treated effluent (nitrogen and phosphorus initial loads of about 45 mg N L^{-1} and 10 mg P L^{-1}, respectively). The authors have concluded that when growing in consortium with *E. asburiae* and *R. ornithinolytica*, *C. vulgaris* growth has significantly increased. Additionally, the three studied consortia have also contributed to higher nutrients removal, since the time required to achieve the limits in discharged effluents (established by European Union legislation) was reduced to at least half of the value determined for the single *C. vulgaris* culture. When comparing ammonium removal from artificial wastewater using a microalgal consortium composed of *Anabena variabilis*, *Chlorella* sp., *Chlorococcus* sp. and *Spirulina* sp. and microalgal–bacterial consortium composed of the same microalgal species and mixed liquor activated sludge from the Harnaschpolder wastewater treatment plant (Delft, The Netherlands), Rada-Ariza et al. (2017) have concluded that the microalgal–bacterial consortium removed ammonium at higher rates (100 mg L^{-1} day^{-1}) than the microalgal consortium (44 mg L^{-1} day^{-1}). More recently, Foladori et al. (2018) have evaluated and optimized total nitrogen removal from municipal wastewater using a mixed microalgal–bacterial consortium spontaneously acclimatized to real wastewater in a photo-sequencing batch reactor. With this study, the authors have reported high chemical oxygen demand (COD) and total Kjeldahl nitrogen (TKN) removal efficiencies: 86 and 97%, respectively. To avoid the requirements for further harvesting of microalgal biomass, some studies have reported the use of these consortia in immobilized growth systems, such as immobilization in solid carriers and biofilms. De-Bashan et al. (2004) have used an immobilized culture of *Chlorella* sp. and *A. brasilense* in alginate beads to treat a municipal wastewater with the following composition (in mg L^{-1}): 0.1–4.3 NH_4-N, 4–5.2 NO_3-N and 4.1 PO_4-P. This study has resulted in an effective removal of nitrogen (both NH_4-N and NO_3-N) and phosphorus, with removal efficiencies ranging between 92 and 100%. When growing *C. sorokiniana* and activated sludge native bacteria in a tubular biofilm PBR treating primary-treated piggery wastewater containing 656 mg L^{-1} of NH_4-N, 117 mg L^{-1} of PO_4-P and 1247 mg L^{-1} of TOC, Godos et al. (2009) have reported the following removal efficiencies: 94–100% for NH_4-N, 70–90% for PO_4-P and 45% for TOC. Posadas et al. (2013) have used a biofilm reactor to promote primary-treated domestic wastewater treatment (TN, PO_4-P and TOC concentrations of 91 mg L^{-1}, 7 mg L^{-1} and 181 mg L^{-1}, respectively) by a centrate wastewater native microalgal–bacterial consortium. This study has revealed TN, PO_4-P and TOC removal efficiencies of 70%, 85% and 90%, respectively. Miranda et al. (2017) have isolated and characterized natural microalgal biofilms from freshwater, saline lakes and marine habitats around Melbourne, Australia. These consortia were composed of several microorganisms, such as cyanobacteria, microalgae, diatoms, bacteria and fungi. From the studied biofilms, Biofilm #52 (composed of two filamentous cyanobacteria, clustered with *Spirulina* and *Oscillatoria* species, one unicellular microalgae, clustered with some *Chlorella* species, and two diatoms, identified as *Nitzschia* sp., and *Fistulifera* sp.) was considered the most promising in terms of nutrients removal from a selenium-rich synthetic wastewater and bioenergy production. Regarding nutrients removal, after 3 days of treatment, the selected biofilm was able to uptake 24% of NH_4-N, 26% of NO_3-N and 17% uptake of PO_4-P from the synthetic wastewater. To reduce time and costs associated with microalgal harvesting, some authors have also described the use of artificial consortia consisting of flocculating microorganisms. Van Den Hende et al. (2011) have reported the use of microalgal–bacterial flocs, mainly composed of *Chlorella* sp., *Pediastrum* sp., *Phormidium* sp., *Scenedesmus* sp. and activated sludge native bacteria, to treat a primary-treated municipal wastewater. The results obtained in this study have shown that these flocs were able to remove 61.2% of TN and 30.2–56.8% of PO_4-P. More recently, Arcila and Buitrón (2017) have evaluated the influence of the solar irradiance level on the formation of microalgal–bacterial aggregates, settling velocity and nutrients removal from a municipal wastewater. With this study, the authors have concluded that the highest irradiance level evaluated (6213 W h m^{-2} day^{-1}) has resulted in a poor wastewater treatment performance, with TN and COD removal efficiencies of 36 and 50%, respectively. In contrast, PO_4-P removal efficiencies obtained in these conditions were 92%. Additionally, these conditions have resulted in low settling velocities and settleability, associated with a poor rate of aggregates formation. On the other hand, low irradiance levels (<3800 W h m^{-2} day^{-1}) have contributed to the formation of microalgal–bacterial aggregates with high settling velocity and settleability and to the increase of TN and COD removal efficiencies: in these conditions, TN and COD removal efficiencies were 60 and 89%, respectively.

10.4.3 Bioenergy Production

In terms of bioenergy production, the potential of both microalgal and microalgal–bacterial consortia has already been evaluated.

Regarding microalgal consortia, Chinnasamy et al. (2010) have evaluated the potential of a microalgal consortium composed of 15 native species from a carpet industry effluent on bioenergy production. In this study, the authors have reported lipid productivities of 6.82% of dry weight. Additionally, the authors have concluded that about 63.9% of algal oil obtained from the consortium could be converted into biodiesel. When evaluating the potential of a microalgal consortium in CO_2 capture using different CO_2 concentrations in the air stream, Boonma et al. (2014) have also studied lipids productivity. Maximum productivity was achieved for a CO_2 concentration of 30% (v/v) and the value obtained was 4.8 mg L^{-1} day^{-1} (approximately 27.6% of dry weight). In the study performed by Qin et al. (2016), four microalgal consortia and respective single cultures were grown in a dairy industry wastewater to assess their potential in both nutrients removal and biomass and bioenergy production. The consortia evaluated in this study were the following: (1) *Chlorella* sp./*Chlorella zofingiensis* (1:1); (2) *Scenedesmus* spp./*C. zofingiensis* (1:1); (3) *Chlorella* sp./*Scenedesmus* spp. (1:1); and (4) *Chlorella* sp./*Scenedesmus* spp./*C. zofingiensis* (1:1:1). Bioenergy production was assessed in terms of lipids productivity, which has shown to be higher in the consortia rather than in each single culture (143.7–150.6 mg L^{-1} day^{-1}). Furthermore, FAME profiles indicated that the lipids produced from microalgal consortia were more suitable for biodiesel production than those produced by single cultured microalgae. More recently, nine native microalgal consortia isolated from rural wastewaters were evaluated in terms of biomass productivities for bioenergy production (Choudhary et al. 2016). Through biochemical characterization of the biomass resulting from the best performing consortium, the authors estimated a lipid content of 31% of dry weight and a theoretical methane potential of 0.79 m^3 kg^{-1} volatile solids, which suggests a good potential of this consortium for biogas generation. In the study performed by Gonçalves et al. (2016a), where single- and dual-species cultures of *S. salina* and *C. vulgaris*, *P. subcapitata* and *M. aeruginosa* were grown in a synthetic medium to evaluate their potential in nutrients removal, fatty acids concentration was also determined. This study has demonstrated higher lipid contents in biomass resulting from microalgal consortia, especially in the consortia *S. salina* + *P. subcapitata* and *S. salina* + *M. aeruginosa*.

Microalgal–bacterial consortia performance in terms of lipids productivity was also studied. For example, in the study performed by Zhao et al. (2014), lipid contents determined in a consortium composed of *Chlorella pyrenoidosa* and native bacteria from a landfill leachate ranged between 14.5 and 20.8% of dry weight, with maximum lipid productivity achieved being 24.1 mg L^{-1} day^{-1}. In the study performed by Miranda et al. (2017), the authors have studied several microalgal–bacterial biofilms, selecting one of them (Biofilm #52) as the most promising in terms of feedstock for bioenergy production, as it was possible to modify the levels and compositions of saturated, monosaturated and polyunsaturated fatty acids through the promotion of the growth of selected individual photosynthetic inhabitants in the consortium.

High lipid contents were also determined in a mixed culture composed of *C. pyrenoidosa* and a yeast (*Rhodotorula mucilaginosa*), both isolated in Nuevo Leon, Mexico (Reyna-Martínez et al. 2015). In this study, the authors have reported lipids contents of about 20% of dry weight, with 94.6% corresponding to triglycerides composed of fatty acid chains between 16 and 18 carbons.

10.5 Research Needs

Although several studies have successfully applied microalgal and microalgal–bacterial consortia in CO_2 capture, nutrients removal and bioenergy production, further research is required for the optimization of culturing parameters, especially for large-scale applications. Additionally, interactions between photosynthetic microorganisms are not fully understood.

Taking into account large-scale applications and the necessity to clearly understand microorganisms' interactions in the consortia, research should be focused on the following topics (Bordel et al. 2009; Fouilland 2012; Padmaperuma et al. 2018):

(1) study of the influence of different environmental conditions (such as light, nutrients availability, pH and temperature) on the consortia behaviour; (2) large-scale outdoor experiments with real environmental conditions; (3) complete understanding of the interactions established between the microorganisms integrating the consortia; and (4) development of reliable mathematical models that correctly describe the behaviour of these consortia, which may be very helpful in process design and operational conditions determination.

To further improve the effectiveness of microalgal consortia in the environmental applications referred in this study, engineering of microalgal consortia should be taken into account (Padmaperuma et al. 2018). For example, microalgal consortia combining microorganisms with different metabolic capacities, specific metal binding abilities, affinities for different nitrogen sources, different affinities to increased CO_2 concentrations and with different lipid contents should be studied. This step will allow the re/introduction or elimination of microorganisms as needed and the complete monitoring of the tasks developed within the consortia.

10.6 Conclusions

The use of microalgae in environmental applications, such as CO_2 capture, nutrients uptake and bioenergy production, has been extensively reported in the literature, with several success cases. Regarding microalgal consortia, various studies have referred their potential in the same applications, especially due to the symbiotic/cooperative interactions that can occur in these combined systems. However, due to the huge number of possible combinations that can be engineered or can naturally occur, the use of microalgal consortia in the referred applications is still under research. The majority of studies have focused on the use of microalgal consortia in nutrients removal from wastewaters, but only a few studies have evaluated their potential in CO_2 capture and bioenergy production.

This chapter provided an overview of current status of research on the use of microalgal consortia, focusing on the main studies conducted regarding CO_2 capture, nutrients removal and bioenergy production. As main conclusion, it is possible to refer that both microalgal and microalgal–bacterial consortia can be effectively applied in wastewater polishing processes. Although further studies are required, it is also possible to conclude that microalgal consortia may be very promising in CO_2 capture and that microalgal and microalgal–bacterial consortia can be an important source of lipids for bioenergy production.

Acknowledgements This work was financially supported by (1) Project UID/EQU/00511/2019—Laboratory for Process Engineering, Environment, Biotechnology and Energy—LEPABE funded by national funds through FCT/MCTES (PIDDAC) and (2) Project POCI-01-0145-FEDER-031736—PIV4Algae—Process Intensification for microalgal production and Valorisation, funded by FEDER funds through COMPETE2020—Programa Operacional Competitividade e Internacionalização (POCI) and by national funds (PIDDAC) through FCT/MCTES. J.C.M. Pires acknowledges the FCT Investigator 2015 Programme (IF/01341/2015).

References

Alcántara C, Domínguez JM, García D et al (2015) Evaluation of wastewater treatment in a novel anoxic-aerobic algal-bacterial photobioreactor with biomass recycling through carbon and nitrogen mass balances. Bioresour Technol 191:173–186

Allen MR, Frame DJ, Huntingford C et al (2009) Warming caused by cumulative carbon emissions towards the trillionth tonne. Nature 458:1163–1166

Andruleviciute V, Makareviciene V, Skorupskaite V et al (2014) Biomass and oil content of *Chlorella* sp., *Haematococcus* sp., *Nannochloris* sp. and *Scenedesmus* sp. under mixotrophic growth conditions in the presence of technical glycerol. J Appl Phycol 26:83–90

Anjos M, Fernandes BD, Vicente AA et al (2013) Optimization of CO_2 bio-mitigation by *Chlorella vulgaris*. Bioresour Technol 139:149–154

Arbib Z, Ruiz J, Álvarez-Díaz P et al (2014) Capability of different microalgae species for phytoremediation processes: wastewater tertiary treatment, CO_2 bio-fixation and low cost biofuels production. Water Res 49:465–474

Arcila JS, Buitrón G (2017) Influence of solar irradiance levels on the formation of microalgae-bacteria aggregates for municipal wastewater treatment. Algal Res 27:190–197

Aslan S, Kapdan IK (2006) Batch kinetics of nitrogen and phosphorus removal from synthetic wastewater by algae. Ecol Eng 28:64–70

Benemann JR, Koopman B, Weissman J et al (1980) Development of microalgae harvesting and high rate pond technologies in California. In: Shelef G, Soeder CJ (eds) Algae biomass production and use. Elsevier/North-Holland Biomedical Press, Amsterdam, pp 457–495

Bhakta JN, Lahiri S, Pittman JK et al (2015) Carbon dioxide sequestration in wastewater by a consortium of elevated carbon dioxide-tolerant microalgae. J CO_2 Util 10:105–112

Bhalamurugan GL, Valerie O, Mark L (2018) Valuable bioproducts obtained from microalgal biomass and their commercial applications: a review. Environ Eng Res 23:229–241

Boonma S, Chaiklangmuang S, Chaiwongsar S et al (2014) Enhanced carbon dioxide fixation and bio-oil production of a microalgal consortium. Clean Soil Air Water 42:1–6

Bordel S, Guieysse B, Muñoz R (2009) Mechanistic model for the reclamation of industrial wastewaters using algal-bacterial photobioreactors. Environ Sci Technol 43:3200–3207

Brennan L, Owende P (2010) Biofuels from microalgae – a review of technologies for production, processing, and extractions of biofuels and co-products. Renew Sust Energy Rev 14:557–577

Cembella AD (2003) Chemical ecology of eukaryotic microalgae in marine ecosystems. Phycologia 42:420–447

Cheah WY, Show PL, Chang J-S et al (2015) Biosequestration of atmospheric CO_2 and flue gas-containing CO_2 by microalgae. Bioresour Technol 184:190–201

Cheng L, Zhang L, Chen H et al (2006) Carbon dioxide removal from air by microalgae cultured in a membrane-photobioreactor. Sep Pur Technol 50:324–329

Chiang C-L, Lee C-M, Chen P-C (2011) Utilization of the cyanobacteria *Anabaena* sp. CH1 in biological carbon dioxide mitigation processes. Bioresour Technol 102:5400–5405

Chinnasamy S, Bhatnagar A, Hunt RW et al (2010) Microalgae cultivation in a wastewater dominated by carpet mill effluents for biofuel applications. Bioresour Technol 101:3097–3105

Chisti Y (2007) Biodiesel from microalgae. Biotechnol Adv 25:294–306

Chojnacka K, Chojnacki A, Górecka H (2004) Trace element removal by *Spirulina* sp. from copper smelter and refinery effluents. Hydrometallurgy 73:147–153

Choudhary P, Prajapati SK, Malik A (2016) Screening native microalgal consortia for biomass production and nutrient removal from rural wastewaters for bioenergy applications. Ecol Eng 91:221–230

Costa JAV, Henrard ASA, Moraes L et al (2017) Use of flue gas as a carbon source. In: Pires JCM (ed) Recent advances in renewable energy, microalgae as a source of bioenergy: products, processes and economics, vol 1. Bentham Science, Sharjah, pp 173–201

Dayananda C, Sarada R, Rani MU et al (2007) Autotrophic cultivation of *Botryococcus braunii* for the production of hydrocarbons and exopolysaccharides in various media. Biomass Bioenergy 31:87–93

De-Bashan LE, Hernandez J-P, Morey T et al (2004) Microalgae growth-promoting bacteria as "helpers" for microalgae: a novel approach for removing ammonium and phosphorus from municipal wastewater. Water Res 38:466–474

Demirbas A (2011) Biodiesel from oilgae, biofixation of carbon dioxide by microalgae: a solution to pollution problems. Appl Energy 88:3541–3547

Doucha J, Straka F, Lívanský K (2005) Utilization of flue gas for cultivation of microalgae (*Chlorella* sp.) in an outdoor open thin-layer photobioreactor. J Appl Phycol 17:403–412

Fergola P, Cerasuolo M, Pollio A et al (2007) Allelopathy and competition between *Chlorella vulgaris* and *Pseudokirchneriella subcapitata*: experiments and mathematical model. Ecol Model 208:205–214

Foess GW, Steinbrecher P, Williams K et al (1998) Cost and performance evaluation of BNR processes. Flo Water Res J 11–13

Foladori P, Petrini S, Nessenzia M et al (2018) Enhanced nitrogen removal and energy saving in a microalgal-bacterial consortium treating real municipal wastewater. Water Sci Technol 78:174–182

Fouilland E (2012) Biodiversity as a tool for waste phycoremediation and biomass production. Rev Environ Sci Biotechnol 11:1–4

Godos I, González C, Becares E et al (2009) Simultaneous nutrients and carbon removal during pretreated swine slurry degradation in a tubular biofilm photobioreactor. Appl Microbiol Biotechnol 82:187–194

Gonçalves AL, Pires JCM, Simões M (2016a) Biotechnological potential of *Synechocystis salina* co-cultures with selected microalgae and cyanobacteria: nutrients removal, biomass and lipid production. Bioresour Technol 200:279–286

Gonçalves AL, Pires JCM, Simões M (2016b) Wastewater polishing by consortia of *Chlorella vulgaris* and activated sludge native bacteria. J Clean Prod 133:348–357

Gonçalves AL, Pires JCM, Simões M (2017) A review on the use of microalgal consortia for wastewater treatment. Algal Res 24:403–415

González-Fernández C, Ballesteros M (2012) Microalgae autoflocculation: an alternative to high-energy consuming harvesting methods. J Appl Phycol 25:991–999

Gouveia L, Graça S, Sousa C et al (2016) Microalgae biomass production using wastewater: treatment and costs: scale-up considerations. Algal Res 16:167–176

Gross EM (2003) Allelopathy of aquatic autotrophs. Crit Rev Plant Sci 22:313–339

He P, Mao B, Lü F et al (2013) The combined effect of bacteria and *Chlorella vulgaris* on the treatment of municipal wastewaters. Bioresour Technol 146:562–568

Hernández D, Riaño B, Coca M et al (2013) Treatment of agro-industrial wastewater using microalgae-bacteria consortium combined with anaerobic digestion of the produced biomass. Bioresour Technol 135:598–603

Ho S-H, Chen W-M, Chang J-S (2010) *Scenedesmus obliquus* CNW-N as a potential candidate for CO_2 mitigation and biodiesel production. Bioresour Technol 101:8725–8730

Ho S-H, Chen C-Y, Lee D-J et al (2011) Perspectives on microalgal CO_2-emission mitigation systems – a review. Biotechnol Adv 29:189–198

Hu Q (2004) Industrial production of microalgal cell-mass and secondary products – major industrial species. In: Richmond A (ed) Handbook of microalgal culture: biotechnology and applied phycology. Blackwell Science, Oxford, pp 268–271

Hulatt CJ, Thomas DN (2011) Productivity, carbon dioxide uptake and net energy return of microalgal bubble column photobioreactors. Bioresour Technol 102:5775–5787

Jagmann N, Philipp P (2014) Reprint of design of synthetic microbial communities for biotechnological production processes. J Biotechnol 192:293–301

Jeyanayagam S (2005) True confessions of the biological nutrient removal process. Flo Water Res J 1:37–46

John RP, Anisha GS, Nampoothiri KM et al (2011) Micro and macroalgal biomass: a renewable source for bioethanol. Bioresour Technol 102:186–193

Johnson MB, Wen Z (2009) Production of biodiesel fuel from the microalga *Schizochytrium limacinum* by direct transesterification of algal biomass. Energy Fuel 23:5179–5183

Kellam SJ, Walker JM (1989) Antibacterial activity from marine microalgae in laboratory culture. Brit Phycol J 24:191–194

Khan MI, Shin JH, Kim JD (2018) The promising future of microalgae: current status, challenges, and optimization of a sustainable and renewable industry for biofuels, feed, and other products. Microb Cell Factories 17:36–56

Kong Q-X, Li L, Martinez B et al (2009) Culture of microalgae *Chlamydomonas reinhardtii* in wastewater for biomass feedstock production. Appl Biochem Biotechnol 160:9–18

Koreivienė J, Valčiukas R, Karosienė J et al (2014) Testing of *Chlorella*/*Scenedesmus* microalgae consortia for remediation of wastewater, CO_2 mitigation and algae biomass feasibility for lipid production. J Environ Eng Landscape Manage 22:105–114

Kumar K, Banerjee D, Das D (2014) Carbon dioxide sequestration from industrial flue gas by *Chlorella sorokiniana*. Bioresour Technol 152:225–233

Lam MK, Lee KT, Mohamed AR (2012) Current status and challenges on microalgae-based carbon capture. Int J Greenhouse Gas Cont 10:456–469

Larsdotter K (2006) Wastewater treatment with microalgae – a literature review. Vatten 62:31–38

Ledda C, Idà A, Allemand D et al (2015) Production of wild *Chlorella* sp. cultivated in digested and membrane-pretreated swine manure derived from a full-scale operation plant. Algal Res 12:68–73

Li F-F, Yang Z-H, Zeng R, Yang G, Chang X, Yan J-B, Hou Y-L (2011) Microalgae capture of CO_2 from actual flue gas discharged from a combustion chamber. Ind Eng Chem Res 50:6496–6502

Lim S-L, Chu W-L, Phang S-M (2010) Use of *Chlorella vulgaris* for bioremediation of textile wastewater. Bioresour Technol 101:7314–7322

Liu J, Danneels B, Vanormelingen P et al (2016) Nutrient removal from horticultural wastewater by benthic filamentous algae *Klebsormidium* sp., *Stigeoclonium* spp. and their communities: from laboratory flask to outdoor Algal Turf Scrubber (ATS). Water Res 92:61–68

López CVG, Fernández FGA, Sevilla JMF et al (2009) Utilization of the cyanobacteria *Anabaena* sp. ATCC 33047 in CO_2 removal processes. Bioresour Technol 100:5904–5910

Maity JP, Bundschuh J, Chen C-Y et al (2014) Microalgae for third generation biofuel production, mitigation of greenhouse gas emissions and wastewater treatment: present and future perspectives – a mini review. Energy 78:104–113

Mandal SK, Singh RP, Patel V (2011) Isolation and characterization of exopolysaccharide secreted by a toxic dinoflagellate, *Amphidinium carterae* Hulburt 1957 and its probable role in harmful algal blooms (HABs). Microbial Ecol 62:518–527

Mathimani T, Baldinelli A, Rajendran K et al (2019) Review on cultivation and thermochemical conversion of microalgae to fuels and chemicals: process evaluation and knowledge gaps. J Clean Prod 208:1053–1064

McGinn PJ, Dickinson KE Bhatti S et al (2011) Integration of microalgae cultivation with industrial waste remediation for biofuel and bioenergy production: opportunities and limitations. Photosyn Res 109:231–247

Meher LC, Vidya SD, Naik SN (2006) Technical aspects of biodiesel production by transesterification – a review. Renew Sust Energy Rev 10:248–268

Mendes LBB, Vermelho AB (2013) Allelopathy as a potential strategy to improve microalgae cultivation. Biotechnol Biofuels 6:152–165

Miao X, Wu Q (2006) Biodiesel production from heterotrophic microalgal oil. Bioresour Technol 97:841–846

Miranda AF, Ramkumar N, Andriotis C et al (2017) Applications of microalgal biofilms for wastewater treatment and bioenergy production. Biotechnol Biofuels 10:120–142

Morais MG, Costa JAV (2007a) Biofixation of carbon dioxide by *Spirulina* sp. and *Scenedesmus obliquus* cultivated in a three-stage serial tubular photobioreactor. J Biotechnol 129:439–445

Morais MG, Costa JAV (2007b) Isolation and selection of microalgae from coal fired thermoelectric power plant for biofixation of carbon dioxide. Energy Convers Manag 48:2169–2173
Moreira D, Pires JCM (2016) Atmospheric CO_2 capture by algae: negative carbon dioxide emission path. Bioresour Technol 215:371–379
Muñoz R, Guieysse B (2006) Algal-bacterial processes for the treatment of hazardous contaminants: a review. Water Res 40:2799–2815
Mustafa E-M, Phang S-M, Chu W-L (2012) Use of an algal consortium of five algae in the treatment of landfill leachate using the high-rate algal pond system. J Appl Phycol 24:953–963
Najdenski HM, Gigova LG, Iliev II et al (2013) Antibacterial and antifungal activities of selected microalgae and cyanobacteria. Int J Food Sci Technol 48:1533–1540
Nath A, Kumar PT, Rai AK et al (2017a) Microalgal consortia differentially modulate progressive adsorption of hexavalent chromium. Physiol Mol Biol Plants 23:269–280
Nath A, Vajpayee G, Dixit K et al (2017b) Micro-algal consortia complexity enhances ecological biomass stability through CO_2 sequestration. J Algal Biomass Utln 8:19–34
Natrah FM, Bossier P, Sorgeloos P et al (2014) Significance of microalgal-bacterial interactions for aquaculture. Rev Aquac 6:48–61
O'Neill BD, Oppenheimer M (2002) Dangerous climate impacts and the Kyoto Protocol. Science 296:1971–1972
Odjadjare EC, Mutanda T, Olaniran AO (2015) Potential biotechnological application of microalgae: a critical review. Crit Rev Biotechnol 37:37–52
Olguín EJ (2012) Dual purpose microalgae-bacteria-based systems that treat wastewater and produce biodiesel and chemical products within a biorefinery. Biotechnol Adv 30:1031–1046
Olguín EJ, Galicia S, Mercado G et al (2003) Annual productivity of *Spirulina* (*Arthrospira*) and nutrient removal in a pig wastewater recycling process under tropical conditions. J Appl Phycol 15:249–257
Padmaperuma G, Kapoore RV, Gilmour DJ et al (2018) Microbial consortia: a critical look at microalgae co-cultures for enhanced biomanufacturing. Crit Rev Biotechnol 38:690–703
Paerl H, Pinckney J (1996) A mini-review of microbial consortia: their roles in aquatic production and biogeochemical cycling. Microbial Ecol 31:225–247
Parmar A, Singh NK, Pandey A et al (2011) Cyanobacteria and microalgae: a positive prospect for biofuels. Bioresour Technol 102:10163–10172
Paul S, Nicholas A, Tseng C et al (2013) Seaweed and microalgae. Aquaculture, 2nd edn. Blackwell, Oxford, pp 268–293
Pielke RA (2009) An idealized assessment of the economics of air capture of carbon dioxide in mitigation policy. Environ Sci Pol 12:216–225
Pires JCM, Martins FG, Alvim-Ferraz MCM et al (2011) Recent developments on carbon capture and storage: an overview. Chem Eng Res Des 89:1446–1460
Posadas E, García-Encina P-A, Soltau A et al (2013) Carbon and nutrient removal from centrates and domestic wastewater using algal-bacterial biofilm bioreactors. Bioresour Technol 139:50–58
Pratt R, Daniels T, Eiler JJ et al (1944) Chlorellin, an antibacterial substance from *Chlorella*. Am Assoc Adv Sci 99:351–352
Pulz O, Gross W (2004) Valuable products from biotechnology of microalgae. Appl Microbiol Biotechnol 65:635–648
Qin L, Wang Z, Sun Y et al (2016) Microalgae consortia cultivation in dairy wastewater to improve the potential of nutrient removal and biodiesel feedstock production. Environ Sci Pollut Res 23:8379–8387
Queiroz MI, Lopes EJ, Zepka LQ et al (2007) The kinetics of the removal of nitrogen and organic matter from parboiled rice effluent by cyanobacteria in a stirred batch reactor. Bioresour Technol 98:2163–2169
Quijano G, Arcila JS, Buitróna G (2017) Microalgal-bacterial aggregates: applications and perspectives for wastewater treatment. Biotechnol Adv 35:772–781

Rada-Ariza AM, Lopez-Vazquez CM, van der Steen NP et al (2017) Nitrification by microalgal-bacterial consortia for ammonium removal in flat panel sequencing batch photo-bioreactors. Bioresour Technol 245:81–89

Radmann EM, Camerini FV, Santos TD et al (2011) Isolation and application of SO_X and NO_X resistant microalgae in biofixation of CO_2 from thermoelectricity plants. Energy Convers Manag 52:3132–3136

Raheem A, Prinsen P, Vuppaladadiyam AK (2018) A review on sustainable microalgae based biofuel and bioenergy production: recent developments. J Clean Prod 181:42–59

Ramanan R, Kim B-H, Cho D-H et al (2016) Algae-bacteria interactions: evolution, ecology and emerging applications. Biotechnol Adv 34:14–29

Ranjan A, Patil C, Moholkar VS (2010) Mechanistic assessment of microalgal lipid extraction. Ind Eng Chem Res 49:2979–2985

Rawat I, Kumar R, Mutanda T et al (2011) Dual role of microalgae: phycoremediation of domestic wastewater and biomass production for sustainable biofuels production. Appl Energy 88:3411–3424

Rawat I, Kumar R, Mutanda T et al (2013) Biodiesel from microalgae: a critical evaluation from laboratory to large scale production. Appl Energy 103:444–467

Renuka N, Sood A, Ratha SK et al (2013) Evaluation of microalgal consortia for treatment of primary treated sewage effluent and biomass production. J Appl Phycol 25:1529–1537

Reyna-Martínez R, Gomez-Flores R, López-Chuken UJ et al (2015) Lipid production by pure and mixed cultures of *Chlorella pyrenoidosa* and *Rhodotorula mucilaginosa* isolated in Nuevo Leon, Mexico. Appl Biochem Biotechnol 175:354–359

Ruiz J, Arbib Z, Álvarez-Díaz P et al (2013) Photobiotreatment model (PhBT): a kinetic model for microalgae biomass growth and nutrient removal in wastewater. Environ Technol 34:979–991

Ryu HJ, Oh KK, Kim YS (2009) Optimization of the influential factors for the improvement of CO_2 utilization efficiency and CO_2 mass transfer rate. J Ind Eng Chem 15:471–475

Safonova E, Kvitko K, Iankevitch M et al (2004) Biotreatment of industrial wastewater by selected algal-bacterial consortia. Eng Life Sci 4:347–353

Shi J, Podola B, Melkonian M (2007) Removal of nitrogen and phosphorus from wastewater using microalgae immobilized on twin layers: an experimental study. J Appl Phycol 19:417–423

Show PL, Tang MSY, Nagarajan D et al (2017) A holistic approach to managing microalgae for biofuel applications. Int J Mol Sci 18:215–248

Silva-Benavides AM, Torzillo G (2012) Nitrogen and phosphorus removal through laboratory batch cultures of microalga *Chlorella vulgaris* and cyanobacterium *Planktothrix isothrix* grown as monoalgal and as co-cultures. J Appl Phycol 24:267–276

Singh G, Thomas PB (2012) Nutrient removal from membrane bioreactor permeate using microalgae and in amicroalgaemembrane photoreactor. Bioresour Technol 117:80–85

Singh S, Kate B, Banerjee U (2005) Bioactive compounds from cyanobacteria and microalgae: an overview. Crit Rev Biotechnol 25:73–95

Solimeno A, García J (2017) Microalgae-bacteria models evolution: from microalgae steady-state to integrated microalgae-bacteria wastewater treatment models – a comparative review. Sci Total Environ 607-608:1136–1150

Spolaore P, Joannis-Cassan C, Duran E et al (2006) Commercial applications of microalgae. J Biosci Bioeng 101:87–96

Subashchandrabose SR, Ramakrishnan B, Megharaj M et al (2011) Consortia of cyanobacteria/microalgae and bacteria: biotechnological potential. Biotechnol Adv 29:896–907

Sydney EB, Sturm W, de Carvalho JC et al (2010) Potential carbon dioxide fixation by industrially important microalgae. Bioresour Technol 101:5892–5896

Tang D, Han W, Li P et al (2011) CO_2 biofixation and fatty acid composition of *Scenedesmus obliquus* and *Chlorella pyrenoidosa* in response to different CO_2 levels. Bioresour Technol 102:3071–3076

Unnithan VV, Unc A, Smith GB (2014) Mini-review: a priori considerations for bacteria-algae interactions in algal biofuel systems receiving municipal wastewaters. Algal Res 4:35–40

Valderrama LT, Del Campo CM, Rodriguez CM et al (2002) Treatment of recalcitrant wastewater from ethanol and citric acid production using the microalga *Chlorella vulgaris* and the macrophyte *Lemna minuscula*. Water Res 36:4185–4192

Van Den Hende S, Vervaeren H, Desmet S et al (2011) Bioflocculation of microalgae and bacteria combined with flue gas to improve sewage treatment. New Biotechnol 29:23–31

Vaz BS, Costa JAV, Morais MG (2016) CO_2 biofixation by the cyanobacterium *Spirulina* sp. LEB 18 and the green alga *Chlorella fusca* LEB 111 grown using gas effluents and solid residues of thermoelectric origin. Appl Biochem Biotechnol 178:418–429

Wang X-J, Xia S-Q, Chen L et al (2006) Nutrients removal from municipal wastewater by chemical precipitation in a moving bed biofilm reactor. Process Biochem 41:824–828

Wilkie AC, Mulbry WW (2002) Recovery of dairy manure nutrients by benthic freshwater algae. Bioresour Technol 84:81–91

Wood A (1987) Simple wastewater treatment system incorporating the selective cultivation of a filamentous algae. Water Sci Technol 19:1251–1254

Xu H, Miao X, Wu Q (2006) High quality biodiesel production from a microalga *Chlorella protothecoides* by heterotrophic growth in fermenters. J Biotechnol 126:499–507

Yeh KL, Chang JS (2011) Nitrogen starvation strategies and photobioreactor design for enhancing lipid content and lipid production of a newly isolated microalga *Chlorella vulgaris* ESP-31: implications for biofuels. Biotechnol J 6:1358–1366

Zhao X, Zhou Y, Huang S et al (2014) Characterization of microalgae-bacteria consortium cultured in landfill leachate for carbon fixation and lipid production. Bioresour Technol 156:322–328

Zhu L (2015) Biorefinery as a promising approach to promote microalgae industry: an innovative framework. Renew Sutain Energy Rev 41:1376–1384

Zhu L, Wang Z, Shu Q et al (2013) Nutrient removal and biodiesel production by integration of freshwater algae cultivation with piggery wastewater treatment. Water Res 47:4294–4302

Chapter 11
Downstream Green Processes for Recovery of Bioactives from Algae

Mónica Bueno, Rocío Gallego, Jose A. Mendiola, and Elena Ibáñez

Abstract Nowadays, macro- and microalgae are being increasingly used as promising raw materials for the food, cosmetic and pharmaceutical industries, thanks to their biodiversity and its variety on valuable bioactive compounds such as carbohydrates, polyunsaturated lipids, proteins and pigments, among others. Furthermore, more efficient and environmentally friendly processes for bioactives' recovery are requested not only by the industry but also by the society. This chapter presents an overview on the use of downstream green processes, mainly based on compressed fluids extraction techniques, in order to recover bioactives from algae that can be lately used in several potential applications for the food, pharmaceutical and cosmetic industries, which is the pillar of algae-based biorefinery.

11.1 Introduction

The increasing knowledge regarding the positive impact of diet on human health has brought about a great interest for seeking new bioactive products of natural origin to be used as functional ingredients for the development of functional foods. The concept of functional food is defined as food that besides the basic nutritional and energetic value provides additional health benefits thanks to the one or more functional ingredients that contains (Merichel Plaza et al. 2009). This definition implies that a functional food must improve well-being or reduce the risk of illness (Diplock et al. 1999).

Micro- and macroalgae have been suggested as a potential natural source of new compounds with biological activity that could be used as functional ingredients, due to their antioxidant (Lv et al. 2015), anti-inflammatory (Caroprese et al. 2012), antidiabetic (Yu Ran et al. 2015), neuroprotective (Pangestuti and Kim 2011), anticancer (Souza et al. 2018), anti-allergic (Thanh-Sang et al. 2012) and antimicrobial activities (Rodriguez-Meizoso et al. 2010), among others.

M. Bueno · R. Gallego · J. A. Mendiola · E. Ibáñez (✉)
Laboratory of Foodomics, Institute of Food Science Research (CIAL, CSIC), Madrid, Spain
e-mail: elena.ibanez@csic.es

A. Hallmann, P. H. Rampelotto (eds.), *Grand Challenges in Algae Biotechnology*, Grand Challenges in Biology and Biotechnology,
https://doi.org/10.1007/978-3-030-25233-5_11

The development and production of these functional ingredients have become of great interest for the food industry, although pharmaceutical and cosmetic industries are also aware of the important bioactive compounds that can be obtained from marine natural sources such as algae and microalgae, thus extending its interest and applicability. In this sense, many algae-derived secondary metabolites are known for their skin benefits, which include protection from UV radiations and prevention of ageing, rough texture, wrinkles, and skin flaccidity (Ariede et al. 2017), of upmost importance for new cosmetics' development. On the other hand, some important secondary metabolites (such as meroterpenoids) have been isolated from marine organisms presenting interesting pharmacological properties, which show cytotoxic activity towards several human cell lines, anti-inflammatory, etc. (García et al. 2018).

At present, the world is not only worried about food and human health but also about the global environmental awareness that continues to be on the rise. This is true in many countries, but especially in Europe and the USA. Climate change, global warming and the realistic threat of a lack of resources in the future for the rapidly growing world population have contributed to push process greenness and sustainability (Herrero and Ibáñez 2018). Sustainability can be understood as a rational way of improving processes to maximize production while minimizing the environmental impact (Herrero and Ibáñez 2015). Considering this framework, the study of the use of solvents that are generally recognized as safe (GRAS) for its use in the food industry, such as water, CO_2 or ethanol, combined with compressed fluids techniques are the most promising engineering approach that offers a fast, cost-effective and environmentally friendly extraction of bioactive compounds from algae. Application of high pressure and moderate-high temperature to the GRAS solvents modifies their properties, contributing to a better extraction process, improving the mass transfer rate and preserving the biological potential of the extracts. In this chapter, green extraction techniques, such as supercritical fluid extraction (SFE), gas expanded liquid (GXL) extraction, pressurized liquid extraction (PLE) and subcritical water extraction (SWE), are presented, and applications to algae bioactives extraction are discussed. Moreover, other important aspects related to upstream processes optimization and biorefinery of algae (achieved through downstream process integration for valorising, in a rational way, all the different algae fractions) are also described.

11.1.1 Marine Resources

Prokaryotic life originated in the oceans about 3.6 billion years ago while eukaryotic life originated between 0.6 and 1 billion years later (Ibáñez and Cifuentes 2013). The long evolution period of marine life compared to terrestrial has generated a huge diversity in terms of number of different species, genes, etc. Furthermore, marine organisms live in hostile environments of light, salinity, and temperature; thus, they must adapt to survive, producing a great variety of secondary (and biologically active) metabolites. This ability, coupled with the immense diversity of species, provides an almost inexhaustible source of natural bioactive compounds from marine resources. Nowadays, the most important source of information for these bioactive compounds is The Dictionary of Marine Natural Products (Blunt and Munro 2008), which lists over 30,000 purified compounds and tends to present a growing number of compounds every year.

Among the marine sources, algae are the most promising due to their easy cultivation and fast growth. Algae are photosynthetic aquatic organisms that possess simple reproductive structures. In general, these can be categorized as unicellular microscopic (microalgae) and multicellular macroscopic organisms (macroalgae). Although the number of different alga species has been estimated to be between one and ten million (Metting 1996), approximately only 40,000 species have been described (Suganya et al. 2016), which involves almost an unlimited field of research.

Macroalgae are classified into groups based on their pigmentation: Chlorophyceae (green algae), Phaeophyceae (brown algae) and Rhodophyceae (red algae) (Oncel 2017). Macroalgae have been extensively utilized as food (or food technological ingredients) for many years, and thus are farmed commercially in several countries (Baghel et al. 2015) (over 30 million tons in 2016) (FAO 2018).

On the other hand, microalgae cultivation is increasing quickly, mainly in large scale, both in outdoor and in indoor production. Microalgae could grow in autotrophic conditions, heterotrophic conditions with enough nutrients but no light availability and even in mixotrophic conditions and, hence, they are able to utilize both inorganic and organic compounds from the medium (Carvalho et al. 2014). Regarding pigment composition, microalgae are classified into nine divisions. Some of the largest groups include Phaeophyceae, Chlorophyceae, Pyrrophyceae (dinoflagellates), Bacillariophyceae (diatoms), Chrysophyceae (golden-brown) and Rhodophyceae (Oncel 2017).

One of the main applications of micro- and macroalgae biomass is biodiesel production (Mata et al. 2010) because of the high level of triglycerides they contain (Yen et al. 2013). Algae as a potential renewable resource is not only used for biofuels (Suganya et al. 2016) but also for food for aquaculture (Suganya et al. 2016), biofertilizer (Marris 2006), environmental applications such as CO_2 mitigation (Bilanovic et al. 2009) or wastewater treatment (Hodaifa et al. 2008), and to obtain high added value foods (Ibáñez and Cifuentes 2013), cosmetics (Ariede et al. 2017) and pharmaceutical products (Thanh-Sang et al. 2012).

In the following section, algae will be presented as a source of different bioactive compounds of interest for the food, cosmetics, and pharmaceutical industries. A revision about the different types of bioactives that have been described in algae is presented, including compounds such as lipids, proteins and peptides, polysaccharides, carotenoids, phenolics, and alkaloids. Table 11.1 presents a summary of potential functional compounds found in different microalgae and macroalgae, together with their possible health effects.

Table 11.1 Functional compounds found in some algae and possible health effects (based on references Ibáñez and Cifuentes 2013; Sathasivam and Ki 2018)

Functional compound	Possible health benefit	Macroalgae	Microalgae and cyanobacteria
PUFAs	Reduce risk of certain heart diseases	*Himanthalia elongate*, *Undaria pinnatifida*, *Porphyra* spp., *Chondrus crispus*	*Dunaliella salina*, *Haematococcus pluvialis*, *Chlorella* spp., *Arthrospira platensis*
Vitamin E	Antioxidant activity		*Porphyridium* spp.
α-Tocopherol	Antioxidant activity	*Himanthalia elongate*	
Folates	Reduce risk of certain types of cancer	*Undaria pinnatifida*	
Sterols	Reduce total and LDL cholesterol and immunosuppressant effects	*Himanthalia elongate*, *Undaria pinnatifida*, *Porphyra* spp., *Chondrus crispus*, *Cystoseira* spp., *Ulva* spp.	*Dunaliella salina*, *Haematococcus pluvialis*, *Chlorella* spp.
Pheophorbide a-, b-like compounds	Inhibition of cytopathic effect of herpes simplex Virus 1		*Dunaliella salina*
Phycobiliproteins	Immunomodulation activity, anti-cancer activity, and hepatoprotective, anti-inflammatory and anti-oxidant properties		*Arthrospira platensis*
Allophycocyanin	Inhibition of cytopathic effect, delay in synthesis of viral RNA of enterovirus		*Cryptomonads*
Soluble fiber	Reduce total and LDL cholesterol	*Himanthalia elongate*, *Undaria pinnatifida*, *Porphyra* spp., *Chondrus crispus*	
Alginic acid, xylofucans	Antiviral activity	*Sargassum vulgare*	
Sulphated polysaccharides	Regulate the bioactivity of growth factors and cytokines, apoptotic, antiviral, antitumour, anti-hyperlipidaemia, and anticoagulant activities	*Undaria pinnatifida*, *Porphyra* spp., *Chondrus crispus*, *Cystoseira* spp., *Ulva* spp.	*Dunaliella salina*, *Haematococcus pluvialis*, *Chlorella* spp., *Arthrospira platensis*, *Porphyridium* spp.
Polysaccharides	Inhibition of hyaluronidase of herpes simplex and influenza A virus and anti-leukaemic activity		*Navicula directa*, *Gymnodinium* sp., *Gyrodinium impudicum*

(continued)

Table 11.1 (continued)

Functional compound	Possible health benefit	Macroalgae	Microalgae and cyanobacteria
Phenolic acids	Antioxidant activity		*Arthrospira platensis*
Terpenes	Valuable curative properties	*Cystoseira* spp.	
Fucoxanthin	Preventive effect on cerebrovascular diseases, increase the metabolism, antioxidant	*Undaria pinnatifida*	*Isochrysis galbana*, *Phaeodactylum tricornutum*
Diadinochrome A, B, diatoxanthin/ cynthiaxanthin	Cytotoxic effect in HeLa cells		*Peridinium bipes*
Carotenoids	Antioxidant, immunomodulation and cancer prevention	*Ulva* spp.	*Haematococcus pluvialis*, *Chlorella* spp., *Muriellopsis* spp. *Scenedesmus* sp., *Porphyridium* sp.
Karatungiols	Antifungal, anti-protozoan		*Amphidinium* spp.

11.2 Algae as Source of Bioactive or Valuable Compounds

11.2.1 Lipids

Algae can produce different kinds of lipids such as glycolipids, phospholipids (polar lipids), glycerolipids with neutral storage lipids, and free fatty acids. Lipid percentages vary within the type of algae, containing 7–16% dry weight for macroalgae and from 1.9% up to 40% for microalgae (Suganya et al. 2016).

Among the lipids, polyunsaturated fatty acids (PUFAs) are the most-studied compounds in algae. PUFA fraction in algae is often higher than in terrestrial vegetables (Kumari et al. 2010). In fact, several microalgae are able to synthesize ω-3 and ω-6 long chain PUFAs, which are essential natural antioxidants for body health, at levels as high as 10–70% of total fatty acids (Kumari et al. 2013), exceeding 20% of their total lipid content (Bellou et al. 2014). However, the amount of PUFAs and the number or position of double bonds on the carbon chain can vary according to the algal species and growing conditions (Villarruel-Lopez et al. 2017). In general, many microalgae have PUFAs such as EPA (eicosapentaenoic acid, ω-3 $C_{20:5}$), DHA (docosahexaenoic acid, ω-3 $C_{22:6}$) and ARA (arachidonic acid, ω-6 $C_{20:4}$).

The specific interest in ω-3 essential PUFAs are their beneficial effects such as the reduction of the risks of heart disease (Chen et al. 2011), depression (Giles et al. 2013), inflammation (Yates et al. 2014), and cancer (Giros et al. 2009; Pottel et al. 2014).

Since humans have difficulty in synthesizing fatty acids with more than 18 carbons, these fatty acids should be obtained from food (Hamed et al. 2015) and in general, algae have low ω-6:ω-3 ratio, as recommended by the WHO. Although fish and seafood are the major source of long-chain PUFAs, it is important to remark that algae have been suggested as a feed for aquaculture with the idea of obtaining the desired fatty acid profile in fish and seafood for consumers.

Other important microalgae-derived lipids are phytosterols, which have been used as additives in many food products such as spread, dairy products and salad dressing (Luo et al. 2015). Phytosterols have been reported to have many beneficial health effects in humans, including immunomodulatory (Caroprese et al. 2012), anti-inflammatory (Ciliberti et al. 2017), anti-hypercholesterolemic (Chen et al. 2014), antioxidant (Lv et al. 2015) and anticancer (Kazlowska et al. 2013).

11.2.2 Proteins and Peptides

Algae can become a potential protein source. The protein content recorded for green and red algae can reach 47% of the dry weight (Ibáñez and Cifuentes 2013) and ranged between 60% and 70% in microalgae such as *Arthrospira platensis*, *Chlorella vulgaris* of *Isochrysis galbana* (Matos et al. 2017). These have been used as a supplement in food, animal feed or aquaculture due to their optimal balance of essential amino acids.

Peptides from protein hydrolysis have been studied due to their bioactivities. Some peptides have potential benefits such as antioxidative (Hu et al. 2015), binding or inhibiting specific receptors (Samarakoon et al. 2014), growth factors, hormones, immunomodulators (de Jesus Raposo et al. 2013), antihypertensive, anticoagulant and antiproliferative (Samarakoon and Jeon 2012).

11.2.3 Polysaccharides

Macroalgae contain large amounts of polysaccharides, mainly cell wall structural polysaccharides such as alginates (brown algae) and carrageenans and agar (red algae) (Ibáñez and Cifuentes 2013); meanwhile, microalgae have a low content (approximately 10% of dry matter) of carbohydrates (Villarruel-Lopez et al. 2017). Nevertheless, macro- and microalgal polysaccharides have health-promoting properties such as anti-inflammatory, antitumour, anti-adhesive, antiviral, antibacterial, immunomodulatory and infection-prevention activities (Gallego et al. 2018). For example, beta glucans are considered immune stimulators while cellulose and starch can act as dietetic fibres, and sulphated polysaccharides have antioxidant and antitumoural activities (Villarruel-Lopez et al. 2017).

11.2.4 Phenolic Compounds

The main bioactivity associated with algal phenolic compounds is their antioxidant effect through scavenging of reactive oxygen species (ROS) or enhancement of intracellular antioxidant defences. For example, extracts from microalgae *Euglena cantabrica* exhibit high antioxidant activity due to their high concentration of phenolic acids, particularly gallic and protocatechuic acids (Jerez-Martel et al. 2017).

Phlorotannins are the major phenolic compounds in brown macroalgae and the most-studied group of phenolic compounds from algae because they constitute an extremely heterogeneous group of molecules, providing a wide range of potential biological activities in addition to antioxidant activity: antiproliferative (Montero et al. 2016), antibiotic (Tanniou et al. 2014), anti-allergic (Kim and Himaya 2011), antidiabetic and anti-inflammatory activities (Catarino et al. 2017).

Other phenolic group with interesting bioactive properties are flavonoids. For instance, it has been reported in microalgae that the synergistic effects of chlorogenic and caffeic acids with 13-*cis*-retinoic acid cannot only prevent lipid peroxidation, but also regress cancer (de Jesus Raposo and Miranda Bernardo de Morais 2015). The flavonoid content in macroalgae has been also studied (Yoshie-Stark et al. 2003).

11.2.5 Alkaloids

Alkaloids present special interest because of their pharmacological activities. Structurally, alkaloids isolated from marine algae mostly belong to the phenylethylamine and indole groups.

Some alkaloids detected in marine macroalgae and microalgae have been associated with relief from depression (phenylethylamine), increased heart rate and blood pressure (tyramine), diuretic effects and inhibition of gut movements (hordenine), treatment of cardiovascular and kidney disorders (dopamine), antitumour, antibacterial and antifungal activity (caulerpin) or antioxidant activities (fragilamide) (Güven et al. 2010).

11.2.6 Carotenoids

Carotenoids are lipophilic compounds that present significant interest as food colorants, feed supplements, nutraceuticals, and for cosmetic and pharmaceutical purposes. Their C_{40} structure is based on isoprene units which can contain oxygen, so they can be classified into two main groups: carotenes and xanthophylls (Gong and Bassi 2016). More than 600 different naturally occurring carotenoids are now known, not including *cis* and *trans* isomers.

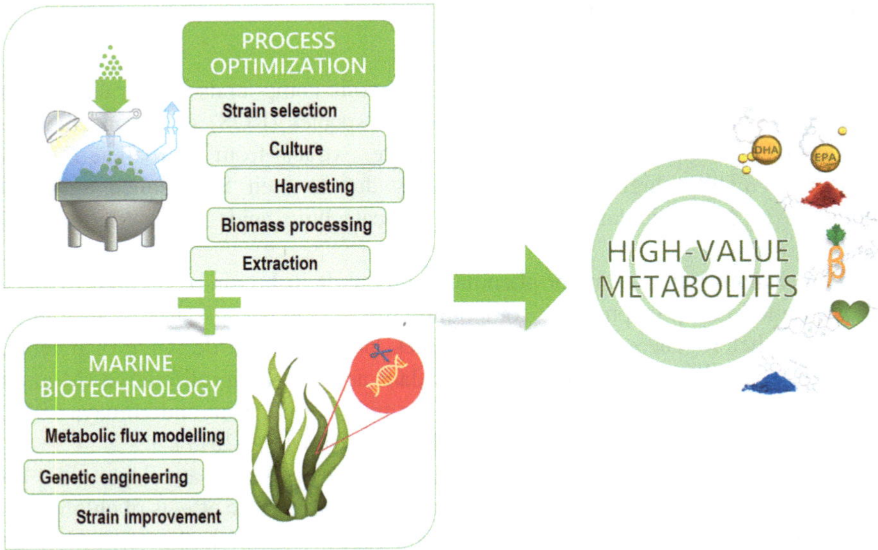

Fig. 11.1 Possibilities of increasing the production of valuable metabolites from algae

Carotenoids from marine macro- and microalgae have been described as powerful antioxidants and their beneficial physiological functions, such as anticancer, anti-obesity, antidiabetic, anti-inflammatory, and cardioprotective activities have also been reported (Hoang Van and Eun 2017). For instance, some of the most-studied carotenoids extracted from algae with beneficial effects on health are fucoxanthin, β-carotene, lutein and zeaxanthin from macro- and microalgae; and astaxanthin, canthaxanthin, capsanthin, α-carotene, crocetin, β-cryptoxanthin, lycopene, neoxanthin and phytoene from microalgae (Christaki et al. 2013; Gallego et al. 2018).

11.3 How to Improve the Production of Bioactive Metabolites

As it was mentioned in the introduction, macro and microalgae have raised an enormous interest, thanks to their potential for being a good source of high added-value compounds that can be used in cosmetic, food and pharmaceutical industries. Furthermore, it is well established that secondary metabolites production can be strongly increased by many factors. Figure 11.1 offers an overview on different ways to increase the production of valuable components from algae: marine biotechnology (through genetic engineering, selection and improvement of strains, metabolic flux modelling, etc.) and optimization of processes including both upstream (strain selection and cultivation conditions) and downstream processes (biomass processing, extraction and purification methods). The main objective would be the integration of these factors in a biorefinery approach, which allows a high production of the bioactives of interest.

11.3.1 Marine Biotechnology

Marine or blue biotechnology can be defined as the application of genetic engineering to marine resources. Thus, by using genetic engineering, it is possible to modify genes and improve algae strains obtaining transgenic algae which are able to overexpress genes and overproduce valuable target compounds.

Marine biotechnology involves the study of the metabolic pathways which lead to the synthesis of bioactive compounds. It is important to consider all biochemical reactions, and their stoichiometry, which occur within the metabolic network. This knowledge will lead to modify or model the metabolomic flux, increasing (or decreasing) the production of selective bioactive metabolites (Ibáñez and Cifuentes 2013).

It is true that genetic manipulation in algae has been limited to a few species due to the complexity and large genome size. Microalgal genome sizes range from 12.6 Mbp for the *Ostreococcus tauri* and 168 Mbp for the *Emiliania huxleyi* to an estimated 10,000 Mbp for the *Karenia brevis* (Cadoret et al. 2012). These large genome sizes can be difficult to sequence and transform.

Furthermore, it is very difficult to obtain new microalgal strains since nuclear transformation has a low efficiency and transgenes expression is not stable (Leon and Fernandez 2007).

Recently, some researchers have proposed new methods to ensure stability and a higher expression of transgenes. For instance, Diaz-Santos et al. (2016) proposed an interesting approach to express transgenes in microalgae using co-transformation with two naked promoterless genes, which are randomly inserted into the nuclear genome. They reached a successful co-transformation of *Chlamydomonas reinhardtii*, concluding that this transformation system could be universally applicable to any microalgal species.

In conclusion, more intense research and the study of new genetic engineering techniques are necessary to better understand, both genetically and metabolically, the complex network involved in the synthesis of bioactive compounds of interest; this way, the full potential of macro and microalgae could be reached.

11.3.2 Optimization of Upstream and Downstream Processes

Upstream and downstream processes involve all stages from the selection of macro and microalgae strains and cultivation to extraction and/or purification of secondary metabolites.

11.3.2.1 Upstream Processes

Of course, depending on the bioactive compound of interest, a specific algae strain must be chosen since metabolite composition is extremely variable among species. Nowadays, there is a huge quantity of compounds obtained from different algae which can be found in many industries. For example, carotenoids such as β-carotene and astaxanthin are obtained from the green microalga *Dunaliella salina* and *Haematococcus pluvialis*, respectively. Another interesting example is the use of *Isochrysis galbana*, which is rich in ω-3, as an ingredient for functional biscuits (Gouveia et al. 2008).

Cultivation conditions are essential in algae biorefinery. The main factors are supply of carbon dioxide (commonly CO_2), nutrient source (i.e. nitrogen and phosphorus) and source and origin of illumination (Vanthoor-Koopmans et al. 2013), and also it is important to take into account other factors such as temperature control, algae concentration, pH, cultivation systems such as ponds or photobioreactors. All these factors are vital for the proper growth of algae. For instance, it is well known that the microalgae *Haematococcus pluvialis* can grow as motile biflagellated green cells when it is subjected to favourable conditions, but under stress conditions (nutrient deficiency, high light intensity or salt stress), the cells lose their motility, their size increases and forms red cysts, allowing its survival for long and stressful periods (Hagen et al. 2002). Thus, in green cells, chlorophylls and carotenoids such as lutein and β-carotene can be found while in red cell phase, astaxanthin and its derivatives (esters, mainly) constitute up to 98% of the total carotenoid content (Boussiba et al. 1999). In terms of light, Aravantinou and Manariotis (Aravantinou and Manariotis 2016) observed a greater growth rate of *Chlorococcum* sp. under artificial light conditions instead of direct sunlight, proving the importance of light intensity and light source on biomass production.

The second main step in upstream processing is harvesting, which has to be optimized for each particular algae strain. In this sense, there are many ways to recover biomass but they are mostly focused in centrifugation and filtration.

Biomass processing is related to the proper disruption of the cell, since most of metabolites of interest are located inside the cell. It includes different techniques such as enzymatic treatments, microwave-assisted processes, pulsed electric fields or high-pressure homogenization. In this sense, Carullo et al. (2018) studied the effect of two different cell disruption techniques in the microalgae *Chlorella vulgaris* and demonstrated that it was possible to selectively recover small-sized cytoplasmic compounds using pulsed electric fields, and high molecular weight intracellular components using high-pressure homogenization.

11.3.2.2 Downstream Processes

Downstream processes involve the extraction and purification methods to isolate the valuable compounds of interest from algae. These procedures can be extremely expensive and can consume a huge quantity of organic solvents, so the optimization of these steps is vital for the global economic viability of the algae biorefinery. As an alternative to conventional processes (such as solid–liquid extraction or Soxhlet extraction), green processes have been proposed as a clean, sustainable and environmentally friendly approach. Table 11.2 shows a list of alternative processes that have been recently used to extract compounds from many sources, including macro and microalgae. These processes are microwave-assisted extraction (MAE), ultrasound-assisted extraction (UAE), enzyme-assisted extraction (EAE), supercritical fluid extraction (SFE), pressurized liquid extraction (PLE) in which subcritical water extraction (SWE) is included, and gas-expanded liquids (GXLs). Even though these techniques are based on different principles, all of them have in common the use of minimal amount of food-grade solvents and its intensification through the employment of microwaves, ultrasound, enzymes or high pressure/temperature (Mendiola et al. 2013) that allow improving the selectivity and the global efficiency of the extraction process.

11.3.2.2.1 Assisted Extraction Techniques

Microwave-assisted extraction is based on the use of microwave radiation that causes heat both inside the matrix and the solvent. In algae, this heat provokes an enormous pressure inside the cells and favours the rupture of the cell wall, thus exposing its constituents to the solvent. Furthermore, the heat helps the solvent to diffuse into the cells, thus improving the transfer of the bioactive compounds between the matrix and the solvent (Tatke and Jaiswal 2011). MAE has been widely used to extract bioactives from algae such as lipids, high-value pigments, proteins, vitamins, carbohydrates and others (Kapoore et al. 2018). Although different organic solvents can be employed, those selected for MAE applications should absorb microwave radiation and, therefore, usually polar and protic solvents are used. Some applications of MAE for the extraction of lipids, pigments and proteins from different algae species such as *Chlorella* sp., *Nannochloropsis salina*, *Phaeodactylum tricornutum* and *Porphyridium purpureum* have been developed using GRAS solvents (Gilbert-Lopez et al. 2017a; Juin et al. 2015; Martinez-Guerra et al. 2014; Patil et al. 2013).

For instance, Martinez-Guerra et al. (2014) studied the extraction of lipids from microalgae using MAE. In this case, algal lipids were extracted from dry *Chlorella* sp. using ethanol as solvent. In comparison to the conventional Bligh and Dyer (BD) method, they obtained an increase in lipid extraction yields (from 13.9% to 20.1%) with a higher fatty acids ethyl esters conversion of the algal lipids (from 78.1% up to 96.2%) under optimum conditions (algae biomass: ethanol molar ratio of 1:250–500 and 2.0–2.5% sodium hydroxide catalyst with reaction times around 6 min).

Another interesting approach was given by Gilbert-Lopez et al. (2017a). They used MAE to obtain high valuable extracts from *Phaeodactylum tricornutum.* Under optimum conditions (30 °C, 100% ethanol and 2 min of extraction), they obtained a

Table 11.2 Advantages and disadvantages of alternative downstream extraction processes (based on references Grosso et al. 2015; Herrero et al. 2017)

	Advantages	Disadvantages
MAE	• Short treatment time and solvent consumption • More efficient than conventional heating • Reduction of extraction temperature using pressurized closed vessels • Organic solvents and water can be used • High extraction yields	• Only solvents with high dielectric properties can be used • Possible thermal degradation of the most thermolabile compounds when using open vessels • High energy consumption
UAE	• Short treatment time and solvent consumption • High efficiency in cell disruption • High extraction yields • Suitable to extract thermolabile compounds • Inexpensive	• Solvents with low surface tension, low viscosity and low vapour pressure are preferable • The presence of a dispersed phase contributes to the ultrasound wave attenuation • Ultrasounds generate heat, being important to accurately control the extraction temperature • Excess of sonication may damage the quality of extracts
EAE	• Water can be used (Green technology) • The enzyme treatment can increase the recovery of bioactive compounds	• The efficiency of enzymatic hydrolysis is very low if materials have low moisture content • Enzyme treatment is usually a slow process, and it may take from hours to days
SFE	• Green technology • Higher selectivity because the solubility of a compound in a supercritical fluid can be manipulated • It is possible to extract more polar compounds with the use of modifiers • Elimination of CO_2 is achieved without residues, yielding a solvent-free extract • Suitable to extract thermolabile compounds	• High costs for the high pressure equipment needed • Can be more time-consuming than the other alternative techniques
PLE/SWE	• Green technology in the case of pressurized water extraction (SWE) • Reduced solvent consumption • Suitable to extract thermolabile compounds	• High costs for the high pressure equipment needed • Extractions performed at high temperatures may lead to degradation of thermolabile compounds
GXL	• Can be considered as a half way from PLE to SFE by increasing the amount of compressed CO_2 • Requires lower working pressures (compared to SCFs) and the subsequent reduction in energy consumption and costs • Suitable to extract compounds with intermediate polarity	• Can be more time-consuming than the other alternative techniques

higher extraction yield (14.51%) and recovered a good amount of lipids such as EPA and carotenoids such as fucoxanthin, even higher than those reported from brown algae.

Ultrasound-assisted extraction also relies on the disruption of cell walls, increasing the contact between solvent and matrix. In this case, the driving force that favours the extraction of the bioactives is the acoustic cavitation produced by the use of high-frequency sounds. Some algae such as *Arthrospira platensis* and *Chlorella* sp. have been used to extract valuable compounds using UAE, with an important increase in the extraction yield. As an example, Zhao et al. (2013) studied different treatments to extract carbohydrates from fresh *Chlorella* sp. UAE treatment showed the best results, reaching the maximum glucose yield (36.94 $\pm$ 2.46 g per 100 g dry cell weight) considering the following extraction conditions: ultrasonic power of 800 w, extraction time of 80 min, flow rate of 1.52 L/min and cell concentration of 0.3 g/L. Another interesting example was given by Hadiyanto and Suttrisnorhadi (2016), who efficiently extracted phycocyanin from *Arthrospira platenis* using UAE. Results showed a significant increase of the extraction yield using UAE (up to 15.7%) in comparison to conventional extraction (11.13 %) under UAE optimal conditions (52.5 °C, 42 min of extraction time and ultrasound frequency of 42 Hz).

One interesting aspect common to both techniques is that it is possible to extract bioactive compounds directly from wet biomass without using any solvent. For instance, Adam et al. (2012) performed a solvent-free ultrasound-assisted extraction from fresh *Nannochloropsis oculata* biomass in order to recover lipids. As the water of the wet alga was used as solvent, lipids were effectively separated into two distinct phases, simplifying the oil recovery. Furthermore, using scanning electron microscopy (SEM), they could observe that after UAE, external structure of cells surface had changed, in contrast to non-treated cells, which appear to be intact. This means that UAE directly from fresh microalgae cells could be an innovative and sustainable option to extract lipids from microalgae.

Passos et al. (2015) studied both pretreatment methods (MAE and UAE) directly from microalgal biomass, finding that all pretreated microalgal biomass had a higher content of all soluble organic macromolecules (proteins, carbohydrates and lipids) than non-pretreated biomass. However, these procedures can damage or degrade thermolabile compounds if extraction conditions are carried out under extremely high temperatures.

Another alternative extraction method relies on the use of enzymes, which are capable of degrading or disrupting cell walls and membranes, thus allowing a better release of bioactives (Munish et al. 2012). In vegetable matrices, pectinases, cellulases and hemicellulases are commonly used. Since algae have a similar cell wall, these enzymes have been also employed for degradation of their cell walls, as many authors have confirmed. For instance, Zuorro et al. (2016) used a multi-enzyme pretreatment based on cellulase and mannanase enzymes for the release of intracellular material, specifically lipids, from the marine microalga *Nannochloropsis* sp, reaching up to 90 % of lipid recovery under optimal conditions. Another interesting example was given by Huo et al. (2015). They applied a mixture of enzymes (cellulase, pectinase and hemicellulase) to extract oil from wet microalgae *Scenedesmus* sp. G4, obtaining up to 86.1% of lipids under optimal conditions and proving the great impact of enzymes on the integrity of microalgae cell. The main problem encountered by using this methodology is the low efficiency of the lysis process and the time required to complete the reaction (that can take from hours to days) (Grosso et al. 2015).

11.3.2.2.2 Compressed Fluids' Extraction Techniques

Compressed fluids' extraction techniques such as SFE, PLE, SWE or GXL are the most innovative methods that have been recently used to obtain high-value compounds from many matrices, including macro and microalgae. The main advantage is that all of them can use green solvents such as CO_2, water or ethanol. Furthermore, the possibility of changing the solvent physicochemical properties and solvating power by changes in pressure and/or temperature of the system provides a great selectivity and efficiency for obtaining a huge range of bioactives with different characteristics.

Despite several differences in the basic principles of SFE, GXL and PLE, they all have in common that they must operate under medium-to-high pressures; for this reason, it is possible to use the same equipment for the three extraction techniques. SFE is based on the use of solvents at temperatures and pressures above their critical points, while PLE operates using liquids at temperatures above their normal boiling points and pressures enough to keep the extracting fluid in the liquid state. GXLs extraction is an intermediate technique between PLE and SFE. GXLs are liquids whose volume has been increased when pressurized with a condensable gas (e.g., CO_2). Under these conditions, at least two fluid phases or a single phase above the bubble point curve but below the critical composition exists (Herrero et al. 2013). Figure 11.2 shows a general scheme of the equipment that can be used for SFE, GXL and PLE. In the following sections, a more detailed explanation on the different configurations employed for each process is included.

11.3.2.2.2.1 Supercritical Fluid Extraction

Briefly, when a fluid is forced to a temperature and pressure above its critical point, it is considered to be a supercritical fluid, and it shares physicochemical characteristics from both liquid and gas states. Some of these properties are low viscosity, high diffusivity and tunable density, which can be easily modified depending on the temperature and pressure applied and, consequently, the solubility of the target compound in the fluid is also modified. Carbon dioxide (CO_2) is the most-employed fluid in SFE, since it has moderate critical temperature and pressure (31.2 °C and 73.8 bar) and it can be recycled, so it can be considered as environmentally friendly. Moreover, a very interesting point is that CO_2 will become a gas at atmospheric conditions, so once the extraction is finished, the CO_2 from the extract is directly evaporated, and the extract is completely solvent-free.

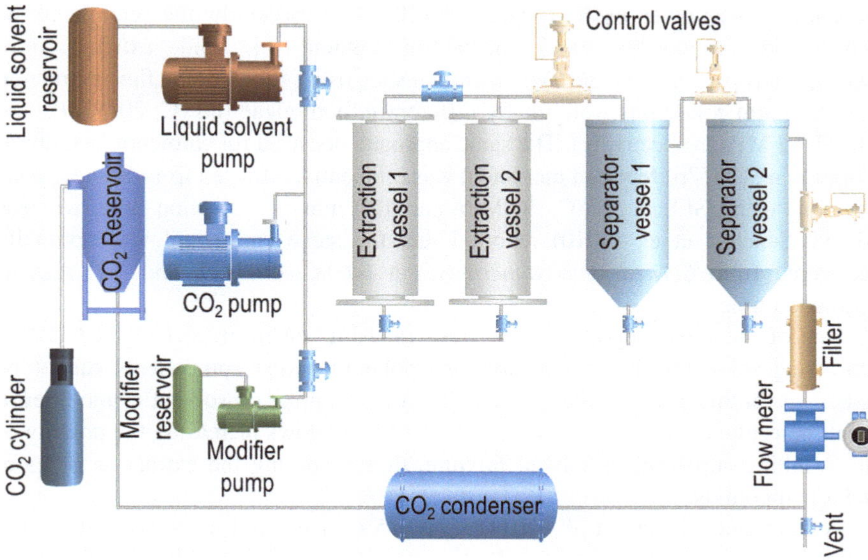

Fig. 11.2 General scheme of equipment used for SFE, CXL and PLE

As mentioned, Fig. 11.2 shows the scheme of a pilot plant that can be used for SFE, GXL extraction and PLE. In SFE configuration, the CO_2 is initially cooled to 0–5 °C in order to be pumped as a liquid; the system includes the possibility of adding a co-solvent as modifier of the polarity of CO_2. Once the mixture is achieved, the fluid is heated to the selected extraction temperature and pumped at the selected pressure into an extraction vessel (E1 and/or E2), kept at working temperature. Algae are placed inside the extraction cell in a basket. If several extraction vessels are used, it is possible to increase productivity since while one is used for extraction, the other can be simultaneously filled with the material. Once the extraction is finished, the pressure is reduced through a control valve (R1) and the extract precipitates and is recovered in the separator vessels (S1 and/or S2). A series of collection vessels at sequentially lower pressures may be employed to trap and fractionate the extract. Flow rate and extraction pressure are controlled by the pumping rate and by the setting of the control valve for a particular pumping rate, respectively. On a pilot and industrial scale, CO_2 is recycled by condensing it, filtering it and sending it back to the reservoir for being pumped in the following extraction.

There are many reviews which summarize the potential of supercritical fluid extraction to obtain bioactives from different natural sources, including algae. For instance, SFE has been used to extract lipids from *Nannochloropsis oculata*, *Tetraselmis suecica*, *Dunaliella salina* and *Crypthecodinium cohnii*, among others; and carotenoids from *Haematococcus pluvialis*, *Chlorococcum littorale*, *Chlorella vulgaris* or *Scenedesmus almeriensis*, among others (Gallego et al. 2018).

As expected, extraction conditions are different depending not only on the compound of interest but also on the algae species. A clear example of this dependence was given by Bong and Loh (2013). In this study, they compared the fatty acid composition and tocopherol content of lipid extracts from *Nannochloropsis oculata* and *Tetraselmis suecica* using supercritical fluid extraction and optimum conditions were totally different in both algae (80 °C, 20.7 MPa and 40 °C, 62 MPa, respectively). The same approach occurred for carotenoids. Gilbert-Lopez et al. (2017b) reported that lutein was efficiently extracted from *Scenedesmus obliquus* using SFE at 50 °C, 36 MPa and 120 min as extraction time, whereas Macías-Sánchez et al. (2010) reported that the same carotenoid was optimally recovered from *Scenedesmus almeriensis* at 60 °C, 40 MPa and 300 min as extraction time.

One of the most important drawbacks of using supercritical CO_2 ($scCO_2$) as extracting solvent is its low polarity, so polar bioactive components cannot be extracted. In this case, an alternative is the use of a polar co-solvent or modifier in small percentages (i.e. ethanol from 1 to 15%) that allows increasing the polarity of the resulting supercritical solvent mixture, thus favouring the extraction of more polar compounds.

For instance, Solana et al. (2014) used a 5% of ethanol as co-solvent for the extraction of α-linolenic acid (αLnA) from *Scenedesmus obliquus*, *Chlorella protothecoides* and *Nannochloropsis salina*. The highest amount of αLnA was reached at 45 °C and 15 MPa after 30 min of extraction.

On the other hand, Ota et al. (2009) extracted β-carotene from *Chlorococcum littorale* comparing SFE with and without ethanol as co-solvent, reaching a high yield (up to 90%) with 10% of ethanol and optimum conditions of 60 °C, 30 MPa and 180 min of extraction time; a yield of 40% was obtained with pure CO_2 as extracting solvent.

Selection of co-solvent is also important for the bioactivity of the obtained extract. For example, Saravana et al. (2017) compared sunflower oil, soybean oil, canola oil, ethanol, and water as co-solvents to support $scCO_2$ extraction of carotenoids, mainly fucoxanthin, and phlorotannins from brown seaweed *Saccharina japonica*. A 2% sunflower oil as co-solvent showed higher carotenoid content and antioxidant activity than the control ($scCO_2$ only).

Regarding microalgae extraction, in general, a drying step prior to $scCO_2$ extraction is required because they are grown in liquid cultures. Reyes al. (2016) studied the direct extraction of carotenoids from *Neochloris oleoabundans* paste (containing around 70–80% water) mixing this paste with adsorbents as supporting media. Results showed that chitosan was the adsorbent with better adsorbent capacities for the recovery of carotenoids. These results are interesting to avoid the drying step, which is energy consuming and could be detrimental for the bioactivity of the extracted compounds.

11.3.2.2.2.2 Gas-Expanded Liquid Extraction

When increasing the amount of polar solvent mixed with CO_2, a different type of solvent is achieved: the so-called "carbon dioxide expanded liquid (CXL)". CXL is a particular case of gas-expanded liquid (GXL) in which carbon dioxide is used as expanding media; CXLs are considered to be half way from pressurized liquids to supercritical fluids (Herrero et al. 2017).

In general terms, GXLs have densities similar to that of organic solvents (without CO_2 added), while their viscosities are between those of supercritical fluids and liquids. GXLs show a wide range of physicochemical properties compared to supercritical fluids, since more diverse properties can be obtained considering the wide variety of different green organic solvents that can be employed (Cunico and Turner 2017). Several physicochemical properties change by changing the pressure and/or temperature in CXL systems; among them are the following properties: density, compressibility, viscosity, mass transfer and dielectric properties. For more in-depth information about GXLs, readers are referred to Sánchez-Camargo et al. (2018).

As shown in Fig. 11.2, the equipment needed to work under CXL conditions is the same as the one required for carrying out SFE; the only difference is that under CXLs conditions, a higher amount of solvent is used and, commonly, lower pressures are employed. In general, the instrumentation consists of two pumps, one for carbon dioxide and another for the solvent, a system for heating the extraction cell (s) (medium–high pressure vessel(s)), valves for controlling the fluid flow path and pressure and a collection device. Operation starts by mixing the liquid solvent with CO_2 at medium–high pressures (CO_2 will expand and the volume of the fluid mixture will increase, depending on the pressure conditions); the fluid is then injected in the medium–high pressure vessel where the extraction takes place (at certain temperature conditions controlled by a heating system); after the extraction time, the outlet valve (R1) is open to control flow/pressure and the extract is continuously collected in a separator vessel (S1).

Some interesting applications of CXLs for the extraction of bioactive compounds from algae have been recently published. For instance, Golmakani et al. (2012) described one of the very first uses of GXL to algal biomass. In this case, two alternative extraction techniques (GXLs and pressurized ethyl lactate: ethanol) were applied to obtain high-value lipids from *Arthrospira platensis*. Results obtained after chemometric optimization allowed understanding the effects of the different factors involved in the studied processes and provide the optimum conditions to get the maximum γ-linolenic acid (γLnA) recovery and lipid yield. GXL (40 °C, 300 atm, 50% ethanol, 90 min extraction time) provided γLnA recovery of 24.7% and total yields of 6.7% (w/w), while PLE (180 °C, 20.7 MPa, ethanol: ethyl lactate 1:1 and 15 min extraction time) provided total yields up to 20.7% (w/w) and γLnA recoveries of 68.3%. In this case, GXL provided lower yields and recoveries than PLE, but gave higher selectivity and demonstrated its performance as intermediate between PLE and supercritical fluids for the extraction of medium-polar compounds.

Reyes et al. (2014) used a Box–Behnken experimental design to examine the effects of mild operating temperature (40–70 °C) and pressure (20–35 MPa), using ethanol in $scCO_2$ (0–13% w/w) on the astaxanthin content, extraction yield, and antioxidant activity of *Haematococcus pluvialis* extract. Since astaxanthin is a carotenoid whose molecular weight and functional groups give low solubility in $scCO_2$, two approaches can be followed to increase its extraction: the first one is to force the extraction by increasing the extraction time and pressure (above 50 MPa), while the other is to employ higher amount of ethanol to increase astaxanthin solubility in $scCO_2$. In the work by Reyes et al. (2014), after demonstrating the important effect of ethanol content in supercritical CO_2 (more significant than pressure and temperature), authors move to the GXL region using higher ethanol content (50–70%, w/w), mild temperature (30–60 °C) and low pressure (7 MPa). Comparing CXE (Carbon Dioxide Expanded Extraction) with $scCO_2$ at optimum extraction conditions (20 MPa, 13% (w/w), 55 °C for $scCO_2$ and 7 MPa, 50% (w/w) ethanol, 45 °C for CXE), CXE showed better results in terms of extraction yield, astaxanthin content and astaxanthin recovery than $scCO_2$ extraction. In fact, these results were better than any previously published manuscript concerning astaxanthin extraction from *H. pluvialis*.

11.3.2.2.2.3 Pressurized Liquid Extraction

Pressurized liquid extraction (PLE) is based on the use of high temperature (below the critical point) and pressures enough to keep the solvent in liquid state. If water is used as extracting solvent, it is called subcritical water extraction (SWE) or pressurized hot water extraction (PHWE) and can be considered as the greenest alternative involving the use of pressurized liquids. Thanks to the high temperatures and pressures, the solvent possesses increased solubility and decreased viscosity, allowing a better mass transfer rates and penetration into the matrix while improving the efficiency of the extraction process.

When working under PLE conditions, instrumentation needed consists of the following: a solvent pump, an extraction vessel (E1), pressure valves, heating systems for controlling temperature and a collection vessel (S1). The solvent is introduced inside the extraction cell by the pump (pressures required range between 35 and 200 bar). Pressure is controlled inside the extraction cell by two on/off valves (or one on/off valve and the restrictor, R1) and the extraction cell is placed inside a heating system, which controls the applied temperature (usually, high-temperature area employed is above the boiling point of the solvent and below its critical point). A collection vessel is needed to recover the extract. It is important to mention that the solvents employed for the extraction should be oxygen-free in order to avoid oxidation of the bioactives as well as to prevent cavitation in the pump; degassing by ultrasounds and helium purge are two systems that can be employed for this purpose.

Depending on the matrix and on the target compound(s), a proper selection of the extracting solvent is needed. Thus, for the extraction of more polar lipids such as short-chain fatty acids and tocopherol or carbohydrates, water can be chosen, whereas less polar lipids such as PUFAs can be extracted using ethanol (Pieber et al. 2012; Rodriguez-Meizoso et al. 2010). In this sense, Otero et al. (2018) studied the selectivity of five solvents of different polarities (hexane, ethyl acetate, acetone, ethanol and ethanol:water 50:50) in the lipid composition of *Fucus vesiculosus* by PLE. Results showed that long-chain fatty acids including oleic acid, arachidonic acid and EPA are selectively extracted using ethyl acetate, producing extracts that at least double the fatty acids quantity in comparison to the other solvents. Nevertheless, the lowest ω-6/ω-3 ratio was achieved with ethanol: water 50:50 (the most polar solvent) with a value of 1.92, much lower than those recommended by FAO (ω-6/ω-3 = 10) (FAO 2010). It is well known that a low ω-6/ω-3 ratio exerts suppressive effects on cardiovascular diseases (Simopoulos 2002).

Several examples can be found in the literature about the use of PLE to extract carotenoids from many different algae species using different solvents such as ethanol, water, acetone and their mixtures. The diversity of solvents that have been employed can be explained by the wide range of polarities of bioactive carotenoids; for example, violaxanthin, neoxanthin and lutein could be effectively extracted from *Chlorella vulgaris* using acetone at 50 °C and 10 MPa (Plaza et al. 2012); fucoxanthin and zeaxanthin could be extracted from *Himanthalia elongata* using ethanol as solvent at 100 °C and 10.3 MPa (Plaza et al. 2010); and also from *Phaeodactylum tricornutum* using ethyl acetate at 100 °C and 10 MPa (Derwenskus et al. 2019); astaxanthin and derivatives could be efficiently extracted from *Haematococcus pluvialis* using pressurized ethanol at 50 °C and 10.3 MPa (Jaime et al. 2010).

11.3.3 Integrated Processes

The integration of processes dealing with the extraction of bioactive compounds from macro- and microalgae is a hot topic. Some interesting approaches have been employed in the literature; for instance, Hernandez et al. (2014) studied the effect of microwave pretreatment prior to $scCO_2$ extraction in different microalgae. Interestingly, the authors reported that the microwave (MW) effect strongly depends on the microalgae tested; whereas in the microalga *Scenedesmus almeriensis*, a positive effect on the yield of lipids was shown, *Nannochloropsis gaditana* seemed to be negatively affected by the microwave-assisted pretreatment. The same approach was studied in *Chlorella vulgaris*, in which Dejoye et al. (2011) concluded that the integration of MAE and $scCO_2$ extraction gives a high quality and yield of recovered lipids.

Among the most promising integration of processes are those involving the coupling of extraction and purification, considering that depending on the extraction conditions and the chemical characteristics of the target compound, sometimes, it would be difficult to obtain pure extracts. Supercritical antisolvent fractionation (SAF), supercritical antisolvent (SAS) or solution-enhanced dispersion (SEDS) by supercritical fluids are processes that could be coupled online to obtain dried encapsulated particles. In general terms, these techniques are based on contacting an organic solution with $scCO_2$. During mixing, the rapid mutual diffusion at the interface of $scCO_2$ and the liquid extract containing the compounds causes the precipitation of solutes, allowing to obtain completely solvent-free products. These processes can also be used to encapsulate or co-precipitate target compounds by super saturation of the polymer/solute, leading to sub-micrometric particles with controlled size. For example, Machado et al. (2016) coupled an enzymatic lysis assisted by ultrasounds, without biomass freezing, for the cell wall disruption of *Haematococcus pluvialis*, with the subsequent encapsulation of carotenoids in the copolymer poly(hydroxybutyrate-co-hydroxyvalerate) (PHBV) using SEDS technique.

11.3.4 Biorefinery

The concept of biorefinery relies on the capability of improving the recovery of different products from a unique biomass. In other words, the main idea consists of the integration of multiple and sequential processes that allow the fractionation of a single biomass into different and isolated compounds of high added value (Subhadra and Grinson 2011).

Therefore, a great effort is being carried out in the development of biorefinery platforms to best exploit available resources. In this sense, a microalgae biorefinery platform was designed in our research group involving the integration of compressed fluids technologies such as SFE, GXLs and PLE in a holistic approach, in which the residue of each extraction is used as a raw material for the next step.

The compressed fluids' biorefinery platform involves the extraction of target compounds of different polarities through the addition/removal of CO_2 and therefore moving from SFE (with neat CO_2) to conventional organic solvents (working under high pressure and temperature) and considering, as intermediate steps, the use of CO_2 plus modifier and/or CXLs. In this approach, working under medium/high pressures, different physicochemical properties can be conveniently modified through the addition of compressed CO_2 (such as polarity, viscosity and diffusivity) (Herrero et al. 2017).

Figure 11.3 shows a scheme of the compressed fluids' platform mentioned above. In this kind of biorefinery platform, the residue of one extraction is the matrix to be treated in the next step; taking into account that all the steps are done in the same equipment, different extraction processes (carried out at medium-high pressure) were sequentially used to extract valuable compounds from algae biomass. Biorefinery started using a dry biomass sample and applying an SFE (CO_2 as solvent) as first step to obtain non-polar bioactives, including carotenoids and lipids; the residue of SFE was subsequently treated with a CXL (and/or PLE with ethanol) to obtain the polar lipids, carotenoids and chlorophylls; and finally, by means of SWE, sugars and proteins were obtained. By this approach, the sample is treated with increasing polarity solvents to provide different extracts enriched in valuable compounds. This was the approach followed by Gilbert-Lopez et al. (2015) in which different compounds were obtained from *Isochrysis galbana*. Thus, the extraction process was partially selective according to the polarity of the solvent/mixture of solvents used. First extracts using $scCO_2$ were rich in triacylglycerides, while extracts obtained using CXL were rich in fucoxanthin, the main carotenoid in

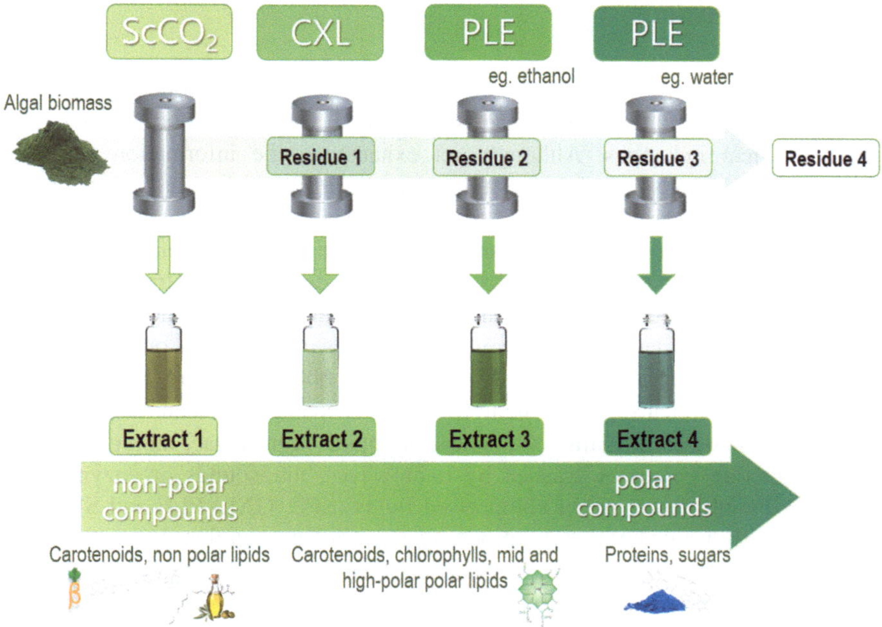

Fig. 11.3 Downstream process for microalgae biorefinery

Isochrysis galbana. The latest extracts obtained were enriched in proteins and carbohydrates.

Similar results were obtained using *Scenedesmus obliquus* as dry biomass. In this case, not fucoxanthin but lutein and β-carotene were extracted in the GXL step (Gilbert-Lopez et al. 2017b).

It is worth mentioning that the same biorefinery approach can be used to extract compounds starting from high polarity to low polarity, by just inverting the order of the processes involved (PLE with water, PLE with ethanol, CXL and SFE with neat CO_2). The viability of this approach has been recently demonstrated considering wet microalgae as starting material (Ibáñez et al. 2017).

It is also important to emphasize that through the integration of green chemistry into biorefineries and the use of low environmental impact technologies such as those based on the use of compressed fluids, future sustainable production chains of biofuels and high-value chemicals from biomass can be established, thus improving the economic viability of the whole biorefinery.

11.4 Conclusions

In this chapter, we presented an overview of the bioactive compounds that can be obtained from macro and microalgae with potential use in the food, cosmetic and pharmaceutical industries. Although not exhaustive, the information has been selected considering some of the most important compounds that can be synthesized by algae and can provide benefits for human health. Some of them are major components such as proteins, lipids and carbohydrates and other minor components (secondary metabolites) generated to protect algal cells against stress conditions. Emphasis has been put on the different possibilities for promoting the enrichment in high-value metabolites, ranging from marine biotechnology to processes (both upstream and downstream) that can be optimized to obtain highly enriched fractions in different components. But the main focus of the chapter has been the description of new technologies to extract valuable compounds from algae, among them are some extraction processes assisted by microwaves, ultrasounds or enzymes and processes based on the use of compressed fluids (SFE, GXL, PLE and SWE). In the framework of this book, these processes have in common that they are greener, more efficient, avoid the use of toxic organic solvents and can be sustainable. Several recent applications of these technologies to the extraction of valuable compounds from algae are described in the text, demonstrating the usefulness and the advantages of such processes compared to conventional ones. Finally, a biorefinery platform based on compressed fluids technology is presented as an example of the possibilities offered by these technologies to completely valorize algae biomass. This platform is intended to be placed in a whole process involving the optimization of the different necessary steps: efficient production of biomass using CO_2 formed by combustion of fossil fuels in thermoelectric power plants, extraction of valuable bioactives using environmentally friendly processes and processing of the oily fraction to produce biofuels; exhausted material can be also used for other purposes (such as fabrication of furniture.). This way, it will be possible to move towards a more sustainable world, in which circular economy will take the lead and sustainable development challenges will start to be met.

References

Adam F, Abert-Vian M, Peltier G et al (2012) "Solvent-free" ultrasound-assisted extraction of lipids from fresh microalgae cells: a green, clean and scalable process. Bioresour Technol 114:457–465

Aravantinou AF, Manariotis ID (2016) Effect of operating conditions on *Chlorococcum* sp. growth and lipid production. J Environ Chem Eng 4(1):1217–1223

Ariede MB, Candido TM, Morocho Jacome AL et al (2017) Cosmetic attributes of algae – a review. Algal Res 25:483–487

Baghel RS, Trivedi N, Gupta V et al (2015) Biorefining of marine macroalgal biomass for production of biofuel and commodity chemicals. Green Chem 17(4):2436–2443

Bellou S, Baeshen MN, Elazzazy AM et al (2014) Microalgal lipids biochemistry and biotechnological perspectives. Biotechnol Adv 32(8):1476–1493

Bilanovic D, Andargatchew A, Kroeger T et al (2009) Freshwater and marine microalgae sequestering of CO_2 at different C and N concentrations – response surface methodology analysis. Energy Convers Manage 50(2):262–267

Blunt JW, Munro MHG (2008) Dictionary of marine natural products, 1st edn. Champmann & Hall/CRC, Boca Raton

Bong SC, Loh SP (2013) A study of fatty acid composition and tocopherol content of lipid extracted from marine microalgae, *Nannochloropsis oculata* and *Tetraselmis suecica*, using solvent extraction and supercritical fluid extraction. Int Food Res J 20(2):721–729

Boussiba S, Bing W, Yuan JP et al (1999) Changes in pigments profile in the green alga *Haeamtococcus pluvialis* exposed to environmental stresses. Biotechnol Lett 21(7):601–604

Cadoret JP, Garnier M, Saint-Jean B (2012) Microalgae, functional genomics and biotechnology. In: Piganeau G (ed) Genomic insights into the biology of algae, vol 64. Academic, Cambridge, pp 285–341

Caroprese M, Albenzio M, Ciliberti MG et al (2012) A mixture of phytosterols from *Dunaliella tertiolecta* affects proliferation of peripheral blood mononuclear cells and cytokine production in sheep. Vet Immunol Immunopathol 150(1–2):27–35

Carullo D, Abera BD, Casazza AA et al (2018) Effect of pulsed electric fields and high pressure homogenization on the aqueous extraction of intracellular compounds from the microalgae *Chlorella vulgaris*. Algal Res 31:60–69

Carvalho JCM, Matsudo MC, Bezerra RP et al (2014) Microalgae bioreactors. In: Bajpai R, Prokop A, Zappi M (eds) Algal biorefineries, cultivation of cells and products, vol 1. Springer, Dordrecht, pp 83–126

Catarino MD, Silva AMS, Cardoso SM (2017) Fucaceae: a source of bioactive phlorotannins. Int J Mol Sci 18(6):1327

Chen J, Shearer GC, Chen Q et al (2011) Omega-3 fatty acids prevent pressure overload-induced cardiac fibrosis through activation of cyclic GMP/protein kinase G signaling in cardiac fibroblasts. Circulation 123(6):584–593

Chen J, Jiao R, Jiang Y et al (2014) Algal sterols are as effective as beta-sitosterol in reducing plasma cholesterol concentration. J Agric Food Chem 62(3):675–681

Christaki E, Bonos E, Giannenas I et al (2013) Functional properties of carotenoids originating from algae. J Sci Food Agric 93(1):5–11

Ciliberti MG, Francavilla M, Intini S et al (2017) Phytosterols from *Dunaliella tertiolecta* reduce cell proliferation in sheep fed flaxseed during post partum. Mar Drugs 15(7):216

Cunico LP, Turner C (2017) Supercritical fluids and gas-expanded liquids. In: Pena-Pereira F, Tobiszewski M (eds) The application of green solvents in separation processes. Elsevier, Atlanta, pp 155–214

de Jesus Raposo MF, Miranda Bernardo de Morais AM (2015) Microalgae for the prevention of cardiovascular disease and stroke. Life Sci 125:32–41

de Jesus Raposo MF, Santos Costa de Morais RM, Miranda Bernardo de Morais AM (2013) Health applications of bioactive compounds from marine microalgae. Life Sci 93(15):479–486

Dejoye C, Vian MA, Lumia G et al (2011) Combined extraction processes of lipid from *Chlorella vulgaris* microalgae: microwave prior to supercritical carbon dioxide extraction. Int J Mol Sci 12 (12):9332–9341

Derwenskus F, Metz F, Gille A et al (2019) Pressurized extraction of unsaturated fatty acids and carotenoids from wet *Chlorella vulgaris* and *Phaeodactylum tricornutum* biomass using subcritical liquids. GCB Bioenergy 11(1):335–344

Diaz-Santos E, Vila M, Vigara J et al (2016) A new approach to express transgenes in microalgae and its use to increase the flocculation ability of *Chlamydomonas reinhardtii*. J Appl Phycol 28 (3):1611–1621

Diplock AT, Agget PJ, Ashwell M et al (1999) Scientific concepts of functional foods in Europe consensus document. Br J Nutr 81(4):S1–S27

FAO (2010) Fats and fatty acids in human nutrition. Report of an expert consulation. FAO Food Nutr Pap 91:1–166

FAO (2018) The state of world fisheries and aquaculture [Online]. http://www.fao.org/state-of-fisheries-aquaculture/en/. Accessed 17 Sept 2018

Gallego R, Montero L, Cifuentes A et al (2018) Green extraction of bioactive compounds from microalgae. J Anal Test 2(2):109–123

García PA, Hernández AP, San Feliciano A et al (2018) Bioactive prenyl- and terpenyl-quinones/hydroquinones of marine origin. Mar Drugs 16(9):292

Gilbert-Lopez B, Mendiola JA, Fontecha J et al (2015) Downstream processing of *Isochrysis galbana*: a step towards microalgal biorefinery. Green Chem 17(9):4599–4609

Gilbert-Lopez B, Barranco A, Herrero M et al (2017a) Development of new green processes for the recovery of bioactives from *Phaeodactylum tricornutum*. Food Res Int 99:1056–1065

Gilbert-Lopez B, Mendiola JA, van den Broek LAM et al (2017b) Green compressed fluid technologies for downstream processing of *Scenedesmus obliquus* in a biorefinery approach. Algal Res 24:111–121

Giles GE, Mahoney CR, Kanarek RB (2013) Omega-3 fatty acids influence mood in healthy and depressed individuals. Nutr Rev 71(11):727–741

Giros A, Grzybowski M, Sohn VR et al (2009) Regulation of colorectal cancer cell apoptosis by the n-3 polyunsaturated fatty acids docosahexaenoic and eicosapentaenoic. Cancer Prev Res 2 (8):732–742

Golmakani MT, Mendiola JA, Rezaei K et al (2012) Expanded ethanol with CO_2 and pressurized ethyl lactate to obtain fractions enriched in gamma-linolenic acid from *Arthrospira platensis* (*Spirulina*). J Supercrit Fluids 62:109–115

Gong M, Bassi A (2016) Carotenoids from microalgae: a review of recent developments. Biotechnol Adv 34(8):1396–1412

Gouveia L, Coutinho C, Mendonca E et al (2008) Functional biscuits with PUFA-omega 3 from *Isochrysis galbana*. J Sci Food Agric 88(5):891–896

Grosso C, Valentao P, Ferreres F et al (2015) Alternative and efficient extraction methods for marine-derived compounds. Mar Drugs 13(5):3182–3230

Güven KC, Percot A, Sezik E (2010) Alkaloids in marine algae. Mar Drugs 8(2):269–284

Hadiyanto H, Suttrisnorhadi (2016) Response surface optimization of ultrasound assisted extraction (UAE) of phycocyanin from microalgae *Spirulina platensis*. Emir J Food Agric 28(4):227–234

Hagen C, Siegmund S, Braune W (2002) Ultrastructural and chemical changes in the cell wall of *Haematococcus pluvialis* (Volvocales, Chlorophyta) during aplanospore formation. Eur J Phycol 37(2):217–226

Hamed I, Ozogul F, Ozogul Y et al (2015) Marine bioactive compounds and their health benefits: a review. Comp Rev Food Sci F 14(4):446–465

Hernandez D, Solana M, Riano B et al (2014) Biofuels from microalgae: lipid extraction and methane production from the residual biomass in a biorefinery approach. Bioresour Technol 170:370–378

Herrero M, Ibáñez E (2015) Green processes and sustainability: an overview on the extraction of high added-value products from seaweeds and microalgae. J Supercrit Fluids 96:211–216

Herrero M, Ibáñez E (2018) Green extraction processes, biorefineries and sustainability: recovery of high added-value products from natural sources. J Supercrit Fluids 134:252–259

Herrero M, Castro-Puyana M, Mendiola JA et al (2013) Compressed fluids for the extraction of bioactive compounds. TrAC-Trends Anal Chem 43:67–83

Herrero M, Mendiola JA, Ibáñez E (2017) Gas expanded liquids and switchable solvents. Curr Opin Green Sustain Chem 5:24–30

Hoang Van C, Eun J-B (2017) Marine carotenoids: bioactivities and potential benefits to human health. Crit Rev Food Sci Nutr 57(12):2600–2610

Hodaifa G, Martinez ME, Sanchez S (2008) Use of industrial wastewater from olive-oil extraction for biomass production of *Scenedesmus obliquus*. Bioresour Technol 99(5):1111–1117

Hu X, Yang XQ, Li LH et al (2015) Antioxidant properties of microalgae protein hydrolysates prepared by neutral protease digestion. Appl Mech Mater 707:149–153
Huo S, Wang Z, Cui F et al (2015) Enzyme-assisted extraction of oil from wet microalgae *Scenedesmus* sp G4. Energies 8(8):8165–8174
Ibáñez E, Cifuentes A (2013) Benefits of using algae as natural sources of functional ingredients. J Sci Food Agric 93(4):703–709
Ibáñez E, Gilbert-López B, Mendiola JA et al (2017) Integrated green extraction processes for downstream processing of microalgae wet biomass (GREENμWETBIO). Spain patent application
Jaime L, Rodriguez-Meizoso I, Cifuentes A et al (2010) Pressurized liquids as an alternative process to antioxidant carotenoids' extraction from *Haematococcus pluvialis* microalgae. Lwt-Food Sci Technol 43(1):105–112
Jerez-Martel I, Garcia-Poza S, Rodriguez-Martel G et al (2017) Phenolic profile and antioxidant activity of crude extracts from microalgae and cyanobacteria strains. J Food Qual 4:1–8
Juin C, Cherouvrier J-R, Thiery V et al (2015) Microwave-assisted extraction of phycobiliproteins from *Porphyridium purpureum*. Appl Biochem Biotechnol 175(1):1–15
Kapoore RV, Butler TO, Pandhal J et al (2018) Microwave-assisted extraction for microalgae: from biofuels to biorefinery. Biology 7(1):18
Kazlowska K, Lin H-TV, Chang S-H et al (2013) In vitro and in vivo anticancer effects of sterol fraction from red algae *Porphyra dentata*. Evid-Based Compl Alt 493869
Kim SK, Himaya SWA (2011) Medicinal effects of phlorotannins from marine brown algae. In: Kim SK (ed) Advances in food and nutrition research, Marine medicinal foods: implications and applications, macro and microalgae, vol 64. Elsevier Academic, Atlanta, pp 97–109
Kumari P, Kumar M, Gupta V et al (2010) Tropical marine macroalgae as potential sources of nutritionally important PUFAs. Food Chem 120(3):749–757
Kumari P, Bijo AJ, Mantri VA et al (2013) Fatty acid profiling of tropical marine macroalgae: an analysis from chemotaxonomic and nutritional perspectives. Phytochemistry 86:44–56
Leon R, Fernandez E (2007) Nuclear transformation of eukaryotic microalgae – historical overview, achievements and problems. Adv Exp Med Biol 616:1–11
Luo X, Su P, Zhang W (2015) Advances in microalgae-derived phytosterols for functional food and pharmaceutical applications. Mar Drugs 13(7):4231–4254
Lv J, Yang X, Ma H et al (2015) The oxidative stability of microalgae oil (*Schizochytrium aggregatum*) and its antioxidant activity after simulated gastrointestinal digestion: relationship with constituents. Eur J Lipid Sci Technol 117(12):1928–1939
Machado FRS Jr, Trevisol TC, Boschetto DL et al (2016) Technological process for cell disruption, extraction and encapsulation of astaxanthin from *Haematococcus pluvialis*. J Biotechnol 218:108–114
Macías-Sánchez MD, Fernandez-Sevilla JM, Acien-Fernandez FG et al (2010) Supercritical fluid extraction of carotenoids from *Scenedesmus almeriensis*. Food Chem 123(3):928–935
Marris E (2006) Black is the new green. Nature 442:624
Martinez-Guerra E, Gude VG, Mondala A et al (2014) Extractive-transesterification of algal lipids under microwave irradiation with hexane as solvent. Bioresour Technol 156:240–247
Mata TM, Martins AA, Caetano NS (2010) Microalgae for biodiesel production and other applications: a review. Renew Sustain Energy Rev 14(1):217–232
Matos J, Cardoso C, Bandarra NM et al (2017) Microalgae as healthy ingredients for functional food: a review. Food Funct 8(8):2672–2685
Mendiola JA, Castro-Puyana M, Herrero M et al (2013) Green foodomics. In: Cifuentes A (ed) Foodomics: advanced mass spectrometry in modern food science and nutrition. Wiley, Hoboken, pp 471–505
Metting FB (1996) Biodiversity and application of microalgae. J Ind Microbiol Biotechnol 17 (5–6):477–489
Montero L, Sanchez-Camargo AP, Garcia-Canas V et al (2016) Anti-proliferative activity and chemical characterization by comprehensive two-dimensional liquid chromatography coupled

to mass spectrometry of phlorotannins from the brown macroalga *Sargassum muticum* collected on North-Atlantic coasts. J Chrom A 1428:115–125

Munish P, Deepika S, Barrow CJ (2012) Enzyme-assisted extraction of bioactives from plants. Trends Biotechnol 30(1):37–44

Oncel SS (2017) Microalgae, taking over the role in the hydrogen future. In: Pires JCM (ed) Recent advances in renewable energy, Microalgae as a source of bioenergy: products, processes and economics, vol 1. Bentham Science, Sharjah, pp 98–149

Ota M, Watanabe H, Kato Y et al (2009) Carotenoid production from *Chlorococcum littorale* in photoautotrophic cultures with downstream supercritical fluid processing. J Sep Sci 32 (13):2327–2335

Otero P, Quintana SE, Reglero G et al (2018) Pressurized liquid extraction (PLE) as an innovative green technology for the effective enrichment of Galician algae extracts with high quality fatty acids and antimicrobial and antioxidant properties. Mar Drugs 16(5):156

Pangestuti R, Kim S-K (2011) Neuroprotective effects of marine algae. Mar Drugs 9(5):803–818

Passos F, Carretero J, Ferrer I (2015) Comparing pretreatment methods for improving microalgae anaerobic digestion: thermal, hydrothermal, microwave and ultrasound. Chem Eng J 279:667–672

Patil PD, Reddy H, Muppaneni T et al (2013) In situ ethyl ester production from wet algal biomass under microwave-mediated supercritical ethanol conditions. Bioresour Technol 139:308–315

Pieber S, Schober S, Mittelbach M (2012) Pressurized fluid extraction of polyunsaturated fatty acids from the microalga *Nannochloropsis oculata*. Biomass Bioenerg 47:474–482

Plaza M, Herrero M, Cifuentes A et al (2009) Innovative natural functional ingredients from microalgae. J Agric Food Chem 57(16):7159–7170

Plaza M, Santoyo S, Jaime L et al (2010) Screening for bioactive compounds from algae. J Pharm Biomed Anal 51(2):450–455

Plaza M, Santoyo S, Jaime L et al (2012) Comprehensive characterization of the functional activities of pressurized liquid and ultrasound-assisted extracts from *Chlorella vulgaris*. Lwt-Food Sci Technol 46(1):245–253

Pottel L, Lycke M, Boterberg T et al (2014) Omega-3 fatty acids: physiology, biological sources and potential applications in supportive cancer care. Phytochem Rev 13(1):223–244

Reyes FA, Mendiola JA, Ibáñez E et al (2014) Astaxanthin extraction from *Haematococcus pluvialis* using CO_2-expanded ethanol. J Supercrit Fluids 92:75–83

Reyes FA, Mendiola JA, Suarez-Alvarez S et al (2016) Adsorbent-assisted supercritical CO_2 extraction of carotenoids from *Neochloris oleoabundans* paste. J Supercrit Fluids 112:7–13

Rodriguez-Meizoso I, Jaime L, Santoyo S et al (2010) Subcritical water extraction and characterization of bioactive compounds from *Haematococcus pluvialis* microalga. J Pharm Biomed Anal 51(2):456–463

Samarakoon K, Jeon Y-J (2012) Bio-functionalities of proteins derived from marine algae – a review. Food Res Int 48(2):948–960

Samarakoon KW, Ko J-Y, Lee J-H et al (2014) Apoptotic anticancer activity of a novel fatty alcohol ester isolated from cultured marine diatom, *Phaeodactylum tricornutum*. J Funct Food 6:231–240

Sánchez-Camargo AdP, Mendiola JA, Ibáñez E (2018) Gas expanded-liquids. In: Hunt AJ, Attard TM (eds) Supercritical and other high-pressure solvent systems: for extraction, reaction and material processing, green chemistry Series No. 57. The Royal Society of Chemistry, London, pp 512–531

Saravana PS, Getachew AT, Yeon-Jin C et al (2017) Influence of co-solvents on fucoxanthin and phlorotannin recovery from brown seaweed using supercritical CO_2. J Supercrit Fluids 120(Part 2):295–303

Sathasivam R, Ki JS (2018) A review of the biological activities of microalgal carotenoids and their potential use in healthcare and cosmetic industries. Mar Drugs 16(1):26

Simopoulos AP (2002) The importance of the ratio of omega-6/omega-3 essential fatty acids. Biomed Pharmacother 56(8):365–379

Solana M, Rizza CS, Bertucco A (2014) Exploiting microalgae as a source of essential fatty acids by supercritical fluid extraction of lipids: comparison between *Scenedesmus obliquus*, *Chlorella protothecoides* and *Nannochloropsis salina*. J Supercrit Fluids 92:311–318
Souza RB, Frota AF, Silva J et al (2018) In vitro activities of kappa-carrageenan isolated from red marine alga *Hypnea musciformis*: antimicrobial, anticancer and neuroprotective potential. Int J Biol Macromol 112:1248–1256
Subhadra B, Grinson G (2011) Algal biorefinery-based industry: an approach to address fuel and food insecurity for a carbon-smart world. J Sci Food Agric 91(1):2–13
Suganya T, Varman M, Masjuki HH et al (2016) Macroalgae and microalgae as a potential source for commercial applications along with biofuels production: a biorefinery approach. Renew Sustain Energy Rev 55:909–941
Tanniou A, Vandanjon L, Incera M et al (2014) Assessment of the spatial variability of phenolic contents and associated bioactivities in the invasive alga *Sargassum muticum* sampled along its European range from Norway to Portugal. J Appl Phycol 26(2):1215–1230
Tatke P, Jaiswal Y (2011) An overview of microwave assisted extraction and its applications in herbal drug research. Res J Med Pla 5(1):21–31
Thanh-Sang V, Dai-Hung N, Kim S-K (2012) Marine algae as a potential pharmaceutical source for anti-allergic therapeutics. Process Biochem 47(3):386–394
Vanthoor-Koopmans M, Wijffels RH, Barbosa MJ et al (2013) Biorefinery of microalgae for food and fuel. Bioresour Technol 135:142–149
Villarruel-Lopez A, Ascencio F, Nuno K (2017) Microalgae, a potential natural functional food source – a review. Polish J Food Nutr Sci 67(4):251–263
Yates CM, Calder PC, Rainger GE (2014) Pharmacology and therapeutics of omega-3 polyunsaturated fatty acids in chronic inflammatory disease. Pharmacol Ther 141(3):272–282
Yen H-W, Hu IC, Chen C-Y et al (2013) Microalgae-based biorefinery – from biofuels to natural products. Bioresour Technol 135:166–174
Yoshie-Stark Y, Hsieh Y-P, Suzuki T (2003) Distribution of flavonoids and related compounds from seaweeds in Japan. J Tokyo Univ Fish 89:1–6
Yu Ran H, Yousof Ali M, Mi-Hee W et al (2015) Anti-diabetic and anti-inflammatory potential of the edible brown alga *Hizikia fusiformis*. J Food Biochem 39(4):417–428
Zhao G, Chen X, Wang L et al (2013) Ultrasound assisted extraction of carbohydrates from microalgae as feedstock for yeast fermentation. Bioresour Technol 128:337–344
Zuorro A, Maffei G, Lavecchia R (2016) Optimization of enzyme-assisted lipid extraction from *Nannochloropsis microalgae*. J Taiwan Inst Chem E 67:106–114

Part IV
Framework and Progress of Practical Applications

Chapter 12
Bioactive Compounds from Microalgae and Their Potential Applications as Pharmaceuticals and Nutraceuticals

Wan-Loy Chu and Siew-Moi Phang

Abstract Microalgae produce a wide array of bioactive compounds with potential applications such as pharmaceuticals and nutraceuticals that are yet to be fully explored. Metabolites from microalgae are known to have various bioactivities including antibacterial, antiviral, anticancer and immunomodulatory activities as well as beneficial effects against metabolic disorders such as obesity, diabetes and hyperlipidaemia. Production of pharmaceuticals and nutraceuticals derived from microalgae is attractive as the photosynthetic organisms can be grown on a large scale under controlled environment. Furthermore, the production platform can be integrated with other applications, especially biofuel production, to improve economic competitiveness based on a biorefinery concept. Microalgae that have been successfully cultured on a large scale for commercial production of pharmaceuticals and nutraceuticals include *Spirulina (Arthrospira)*, *Dunaliella*, *Chlorella* and *Haematococcus*. While various bioactive compounds have been discovered from microalgae, most studies on those compounds are still at the preclinical stage, confined to laboratory experiments. Overall, only very few microalgal products have been successfully brought to the market. This chapter aims to give an overview of the wide range of potential pharmaceuticals and nutraceuticals from microalgae and current status of the production of such compounds as well as to highlight the future directions of research in this field.

W.-L. Chu (✉)
School of Postgraduate Studies, International Medical University, Kuala Lumpur, Malaysia
e-mail: wanloy_chu@imu.edu.my

S.-M. Phang
Institute of Biological Sciences, University of Malaya, Kuala Lumpur, Malaysia

Institute of Ocean and Earth Sciences (IOES), University of Malaya, Kuala Lumpur, Malaysia

A. Hallmann, P. H. Rampelotto (eds.), *Grand Challenges in Algae Biotechnology*, Grand Challenges in Biology and Biotechnology,
https://doi.org/10.1007/978-3-030-25233-5_12

12.1 Introduction

Microalgae comprise a diverse group of organisms with an estimated number of species of 72,500, but only about 44,000 have been described (de Clerck et al. 2013). This group of organisms consists of the eukaryotic (microalgae) and prokaryotic taxa (cyanobacteria). Apart from the rich biological diversity, microalgae are also known to produce a myriad of metabolites, many of which have potential biomedical applications that have yet to be fully explored (Coates et al. 2013). The breadth of chemical diversity of secondary metabolites from microalgae is related to the vast number of environments they inhabit. While microalgae are mainly found in aquatic environments (49.78% in freshwater and 44.48% in seawater), some species also grow in contaminated and extreme environments including thermal and glacial lakes (Cezare-Gomes et al. 2019). The production of secondary metabolites is related to their possible ecological roles including for defence of the microalgae against predators, protection against infection by various microorganisms and sexual communication (Coates et al. 2013). Interestingly, some of the bioactive compounds that were first isolated from invertebrates were later found to be produced by symbiotic microalgae in those animals. For instance, the anticancer compound dolastatin was originally isolated from the sea hare (*Dolabella auricularia*) but was later discovered to be synthesized by the cyanobacteria *Lyngbya* and *Symploca* (Luesch et al. 2002).

Microalgae are a potential and attractive source of bioactive compounds, as they can be easily cultured, and have short generation times, with doubling time of 3.9–8.1 h for some species (Table 12.1). The doubling time of microalgae can be further shortened by manipulating the culture conditions (de Swaaf et al. 2003; Chen et al. 2013; Table 12.1). Furthermore, microalgae can synthesize structurally complex molecules, which are difficult or impossible to produce by chemical synthesis (Dewi et al. 2018). The possibility of culturing those microalgae also allows an environmentally friendly way to drug discovery, overcoming problems such as overutilization of natural bioresources and use of destructive collection practices (Lauritano et al. 2016). In addition, a "milking" approach can be used where extraction of the high-value products (e.g. use of biocompatible organic solvents and pulsed electric field) does not kill the algae and thus, the process is more environmentally and economically friendly (Vinayak et al. 2015). Large-scale cultivation of microalgae in photobioreactors is possible to produce large volumes of cultures, ranging from 1000 to 6000 L of microalgae such as *Chlorella* and *Spirulina*, for the extraction of bioactive compounds (Borowitzka 1999). In addition, microalgae are potential candidates for exploration of new compounds due to their metabolic plasticity, which can be exploited for the production of several compounds with applications in various biotechnology sectors including food, energy, health, environment and biomaterials (Martinez Andrade et al. 2018).

Regulatory bodies such as the US Food and Drug Administration (FDA) set the standards that need to be met before any microalgal product is certified safe for human consumption (Gantar and Svircev 2008). Some of the microalgal products have attained GRAS (generally recognized as safe) status and are marketed as

Table 12.1 Specific growth rates (μ, day^{-1}) and doubling time (h) of selected microalgae grown in different media and culture conditions

Microalgal species	Culture medium	Culture condition	Specific growth rate (μ, day^{-1})	Doubling time (h)	References
Crypthecodinium cohnii	Medium containing glucose, yeast extract and sea salt	2 L bioreactor (Fed-batch, glucose levels maintained at between 5 and 20 g/L)	2.06	8.1	de Swaaf et al. (2003)
Nannochloropsis sp. *Isochrysis* sp. *Tetraselmis* sp. *Rhodomonas* sp.	L1, f/2 and K-medium	15 L carboys (Aerated batch cultures)	0.32–0.41 0.10–0.25 0.14–0.24 0.26–0.30	40.6–52.0 66.5–166.3 69.3–118.8 55.4–64.0	Huerlimann et al. (2010)
Spirulina platensis	Defined inorganic medium (16.8 g $NaHCO_3$/L)	Flat-type photobioreactor (Irradiance up to 1300 μmol/m^2/s)	4.31 (μ_{max})	3.9	Chen et al. (2013)
Ankistrodesmus falcatus	COMBO medium	1 L flasks (Manipulation of nitrogen and phosphorus levels, and irradiance)	1.03–1.56 (μ_{max})	10.7–16.1	Alvarez-Diaz et al. (2014)
Chlorella vulgaris	Bold Basal Medium (BBM)	250 mL flasks (Aerated cultures with different medium concentrations—25, 50 and 100%)	0.37–0.38 (μ_{max})	43.8–45.0	Blair et al. (2014)
Scenedemus obliquus	Final effluent of wastewater treatment plant	High Rate Algae Pond (Working volume—266 vs. 533 L)	0.37–0.65 (μ_{max})	25.6–45.0	Arbib et al. (2017)
Amphidinium carterae	Modified K-medium	T-flasks (60 mL) (Daily mean irradiance > 50 μE/m^2/s)	0.60 (μ_{max})	27.7	Molina-Miras et al. (2018)

functional food (Gantar and Svircev 2008; Jha et al. 2017a). The most widely sold microalgal products as functional food include *Spirulina* (*Arthrospira*) and *Chlorella* (Andrade 2018). In addition, *Haematococcus pluvialis* has been approved for human consumption as dietary supplement in several European countries (Lorenz and Cysewski 2000). The food-grade microalgae are a rich source of nutraceuticals such as carotenoids (β-carotene and astaxanthin), C-phycocyanin, long-chain polyunsaturated fatty acids (LC-PUFA), vitamins and polysaccharides (Lorenz and Cysewski 2000; Andrade 2018).

Microalgae produce a wide array of metabolites with diverse bioactivities including anticancer, antibacterial, antiviral and immunomodulatory activities and ameliorating effects against metabolic disorders such as diabetes and hyperlipidaemia (Nagaoka et al. 2005; Renju et al. 2014a; Aboulthana et al. 2018; Dewi et al. 2018). Apart from the limited number of microalgal species that have reached commercial status, the rich microalgal resources with the mentioned bioactivities have not been fully exploited as sources of pharmaceuticals and nutraceuticals (Jha et al. 2017a). There are several reports on large-scale screening of microalgae for bioactive compounds. For instance, in an early study, screening of 1000 cyanobacterial strains from diverse habitats for novel antineoplastic agents identified the families Scytonemataceae and Stigonemataceae as prolific producers of novel cytotoxic compounds (Patterson et al. 1991). Recently, Caires et al. (2018) screened various strains of the benthic filamentous cyanobacteria *Neolyngbya* for antimicrobial and antioxidant activities. Solvent extracts of the cyanobacteria showed inhibitory activity against the tested bacteria and *Candida albicans*, and the methanolic extract showed the greatest antioxidant activity. In another study, Lauritano et al. (2016) screened 21 diatoms, seven dinoflagellates, and four flagellate species for antioxidant, anti-inflammatory, anticancer, antidiabetic, antibacterial, and antibiofilm activities.

The aim of this chapter is to give an overview of the wide range of potential pharmaceuticals and nutraceuticals derived from microalgae. The aspect on mass-culture systems, particularly photobioreactors, currently used to produce microalgal biomass for the extraction of pharmaceuticals and nutraceuticals is also discussed. This chapter also aims to highlight the challenges faced in the commercialization of microalgal products as pharmaceuticals and nutraceuticals, and to propose some directions for future research.

12.2 Bioactivities of Microalgal Compounds and Their Potential Applications as Pharmaceuticals and Nutraceuticals

Microalgal compounds with potential for development into pharmaceuticals and nutraceuticals include astaxanthin, fucoxanthin, C-phycocyanin, sulphated polysaccharides and novel secondary metabolites (Figs. 12.1 and 12.2). In the subsequent sections, the following bioactivities of microalgal compounds will be covered in detail: antibacterial, antiviral, anticancer and immunomodulatory activities, and beneficial effects against metabolic disorders such as obesity, diabetes and hyperlipidaemia. A summary of the microalgal compounds with antibacterial,

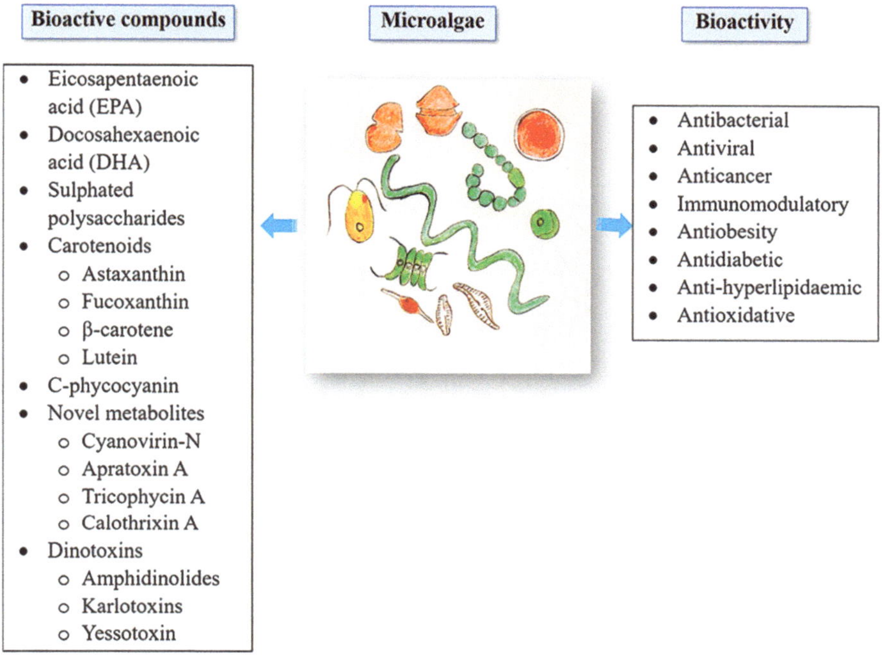

Fig. 12.1 Schematic overview of the bioactive compounds from microalgae

antiviral and immunomodulatory activities is given in Table 12.2 while those with anticancer activity are listed in Table 12.3.

12.2.1 Antibacterial Activity

There has been a long history in attempts to isolate antibacterial compounds from microalgae since the discovery of chlorellin, which is a mixture of fatty acids, from *Chlorella vulgaris* and *Chlorella pyrenoidosa* for its activity against various Gram-positive and Gram-negative bacteria (Pratt et al. 1944). Since then, a variety of antibacterial compounds from microalgae have been characterized, including fatty acids (Desbois et al. 2009), phenolic compounds (Kuntzler et al. 2018), biphenyls (Volk and Furkert 2006), polysaccharides (de Jesus Raposo et al. 2015) and nanoparticles (Patel et al. 2015).

Amongst the fatty acids from microalgae which display antibacterial activity include eicosapentaenoic acid (EPA) and palmitic acid. For instance, Desbois et al. (2009) showed that EPA from *Phaeodactylum tricornutum* displayed potent antibacterial activity against diverse bacteria including multi-resistant *Staphylococcus aureus* (MRSA). Recently, Wang et al. (2018a) obtained two active fractions from *Phaeodactylum tricornutum*, a pure fraction of EPA and another (fraction B)

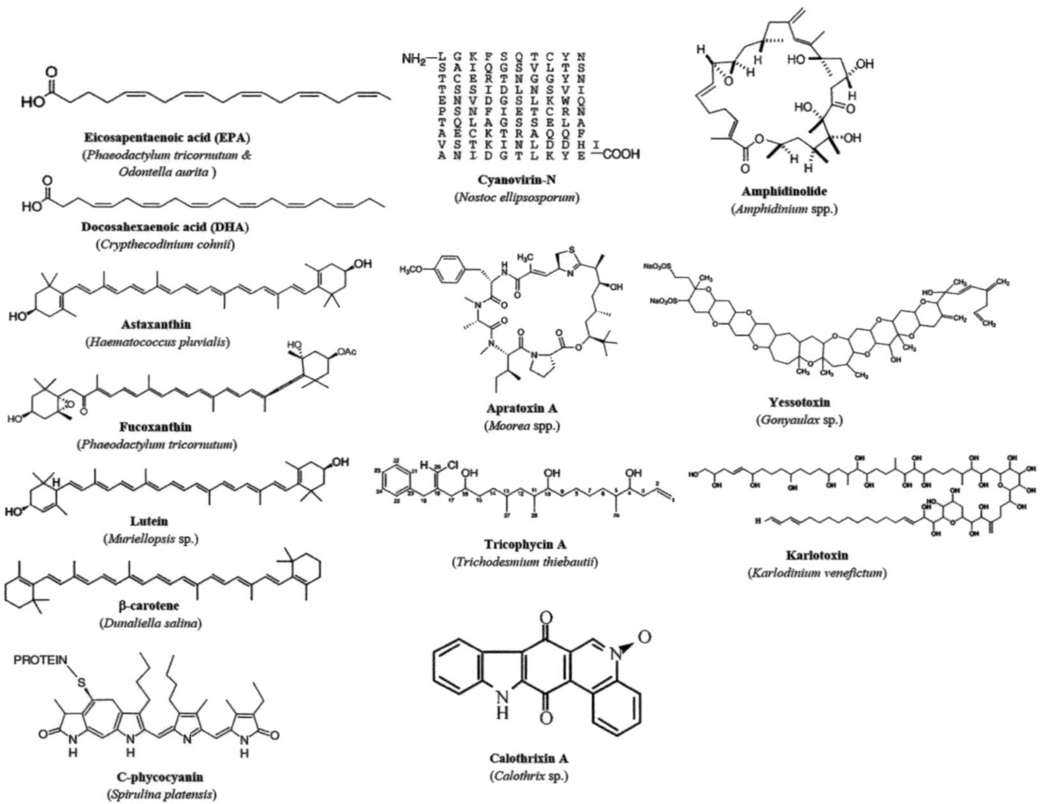

Fig. 12.2 Selected bioactive compounds derived from microalgae with potential for the development of pharmaceuticals and nutraceuticals. Example of microalgal producer for each compound is given in parenthesis (Burja et al. 2001; Chen et al. 2003; Mori et al. (2004); Mimouni et al. 2012; Coates et al. 2013; Rodrigues et al. 2015; Assunção et al. 2017; Bertin et al. 2017)

Table 12.2 Summary of findings on antibacterial, antiviral and immunomodulatory activities of selected compounds from microalgae

Compound	Class of compound	Microalgal species	Findings	References
Antibacterial activity				
Eicosapentaenoic acid (EPA)	Polyunsaturated fatty acids (PUFA)	*Phaeodactylum tricornutum*	Active against a range of both Gram-positive and Gram-negative bacteria, including multidrug-resistant *Staphylococcus aureus* (MRSA); IC_{50} = 10–20 μM	Desbois et al. (2009)
Hapalindoles	Indole alkaloids	*Westiellopsis* sp. *Fischerella muscicola*	MIC values against *Mycobacterium tuberculosis*: Haplindole X = 2.5 μM Haplindole I = 2.0 μM Haplindole A =< 0.6 μM	Kim et al. (2012a)
Silver chloride nanoparticles (AgCl-NP)	Nanoparticles (synthesized using microalgae)	*Chlorella vulgaris*	AgCl-NP (from 10 μg/mL) reduced 98% of the growth of *Staphylococcus aureus* (Gram-positive) and *Klebsiella pneumoniae* (Gram-negative)	da Silva Ferreira et al. (2017)
Palmitic acid methyl ester (C16:0)	Fatty acid methyl esters (FAME)	*Scenedesmus intermedius*	Minimum inhibitory concentration (MIC): Gram-positive bacteria = 24–48 μg/mL Gram-negative bacteria = 12–24 μg/mL	Davoodbasha et al. (2018)
Eicosapentaenoic acid (EPA)	Polyunsaturated fatty acids (PUFA)	*Phaeodactylum tricornutum*	Two fractions, EPA and fraction B, displayed antibacterial activities against *Vibrio vulnificus*, *V. parahae molyticus*, and *V. cholerae*	Wang et al. (2018a)
Antiviral activity				
Calcium spirulan	Sulphated polysaccharides	*Spirulina platensis*	IC_{50} against: HIV = 9.3 ± 1.7 μg/mL Herpes simplex virus (HSV) = 9.7 ± 0.8 μg/mL Activity against HIV—complete inhibition of syncytium formation at >25 μg/mL; IC_{50} = 7.3 μg/mL	Hayashi et al. (1996)

(continued)

Table 12.2 (continued)

Compound	Class of compound	Microalgal species	Findings	References
Cell-wall sulphated polysaccharide	Sulphated polysaccharide	*Porphyridium* sp.	Active against HSV 1 and 2, and *Varicella zoster* virus (VZV)—50% inhibition at 1 μg/mL	Huleihel et al. (2001)
Scytovirin	Proteins	*Scytonema varium*	Potent anticytopathic activity against HIV-1: $EC_{50} = 0.3–22$ nM	Bokesch et al. (2003)
Cyanovirin-N	Proteins	*Nostoc ellipsosporum* (recombinant protein produced in *E. coli*)	Potent antiviral activity against influenza A and B virus—$EC_{50} = 0.004–0.04$ μg/mL	O'Keefe et al. (2003)
Naviculan	Sulphated polysaccharide	*Navicula directa*	Active against HSV type 1 and 2, and influenza A virus—selectivity indices (CC_{50}/IC_{50}) = 270, 510 and 32, respectively	Lee et al. (2006)
p-KG03	Sulphated polysaccharides	*Gyrodinium impudium*	Activity against influenza type A virus—$EC_{50} = 0.19–0.48$ μg/mL; selective index >200	Kim et al. (2012b)
Protein hydrolysates (rich in Arg, Lys, Ala and His)	Proteins	*Scenedesmus obliquus*	Activity against Coxsackie B_3 virus—57.6–66.2% inhibition at 100 μg/mL	Afify et al. (2018)
Immunomodulatory activity				
C-phycocyanin (C-PC)	Phycobiliproteins	*Spirulina maxima*	Inhibition of allergic inflammatory response and histamine release from isolated rat mast cells	Remirez et al. (2002)
Sulphated polysaccharide	Polysaccharides	*Porphyridium* sp.	Topical application of the polysaccharide in human subjects inhibited cutaneous erythema induced by a known irritant In vitro study—the polysaccharide inhibited migration of polymorphonuclear leukocytes (PMN) toward a standard chemoattractant and blocked adhesion of PMN to endothelial cells	Matsui et al. (2003)

Violaxanthin	Carotenoids	*Chlorella ellipsoidea*	Anti-inflammatory effect—significantly inhibited nitric oxide (NO) and prostaglandin E2 (PGE2), and NK-κB pathway in LPS-stimulated RAW 264.7 macrophage cells	Soontornchaiboon et al. (2012)
Lycopene	Carotenoids	*Chlorella marina*	Significant anti-inflammatory effect in high-cholesterol-fed rats—activities of inflammatory marker enzymes decreased	Renju et al. (2014a)
ROQUETTE *Schizochytrium* sp. (SC) (marine microalgal formulation)	PUFA, especially DHA	*Schizochytrium* sp.	Oral (250 and 500 mg/kg) and topical (5.0 and 10.0%) administration—significant effects on reducing macroscopic score of skin inflammation in mice	Hidalgo-Lucas et al. (2014b)
Protein hydrolysate	Proteins	*Schizochytrium* sp.	Ameliorated bowel inflammation in mice-marked reductions in body weight loss, clinical colitis scores, colonic histological damage, and colonic inflammation Attenuated the induction of pro-inflammatory cytokines and increased the induction of anti-inflammatory cytokines	Wang et al. (2018b)

Table 12.3 Summary of findings on anticancer activity of selected compounds from microalgae

Compound	Class of compound	Microalgal species	Findings	References
Apratoxin A	Cyclic depsipeptides	*Lyngbya* spp. (*Moorea* spp.)	Mediated tumour cytotoxicity through the induction of cell cycle arrest and of apoptosis; interfered with tyrosine kinase signalling and ultimately inhibited the phosphorylation of transcription factor STAT3	Luesch et al. (2006)
Trichophycin A	Polyketides	*Trichodesmium thiebautii*	Anticancer activity against neuroblastoma cell line, Neuro-2A ($EC_{50} = 6.5$ μM) HCT-116 cells (EC_{50}: 11.7 μM)	Bertin et al. (2017)
Calothrixin A	Alkaloids	*Calothrix* sp.	Antiproliferative effect against Jurkat cancer cells ($IC_{50} = 1.6$ μM) via induction of apoptosis	Chen et al. (2003)
Curacin A	Lipids	*Lyngbya majuscula*	Antimitotic agent—inhibited microtubule assembly and binding of colchicine to tubulin IC_{50}: L1210 leukaemia cells $= 9 \times 10^{-9}$M Burkitt lymphoma cells $= 2 \times 10^{-7}$M	Gerwick et al. (1994)
Haplindole H	Alkaloids	*Fischerella muscicola*	Potent inhibitor of NF-κB (0.76 μM) with selective cytotoxicity against PC-3 prostate cancer cell line (0.02 μM)	Acuna et al. (2018)
Scytonemin	Pigments	*Stigonema* sp.	Inhibitor of *polo*-like kinase 1 activity ($IC_{50} = 2$ μM); induced apoptosis in Jurkat T cells	Stevenson et al. (2002)
C-phycocyanin	Phycobiliproteins	Not stated	Active against MDA-MB-231 cells ($IC_{50} = 5.98 \pm 0.95$ μM); reduced cell proliferation and colony formation ability; caused G1 arrest due to decreased expression of Cyclin-E and CDK-2 and increased p21 levels	Ravi et al. (2015)
Selenium-containing phycocyanin (Se-PC)	Phycobiliproteins	*Spirulina platensis*	Photodynamic therapy (PDT) with Se-PC showed higher tumour inhibition (75.4%) compared to PC PDT group (52.6%) in mice; Se-PC PDT induced cell death through free radical production of PDT in tumours and enhanced the activity of antioxidant enzymes with Se in vivo	Liu et al. (2018b)

Lycopene	Carotenoids	*Chlorella marina*	Inhibited the growth and colony formation of prostate cancer cell lines PC-3 and DU-145 at 20 and 50 μM respectively	Renju et al. (2014b)
Astaxanthin	Carotenoids	*Haematococcus pluvialis*	Significantly reduced proliferation rates and inhibited breast cancer cell migration compared to normal breast epithelial cells	McCall et al. (2018)
Astaxanthin-rich extract	Carotenoids	*Haematococcus pluvialis*	Astaxanthin-rich extract at 5–25 μg/mL inhibited cell growth of HCT-116 colon cancer cells by arresting cell cycle progression and by promoting apoptosis	Palozza et al. (2009)
Fucoxanthin	Carotenoids	*Undaria pinnatifida* (Wakame seaweed)	Significantly inhibited the growth of sarcoma at the dose of 50 or 100 mg/kg in xenografted sarcoma (S180) mice; suppressed expression of bcl-2 and enhanced expression of cleaved caspase-3	Wang et al. (2012b)
Amphidinol-2	Polyketides (dino-flagellate toxins)	*Amphidinium klebsii*	Cytotoxic against HCT-116, HT-29 and MCF-7 cancer cells: IC_{50} = 1.2–8.5 μg/mL; upregulation of the pre-apoptosis markers *cfos* and *cjun*	Espiritu et al. (2017)
D-galactan sulphate (GA3P)	Extracellular polysaccharides	*Gymnodinium* sp. A3	Potent inhibitor of DNA topoisomerase (topo) I and topo II; exhibited significant cytotoxicity against a variety of cancer cells	Umemura et al. (2003)
Cell wall sulphated polysaccharide	Polysaccharides	*Porphyridium cruentum*	In vivo—growth of Graffi myeloid tumour in hamsters retarded and the mean survival time prolonged by 10–16 days In vitro-induced apoptosis in primary Graffi tumour cells	Gardeva et al. (2014)
Nonyl 8-acetoxy-6-methyl-octanoate (NAMO)	Fatty alcohol esters	*Phaeodactylum tricornutum*	Strong suppression against human promyelocytic leukaemia (HL-60) cells—IC_{50} = 65.15 μM; induced apoptosis by activating Bax and suppressing Bcl-xL and upregulating caspase-3 and p53	Samarakoon et al. (2014)

containing new antibacterial substances that displayed antibacterial activities against human pathogenic bacteria. In addition, fatty acid methyl esters (FAME), particularly palmitic acid methyl ester (C16:0), extracted from *Scenedesmus intermedius* have also been found to display antibacterial activities against both Gram-positive and Gram-negative bacteria (Davoodbasha et al. 2018).

Recently, Senhorinho et al. (2018) screened 40 freshwater green microalgae from water bodies near abandoned mine sites in northern Canada. Of the 40 methanol extracts from the microalgae tested, 37.5% displayed strong inhibitory activity against the human pathogen *Staphylococcus aureus*, with Minimum Inhibitory Concentrations (MIC) ranging from 16 to >1024 μg/mL. The metabolites identified in the active extracts include amino acids, glycerolipids, sphingolipids, acylcarnitines and biogenic amines. However, there was no significant correlation between the metabolites analysed and antibacterial activity observed.

Nanoparticles (NP) such as silver and gold NP are well known to possess antibacterial activity. There has been increasing interest in using microalgae as bio-factory for the synthesis of such NP (Shankar et al. 2016). The green synthesis approach that uses cell-free extracts from the microalgae is more attractive compared to conventional physical and chemical methods, which are costly and involve the use of hazardous chemicals (Narayanan and Sakthivel 2011). Synthesis of NP may involve extracellular process, where the metal ions are trapped on the cell surface and reduced to elemental metal atoms by specific enzymes. On the other hand, in intracellular synthesis of NP, the ions are transported into the algal cell to form NP in the presence of enzymes.

Da Silva Ferreira et al. (2017) synthesized silver chloride nanoparticles (AgCl-NP) by incubating conditioned medium of *Chlorella vulgaris* with $AgNO_3$. The AgCl-NP formed was found to have potent antibacterial activity, reducing 98% the growth of Gram-positive *Staphylococcus aureus* and Gram-negative *Klebsiella pneumoniae* bacterial pathogens. In another study, Patel et al. (2015) evaluated the ability of 16 strains of cyanobacteria and microalgae to synthesize silver nanoparticles (Ag-NP). The Ag-NP were formed in the presence of the biomass as well as in the cell-free culture medium and might involve extracellular polysaccharide. In the same study, C-phycocyanin from *Spirulina* and *Limnothrix* were successfully used for the biosynthesis of Ag-NP. Further, the Ag-NP from all except one strain was found to display antibacterial activity against bacteria such as *Staphylococcus aureus* and *Pseudomonas aeruginosa*.

Some cyanotoxins have also been shown to be a promising source of novel antibiotics against multi-resistant bacteria. For instance, microcystin-LR displayed antimicrobial activity against four highly resistant species of *Mycobacterium*, with MIC ranging from 0.42 to 0.53 μM (Ramos et al. 2015). In addition, phenolic compounds from *Spirulina* LEB 18 were found to have antibacterial activity against *Escherichia coli* and *Staphylococcus aureus* when tested alone or incorporated into nanofibers (Kuntzler et al. 2018). The phenolic compounds in *Spirulina* contain organic acids such as caffeic, gallic, salicylic and *trans*-cinnamic acids (de Souza et al. 2015). In another study, simple phenols such as ferulic acid from the chloroform extracts of *Dunaliella* sp. accounted for the antibacterial activity against

Bacillus subtilis ATCC 6633 and *E. coli* 0157:H7 ATCC 35150 (Kim et al. 2012a). In addition, both purified β-carotene and chlorophyll pigments from the organic solvent extracts of *Chlorococcum humicola* were shown to display antibacterial activity against *Staphylococcus aureus*, *Vibrio cholerae* and *Klebsiella pneumoniae* (Bhagavathy et al. 2011).

12.2.2 Antiviral Activity

As viral infection is one of the main threats to human health, there has been great interest in looking for safe, effective and specific antiviral agents. Screening of water, methanolic or ethanolic extracts from microalgae for antiviral activity has been reported by several authors. For instance, in an early study, Lau et al. (1993) reported that only 2% of approximately 900 strains of cyanobacteria were able to inhibit reverse transcriptase activity of the human immunodeficiency virus (HIV) and avian myeloblastosis virus. In another study, Fabregas et al. (1999) screened extracts from ten microalgae for antiviral activity against the viral haemorrhagic septicaemia virus (VHSV) of salmonid fish and the African swine fever virus (ASFV). The study showed that aqueous extracts from *Porphyridium cruentum*, *Chlorella autotrophica* and *Ellipsoidon* sp. displayed significant inhibition of the replication of both viruses, which could be attributed to sulphated polysaccharides. In addition, Abdo et al. (2012) tested methanol and water extracts of five freshwater microalgae against Adenovirus Type 40 and found that only extracts from *Spirulina platensis* showed significant activity at non-cytotoxic concentrations. In another study, Kok et al. (2011) assessed the inhibitory activities of methanol extracts from *Ankistrodesmus convolutus*, *Synechococcus elongatus* and *Spirulina platensis* against Epstein–Barr virus (EBV) in three Burkitt's lymphoma (BL) cell lines. The methanol extracts were obtained by homogenizing the dried algal biomass in methanol and after centrifugation, the supernatant was evaporated and the dried extract was reconstituted in the solvent before use for the testing. The study found that the HPLC sub-fraction of the methanol extracts from *Synechococcus elongatus* (SEF1′) was most effective in reducing cell-free EBV DNA ($EC_{50} = 1.38$ μg/mL, therapeutic index >69). In addition, Komatsu et al. (2013) found that acidic polysaccharide fraction from *Coccomyxa gleobotrydiformi* displayed antiviral activity against influenza A virus through preventing the cell attachment and/or penetration of the virus.

Recently, Afify et al. (2018) found that protein hydrolysates obtained from *Scenedesmus obliquus* exhibited antiviral activity against Coxsackie B_3 virus (CVB_3). The inhibitory action of the hydrolysates was at the steps involving attachment, penetration, and adsorption of the viral particles. In addition, the antiviral activity of the hydrolysates was found to be positively correlated with the antioxidant activity of the extracts.

Polysaccharides are amongst the compounds that are known to possess antiviral activity and have great potential for pharmaceutical applications (Yu et al. 2018). For instance, a sulphated polysaccharide named calcium spirulan isolated from *Spirulina platensis* was shown to have potent antiviral activity against both HIV-1 and anti-herpes simplex virus type 1 (HSV-1) (Hayashi et al. 1996). In another study, Rechter et al. (2006) assessed the antiviral activity of intracellular and extracellular fractions containing spirulan-like molecules from *Spirulina*. The study showed that the spirulan-like fractions displayed strong inhibition against human cytomegalovirus, HSV-1, human herpesvirus type 6 and HIV-1, but only weak or no inhibition against EBV and influenza A virus. While for herpesviruses, the antiviral action of the fractions mainly targeted at viral entry, the action against HIV-1 occurred at a stage later than viral entry.

The sulphated polysaccharide p-KG03 purified from the dinoflagellate *Gyrodinium impudium* was found to exhibit antiviral activity against influenza type A virus (Kim et al. 2012b). The study further showed that the antiviral activity of the sulphated polysaccharide was associated with its interaction with the viral particles, particularly in the adsorption and internalization steps. In a subsequent study, Yim et al. (2004) showed that p-KG03 also displayed potent antiviral activity (EC_{50} = 26.9 μg/mL) against encephalomyocarditis virus (EMCV) without any cytotoxic effect. In addition, cell wall sulphated polysaccharide from the red microalga *Porphyridium* sp. was found to display strong antiviral activity against HSV-1 and 2 and *Varicella zoster* virus (VZV) (Huleihel et al. 2001). Another sulphated polysaccharide isolated from *Navicula directa*, named naviculan, was found to have antiviral activities against HSV-1 and 2, and influenza A virus (Lee et al. 2006). In addition, naviculan displayed inhibitory effect against cell–cell fusion between CD4-expressing and HIV gp160-expressing cells.

Amongst the cyanobacterial compounds, cyanovirin-N (CVN) is a potent antiviral agent against HIV that acts by inhibiting the virus cell entry in a highly specific manner (Lotfi et al. 2018). Cyanovirin-N (CVN) is a type of lectins, which are proteins of non-immune origin that are capable of recognizing and binding to glycoconjugate moieties non-covalently (Sharon and Lis 1989) There have been attempts to develop different expression systems for the recombinant production of CVN for use as an effective anti-HIV microbicide. Formulation of the microbicide can be in the form of vagina gels/rings, creams, lubricants or suppositories, delivering the active ingredients slowly during coitus or over extended periods of time.

Cyanovirin-N (CVN) was first isolated from the cultures of *Nostoc ellipsosporum* (Boyd et al. 1997). The protein was successfully produced recombinantly by expression of a corresponding DNA sequence in *E. coli*. The antiviral activity of CVN was partly due to its high-affinity interactions with the viral surface envelope glycoprotein gp120. Cyanovirin-N (CVN) has also been found to display antiviral activity against Ebola (Barrientos et al. 2003) and influenza virus (O'Keefe et al. 2003). The potency of CVN against viruses is attributed to the N-linked oligosaccharides with high mannose content that interact with the viral glycoproteins (Barrientos et al. 2006). This shows that the antiviral activity of CVN is not due to its induction of viral particle agglutination, but via inhibition of viral entry.

Other lectins from cyanobacteria which have been shown to exhibit anti-HIV activity include scytovirin (SVN) from *Scytonema varium* (Bokesch et al. 2003) and agglutinin from *Oscillatoria agardhii* (OAA) (Sato et al. 2007). O'keefe et al. (2011) have filed a patent on the use scytovirin for treating infections caused by high mannose-enveloped viruses including hepatitis C virus (HCV). Recently, Siqueira et al. (2018) performed a genomic screening of new putative antiviral lectins from the genomic sequences of Amazonian cyanobacteria using a bioinformatics approach. Amongst the seven strains of cyanobacteria screened, 75 unique coding sequences for one or more lectin domains were identified. Using homology modelling and molecular dynamics simulations, the study discovered that *Nostoc* sp. CACIAM 19 and *Tolypothrix* sp. CCAM 22 strains presented cyanovirin-N homologues.

12.2.3 Immunomodulatory Activity

Several bioactive compounds from microalgae have been shown to possess immunomodulatory activity, especially anti-inflammatory effect. For instance, lycopene (*cis*/all-*trans* 40:60) from *Chlorella marina* was found to show significant antioxidant and anti-inflammatory effect in high-cholesterol fed rats (Renju et al. 2014a). Inflammatory biomarkers such as cyclooxygenase, 15-lipoxygenase and myeloperoxidase activity, and C-reactive protein and ceruloplasmin levels in serum were found to decrease significantly in rats fed algal lycopene. In another study, lycopene from *Chlorella marina* was found to reduce significantly inflammatory marker enzymes such as cyclooxygenase, lipoxygenase and myeloperoxidase in an arthritis rat model, compared to tomato lycopene and the anti-inflammatory drug indomethacin (Renju et al. 2013). There was also a marked reduction in oedema of paw and joint tissues in the rats supplemented with the algal lycopene. The authors further suggest that algal lycopene could be a potential agent for the treatment of anti-inflammatory diseases such as arthritis. In addition, violaxanthin isolated from *Chlorella ellipsoidea* was found to significantly inhibit nuclear factor-κB (NF-κB) pathways in exerting its anti-inflammatory effect against lipopolysaccharide (LPS)-stimulated RAW 264.7 mouse macrophage cells (Soontornchaiboon et al. 2012).

Another pigment which has been shown to have immunomodulatory effect is C-phycocyanin, which is one of the major pigments in cyanobacteria, especially *Spirulina*. C-phycocyanin has been shown to display inhibitory activity against allergic responses, such as ear swelling, skin reactions and histamine release from mast cells in a rat model (Remirez et al. 2002). In addition, C-phycocyanin was found to reduce allergic inflammation by suppressing antigen-specific IgE response in a mouse model (Nemoto-Kawamura et al. 2004). In another study, water extract from *Botryococcus braunii* exerted anti-inflammatory effect by inhibiting the expression of nitric oxide synthase (iNOS) gene and the consequent production of nitric oxide (NO) under oxidative stress in RAW 264.7 murine macrophages (Buono et al. 2012). Recently, Wang et al. (2018b) showed that protein-rich materials after extraction of DHA from bioengineered *Schizochytrium* sp. ameliorated bowel inflammation in mice. The protein hydrolysate attenuated the induction of pro-inflammatory cytokines and increased the induction of anti-inflammatory cytokines. In addition, the protein hydrolysate promoted the proliferation of colonic crypt stem cells and progenitor cells required for crypt repair. In another study, feeding of *Schizochytrium* oil rich in DHA has been shown to increase plasma level of the anti-inflammatory cytokine TGF-β1 (Komprda et al. 2016). In another study, supplementation of *Spirulina* was found to enhance primary immune response in terms of antibody production following tetanus toxoid vaccination in a mouse model (Chu et al. 2013).

There has also been interest in screening of microalgae for compounds for skin health, including anti-inflammatory, wound-healing and antioxidative effects (Kim et al. 2018). For instance, Hidalgo-Lucas et al. (2014a) assessed the beneficial effect of a formulation based on *Chlorella sorokiniana* (ROQUETTE *Chlorella* sp., RC) on two dermatological disorder models in mice, namely skin inflammation and wound healing. Oral and topical administration of RC at high doses showed significant effects in reducing skin inflammation and had efficient effect on wound-healing process. Similarly, topical and oral administration of formulation based on *Schizochytrium* (ROQUETTE *Schizochytrium* sp.) had significant effects on macroscopic score of skin inflammation and efficient effect on healing process, especially at high doses (Hidalgo-Lucas et al. 2014b). In addition, supplementation of *Chlorella vulgaris* has been able to reduce atopic dermatitis-like symptoms, including dermatitis scores, epidermal thickness and skin hydration, in a mouse model (Kang et al. 2015). In the same study, treatment of *Chlorella vulgaris* was found to downregulate serum levels of thymus- and activation-regulated chemokine (TARC) and macrophage-derived chemokine (MDC). Also, the treatment downregulated mRNA expression levels of IL-4 and IFN-γ. In another study, sulphated polysaccharide of *Porphyridium* was found to be an effective anti-inflammatory agent for topical use. Topical application of the polysaccharide on human subjects was found to inhibit the development of erythema induced by a powerful irritant (Matsui et al. 2003).

12.2.4 Anticancer Activity

Cancer is the second leading cause of death globally, accounting for an estimate of 9.6 million deaths in 2018 (WHO 2018). There has been much interest in exploring for new therapeutic agents for cancer, particularly from natural products, in view of the side effects posed by the current treatment strategies (Pádua et al. 2015) and the development of resistance of tumour cells against currently available drugs (Li et al. 2018). Microalgae are a potential source of anticancer compounds that have yet to be fully explored (Abd El-Hack et al. 2018). Potential anticancer compounds from microalgae include pigments (C-phycocyanin, astaxanthin and fucoxanthin), bioactive peptides, lipid compounds, alkaloids and dinoflagellate toxins (Table 12.3). The anticancer effects of the microalgal compounds are mediated through their action in

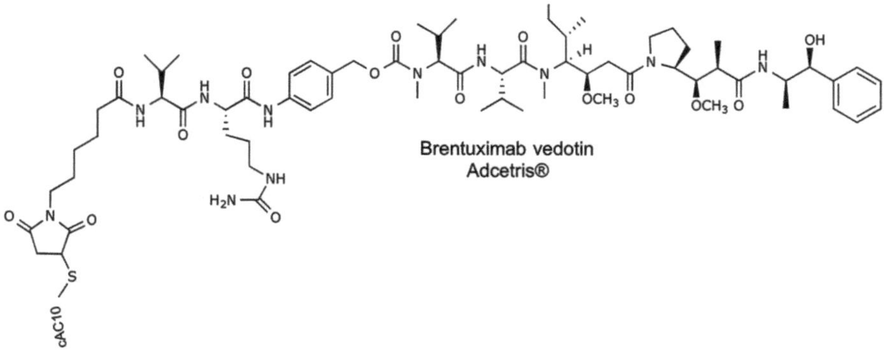

Fig. 12.3 Brentuximab vedotin, conjugation of antibody with dolastatin 10 (cyanobacterial metabolite), is a drug that has been approved by the US FDA for the treatment of Hodgkin's lymphoma (Alves et al. 2018)

inducing cytotoxicity, downregulating invasion of tumour cells, and enhancing cancer cell apoptosis.

A wide range of novel metabolites, especially bioactive peptides, from cyanobacteria have been shown to display anticancer activity (Gerwick and Moore 2012). For instance, symplopastin 1, a linear peptide isolated from *Symploca hydroides*, was found to be a potent microtubule inhibitor that was effective against drug-insensitive mammary tumour and drug-insensitive colon tumour (Luesch et al. 2001). In addition, TZT-1027 (auristatin PE), an analogue of dolastatin (linear peptide), was tested in phase 1 clinical trials due to its good preclinical activity, particularly its antitumour activity against breast carcinoma and lung carcinoma in mice (Kobayashi et al. 1997). In another development, a drug named brentuximab vedotin (Fig. 12.3) developed based on antibody conjugated to dolastatin 10 has been approved by the US FDA for treating Hodgkin's lymphoma and anaplastic large-cell lymphoma (Gerwick and Moore 2012). Recently, a patent on novel derivatives of dolastatin 10 and auristatins, particularly on their methods of production and use as medicinal products in the treatment of cancer, was filed (Perez et al. 2018).

Cryptophycin-52 (LY355703), an analogue of the cyanobacterial depsipeptide cryptophycin, is another compound that has been brought into phase 1 clinical trial (Sessa et al. 2002). The trial was on patients with advanced solid tumour; however, adverse effects related to neuropathy and myalgia were reported. This was followed by a multi-centre phase 2 clinical trial on patients with non-small-cell-lung cancer (NSCLC) (Edelman et al. 2003). However, the results showed that cryptophycin-52 only has limited activity in NSCLC cells.

Other metabolites from cyanobacteria that have been found to display anticancer activity include apratoxin A, tricophycin A, calothrixin A and hapalindole H (Table 12.3). For instance, apratoxin A, a cyclic depsipeptide isolated from *Lyngbya* spp. (=*Moorea* spp.), was found to display potent inhibitory activity against cancer cell growth by inducing apoptosis and cell cycle arrest at G1 phase (Luesch et al.

2006). In addition, apratoxin A was found to show potent antitumour activity against U2OS osteosarcoma cells and this was attributed to its inhibitory action on the cellular secretory pathways (Liu et al. 2009). Recently, Cai et al. (2018) synthesized an analogue of apratoxin, named apratoxin 10 (Apra S10), and demonstrated its potential as an antipancreatic cancer agent using an orthotopic pancreatic patient-derived xenograft mouse model. The compound inhibited growth of pancreatic cancer cells via downregulation of multiple receptor tyrosine kinase and inhibition of growth factor and cytokine secretion.

Trichophycin A, a polycavernoside (polyketide) isolated from *Trichodesmium thiebautii*, was found to exhibit cytotoxic activity against neuro-2a-neuroblastoma and human colon cancer cell lines (Bertin et al. 2017). In addition, calothrixin A, a phenanthridine alkaloid isolated from *Calothrix* sp., was found to induce apoptosis in human Jurkat cancer cells and caused G2/M cell cycle arrest (Chen et al. 2003). Recently, Acuna et al. (2018) found that hapalindole H, an alkaloid compound isolated from *Fischerella muscicola*, displayed selective cytotoxicity against PC-3 prostate cancer cells.

The cyanobacterial pigment C-phycocyanin is another potential compound for the development of anticancer agent (Beata and Katalin 2017). The cell targets of C-phycocyanin include MDR1 gene, cytoskeletal proteins and COX-2 enzymes, making it capable of killing drug-resistant cancer cells (Fernandes et al. 2018). C-phycocyanin has been shown to induce apoptosis and cell cycle arrest as well as suppress cell migration, proliferation and colony formation ability of NSCLC cells by regulating multiple key genes (Hao et al. 2018). In addition, C-phycocyanin was found to show potent anticancer activity against triple-negative MDA-MB-231 breast cancer cells (Ravi et al. 2015). In another study, C-phycocyanin from *Spirulina platensis* was found to display antiangiogenesis effect on B16-F10 melanoma tumours in a C57BL/6 mouse model (Dibaei et al. 2018). Recently, Liu et al. (2018b) demonstrated that selenium-enriched phycocyanin (Se-PC), when applied via photodynamic therapy approach, was effective in inhibiting tumour development in a mouse liver cancer model.

There have been many studies that evaluated the anticancer activity of algal carotenoids in extract or purified form (Gateau et al. 2017; Cha et al. 2008). For instance, Cha et al. (2008) demonstrated that extracts from *Chlorella vulgaris* and *Chlorella ellipsoidea*, containing mainly violaxanthin and lutein, inhibited the growth of HCT116 human colon cells and enhanced apoptosis. In another study, lycopene from *Chlorella marina* was found to display stronger inhibitory effect against the growth and colony formation of human prostate cancer cells than the carotene from tomatoes (Renju et al. 2014b).

Astaxanthin and fucoxanthin are amongst the major algal carotenoids that have been widely studied for their anticancer activity. Astaxanthin is a keto-carotenoid produced by *Haematococcus pluvialis*, especially when it is under unfavourable growth conditions. The carotenoid is accumulated outside the chloroplast and the content can reach up to 4.5% dry weight in *Haematococcus pluvialis* (Boussiba et al. 1999). The anticancer effect of astaxanthin is suggested to be due to its inflammation and oxidative stress-reducing properties (McCall et al. 2018). In a recent study,

McCall et al. (2018) demonstrated that treatment of astaxanthin significantly reduced proliferation rates and migration of breast cancer cells. In another study, Rao et al. (2013) demonstrated that mono- and diester forms of astaxanthin were more potent than the unesterified form of the xanthophyll in reducing UV-DBMA-induced skin cancer in a rat model. In addition, astaxanthin was found to be able to resensitize gemcitabine-resistant human pancreatic cancer cells (GR-HPCC) to gemcitabine (Yan et al. 2017). In addition, Palozza et al. (2009) showed that an astaxanthin-rich lipid CO_2 extract from *Haematococcus pluvialis* inhibited the growth of human colon cancer cell lines. The use of *Haematococcus* extract instead of purified astaxanthin is attractive, as it reduces the high costs involved in the production, isolation and purification of the carotenoid, which may limit its application in cancer therapy (Palozza et al. 2009).

Fucoxanthin is another algal carotenoid that has been shown to display anticancer activity (Nakazawa et al. 2009; Kim et al. 2010, 2013; Jin et al. 2018; Wang et al. 2018a). While brown seaweeds are the major producers of fucoxanthin, marine microalgae such as *Phaeodactylum tricornutum* and *Isochrysis galbana* have also been shown to have the potential for commercial production of the carotenoid (Kim et al. 2012c, d). Fucoxanthin was found to induce reactive oxygen species (ROS) production, promoting apoptosis in HL-60 human leukaemia cells (Kim et al. 2010). In addition, the carotenoid was shown to inhibit growth of other cancerous cells including colon cancer (Caco-2) and prostate cancer (PC-3 and LNCaP) cells by inducing apoptosis (Nakazawa et al. 2009). Recently, Jin et al. (2018) demonstrated that combination of fucoxanthin with tumour necrosis factor-related apoptosis-inducing ligand (TRAIL) exerted a strong synergistic effect on apoptosis in human cervical cancer cells by targeting the P13K/Akt/NF-κβ signalling pathway. In another study, Wang et al. (2012b) demonstrated that fucoxanthin could significantly inhibit the growth of sarcoma in xenografted sarcoma 180 (S180) mice. Another in vivo model study showed that the administration of fucoxanthin could significantly inhibit the growth of tumour mass in mice implanted with melanoma B16F10 cells (Kim et al. 2013).

Several toxins produced by dinoflagellates including amphidinolides (APDN), amphidinols (APDL), karlotoxins (KTX), pectenotoxins (PTX) and yessotoxin (YTX) have been shown to display anticancer activity (Leira et al. 2002; Kobayashi and Tsuda 2004; Chae et al. 2005; Waters et al. 2010; Espiritu et al. 2017). For instance, APDN produced by dinoflagellates of the genus *Amphidinium* displayed strong cytotoxicity against murine lymphoma L1210 and human epidermoid carcinoma KB cell lines (Kobayashi and Tsuda 2004). A study on amphidinol-2 (AM-2) showed that the compound displayed cytotoxicity in HCT-116, HT-29 and MCF-7 cancer cell lines and upregulated the pre-apoptosis markers *cfos* and *cjun* in those cell lines (Espiritu et al. 2017). Karlotoxins (KTX), produced by *Karlodinium venefictum*, are potentially useful for lowering cholesterol or targeting cancer cells high in cholesterol (Waters et al. 2010). Pectenotoxins (PTX) are a group of toxins produced by dinoflagellates of the genus *Dinophysis* that cause diarrhetic shellfish poisoning (PSP) (Miles et al. 2006). The toxin was found to be highly effective in activating an intrinsic apoptosis pathway in p53-deficient tumour cells (Chae et al.

2005). Yessotoxin (YTX), a polyether toxin produced by *Prorocentrum* and *Gonnyaulax*, was found to induce apoptotic changes in BE(2)-M17 neuroblastoma cells (Leira et al. 2002).

Some of the microalgal polysaccharides have also been found to have anticancer activity. For instance, an extracellular polysaccharide (D-galactan sulphate) isolated from the toxic dinoflagellate *Gymnodinium* sp. A3 (GA3), exhibited cytotoxic effect against a panel of cancer cells (Umemura et al. 2003). The polysaccharide was found to exert its effect by inhibiting DNA topoisomerase I and II. In addition, cell wall sulphated polysaccharide from *Porphyridium cruentum* was found to exhibit strong antitumour activity against Graffi myeloid tumour in hamsters (Gardeva et al. 2014). The anticancer activity of the sulphated polysaccharide could be associated with its immunostimulating effect as well as with its direct cytotoxic properties.

Several lipid compounds from microalgae have been shown to display anticancer activity. For instance, nonyl 8-acetoxy-6-methyloctanate (NAMC), which is a novel fatty alcohol ester isolated from *Phaeodactylum tricornutum*, displayed strong suppression on the growth of HL-60 leukaemia cells by upregulating apoptotic pathways (Samarakoon et al. 2014). In addition, curacin A, a lipid compound isolated from *Lyngbya masjucula*, is another potent agent found to have selective antiproliferative activity against colon, renal and breast cancer cell lines (Gerwick et al. 1994). The compound acts by suppressing the binding of tubulin to colchicine. Another lipid compound, sigmasterol, isolated from *Navicula incerta* showed potent apoptotic effect against HepG2 liver cancer cells (Kim et al. 2014). Further, polyunsaturated aldehydyes (PUA), an allelopathic agent produced by diatoms, were reported to display cytotoxic effect against A549 lung cancer and COLO 205 colon cancer cell lines (Sansone et al. 2014). In a recent study, supplementation of algal oil rich in omega-3 PUFA was found to significantly suppress pulmonary metastases and outgrowth of melanoma cells in a mouse model (C57BL/6) (Tan et al. 2018).

12.2.5 Beneficial Effects Against Metabolic Disorders and Other Diseases

Several reports have highlighted the beneficial effects of consuming microalgae as a functional food in ameliorating metabolic diseases such as hyperlipidaemia, diabetes and obesity. For instance, feeding of rats with diets containing dietary fibre from pelleted red microalgal biomass (*Porphyridium* sp.) or their sulphated polysaccharides was found to significantly lower serum cholesterol levels (Dvir et al. 2000). In addition, the sulphated polysaccharides increased mucosa and muscularis cross-sectional area of the jejunum and caused hypertrophy in the muscularis layer. Similarly, feeding of *Spirulina platensis* concentrate was found to have hypocholesterolaemic effect in rats and the action was suggested to be due to the action of C-phycocyanin in inhibiting both jejunal cholesterol absorption and ileal bile acid reabsorption (Nagaoka et al. 2005).

There have also been animal model studies which showed that consumption of microalgae (whole biomass or extract) could have antidiabetic effect. For instance, algal extract prepared from *Nannochloropsis oculata* was found to lower glucose and C-reactive protein levels and restore insulin to normal level in streptozotocin-induced diabetic rats (Aboulthana et al. 2018). A study on diabetic rats showed that oral supplementation of *Spirulina platensis* could lead to an increase in trace minerals and antioxidant enzymes and lowering of plasma concentrations of glucose and inflammatory markers such as TNF-α and IL-6 (Nasirian et al. 2018). Another study showed that supplementation of omega-3 fatty acids extracted from microalgae could prevent the appearance of health complications caused by inflammatory states in diabetic rats (Gutierrez-Pliego et al. 2018). Recently, Zhang et al. (2018) demonstrated that administration of fucoxanthin significantly improved glucose/lipid metabolism and insulin resistance and prevented pancreatic histological changes in a diabetic mouse model. In addition, administration of astaxanthin has been shown to prevent the progression of diabetic nephropathy induced by oxidative stress in a mouse model (Naito et al. 2004).

Another metabolic disorder that is of great concern is obesity, as it has become a global threat to public health and imposes a huge expenditure on healthcare costs (Chu and Phang 2016). Amongst the algal compounds, fucoxanthin is one of the most well studied in terms of antiobesity effect (Gammone and D'Orazio 2015). Feeding of fucoxanthin was found to reduce the levels of inflammatory markers such as interleukin-18 (IL-18), TNF-α, inducible nitric oxide synthase (iNOS) and cyclooxygenase-2 (COX-2) in an obese mouse model (Tan and Hou 2014). In addition, a combination of fucoxanthin and conjugated linoleic acid was found to reduce serum levels of triacylglycerols, glucose and leptin in diet-induced obese rats (Hu et al. 2012). Fucoxanthin also affects hepatic lipid contents by regulating metabolic enzyme activities and stimulating fatty acid oxidation activity (Woo et al. 2010). In addition, fucoxanthin exerts its antiobesity effect by inducing uncoupling protein 1 (UCP-1) expression in white adipose tissue (WAT), which enhances energy dissipation through fatty acid oxidation and heat production (Maeda 2015).

Long-chain omega-3 PUFA such as EPA and docosahexaenoic acid (DHA) has long been recognized to provide significant health benefits, particularly in reducing cardiac diseases such as arrythmia, stroke and hypertension (Romieu et al. 2005; von Schacky 2008; Ulmann et al. 2017). Dietary DHA has also been shown to be important for proper development of the brain and eye (Ward and Singh 2005). While fish oil is the major dietary source of EPA and DHA, it is not suitable for vegetarians and the odour makes it unattractive. Furthermore, there is concern over the decreasing global stocks of fish as sustainable sources of EPA and DHA for human nutrition (Ulmann et al. 2017). There are a variety of alternative sources of EPA and DHA including microalgae that are being explored for commercial production (Adarme-Vega et al. 2012). Amongst the microalgae, species of *Thraustochytrium*, *Schizochytrium* and *Crypthecodinium* are rich in DHA, while those of *Phaeodactylum* and *Monodus* are good sources of EPA (Ward and Singh 2005). In addition, the diatom *Odontella aurita* is another source of EPA, which has been commercialized as a dietary supplement (Mimouni et al. 2012). Algal DHA from *Crypthecodinium cohnii* produced by Martek Inc. has received approval for inclusion in infant formulae in the US (Ward and Singh 2005).

12.2.6 Other Bioactivities

Microalgae are known to produce a variety of compounds with antioxidative activity, particularly carotenoids such as astaxanthin, β-carotene and lutein (Chu 2011; Gateau et al. 2017). Such carotenoids are not only consumed as dietary supplement but also as natural food additives, which are better accepted by consumers than synthetic ones (Gateau et al. 2017). Amongst the carotenoids, astaxanthin is a very powerful antioxidant, as its activity is ten times more than other carotenoids such as zeaxanthin, lutein, canthaxanthin and β-carotene, and 100 times higher than α-tocopherol (Higuera-Ciapara et al. 2006). The protective effects of astaxanthin against diseases such as cancer, inflammatory diseases, diabetes and cardiovascular disease involve antioxidative mechanisms that prevent oxidative damage to cells (Ambati et al. 2014).

Among the carotenoids, consumption of lutein and zeaxanthin has been shown to be beneficial in reducing the development and progression of age-related macular degeneration (AMD) (Carpentier et al. 2009). Lutein and zeaxanthin are the two major macular pigments in the retina, which function as blue light filters and antioxidants. Increasing age has been proven to be a risk factor for the development and progression of AMD (Carpentier et al. 2009). Lutein is currently produced from marigold oleoresin, but several microalgae have been shown to contain higher amounts of the pigment than most marigold cultivars (Fernandez-Sevilla et al. 2010). The potential microalgal producers of lutein include *Muriellopsis* sp. (Del Campo et al. 2001), *Scenedesmus almeriensis* (Sánchez et al. 2008) and *Chlorella protothecoides* (Shi and Chen 2002).

In another development, there have been several reports on the beneficial effects of microalgae consumption in enhancing cognitive activity and ameliorating memory loss. For instance, in a recent study, Choi et al. (2018a) showed that extract from *Spirulina maxima* fermented with *Lactobacillus planterium* HY-08 could ameliorate scopolamide-induced memory impairment in mice. The efficacy of the fermented extract was suggested to be due to the synergistic effects of β-carotene and other bioactive substances, particularly in increasing extracellular signal-regulated kinases (ERK)-signalling and inducing expression of p-cAMP response element-binding protein (p-CREB) and brain-derived neurotrophic factor (BDNF). In another study, feeding of *Chlorella*-supplemented diet was found to reduce oxidative stress and prevent the decline in memory loss in a transgenic mouse model of age-dependent dementia (Nakashima et al. 2009). The studies by both Choi et al. (2018a) and Nakashima et al. (2009) showed the beneficial effects of *Chlorella* and *Spirulina* as functional food in enhancing cognitive activities. In both studies, algal extract or whole algal meal was tested rather than pure compounds.

12.3 Mass Culture of Microalgae for the Production of Pharmaceuticals and Nutraceuticals

Scaling up of microalgal cultures is an important aspect that needs to be considered in commercial production of pharmaceuticals and nutraceuticals from microalgae, as critically reviewed by Borowitzka and Vonshak (2017). It has been estimated that to produce 1 tonne/year of protein, about 18,300 L of culture are required, while production of β-carotene requires about 182,600 L of culture and fucoxanthin would require about 548,000 L. Most commercial production of microalgae is operated using outdoor open ponds due to its lower costs rather than closed systems, although the processing conditions are far from being optimal (Guedes et al. 2011). However, open pond systems have numerous issues including the difficulty in controlling temperature, light distribution and CO_2 concentration, possibility of microbial contamination, and low biomass productivity (Guedes et al. 2011; Borowitzka and Vonshak 2017). Of great concern is microbial contamination, as it may compromise the quality of the final products. Thus, the production of high-value pharmaceuticals and nutraceuticals needs to be operated using photobioreactors (PBR). Closed systems using PBR allow the culture to be well controlled with the benefits of low contamination risk, high biomass productivity and high CO_2 fixation efficiency (Wang et al. 2012a).

Many different types of PBR have been developed since the first flat panel PBR described by Burlew (1953). There are a variety of reactor designs including vertical column, tubular and flat panel PBR. Some examples of PBR used for the culturing of microalgae to produce pharmaceuticals and nutraceuticals are given in Table 12.4. Tubular PBR, which consist of an array of transparent tubes built in straight, bent or spiral patterns are amongst the most popular configuration (Wang et al. 2011). Using a Computational Fluid Dynamics (CFD) simulation approach, Guler et al. (2019) compared the performance of three types of PBR, namely flat plate, airlift and stirred tank for biomass and fucoxanthin production from *Phaeodactylum tricornutum*. The authors found that the highest biomass and fucoxanthin production was from the cultures grown in the flat-plate PBR.

There are also instances where both open and closed systems are integrated for commercial production of high-value chemicals from microalgae. For instance, an outdoor culture system consisting of both PBR and open pond was successfully developed to generate dense biomass of *Haematococcus pluvialis* for the extraction of astaxanthin (Olaizola 2000). In this system, the microalga was first grown in an outdoor PBR (25,000 L) before being transferred daily to an open pond system, where astaxanthin accumulation was induced. In a recent study, a hybrid system consisting of open pond and PBR, with culture medium circulating between both, was used to grow the oleaginous microalga *Scenedesmus dimorphus* under indoor and outdoor conditions (Liu et al. 2018a). The algal biomass productivity of the hybrid system increased 46.3–74.3% compared with the open pond and was 12.5% higher than that of PBR. Recently, Wang et al. (2018a) set up a pilot plant of 20 specially designed polyethylene phytobags (volume 150 L each) for the culturing

Table 12.4 Mass culture systems of microalgae for the production of selected pharmaceuticals and nutraceuticals

Algal product	Microalgal species	Culture system	Concentration/Productivity	References
Eicosapentaenoic acid (EPA)	*Phaeodactylum tricornutum*	Polyethylene phytobags (20 units, 4 m long, 1.5 m high, 150 L each)	Biomass = 7.62 g/m^2/day dry biomass EPA = 3.68% dw; 0.28 g/m^2/day	Wang et al. (2018a)
Astaxanthin	*Haematococcus pluvialis*	Bubble column reactors (diameter = 10–20 cm; height/diameter ratios = 1:1–9:1)	Biomass: 0.052–0.053 g/L/day Astaxanthin: 1.47–1.48 mg/L/day	Choi et al. (2018b)
Fucoxanthin	*Phaeodactylum tricornutum*	Airlift PBR Flat plate PBR Stirred tank	Biomass (g dry wt/L): Airlift = 1.93 ± 0.09 Flat plate = 3.03 ± 0.30 Stirred tank = 2.09 ± 0.19 Fucoxanthin (mg/L/day): Airlift = 0.387 ± 0.12 Flat plate = 1.05 ± 0.23 Stirred tank = 0.245 ± 0.08	Guler et al. (2019)
Lutein	*Scenedesmus almeriensis*	Bubble column PBR (4 units; 2.0 L capacity, 0.06 m diameter and 0.50 m height)	Max. productivity: Biomass = 0.87 g/L/day Lutein = 4.77 mg/L/day	Sánchez et al. (2008)
Lutein	*Muriellopsis* sp.	Methyl polymetacylate tubular PBR (55 L; tubes of 90 m long, 2.4 cm inner diameter and 2.2 m^2 surface)	Max. productivity: Biomass = 40 g dry wt/m^2/day; Lutein = 180 mg/m^2/day	Del Campo et al. (2001)
Karlotoxins (KTX)	*Karlodinium veneficum* (strain K10)	Bubble column PBR (internal diameter = 0.242 m; height = 1.86 m; vol = 80 L) Flat-panel PBR (width = 0.09 m; length = 2.4 m; vol = 281 L)	Max. cell productivity: Semi-continuous: Bubble column = 58 × 10^3 cells/mL/day Batch: Bubble column and flat-panel = 57 × 10^3 cells/mL/day Three congeners of KTX were	Lopez-Rosales et al. (2018)

			recovered: KmTx-10, sulfo-KmTx-10 and KmTx-12	
Amphidinols (APDN)	*Amphidinium carterae*	Bubble column PBR (80 L)	Biomass: 0.540 g/dw/L APDN (excreted): 49 mg/L supernatant	Molina-Miras et al. (2018)
C-phycocyanin (C-PC)	*Spirulina platensis*	Flat-plate PBR	Max biomass productivity = 0.74 g/L/day Max CO_2 consumption rate = 1.53 g/L/day Max C-PC productivity = 0.11 g/L/day	Chen et al. (2013)
Carotenoids (Astaxanthin, lutein/zeaxanthin, canthaxanthin and β-carotene)	*Nannochloropsis* sp.	Polyethylene bags (10 L)	Carotenoids = 1.47 ± 0.10 mg/g dry biomass Lipids = 45 g/100 g dry biomass	Nobre et al. (2013)
β-carotene	*Dunaliella salina*	Floating PBR (75 L)	300 g algal biomass containing 14.3 g β-carotene can be produced from 1 m^3 desalination concentrate	Zhu et al. (2018)

of *Phaeodactylum tricornutum* to produce EPA. The EPA was shown to have potent antibacterial activity and have potential use as a feed supplement. In a recent development, *Dunaliella salina* was successfully grown on seawater desalination concentrate in floating PBR anchored on the ocean field of Lingshui Bay in China for the production of β-carotene (Zhu et al. 2018). The floating PBR system is an innovative design as the ocean provides a natural habitat with large cultivable area for microalgae and it facilitates temperature control due to the high specific heat capacity of seawater (Kim et al. 2016). In addition, the energy from waves can be used to mix the culture.

There have been reports on the successful cultivation of microalgae for products such as carotenoids and DHA under heterotrophic condition. For instance, *Chlorella protothecoides* grown heterotrophically on glucose and urea in fermenters scaled up to 30 L produced high biomass concentration (45.8 g/L) and attained high lutein productivity (49.18 mg/L/day) (Shi and Chen 2002). Similarly, *Crypthecodinium cohnii* grown on acetic acid (feeding rate of 50% w/w) on a fed-batch mode under heterotrophic condition afforded final concentrations of 109 g/L dry biomass, 61 g/L lipids and 19 g/L DHA (de Swaaf et al. 2003). The use of fermentation strategy under heterotrophic mode is an attractive strategy to reduce costs in producing high-density algal biomass without the need of light. However, not all microalgae or microalgal products can be produced this way (Borowitzka 1999).

Dinoflagellate toxins have great potential in pharmaceutical applications, as discussed in the previous sections. However, only few of those compounds have led to commercial products. The main challenge faced is to generate sufficient biomass for the extraction of the bioactive compounds (Camacho et al. 2007). While producing large quantities of other microalgae such as *Chlorella*, *Spirulina*, *Dunaliella salina* and *Haematococcus pluvialis* has proved to be successful (Borowitzka 1999; Ben-Amotz 2004; Belay 2013), culturing of dinoflagellates poses new challenges. The growth rates of dinoflagellates are lower than typical microalgae and the biomass attained are usually low. As the toxin concentrations are very low, extremely large volumes of cultures are required for the extraction process (Camacho et al. 2007).

The design of PBR for the culturing of dinoflagellates needs to consider the extreme sensitivity of such microalgae to shear stress and hydrodynamic forces (Cao et al. 2015; Gallardo-Rodriguez et al. 2016). The levels of shear stress that dinoflagellates can withstand are two orders of magnitude lower than most animal and plant cells (Chisti 1999; Juhl et al. 2000). In one study, García Camacho et al. (2007) showed that the tolerance of *Protoceratium reticulatum*, a producer of yessotoxin, to shear stress was very low compared to other dinoflagellates and animal cells. The dinoflagellate cells were damaged to varying degrees under low shear stress of >0.12 to 0.20 mN/m^2. Turbulence due to agitation, shaking, aeration and stirring has been reported to inhibit the growth of dinoflagellate cultures (Pollingher and Zemel 1981; Berdalet 1992; Juhl et al. 2001).

Despite the challenges, there have been several successful ventures in mass culturing of dinoflagellates to produce bioactive compounds. For instance, a pilot-scale bioprocess was developed for the production of karlotoxin-enriched extracts of

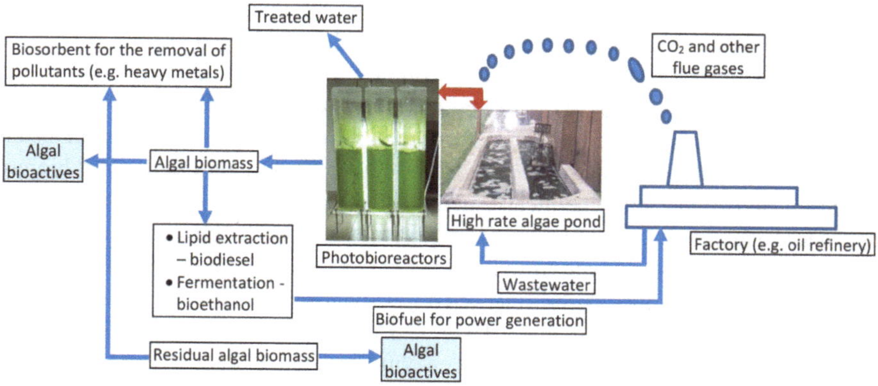

Fig. 12.4 Integration of bioactive compound production with other applications of microalgae, including biofuel production, based on a biorefinery concept

Karlodinium veneficum using a bubble column and a flat-panel PBR (80–281 L) under outdoor conditions (Lopez-Rosales et al. 2018). The bubble column PBR proved to be a better culture system than the flat panel PBR based on the maximum cell productivity attained (58×10^3 cells/mL/day). In another pilot-scale study, Molina-Miras et al. (2018) developed a LED-illuminated bubble column PBR (80 L) to culture *Amphidinium carterae* for the production of amphidinols (APDN). As APDN were excreted by cells into the medium, a dry APDN-enriched extract (49 mg/L) was recovered from the supernatant.

Integration of bioprocessing of the algal biomass for bioactive compound production with other applications such as biofuel production, CO_2 fixation and bioremediation based on a biorefinery concept is an attractive approach to improve economic competitiveness (Giraldo-Calderón et al. 2018; Costa et al. 2019). Figure 12.4 illustrates how the various applications of microalgae can be integrated using a biorefinery approach. An example of how this concept is applied is the culturing of *Haematococcus pluvialis* in bubble column reactors to remove flue gas, attaining high biomass and astaxanthin productivity (Choi et al. 2018b). In another set-up, *Spirulina platensis* grown in a flat-plate PBR was used to produce C-phycocyanin and simultaneously mitigate CO_2 emissions during its growth (Chen et al. 2013).

Another example of biorefinery process, focusing on the downstream processing aspect, was reported for the production of fucoxanthin by *Isochrysis galbana* (Gilbert-López et al. 2015). A green chemistry approach was used to extract a variety of products from the microalgal biomass. The process involved first the extraction of fucoxanthin using supercritical CO_2, followed by recovering other fractions with high antioxidant activity and, eventually, carbohydrates and proteins from the residual biomass. In another biorefinery process, lipids and carotenoids were extracted from *Nannochloropsis* sp. using supercritical CO_2 and the residual biomass was subjected to dark fermentation for production of biohydrogen (Nobre et al. 2013). Details of various aspects of biorefinery for *Nannochloropsis* in relation to induction, harvesting and extraction of EPA-oil and other products such as high-value proteins were recently reviewed by Chua and Schenk (2017).

The use of wastewater as a low-cost substrate for the culturing of microalgae can be an important component of the biorefinery approach. For instance, Olguin (2012) highlighted the potential of growing *Spirulina* in seawater supplemented with anaerobically digested piggery waste for waste treatment and production of biogas, biodiesel and high-value chemicals such as C-phycocyanin using the biorefinery approach. However, strict regulatory requirements need to be complied with in ensuring that products derived from microalgae in wastewater are free from potential impurities if they are to be marketed as pharmaceuticals or nutraceuticals (Borowitzka and Vonshak 2017). Efficient extraction technologies (e.g. supercritical fluid extraction and enzyme-assisted extraction) are crucial in order to attain high yields and to obtain products of high purity (Sosa-Hernandez et al. 2018).

12.4 Future Directions of Research

The pharmaceutical and nutraceutical industries based on microalgae are still confined to a few well-established products such as astaxanthin and C-phycocyanin. Also, commercial production of microalgae as functional food is based on only a few species, notably *Spirulina platensis*, *Haematococcus pluvialis*, *Chlorella* spp. and *Dunaliella salina*. As highlighted in this chapter, there is a diverse range of potential bioactive compounds that have been discovered from microalgae. However, there are many hurdles that need to be overcome before such compounds can be further developed for the pharmaceutical and nutraceutical industries.

Despite the many reported studies on bioactive compounds from microalgae, most have been exploratory, focusing mainly on mechanistic aspects confining to laboratory investigations. Most studies on the bioactivities of microalgal compounds are still at the pre-clinical stage based on cell lines or animal models. Very few clinical trials have been conducted to validate the efficacy of such compounds in human subjects. Dolastatin 10 (Vaishampayan et al. 2000) and cryptophycin 52 (Edelman et al. 2003) are amongst the microalgal compounds that have been brought to clinical trials to evaluate their anticancer activity. There have been more clinical trials on astaxanthin compared to other microalgal compounds. The clinical trials on astaxanthin have been conducted on human subjects to assess its hypolipidaemic, hypotensive and antioxidant effects (Ambati et al. 2014). Despite the potential anticancer activity of astaxanthin, as highlighted in this chapter, there have been not been any clinical trials of astaxanthin conducted on cancer patients. So far, there has been only one drug developed from microalgal compound, brentuximab vedotin (dolastatin conjugated with antibody), which has been approved as a cancer therapeutic agent (Gerwick and Moore 2012). More clinical trials on microalgal compounds are urgently required before new therapeutic agents can be brought to the market.

Scaling up of microalgal cultures to ensure sufficient supply of biomass for the extraction of bioactive compounds is a crucial part in the development of the pharmaceutical and nutraceutical industries based on microalgae (Borowitzka and Vonshak 2017). Although many microalgae have been identified as a source of bioactive compounds, very few have been evaluated beyond laboratory culture conditions. The potential species should be assessed for their performance when grown on a large scale under outdoor conditions. An example of such large-scale study using 10 pilot-scale trials involving high-rate algae ponds (HRAP) and PBR (flat-panel and tubular forms) was reported by Wolf et al. (2016). The aim was to assess the interaction between solar energy input, ambient temperature and system surface area-to-volume ratio in influencing yield and areal and volumetric productivities of *Chlorella sorokiniana* and *Chlorella* sp. cultured in those systems. Moving forward, new tools such as artificial intelligence (AI), particularly neural learning and statistical and evolutionary learning-based techniques, should be applied in optimizing productivity and costs in mass culture of microalgae (Jha et al. 2017b). The AI tools that can be applied involve statistical models such as Bayesian and naïve Bayes, clustering, hidden Markov and nearest-neighbour models.

The use of molecular approaches, particularly "omics"' tools, to enhance production of the desired microalgal metabolites is another area that has seen much advances in recent years (Guarnieri and Pienkos 2015). Microalgae are now regarded as highly attractive candidates for development as microbial factories. More studies are needed to unravel the biosynthetic pathways for the high-value metabolites in microalgae. Metabolomics approach will be useful for a comprehensive analysis of the secondary metabolites and elucidation of the pathways involved in the synthesis of the novel compounds in microalgae (Prakash et al. 2018). Metabolic and genetic engineering can then be applied to manipulate production of the microalgal metabolites. For instance, modification of the carotenoid biosynthesis pathway by genetic engineering has resulted in an increase of 67% astaxanthin production in the transformed strains of *Haematococcus pluvialis* compared to the wild-type (Galarza et al. 2018). Genome mining is another approach that is worth further exploring in attempts to discover novel metabolites in microalgae for the development of new drugs. Using this approach, instead of whole-genome sequencing, it allows rapid identification of gene clusters that code for specific compounds (Shih et al. 2013). The chemical structure of the compound can then be predicted based on the sequence information using in silico modelling.

Improving the delivery and bioavailability of bioactive compounds from microalgae is another area that warrants further investigations. This area is particularly relevant to microalgal products such as astaxanthin, which is already in the market, and potential new drugs intended for commercialization. For instance, the low bioavailability of astaxanthin may result in low assimilation into the body system, reducing its effectiveness as a supplement (Anarjan et al. 2013). There have been some recent studies that aimed to develop astaxanthin formulations in encapsulated forms to enhance its bioavailability (Khalid and Barrow 2018; Niizawa et al. 2018). In another line of development, nanoformulations using a wide variety of nanoparticles and their polymerized forms is an emerging approach in the development of new cancer drugs (Bajpai et al. 2018). For instance, a carbon-based nano-delivery system for anticancer drugs, in combination with genetically bioengineered biosilica from *Thalassiosira pseudonana,* was developed to kill targeted cancer cells without any adverse effect on normal and healthy cells (Delalat et al. 2015).

12.5 Concluding Remarks

There are indeed a diverse range of bioactive compounds from microalgae that have great potential to be developed for pharmaceuticals and nutraceutical applications. The bioactivities of microalgal metabolites that have attracted much interest include anticancer, antibacterial, antiviral and immunomodulatory activities, and their beneficial effects in ameliorating metabolic disorders such as diabetes, obesity and hyperlipidaemia. However, there are only a few pharmaceuticals and nutraceuticals derived from microalgae that have been successfully brought to commercialization. Amongst the well-known commercial microalgal products include astaxanthin, fucoxanthin, C-phycocyanin, β-carotene and PUFA. Clinical trials to evaluate the efficacy and safety of other bioactive compounds from microalgae are urgently needed before the compounds can be considered for the development of new drugs. Also, the feasibility of mass culturing of the selected microalgae under outdoor conditions needs to be critically evaluated. Production of bioactive compounds integrated with other applications such as biofuel production and CO_2 biofixation based on a biorefinery concept is an attractive strategy to enhance economic feasibility.

Acknowledgements The first author would like to acknowledge the funding and support from the International Medical University for algal biotechnology research. Phang S.M. would like to acknowledge the following grants: UM Algae (GA003-2012); MOHE-HiCoE Grant Phase 2 (2019-2021); UM Grand Challenge-SBS No.GC002B-15SBS.

References

Abd El-Hack ME, Abdelnour S, Alagawany M, Abdo M, Sakr MA, Khafaga AF, Mahgoub SA, Elnesr SS, Gebriel MG (2018) Microalgae in modern cancer therapy: current knowledge. Biomed Pharmacother 111:42–50

Abdo SM, Hetta MH, El-Senousy WM, El RAS, Ali GH (2012) Antiviral activity of freshwater algae. J Appl Pharm Sci 2:21

Aboulthana WM, El-Feky AM, Ibrahim NE, Sahu RK, El-Sayed AEB (2018) Evaluation of the pancreatoprotective effect of *Nannochloropsis oculata* extract against streptozotocin-induced diabetes in rats. J Appl Pharm Sci. 8:46–58

Acuna UM, Mo S, Zi J, Orjala J, De Blanco EJ (2018) Hapalindole H induces apoptosis as an inhibitor of NF-kB and affects the intrinsic mitochondrial pathway in PC-3 androgen-insensitive prostate cancer cells. Anticancer Res 38:3299–3307

Adarme-Vega TC, Lim DK, Timmins M, Vernen F, Li Y, Schenk PM (2012) Microalgal biofactories: a promising approach towards sustainable omega-3 fatty acid production. Microb Cell Factories 11:96

Afify AE, El Baroty GS, El Baz FK, El Baky HH, Murad SA (2018) *Scenedesmus obliquus*: antioxidant and antiviral activity of proteins hydrolyzed by three enzymes. J Genet Eng Biotechnol 16:399–408

Alvarez-Diaz PD, Ruiz J, Arbib Z, Barragan J, Garrido-Perez C, Perales JA (2014) Lipid production of microalga *Ankistrodesmus falcatus* increased by nutrient and light starvation in a two-stage cultivation process. Appl Biochem Biotechnol 174:1471–1483

Alves C, Silva J, Pinteus S, Gaspar H, Alpoim MC, Botana LM, Pedrosa R (2018) From marine origin to therapeutics: the antitumor potential of marine algae-derived compounds. Front Pharmacol 9:777

Ambati RR, Phang SM, Ravi S, Aswathanarayana RG (2014) Astaxanthin: sources, extraction, stability, biological activities and its commercial applications—a review. Mar Drugs 12:128–152

Anarjan N, Nehdi IA, Tan CP (2013) Influence of astaxanthin, emulsifier and organic phase concentration on physicochemical properties of astaxanthin nanodispersions. Chem Central J 7:127

Andrade LM (2018) *Chlorella* and *Spirulina* microalgae as sources of functional foods, nutraceuticals, and food supplements: an overview. MOJ Food Process Technol 6:45–58

Arbib Z, de Godos I, Ruiz J, Perales JA (2017) Optimization of pilot high rate algal ponds for simultaneous nutrient removal and lipids production. Sci Total Environ 589:66–72

Assunção J, Guedes A, Malcata F (2017) Biotechnological and pharmacological applications of biotoxins and other bioactive molecules from Dinoflagellates. Mar Drugs 15(12):393

Bajpai VK, Shukla S, Kang SM, Hwang SK, Song X, Huh YS, Han YK (2018) Developments of cyanobacteria for nano-marine drugs: relevance of nanoformulations in cancer therapies. Mar Drugs 16:179

Barrientos LG, O'Keefe BR, Bray M, Sanchez A, Gronenborn AM, Boyd MR (2003) Cyanovirin-N binds to the viral surface glycoprotein, GP1,2 and inhibits infectivity of Ebola virus. Antivir Res 58:47–56

Barrientos LG, Matei E, Lasala F, Delgado R, Gronenborn AM (2006) Dissecting carbohydrate-cyanovirin-N binding by structure-guided mutagenesis: functional implications for viral entry inhibition. Protein Eng Des Sel 19:525–535

Beata MK, Katalin S (2017) Phycobilins and phycobiliproteins used in food industry and medicine. Mini Rev Med Chem 17:1173–1193

Belay A (2013) Biology and industrial production of *Arthrospira* (*Spirulina*). In: Richmond A, Qiang H (eds) Handbook of microalgal culture: applied phycology and biotechnology. Wiley, Hoboken, pp 339–358

Ben-Amotz A (2004) Industrial production of microalgal cell-mass and secondary products-major industrial species: *Dunaliella*. In: Richmond A (ed) Handbook of microalgal culture: biotechnology and applied phycology. Wiley, Hoboken, pp 273–288

Berdalet E (1992) Effects of turbulence on the marine dinoflagellate *Gymnodinium nelsonii* 1. J Phycol 28:267–272

Bertin M, Wahome P, Zimba P, He H, Moeller P (2017) Trichophycin A, a cytotoxic linear polyketide isolated from a *Trichodesmium thiebautii* bloom. Mar Drugs 15:10

Bhagavathy S, Sumathi P, Bell IJ (2011) Green algae *Chlorococcum humicola* – a new source of bioactive compounds with antimicrobial activity. Asian Pac J Trop Biomed 1:S1–S7

Blair MF, Kokabian B, Gude VG (2014) Light and growth medium effect on *Chlorella vulgaris* biomass production. J Environ Chem Eng 2:665–674

Bokesch HR, O'Keefe BR, McKee TC, Pannell LK, Patterson GM, Gardella RS, Sowder RC, Turpin J, Watson K, Buckheit RW (2003) A potent novel anti-HIV protein from the cultured cyanobacterium *Scytonema varium*. Biochemistry 42:2578–2584

Borowitzka MA (1999) Commercial production of microalgae: ponds, tanks, and fermenters. Prog Ind Microbiol 35:313–321

Borowitzka MA, Vonshak A (2017) Scaling up microalgal cultures to commercial scale. Eur J Phycol 52:407–418

Boussiba S, Bing W, Yuan JP, Zarka A, Chen F (1999) Changes in pigments profile in the green alga *Haeamtococcus pluvialis* exposed to environmental stresses. Biotechnol Lett 21:601–604

Boyd MR, Gustafson KR, McMahon JB, Shoemaker RH, O'Keefe BR, Mori T, Gulakowski RJ, Wu L, Rivera MI, Laurencot CM (1997) Discovery of cyanovirin-N, a novel human immunodeficiency virus-inactivating protein that binds viral surface envelope glycoprotein gp120: potential applications to microbicide development. Antimicrob Agents Chemother 41:1521–1530

Buono S, Langellotti AL, Martello A, Bimonte M, Tito A, Carola A, Apone F, Colucci G, Fogliano V (2012) Biological activities of dermatological interest by the water extract of the microalga *Botryococcus braunii*. Arch Dermatol Res 304:755–764

Burja AM, Banaigs B, Abou-Mansour E, Grant Burgess J, Wright PC (2001) Marine cyanobacteria—a prolific source of natural products. Tetrahedron 57:9347–9377

Burlew J (1953) Algal culture from laboratory to pilot plant. Carnegie Institute of Washington, Washington, DC

Cai W, Ratnayake R, Gerber MH, Chen QY, Yu Y, Derendorf H, Trevino JG, Luesch H (2018) Development of apratoxin S10 (Apra S10) as an anti-pancreatic cancer agent and its preliminary evaluation in an orthotopic patient-derived xenograft (PDX) model. Investig New Drugs 37:364–374

Caires TA, da Silva AMS, Vasconcelos VM, Affe HMJ, de Souza Neta LC, Boness HVM, Sant'Anna CL, Nunes JMC (2018) Biotechnological potential of *Neolyngbya* (Cyanobacteria), a new marine benthic filamentous genus from Brazil. Algal Res 36:1–9

Camacho FG, Rodriguez JG, Miron AS, Garcia MC, Belarbi EH, Chisti Y, Grima EM (2007) Biotechnological significance of toxic marine dinoflagellates. Biotechnol Adv 25:176–194

Cao Z, Li X, Zou X, Greenwood M, Gerwick WH, Murray TF (2015) Involvement of JNK and caspase activation in hoiamide A-induced neurotoxicity in neocortical neurons. Mar Drugs 13:903–919

Carpentier S, Knaus M, Suh M (2009) Associations between lutein, zeaxanthin, and age-related macular degeneration: an overview. Crit Rev Food Sci Nutr 49:313–326

Cezare-Gomes EA, Mejia-da-Silva LDC, Perez-Mora LS, Matsudo MC, Ferreira-Camargo LS, Singh AK, de Carvalho JCM (2019) Potential of microalgae carotenoids for industrial application. Appl Biochem Biotechnol 188(3):602–634

Cha KH, Koo SY, Lee DU (2008) Antiproliferative effects of carotenoids extracted from *Chlorella ellipsoidea* and *Chlorella vulgaris* on human colon cancer cells. J Agri Food Chem 56:10521–10526

Chae HD, Choi TS, Kim BM, Jung JH, Bang YJ, Shin DY (2005) Oocyte-based screening of cytokinesis inhibitors and identification of pectenotoxin-2 that induces Bim/Bax-mediated apoptosis in p53-deficient tumors. Oncogene 24:4813

Chen X, Smith GD, Waring P (2003) Human cancer cell (Jurkat) killing by the cyanobacterial metabolite calothrixin A. J Appl Phycol 15:269–277

Chen CY, Kao PC, Tsai CJ, Lee DJ, Chang JS (2013) Engineering strategies for simultaneous enhancement of C-phycocyanin production and CO_2 fixation with *Spirulina platensis*. Bioresour Technol 145:307–312

Chisti Y (1999) Shear sensitivity. In: Flickinger MC, Drew SW (eds) Encyclopedia of bioprocess technology: fermentation, biocatalysis, and bioseparation, vol 5. Wiley, New York, pp 2379–2406

Choi WY, Kang DH, Lee HY (2018a) Effect of fermented *Spirulina maxima* extract on cognitive-enhancing activities in mice with scopolamine-induced dementia. Evid Based Complement Alternat Med Vol 2018, Article FD7218504, 9 p

Choi YY, Hong ME, Jin ES, Woo HM, Sim SJ (2018b) Improvement in modular scalability of polymeric thin-film photobioreactor for autotrophic culturing of *Haematococcus pluvialis* using industrial flue gas. Bioresour Technol 249:519–526

Chu WL (2011) Potential applications of antioxidant compounds derived from algae. Curr Top Nutra Res 9:83

Chu WL, Phang SM (2016) Marine algae as a potential source for anti-obesity agents. Mar Drugs 14:222

Chu WL, Quynh lV, Radhakrishnan AK (2013) Effect of *Spirulina* (*Arthrospira*) supplementation on the immune response to tetanus toxoid vaccination in a mouse model. J Diet Suppl 10:229–240

Chua ET, Schenk PM (2017) A biorefinery for *Nannochloropsis*: induction, harvesting, and extraction of EPA-rich oil and high-value protein. Bioresour Technol 244:1416–1424

Coates RC, Trentacoste E, Gerwick WH (2013) Bioactive and novel chemicals from microalgae. In: Richmond A, Qiang H (eds) Handbook of microalgal culture: applied phycology and biotechnology. Wiley, Hoboken, pp 504–531

Costa JAV, Freitas BCBD, Lisboa CR, Santos TD, Brusch LRDF, de Morais MG (2019) Microalgal biorefinery from CO_2 and the effects under the Blue Economy. Renew Sustain Energy Rev 99:58–65

da Silva Ferreira V, ConzFerreira ME, Lima LM, Frases S, de Souza W, Sant'Anna C (2017) Green production of microalgae-based silver chloride nanoparticles with antimicrobial activity against pathogenic bacteria. Enzym Microb Technol 97:114–121

Davoodbasha M, Edachery B, Nooruddin T, Lee SY, Kim JW (2018) An evidence of C16 fatty acid methyl esters extracted from microalga for effective antimicrobial and antioxidant property. Microb Pathog 115:233–238

de Clerck O, Guiry MD, Leliaert F, Samyn Y, Verbruggen H (2013) Algal taxonomy: a road to nowhere? J Phycol 49:215–225

de Jesus Raposo MF, de Morais AM, de Morais RM (2015) Marine polysaccharides from algae with potential biomedical applications. Mar Drugs 13:2967–3028

de Souza TD, Prietto L, de Souza MM, Furlong EB (2015) Profile, antioxidant potential, and applicability of phenolic compounds extracted from *Spirulina platensis*. Afr J Biotechnol 14:2903–2909

de Swaaf ME, Sijtsma L, Pronk JT (2003) High-cell-density fed-batch cultivation of the docosahexaenoic acid producing marine alga *Crypthecodinium cohnii*. Biotechnol Bioeng 81:666–672

Del Campo JA, Rodnguez H, Moreno J, Vargas MA, Rivas J, Guerrero MG (2001) Lutein production by *Muriellopsis* sp. in an outdoor tubular photobioreactor. J Biotechnol 85:289–295

Delalat B, Sheppard VC, Rasi Ghaemi S, Rao S, Prestidge CA, McPhee G, Rogers ML, Donoghue JF, Pillay V, Johns TG, Kroger N, Voelcker NH (2015) Targeted drug delivery using genetically engineered diatom biosilica. Nat Commun 6:8791

Desbois AP, Mearns-Spragg A, Smith VJ (2009) A fatty acid from the diatom *Phaeodactylum tricornutum* is antibacterial against diverse bacteria including multi-resistant *Staphylococcus aureus* (MRSA). Mar Biotechnol 11:45–52

Dewi IC, Falaise C, Hellio C, Bourgougnon N, Mouget JL (2018) Anticancer, antiviral, antibacterial, and antifungal properties in microalgae. In: Levine IA, Fleurence J (eds) Microalgae in health and disease prevention. Academic, Cambridge, pp 235–261

Dibaei F, Fazilati M, Moenzadeh F, Kafayat A, Jazayeri N, Talebi A (2018) Anti-angiogenesis effect of C-phycocyanin of *Spirulina platensis* on B16-F10 melanoma tumors in C57BL/6 mouse. Pathobiol Res 21:141–146

Dvir I, Chayoth R, Sod-Moriah U, Shany S, Nyska A, Stark AH, Madar Z, Arad SM (2000) Soluble polysaccharide and biomass of red microalga *Porphyridium* sp. alter intestinal morphology and reduce serum cholesterol in rats. Br J Nutr 84:469–476

Edelman MJ, Gandara DR, Hausner P, Israel V, Thornton D, DeSanto J, Doyle LA (2003) Phase 2 study of cryptophycin 52 (LY355703) in patients previously treated with platinum based chemotherapy for advanced non-small cell lung cancer. Lung Cancer 39:197–199

Espiritu RA, Tan MCS, Oyong GG (2017) Evaluation of the anti-cancer potential of amphidinol 2, a polyketide metabolite from the marine dinoflagellate *Amphidinium klebsii*. Jordan J Biol Sci 10:297–302

Fabregas J, Garcia D, Fernandez-Alonso M, Rocha AI, Gomez-Puertas P, Escribano JM, Otero A, Coll JM (1999) In vitro inhibition of the replication of haemorrhagic septicaemia virus (VHSV) and African swine fever virus (ASFV) by extracts from marine microalgae. Antivir Res 44:67–73

Fernandes ESE, Figueira FDS, Lettnin AP, Carrett-Dias M, Filgueira D, Kalil S, Trindade GS, Votto APS (2018) C-Phycocyanin: cellular targets, mechanisms of action and multi drug resistance in cancer. Pharmacol Rep 70:75–80

Fernandez-Sevilla JM, Acien Fernandez FG, Molina Grima E (2010) Biotechnological production of lutein and its applications. Appl Microbiol Biotechnol 86:27–40

Galarza JI, Gimpel JA, Rojas V, Arredondo-Vega BO, Henríquez V (2018) Over-accumulation of astaxanthin in *Haematococcus pluvialis* through chloroplast genetic engineering. Algal Res 31:291–297

Gallardo-Rodriguez JJ, Lopez-Rosales L, Sanchez-Miron A, Garcia-Camacho F, Molina-Grima E, Chalmers JJ (2016) New insights into shear-sensitivity in dinoflagellate microalgae. Bioresour Technol 200:699–705

Gammone MA, D'Orazio N (2015) Anti-obesity activity of the marine carotenoid fucoxanthin. Mar Drugs 13:2196–2214

Gantar M, Svircev Z (2008) Microalgae and cyanobacteria: food for thought. J Phycol 44:260–268

García Camacho F, Gallardo Rodríguez JJ, Sánchez Mirón A, Cerón García MC, Belarbi EH, Molina Grima E (2007) Determination of shear stress thresholds in toxic dinoflagellates cultured in shaken flasks. Process Biochem 42:1506–1515

Gardeva E, Toshkova R, Minkova K, Gigova L (2014) Cancer protective action of polysaccharide, derived from red microalga *Porphyridium cruentum* – biological background. Biotechnol Biotechnolog Equip 23:783–787

Gateau H, Solymosi K, Marchand J, Schoefs B (2017) Carotenoids of microalgae used in food industry and medicine. Mini Rev Med Chem 17:1140–1172

Gerwick WH, Moore BS (2012) Lessons from the past and charting the future of marine natural products drug discovery and chemical biology. Chem Biol 19:85–98

Gerwick WH, Proteau PJ, Nagle DG, Hamel E, Blokhin A, Slate DL (1994) Structure of curacin A, a novel antimitotic, antiproliferative and brine shrimp toxic natural product from the marine cyanobacterium *Lyngbya majuscula*. J Org Chem 59:1243–1245

Gilbert-López B, Mendiola JA, Fontecha J, van den Broek LAM, Sijtsma L, Cifuentes A, Herrero M, Ibáñez E (2015) Downstream processing of *Isochrysis galbana*: a step towards microalgal biorefinery. Green Chem 17:4599–4609

Giraldo-Calderón ND, Romo-Buchelly RJ, Arbeláez-Pérez AA, Echeverri-Hincapié D, Atehortúa-Garcés L (2018) Microalgae biorefineries: applications and emerging technologies. Dyna 85:219–233

Guarnieri MT, Pienkos PT (2015) Algal omics: unlocking bioproduct diversity in algae cell factories. Photosynth Res 123:255–263

Guedes AC, Amaro HM, Malcata FX (2011) Microalgae as sources of high added-value compounds—a brief review of recent work. Biotechnol Prog 27:597–613

Guler BA, Deniz I, Demirel Z, Oncel SS, Imamoglu E (2019) Comparison of different photobioreactor configurations and empirical computational fluid dynamics simulation for fucoxanthin production. Algal Res 37:195–204

Gutierrez-Pliego LE, Martinez-Carrillo BE, Resendiz-Albor AA, Arciniega-Martinez IM, Escoto-Herrera JA, Rosales-Gomez CA, Valdes-Ramos R (2018) Effect of supplementation with n-3 fatty acids extracted from microalgae on inflammation biomarkers from two different strains of mice. J Lipids Vol 2018, Article ID 4765358, 10 p, https://doi.org/10.1155/2018/4765358

Hao S, Yan Y, Li S, Zhao L, Zhang C, Liu L, Wang C (2018) The in vitro anti-tumor activity of phycocyanin against non-small cell lung cancer cells. Mar Drugs 16:178

Hayashi K, Hayashi T, Kojima I (1996) A natural sulfated polysaccharide, calcium spirulan, isolated from *Spirulina platensis*: in vitro and ex vivo evaluation of anti-herpes simplex virus and anti-human immunodeficiency virus activities. AIDS Res Hum Retroviruses 12:1463–1471

Hidalgo-Lucas S, Bisson JF, Duffaud A, Nejdi A, Guerin-Deremaux L, Baert B, Saniez-Degrave MH, Rozan P (2014a) Benefits of oral and topical administration of ROQUETTE *Chlorella* sp. on skin inflammation and wound healing in mice. Anti-Inflamm Anti-Allergy Agents Med Chem 13:93–102

Hidalgo-Lucas S, Rozan P, Guerin-Deremaux L, Violle N, Baert B, Saniez-Degrave MH, Bisson JF (2014b) Oral and topical administration of ROQUETTE *Schizochytrium* sp. alleviate skin inflammation and improve wound healing in mice. Anti-Inflamm Anti-Allergy Agents Med Chem 13:154–164

Higuera-Ciapara I, Felix-Valenzuela L, Goycoolea FM (2006) Astaxanthin: a review of its chemistry and applications. Crit Rev Food Sci Nutr 46:185–196

Hu X, Li Y, Li C, Fu Y, Cai F, Chen Q, Li D (2012) Combination of fucoxanthin and conjugated linoleic acid attenuates body weight gain and improves lipid metabolism in high-fat diet-induced obese rats. Arch Biochem Biophys 519:59–65

Huerlimann R, De Nys R, Heimann K (2010) Growth, lipid content, productivity, and fatty acid composition of tropical microalgae for scale-up production. Biotechnol Bioeng 107:245–257

Huleihel M, Ishanu V, Tal J, Arad SM (2001) Antiviral effect of red microalgal polysaccharides on *Herpes simplex* and *Varicella zoster* viruses. J Appl Phycol 13:127–134

Jha D, Jain V, Sharma B, Kant A, Garlapati VK (2017a) Microalgae-based pharmaceuticals and nutraceuticals: an emerging field with immense market potential. ChemBioEng Rev 4:257–272

Jha SK, Bilalovic J, Jha A, Patel N, Zhang H (2017b) Renewable energy: present research and future scope of artificial intelligence. Renew Sust Energ Rev 77:297–317

Jin Y, Qiu S, Shao N, Zheng J (2018) Fucoxanthin and tumor necrosis factor-related apoptosis-inducing ligand (TRAIL) synergistically promotes apoptosis of human cervical cancer cells by targeting PI3K/Akt/NF-κB signaling pathway. Med Sci Monitor 24:11–18

Juhl AR, Velazquez V, Latz MI (2000) Effect of growth conditions on flow-induced inhibition of population growth of a red-tide dinoflagellate. Limnol Oceanogr 45:905–915

Juhl AR, Trainer VL, Latz MI (2001) Effect of fluid shear and irradiance on population growth and cellular toxin content of the dinoflagellate *Alexandrium fundyense*. Limnol Oceangr 46:758–764

Kang H, Lee CH, Kim JR, Kwon JY, Seo SG, Han JG, Kim BG, Kim JE, Lee KW (2015) *Chlorella vulgaris* attenuates *Dermatophagoides farinae*-induced atopic dermatitis-like symptoms in NC/Nga mice. Int J Mol Sci 16:21021–21034

Khalid N, Barrow CJ (2018) Critical review of encapsulation methods for stabilization and delivery of astaxanthin. J Food Bioactives 1:104–115–104–115

Kim KN, Heo SJ, Kang SM, Ahn G, Jeon YJ (2010) Fucoxanthin induces apoptosis in human leukemia HL-60 cells through a ROS-mediated Bcl-xL pathway. Toxicol In Vitro 24:1648–1654

Kim H, Lantvit D, Hwang CH, Kroll DJ, Swanson SM, Franzblau SG, Orjala J (2012a) Indole alkaloids from two cultured cyanobacteria, *Westiellopsis* sp. and *Fischerella muscicola*. Bioorg Med Chem 20:5290–5295

Kim M, Yim JH, Kim SY, Kim HS, Lee WG, Kim SJ, Kang PS, Lee CK (2012b) In vitro inhibition of influenza A virus infection by marine microalga-derived sulfated polysaccharide p-KG03. Antivir Res 93:253–259

Kim SM, Jung YJ, Kwon ON, Cha KH, Um BH, Chung D, Pan CH (2012c) A potential commercial source of fucoxanthin extracted from the microalga *Phaeodactylum tricornutum*. Appl Biochem Biotechnol 166:1843–1855

Kim SM, Kang SW, Kwon ON, Chung D, Pan CH (2012d) Fucoxanthin as a major carotenoid in *Isochrysis* aff. *galbana:* characterization of extraction for commercial application. J Korean Soc Appl Biol Chem 55:477–483

Kim KN, Ahn G, Heo SJ, Kang SM, Kang MC, Yang HM, Kim D, Roh SW, Kim SK, Jeon BT, Park PJ, Jung WK, Jeon YJ (2013) Inhibition of tumor growth in vitro and in vivo by fucoxanthin against melanoma B16F10 cells. Environ Toxicol Pharmacol 35:39–46

Kim YS, Li XF, Kang KH, Ryu B, Kim SK (2014) Stigmasterol isolated from marine microalgae *Navicula incerta* induces apoptosis in human hepatoma HepG2 cells. BMB Rep 47:433–438

Kim ZH, Park H, Hong SJ, Lim SM, Lee CG (2016) Development of a floating photobioreactor with internal partitions for efficient utilization of ocean wave into improved mass transfer and algal culture mixing. Bioprocess Biosyst Eng 39:713–723

Kim JH, Lee JE, Kim KH, Kang NJ (2018) Beneficial effects of marine algae-derived carbohydrates for skin health. Mar Drugs 16(11):459

Kobayashi JI, Tsuda M (2004) Amphidinolides, bioactive macrolides from symbiotic marine dinoflagellates. Nat Prod Rep 21:77–93

Kobayashi M, Natsume T, Tamaoki S, Watanabe J, Asano H, Mikami T, Miyasaka K, Miyazaki K, Gondo M, Sakakibara K, Tsukagoshi S (1997) Antitumor activity of TZT-1027, a novel dolastatin 10 derivative. Jpn J Cancer Res 88:316–327

Kok YY, Chu WL, Phang SM, Mohamed SM, Naidu R, Lai PJ, Ling SN, Mak JW, Lim PK, Balraj P, Khoo AS (2011) Inhibitory activities of microalgal extracts against Epstein-Barr virus DNA release from lymphoblastoid cells. J Zhejiang Univ Sci B 12:335–345

Komatsu T, Kido N, Sugiyama T, Yokochi T (2013) Antiviral activity of acidic polysaccharides from *Coccomyxa gloeobotrydiformi*, a green alga, against an in vitro human influenza A virus infection. Immunopharmacol Immunotoxicol 35:1–7

Komprda T, Sladek Z, Skultety O, Krizkova S, Rozikova V, Nemcova B, Sustrova T, Valova M (2016) Effect of dietary *Schizochytrium* microalga oil on selected markers of low-grade inflammation in rats. J Anim Physiol Anim Nutr 100:1169–1178

Kuntzler SG, Costa JAV, Morais MG (2018) Development of electrospun nanofibers containing chitosan/PEO blend and phenolic compounds with antibacterial activity. Int J Biol Macromol 117:800–806

Lau AF, Siedlecki J, Anleitner J, Patterson GM, Caplan FR, Moore RE (1993) Inhibition of reverse transcriptase activity by extracts of cultured blue-green algae (Cyanophyta). Planta Med 59:148–151

Lauritano C, Andersen JH, Hansen E, Albrigtsen M, Escalera L, Esposito F, Helland K, Hanssen KØ, Romano G, Ianora A (2016) Bioactivity screening of microalgae for antioxidant, anti-inflammatory, anticancer, anti-diabetes, and antibacterial activities. Front Mar Sci 3:68

Lee JB, Hayashi K, Hirata M, Kuroda E, Suzuki E, Kubo Y, Hayashi T (2006) Antiviral sulfated polysaccharide from *Navicula directa*, a diatom collected from deep-sea water in Toyama Bay. Biol Pharm Bull 29:2135–2139

Leira F, Alvarez C, Vieites J, Vieytes M, Botana L (2002) Characterization of distinct apoptotic changes induced by okadaic acid and yessotoxin in the BE (2)-M17 neuroblastoma cell line. Toxicol In Vitro 16:23–31

Li Y, Gao X, Yu Z, Liu B, Pan W, Li N, Tang B (2018) Reversing multidrug resistance by multiplexed gene silencing for enhanced breast cancer chemotherapy. ACS Appl Mat Interfaces 10:15461–15466

Liu Y, Law BK, Luesch H (2009) Apratoxin a reversibly inhibits the secretory pathway by preventing cotranslational translocation. Mol Pharmacol 76:91–104

Liu W, Chen Y, Wang J, Liu T (2018a) Biomass productivity of *Scenedesmus dimorphus* (Chlorophyceae) was improved by using an open pond–photobioreactor hybrid system. Eur J Phycol 54:1–8. https://doi.org/10.1080/09670262.2018.1519601

Liu Z, Fu X, Huang W, Li C, Wang X, Huang B (2018b) Photodynamic effect and mechanism study of selenium-enriched phycocyanin from *Spirulina platensis* against liver tumours. J Photochem Photobiol B 180:89–97

Lopez-Rosales L, Sanchez-Miron A, Garcia-Camacho F, Place AR, Chisti Y, Molina-Grima E (2018) Pilot-scale outdoor photobioreactor culture of the marine dinoflagellate *Karlodinium veneficum*: production of a karlotoxins-rich extract. Bioresour Technol 253:94–104

Lorenz RT, Cysewski GR (2000) Commercial potential for *Haematococcus* microalgae as a natural source of astaxanthin. Trends Biotechnol 18:160–167

Lotfi H, Sheervalilou R, Zarghami N (2018) An update of the recombinant protein expression systems of Cyanovirin-N and challenges of preclinical development. Bioimpacts 8:139–151

Luesch H, Moore RE, Paul VJ, Mooberry SL, Corbett TH (2001) Isolation of dolastatin 10 from the marine cyanobacterium *Symploca* species VP642 and total stereochemistry and biological evaluation of its analogue symplostatin 1. J Nat Prod 64:907–910

Luesch H, Harrigan GG, Goetz G, Horgen FD (2002) The cyanobacterial origin of potent anticancer agents originally isolated from sea hares. Curr Med Chem 9:1791–1806

Luesch H, Chanda SK, Raya RM, de Jesus PD, Orth AP, Walker JR, Izpisua Belmonte JC, Schultz PG (2006) A functional genomics approach to the mode of action of apratoxin A. Nat Chem Biol 2:158–167

Maeda H (2015) Nutraceutical effects of fucoxanthin for obesity and diabetes therapy: a review. J Oleo Sci 64:125–132

Martinez Andrade KA, Lauritano C, Romano G, Ianora A (2018) Marine microalgae with anti-cancer properties. Mar Drugs 16:165

Matsui MS, Muizzuddin N, Arad S, Marenus K (2003) Sulfated polysaccharides from red microalgae have antiinflammatory properties in vitro and in vivo. Appl Biochem Biotechnol 104:13–22

McCall B, McPartland CK, Moore R, Frank-Kamenetskii A, Booth BW (2018) Effects of astaxanthin on the proliferation and migration of breast cancer cells in vitro. Antioxidants 7:135

Miles CO, Wilkins AL, Munday JS, Munday R, Hawkes AD, Jensen DJ, Cooney JM, Beuzenberg V (2006) Production of 7-epi-pectenotoxin-2 seco acid and assessment of its acute toxicity to mice. J Agric Food Chem 54:1530–1534

Mimouni V, Ulmann L, Pasquet V, Mathieu M, Picot L, Bougaran G, Cadoret JP, Morant-Manceau A, Schoefs B (2012) The potential of microalgae for the production of bioactive molecules of pharmaceutical interest. Curr Pharm Biotechnol 13:2733–2750

Molina-Miras A, Morales-Amador A, de Vera CR, López-Rosales L, Sánchez-Mirón A, Souto ML, Fernández JJ, Norte M, García-Camacho F, Molina-Grima E (2018) A pilot-scale bioprocess to produce amphidinols from the marine microalga *Amphidinium carterae*: isolation of a novel analogue. Algal Res 31:87–98

Mori K, Ooi T, Hiraoka M, Oka N, Hamada H, Tamura M, Kusumi T (2004) Fucoxanthin and its metabolites in edible brown algae cultivated in deep seawater. Mar Drugs 2:63–72

Nagaoka S, Shimizu K, Kaneko H, Shibayama F, Morikawa K, Kanamaru Y, Otsuka A, Hirahashi T, Kato T (2005) A novel protein C-phycocyanin plays a crucial role in the hypocholesterolemic action of *Spirulina platensis* concentrate in rats. J Nutr 135:2425–2430

Naito Y, Uchiyama K, Aoi W, Hasegawa G, Nakamura N, Yoshida N, Maoka T, Takahashi J, Yoshikawa T (2004) Prevention of diabetic nephropathy by treatment with astaxanthin in diabetic db/db mice. Biofactors 20:49–59

Nakashima Y, Ohsawa I, Konishi F, Hasegawa T, Kumamoto S, Suzuki Y, Ohta S (2009) Preventive effects of *Chlorella* on cognitive decline in age-dependent dementia model mice. Neurosci Lett 464:193–198

Nakazawa Y, Sashima T, Hosokawa M, Miyashita K (2009) Comparative evaluation of growth inhibitory effect of stereoisomers of fucoxanthin in human cancer cell lines. J Funct Foods 1:88–97

Narayanan KB, Sakthivel N (2011) Green synthesis of biogenic metal nanoparticles by terrestrial and aquatic phototrophic and heterotrophic eukaryotes and biocompatible agents. Adv Colloid Interf Sci 169:59–79

Nasirian F, Dadkhah M, Moradi-Kor N, Obeidavi Z (2018) Effects of *Spirulina platensis* microalgae on antioxidant and anti-inflammatory factors in diabetic rats. Diabetes Metab Syndr Obes 11:375–380

Nemoto-Kawamura C, Hirahashi T, Nagai T, Yamada H, Katoh T, Hayashi O (2004) Phycocyanin enhances secretary IgA antibody response and suppresses allergic IgE antibody response in mice immunized with antigen-entrapped biodegradable microparticles. J Nutr Sci Vitaminol 50:129–136

Niizawa I, Espinaco BY, Zorrilla SE, Sihufe GA (2018) Natural astaxanthin encapsulation: use of response surface methodology for the design of alginate beads. Int J Biol Macromol 121:601–608

Nobre BP, Villalobos F, Barragan BE, Oliveira AC, Batista AP, Marques PA, Mendes RL, Sovova H, Palavra AF, Gouveia L (2013) A biorefinery from *Nannochloropsis* sp. microalga—extraction of oils and pigments. Production of biohydrogen from the leftover biomass. Bioresour Technol 135:128–136

O'keefe B, McMahon JB, Gustafson K, Takebe Y (2011) Antiviral activity of the protein scytovirin and methods of use. Google Patents

O'Keefe BR, Smee DF, Turpin JA, Saucedo CJ, Gustafson KR, Mori T, Blakeslee D, Buckheit R, Boyd MR (2003) Potent anti-influenza activity of cyanovirin-N and interactions with viral hemagglutinin. Antimicrob Agents Chemother 47:2518–2525

Olaizola M (2000) Commercial production of astaxanthin from *Haematococcus pluvialis* using 25,000-liter outdoor photobioreactors. J Appl Phycol 12:499–506

Olguin EJ (2012) Dual purpose microalgae-bacteria-based systems that treat wastewater and produce biodiesel and chemical products within a biorefinery. Biotechnol Adv 30:1031–1046

Pádua D, Rocha E, Gargiulo D, Ramos AA (2015) Bioactive compounds from brown seaweeds: phloroglucinol, fucoxanthin and fucoidan as promising therapeutic agents against breast cancer. Phytochem Lett 14:91–98

Palozza P, Torelli C, Boninsegna A, Simone R, Catalano A, Mele MC, Picci N (2009) Growth-inhibitory effects of the astaxanthin-rich alga *Haematococcus pluvialis* in human colon cancer cells. Cancer Lett 283:108–117

Patel V, Berthold D, Puranik P, Gantar M (2015) Screening of cyanobacteria and microalgae for their ability to synthesize silver nanoparticles with antibacterial activity. Biotechnol Rep 5:112–119

Patterson GML, Baldwin CL, Bolis CM, Caplan FR, Karuso H, Larsen LK, Levine IA, Moore RE, Nelson CS, Tschappat KD, Tuang GD, Furusawa E, Furusawa S, Norton TR, Raybourne RB (1991) Antineoplastic activity of cultured blue-green algae (Cyanophyta). J Phycol 27:530–536

Perez M, Rilatt I, Lamothe M (2018) Derivatives of dolastatin 10 and auristatins. Google Patents

Pollingher U, Zemel E (1981) In situ and experimental evidence of the influence of turbulence on cell division processes of *Peridinium cinctum* forma *westii* (Lemm.) Lefevre. Br Phycol J 16:281–287

Prakash B, Kujur A, Yadav A (2018) Drug synthesis from natural products: a historical overview and future perspective. In: Tewari A, Tiwari S (eds) Synthesis of medicinal agents from plants. Elsevier, Amsterdam, pp 25–46

Pratt R, Daniels T, Eiler JJ, Gunnison J, Kumler W, Oneto JF, Strait LA, Spoehr H, Hardin G, Milner H (1944) Chlorellin, an antibacterial substance from *Chlorella*. Science 99:351–352

Ramos DF, Matthiensen A, Colvara W, de Votto AP, Trindade GS, da Silva PE, Yunes JS (2015) Antimycobacterial and cytotoxicity activity of microcystins. J Venom Anim Toxins Incl Trop Dis 21:9

Rao AR, Sindhuja H, Dharmesh SM, Sankar KU, Sarada R, Ravishankar GA (2013) Effective inhibition of skin cancer, tyrosinase, and antioxidative properties by astaxanthin and astaxanthin esters from the green alga *Haematococcus pluvialis*. J Agric Food Chem 61:3842–3851

Ravi M, Tentu S, Baskar G, Rohan Prasad S, Raghavan S, Jayaprakash P, Jeyakanthan J, Rayala SK, Venkatraman G (2015) Molecular mechanism of anti-cancer activity of phycocyanin in triple-negative breast cancer cells. BMC Cancer 15:768

Rechter S, Konig T, Auerochs S, Thulke S, Walter H, Dornenburg H, Walter C, Marschall M (2006) Antiviral activity of *Arthrospira*-derived spirulan-like substances. Antivir Res 72:197–206
Remirez D, Ledon N, Gonzalez R (2002) Role of histamine in the inhibitory effects of phycocyanin in experimental models of allergic inflammatory response. Mediat Inflamm 11:81–85
Renju GL, Muraleedhara Kurup G, Saritha Kumari CH (2013) Anti-inflammatory activity of lycopene isolated from *Chlorella marina* on type II collagen induced arthritis in Sprague Dawley rats. Immunopharmacol Immunotoxicol 35:282–291
Renju GL, Kurup GM, Saritha Kumari CH (2014a) Effect of lycopene from *Chlorella marina* on high cholesterol-induced oxidative damage and inflammation in rats. Inflammopharmacology 22:45–54
Renju GL, Muraleedhara Kurup G, Bandugula VR (2014b) Effect of lycopene isolated from *Chlorella marina* on proliferation and apoptosis in human prostate cancer cell line PC-3. Tumour Biol 35:10747–10758
Rodrigues DB, Menezes CR, Mercadante AZ, Jacob-Lopes E, Zepka LQ (2015) Bioactive pigments from microalgae Phormidium autumnale. Food Res Int 77:273–279
Romieu I, Téllez-Rojo MM, Lazo M, Manzano-Patiño A, Cortez-Lugo M, Julien P, Bélanger MC, Hernandez-Avila M, Holguin F (2005) Omega-3 fatty acid prevents heart rate variability reductions associated with particulate matter. Am J Resp Crit Care Med 172:1534–1540
Samarakoon KW, Ko JY, Lee JH, Kwon ON, Kim SW, Jeon YJ (2014) Apoptotic anticancer activity of a novel fatty alcohol ester isolated from cultured marine diatom, *Phaeodactylum tricornutum*. J Funct Foods 6:231–240
Sánchez JF, Fernández-Sevilla JM, Acién FG, Cerón MC, Pérez-Parra J, Molina-Grima E (2008) Biomass and lutein productivity of *Scenedesmus almeriensis*: influence of irradiance, dilution rate and temperature. Appl Microbiol Biotechnol 79:719–729
Sansone C, Braca A, Ercolesi E, Romano G, Palumbo A, Casotti R, Francone M, Ianora A (2014) Diatom-derived polyunsaturated aldehydes activate cell death in human cancer cell lines but not normal cells. PLoS One 9:e101220
Sato Y, Okuyama S, Hori K (2007) Primary structure and carbohydrate-binding specificity of a potent anti-HIV lectin isolated from the filamentous cyanobacterium, *Oscillatoria agardhii*. J Biol Chem 282(15):11021–11029
Senhorinho GNA, Laamanen CA, Scott JA (2018) Bioprospecting freshwater microalgae for antibacterial activity from water bodies associated with abandoned mine sites. Phycologia 57:432–439
Sessa C, Weigang-Köhler K, Pagani O, Greim G, Mora O, de Pas T, Burgess M, Weimer I, Johnson R (2002) Phase I and pharmacological studies of the cryptophycin analogue LY355703 administered on a single intermittent or weekly schedule. Eur J Cancer 38:2388–2396
Shankar PD, Shobana S, Karuppusamy I, Pugazhendhi A, Ramkumar VS, Arvindnarayan S, Kumar G (2016) A review on the biosynthesis of metallic nanoparticles (gold and silver) using bio-components of microalgae: formation mechanism and applications. Enzym Microb Technol 95:28–44
Sharon N, Lis H (1989) Lectins as cell recognition molecules. Science 246:227–234
Shi XM, Chen F (2002) High-yield production of lutein by the green microalga *Chlorella protothecoides* in heterotrophic fed-batch culture. Biotechnol Prog 18:723–727
Shih PM, Wu D, Latifi A, Axen SD, Fewer DP, Talla E, Calteau A, Cai F, Tandeau de Marsac N, Rippka R, Herdman M, Sivonen K, Coursin T, Laurent T, Goodwin L, Nolan M, Davenport KW, Han CS, Rubin EM, Eisen JA, Woyke T, Gugger M, Kerfeld CA (2013) Improving the coverage of the cyanobacterial phylum using diversity-driven genome sequencing. Proc Natl Acad Sci USA 110:1053–1058
Siqueira AS, Lima ARJ, Aguiar DCF, Santos AS, Vianez Júnior JLDSG, Gonçalves EC (2018) Genomic screening of new putative antiviral lectins from Amazonian cyanobacteria based on a bioinformatics approach. Proteins 86:1047–1054

Soontornchaiboon W, Joo SS, Kim SM (2012) Anti-inflammatory effects of violaxanthin isolated from microalga *Chlorella ellipsoidea* in RAW 264.7 macrophages. Biol Pharm Bull 35:1137–1144

Sosa-Hernandez JE, Escobedo-Avellaneda Z, Iqbal HMN, Welti-Chanes J (2018) State-of-the-art extraction methodologies for bioactive compounds from algal biome to meet bio-economy challenges and opportunities. Molecules 23:2953

Stevenson CS, Capper EA, Roshak AK, Marquez B, Eichman C, Jackson JR, Mattern M, Gerwick WH, Jacobs RS, Marshall LA (2002) The identification and characterization of the marine natural product Scytonemin as a novel Antiproliferative Pharmacophore. J Pharmacol Exp Ther 303(2):858–866

Tan CP, Hou YH (2014) First evidence for the anti-inflammatory activity of fucoxanthin in high-fat-diet-induced obesity in mice and the antioxidant functions in PC12 cells. Inflammation 37:443–450

Tan RH, Wang F, Fan CL, Zhang XH, Zhao JS, Zhang JJ, Yang Y, Xi Y, Zou ZQ, Bu SZ (2018) Algal oil rich in n-3 polyunsaturated fatty acids suppresses B16F10 melanoma lung metastasis by autophagy induction. Food Funct 9:6179–6186

Ulmann L, Blanckaert V, Mimouni VX, Andersson M, Schoefs B, Chenais B (2017) Microalgal fatty acids and their implication in health and disease. Mini Rev Med Chem 17:1112–1123

Umemura K, Yanase K, Suzuki M, Okutani K, Yamori T, Andoh T (2003) Inhibition of DNA topoisomerases I and II, and growth inhibition of human cancer cell lines by a marine microalgal polysaccharide. Biochem Pharmacol 66:481–487

Vaishampayan U, Glode M, Du W, Kraft A, Hudes G, Wright J, Hussain M (2000) Phase II study of dolastatin-10 in patients with hormone-refractory metastatic prostate adenocarcinoma. Clin Cancer Res 6:4205–4208

Vinayak V, Manoylov KM, Gateau H, Blanckaert V, Herault J, Pencreac'h G, Marchand J, Gordon R, Schoefs B (2015) Diatom milking: a review and new approaches. Mar Drugs 13:2629–2665

Volk RB, Furkert FH (2006) Antialgal, antibacterial and antifungal activity of two metabolites produced and excreted by cyanobacteria during growth. Microbiol Res 161:180–186

von Schacky C (2008) Omega-3 fatty acids: antiarrhythmic, proarrhythmic or both? Curr Opin Clin Nutr Metab Care 11:94–99

Wang H, Fewer DP, Sivonen K (2011) Genome mining demonstrates the widespread occurrence of gene clusters encoding bacteriocins in cyanobacteria. PLoS One 6:e22384

Wang B, Lan CQ, Horsman M (2012a) Closed photobioreactors for production of microalgal biomasses. Biotechnol Adv 30:904–912

Wang J, Chen S, Xu S, Yu X, Ma D, Hu X, Cao X (2012b) In vivo induction of apoptosis by fucoxanthin, a marine carotenoid, associated with down-regulating STAT3/EGFR signaling in sarcoma 180 (S180) xenografts-bearing mice. Mar Drugs 10:2055–2068

Wang S, Said IH, Thorstenson C, Thomsen C, Ullrich MS, Kuhnert N, Thomsen L (2018a) Pilot-scale production of antibacterial substances by the marine diatom *Phaeodactylum tricornutum* Bohlin. Algal Res 32:113–120

Wang X, Wang H, Pierre JF, Wang S, Huang H, Zhang J, Liang S, Zeng Q, Zhang C, Huang M, Ruan C, Lin J, Li H (2018b) Marine microalgae bioengineered *Schizochytrium* sp. meal hydrolysates inhibits acute inflammation. Sci Rep 8:9848

Ward OP, Singh A (2005) Omega-3/6 fatty acids: alternative sources of production. Process Biochem 40:3627–3652

Waters AL, Hill RT, Place AR, Hamann MT (2010) The expanding role of marine microbes in pharmaceutical development. Curr Opin Biotechnol 21:780–786

WHO (World Health Organization) Cancer (2018) http://www.who.int/en/news-room/fact-sheets/detail/cancer. Accessed 19 Jan 2019

Wolf J, Stephens E, Steinbusch S, Yarnold J, Ross IL, Steinweg C, Doebbe A, Krolovitsch C, Müller S, Jakob G, Kruse O, Posten C, Hankamer B (2016) Multifactorial comparison of photobioreactor geometries in parallel microalgae cultivations. Algal Res 15:187–201

Woo MN, Jeon SM, Kim HJ, Lee MK, Shin SK, Shin YC, Park YB, Choi MS (2010) Fucoxanthin supplementation improves plasma and hepatic lipid metabolism and blood glucose concentration in high-fat fed C57BL/6N mice. Chem Biol Interact 186:316–322

Yan T, Li HY, Wu JS, Niu Q, Duan WH, Han QZ, Ji WM, Zhang T, Lv W (2017) Astaxanthin inhibits gemcitabine-resistant human pancreatic cancer progression through EMT inhibition and gemcitabine resensitization. Oncol Lett 14:5400–5408

Yim JH, Kim SJ, Ahn SH, Lee CK, Rhie KT, Lee HK (2004) Antiviral effects of sulfated exopolysaccharide from the marine microalga *Gyrodinium impudicum* strain KG03. Mar Biotechnol 6:17–25

Yu Y, Shen M, Song Q, Xie J (2018) Biological activities and pharmaceutical applications of polysaccharide from natural resources: a review. Carbohydr Polym 183:91–101

Zhang Y, Xu W, Huang X, Zhao Y, Ren Q, Hong Z, Huang M, Xing X (2018) Fucoxanthin ameliorates hyperglycemia, hyperlipidemia and insulin resistance in diabetic mice partially through IRS-1/PI3K/Akt and AMPK pathways. J Funct Foods 48:515–524

Zhu C, Zhai X, Jia J, Wang J, Han D, Li Y, Tang Y, Chi Z (2018) Seawater desalination concentrate for cultivation of *Dunaliella salina* with floating photobioreactor to produce β-carotene. Algal Res 35:319–324

Chapter 13
Metal Pollution in Water: Toxicity, Tolerance and Use of Algae as a Potential Remediation Solution

Rossella Pistocchi, Ly Thi Hai Dao, Paulina Mikulic, and John Beardall

Abstract Metals are persistent pollutants which, in consequence of their accumulation in water bodies, can exert dangerous effects on human health and on the ecosystem. Conventional techniques used for their removal are expensive, energy-consuming and can release additional noxious chemicals, all aspects that could be overcome through the use of living organisms for bioremediation purposes. Algae are naturally exposed to complex and stressful environments so that they possess a certain degree of resistance to metal toxicity, thanks to physiological defence mechanisms that are at the basis of the utility for bioremediation. Both living and dead micro- and macroalgae were successfully demonstrated to sequester different metals from aqueous solutions, the main advantages being the high selectivity and affinity also at low metal concentrations and the possibility to interact with multiple ions. On the other hand, the process is species-specific, as algae are very diverse in composition, growth requirements, resistance to pre- and post-treatments and their exploitation usually requires tailored processes. This chapter summarises the basic aspects of cells' interactions with the metals and the evidence of the great utility these organisms can have in metal removal. It analyses the applicative achievements obtained until now which are still at a base level due to the complex nature of real wastewaters and to the difficulty in scaling up the process at industrial level.

13.1 Introduction

Many metals (e.g. zinc, copper and iron) play an essential role in biology. Others such as cadmium and lead, and metalloids such as arsenic, are not involved in metabolism, but can interfere with biological processes and pose problems of

R. Pistocchi
Department of Biological, Geological, and Environmental Sciences, University of Bologna, Ravenna, Italy

L. T. H. Dao · P. Mikulic · J. Beardall (✉)
School of Biological Sciences, Monash University, Clayton, VIC, Australia
e-mail: john.beardall@monash.edu

A. Hallmann, P. H. Rampelotto (eds.), *Grand Challenges in Algae Biotechnology*, Grand Challenges in Biology and Biotechnology,
https://doi.org/10.1007/978-3-030-25233-5_13

Table 13.1 Metals commonly found as pollutants, sources and the US EPA guideline concentrations for levels acceptable for aquatic (freshwater) life

Metal	Source	Recommended limits for aquatic life in freshwater USEPA CCC[a] μg/L
Arsenic	Biogeochemical processes, smelting operations for Cu, Zn and Pb, wood preservatives, pest controls	150
Cadmium[b]	Zinc and lead smelting, waste batteries, sludge, waste incineration and fuel combustion	0.72
Chromium[b]	Mining, industrial coolants, leather tanning	74 Trivalent 11 Hexavalent
Copper	Mining, electroplating, smelting operations	1.45
Lead[b]	Lead acid batteries, older paints, E-waste, smelting operations, coal-based thermal power plants, ceramics	2.5
Mercury	Chlor-alkali plants, thermal power plants, fluorescent lamps, hospital waste (damaged thermometers, barometers, sphygmomanometers), electrical appliances etc.	0.77
Nickel[b]	Smelting operations, thermal power plants, battery industry	52
Zinc[b]	Smelting, electroplating, urban run-off from roofs (galvanised roofs) and roads (tyre wear)	120

Modified from Gautam et al. (2015)
Data on chronic exposure levels taken from National Recommended Water Quality Criteria https://www.epa.gov/wqc/national-recommended-water-quality-criteria-aquatic-life-criteria-table
[a]*CCC* Criterion Continuous Concentration, i.e. chronic exposure levels
[b]Values are dependent on the hardness of the water, those cited here are for a hardness of 100 mg/L for ease of comparison

toxicity. Even essential metals such as copper and zinc can be toxic at elevated concentrations. Although there are some environments which have naturally elevated levels of metals, human activities such as mining and industrialisation have greatly exacerbated the problems of metal toxicity.

13.1.1 Metal Pollution in Aquatic Environments

Metals generated by industry and other human activities often find their way into water bodies, including importantly, drinking water sources for human consumption. In aquatic ecosystems, metal pollution can have a range of effects on organisms and severely disrupt food chain function (Baby et al. 2010; Klerks and Levinton 1989; Rai 2008). A range of metals are commonly found as pollutants and the most common, along with the sources and the US EPA guideline concentrations for levels acceptable for aquatic (freshwater) life, are summarised in Table 13.1.

Table 13.2 Concentrations of heavy metals in some freshwaters

	Metal concentration (μg/L)					
Site	Cd	Cr	Cu	Pb	Ni	Zn
Unpolluted lake Parma Valley, Italy	0.37	0.4	**14.4**	**3.45**	–	–
Polluted river downstream Parma Valley, Italy	**1.82**	4.42	**15.3**	**7.3**	–	–
Sunguncha y stream, Iran	–	**28.3**	1.9	**5.5**	13.2	–
Electroplating effluent, India	**2000**	**56,000**	**4000**	**4000**	**2 × 10^5**	**5000**
Lake Chapala, Mexico	**2.5**	2.80	**16.5**	**12.8**	18.3	32
"Drinking water" Ethiopia (max found)	**6.41**	**21.3**	**27**	**46**	11.2	**5140**
Chari River, Chad		**130**	**190**		**650**	**250**
Rio Tinto, Spain	**700–43,000**		**12,000–132,000**			**14,000–118,000**

Values in bold are in excess of the US EPA chronic exposure guidelines shown in Table 13.1. Data abstracted from De Filippis and Pallaghy (1994), Moore et al. (2011), Reimann et al. (2003), Ford and Ryan (1995), Nambatingar et al. (2017), Aguilera (2013)

Concentrations of heavy metals are often found to be in excess of guidelines and a small set of examples is given, by way of illustration, in Table 13.2. Most of the situations where metal concentrations exceed safe limits for ecosystem function or human health are associated with industrial waste, including acid mine drainage and smelting operations.

Low pH sites are often accompanied by high metal levels, as metal ions readily dissolve in acid. Not surprisingly, many metal-enriched sites are also rich with metal-tolerant extremophiles, and acidophilic algae are described as extremely tolerant of high metal levels in their environment (Mikulic and Beardall 2014; Varshney et al. 2015; Whitton 1970). This tolerance of some algae to high levels of metals makes them of interest in terms of possible approaches to bioremediation.

13.1.2 Mechanisms of Metal Toxicity to Algae

Metals, both essential and non-essential, are toxic when present at a level that will induce a deleterious response in an organism (Påhlsson 1989). To have a toxic effect, a metal must thus be present at a level that causes stress responses that cannot be overcome by the cell's physiological defences. The concentration of metal that will cause toxicity symptoms is largely determined by its bioavailability which, in turn, is determined by the metal's chemical and physical properties. The metal's ionic radius, charge and coordination geometry determine its interaction with other metals and nutrients and its mass transfer at the biological interface, which in turn determines the path of the metal from the growth medium to the cell. Finally, in addition to the common effects of toxic metal levels on cells, there are many specific differential effects on different genotypes, resulting in different physiological responses (Foy et al. 1978).

Generally, metal-coordinating groups such as amine (of amino acids and nucleotides), imine (of histidine and nucleotides), sulfhydryl (of cysteine) and carboxylate groups are important for activities of biological functional molecules like proteins, enzymes and polynucleotides (Ochiai 1987). Consequently, a metal may displace/substitute for an existing essential element bound to an active site, or simply bind itself to the active site, or interact with the groups, resulting in an inhibition of the biomolecule's activity (Stauber and Florence 1987). Therefore, general metal toxicity mechanisms can be grouped as follows:

1. Interactions with Essential Functional Groups of Biomolecules

Metals will bind to many functional groups in the cell's macromolecules, including lipids, polysaccharides and proteins that form important cell structures, including those involved in respiration and photosynthesis, resulting in their impaired functionality (Eichhorn 1975). Metals also cause damage to essential functional groups through the formation of reactive oxygen species, which can then cause oxidative damage to important cell structures (Pinto et al. 2003; Stohs and Bagchi 1995). The metal can also displace essential elements used in processes through molecular mimicry, resulting in deficiency symptoms of the displaced element (Clijsters and Van Assche 1985; De Filippis and Ziegler 1993; Singh and Singh 1987; Van Assche and Clijsters 1986).

Through their interactions with thiol (–SH) groups of proteins, metals can inhibit the functioning of many proteins, including those involved in PSII (Clijsters and Van Assche 1985; Gimmler 2001). Many essential metals are bivalent, including Mn, Fe, Cu, Mg and Zn; thus, they share a similar size atomic radius. In conditions of excess zinc, molecular mimicry results in competition at different sites in the cell and a change in physiological equilibria (De Filippis and Ziegler 1993; Singh and Singh 1987). Thus, in photosynthesis, Zn has been found to displace Mg in the active ternary Rubisco-CO_2 metal complex. Another example of molecular mimicry is substitution of Mn by Zn in the water splitting enzyme (Clijsters and Van Assche 1985; Van Assche and Clijsters 1986).

Metal interactions with macromolecules generally cause conformational change to proteins either by directly binding to functional groups of the active and non-active sites of the proteins (Eichhorn 1975), or by altering lipids in the cell membranes to which proteins are bound (Devi and Prasad 2013).

Lipids can be altered by metals either directly through metal-lipid binding, indirectly by lipid peroxidation resulting from metal oxidative stress or by alterations in the metabolism of the lipids (Akeson et al. 1989; Jones and Kochian 1997; Maksymiec et al. 1992; Skórzyńska et al. 1991; Stefanov et al. 1993; Verma and Dubey 2003; Zhao et al. 1987). Metals have been shown to cause damage to nucleic acids by binding to their heterocyclic bases, ribose, hydroxyls, and phosphate groups (Eichhorn 1975). A metal bound to the base will disrupt base-pair hydrogen bonds, and thus the double helix structure of DNA (Eichhorn 1962, 1973). Metals can result in cross links in the double-helix, interfering with the unwinding of the two strands in transcription (Eichhorn and Butzow 1965; Eichhorn and Shin 1968; Shin and Eichhorn 1968; Yamane and Davidson 1961). Metals can neutralise the negatively charged DNA molecule by binding to the phosphate backbone, thus resulting in an increase in base-pair mismatching (Eichhorn 1975; Szer and Ochoa 1964). Depolymerisation of nucleic acids has also been observed in cases of metal–phosphate interactions (Butzow and Eichhorn 1965; Eichhorn and Butzow 1965).

2. Disruption of the Integrity of Biomembranes

Uptake of metals into the cell and cell organelles can result in membrane depolarisation, damage to membrane lipids via the various mechanisms mentioned above, acidification of the cytoplasm and the onset of disruption of homeostasis (Conner and Schmid 2003; Pinto et al. 2003; Cardozo et al. 2002).

3. Production of Reactive Oxygen Species

A major mechanism of metal-induced damage to biological systems involves generation of reactive oxygen species (ROS). The mechanisms by which metals cause ROS production are complex, involving many different sources. Metals can be either redox active or non-active, which determines the mechanism by which they increase ROS levels (Ercal et al. 2001). Redox active metals will partake in redox reactions resulting in free radicals and ROS. These include Fe^{3+} and Cu^{2+}, involved in the Fenton type reactions (Eq. 13.1–13.2) and overall Haber-Weiss reaction (Eqs. 13.3), (Pinto et al. 2003; Stohs and Bagchi 1995).

$$M^n + O_2^{\bullet -} \rightarrow M^{n-1} + O_2 \tag{13.1}$$

$$M^{n-1} + H_2O_2 \rightarrow M^n + {}^{\bullet}OH + OH^- \tag{13.2}$$

The overall reaction is thus:

$$O_2^{\bullet -} + H_2O_2 \xrightarrow{M^n/M^{n-1}} O_2 + {}^{\bullet}OH + OH^- \tag{13.3}$$

Chromium will undergo reactions analogous to the Fenton reactions to produce ${}^{\bullet}OH$ (Shi and Dalal 1990), and other metals (cobalt, nickel and vanadium) have been shown to be involved in similar, Fenton-like, reactions (Leonard et al. 2004). Fenton-like reactions are most commonly associated with membrane fractions in the cell (Stohs and Bagchi 1995).

Redox non-active metals, for example, Cd, As, Pb, Mn, Zn, Ni and Al, increase ROS generation indirectly by reducing antioxidative capacity, such as by reducing the antioxidant glutathione (GSH) pool (Piotrowska-Niczyporuk et al. 2015; Malea et al. 2006) or inhibiting the antioxidative enzymes such as catalase, ascorbate peroxidase (APX), glutathione peroxidase (GPX) and superoxide dismutase (SOD) by interacting with the –SH groups on the enzymes/displacing essential metal ions from the enzymes (Leão et al. 2014).

Although thiol-containing molecules are generally associated with the cells' antioxidative capacity (see below), Shi and Dalal (1990) showed that Cr (V) induced the production of thyil, a glutathione radical, in the presence of high levels of glutathione. Cr(V) and vanadium also induced the production of a thyil radical from biologically relevant thiol-containing molecules, for example, glutathione, cysteine and pencillamine (Shi and Dalal 1990; Shi et al. 1994). Thyils produced through Cr(V) toxicity in the presence of H_2O_2, yielded $^{\bullet}OH$ and, in the presence of thiols, lipid hydroperoxide free radicals (Shi et al. 1994).

13.1.3 Strategies Used by Algae to Cope with Metals in Their Environment

Many algae possess a degree of tolerance to metal toxicity, which can be phenotypic or genotypic. Table 13.3 shows EC_{50} values for growth (the concentration of metal required to inhibit growth rate by 50%) for a range of microalgae with varying levels of tolerance to different metals. The underlying defence mechanisms providing metal tolerance form the basis of the utility of algae for bioremediation. In essence, these can be summarised as (a) reduced uptake or exclusion from the cell, (b) efflux mechanisms and biotransformations converting the toxic metal into a less toxic form and (c) sequestration (Fig. 13.1). Other mechanisms involve detoxification, or prevention of synthesis, of ROS and involvement of heat shock proteins.

13.1.3.1 Cellular Defence Through Metal Uptake and Sequestration Mechanisms

Defence against metal toxicity begins at the cell surface, with cells achieving metal tolerance through *exclusion* by blocking transport via ionophores and other membrane transporters, or by binding the metal ions to functional groups on the cell wall (Cumming and Taylor 1990; Hall 2002; Macfie and Welbourn 2000). Acidophiles in particular are often thought to employ the exclusion mechanism through reduced metal uptake, as a result of the increased proton concentration in the low pH environment they inhabit (Beardall and Entwisle 1984; Nalewajko et al. 1997; Olaveson and Stokes 1989). Many macroalgae, on the other hand, possess cell wall materials with negatively charged functional groups (e.g. alginates in brown algae) that attract and readily bind the positively charged metal ions (Moenne et al. 2016) (see Sect. 13.2.1). Exclusion mechanisms in green and red algae can be also based on metal chelation by epibionts, as well as on permeability of the plasmatic membrane to metal ions (Andrade et al. 2004; dos Santos et al. 2014).

The stimulation of *efflux mechanisms* can help maintain homoeostasis. In this mode of protection, cells take up metals through passive or active transport. The metals are often biotransformed to a less toxic form then sent back out of the cell

Table 13.3 EC_{50} values reported in the literature for a range of different algae and metals

Metal	Algal strain	IC50/EC50 (mg/L)	References
Pb	*Chlorella* sp.	0.4×10^{-3a}	Dao and Beardall (2016)
	Scenedesmus sp.	0.83×10^{-4a}	
	Pseudokirchneriella subcapitata	0.66^{b}	De Schamphelaere et al. (2014)
	Chlamydomonas reinhardtii	0.02	Scheidegger et al. (2011)
	Isochrysis galbana	7.83	Liu et al. (2011)
	Tetraselmis chui	2.26	
	Nannochloropsis gaditana	0.75	Debelius et al. (2009)
	Isochrysis galbana (T-iso)	1.346	
	Chaetoceros sp.	0.103	
	Rhodomonas salina	0.89	
	Tetraselmis chuii	2.69	
	Chlorella kessleri	3.025	Fujiwara et al. (2008)
	Phaeocystis antarctica	0.57	Gissi et al. (2015)
Cu	*Pavlova* sp.	0.05	Purbonegoro et al. (2018)
	Planothidium lanceolatum	0.62	Sbihi et al. (2012)
	Isochrysis galbana	1.99	Liu et al. (2011)
	Tetraselmis chui	0.082	
	Tetraselmis chuii	0.33	Debelius et al. (2009)
	Nannochloropsis gaditana	0.14	
	Isochrysis galbana (T-iso)	0.057	
	Chaetoceros sp.	0.089	
	Rhodomonas salina	0.045	
	Minutocellus polymorphus	0.6×10^{-3}	Levy et al. (2007)
	Dunaliella tertiolecta	0.53	
	Gephyrocapsa oceanica	0.017	
	Scenedesmus obliquus	0.05	Yan and Pan (2002)
	Chlorella pyrenoidosa	0.068	
	Closterium lunula	0.2	
	Chlamydomonas acidophila	8.96	Nishikawa and Tominaga (2001)
	Isochrysis galbana	0.018	Trenfield et al. (2015)
	Phaeocystis antarctica	0.59×10^{-2}	Gissi et al. (2015)
Cd	*Planothidium lanceolatum*	0.25	Sbihi et al. (2012)
	Pseudokirchneriella subcapitata	0.67	Rodgher et al. (2012)
	Microcystis aeruginosa	1.12×10^{-3}	
	Isochrysis galbana	13.67	Liu et al. (2011)
	Tetraselmis chui	4.25	
	Chlamydomonas acidophila	1.62	Nishikawa and Tominaga (2001)
	Scenedesmus armatus	0.54	Baścik-Remisiewicz and Tukaj (2002)

(continued)

Table 13.3 (continued)

Metal	Algal strain	IC50/EC50 (mg/L)	References
	Phaeocystis antarctica	1.5	Gissi et al. (2015)
	Scenedesmus obliquus	0.058	Monteiro et al. (2011)
	Desmodesmus pleiomorphus	1.92	
Zn	*Planothidium lanceolatum*	0.35	Sbihi et al. (2012)
	Cyanidium caldarium	4.87	Mikulic and Beardall (2014)
	Chlamydomonas reinhardtii	0.038	
	Chlamydomonas acidophila	0.13	Nishikawa and Tominaga (2001)
	Phaeocystis antarctica	1.11	Gissi et al. (2015)
	Scenedesmus obliquus	16.99	Monteiro et al. (2011)
	Desmodesmus pleiomorphus	4.87	
Cr	*Pseudokirchneriella subcapitata*	1.04[a]	Rodgher et al. (2012)
	Microcystis aeruginosa	0.58[a]	
Co	*Chlamydomonas acidophila*	4.79	Nishikawa and Tominaga (2001)
Ag	*Chlorella* sp.	0.39	Yoo-iam et al. (2014)
Al	*Isochrysis galbana*	3.05	Trenfield et al. (2015)
As	*Chlorella* sp. strain CE-35	0.57	Rahman et al. (2014)
	Chlorella sp.	25	Levy et al. (2005)
	Monoraphidium arcuatum	0.25	

Unless indicated otherwise, values are for total metal concentrations rather than free ion concentrations

[a]Calculated as free metal ion concentrations

[b]Calculated as filtered Pb concentration (measured at the start of the exposure)

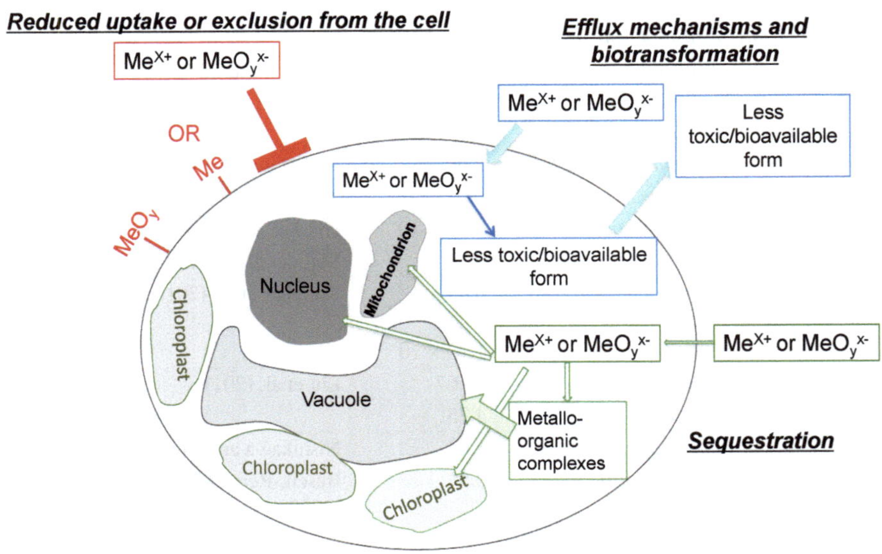

Fig. 13.1 Mechanisms involved in metal tolerance in algae

(Rosen 1996). Transporters in the cell membrane often have a lower affinity for these new biotransformants and thus the cell has ensured protection from accumulation of the metal at sensitive sites (Hall 2002). In some species, for example, arsenate is transformed to arsenite, and then to other arsenical compounds, thus altering it, so it is no longer a substrate for cell membrane transporters (Qin et al. 2009). In *Ectocarpus siliculosus* (Roncarati et al. 2015) and *Ulva compressa* (Navarrete et al. 2019) some evidence was obtained that a release of copper from cells to the external medium could be part of a detoxification mechanism.

Sequestration involves the metal ion entering the cell, and then being incorporated into cellular processes/structures/compartments without compromising these processes and structures greatly (Hall 2002; Pinto et al. 2003). Complexation of the metal with either an organic acid or a metal-binding protein, and then the storage of these complexes in the vacuole, is another common mode of sequestration. This is, for instance, a common tolerance method for zinc and cadmium in photosynthetic organisms (Brune et al. 1994; Cobbett and Goldsbrough 2002; Goldsbrough 1999; Howden et al. 1995a, b; Robinson et al. 1993).

13.1.3.2 Maintenance of Cellular Redox Balance Under Metal Stress

Mechanisms underlying the maintenance of cellular redox balance in the face of metal toxicity is crucial to metal tolerance in many species, including algae (Bielen et al. 2013; Siripornadulsil et al. 2002).

Thiols, which are organic sulphur derivatives known as mercaptans, and also referred to as sulfhydryl groups (–SH), have a central role in maintaining the antioxidant capacity of a cell. Thiols can be classified into low-molecular-weight and high-molecular-weight thiol-containing proteins. The overall thiol redox state reflects the oxidative state of the cell, and its capacity to defend the cell against oxidative stress, and scavenge for ROS. ROS scavenging mechanisms can be non-enzymatic and enzymatic (Apel and Hirt 2004; Halliwell 2006).

Enzymatic mechanisms include enzymes which catalyse the reduction of ROS. These include enzymes such as superoxide dismutase (SOD), ascorbate peroxidase (APX), glutathione peroxidase (GPX) thioredoxin, peroxiredoxin and lipid peroxidase (Apel and Hirt 2004).

Non-enzymatic mechanisms involve molecules with high reducing potential. Two major cellular redox buffers are ascorbate and the highly important, low-molecular-weight thiol, glutathione (GSH) (Apel and Hirt 2004; Lu 2009). Other important antioxidant molecules in the cell include tocopherol and flavonoids (Noctor and Foyer 1998).

GSH is a sulfhydryl tripeptide formed from glutamate, cysteine and glycine. In addition to acting as an antioxidant which can directly reduce ROS (Lu 2009; Meister and Anderson 1983), GSH is also a coenzyme for many enzymes that act in ROS reduction, for example, glutathione peroxidase. GSH is used as a substrate in the biosynthesis of phytochelatins, but will also bind spontaneously, and non-enzymatically, to toxic metals, forming metal–thiol conjugates, and also enzymatically, through the action of glutathione S-transferases, forming conjugates with toxic electrophiles (Cobbett and Goldsbrough 2002; Dalle-Donne et al. 2009). Due to these detoxifying roles, GSH levels are crucial in determining the redox state of a cell, which in turn modulates many cellular processes including cell division, differentiation and enzyme regulation (Dalle-Donne et al. 2009).

Furthermore, photosynthetic organisms such as algae also have a broad range of accessory pigments that have the primary role of extending the spectral absorption for PAR, but some of these also act as antioxidants to protect the light-harvesting complexes from ROS molecules produced in the light phase of photosynthesis and elsewhere in the cell (Pinto et al. 2003). These include fucoxanthin, peridinin and β-carotene which have a role in reducing singlet oxygen (Mallick and Mohn 2000; Noctor et al. 2015).

13.1.3.3 Heat Shock Protein Response to Metal Stress

The heat shock protein (HSP) response is commonly associated with environmental stress (Feder and Hofmann 1999; Sanità di Toppi et al. 2008). A study of the metal tolerance in *Chlamydomonas acidophila* isolated from acidic mining lakes showed increases in heat shock proteins (HSPs) (Spijkerman et al. 2007), much like an increased HSPs response that was seen in the extremophilic *Chlamydomonas acidophila* compared to *Chlamydomonas reinhardtii* when subjected to heat and pH stress (Gerloff-Elias et al. 2006). HSPs, which prevent permanent membrane damage by maintaining proteins in a folding-competent state (Parsell and Lindquist 1993), have been proven to protect cells against metal stress in tomato cell cultures (Neumann et al. 1994) and in moss (Esposito et al. 2012). Bierkens et al. (1998) showed a dose-dependent induction of heat shock protein 70 (HSP70) for many pollutants, including Zn and Se, in the green alga *Raphidocelis subcapitata*, suggesting that HSP70 is a potential bio-marker for pollution in algae.

13.1.3.4 Tolerance Acquired Through Physiological Acclimation and Genetic Adaptation to Metal Stress

Algae can show both physiological acclimation and genetic adaptation to high levels of metals. For instance, physiological adaptations to metal stress were found by Bossuyt and Janssen (2004) who found increased tolerance in *Pseudokirchneriella subcapitata* after a 3-month acclimation period to higher concentrations of copper (from 0.5 to 100 μg L^{-1}) and Kobayashi et al. (2006) showed that pre-exposure of *Chlamydomonas reinhardtii* to cadmium-stimulated production of phytochelatins, which also resulted in a higher tolerance of the alga to arsenic. There are also a number of studies showing differential tolerance acquired by macroalgae previously exposed to metal pollution, such as those reviewed by Moenne et al. (2016) for copper. Complication is added by the observation that the same species can acquire resistance in certain areas, but it did not seem to display differential tolerance to copper excess in others (Moenne et al. 2016) and by the finding that two Cu-tolerant populations of *E. siliculosus*, from Chile and England, could display distinctive defence strategies as, for example, differential antioxidant responses to elevated Cu concentrations (Sáez et al. 2015).

Klerks and Weis (1987) distinguished between resistance to pollutants acquired through physiological acclimation during exposure to sub-toxic concentrations, which is not passed down to subsequent generations, and populations that have evolved genetically based resistance through the act of natural selection at locations with high levels of toxicants, referred to as adaptation.

Reports of genetic adaptation of species to toxicants were based on the algae displaying tolerance upon return to standard media, and passing on the tolerance to subsequent generations (Klerks and Weis 1987). For example, Shehata and Whitton (1982) selected strains of *Anacystis nidulans* that were genetically adapted to Zn by growing cells at increasing inhibitory concentrations of Zn. Genetic adaptation to selenium was also observed in cultures of *Chlorella vulgaris*, with resistance remaining after sub-culturing in Se-free media (Shrift 1954). Harding and Whitton (1976) collected *Stigeoclonium tenue* from sites that varied in environmentally relevant Zn levels, and found increased Zn resistance in strains that grew at sites with more than 0.2 mg L^{-1} Zn. Interestingly, Harding and Whitton (1976) also found that this resistance was neither lost nor increased after culturing the strains for 6 months in clean water and in water with high levels of Zn. Other authors have reported successful selection of metal-tolerant algal species, which retained their metal tolerance upon culturing in non-toxic levels of metals (Foster 1977; Jensen et al. 1974; Kellner 1955; Russell and Morris 1970; Say and Whitton 1977).

In contrast, resistance acquired through physiological acclimation is lost upon cells' return to media with non-toxic levels of metal. Thus, algal acclimation to elevated Zn levels for a period of 100 days resulted in a threefold increase in Zn tolerance, with tolerance being lost upon the algae being returned to standard culture media (Muyssen and Janssen 2001).

An integrated "omics" approach could provide useful insight into the underlying mechanisms of physiological acclimation and genetic adaptation, which could uncover physiological mechanisms useful in bioremediation (Malla et al. 2018).

13.2 Can Algae Be Used for Bioremediation of Metal Pollution?

Current physicochemical approaches to removal of heavy metals from polluted effluent concentrate on techniques such as ion exchange and membrane transfer. However, ion exchange can release noxious chemicals from the resins involved and both this and membrane transfer are expensive. Membrane transfer is also energy consuming. As a consequence, biological approaches to remediation of metal pollution are of considerable interest, especially for countries where the high cost of technological approaches is precluded because of economic constraints.

Conversely, bioremediation involves the use of living organisms to clean polluted aquatic or terrestrial habitats as a cost-effective and eco-friendly technology. It is aimed at the elimination of pollutants such as nutrients (e.g. nitrate and phosphate) as well as xenobiotic compounds (polycyclic aromatic hydrocarbons, chlorophenols and pesticides). In addition to micro- and macroalgae which are mainly suited for treating contaminated water bodies, many other organisms such as bacteria, fungi and vascular plants are also considered for bioremediation applications in terrestrial or aquatic environments (Pacheco et al. 2015).

With regard to heavy metals remediation, the advantage of using algae to clean wastewaters lies in the lower cost when compared with conventional techniques (e.g. chemical precipitation, flocculation, ion exchange, electrochemical methods and membrane filtration); the economic convenience is determined by the lower energy demand and lower consumption of expensive chemicals, but other advantages include a better performance of the biological methods, especially at low metal concentrations, and the algal selectivity towards certain metals. Generally, algae are considered as a good source of biomass because they have a highly efficient photosynthesis and can grow faster than vascular plants and, furthermore, they can grow in wastewaters or in seawater and thus do not compete with freshwater use for other purposes.

Removal of metals by algae can occur through two mechanisms: biosorption or bioaccumulation, where the former consists of the binding of the pollutant to the cell surface, a passive process that does not require energy from metabolic activity, while the second involves transport into the cell and is an active metabolic process. Nevertheless, it has been ascertained that the uptake of metal ions into cells occurs secondarily after their adhesion to the cell wall (Chojnacka 2010) and thus the whole mechanism of bioaccumulation occurs in two steps with different velocity: a rapid adsorption of the metal to the cell wall followed by a slower uptake into the cell (see Fig. 13.2 for an example). It has also been shown that the biosorption process can be performed by both living and dead biomass.

Recently, biosorption has been increasingly considered for bioremediation with algae, as it is a very rapid process. In addition, the necessary biomass can be found in nature or can be a waste product from other processes, requires little treatment and can be reused. Nevertheless, the choice of the process as well as of the organism to be utilised depends on several factors, such as location of the decontamination plant and local availability of organisms/biomass, properties of the wastewater and whether there are additional aims for final biomass utilisation, including the possibility of metal recovery, and all these issues have to be carefully considered for each situation. For example, the use of living biomass, growing in the wastewater, has often been considered because of the presence of additional advantages such as the possibility to couple this process to the utilisation of waste CO_2 or to the reduction of nitrates and phosphates, as well as for the potential use of the biomass in biofuel production. It is highly important to consider that "algae" are very different organisms with many differences in cell composition, including the cell walls (i.e. the first

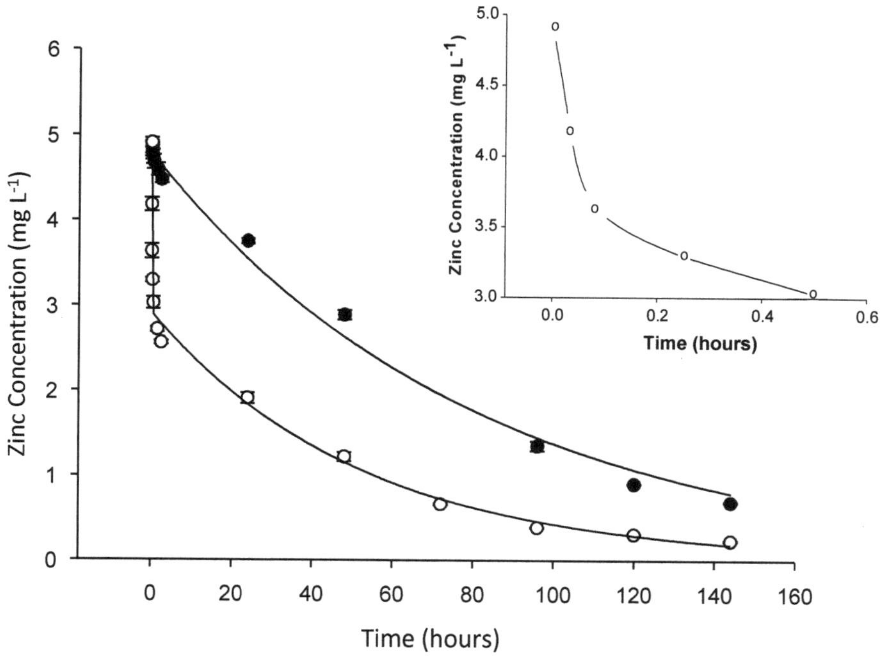

Fig. 13.2 Uptake of zinc by biofilms of the algae *Oocystis* sp. (closed symbols) and *Nitzschia* sp. (open symbols). Note the very marked rapid initial uptake by *Nitzschia*, associated with biosorption, followed by a slower, but high affinity uptake that results in effective removal of a large proportion of the metal within 2.5 h. Inset shows the rapid phase of biosorption by *Nitzschia*. *Oocystis* shows a much less marked biosorption phase. Data from the BSc Honours thesis of Katherina Bilski, School of Biological Sciences, Monash University

area of contact with the metals), but also in growth rates and in their adaptation to environmental parameters.

Below, we discuss separately the use of macroalgae and microalgae in the bioremediation process; however, there are some common aspects that have been extensively discussed in the review by Zeraatkar et al. (2016). For example, the interaction of metals with a surface of biological origin, even if it is a dead biomass, is influenced by several factors which are pH, temperature, organic matter, initial metal concentration, biomass concentration and the co-presence of a mixture of anions and cations. If the process is aimed at using living biomass, additional aspects should be taken into account such as the metal concentration (which should be at a level that does not inhibit growth), the effect of nutrient concentration, the influence of the algal growth phase on metal absorption and the presence of optimal temperature and light. Accordingly, a general knowledge of the physiological characteristics of the employed species is required.

13.2.1 Macroalgae

Macroalgae represent an important source of biomass, most of which comes from the harvesting of naturally growing seaweeds. Due to an increasing demand for biomass for a range of different purposes (food, phycocolloid extraction, plant stimulant formulations, cosmetics and pigment extraction), world harvesting is increasing, giving rise to some environment-related concerns; however, seaweed farming is also increasing (FAO 2016) so that an increasing chance to have access to waste biomass can be considered.

Studies on the use of macroalgae in metal bioremediation have been mainly addressed to dried biomass; nevertheless, cultured species have also been considered for applicative purposes.

Macroalgae belong to a range of different phylogenetic groups, characterised by large differences in cell constituents, including the cell walls, which are made of different arrays of polysaccharides, and whose structure could strongly influence the binding efficiency of metals. Metal adsorption is usually mediated by the negatively charged functional groups present on the cell surface, such as carboxyl, sulfhydryl, hydroxyl, sulphate and amino groups (Davis et al. 2003). Brown algae attracted the early attention of researchers in the bioremediation field because of the abundance in the cell walls (up to 40%) of alginic acid, a polysaccharide with several negatively charged groups which were considered to play an important role in metal binding (Davis et al. 2003). A comparison among different brown, red and green seaweeds (two of each group) made with dried biomass mixed with synthetic solutions of five metals (Romera et al. 2007) indicated that brown macroalgae, in particular *Fucus spiralis,* showed higher sorption capacity than any other algae, at a level that was at least twice that observed with the other species examined. The relative levels of efficiency were brown > red > green; nevertheless, all the algae showed similar affinities for each metal tested, with the following sequence: lead > cadmium ≥ copper > nickel > zinc. A more extensive comparison of the biosorption capacity of 14 different metal ions by different micro- and macroalgal strains is reported by Zeraatkar et al. (2016). Among macroalgae, the brown algae (in particular, *Laminaria japonica* and *Fucus vesiculosus*) show maximum absorption values, and algae of this group have been intensively studied, especially as dead biomass, because of the advantages of their large size/biomass and the high sorption efficiency. Recently, however, two red macroalgae, *Gracilaria caudata* and *Gracilaria cervicornis*, were also used as cation exchangers for remediation of a synthetic petrochemical wastewater rich in transition metals (copper, nickel and zinc) (Cechinel et al. 2018). The rationale for this choice was based on the fact that these members of the Rhodophyta (red algae) are rich in sulphated polysaccharides (agarans and carrageenans), having several types of binding site which could display cation exchange capacity, and on the fact that this genus is present in several coastal areas of the world and has also industrial application for agar production. The ion exchange capacities of the two algae were shown to correspond to the total amount of metals present on the surface of living raw algae. The observed differences in ion exchange capacity between the two species were compared and attributed to the different thallus surface microscopic conformation, which displayed different porosity, and to their array of functional groups which were investigated through Fourier transform infra-red spectroscopic analysis. A model was then developed in order to enable the identification and quantification of the functional groups present on the biomass surface and the results obtained were in agreement with the total amount of metals measured by biomass digestion.

Dead algae have additional advantages as, in contrast to live algae, the dead biomass will tolerate chemical treatments that have the effect of increasing the metal sorption capacity, for example, pretreatments with $CaCl_2$ or HCl, which have been demonstrated (Kumar and Gaur 2011; Mehta et al. 2002; Kalyani et al. 2004) to enhance accumulation of metals, presumably by removing cations that were bound to the negatively charged groups of the cell walls; dead algae also represent a useful biomass if there is the aim to remove the bound metals and reuse the algal biomass for more absorption cycles.

Studies on bioremediation performed with macroalgae in real wastewaters are far fewer than the number of studies performed with dissolved metal solutions; comparative studies have reported a lesser efficiency of absorption in wastewaters than in pure metal solutions. Wastewater composition is indeed another factor to be considered in the process design; the presence of a complex matrix, due for example to the presence of organic matter which binds the metals, or the presence of multiple ions which compete with each other for the binding sites, are factors that can decrease the process efficiency. Concerning the ions, competition was observed with both cations, especially Ca^{2+} which is often present in wastewater (Cechinel et al. 2016), and anions (Vijayaraghavan et al. 2005).

Complete trials of metal remediation performed with packed biomass are still few; those using brown algae have been reviewed by Mazur et al. (2018), while the use of red algae was reported by Cechinel et al. (2018). Both these authors reported that in order to increase efficiency of wastewater purification in the presence of multiple ions, the utilisation of a two-stage process, with in-series columns or the use of multiple columns filled with differently pre-treated algae, could represent a method to improve the removal of metals (Castro et al. 2017; Park et al. 2006). Elution trials also showed the feasibility of recovering each metal separately.

Other comparative studies have shown that green macroalgae can also perform quite well in metal accumulation (e.g. Gosavi et al. 2004; Kumar et al. 2009; Oberholster et al. 2014), so applied studies have also considered this group, with the advantages that green algae are ubiquitous, quite robust in facing adverse environmental conditions and that it is possible to culture them in wastewaters, which are not usually saline, where they can display high growth rates. The above-cited studies have indicated that filamentous or laminar algae, such as those of the genera *Cladophora*, *Chaetomorpha*, *Ulva* and *Oedogonium*, were the most efficient. A good absorption performance was also demonstrated by two charophytes, *Chara aculeolata* and *Nitella opaca*, reaching high bioconcentration factor values (>1000 for Cd and Pb), but with the former species being better suited to bioremediation because of a higher tolerance to the metals (Sooksawat et al. 2013).

In view of industrial applications, living macroalgae should be grown in open ponds and *Oedogonium*, a cosmopolitan genus characterised by high growth rate and frequent dominance in the natural environment, appears as highly promising. In a laboratory study, it was shown to significantly improve wastewater quality by reducing the levels of five metals (Al, As, Cd, Ni and Zn) initially present (Roberts et al. 2013); in another study, different species of *Oedogonium* displayed significant differences in their growth rates in "clean" water; nevertheless, all species bioaccumulated the same metals to a similar extent (Ellison et al. 2014). *Oedogonium* was then utilised for the first successful example of macroalgae cultivation in a real contaminated wastewater produced from a coal-fired station, at a scale of 15,000 L: despite the presence of multiple contaminants, *Oedogonium* displayed a 50% increase in dry weight (DW) over 3–4-day trials and the rate of remediation of eight elements correlated with productivity. Among the elements, Al and Zn had the greatest bioremediation rates being decreased to below regulatory criteria in a single 3-day harvest cycle; the others had intermediate (As, Cu and Ni) or low (Cd, Cr and Se) rates of bioremediation; however, final concentrations of each of these elements were lower than initial concentrations in wastewater.

13.2.2 Microalgae

Microalgae also show a strong capacity for absorption and uptake of metals. This is especially true of extremophiles (Mikulic and Beardall 2014; Varshney et al. 2015), especially acidophiles tolerant of low pH conditions with, consequently, frequently high concentrations of dissolved metals. Nonetheless, even non-extremophilic microalgae can be tolerant of high metal concentrations. This, together with their relatively fast growth rates, has resulted in microalgae having attracted attention as possible bioremediation species.

As is the case with macroalgae, metal removal from water by microalgae comprises two mechanisms: an initial rapid phase due to adsorption to charged groups on the cell surface and a second, active, component due to uptake into the cell (Mantzorou et al. 2018) (see Fig. 13.2, for an example).

The capacity of microalgae (as well as macroalgae) for metal bioremediation has been reviewed recently by Zeraatkar et al. (2016) and more generally for a range of wastewater types (nutrients, pharmaceuticals, dyes and metals) by Wang et al. (2016).

In contrast to those macroalgae investigated for bioremediation, which include red, green and brown seaweeds, most studies of metal removal by microalgae have concentrated on freshwater species, notably green algae such as *Chlorella*, *Chlamydomonas* and *Scenedesmus*.

From the studies carried out so far, it is apparent that the capacity to strip metals from the surrounding medium is dependent on both the metal involved and the algal species. For instance, Al-Rub et al. (2004) and Mehta and Gaur (2001) showed that *Chlorella vulgaris* can bind nickel strongly, while Wong et al. (2000) reported that

Table 13.4 Cell-associated concentrations of metals in a selection of green algae and a cyanobacterium

Metal	Species	Accumulation in algae (mg g^{-1} alga)	References
Ni	*Chlorella vulgaris*	15–24	Al-Rub et al. (2004), Mehta and Gaur (2001)
Ni	*C. miniata* live	1.4	Wong et al. (2000)
Ni	*C. miniata* dead	20.4	Wong et al. (2000)
Cr (III)	*C. miniata*	41	Han et al. (2006)
U	*C. vulgaris*	14–26.6	Vogel et al. (2010)
Cd	*Chlamydomonas reinhardtii*	42.6	Tüzün et al. (2005)
Hg	*C. reinhardtii*	72.2	Tüzün et al. (2005)
Pb	*C. reinhardtii*	96.3	Tüzün et al. (2005)
Zn	*Scenedesmus obliquus*	210–837	Monteiro et al. (2011)
Cr (VI)	*S. incrassatulus*	4.4	Jácome-Pilco et al. (2009)
Cu (II)	*Nostoc muscorum*	42.1	Hazarika et al. (2014)
Pb (II)	*N. muscorum*	27.1	Hazarika et al. (2014)
Cd (II)	*N. muscorum*	7.0	Hazarika et al. (2014)
Zn (II)	*N. muscorum*	2.1	Hazarika et al. (2014)

the related species *C. miniata* would only do so to a limited extent. On the other hand, Han et al. (2006) report that *C. miniata* will accumulate Cr(III) to a much greater extent than it does Ni (Table 13.4).

The capacity of green algae to absorb a range of other metals has also been shown, with Vogel et al. (2010) reporting high cellular U concentrations for *C. vulgaris* and Tüzün et al. (2005) showing that *Chlamydomonas reinhardtii* cells absorbed Cd, Hg and Pb (Table 13.4). Zn is taken up to higher levels than other metals by *Scenedesmus obliquus* (Monteiro et al. 2011), though another *Scenedesmus* species, *S. incrassatulus*, only accumulated small amounts of Cr (VI) (Jácome-Pilco et al. 2009). *Scenedesmus abundans* also shows a marked ability to remove Cu and Cd from solution with initial concentrations of 10 mg L^{-1} of these metals being decreased to 0.09 (99% removal) and 0.26 mg L^{-1} (97% removal), respectively, over 30 h in batch culture experiments. Lower, but still significant (25–78%), removal of Cu, Cd and Cr has been reported for continuous cultures of *Scenedesmus incrassatulus* (Peña-Castro et al. 2004). In simulated industrial effluents, *Desmodesmus* sp. (=*Scenedesmus*) removed 94% of Cu and 85% of Ni within 2 days (Rugnini et al. 2018).

Cyanobacteria have also been shown to have a capacity to remove metals from solution. Cu(II), Pb(II), Cd(II) and Zn(II) are accumulated by *Nostoc muscorum* over 30 h (Hazarika et al. 2014), representing removal of 60–98% depending on the metal and its concentration (removal efficiency dropping with increased metal concentration). *Synechocystis* strains removed 51–95% of 15 mg L^{-1} Ni over 168 h exposure (Yilmaz et al. 2012).

There has, in addition, been a limited amount of work on metal removal using marine species. Pistocchi et al. (2000) examined Cd removal by cultures of the diatoms *Cylindrotheca fusiformis* and *Achnanthes brevipes* and of the dinoflagellate *Prorocentrum micans* and reported that these species removed 76%, 46% and 37% of Cd supplied at 0.2 mg L^{-1}, with lower percentage removal at higher concentrations.

Several studies have examined metal removal by dead microalgal cells (see Kumar et al. 2015 for an extensive review). *Desmodesmus pleiomorphus* was shown to have similar uptake of Cd in live or dead cells and Monteiro et al. (2010, 2011) and Al-Rub et al. (2004) reported almost identical Ni uptake by live and dead cells of *Chlorella vulgaris*. On the other hand, uptake of Ni by *C. miniata* live cells was reported as only 1.3 mg g^{-1} by Lau et al. (1999) compared to values for dead cells of 20.4 mg g^{-1} (Wong et al. 2000), though the pH values for the two sets of measurements differed (pH 7.4 and 6 respectively). Tien et al. (2005) also showed twofold greater accumulation by dead versus living cells of the dinoflagellate *Ceratium hirundinella*, though the reverse was true for the diatom *Asterionella formosa*.

One of the problems in using microalgae to remove metals from wastewater involves the separation of the algae after treatment. In free-living suspensions, this can involve chemical treatments to induce flocculation or energy-expensive processes such as centrifugation. An alternative approach has been to immobilise the microalgae; this can be achieved through various methods, mostly designed to keep the algae alive, and the most common forms are spherical beads made of natural polymers such as alginate, carragenaan or agar (de-Bashan and Bashan 2010). Mehta and Gaur (2001) compared Cu and Ni uptake by free and immobilised *Chlorella vulgaris* and showed that algal beads achieved a higher removal (70% Ni and >90% of Cu within an hour) than free cells (53% Ni and 60% Cu within 45 min), though the alginate of the beads was responsible for a proportion of metal removal. In the same study, maximum quota of Cu was similar in free and immobilised cells, but twofold higher in immobilised cell in the case of Ni. Al-Rub et al. (2004) examined Ni uptake by free living *C. vulgaris* cells, suspended dead algal cells, immobilised powdered algal cells and alginate beads alone. Immobilised dead cells showed the greatest uptake, much higher than free living cells, but a large proportion of the uptake by the immobilised cells could be contributed to the alginate alone. The advantage of immobilised systems is that the support (alginate in the case of Al-Rub et al. 2004) can be recycled and used for several cycles of metal removal. Other substrates have been used: Akhtar et al. (2003) used a matrix formed from the vegetable sponge *Luffa cylindrica* and showed that cadmium sorption capacity is rapid, the amount removed equivalent in the immobilised and free living cells and that the immobilised system operating in continuous flow could remove 73% of cadmium supplied at 5 mg L^{-1}. As is the case with algae immobilised in alginate beads, Akhtar et al. (2003) demonstrated that their system was reusable after removal of all Cd from the immobilised cells.

An alternative approach to immobilising microalgae in beads is to make use of the ability of algae to grow as biofilms. Since this often involves growth as part of a complex community, biofilms are dealt with separately below.

Microalgal cultures are increasingly considered in bioremediation processes due to the possibility that they incorporate CO_2 from the atmosphere, thereby contributing to amelioration of anthropogenic CO_2 emissions, and can be employed for different purposes as, for example, in biofuel production (Olguìn 2012; Palma et al. 2017). Microalgae also have numerous additional uses, although their use in food and feed applications or in the extraction of bioactive molecules for nutraceutical and pharmaceutical purposes can be constrained by the high content of accumulated metals. Living and actively growing cultures need additional nutrients for biomass growth and both growth and metal removal are influenced by the variety of factors reported above. A process based on living algae has the advantage that it could be run continuously for extended periods, although problems related to metal toxicity to the growing cells must be considered. Another positive and interesting aspect could involve the possibility of achieving a metal transformation/detoxification process by live algae; this was observed, for instance, in *Chlorella vulgaris* that was shown to be able to convert Cr(VI) into the less toxic form Cr(III) (Shen et al. 2013) or in the marine microalga *Ostreococcus tauri* that was able to transform inorganic arsenic (As) into a less toxic organic form and promote its volatilisation through biomethylation (Zhang et al. 2013).

Despite the high number of studies showing efficient removal of metals by growing microalgae, application trials were mostly limited to cultures operating in batch mode and very few have considered a semicontinuous system (Aoyama and Okamura 1993; Jácome-Pilco et al. 2009; Muñoz et al. 2006; Peña-Castro et al. 2004). Jácome-Pilco et al. (2009) cultivated *S. incrassatulus* grown in continuous culture in an airlift photobioreactor in the presence of Cr(VI) for 29 days and found that at steady state, Cr(VI)-removal efficiency of 43.5% and Cr(VI) uptake of 1.7 mg Cr(VI) g^{-1} dry biomass could be obtained and showed that, after reaching a defined physiological state, removal efficiency by the algae could remain constant for a long period of time.

Photobioreactors generally provide good conditions for microalgae growth. However, their cost is high and their use imposes high energy consumption rates, so they are generally disregarded for remediation (Acién et al. 2016). For this reason, algae-based biotechnologies that can be used for the removal of pollutants (Adey et al. 1996; Hoffmann 1998; Perales-Vela et al. 2006; Toumi et al. 2000) are based largely on High Rate Algal Ponds (HRAP). These technologies are reported to have a high metal removal capacity due to the high pH, achieved as a result of algal photosynthesis, that enhances metal precipitation (Toumi et al. 2000).

There are already commercially available technologies based on the use of microalgae either solely or in association with other organisms, for example, the

AlgaSORB sorption process (Bio-recovery System Inc., Cincinnati, OH) which consists of dead cells (mainly *Scytonema* and *Chlorella vulgaris*) immobilised in a silica gel polymer (Kuyucak and Volesky 1988), the biosorbent BIO-FIX, made up of algae and other biomass immobilised in high-density polysulfone, and the biosorbents of the company B.V. SORBEX, Inc., Canada, based on different types of biomaterials which include microalgae (*C. vulgaris*) and macroalgae.

13.2.3 Bacterial–Periphyton Interactions in Biofilms

It is well known that microalgae and bacteria can interact in a number of ways with positive or negative consequences for both partners (de-Bashan and Bashan 2010). Importantly to bioremediation, such interactions often allow algae to better tolerate or adapt to the stress caused by contaminants. For example, Fouilland et al. (2018) showed that although bacteria may have a general negative effect on algal growth, in the presence of pollutants, such as a mixture of pesticides or metals, the two microalgae species used were less affected by levels of these compounds that had been shown previously to be toxic to algae grown alone. Other reported effects are the observation that Cu(II) adsorbed faster onto the algal-bacterial biomass than *C. sorokiniana* alone and that this interaction offers the possibility to combine BOD and heavy metals removal in a single treatment step (Muñoz et al. 2006).

Phototrophic microbial communities often exist in nature in the form of biofilms, usually with one dominating species due to a reduced diversity in such environments (Aguilera et al. 2007; Garcia-Meza et al. 2005; Souza-Egipsy et al. 2011). These microorganisms form biofilms by extruding extracellular polymeric substances (EPS), with high levels of negatively charged functional groups, and as such can double as a detoxifying agent, reducing the bioavailability of toxicants (Gaur and Rai 2001; Hall 2002). Biofilms of metal-tolerant algae may thus have an interesting place in bioengineering solutions for water remediation of metal polluted sites (Ciniglia et al. 2004; Morozkina et al. 2010).

The term "algal biofilms" refers to biofilms dominated by microalgae and cyanobacteria which are formed on a variety of substrata in the presence of sufficient light and moisture. Algae provide nutrients and oxygen to the bacteria which in turn provide vitamins and inorganic carbon to the algae (Keesano and Sims 2014). As biofilms consist of living organisms, they have received attention in the remediation field (Roeselers et al. 2008), the aim being that of coupling wastewater treatment with the production of biomass for other purposes, including biofuel (Miranda et al. 2017). Furthermore, another significant advantage of biofilms could be that, by using immobilised phototrophic communities, the problem of separation of suspended algal biomass from water can be avoided.

Early studies on metal removal from phototrophic biofilms, as those reported in Roeselers et al. (2008), indicated that the EPS, in the presence of the elevated pH levels generated via CO_2 removal during oxygenic photosynthesis, could account for the metal-binding properties of phototrophic biofilms.

Applied studies with living biofilms were at first developed for bioremediation of nutrients in wastewaters and were only later applied to metal removal; thus, studies in this field are fewer, but could take advantage of technologies already developed (reviewed in Keesano and Sims 2014) consisting, for example, in specific photobioreactors that could allow a high surface area of contact between water and the biofilm. Travieso et al. (2002) designed a pilot rotary reactor (BIOALGA) where *Scenedesmus obliquus* was attached and able to grow, being able to remove 82% of supplied Co in 3 days, increasing to 93% over 7 days. Orandi et al. (2012) in a different design of rotating photobioreactor demonstrated good removal of multiple ions (such as Cu, Zn, Ni and Mn) by an indigenous microalgal consortium dominated by *Ulothrix* sp., although the authors affirmed that an appropriate immobilisation technique is required to maintain and exploit microbial biomass for metal absorption during a continuous industrial process.

Ma et al. (2018) used a spiral tubular photobioreactor to investigate Cu removal by a periphyton community that was allowed to develop before the experiments. The community comprised algae (mostly green algae and diatoms), bacteria, fungi and protozoa and could remove 2 and 10 mg L^{-1} Cu by 99% and 98.2%, respectively, at a retention time of 12 h.

These results demonstrate that biofilm-based technologies could be utilised for wastewater remediation from metals; these studies are still in their infancy and the applications are currently limited. Future development is however expected, as these systems are becoming increasingly popular.

13.3 Conclusions and Suggestions for Further Work

Algae clearly have the potential to be used effectively in bioremediation of metal pollution. Currently there is considerable variability in the available data, due in part to the range of environmental factors (pH, ionic strength, temperature, interactions between different metals, etc.) employed in the various studies, and also to the different composition and physiological characteristics of the numerous algal species employed. These aspects are described in detail in the recent reviews by Kumar et al. (2015), Zeraatkar et al. (2016) and Wang et al. (2016). Nonetheless, it seems likely that different species show specificity for different metals and this could be exploited to 'tailor' approaches to bioremediation.

Free living microalgae clearly have a capacity to take up metals from solution. Positive results are mainly linked to the mechanism(s) they use for coping with the presence of metals, to the successful achievement of optimal growth capacity and to the general composition of the wastewater utilised. Detailed studies on the species physiology are thus important for the use of living biomass; such knowledge is important also because cellular aspects, such as transporter genes, ligand synthesis and metal transformation to less toxic species, could represent the target of genetic engineering approaches to enhance tolerance and accumulation of toxic metals (Mosa et al. 2016). Alternatively, there could be the chance to discover new microorganisms having a higher resistance to metals and isolating them from extreme environments.

The problems associated with harvesting the cells to remove metals from the environment mean that use of macroalgae biomass (live or dead) or immobilised/ biofilm microalgae has the greater potential at present.

To date, most of the studies carried out have been at the controlled laboratory scale, with well-defined media, and in spite of the great utility of a phytoremediation process, studies on operating modes, such as those in continuous mode or in large volumes, are still few (Ghosh et al. 2016; Roberts et al. 2015); algal-based bioremediation has therefore not yet been developed for larger scale application in industries and commercialisation of these technologies is still poor. It will be a challenge for the future to scale up to an industrially relevant process and to show that algal approaches will cope with the more complex chemical conditions in real-life wastewater effluent, which contains a suite of organic and inorganic materials in addition to the metals of interest. Nonetheless, the use of algae is a promising and sustainable approach to bioremediation either alone or in combination with other organisms.

References

Acién FG, Gómez-Serrano C, Morales-Amaral MM et al (2016) Wastewater treatment using microalgae: how realistic a contribution might it be to significant urban wastewater treatment? Appl Microbiol Biotechnol 100:9013–9022

Adey WH, Luckett C, Smith M (1996) Purification of industrially contaminated groundwaters using controlled ecosystems. Ecol Eng 7:191–212

Aguilera A (2013) Eukaryotic organisms in extreme acidic environments, the Río Tinto Case. Life 3:363–374

Aguilera A, Souza-Egipsy V, Gomez F et al (2007) Development and structure of eukaryotic biofilms in an extreme acidic environment, Rio Tinto (SW, Spain). Microb Ecol 53:294–305

Akeson MA, Munns DN, Burau RG (1989) Adsorption of Al^{3+} to phosphatidylcholine vesicles. Biochim Biophys Acta 986:33–40

Akhtar N, Saeed A, Iqbal M (2003) *Chlorella sorokiniana* immobilized on the bio-matrix of vegetable sponge of *Luffa cylindrica*: a new system to remove cadmium from contaminated aqueous medium. Bioresour Technol 88:163–165

Al-Rub FA, El-Naas MH, Benyahia F et al (2004) Biosorption of nickel on blank alginate beads, free and immobilized algal cells. Process Biochem 39:1767–1773

Andrade LR, Farina M, Amado GM (2004) Effects of copper on *Enteromorpha flexuosa* (Chlorophyta) *in vitro*. Ecotoxicol Environ Saf 58:117–125

Aoyama I, Okamura H (1993) Interactive toxic effect and bioconcentration between cadmium and chromium using continuous algal culture. Environ Toxicol 8:255–269

Apel K, Hirt H (2004) Reactive oxygen species: metabolism, oxidative stress, and signal transduction. Annu Rev Plant Biol 55(1):373–399

Baby J, Rai JS, Biby ET et al (2010) Toxic effect of heavy metals on aquatic environment. Int J Biol Chem Sci 4(4):939–952

Baścik-Remisiewicz A, Tukaj Z (2002) Toxicity of inorganic cadmium salts to the microalga *Scenedesmus armatus* (Chlorophyta) with respect to medium composition, pH and CO_2 concentration. Acta Physiol Plant 24(1):59–65

Beardall J, Entwisle L (1984) Internal pH of the obligate acidophile *Cyanidium caldarium* Geitler (Rhodophyta). Phycologia 23(3):397–399
Bielen A, Remans T, Vangronsveld J et al (2013) The influence of metal stress on the availability and redox state of ascorbate, and possible interference with its cellular functions. Int J Mol Sci 14(3):6382–6413
Bierkens J, Maes J, Plaetse FV (1998) Dose-dependent induction of heat shock protein 70 synthesis in *Raphidocelis subcapitata* following exposure to different classes of environmental pollutants. Environ Pollut 101:91–97
Bossuyt BTA, Janssen CR (2004) Long-term acclimation of *Pseudokirchneriella subcapitata* (Korshikov) Hindak to different copper concentrations: changes in tolerance and physiology. Aquat Toxicol 68:61–74
Brune A, Urbach W, Dietz KJ (1994) Compartmentation and transport of zinc in barley primary leaves as basic mechanisms involved in zinc tolerance. Plant Cell Environ 17:153–162
Butzow JJ, Eichhorn GL (1965) Interactions of metal ions with polynucleotides and related compounds. IV. Degradation of polyribonucleotides by zinc and other divalent metal ions. Biopolymers 3:95–107
Cardozo KHM, Oliveira MAL, Tavares MFM et al (2002) Daily oscillation of fatty acids and malondialdehyde in the dinoflagellate *Lingulodinium polyedrum*. Biol Rhythm Res 33 (4):371–382
Castro L, Blázquez ML, González F et al (2017) Biosorption of Zn (II) from industrial effluents using sugar beet pulp and *F. vesiculosus*: from laboratory tests to a pilot approach. Sci Total Environ 598:856–866
Cechinel MAP, Mayer DA, Pozdniakova TA et al (2016) Removal of metal ions from a petrochemical wastewater using brown macro-algae as natural cation-exchangers. Chem Eng J 286:1–15
Cechinel MAP, Mayer DA, Mazur LP (2018) Application of ecofriendly cation exchangers (*Gracilaria caudata* and *Gracilaria cervicornis*) for metal ions separation and recovery from a synthetic petrochemical wastewater: batch and fixed bed studies. J Clean Prod 172:1928e1945
Chojnacka K (2010) Biosorption and bioaccumulation – the prospects for practical applications. Environ Int 36:299–307
Ciniglia C, Yoon HS, Pollio A et al (2004) Hidden biodiversity of the extremophilic cyanidiales red algae. Mol Ecol 13:1827–1838
Clijsters H, Van Assche F (1985) Inhibition of photosynthesis by heavy metals. Photosynth Res 7:31–40
Cobbett C, Goldsbrough P (2002) Phytochelatins and metallothioneins: roles in heavy metal detoxification and homeostasis. Annu Rev Plant Biol 53:159–182
Conner SD, Schmid SL (2003) Regulated portals of entry into the cell. Nature 422:37–44
Cumming JR, Taylor GJ (1990) Mechanisms of metal tolerance in plants: physiological adaptations for exclusion of metal ions from the cytoplasm. In: Alscher RC, Cumming JR (eds) Stress responses in plants: adaptation and acclimation mechanisms. Wiley-Liss, New York, pp 329–359
Dalle-Donne I, Rossi R, Colombo G et al (2009) Protein S-glutathionylation: a regulatory device from bacteria to humans. Trends Biochem Sci 34(2):85–96
Dao LHT, Beardall J (2016) Effects of lead on growth, photosynthetic characteristics and production of reactive oxygen species of two freshwater green algae. Chemosphere 147:420–429
Davis TA, Volesky B, Mucci A (2003) A review of the biochemistry of heavy metal biosorption by brown algae. Water Res 37:4311–4330
De Filippis LF, Pallaghy CK (1994) Heavy metals: sources and biological effects. In: Rai LC, Caur JP, Soeder CJ (eds) Algae and water pollution: advances in limnology series. Schweizerbart, Stuttgart, pp 32–77
De Filippis LF, Ziegler H (1993) Effect of sublethal concentrations of zinc, cadmium and mercury on the photosynthetic carbon reduction cycle of *Euglena*. J Plant Physiol 142(2):167–172

De Schamphelaere K, Nys C, Janssen C (2014) Toxicity of lead (Pb) to freshwater green algae: development and validation of a bioavailability model and inter-species sensitivity comparison. Aquat Toxicol 155:348–359

de-Bashan LE, Bashan Y (2010) Immobilized microalgae for removing pollutants: review of practical aspects. Bioresour Technol 101:1611–1627

Debelius B, Forja JM, DelValls Á et al (2009) Toxicity and bioaccumulation of copper and lead in five marine microalgae. Ecotoxicol Environ Saf 72:1503–1513

Devi SR, Prasad MNV (2013) Membrane lipid alterations in heavy metal exposed plants. In: Prasad MNV, Hagemeyer J (eds) Heavy metal stress in plants: from molecules to ecosystems. Springer, Heidelberg, pp 99–116

dos Santos RW, Schmidt EC, Felix MRD et al (2014) Bioabsorption of cadmium, copper and lead by the red macroalga *Gelidium floridanum*: physiological responses and ultrastructure features. Ecotoxicol Environ Saf 105:80–89

Eichhorn GL (1962) Metal ions as stabilizers or destabilizers of the deoxyribonucelic acid structure. Nature 194:474–475

Eichhorn GL (1973) Inorganic biochemistry. Elsevier, Amsterdam

Eichhorn GL (1975) Active sites of biological macromolecules and their interaction with heavy metals. In: McIntyre AD, Mills CF (eds) Ecological toxicology research: effects of heavy metal and organohalogen compounds. Plenum, New York, pp 123–142

Eichhorn GL, Butzow JJ (1965) Interactions of metal ions with polynucleotides and related compounds. III. Degradation of polyribonucleotides by lanthanum ions. Biopolymers 3 (1):79–94

Eichhorn GL, Shin YA (1968) Interaction of metal ions with polynucleotides and related compounds. XII. The relative effect of various metal ions on DNA helicity. J Am Chem Soc 90 (26):7323–7328

Ellison MB, de Nys R, Paul NA, Roberts DA (2014) Growth and metal bioconcentration by conspecific freshwater macroalgae cultured in industrial waste water. Peer J 2:e401. https://doi.org/10.7717/peerj.401

Ercal N, Gurer-Orhan H, Aykin-Burns N (2001) Toxic metals and oxidative stress. Part I: Mechanisms involved in metal induced oxidative damage. Curr Top Med Chem 1(6):529–539

Esposito S, Sorbo S, Conte B et al (2012) Effects of heavy metals on ultrastructure and HSP70s induction in the aquatic moss *Leptodictyum riparium* Hedw. Int J Phytoremediation 14 (4):443–455

FAO (2016) The state of world fisheries and aquaculture. Contributing to food security and nutrition for all. Food and Agriculture Organization of the United Nations, Rome

Feder ME, Hofmann GE (1999) Heat-shock proteins, molecular chaperones, and the stress response: evolutionary and ecological physiology. Annu Rev Physiol 61:243–282

Ford T, Ryan D (1995) Toxic metals in aquatic ecosystems: a microbiological perspective. Environ Health Perspect 103(Suppl1):25–28

Foster PL (1977) Copper exclusion as a mechanism of heavy metal tolerance in a green alga. Nature 269:322–323

Fouilland E, Galès A, Beaugelin I, Lanouguère E, Pringault O, Leboulanger C (2018) Influence of bacteria on the response of microalgae to contaminant mixtures. Chemosphere 211:449–455

Foy CD, Chaney RL, White MC (1978) The physiology of metal toxicity in plants. Ann Rev Plant Physiol 29:511–566

Fujiwara K, Matsumoto Y, Kawakami H et al (2008) Evaluation of metal toxicity in *Chlorella kessleri* from the perspective of the periodic table. Bull Chem Soc Jpn 81:478–488

Garcia-Meza JV, Barrangue C, Admiraal W (2005) Biofilm formation by algae as a mechanism for surviving on mine tailings. Environ Toxicol Chem 24:573–581

Gaur JP, Rai LC (2001) Heavy metal tolerance in algae. In: Rai LC, Gaur JP (eds) Algal adaptation to environmental stresses: physiological, biochemical and molecular mechanisms. Springer, Berlin, pp 363–388

Gautam RK, Sharma SK, Mahiya S et al (2015) Contamination of heavy metals in aquatic media: transport, toxicity and technologies for remediation. In: Sanjay K, Sharma R (eds) Heavy metals in water: presence, removal and safety. The Royal Society of Chemistry. www.rsc.org
Gerloff-Elias A, Barua D, Mölich A et al (2006) Temperature and pH dependent accumulation of heat-shock proteins in the acidophilic green alga *Chlamydomonas acidophila*. FEMS Microbiol Ecol 56(3):345–354
Ghosh A, Dastidar MG, Sreekrishnan TR (2016) Recent advances in bioremediation of heavy metals and metal complex dyes: review. J Environ Eng 142(9):C4015003
Gimmler H (2001) Acidophilic and acidotolerant algae. In: Rai LC, Gaur JP (eds) Algal adaptation to environmental stresses: physiological, biochemical and molecular mechanisms. Springer, Berlin, pp 259–290
Gissi F, Adams MS, King CK et al (2015) A robust bioassay to assess the toxicity of metals to the Antarctic marine microalga *Phaeocystis antarctica*. Environ Toxicol Chem 34(7):1578–1587
Goldsbrough P (1999) Metal tolerance in plants: the role of phytochelatins and metallothioneins. In: Terry NG, Bañuelos G (eds) Phytoremediation of contaminated soil and water. CRC Press LLC, Boca Raton, pp 221–233
Gosavi K, Sammut J, Gifford S, Jankowski J (2004) Macroalgal biomonitors of trace metal contamination in acid sulfate soil aquaculture ponds. Sci Total Environ 324:25–39
Hall JL (2002) Cellular mechanisms for heavy metal detoxification and tolerance. J Exp Bot 53 (366):1–11
Halliwell B (2006) Reactive species and antioxidants. Redox biology is a fundamental theme of aerobic life. Plant Physiol 141(2):312–322
Han X, Wong YS, Tam NFY (2006) Surface complexation mechanism and modeling in Cr(III) biosorption by a microalgal isolate, *Chlorella miniata*. J Colloid Interface Sci 303:365–371
Harding JPC, Whitton PA (1976) Resistance to zinc of *Stigeoclonium tenue* in the field and the laboratory. Br Phycol J 11(4):417–426
Hazarika J, Pakshirajan K, Sinharoy A et al (2014) Bioremoval of Cu(II), Zn(II), Pb(II) and Cd (II) by *Nostoc muscorum* isolated from a coal mining site. J Appl Phycol. https://doi.org/10.1007/s108110140475-3
Hoffmann JP (1998) Wastewater treatment with suspended and nonsuspended algae. J Phycol 34:757–763
Howden R, Andersen CR, Goldsbrough PB et al (1995a) A cadmium sensitive, glutathione-deficient mutant of *Arabidopsis thaliana*. Plant Physiol 107(4):1067–1073
Howden R, Goldsbrough PB, Andersen CR et al (1995b) Cadmium sensitive, cad1 mutants of *Arabidopsis thaliana* are phytochelatin deficient. Plant Physiol 107(4):1059–1066
Jácome-Pilco CR, Cristiani-Urbina E, Flores-Cotera LB et al (2009) Continuous Cr(VI) removal by *Scenedesmus incrassatulus* in an airlift photobioreactor. Bioresour Technol 100:2388–2391
Jensen A, Rystad B, Melsom S (1974) Heavy metal tolerance of marine phytoplankton. I. The tolerance of three algal species to zinc in coastal sea water. J Exp Mar Biol Ecol 15(2):145–157
Jones DL, Kochian LV (1997) Aluminum interaction with plasma membrane lipids and enzyme metal binding sites and its potential role in Al cytotoxicity. FEBS Lett 400(1):51–57
Kalyani S, Rao PS, Krishnaiah A (2004) Removal of nickel(II) from aqueous solutions using marine macroalgae as the sorbing biomass. Chemosphere 57:1225–1229
Keesano M, Sims RC (2014) Algal biofilm based technology for wastewater treatment. Algal Res 5:231–240
Kellner K (1955) Die adaptation von *Ankistrodesmus braunii* an rubidium und kupfer. Biol Zbl 74:662–691
Klerks PL, Levinton JS (1989) Effects of heavy metals in a polluted aquatic ecosystem. In: Levin SA, Kelly JR, Harwell MA, Kimball KD (eds) Ecotoxicology: problems and approaches. Springer advanced text in life sciences. Springer, New York
Klerks PL, Weis JS (1987) Genetic adaptation to heavy metals in aquatic organisms: a review. Environ Pollut 45(3):173–205

Kobayashi I, Fujiwara S, Saegusa H et al (2006) Relief of arsenate toxicity by Cd-stimulated phytochelatin synthesis in the green alga *Chlamydomonas reinhardtii*. Mar Biotechnol 8 (1):94–101

Kumar D, Gaur JP (2011) Metal biosorption by two cyanobacterial mats in relation to pH, biomass concentration, pretreatment and reuse. Bioresour Technol 102:2529–2535

Kumar JIN, Oommen C, Kumar RN (2009) Biosorption of heavy metals from aqueous solution by green marine macroalgae from Okha Port, Gulf of Kutch, India. Am-Eurasia J Agric Environ Sci 6:317–323

Kumar KS, Dahms H-U, Won E-J et al (2015) Microalgae – a promising tool for heavy metal remediation. Ecotox Environ Safe 113:329–352

Kuyucak N, Volesky B (1988) Biosorbents for recovery of metals from industrial solutions. Biotechnol Lett 10(2):137–142

Lau S, Lee HY, Tsang CCK et al (1999) Effect of metal interference, pH and temperature on Cu and Ni biosorption by *Chlorella vulgaris* and *Chlorella miniata*. Environ Technol 20:953–961

Leão GA, de Oliveira JA, Felipe RTA et al (2014) Anthocyanins, thiols, and antioxidant scavenging enzymes are involved in *Lemna gibba* tolerance to arsenic. J Plant Interact 9:143–151

Leonard SS, Harris GK, Shi X (2004) Metal-induced oxidative stress and signal transduction. Free Radic Biol Med 37(12):1921–1942

Levy JL, Stauber JL, Adams MS, Maher WA, Kirby JK, Jolley DF (2005) Toxicity, biotransformation, and mode of action of arsenic in two freshwater microalgae (*Chlorella* sp. and *Monoraphidium arcuatum*). Environ Toxicol Chem 24(10):2630–2639

Levy J, Stauber JL, Jolley DF (2007) Sensitivity of marine microalgae to copper: the effect of biotic factors on copper adsorption and toxicity. Sci Total Environ 387(1–3):141–154

Liu G, Chai X, Shao Y et al (2011) Toxicity of copper, lead, and cadmium on the motility of two marine microalgae *Isochrysis galbana* and *Tetraselmis chui*. J Environ Sci 23(2):330–335

Lu SC (2009) Regulation of glutathione synthesis. Mol Aspects Med 30(1-2):42–59

Ma L, Wang F, Yu Y et al (2018) Cu removal and response mechanisms of periphytic biofilms in a tubular bioreactor. Bioresour Technol 248:61–67

Macfie SM, Welbourn PM (2000) The cell wall as a barrier to uptake of metal ions in the unicellular green alga *Chlamydomonas reinhardtii* (Chlorophyceae). Arch Environ Contam Toxicol 39 (4):413–419

Maksymiec W, Russa R, Urbanik-Sypniewska T et al (1992) Changes in acyl lipid and fatty acid composition in thylakoids of copper non-tolerant spinach exposed to excess copper. J Plant Physiol 140(1):52–55

Malea P, Rijstenbil J, Haritonidis S (2006) Effects of cadmium, zinc and nitrogen status on non-protein thiols in the macroalgae *Enteromorpha* spp. from the Scheldt Estuary (SW Netherlands, Belgium) and Thermaikos Gulf (N Aegean Sea, Greece). Mar Environ Res 62:45–60

Malla MA, Dubey A, Yadev S et al (2018) Understanding and designing the strategies for the microbe-mediated remediation of environmental contaminants using omics approaches. Front Microbiol 9:1132

Mallick N, Mohn FH (2000) Reactive oxygen species: response of algal cells. J Plant Physiol 157 (2):183–193

Mantzorou A, Navakoudis E, Paschalidis K et al (2018) Microalgae: a potential tool for remediating aquatic environments from toxic metals. Int J Environ Sci Technol 15:1815–1830

Mazur LP, Cechinel MAP, Guelli SMA, de Souza U, Boaventura RAR, Vilar VJP (2018) Brown marine macroalgae as natural cation exchangers for toxic metal removal from industrial wastewaters: a review. J Environ Manag 223:215–253

Mehta SK, Gaur JP (2001) Removal of Ni and Cu from single and binary metal solutions by free and immobilized *Chlorella vulgaris*. Eur J Protistol 37:261–271

Mehta SK, Tripathi BN, Gaur JP (2002) Enhanced sorption of Cu2+ and Ni2+ by acid-pretreated *Chlorella vulgaris* from single and binarymetal solutions. J Appl Phycol 14:267–273

Meister A, Anderson ME (1983) Glutathione. Annu Rev Biochem 52:711–760

Mikulic P, Beardall J (2014) Contrasting toxicity effects of zinc on growth and photosynthesis in a neutrophilic alga (*Chlamydomonas reinhardtii*) and an extremophilic alga (*Cyanidium caldarium*). Chemosphere 112:402–411

Miranda AF, Ramkumar N, Andriotis C et al (2017) Applications of microalgal biofilms for wastewater treatment and bioenergy production. Biotechnol Biofuels 10:120. https://doi.org/10.1186/s1306801707989

Moenne A, González A, Sáez CA et al (2016) Mechanisms of metal tolerance in marine macroalgae, with emphasis on copper tolerance in Chlorophyta and Rhodophyta. Aquat Toxicol 176:30–37

Monteiro CM, Castro PML, Malcata FX (2010) Cadmium removal by two strains of *Desmodesmus pleiomorphus* cells. Water Air Soil Pollut 208:17–27

Monteiro CM, Fonseca SC, Castro PML et al (2011) Toxicity of cadmium and zinc on two microalgae, *Scenedesmus obliquus* and *Desmodesmus pleiomorphus*, from Northern Portugal. J Appl Phycol 23(23):97–103

Moore F, Esmaeili K, Keshavarzi B (2011) Assessment of heavy metals contamination in stream water and sediments affected by the sungun porphyry copper deposit, East Azerbaijan Province, Northwest Iran. Water Qual Expo Health 3:37–49

Morozkina EV, Slutskaya ES, Fedorova TV et al (2010) Extremophilic microorganisms: biochemical adaptation and biotechnological application (review). Appl Biochem Microbiol 46:1. https://doi.org/10.1134/s0003683810010011

Mosa KA, Saadoun I, Kumar K et al (2016) Potential biotechnological strategies for the cleanup of heavy metals and metalloids. Front Plant Sc 7:303. https://doi.org/10.3389/fpls.2016.00303

Muñoz R, Alvarez MT, Muñoz A et al (2006) Sequential removal of heavy metals ions and organic pollutants using an algal–bacterial consortium. Chemosphere 63:903–911

Muyssen BTA, Janssen CR (2001) Zinc acclimation and its effect on the zinc tolerance of *Raphidocelis subcapitata* and *Chlorella vulgaris* in laboratory experiments. Chemosphere 45:507–514

Nalewajko C, Colman B, Olaveson M (1997) Effects of pH on growth, photosynthesis, respiration, and copper tolerance of three *Scenedesmus* strains. Environ Exp Bot 37:153–160

Nambatingar N, Clement Y, Merle A et al (2017) Heavy metal pollution of Chari river water during the crossing of N'Djamena (Chad). Toxics 5:26. https://doi.org/10.3390/toxics5040026

Navarrete A, González A, Gómez M et al (2019) Copper excess detoxification is mediated by a coordinated and complementary induction of glutathione, phytochelatins and metallothioneins in the green seaweed *Ulva compressa*. Plant Physiol Biochem 135:423–431

Neumann D, Lichtenberger O, Günther D et al (1994) Heatshock proteins induce heavy-metal tolerance in higher plants. Planta 194:360–367

Nishikawa K, Tominaga N (2001) Isolation, growth, ultrastructure, and metal tolerance of the green alga, *Chlamydomonas acidophila* (Chlorophyta). Biosci Biotechnol Biochem 65 (12):2650–2656. https://doi.org/10.1271/bbb.65.2650

Noctor G, Foyer CH (1998) Ascorbate and glutathione: keeping active oxygen under control. Annu Rev Plant Physiol Plant Mol Biol 49:249–279

Noctor G, Lelarge-Trouverie C, Mhamdi A (2015) The metabolomics of oxidative stress. Phytochemistry 112:33–53

Oberholster PJ, Cheng PH, Botha AM et al (2014) The potential of selected macroalgal species for treatment of AMD at different pH ranges in temperate regions. Water Res 60:82–92

Ochiai E-I (1987) General principles of biochemistry of the elements. Plenum, New York

Olaveson MM, Stokes PM (1989) Responses of the acidophilic alga *Euglena mutabilis* (Euglenophyceae) to carbon enrichment at pH 3. J Phycol 25:529–539

Olguìn EG (2012) Dual purpose microalgae–bacteria-based systems that treat wastewater and produce biodiesel and chemical products within a biorefinery. Biotechnol Adv 30:1031–1046

Orandi S, Lewis DM, Moheimani NR (2012) Biofilm establishment and heavy metal removal capacity of an indigenous mining algal-microbial consortium in a photo-rotating biological

contactor. J Ind Microbiol Biotechnol 39(9):1321–1331. https://doi.org/10.1007/s1029501211429
Pacheco MM, Hoeltz M, Moraes MSA, Schneider RCS (2015) Microalgae: cultivation techniques and wastewater phycoremediation. J Environ Sci Health A 50:573–589
Påhlsson A-MB (1989) Toxicity of heavy metals (Zn, Cu, Cd, Pb) to vascular plants. Water Air Soil Poll 47(3):287–319
Palma H, Killoran E, Sheehan M et al (2017) Assessment of microalga biofilms for simultaneous remediation and biofuel generation in mine tailings water. Bioresour Technol 234:327–335
Park D, Yun Y-S, Jo JH et al (2006) Biosorption process for treatment of electroplating wastewater containing Cr(VI): laboratory-scale feasibility test. Ind Eng Chem Res 45:5059–5065
Parsell DA, Lindquist S (1993) The function of heat-shock proteins in stress tolerance: degradation and reactivation of damaged proteins. Annu Rev Genet 27:437–496
Peña-Castro JM, Martinez-Jeronimo F, Esparza-Garcia F et al (2004) Phenotypic plasticity in *Scenedesmus incrassatulus* (Chlorophyceae) in response to heavy metals stress. Chemosphere 57:1629–1636
Perales-Vela HV, Peña-Castro JM, Cañizares-Villanueva RO (2006) Heavy metal detoxification in eukaryotic microalgae. Chemosphere 64:1–10
Pinto E, Sigaud-Kutner TCS, Leitão MAS et al (2003) Heavy metal-induced oxidative stress in algae. J Phycol 39(6):1008–1018
Piotrowska-Niczyporuk A, Bajguz A, Talarek M et al (2015) The effect of lead on the growth, content of primary metabolites, and antioxidant response of green alga *Acutodesmus obliquus* (Chlorophyceae). Environ Sci Poll Res Int 22:19112–19123
Pistocchi R, Mormile M, Guerrini F et al (2000) Increased production of extra-and intracellular metal-ligands in phytoplankton exposed to copper and cadmium. J Appl Phycol 12:469–477
Purbonegoro T, Suratno, Puspitasari R et al (2018) Toxicity of copper on the growth of marine microalgae *Pavlova* sp. and its chlorophyll-*a*. IOP Conf Ser Earth Environ Sci 118:012060. https://doi.org/10.1088/1755-1315/118/1/012060
Qin J, Lehr CR, Yuan C et al (2009) Biotransformation of arsenic by a Yellowstone thermoacidophilic eukaryotic alga. Proc Natl Acad Sci USA 106(13):5213–5217
Rahman MA, Hogan B, Duncan E et al (2014) Toxicity of arsenic species to three freshwater organisms and biotransformation of inorganic arsenic by freshwater phytoplankton (*Chlorella* sp. CE-35). Ecotoxicol Environ Saf 106:126–135
Rai PK (2008) heavy metal pollution in aquatic ecosystems and its phytoremediation using wetland plants: an ecosustainable approach. Int J Phytoremediat 10:133–160
Reimann C, Bjorvatn K Frengstad B et al (2003) Drinking water quality in the Ethiopian section of the East African Rift Valley I—data and health aspects. Sci Total Environ 311:65–80
Roberts DA, de Nys R, Paul NA (2013) The effect of CO_2 on algal growth in industrial wastewater for bioenergy and bioremediation applications. PLoS One 8:e81631. https://doi.org/10.1371/journal.pone.0081631
Roberts DA, Paul NA, Bird MI (2015) Bioremediation for coal-fired power stations using macroalgae. J Environ Manag 153:25e32
Robinson NJ, Tommey AM, Kuske C et al (1993) Plant metallothioneins. Biochem J 295:1–10
Rodgher S, Espíndola ELG, Simões FCF et al (2012) Cadmium and chromium toxicity to *Pseudokirchneriella subcapitata* and *Microcystis aeruginosa*. Braz Arch Biol Technol 55:161–169
Roeselers G, van Loosdrecht MCM, Muyzer G (2008) Phototrophic biofilms and their potential applications. J Appl Phycol 20:227–235
Romera E, González F, Ballester A et al (2007) Comparative study of biosorption of heavy metals using different types of algae. Bioresour Technol 98:3344–3353
Roncarati F, Sáez CA, Greco M et al (2015) Response differences between *Ectocarpus siliculosus* population to copper stress involve cellular exclusion and induction of the phytochelatin biosynthethic pathway. Aquat Toxicol 159:167–175

Rosen BP (1996) Bacterial resistance to heavy metals and metalloids. J Biol Inorg Chem 1 (4):273–277

Rugnini L, Costa G, Congestri R et al (2018) Phosphorus and metal removal combined with lipid production by the green microalga *Desmodesmus* sp.: an integrated approach. Plant Physiol Biochem 125:45–51

Russell G, Morris OP (1970) Copper tolerance in the marine fouling alga *Ectocarpus siliculosus*. Nature 228:288–289

Sáez CA, Roncarati F, Moenne A et al (2015) Copper-induced intra-specific oxidative damage and antioxidant responses in strains of the brown alga *Ectocarpus siliculosus* with different pollution histories. Aquat Toxicol 159:81–89

Sanità di Toppi L, Pawlik-Skowrońska B, Vurro E et al (2008) First and second line mechanisms of cadmium detoxification in the lichen photobiont *Trebouxia impressa* (Chlorophyta). Environ Pollut 151(2):280–286

Say PJ, Whitton BA (1977) Influence of zinc on lotic plants. II. Environmental effects on toxicity of zinc to *Hormidium rivulare*. Freshwater Biol 7(4):377–384

Sbihi K, Cherifi O, El Gharmali A et al (2012) Accumulation and toxicological effects of cadmium, copper and zinc on the growth and photosynthesis of the freshwater diatom *Planothidium lanceolatum* (Brébisson) Lange-Bertalot: a laboratory study. J Mater Environ Sci 3(3):497–506

Scheidegger C, Behra R, Sigg L (2011) Phytochelatin formation kinetics and toxic effects in the freshwater alga *Chlamydomonas reinhardtii* upon short-and long-term exposure to lead (II). Aquat Toxicol 101(2):423–429

Shehata FHA, Whitton BA (1982) Zinc tolerance in strains of the blue-green alga *Anacystis nidulans*. Br Phycol J 17:5–12

Shen Q-H, Zhi T-T, Cheng L-H, Xu X-H, Chen H-L (2013) Hexavalent chromium detoxification by nonliving *Chlorella vulgaris* cultivated under tuned conditions. Chem Eng J 228:993–1002

Shi XL, Dalal NS (1990) Evidence for a fenton-type mechanism for the generation of •OH radicals in the reduction of Cr(VI) in cellular media. Arch Biochem Biophys 281(1):90–95

Shi XL, Dong Z, Dalal NS et al (1994) Chromate-mediated free radical generation from cysteine, penicillamine, hydrogen peroxide, and lipid hydroperoxides. Biochim Biophys Acta 1226:65–72

Shin YA, Eichhorn GL (1968) Interactions of metal ions with polynucleotides and related compounds. XI. The reversible unwinding and rewinding of deoxyribonucleicacid by zinc (II) ions through temperature manipulation. Biochemistry 7(3):1026–1032

Shrift A (1954) Sulfur-selenium antagonism. II. Antimetabolite action of selenomethionine on the growth of *Chlorella vulgaris*. Am J Bot 41(4):345–352

Singh DP, Singh SP (1987) Action of heavy metals on Hill activity and O_2 evolution in *Anacystis nidulans*. Plant Physiol 83:12–14

Siripornadulsil S, Traina S, Verma DP, Sayre RT (2002) Molecular mechanisms of proline-mediated tolerance to toxic heavy metals in transgenic microalgae. Plant Cell 14(11):2837–2847

Skórzyńska E, Urbanik-Sypniewska T, Russa R et al (1991) Galactolipase activity of chloroplasts in cadmium-treated runner bean plants. J Plant Physiol 138(4):454–459

Sooksawat N, Meetam M, Kruatrachue M et al (2013) Phytoremediation potential of charophytes: bioaccumulation and toxicity studies of cadmium, lead and zinc. J Environ Sci 25(3):596–604

Souza-Egipsy V, Altamirano M, Amils R, Aguilera A (2011) Photosynthetic performance of phototrophic biofilms in extreme acidic environments. Environ Microbiol 13:8

Spijkerman E, Barua D, Gerloff-Elias A et al (2007) Stress responses and metal tolerance of *Chlamydomonas acidophila* in metal-enriched lake water and artificial medium. Extremophiles 11(4):551–562

Stauber J, Florence T (1987) Mechanism of toxicity of ionic copper and copper complexes to algae. Mar Biol 94:511–519

Stefanov K, Popova I, Kamburova E et al (1993) Lipid and sterol changes in *Zea mays* caused by lead ions. Phytochemistry 33(1):47–51

Stohs SJ, Bagchi D (1995) Oxidative mechanisms in the toxicity of metal ions. Free Radic Biol Med 18(2):321–336

Szer W, Ochoa S (1964) Complexing ability and coding properties of synthetic polynucleotides. J Mol Biol 8(6):823–834

Tien CJ, Sigee DC, White KN (2005) Copper adsorption kinetics of cultured algal cells and freshwater phytoplankton with emphasis on cell surface characteristics. J Appl Phycol 17:379–389

Toumi A, Nejmeddine A, ElHamouri B (2000) Heavy metal removal in waste stabilization ponds and high rate ponds. Water Sci Technol 42:17–21

Travieso L, Pellon A, Benitez F, Sanchez E, Borja R, O'Farrill N, Weiland P (2002) BIOALGA reactor: preliminary studies for heavy metals removal. Biochem Eng J 12:87–91

Trenfield MA, van Dam JW, Harford AJ et al (2015) Aluminium, gallium, and molybdenum toxicity to the tropical marine microalga *Isochrysis galbana*. Environ Toxicol Chem 34 (8):1833–1840

Tüzün İ, Bayramoğlu G, Yalçın E et al (2005) Equilibrium and kinetic studies on biosorption of Hg (II), Cd(II) and Pb(II) ions onto microalgae *Chlamydomonas reinhardtii*. J Environ Manag 77:85–92

Van Assche F, Clijsters H (1986) Inhibition of photosynthesis in *Phaseolus vulgaris* by treatment with toxic concentrations of zinc: effects on electron transport and photophosphorylation. Physiol Plant 66:717–721

Varshney P, Mikulic P, Vonshak A et al (2015) Extremophilic micro-algae and their potential contribution in biotechnology. Bioresour Technol 184:363–372

Verma S, Dubey RS (2003) Lead toxicity induces lipid peroxidation and alters the activities of antioxidant enzymes in growing rice plants. Plant Sci 164(4):645–655

Vijayaraghavan K, Jegan J, Palanivelu K et al (2005) Batch and column removal of copper from aqueous solution using a brown marine alga *Turbinaria ornata*. Chem Eng J 106:177–184

Vogel M, Gunther A, Rossberg A et al (2010) Biosorption of U(VI) by the green algae *Chlorella vulgaris* in dependence of pH value and cell activity. Sci Total Environ 409(2):384e395

Wang Y, Ho S-H, Cheng C-L et al (2016) Perspectives on the feasibility of using microalgae for industrial wastewater treatment. Bioresour Technol 222:485–497

Whitton BA (1970) Ecotoxicity of heavy metals to freshwater algae. Phykos 9:116

Wong JPK, Wong YS, Tam NFY (2000) Nickel biosorption by two *Chlorella* species, *C. vulgaris* (a commercial species) and *C. miniata* (a local isolate). Bioresour Technol 73:133–137

Yamane T, Davidson N (1961) On the complexing of desoxyribonucleic acid (DNA) by mercuric ion. J Am Chem Soc 83(12):2599–2607

Yan H, Pan G (2002) Toxicity and bioaccumulation of copper in three green microalgal species. Chemosphere 49(5):471–476

Yilmaz ES, Aslim B, Cansunar E (2012) Potential of nickel(ii) removal by *Synechocystis* sp. isolates. Fresen Environ Bull 21(4):853–859

Yoo-iam M, Chaichana R, Satapanajaru T (2014) Toxicity, bioaccumulation and biomagnification of silver nanoparticles in green algae (*Chlorella* sp.), water flea (*Moinamacrocopa*), blood worm (*Chironomus* spp.) and silver barb (*Barbonymusgonionotus*). Chem Spec Bioavail 26 (4):257–265

Zeraatkar AK, Ahmadzadeh H, Talebi AF et al (2016) Potential use of algae for heavy metal bioremediation, a critical review. J Environ Manag 181:817–831

Zhang S-Y, Sun G-X, Yin X-X et al (2013) Biomethylation and volatilization of arsenic by the marine microalgae *Ostreococcus tauri*. Chemosphere 93:47–53

Zhao X-J, Sucoff E, Stadelmann EJ (1987) Al^{3+} and Ca^{2+} alteration of membrane permeability of *Quercus rubra* root cortex cells. Plant Physiol 83(1):159–162

Chapter 14
Benefits of Algal Extracts in Sustainable Agriculture

Sharadwata Pan, Jaison Jeevanandam, and Michael K. Danquah

Abstract Algae possess inherent complex physiological photosynthetic mechanisms, which enable beneficial transformation of solar energy into other energy forms, for food and active metabolite synthesis. A number of active metabolites derived from algae, many of which demonstrate bioactive properties, have found profound, multifunctional applications in biofuels, nutraceuticals and functional foods, pharmaceuticals, and cosmetics industries. In spite of the evolving global interests and market demand of algal biomass and metabolites, studies and applications pertaining to sustainable agriculture challenges, such as soil nutrient deficiency, drought stress, soil toxicity, leaf discoloration and plant growth stunts, are limited. The generation and functional determination of novel bioactive compounds from algal biomass may offer innovative opportunities to address some of the aforementioned challenges. This chapter profiles and discusses the prospects of key algal metabolites in addressing plant growth challenges. Additionally, research findings from specific studies based on the use of algal metabolites and phytohormones as biostimulants, their influence in host animal physiology, and protective mechanisms against adventitious organisms or foreign pathogens, have been discussed. The chapter lays down progressive perspectives for optimal exploitation of algal metabolites and phytohormones in enhancing agricultural outputs.

S. Pan
School of Life Sciences Weihenstephan, Technical University of Munich, Freising, Germany

J. Jeevanandam
Department of Chemical and Petroleum Engineering, Curtin University of Technology, Sarawak, Malaysia

M. K. Danquah (✉)
Department of Chemical Engineering, University of Tennessee, Chattanooga, TN, USA
e-mail: michael-danquah@utc.edu

A. Hallmann, P. H. Rampelotto (eds.), *Grand Challenges in Algae Biotechnology*, Grand Challenges in Biology and Biotechnology,
https://doi.org/10.1007/978-3-030-25233-5_14

14.1 Introduction

Since the last few decades, agro-based products are accepted by the market and common citizens across the globe, with certain degrees of apprehension, mainly with concerns pertaining to deleterious health effects, posed by over-usage of chemical stimulants or fertilizers in the field (Chatzissavvidis and Therios 2014). Whilst it directly affects the productivity of crops and plants in general, indirect effects are no less stern, with grave implications towards the overall human and livestock fitness and development (Singh et al. 2017). Consequently, the focus has cautiously shifted to more 'green' or natural solutions, including concomitant use of biofertilizers and related solicitations, which would ensure a more sustainable and organic agricultural output (Sharma et al. 2012). On this front, extraordinary measures have been taken, with specific, domain-based applications. For instance, the EU region has tried to minimalize or eradicate unwarranted GMO applications. Particularly in this context, irrespective of country or the global constituency, applications of algal metabolites and phytohormones towards manifestation of viable agricultural practices, have taken a centre stage. For instance, the global algal market is touted to attain a staggering US$45 billion by 2023 (Cosgrove 2017). Solicitations of algal biofertilizers are a prime focus in the EU region, with an eye towards green farming of vegetables ('VegaAlga—Result In Brief' 2018). Additionally, a very recent US Bill is currently underway, with an objective to boost algal applications, both as food and feed (Einstein-Curtin 2018). All these are primarily ascribing to the extraordinary ranges of benefits, associated with intrinsic metabolites, derived from both microalgae and macroalgae, including the perennial favourites, cyanobacteria and marine seaweeds (Chatzissavvidis and Therios 2014; Singh et al. 2017; Sharma et al. 2012; references therein). It may suffice to note that, the utilization of algal metabolites has received widespread attention, including critical acclaim, in diverse domains transcending medicines, agriculture and in the food sectors. However, much of these may as well be a consequence of the foremost focus on algal employment in the biofuel sector, including assessments of their economic prospects (Peng et al. 2018; Yeong et al. 2018; Chye et al. 2018).

Several studies in the past have outlined the impeccable benefits associated with administration of algal extracts, directly or indirectly, in order to boost agricultural yield and productivity. An overview of annual publications concerning associated domains in recent years can be seen in Fig. 14.1. Interested readers may refer to the insightful reviews by Sharma et al. (2012), Singh et al. (2017), Pathak et al. (2018), Han et al. (2018), Trentacoste et al. (2015) and Chatzissavvidis and Therios (2014), in this context. These studies have validated the claim that much of the advantages pertaining to algal extract employments, cater to the constructive attributes associated with the underlying metabolites and phytohormones. The range of benefits, as revealed by these studies, is truly awe-inspiring, ranging from their abilities to transform the solar energy to readily absorbable forms; ability to manifest themselves as omnipresent organisms, without the necessity of cultivable lands; ability to control and minimize the loss of important nutrients such as phosphorus and

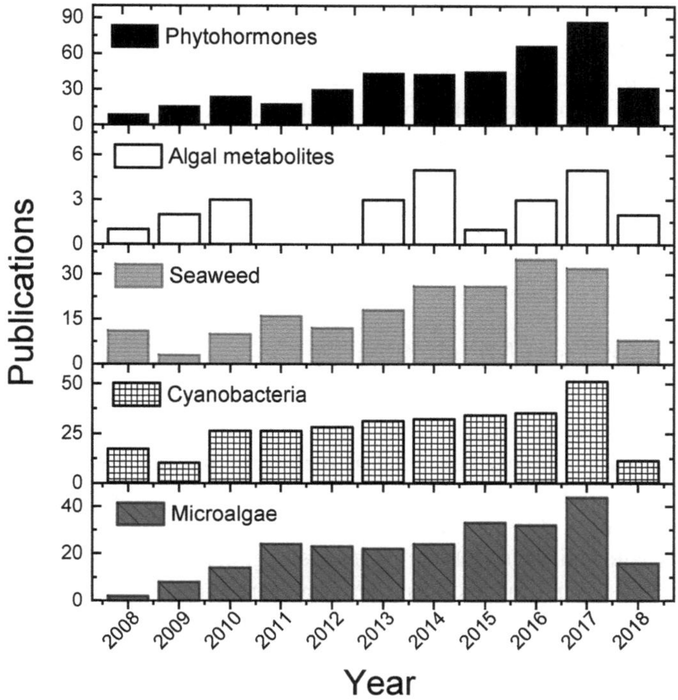

Fig. 14.1 Archival and peer-reviewed publications in the database SCOPUS (http://www.scopus.com), from 2008 to 2017, in the technical literature. A comprehensive examination of the database was carried out employing the expressions 'sustainable agriculture' and the individual categories, including algal metabolites or phytohormones, as outlined in the figure legends in each stack with sufficient terminologies in all domains

nitrogen; contributions to wastewater management, including treatments of replenished water; and appreciable influences in diverse processes of soil treatment and enrichment. These, in addition to their noteworthy influences towards plant and animal physiology, including astonishing health benefits, which are perceived to be the expressions of their intrinsic bioactive components. Despite all these, there are only a few studies that target a holistic discussion on the cumulative impacts of algal metabolites and phytohormones with a targeted approach towards the realization of an organic and viable agriculture system and subsequent boost to the agricultural yield.

The current chapter lays down rational viewpoints for optimum management of algal metabolites and phytohormones in ensuring sustainable agricultural outcomes. Several outlooks have been considered. Major algal metabolites are categorized, and their influences in realization of a viable agro-economy are discussed in the next section. Section 14.3 deals with the active solicitations of algae-based metabolites as biostimulants, including soil amendments, and advancements from the purview of genetic engineering and manipulations. Section 14.4 discusses various strategies and benefits of the algal metabolites, in host defence against attacks from foreign pathogens, including adventitious agents, nematodes and insect pests. Section 14.5 captures the perspectives on the influence of algal metabolites from micro- and macroalgae on animal host physiology, including isolated symbiotic associations for metabolite transfer. The major classes of phytohormones and their effects on viable agriculture are elaborated in Sect. 14.6. Finally, the central conclusions are summarized in Sect. 14.7 along with an eye on probable future endeavours in the domain.

14.2 Major Algal Metabolites

There are diverse algae groups, from which, a variety of chemical compounds are extracted, which are highly beneficial as medicines and facilitates the synthesis of other economically important products. Algae possess the ability to survive even in harsh environments due to their enhanced defence strategies, and yields novel metabolites with unique metabolic pathways, significant diversity in structure and chemistry. These algal metabolites gained enormous implicational significance, as they can be utilized as essential supplements for human nutrition and novel bioactive substances (Cardozo et al. 2007). Major metabolites that are commonly extracted from algae are phenols, terpenoids, free fatty acids, polysaccharides and carotenoids.

14.2.1 Phenols

A hydroxyl class of chemical compounds that are directly bonded to a group of aromatic hydrocarbons are called phenols (Khoddami et al. 2013). Several algae release phenols as a consequence of their secondary metabolism, and these phenolic compounds are isolated from the crude extract to use them as algal products. Generally, phenolic compounds are classified into simple phenols and polyphenols, whereas phenolic acids, flavonoids and phenylpropanoids are the sub-classes of phenols, which are obtained from algae (Bravo 1998). The polyphenols play a significant role in the growth, reproduction, adhesion and cell wall formation (Schoenwaelder 2002). It is noteworthy that the properties of algal extracts depend on the quantity of metabolites, especially phenols, whilst the quality of phenols in algae is determined by several physical, chemical and environmental factors (Jormalainen and Honkanen 2008). In a recent study with six microalgal species, it was reported that *Chaetoceros calcitrans* and *Isochrysis galbana* possess higher antioxidant activity followed by *Ophiocordyceps sinenis*, *Skeletonema costatum*, *Phaeodactylum tricornutum* and *Saccharina japonica*. The antioxidant property of these species was attributed to their phenolic content, in the form of gallic acid (Foo et al. 2017). Similarly, *Sargassum fluitans* brown algae and microalgal species from industrial wastewater were also shown to possess high quantity of polyphenols

(Gutiérrez et al. 2017). Algal groups, namely Rhodophyceae, contain phenolic compounds, such as mycosporine-like amino acids (MAAs) and bromophenols; Phaeophyceae contains Colpol, bromophenols and phlorotannins; whereas Chlorophyceae algal group contains coumarins and vanillic acid, along with MAAs and bromophenols (Freile-Pelegrín and Robledo 2014). These algal phenolic compounds are highly useful in agricultural practices as antimicrobials, fungicides, antioxidants and possess antiproliferative properties against certain cancer types (Esquivel-Hernández et al. 2017).

14.2.2 Terpenoids

Terpenoid is a group of natural organic chemicals that are derived from terpenes and contains functional groups of oxygen with multicyclic structures (Cseke et al. 2016). These compounds are also known as isoprenoids, and are similar to terpenes, with an additional oxygen-based or methyl functional groups (Kuzuyama 2017). The number of isoprenes that are attached to the parent terpene, and the number of cyclic structures in forming terpenoids, are used to classify terpenoids (Ayoola et al. 2008). Hemiterpenoids, monoterpenoids, sesquiterpenoids, diterpenoids, sesterterpenoids, triterpenoids, tetraterpenoids and polyterpenoids are the different types of terpenoids that are classified depending on the isoprene numbers (Ashour et al. 2010). Similarly, monocyclic, bicyclic, rearranged and linear, constitute another set of terpenoid classification, based on the existence of cyclic structures (Gouveia et al. 2013). Amongst algae, it is noteworthy that the red algae of Rhodophyta family possess the capability to produce bromine halogenated terpenoids (Fenical and Paul 1984). Terpenoids are also reported to be present and secreted as secondary metabolites in benthic marine algae, Caribbean marine algae, Caulerpale algae, tropical marine algae, *Halimeda* algal species, algal class of Phaeophyceae under the family Dictyotaceae, *Cystoseira barbata*, *Cystoseira crinita* from black sea, *Laurencia pinnatifida* and certain blue-green as well as green algae (Zbakh et al. 2012; Freile-Pelegrín et al. 2008; Commeiras et al. 2006; Paula et al. 2011).

Algal terpenoids are highly beneficial as agricultural, medicinal and cosmetic entities, due to their enhanced bioactivity and bioavailability (Cocchietto et al. 2002). Subsequently, several novel techniques have been used recently to elevate the production of algal terpenoids and their effective isolation procedures. The algal metabolic alterations, via genetic engineering coupled with high-throughput protocols, are gaining wide attention to produce large quantities of terpenoids from algae (Kempinski et al. 2015). Besides these complex procedures, it was recently reported that sargassacean algal species are capable of producing high quantity of terpenoids, particularly mero and linear diterpenes. Likewise, recent literature proved that Dictyotacean algal species are able to produce cyclic diterpenoids, such as prenylated guaianes, dolabellanes and xenicanes, sesquiterpenes, as well as several monoterpenes (Vallim et al. 2005). Another study revealed that *Cystoseira amentacea* var. *stricta*, a Mediterranean brown algae, has the ability to synthesize high quantities of monocyclic, bicyclic, rearranged meroditerpenoids and methoxybifurcarenone (Stengel and Connan 2015). These complex terpenoid compounds are extracted and isolated from algae using novel and latest techniques, to use them either as potential drugs, or in other biological applications (de Morais et al. 2015). Terpenoids possess unique properties, such as being an anti-angiogenic, a curative agent against the Alzheimer's disease, demonstrate specific pharmacological properties, possess radical scavenging abilities, act as an anticancer, antimicrobial, antioxidant as well as a sweetening agent, act as a cardio and hormonal stimulant, fragrances and spices (Awasthi et al. 2018; Ngo et al. 2017; Rodriguez-Garcia et al. 2017). In agriculture, terpenoids are beneficial in controlling insect pests, post-harvest disease management, biopesticide formulations, ecosystem intensification, oviposition enhancement and allelopathy (Sarwar 2015; Seethapathy et al. 2016; Kevan and Shipp 2017; Pavela and Benelli 2016).

14.2.3 Free Fatty Acids

Free or nonesterified fatty acids are present in the crude secondary metabolites of algae, which are more susceptible towards auto-oxidation than the esterified fatty acids, and act as pro-oxidant metabolites (Tena et al. 2018). The presence of free fatty acids in algae is a significant ecological and physiological marker, which is effectively utilized in taxonomy, and to identify organic matter from soil and aquatic ecosystems (Sushchik et al. 2001). These fatty acids are either classified based on their length, such as short, medium, long and very long chain; saturated, including lignoceric, palmitic, capric, behenic, cerotic, caprylic, lauric, stearic, myristic, arachidic acids; and unsaturated fatty acid compounds, which are broadly sub-classified into *cis*- and *trans*-fatty acids (Tvrzicka et al. 2011). Investigations on free fatty acids have been reported in microalgal oils, as well as algal species such as marine Haptophycean, fresh water eustigamatophyte *Monodus subterraneus*, red algae, blue-green algae including *Anabaena variabilis* and *Anacystis nidulans*, and certain green algae including *Chlorella* species and *Haematococcus pluvialis* (Cherng and Shih 2006; Zhekisheva et al. 2005; Khozin-Goldberg and Cohen 2006). Recently, non-esterified fatty acids have been extracted from a variety of algal species, due to their high demands in biodiesel production. In a recent study, free fatty acids were extracted from microalgae, mixed with methanol to esterify and use them as heterogenous biodiesel catalysts (Veillette et al. 2017). Similarly, it has been reported that the nitrogen-depleted *Chlorella zofingiensis*, *Haematococcus pluvialis*, *Nannochloropsis gaditana*, *Nannochloropsis oceania* CCALA 804, *Scenedesmus obliquus*, *Botryococcus braunii*, *Scenedesmus dimosphus* and *Scenedesmus acutus* produce high quantity of non-esterified fatty acids (Mulders et al. 2015; Solovchenko et al. 2014; Abomohra et al. 2016; Avula et al. 2017). Several algal growth parameters, such as increasing carbon dioxide concentration, treatment with residual nitrogen, depleting nitrogen, mixed growth of microalgal consortium and increasing myristic acid content, also help to increase their free fatty acid content (Aslam et al. 2018; Ryu et al. 2015; Saito et al. 2018). These free fatty acids from algae are extensively utilized in biodiesel production, as antibacterial agents and antioxidants, and to enhance plant growth, all of which make these highly advantageous towards agricultural applications (Desbois and Smith 2010; Lin et al. 2016).

14.2.4 Polysaccharides

Polymers of carbohydrate molecules with long monosaccharide chains, which are bound together by glycosidic linkages, are known as polysaccharides. These polysaccharides are always heterogenous, characterized by a few alterations in their polymeric chain, with exclusive features depending on the building blocks, the monosaccharides (Stick 2001). Polysaccharides are broadly classified based on their structures, such as arabinoxylans, cellulose, chitin and pectins; whereas starch and glycogen are the storage-based classes of polysaccharides (Gopinath et al. 2018). Acidic, bacterial capsules and nano-sized polymers are other minor classifications of polysaccharides (Wang et al. 2015; Johnson et al. 2014). Green and blue-green algae are the highest polysaccharide yielding algal species, which are made with monosaccharides such as hexose, pentose, uronic acid and methyl pentose (Moore and Tischer 1964). Likewise, marine algae are reported to produce high quantities of sulphated polysaccharides, arabinoxylans by gums and mucilages of marine green and brown algae, celluloses by cladophorales green algae, cell walls of blue and red algae, chitin by *Chlorella* species and calcified coralline algae *Clathromorphum compactum*, and pectins by green algae *Micrasterias*, *Penium margaritaceum* and *Netrium digitus* (Eder and Lütz-Meindl 2010; Domozych et al. 2014; Mihranyan 2011; Athukorala et al. 2007; Jiao et al. 2011). Similarly, starch is strongly yielded by red algae, green algae and microalgae, whereas glycogen is largely synthesized in *Porphyridium*, blue-green algae and certain algal species that grow in wastewater (Ellis et al. 2012; Boeckaert et al. 2008; Shimonaga et al. 2007). In recent times, marine algae, Pacific red and brown algae, Kamchatka brown algae and fucoidans of brown and green algae were explored for the production of polysaccharides (Vaskovsky et al. 2015; Bilan et al. 2016; Skriptsova 2015; Berri et al. 2016). These polysaccharides are used in agriculture as soil conditioners, nutrient carriers, immunostimulatory and anti-metastatic agents, and as safe release systems for agrochemicals and sustainable agriculture (Guilherme et al. 2015; Jung et al. 2017; Campos et al. 2015).

14.2.5 Carotenoids

Carotenoids are natural derivatives of five-carbon isoprene units that are formed into pigment with highly conjugated, regular structures of 40-carbon via enzymatic polymerization. Bacteria, fungi, algae and plants are reported to possess and secrete about 600 different types of carotenoids as secondary metabolites (Cardozo et al. 2007). Carotenoids are essential in algae, specifically in the photosynthetic variants, as they act as accessory pigments in light-harvesting processes, protein structure assembly stabilizers in photosystems, and excess light exposure mediated photo or free radical oxidation inhibitors (Zhang et al. 1999). Alpha- and beta-carotene, lutein, lycopene, beta-cryptoxanthin and zeaxanthin are the common types of algal carotenoids, which are extensively studied for agricultural purposes (Takaichi 2011). Usually, microalgae produce high quantities of carotenoids that are used in food, cosmetics, feed, pharmaceutical and nutraceutical applications (Henríquez et al. 2016). Recently, the yield of carotenoids from microalgae, namely *Phormidium autumnale*, was shown to be upgraded by cultivating those using agricultural wastes (Rodrigues et al. 2014). Other algae, such as *Chlamydomonas reinhardtii*, *Dunaliella salina*, red algae, *Porphyra umbilicalis* and *Chlorella vulgaris*, have also been used to yield considerable quantities of carotenoids (Gille et al. 2016; Lohr 2017). Specifically, the algal metabolites, which belongs to carotenoid groups, such as alpha carotenes, are extracted from *Spirulina* species, beta-carotenes from *Dunaliella* species, including several marine algae, lutein from *Chlorella protothecoids* and *Muriellopsis* species, lycopene from several green algae and macrophytic red algae, beta-cryptoxanthin from green and macroalgae, and zeaxanthin from red, green and mutant algal species (Huang et al. 2018; Dautermann and Lohr 2017; Hoang et al. 2016; Chang et al. 2015). These carotenoids are extensively used in agricultural practices to increase provitamin A in biofortified crop plants, as antioxidants, as fertilizers for petroleum-contaminated soil remediation and as pesticides, which also increase resistivity towards various phyto-ailments (Sakamoto et al. 2017; Han et al. 2016). Table 14.1 summarizes major algal metabolites and their agricultural properties.

14.3 Algal Metabolites as Biostimulants or Biofertilizers

A host of metabolites from seaweed extracts, including marine algae, have received particular attention, due to their beneficial attributes towards plant nourishment (Tuchy et al. 2013). Additionally, red algae (*Corralina mediterranea, Jania rubens*), brown algae (*Ecklonia maxima, Ascophyllum nodosum*) and green algae (*Ulva lactuca, Cladophora dalmatica*) have found profound usage in viable agriculture, pertaining to their noteworthy biostimulant activities. Algal metabolites have long been utilized as potent biostimulants of overall plant development (Tuchy et al. 2013; Chatterjee et al. 2017; references therein). Based on the wealth of existing literature, majority of the investigations carried out so far have focussed on the influence of algal metabolites on soil agglomeration, and consequential coalescence and/or flocculence, whilst several groups have tried to relate the metabolite effects with the manifestation of the soil micro- and macro-ecology. Precisely, the effects of

Table 14.1 Major algal metabolites and their agricultural properties

Metabolites	Algal groups	Agricultural properties or uses
Phenols	*Chaetoceros calcitrans*, *Isochrysis galbana*, *Ophiocordyceps sinenis*, *Skeletonema costatum*, *Phaeodactylum tricornutum*, *Saccharina japonica*, *Sargassum fluitans* brown algae, rhodophyceae phaeophyceae chlorophyceae class and microalgal species	Antimicrobials, fungicides, and antioxidants
Terpenoids	Red algae of Rhodophyta family, benthic marine algae, Caribbean marine algae, caulerpale algae, tropical marine algae, algal class of Phaeophyceae, *Cystoseira barbata*, *Cystoseira crinita* from Black Sea, *Laurencia pinnatifida*, blue-green and green algae, sargassacean, Dictyotacean algal species and Mediterranean brown algae	Anti-angiogenic, radical scavenging ability, antimicrobial, antioxidant, plant hormone stimulation, controls insect pest, post-harvest plant disease management, biopesticide formulation, ecosystem intensification, oviposition enhancement and allelopathy
Free fatty acids	Marine Haptophycean, microalgal oils, *Monodus subterraneus*, red algae, *Anabaena variabilis*, *Anacystis nidulans*, *Chlorella* species, *Haematococcus pluvialis*, nitrogen-depleted *Chlorella zofingiensis*, *Nannochloropsis gaditana*, *Nannochloropsis oceania* CCALA 804, *Scenedesmus obliquus*, *Botryococcus braunii*, *Scenedesmus dimosphus* and *Scenedesmus acutus*	Biodiesel production, antibacterial agents, antioxidants and to enhance plant growth
Polysaccharides	Green, blue-green algae, marine green and brown algae, cladophorales green algae, blue and red algae, *Chlorella* species, *Clathromorphum compactum*, green alga *Micrasterias*, *Penium margaritaceum*, *Netrium digitus*, Microalgaes, *Porphyridium*, blue-green algae from wastewater, Pacific red and brown algae, Kamchatka brown algae and fucoidans of brown algae	Soil conditioner, nutrient carrier, immunostimulatory, anti-metastatic agent, safe release systems for agrochemicals and sustainable agriculture
Carotenoids	Microalgae *Phormidium autumnale*, *Chlamydomonas reinhardtii*, *Dunaliella salina*, *Porphyra umbilicalis*, *Chlorella vulgaris*, *Spirulina*, *Dunaliella*, *Muriellopsis* species, *Chlorella protothecoids*, green algae, macrophytic red algae and mutant algal species	Elevate provitamin A in biofortified crop plants, antioxidant, fertilizer for petroleum-contaminated soil remediation, pesticide to increase resistivity towards various diseases in plants

algal metabolites on the physical, chemical and biochemical aspects of the soil are discussed in the following subsections.

14.3.1 Impact on Soil Aggregation and Porosity

Major classes of algal species have been reported to be associated with significant biofertilizer characteristics, including brown macroalgae (*Stoechospermum marginatum*, *Fucus vesiculosus*), blue-green algae (*Aulosira*, *Wollea*, *Anabaena*, *Calothrix*), and red macroalgae (*Lithothamnion corallioides*, *Phymatolithon caldarium*). For a detailed list, see Table 10.1 in Chatterjee et al. (2017). These algae have been reported to be associated with a wide range of assistances to soil accretion and permeability from enriching the soil with important minerals and nutrients, such as nitrogen, phosphorus and potassium, stimulating plant development including conferment of abiotic stress tolerance, facilitating photosynthesis, in vivo manufacturing of growth regulators, resistance from pathogen attacks, and improving ventilation and soil assembly, including holding soil humidity (Ramya et al. 2015; Sharma et al. 2014; Stadnik and De Freitas 2014; Craigie 2011; Chatterjee et al. 2017). The aggregation, which is thought to be a direct manifestation of the adhesive properties of the algal proteoglycans (Flaibani et al. 1989) has a direct influence on the soil physical properties, such as ventilation and temperature, which secondarily progresses the material milieu of the crop (Falchini et al. 1996).

In addition to soil aggregation, the permeability or porosity of the soil is an important consideration. As aforementioned, several classes of algae have been positively linked to improve the soil porosity by increasing the assimilation of nutrients, such as nitrogen and phosphorous. It has been reported that soil permeability may be positively invigorated by countering the deleterious influences of accumulation of water (Falchini et al. 1996). Furthermore, a few studies have focused on the forceful solubilization of unsolvable conformations of inorganic phosphate (Kleiner and Harper 1977; Cameron and Julian 1988; Roychoudhury et al. 1979). In fact, an increase in crop productivity has been reported by the administration of nitrogen, thereby making the soil more spongy and bountiful (Chatterjee et al. 2017).

14.3.2 Impact on Soil Macro- and Micro-environment

Algal biofertilizers layout effective biostimulant strategies by manipulating the soil macro-ecology. This, by directly influencing the soil acidity or basic nature by regulating the pH, and by enforcing activities analogous to bioremediation, by sequestration of metal ions in the soil (Chatterjee et al. 2017). However, contradictory accounts of algal species administration on soil pH have been documented. For instance, whilst the utilization of algal biostimulants has been linked to an enhancement of soil pH (Saha and Mandal 1979), Subhashini and Kaushik (1981) documented a reduction in soil pH, agglomeration, and conductivity. Whilst the influence on soil pH is noteworthy, the biosorption activities of the algal metabolites are more encouraging. Past studies have reported on the capabilities of cyanobacterial species to sequester trace elements, such as iron, zinc, copper, manganese and zinc, from the unsolvable entities (Das et al. 1991; Lange 1976), which also provide supplementary benefits in terms of evading soil erosion and adsorption of cations (Whitton 2000).

Just as the macro-environment, the micro-ecology of soil gets considerably affected through employment of algal metabolites. In spite of the wealth of data on the context, precise interpretations of the related transformations in the microbial community, following the administration of certain algal species, have not been rigorously investigated (Chatterjee et al. 2017). Past studies have indicated species-dependent (bacteria, fungi, etc.) enhancement of microbial population post algal treatment (Rao and Burns 1990; Acea et al. 2001; Rogers and Burns 1994). Mostly, this extraordinary increase in the microbial population has been tackled with assorted explanations, including an introduction of supplementary energy reservoirs in the form of algal polysaccharides, enhancement in the absolute nitrogen matter, as well as the mineralization of accessible carbon, which influences the thriving of the microbes (Chatterjee et al. 2017; Anderson and Gray 1991).

14.3.3 Genetically Modified Algae in Sustainable Agriculture

Not only the native species, but instances of algal species with genetically altered genomes in viable agriculture, have been documented. However, this is species dependent. For instance, in the context of genetics research concerning seaweeds, active solicitations in high throughput genetic characterizations, alteration and modifications, have only been realized since the beginning of the new millennium, although early classical genetic studies have been carried out since the last 30 years or so (Chatzissavvidis and Therios 2014). The premise of genetic modifications encompasses diverse motivations, including conferring higher biotic and abiotic stress tolerance to improve nitrogen and other mineral assimilations, all of which are in sync with the vision of enforcement of a strict, sustainable and agronomic culture (Chatterjee et al. 2017). For instance, Singh et al. (1987) reported successful transfer of herbicide lenient genes from the diazotrophic *Gloeocapsa* species, to the genome of a *Nostoc* species, thereby conferring herbicide defiance with subsequent beneficial biofertilizer attributes. Additionally, enhanced nitrogen fixation has been reported by Chaurasia et al. (2008) through the transfer and overexpression of the *hetR* gene. This was mediated by the manifestation of numerous heterocysts, being subjected to the heterocyst proliferation abilities of the concerned gene. There is no doubt that the introduction of novel genes into the existing algal genomes, helps to create better strains with affiliated effects on metabolic rates and higher productivities (Dhargalkar and Pereira 2005). However, in global niches, where GMOs are dealt with seriously, and with strictest possible scrutiny, for instance in the EU region, the efficacy of the genetically modified algae in sustainable agriculture is ambiguous. Nonetheless, advances in genetically modified algal strains will additionally act as forerunners to a 'greener' sustainable approach to the growth of marine resources.

14.4 Algal Metabolites in Plant Protection and Development

The effects of algal metabolites on host plant defence systems against the attacks from foreign pathogens, predominantly microbes, nematodes and insects, have been recently reviewed (Chatzissavvidis and Therios 2014). More importantly, a very recent study by Hamed et al. (2018) critically evaluated the roles of marine macroalgae in viable agriculture by reviewing the influences of the same on host plant fortification, against a host of pathogens including bacteria, fungi, nematodes, insects and viruses. Evidences of enhanced resistance exist towards pests and ailments, following seaweed extract administration (Pardee et al. 2004; Mercier et al. 2001), which is attributed to the induced resistance capabilities conferred by algal polysaccharides, such as carrageenan and laminarin. The algal metabolites levy a wide range of influence on the host plant physiology, ranging from regulation of signalling pathways, gene expressions targeting defence mechanisms, metabolism, and a holistic impact on the rhizosphere. The following subsections discuss these aspects in the light of existing literature.

14.4.1 Antimicrobial Activity

Past studies have focused on the impacts of critical algal metabolites, such as alkaloids, cyclic peptides, sterols, lipids and polysaccharides, amongst others, on conferring resistances against microbial attacks (Al-Saif et al. 2014; Abdel-Raouf et al. 2015). For a detailed accounts of algal species-dependent antimicrobial effects on various host plants or crops, see the insightful reviews by Hamed et al. (2018) and Chatzissavvidis and Therios (2014). Identical assertions have been reported concerning marine macroalgae, with antimicrobial, cytotoxic and anti-free radical characteristics (Moubayed et al. 2017). Much of the studies have been based on findings carried out in diverse solvents, which may act as an efficient strategy to investigate the influence of the constituent biologically active peptides and other compounds, towards these beneficial attributes (Michalak and Chojnacka 2015). Specifically, in marine macroalgal systems, several antimicrobial features have been reported to be a direct manifestation of the kind and proportion of the underlying bioactive saturated and unsaturated fatty acids, the algal polysaccharides and the derived oligosaccharides, and the phenolic compounds (Ibraheem et al. 2017; Benkendorff et al. 2005; Lee and Jeon 2013; Hamed et al. 2018; references therein).

Several recent studies have reported on the antifungal activities of the crude algal extracts (Galal et al. 2011; Paulert et al. 2010; Baloch et al. 2013; Ibraheem et al. 2017). Several mechanisms could be identified to trace the influence, including protection from root fungal infections, earlier fruiting, better shoot development, and increased root depths. Biologically active constituents in the algal extracts, such as flavonoids and phenolic acids, seem to act as robust antifungal entities (Ammar et al. 2017), as is the case with other components, like reserve polysaccharides and carbohydrates, which influence energy storage, enzyme and moisture control, and an overall endurance towards fungal attacks (Prabhavathi and Rajam 2007; Hamed et al. 2018). On the other hand, the protective actions of algal extracts against plant viruses, a significant deterrent towards high agricultural yield, have been associated with biological antiviral ingredients, such as alginates, polysaccharides, alkaloids, polyphenols, vitamin C, peptides, omega-3 fatty acids, and carbohydrate-binding proteins, such as lectins (Zhao et al. 2017; Manzo et al. 2009; Pardee et al. 2004; Matanjun et al. 2009; Wang et al. 2004).

14.4.2 Antinematodal Activity

The protection against scrounging nematodes is necessary, considering the staggering damages (in excess of US$100 billion) they cause to the floral agriculture (Saifullah et al. 2007). Algal metabolites and extracts have shown considerable antinematodal activities in host plants, either by modifying the ratio of plant growth regulators, by diminishing the plant root infiltration by the nematodes, by ensuring higher shoot progression, or by alleviating the overall root damages, all resulting in an increase in the inclusive crop yield in agricultural fields (Khan et al. 2009; Sultana et al. 2011; Chatzissavvidis and Therios 2014). Particularly, macroalgae from marine resources have been shown to possess noteworthy antinematodal characteristics, both in a controlled greenhouse environment and under open field conditions (Hamed et al. 2018; Sultana et al. 2011). Additionally, the antinematodal actions on host plants, by administration of seaweed extracts, have been shown to be effected either through reduced contagion by diminishing gall manifestations (Sultana et al. 2011), or by employing a pre-treatment of the plant seedlings with a betaine mixture (Wu et al. 1998). Furthermore, the presence of growth promoting substances in macroalgae, like cytokinins, and predecessors of ethylene production, are believed to be involved in bestowing shielding actions against the nematodes harbouring in plant root knots (Glazer et al. 1985). An interesting observation from a recent in vivo study by Ngala et al. (2016) is that the commercially procurable seaweed extracts demonstrate inhibiting action mechanisms towards root-knot nematodes by adversely affecting their egg producing capabilities and physical discernments.

14.4.3 Bioinsecticidal Activity

As in the case with microbes and nematodes, algal extracts also demonstrate protection against pest (insect) attacks, predominantly mediated by the constituent secondary metabolites (Ara et al. 2005). This has especially been established with targeted studies with seaweed extracts. Although the precise mechanisms are yet to be convincingly established, presence of chelated metal ions and enhanced proportions of phenolic compounds and/or anthocyanins, may act in favour of the affirmative confrontation of the leaves against the insect bouts (Khan et al. 2009; Chatzissavvidis and Therios 2014; references therein). The insect repelling and insecticidal features of oceanic macroalgal species have been well documented (Bantoto and Danilo 2013; Ali et al. 2013; Cetin et al. 2010). Several past studies have focused on the bioinsecticidal effects of chloroform, and aqueous and methanolic extracts of algal species, such as *Sargassum swartzii, Padina pavonica, Caulerpa scalpelliformis, Ulva fasciata* on the insect pest *Dysdercus cingulatus* (Kombiah and Sahayaraj 2012; Sahayaraj and Kalidas 2011; Asha et al. 2012; Asharaja and Sahayaraj 2013). A range of activities against the pest species have been documented, including nymphal death, prolonged breeding times, limitation of endurance, as well as diminished fertility and hatchability. The observations made by Asharaja and Sahayaraj (2013) are particularly significant, who attributed the insecticidal activities of the chloroform extracts from *P. pavonica* and *S. swartzii* to the existence of hexadecanoic acid methyl ester and stigmastan-6, 22-dien,3,5-dedihydro, correspondingly.

14.5 Influence of Algal Metabolites in Animal Host Physiology

14.5.1 Macroalgal Impacts

Not only plants, but past studies have also reported on the noteworthy impacts of seaweed extracts on elucidating bioactive reactions in animals. These include a wide range of responses, ranging from a general health enhancement to particular benefits, such as increased milk productivity, enhanced fertility and facilitating reproduction, improvement in the colour of egg yolks, better ability to withstand heat trauma, modifications of fat accumulation configurations, prolonged shelf life, and ability to better counter the attacks by foreign pathogens, including adventitious agents (Chatzissavvidis and Therios 2014; Craigie 2011; Chapman and Chapman 1980; He et al. 2009). Incidentally, these trials are not entirely novel. For instance, in the EU region, the regular usage of seaweed as a feed to the livestock, rather date back to ancient times, stretching as early as the Roman civilization (Craigie 2011). Additional benefits of the utilization of algal extracts include better iodine supplements in animal milk (Abdel-Raouf et al. 2012). One of the most laudable advantages of algal metabolites, in the animal host physiology, is the transformation of the substandard algal protein to an above-average animal meat protein, devoid of the intermediate pre-handling steps (Venkataraman 1978). In a later study, further validation of this was demonstrated by He et al. (2009), whilst investigating the impacts of a phlorotannin supplemented algal extract on fatty acid deposition, who reported that the ability of phlorotannin to modify the metabolism of fats, is probably responsible for the amendment of the meat harvests. The utilization of flour, supplemented with ~30% algal extracts, has additionally been shown to be favourable as a feed to goats, without any noticeable performance reduction (Castro et al. 2009). Furthermore, administration of algal metabolites in the form of commercially procurable extracts, such as Tasco, have been reported to lay a wide range of positive influences, including alleviation of the enterohemorrhagic bacterial community in cattles (Braden et al. 2007), as well as manifesting better colour and textural characteristics of cattle meat products (Montgomery et al. 2001).

14.5.2 Symbiosis Through Metabolite Transfer

Evidences of transmission of metabolites, from symbiotic algal species to host animals, have been documented in literature (Whitehead and Douglas 2003; references therein). This is both interesting and noteworthy, since there is a fast rate of photosynthesis attributed to the algal cells, which allows majority of the photosynthetically secured carbon to be transferred to the animal host tissues, thereby permitting a considerable involvement to the energy (carbon) resources of the animal cohort (Muscatine et al. 1984; Edmunds and Davies 1986). A wide range of strategies have been devised to study metabolite transfer between the algal symbionts and the animal tissues, including detection of radioactive photosynthetic components, detection of constituents specific to organisms, and respiratory proportion quantification (Johnston et al. 1995; Gattuso and Jaubert 1990; Harland et al. 1991). These symbiotic associations also serve as nutritional foundations, as has been shown by Whitehead and Douglas (2003), through a series of experiments carried out using the dinoflagellate algal species of the genus *Symbiodinium* and *Anemonia viridis*, a sea anemone. The authors reported that predominantly photosynthetic components, such as fumarate/succinate and glucose, are transported to the tissues of the animal cells from the symbiotic algal cells. Moreover, the authors assert that assessment of metabolites (or 'mobile' compounds) is a reasonable strategy to investigate the nutritional relations between the symbiotic algae and the hosts, with suitable amendments.

14.5.3 Microalgal Impacts

In addition to macroalgae, several microalgal species have been routinely employed as a dominant constituent of animal feedstock, primarily attributing to the extraordinary ranges of health benefits that they cater to (Sathasivam et al. 2017; references therein). For instance, the utilization of extracts rich in *Chlorella* sp., has been linked with an enhanced plasma haemoglobin, alleviating effects on serum cholesterol and plasma glucose, including protecting the liver against malnutrition effects, most of which may be credited to the presence of β-1,3-glucan, which is thought to participate in lowering plasma lipids, shows antioxidant activities, and stimulates the animal host immune systems (Barrow and Shahidi 2008). The health situation of animals, involved in viable agriculture, are directly or indirectly influenced by the feed, which affects growth and development, reproduction and even existence. Beneficial properties of microalgal metabolites also confer distinct advantages to the animal livestock, in the form of resistance against stress, improved metabolism, hatchability, all of which have made it an attractive alternative for the conventional protein supplements (Shields and Lupatsch 2012; Svircev 2005; Spolaore et al. 2006). The advantages of utilization of algal species as an active feed component have also been extended to animal husbandry, such as farming of pigs and in the poultry, post elaborative toxicological and dietetic assessments (Becker 2004). Similar health benefits, as mentioned earlier, are conferred to the farmed pigs and poultry animals, by the bioactive metabolites present in the algal extracts, owing to the presence of nutrients such as essential fatty acids, minerals and vitamins (Certik and Shimizu 1999). It has even been postulated that administration of algal metabolites to the poultry animals is by far the best possible outlook for its commercialization (Becker 2007). Ginzberg et al. (2000) additionally reported on the positive impact of the algal administration to poultry animals in the form of reduced level of cholesterol and darker colours of the egg yolks. Furthermore, the effects of microalgae on the farming of aquatic animals have also been well documented (Sathasivam et al. 2017; references therein). Although a wide range of algal species have been employed as feed components, including *Chlorella*, *Pavlova*, *Dunaliella* sp. *Chaetoceros*, *Thalassiosira*, amongst others, *Spirulina* stands by far with the most exhaustive investigations and solicitations (Richmond 1988), catering to the wide range of health advantages it offers, including blood cell manifestations, particularly the synthesis of leucocytes, and regulation of hormones (Vonshak 1986; Ciferri and Tiboni 1985; Del Campo et al. 2000). These effects are also direct and indirect manifestations of the underlying biologically active entities, including astaxanthin, phycocyanin, carotenoids such as beta-carotene, and polyunsaturated fatty acids or PUFAs, such as eicosapentaenoic acid (EPA) and docosahexaenoic acid (DHA) (Lakeh et al. 2010; Skjanes et al. 2013; Gouveia et al. 2008; Vonshak 1986).

14.6 Algal Phytohormones in Sustainable Agriculture

Phytohormones are produced within algae in low concentrations as signal molecules, which control the general algal development. Algal phytohormones are produced by green algae, which derive energy via photosynthesis, similar to plants (Tarakhovskaya et al. 2007), and are gaining unique research interests in recent times. Auxins, gibberellins, cytokinins, abscisic acid and ethylene are the major phytohormones, which are widely extracted and explored for agricultural applications.

14.6.1 Auxins

Auxins represent a hormonal class, which possesses a cardinal role in coordinating several behavioural and growth processes in the life cycle and development of an organism, with morphogen-like characteristics. Auxins are found to be produced and secreted by algae only in the late 1950s, and algae such as *Laminaria*, *Macrocystis* brown algae, *Chlorella*, *Cladophora*, *Enteromorpha* green algae and *Botryocladia* red algae were reported to possess auxins and their inactive analogues. The developments, in the modern methods of extraction, illustrated that auxins are highly yielded by green, characean and brown algae (Tarakhovskaya et al. 2007). Currently, algal auxins are broadly extracted from *Klebsormidium nitens*, *Acutodesmus obliquus*, *Chlorella vulgaris*, *Chlorella pyrenoidosa*, *Scenedesmus quadricauda* and charophyte algae (Ohtaka et al. 2017; Piotrowska-Niczyporuk et al. 2018; Liu et al. 2016). Algal auxins are widely utilized in agriculture due to unique properties and their ability to facilitate the growth and development of plants. Several current studies proved that the auxins possess enhanced herbicidal properties (Quareshy et al. 2017). It has been reported that *Chlorella*-based green algal cell suspension is highly beneficial in screening herbicides. Inhibitory effects of 39 herbicides, spanning 19 chemical classes with 9 distinct modes of action, were studied on liquid media cultured green algae, and the result showed that green algae are beneficial in high-throughput in vitro herbicidal screening (Ma et al. 2002). Likewise, auxins are also used to enhance the budding and growth of *Aegle marmelos* Correa medicinal plant, and rooting of various Fig cuttings (Patel and Patel 2018). Algal auxins are also extensively utilized in the micropropagation of Musa variety of Yangambi, in vitro regeneration procedures in sugarcane soma clones, rooting of rosemary, elderberry and sage shown to enhance floral attributes and yield of essential oil in *Matricaria chamomilla* (Keshari et al. 2016; Solangi et al. 2018). It is used to develop sustainable agriculture by acting as a pesticide, an antibacterial, antifungal and antiviral agent, and helps in enhancing plant growth as well as soil remediation, which improves agricultural productivity (Kumar et al. 2016; Munira et al. 2018; Napier 2017).

14.6.2 Gibberellins

Fungal ascomycetes, namely *Gibberella fujikuroi*, were used to discover gibberellins or gibberellic acid, which was later illustrated to be present and extracted from marine algae (Mowat 1965). Gibberellins are broadly classified into gibberellins A and B, which were further sub-classified into GA1, GA3, GA4, GA5, GA6 and GA7 categories (Takahashi et al. 2012). The liquid chromatography–mass spectroscopy (LC-MS) investigations amongst 31 microalgal species revealed that *Chlamydomonas reinhardtii* algal species contains 19 different types of gibberellins (Stirk et al. 2013). Likewise, brown algae, tissue extracts of *Fucus vesiculosus* and *Fucus spiralis* were investigated and revealed that these algae possess higher gibberellin contents (Tarakhovskaya et al. 2007). Other algae such as *Microcystis aeruginosa*, *Haematococcus pluvialis*, *Chlorella vulgaris*, *Cyamopsis tetragonolaba* and *Chaetomorpha litorea* were also illustrated to produce large quantities of various types of gibberellins (Gao et al. 2013; Falkowska et al. 2011; Sangeetha et al. 2011). These phytohormones promote algal growth by shortening the log phase, stimulating pigments and proteins accumulation, decreasing heavy metals toxicity, activating cell division and elevating growth rate in log phase during the algal growth (Han et al. 2018). Gibberellins were discovered in 1930s due to excessive rice stem elongation, which is caused by these phytohormones secreted by fungi. Thus, these phytohormones are explored to develop agriculture, exclusively by enhancing plant growth (Silverstone and Sun 2000). It is noteworthy that gibberellins assist in elevating plant growth, enhancing length of xylem fibre, biomass production, sinking demand increment during rapid fruit growth of Japanese pear fruit, and as unique biofertilizers (Zhang et al. 2007). Overloading gibberellins during plant growth, inhibits the growth of other predatory herbs, and acts as the enhanced, new generation herbicides (Dayan and Duke 2014). These are also used as pesticides, enhance seed germination, and as antibacterial, antifungal and antiviral agents to protect plants from microbes, as well as to elevate nutrient uptake in soil and in soil remediation (Liu et al. 2013; Altunok et al. 2015; Rasoulpour et al. 2018).

14.6.3 Cytokinins

Cytokinins embody a phytohormone group, which possesses similar growth functions as those of gibberellins, and promotes cytokinesis during cell division and differentiation (Kieber 2002). Cytokinins are broadly classified into adenine-type, which includes zeatin, kinetin and 6-benzylaminopurine and phenylurea-type, such as tridiazutal and diphenylurea (Aina et al. 2012). In other cases, they are classified into isoprenoid and aromatic cytokinins (Sakakibara 2006). Amongst algae, cytokinins are initially identified in marine algae, such as chloroplast of *Euglena gracilis* and green algae of *Chlorella*, *Protococcus* and *Scenedesmus* algal classes (Tarakhovskaya et al. 2007). Phytohormones that belong to the cytokinin class, including isopentenyladenosine (IPA), zeatin, aromatic cytokinins, zeatin riboside and topolin conjugates, were identified and extracted from green macroalgae, microalgae and characean (Ördög et al. 2004). Until 1980s, there were about 18 algal species, which were reported to possess cytokinin activities, and all these 18 species could be utilized for cytokinin extraction procedures (Mooney and van Staden 1986). Later, certain macroalgae, brown algae, Brazilian red algae, green algae, *Cyanophyta* and microalgae were also demonstrated to yield high quantities of cytokinins, which promote the algal growth (Stirk et al. 2003; Yokoya et al. 2010). Recently, *Acutodesmus obliquus*, *Chlorella vulgaris*, *Gracilaria caudata*, *Klebsormidium* and *Clorella sorokiniana*, were utilized for the extraction of algal cytokinins, as they produce large quantities of cytokinins during their growth phase (Ozioko et al. 2015; Souza and Yokoya 2016; Piotrowska-Niczyporuk et al. 2018). In agriculture, algal cytokinins are highly significant for farmers to elevate plant growth and cell division (Basra and Lovatt 2016). It was reported that cytokinins help to increase the growth of cotton seedlings by 10% during drought conditions (Burke 2010). Thus, algal cytokinins are utilized as biofertilizers and stimulants of plant growth for sustainable agriculture (Win et al. 2018). Additionally, algal cytokinins act as antibacterial and antifungal agents against diseases in tomato, as antiviral agents, fertilizers, biostimulants and antioxidants (Esserti et al. 2017; Dmytryk and Chojnacka 2018).

14.6.4 Abscisic Acid

Abscisic acid is a 15-C sesquiterpenoid plant hormone, which is crucial in plant growth, particularly in controlling the stomatal closure, seed and bud dormancy. It is an essential hormone in plants, which gives them the ability to tolerate soil salinity, heat, cold and other environmental stresses, as well as heavy metal ion concentrations (Ruggiero et al. 2004). Abscisic acid is significant in leaf abscission, and are also produced in roots, as a response to decrement in the soil water potential and other stresses. As this hormone is involved in maintaining the functions of plant leaves, which is essential for photosynthesis to produce energy for the whole plant, abscisic acid is a unique and significant hormone for plant growth (Boyle et al. 2016). Abscisic acid is classified based on the isomerism of the carboxyl group at the second carbon, such as *cis*- and *trans*-abscisic acid (Li et al. 2017). Several algal groups also produce abscisic acid, which can be extracted and used as biofertilizers, to enhance plant growth amidst the stress environment. Conventionally, *Chlorella* species, *Haematococcus pluvialis*, *Dunaliella salina*, certain algae from the *Ascophyllum* genus and *Laminaria* species, were identified to yield high quantities of abscisic acid, and to extract them for various applications (Nimura and Mizuta 2002). Also, algae, such as *Chlamydomonas reinhardtii*, Brazilian red algae, *Chlorella vulgaris* and *Laminaria japonica*, were proved to possess high abscisic acid contents (Yoshida et al. 2003; Yokoya et al. 2010). Currently, *Cyanidioschyzon merolae*, *Scenedesmus quadricauda*, *Ecklonia maxima*, *Nannochloropsis oceanica* and *Chlorella sorokiniana* were used for efficient extraction of abscisic acid for pharmaceutical applications (Khasin et al. 2017; Kobayashi et al. 2016). In agriculture, abscisic acid is highly essential for the plant growth, which helps in enhancing crop production (Forchetti et al. 2007). In recent times, they are used in efficient water utilization in *Arabidopsis*, controlling agrochemical usage, improving drought tolerance, regulating gas exchange and biostimulants in plants, which eventually enhances agricultural sustainability (Valerio et al. 2017; Yang et al. 2016).

14.6.5 Ethylene

Ethylene is also a plant-specific hormone, which regulates fruit ripening, leaf abscission and opening of flowers (Lin et al. 2009). Ethylene has been conserved for over 450 million years of plant evolution as hormones, and are used for ripening of fruits even by the ancient Egyptians and Chinese (Tsai et al. 2008). In 1934, the synthesis of ethylene by plants was proved, and it was reported that the secretion of ethylene phytohormone is responsible for the ripening of fruits and senescence of vegetative tissues in plants (Ju et al. 2015). Ethylene, which is extracted and isolated from algae, is gaining implicational importance in various agricultural practices (Arshad and Frankenberger 2012). Ethylene hormones are conventionally extracted from nitrogen-fixing blue-green algae, marine red algae, *Scenedesmus obliquus*, *Chlorella pyrenoidosa* and coral reef algae (Ma 2002). Recently, *Klebsormidium*, *Spirogyra pratensis*, *Pyropia yezoensis*, *Chlorella protothecoides* and *Grateloupia imbricata* were used for ethylene extraction to suit various applications (Pilar et al. 2016; Uji et al. 2016; Kishi et al. 2015). These algal ethylene phytohormones are used in several agricultural purposes, such as seedling, pollination, senescence of xylem cells, enhancement of seed germination, elevation of auxin transport, advancement of root hair growth for efficient mineral and water absorption, induction of adventitious root growth and assistance in fruit storage (Mao et al. 2016; Esashi 2017; Woodson 2018). Other algal phytohormones, such as jasmonic acid, polyamines, brassinosteroids and rhodomorphin, were used to implement agricultural sustainability (Tarakhovskaya et al. 2007). Table 14.2 lists significant algal phytohormones, which are extensively used in several agricultural practices to promote sustainable agriculture.

14.7 Summary and Future Perspectives

The discussions, guidelines, findings and the corresponding arguments pertaining to the existing literature lay claim to the proclamation that algal metabolites and phytohormones are indispensable for the realization of a sustainable agricultural tradition. Figure 14.2 is an illustrative schematic in the light of these opinions. From both the perspectives of overall growth and development of the host plants, as well as an advancement in agricultural productivity, the shift in focus towards a more rational and green approach through administration of algal extracts, is thoroughly justified. The enormous flow of funds, both from public and private sectors, only glorify this further. As mentioned in Sect. 14.1, there is a global effort to boost the agricultural yield, mediated by sustainable strategies. For instance, two US states,

Table 14.2 Major phytohormones extracted from algae and their applications in agriculture

Phytohormones	Algal groups	Agricultural applications
Auxins	*Laminaria*, *Macrocystis* brown algae, *Chlorella*, *Cladophora*, *Enteromorpha* green algae, *Botryocladia* red algae, characean algae, *Klebsormidium nitens*, *Acutodesmus obliquus*, *Chlorella vulgaris*, *Chlorella pyrenoidosa*, *Scenedesmus quadricauda* and charophyte algae	Budding and growth of *Aegle marmelos* Correa medicinal plant, in vitro herbicidal screening, rooting of fig cuttings, micropropagation of *Musa* Yangambi variety, in vitro regeneration of sugarcane soma clones, rooting of rosemary, elderberry and sage, enhances floral attributes and yield of essential oil in *Matricaria chamomilla*, acts as pesticides, antibacterial, antifungal, antiviral agent, enhances plant growth, sustainability in agriculture and helps in soil remediation
Gibberellins	*Chlamydomonas reinhardtii*, *Fucus vesiculosus*, *Fucus spiralis*, *Microcystis aeruginosa*, *Haematococcus pluvialis*, *Chlorella vulgaris*, *Cyamopsis tetragonolaba* and *Chaetomorpha litorea*	Elevates plant growth, enhances xylem fibre length, biomass production, sink demand increment during rapid Japanese pear fruit growth, unique biofertilizers, new generation herbicides, pesticides, enhances seed germination, increase soil nutrients, antibacterial, antifungal, antiviral agents and soil remediation
Cytokinins	*Euglena gracilis Chlorella*, *Protococcus*, *Scenedesmus* algal classes green macro- and microalgae, characean, Brazilian red algae, *Cyanophyta*, brown algae, *Acutodesmus obliquus*, *Chlorella vulgaris*, *Gracilaria caudata* and *Clorella sorokiniana*	Elevate plant growth and cell division, biofertilizers, plant growth biostimulants, antibacterial, antifungal, antiviral agents and antioxidants
Abscisic acid	*Chlorella* species, *Haematococcus pluvialis*, *Dunaliella salina*, *Ascophyllum* genus, *Laminaria* species, *Chlamydomonas reinhardtii*, Brazilian red algaes, *Chlorella vulgaris*, *Laminaria japonica*, *Cyanidioschyzon merolae*, *Scenedesmus quadricauda*, *Ecklonia maxima*, *Nannochloropsis oceanica* and *Chlorella sorokiniana*	Enhancing crop production, efficient water utilization in *Arabidopsis*, agrochemical usage control, improves drought tolerance, regulates gas exchange and biostimulants
Ethylene	Nitrogen-fixing blue-green algae, marine red algae, *Scenedesmus obliquus*, *Chlorella pyrenoidosa*, coral reef algae, *Klebsormidium*, *Spirogyra pratensis*, *Pyropia yezoensis*, *Chlorella protothecoides* and *Grateloupia imbricata*	Improves seedling, pollination, xylem cell senescence, enhances seed germination, auxin transport, mineral and water absorption via root hair growth, induces adventitious root growth and helps in fruit storage

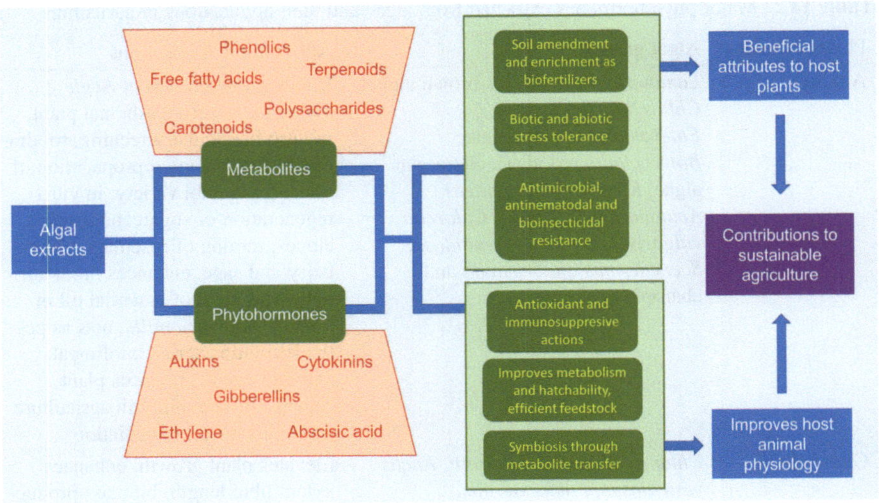

Fig. 14.2 The two predominant categories of algal extracts and their contributions to viable agriculture

Ohio and Arizona, included algal culture as an integral subset of agriculture (Trentacoste et al. 2015). This is also to attract and increase the flow of investments in algal cultivation from the industrial sector, as much as to keep constraints, such as taxes, within acceptable boundaries. However, saying that, large-scale production of algal biomass and subsequent procedures to extract and realize the optimum extraction of nutrients from algal extracts, are far from being optimal. This aspect demands significant attention, aided by both industrial and government ventures.

Works are underway, focusing on alternative protocols of algal biomass culture, which would ensure high efficiency, and would alleviate the prohibitively high concurrent costs. One such effort is the proposal for eco-friendly and viable agricultural practices for the future, mainly involving cyanobacterial systems, including applications of hybrid reactors. This encompasses the benefits of both conventional photo-bioreactor and raceway pond prototypes (Pathak et al. 2018). Traditional and recent advancements, facilitating manufacturing and harvesting of algal biomass, are shown in Fig. 14.3. Additionally, inherent and associated challenges need to be addressed, including design of next generation of algal species that would better withstand biotic and abiotic stress, lower prerequisite of water and other nutrients for harvest, regulating ecological instabilities to facilitate open-air farming, and focusing on algal systems with low-protein and high-fat contents. In this context, combinatorial approaches are advised, including genetic engineering, genomics, proteomics and metabolomics, in addition to high-throughput characterization techniques. The genetic transformations could as well boost the algal applications as biostimulants and biofertilizers, by improving their stress tolerance and sustenance, as well as resistance against microbes, insects and nematodes.

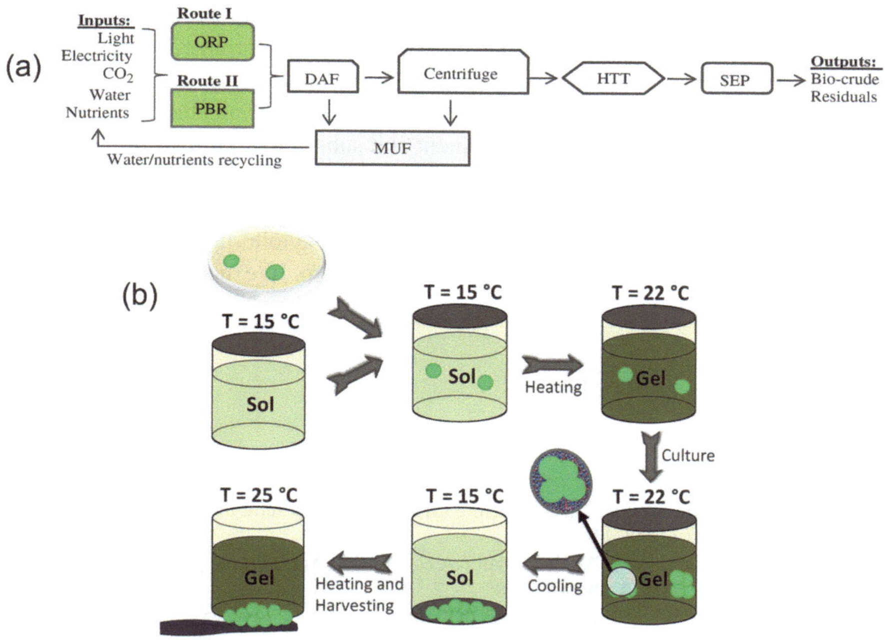

Fig. 14.3 (**a**) Process flow diagram of open pond and closed photobioreactor-based microalgal feedstock production for biofuels. *ORP* open raceway pond, *PBR* photobioreactor, *DAF* dissolved air flotation, *HTT* hydrothermal treatment, *SEP* solvent-based extraction and purification, *MUF* membrane ultrafiltration. Reproduced from Richardson et al. (2014) [Open Access]. Copyright 2013 Richardson et al. (**b**) Schematic of microalgal cultivation and harvesting process using thermoreversible sol–gel transition. Microalgal cells are seeded in the TAPP medium in solution phase at 15 °C. Then, the temperature is raised at 22 °C for gelation of the medium and entrapped microalgal cultivation. After the cultivation period, the temperature is decreased to 15 °C allowing microalgal clusters to gravimetrically settle at the bottom. The temperature is finally raised to 25 °C and settled microalgal clusters are scraped off the TAPP surface. Reproduced from Estime et al. (2017) [Open Access]. Copyright 2017 Estime et al.

Predominantly, standardization of two principal and yet conflicting approaches, is a *quid pro quo*, as effectively postulated by du Jardin (2015). First is the transition from the laboratory to the open field, mainly to acclimatize the lab-scale standard model parameters with the variations in the field, including access to a wide range of cultivars and host plant species. Second is the reverse strategy: a shift from field to the lab. The latter is somewhat unconventional, but stands extraordinarily scientific, concerning how far the genotype-specific alterations, as seen in the root environment, influence the interactions between the plant health and growth-promoting substances. Consequences will lie on both macroalgal usage, which stands a better alternative as bioinoculants for viable agricultural strategies, subject to additional investigations to recover their full potential as well as on the employment of microalgal inoculums or extracts, which in spite of possessing distinct advantages in the form of being a far superior candidate for greater biofuel yield compared to the energy crops, faces challenges in the domains of scale-up and hormonal manipulations. Furthermore, significant further investigations are warranted for uncovering the full-scale benefits of the algal secondary metabolites on plant and animal physiology. Finally, several outlooks should be critically evaluated, the aptness of the shift in focus towards algal involvement and subsequent inflow of investments, long-term proficiency and effects on the biofuel, food and pharmaceutical markets, and desired rate of metabolite production and screening to meet the needs of the industry and the economy. Algal metabolites and phytohormones hold immense promise to ensure a sustainable mode of agriculture. We need to work on the promise.

References

Abdel-Raouf N, Al-Homaidan AA, Ibraheem IBM (2012) Agricultural importance of algae. Afr J Biotechnol 11(54):11648–11658

Abdel-Raouf N, Al-Enazi NM, Al-Homaidan AA et al (2015) Antibacterial β-amyrin isolated from *Laurencia microcladia*. Arab J Chem 8:32–37

Abomohra AE-F, Jin W, El-Sheekh M (2016) Enhancement of lipid extraction for improved biodiesel recovery from the biodiesel promising microalga *Scenedesmus obliquus*. Energy Convers Manag 108:23–29

Acea MJ, Diz N, Prieto-Fernández A (2001) Microbial populations in heated soils inoculated with cyanobacteria. Biol Fert Soils 33:118–125

Aina O, Quesenberry K, Gallo M (2012) Thidiazuron-induced tissue culture regeneration from quartered-seed explants of *Arachis paraguariensis*. Crop Sci 52(3):1076–1083

Ali SMY, Ravikumar S, Beula JM (2013) Mosquito larvicidal activity of seaweeds extracts against *Anopheles stephensi*, *Aedes aegypti* and *Culex quinquefasciatus*. Asian Pac J Trop Dis 3:196–201

Al-Saif SS, Abdel-Raouf N, El-Wazanani HA et al (2014) Antibacterial substances from marine algae isolated from Jeddah coast of Red Sea, Saudi Arabia. Saudi J Biol Sci 21:57–64

Altunok M, Ozkaya FC, Engin S et al (2015) In vitro antibacterial activity of sponge-associated fungi against bacterial aquaculture pathogens. Fresenius Environ Bull 24(6A):2158–2166

Ammar N, Jabnoun-Khiareddine H, Mejdoub-Trabelsibi B et al (2017) Pythium leak control in potato using aqueous and organic extracts from the brown alga *Sargassum vulgare* (C. Agardh, 1820). Postharvest Biol Technol 130:81–93

Anderson TH, Gray TRG (1991) The influence of soil organic carbon on microbial growth and survival. In: Wilson WS (ed) Advances in soil organic matter research: the impact on agriculture and the environment. Royal Society of Chemistry Cambridge, Cambridge, pp 253–266

Ara J, Sultana V, Qasim R et al (2005) Biological activity of *Spatoglossum asperum*: a brown alga. Phytother Res 19:618–623

Arshad M, Frankenberger WT Jr (2012) Ethylene: agricultural sources and applications. Springer Science & Business Media, New York

Asha A, Rathi JM, Patric Raja D et al (2012) Biocidal activity of two marine green algal extracts against third instar nymph of *Dysdercus cingulatus* (Fab.) (Hemiptera: Pyrrhocoridae). J Biopest 5:129–134

Asharaja A, Sahayaraj K (2013) Screening of insecticidal activity of brown macroalgal extracts against *Dysdercus cingulatus* (Fab.) (Hemiptera: Pyrrhocoridae). J Biopest 6:193–203

Ashour M, Wink M, Gershenzon J (2010) Biochemistry of terpenoids: monoterpenes, sesquiterpenes and diterpenes. Annu Plant Rev 2:258–303

Aslam A, Thomas-Hall SR, Manzoor M et al (2018) Mixed microalgae consortia growth under higher concentration of CO_2 from unfiltered coal fired flue gas: fatty acid profiling and biodiesel production. J Photochem Photobiol B 179:126–133
Athukorala Y, Lee K-W, Kim S-K et al (2007) Anticoagulant activity of marine green and brown algae collected from Jeju Island in Korea. Bioresour Technol 98(9):1711–1716
Avula SGC, Belovich JM, Xu Y (2017) Determination of fatty acid methyl esters derived from algae *Scenedesmus dimorphus* biomass by GC–MS with one-step esterification of free fatty acids and transesterification of glycerolipids. J Sep Sci 40(10):2214–2227
Awasthi M, Upadhyay AK, Singh S et al (2018) Terpenoids as promising therapeutic molecules against Alzheimer's disease: amyloid beta-and acetylcholinesterase-directed pharmacokinetic and molecular docking analyses. Mol Simul 44(1):1–11
Ayoola G, Coker H, Adesegun S et al (2008) Phytochemical screening and antioxidant activities of some selected medicinal plants used for malaria therapy in Southwestern Nigeria. Trop J Pharm Res 7(3):1019–1024
Baloch GN, Tariq S, Ehteshamul-Haque S (2013) Management of root diseases of eggplant and watermelon with the application of asafoetida and seaweeds. J Appl Bot Food Qual 86:138–142
Bantoto V, Danilo DY (2013) The larvicidal activity of brown algae *Padina minor* (Yamada) and *Dicyota linearis* (Greville) against the dengue vector, *Aedes aegypti* (Linn) (Diptera: Culicidae). J Vector Borne Dis 50:68–70
Barrow C, Shahidi F (2008) Marine nutraceuticals and functional foods. CRC, London
Basra SM, Lovatt CJ (2016) Exogenous applications of moringa leaf extract and cytokinins improve plant growth, yield, and fruit quality of cherry tomato. Hortic Technol 26(3):327–337
Becker W (2004) Microalgae in human and animal nutrition. In: Richmond A (ed) Handbook of Microalgal Culture. Blackwell, Oxford, pp 312–351
Becker EW (2007) Microalgae as a source of protein. Biotechnol Adv 25(2):207–210
Benkendorff K, Davis AR, Rogers CN et al (2005) Free fatty acids and steroids in the benthic spawn of aquatic molluscs, and their associated antimicrobial properties. J Exp Mar Biol Ecol 316:29–44
Berri M, Slugocki C, Olivier M et al (2016) Marine-sulfated polysaccharides extract of *Ulva armoricana* green algae exhibits an antimicrobial activity and stimulates cytokine expression by intestinal epithelial cells. J Appl Phycol 28(5):2999–3008
Bilan M, Klochkova N, Ustyuzhanina N et al (2016) Polysaccharides of algae 68. Sulfated polysaccharides from the Kamchatka brown alga *Laminaria bongardiana*. Russ Chem Bull 65 (11):2729–2736
Boeckaert C, Vlaeminck B, Dijkstra J et al (2008) Effect of dietary starch or micro algae supplementation on rumen fermentation and milk fatty acid composition of dairy cows. J Dairy Sci 91(12):4714–4727
Boyle RK, McAinsh M, Dodd IC (2016) Daily irrigation attenuates xylem abscisic acid concentration and increases leaf water potential of Pelargonium × hortorum compared with infrequent irrigation. Physiol Plant 158(1):23–33
Braden KW, Blanton JR Jr, Montgomery JL et al (2007) Tasco supplementation: effects on carcass characteristics, sensory attributes, and retail display shelf-life. J Anim Sci 85:754–768
Bravo L (1998) Polyphenols: chemistry, dietary sources, metabolism, and nutritional significance. Nutr Rev 56(11):317–333
Burke J (2010) Plant hormone increases cotton yields in drought conditions. Agric Res 58(7):31
Cameron HJ, Julian GR (1988) Utilization of hydroxyapatite by cyanobacteria as their sole source of phosphate and calcium. Plant Soil 109:123–124
Campos EVR, De Oliveira JL, Fraceto LF et al (2015) Polysaccharides as safer release systems for agrochemicals. Agron Sustain Dev 35(1):47–66
Cardozo KH, Guaratini T, Barros MP et al (2007) Metabolites from algae with economical impact. Comp Biochem Physiol C Toxicol Pharmacol 146(1–2):60–78
Castro NM, Valdez MC, Álvarez AM et al (2009) The kelp *Macrocystis pyrifera* as nutritional supplement for goats. Rev Cien 19(1):63–70

Certik M, Shimizu S (1999) Biosynthesis and regulation of microbial polyunsaturated fatty acid production. J Biosci Bioeng 87:1–14
Cetin H, Gokoglu M, Oz E (2010) Larvicidal activity of the extract of seaweed, *Caulerpa scalpelliformis*, against *Culex pipiens*. J Am Mosq Control Assoc 26:433–435
Chang J-J, Thia C, Lin H-Y et al (2015) Integrating an algal β-carotene hydroxylase gene into a designed carotenoid-biosynthesis pathway increases carotenoid production in yeast. Bioresour Technol 184:2–8
Chapman VJ, Chapman DJ (1980) Seaweed and their uses, 3rd edn. Chapman & Hall, New York, pp 63–85
Chatterjee A, Singh S, Agrawal C et al (2017) Role of algae as a biofertilizer. In: Rastogi RP, Madamwar D, Pandey A (eds) Algal green chemistry: recent progress in biotechnology. Elsevier, Amsterdam, pp 189–200
Chatzissavvidis C, Therios I (2014) Role of algae in agriculture. In: Pomin VH (ed) Seaweeds: agricultural uses, biological and antioxidant agents. Nova Science, New York, pp 1–37
Chaurasia AK, Parasnis A, Apte SK (2008) An integrative expression vector for strain improvement and environmental applications of the nitrogen fixing cyanobacterium, *Anabaena* sp. strain PCC7120. J Microbiol Methods 73:133–141
Cherng J-Y, Shih M-F (2006) Improving glycogenesis in Streptozocin (STZ) diabetic mice after administration of green algae *Chlorella*. Life Sci 78(11):1181–1186
Chye JTT, Jun LY, Yon LS et al (2018) Biofuel production from algal biomass. In: Konur O (ed) Bioenergy and biofuels, 1st edn. CRC Press, Taylor and Francis, Boca Raton, pp 87–117
Ciferri O, Tiboni O (1985) The biochemistry and industrial potential of spirulina. Ann Rev Microbiol 39:503–526
Cocchietto M, Skert N, Nimis P et al (2002) A review on usnic acid, an interesting natural compound. Naturwissenschaften 89(4):137–146
Commeiras L, Bourdron J, Douillard S et al (2006) Total synthesis of terpenoids isolated from caulerpale algae and their inhibition of tubulin assembly. Synthesis 2006(01):166–181
Cosgrove E (2017) Report: algae market to reach $45bn by 2023. https://agfundernews.com/report-algae-market-could-to-reach-45b-by-2023.html. Accessed 11 Aug 2018
Craigie JS (2011) Seaweed extract stimuli in plant science and agriculture. J Appl Phycol 23:371–393
Cseke LJ, Kirakosyan A, Kaufman PB et al (2016) Natural products from plants. CRC, London
Das SC, Mandal M, Mandal LN et al (1991) Effect of growth and subsequent decomposition of blue-green algae on the transformation of iron and manganese in submerged soils. Plant Soil 138:75–84
Dautermann O, Lohr M (2017) A functional zeaxanthin epoxidase from red algae shedding light on the evolution of light-harvesting carotenoids and the xanthophyll cycle in photosynthetic eukaryotes. Plant J 92(5):879–891
Dayan FE, Duke SO (2014) Natural compounds as next generation herbicides. Plant Physiol 114 (239061):1090–1105
de Morais MG, Vaz BS, de Morais EG et al (2015) Biologically active metabolites synthesized by microalgae. BioMed Res Int 2015:835761
Del Campo JA, Moreno J, Rodríguez H et al (2000) Carotenoid content of chlorophycean microalgae: factors determining lutein accumulation in Muriellopsis sp. (Chlorophyta). J Biotechnol 76:51–59
Desbois AP, Smith VJ (2010) Antibacterial free fatty acids: activities, mechanisms of action and biotechnological potential. Appl Microbiol Biotechnol 85(6):1629–1642
Dhargalkar VK, Pereira N (2005) Seaweed: promising plant of millennium. Sci Cult 71(3-4):60–66
Dmytryk A, Chojnacka K (2018) Algae as fertilizers, biostimulants, and regulators of plant growth. Algae biomass, characteristics and applications. Springer, Cham, pp 115–122
Domozych DS, Sørensen I, Popper ZA et al (2014) Pectin metabolism and assembly in the cell wall of the charophyte green alga *Penium margaritaceum*. Plant Physiol 114(236257):105–118

du Jardin P (2015) Plant biostimulants: definitions, concept, main categories and regulation. Sci Hort 196:3–14
Eder M, Lütz-Meindl U (2010) Analyses and localization of pectin-like carbohydrates in cell wall and mucilage of the green alga *Netrium digitus*. Protoplasma 243(1–4):25–38
Edmunds PJ, Davies PS (1986) An energy budget for *Porites porites* (Scleractinia). Mar Biol 92:339–347
Einstein-Curtin A (2018) A more robust algae farming industry in the US will provide us with new and sustainable agricultural products. https://www.feednavigator.com/Article/2018/04/19/A-more-robust-algae-farming-industry-in-the-US-will-provide-us-with-new-and-sustainable-agricultural-products?utm_source=copyright&utm_medium=OnSite&utm_campaign=copyright. Accessed 11 Aug 2018
Ellis JT, Hengge NN, Sims RC et al (2012) Acetone, butanol, and ethanol production from wastewater algae. Bioresour Technol 111:491–495
Esashi Y (2017) Ethylene and seed germination. The plant hormone ethylene. CRC, London, pp 133–157
Esquivel-Hernández DA, Ibarra-Garza IP, Rodríguez-Rodríguez J et al (2017) Green extraction technologies for high-value metabolites from algae: a review. Biofuels Bioprod Biorefin 11 (1):215–231
Esserti S, Smaili A, Rifai LA et al (2017) Protective effect of three brown seaweed extracts against fungal and bacterial diseases of tomato. J Appl Phycol 29(2):1081–1093
Estime B, Ren D, Sureshkumar R (2017) Cultivation and energy efficient harvesting of microalgae using thermoreversible sol-gel transition. Sci Rep 7:40725
Falchini L, Sparvoli E, Tomaselli L (1996) Effect of *Nostoc* (cyanobacteria) inoculation on the structure and stability of clay soils. Biol Fert Soils 23:246–252
Falkowska M, Pietryczuk A, Piotrowska A et al (2011) The effect of gibberellic acid (GA3) on growth, metal biosorption and metabolism of the green algae *Chlorella vulgaris* (Chlorophyceae) Beijerinck exposed to cadmium and lead stress. Pol J Environ Stud 20:53–59
Fenical W, Paul VJ (1984) Antimicrobial and cytotoxic terpenoids from tropical green algae of the family *Udoteaceae*. In: 11th International seaweed symposium. Springer, Dordrecht
Flaibani A, Olsen Y, Painter TJ (1989) Polysaccharides in desert reclamation: composition of exocellular proteoglycan complexes produced by filamentous blue-green and unicellular green edaphic algae. Carbohydr Res 190:235–248
Foo SC, Yusoff FM, Ismail M et al (2017) Antioxidant capacities of fucoxanthin-producing algae as influenced by their carotenoid and phenolic contents. J Biotechnol 241:175–183
Forchetti G, Masciarelli O, Alemano S et al (2007) Endophytic bacteria in sunflower (*Helianthus annuus* L.): isolation, characterization, and production of jasmonates and abscisic acid in culture medium. Appl Microbiol Biotechnol 76(5):1145–1152
Freile-Pelegrín Y, Robledo D (2014) Bioactive phenolic compounds from algae. In: Hernández-Ledesma B, Herrero M (eds) Bioactive compounds from marine foods: plant and animal sources. Wiley, Chichester
Freile-Pelegrín Y, Robledo D, Chan-Bacab M et al (2008) Antileishmanial properties of tropical marine algae extracts. Fitoterapia 79(5):374–377
Galal HRM, Salem WM, Naser F et al (2011) Biological control of some pathogenic fungi using marine algae extracts. Res J Microbiol 6:645–657
Gao Z, Meng C, Gao H et al (2013) Carotenoid genes transcriptional regulation for astaxanthin accumulation in fresh water unicellular alga *Haematococcus pluvialis* by gibberellin a3 (GA3). Indian J Biochem Biophys 50:548–553
Gattuso J-P, Jaubert J (1990) Effect of light on oxygen and carbon dioxide fluxes and on metabolic quotients measured in situ in a zooxanthellate coral. Limnol Oceanogr 35:1796–1804
Gille A, Trautmann A, Posten C et al (2016) Bioaccessibility of carotenoids from *Chlorella vulgaris* and *Chlamydomonas reinhardtii*. Int J Food Sci Nutr 67(5):507–513
Ginzberg A, Cohen M, Sod-Moriah UA et al (2000) Chickens fed with biomass of the red microalga *Porphyridium* sp. have reduced blood cholesterol level and modified fatty acid composition in egg yolk. J Appl Phycol 12:325–330

Glazer I, Apelbaum A, Orion D (1985) Effect of inhibitors and stimulators of ethylene production on gall development in *Meloidogyne javanica*-infected tomato roots. J Nematol 17:145–149

Gopinath V, Saravanan S, Al-Maleki A et al (2018) A review of natural polysaccharides for drug delivery applications: special focus on cellulose, starch and glycogen. Biomed Pharmacother 107:96–108

Gouveia L, Batista AP, Sousa I (2008) Microalgae in novel food products. In: Papadoupoulos K (ed) Food chemistry research developments. Nova Science, New York, pp 75–112

Gouveia V, Seca AM, Barreto MC et al (2013) Di-and sesquiterpenoids from *Cystoseira* genus: structure, intra-molecular transformations and biological activity. Mini Rev Med Chem 13 (8):1150–1159

Guilherme MR, Aouada FA, Fajardo AR et al (2015) Superabsorbent hydrogels based on polysaccharides for application in agriculture as soil conditioner and nutrient carrier: a review. Eur Polym J 72:365–385

Gutiérrez R, Núñez R, Quintana L et al (2017) Optimization of the extraction process of phenolic compounds from the brown algae *Sargassum fluitans* Børgesen (Børgesen). Biotecnol Apl 34 (3):3301–3304

Hamed SM, Abd El-Rhman AA, Abdel-Raouf N et al (2018) Role of marine macroalgae in plant protection & improvement for sustainable agriculture technology. Beni-Seuf Univ J Appl Sci 7:104–110

Han T, Zhao Z, Wang Y (2016) The effect of ryegrass and fertilizer on the petroleum contaminated soil remediation. Feb Fresenius Environ Bull 25(6):2243–2250

Han X, Zeng H, Bartocci P et al (2018) Phytohormones and effects on growth and metabolites of microalgae: a review. Fermentation 4:25

Harland AD, Fixter LM, Davies PS et al (1991) Distribution of lipids between the zooxanthellae and animal compartment in the symbiotic sea anemone *Anemonia viridis*: wax esters, triglycerides and fatty acids. Mar Biol 110:13–19

He ML, Wang Y, You JS et al (2009) Effect of a seaweed extract on fatty acid accumulation and glycerol-3-phosphate dehydrogenase activity in 3T3-L1 adipocytes. Lipids 44:125–132

Henríquez V, Escobar C, Galarza J et al (2016) Carotenoids in microalgae. Carotenoids in nature. Springer, Cham, pp 219–237

Hoang TH, Qin JG, Stone DA et al (2016) Colour changes of greenlip abalone (*Haliotis laevigata* Donovan) fed fresh macroalgae and dried algal supplement. Aquaculture 456:16–23

Huang W, Lin Y, He M et al (2018) Induced high-yield production of zeaxanthin, lutein, and β-carotene by a mutant of *Chlorella zofingiensis*. J Agric Food Chem 66(4):891–897

Ibraheem BMI, Hamed SM, Abd Elrhman AA et al (2017) Antimicrobial activities of some brown macroalgae against some soil borne plant pathogens and in vivo management of *Solanum melongena* root diseases. Aust J Basic Appl Sci 11:157–168

Jiao G, Yu G, Zhang J et al (2011) Chemical structures and bioactivities of sulfated polysaccharides from marine algae. Mar Drugs 9(2):196–223

Johnson JL, Jones MB, Cobb BA (2014) Bacterial capsular polysaccharide prevents the onset of asthma through T-cell activation. Glycobiology 25(4):368–375

Johnston M, Yellowlees D, Gilmour I (1995) Carbon isotopic analysis of the free fatty acids in a tridacnid–algal symbiosis: interpretation and implications for the symbiotic association. Proc R Soc Lond B 260:293–297

Jormalainen V, Honkanen T (2008) Macroalgal chemical defenses and their roles in structuring temperate marine communities. Algal chemical ecology. Springer, Cham, pp 57–89

Ju C, Van de Poel B, Cooper ED et al (2015) Conservation of ethylene as a plant hormone over 450 million years of evolution. Nat Plants 1(1):14004

Jung S, Sung H, Young S (2017) Immunostimulatory and anti-metastatic activity of polysaccharides isolated from byproducts of the corn starch industry. Carbohydr Polym 181:911–917

Kempinski C, Jiang Z, Bell S et al (2015) Metabolic engineering of higher plants and algae for isoprenoid production. In: Schrader J, Bohlmann J (eds) Biotechnology of isoprenoids. Springer, Cham, pp 161–199

Keshari B, Pradhan B, Deo B (2016) Effects of cytokinins and auxins on micropropagation of *Musa* spp. cv. Yangambi. IJOER 2(5):156–158
Kevan PG, Shipp L (2017) Biological control as biotechnological amelioration and ecosystem intensification in managed ecosystems. In: Reference module in life sciences. Elsevier, Amsterdam. https://doi.org/10.1016/B978-0-12-809633-8.09246-3. ISBN 978-0-12-809633-8
Khan W, Rayirath UP, Subramanian S et al (2009) Seaweed extracts as biostimulants of plant growth and development. Plant Growth Regul 28:386–399
Khasin M, Cahoon RE, Alvarez S et al (2017) Synthesis, secretion, and perception of abscisic acid regulates stress responses in *Chlorella sorokiniana*. bioRxiv 180547
Khoddami A, Wilkes MA, Roberts TH (2013) Techniques for analysis of plant phenolic compounds. Molecules 18(2):2328–2375
Khozin-Goldberg I, Cohen Z (2006) The effect of phosphate starvation on the lipid and fatty acid composition of the fresh water eustigmatophyte Monodus subterraneus. Phytochemistry 67 (7):696–701
Kieber JJ (2002) Tribute to Folke Skoog: recent advances in our understanding of cytokinin biology. J Plant Growth Regul 21(1):1
Kishi M, Kawai M, Toda T (2015) Heterotrophic utilization of ethylene glycol and propylene glycol by *Chlorella protothecoides*. Algal Res 11:428–434
Kleiner KT, Harper KT (1977) Soil properties in relation to cryptogamic ground cover in Canyonlands National Park. J Range Manag 30:202–205
Kobayashi Y, Ando H, Hanaoka M et al (2016) Abscisic acid participates in the control of cell cycle initiation through heme homeostasis in the unicellular red alga *Cyanidioschyzon merolae*. Plant Cell Physiol 57(5):953–960
Kombiah P, Sahayaraj K (2012) Repellent activity of *Caulerpa scalpelliformis* extracts and its formulations against *Spodoptera litura* and *Dysdercus cingulatus* (fab.). J Biopest 5:145–150
Kumar M, Tomar RS, Lade H et al (2016) Methylotrophic bacteria in sustainable agriculture. World J Microbiol Biotechnol 32(7):120
Kuzuyama T (2017) Biosynthetic studies on terpenoids produced by Streptomyces. J Antibiot 70 (7):811
Lakeh AAB, Ahmadi MR, Safi S (2010) Growth performance, mortality and carotenoid pigmentation of fry offspring as affected by dietary supplementation of astaxanthin to female rainbow trout (*Oncorhynchus mykiss*) broodstock. J Appl Ichthyol 26:35–39
Lange W (1976) Speculations on a possible essential function of the gelatinous sheath of blue-green algae. Can J Microbiol 22:1181–1185
Lee SH, Jeon YJ (2013) Anti-diabetic effects of brown algae derived phlorotannins, marine polyphenols through diverse mechanisms. Fitoterapia 86:129–136
Li J, Wu Y, Xie Q et al (2017) 5-abscisic acid. In: Li J, Li C, Smith SM (eds) Hormone metabolism and signaling in plants. Academic, San Diego, pp 161–202
Lin Z, Zhong S, Grierson D (2009) Recent advances in ethylene research. J Exp Bot 60 (12):3311–3336
Lin J-T, Liu S-C, Hu C-C et al (2016) Effects of roasting temperature and duration on fatty acid composition, phenolic composition, Maillard reaction degree and antioxidant attribute of almond (*Prunus dulcis*) kernel. Food Chem 190:520–528
Liu Y, Sun Y, He S et al (2013) Synthesis and characterization of gibberellin–chitosan conjugate for controlled-release applications. Int J Biol Macromol 57:213–217
Liu J, Qiu W, Song Y (2016) Stimulatory effect of auxins on the growth and lipid productivity of *Chlorella pyrenoidosa* and *Scenedesmus quadricauda*. Algal Res 18:273–280
Lohr M (2017) Genomic insights into the biosynthesis of photoprotective pigments in *Porphyra umbilicalis*: carotenoids and mycosporine-like amino acids. Phycologia 56(4):121
Ma J (2002) Differential sensitivity to 30 herbicides among populations of two green algae *Scenedesmus obliquus* and *Chlorella pyrenoidosa*. Bull Environ Contam Toxicol 68(2):275–281

Ma J, Xu L, Wang S (2002) A quick, simple, and accurate method of screening herbicide activity using green algae cell suspension cultures. Weed Sci 50(5):555–559

Manzo E, Ciavatta ML, Bakkas S et al (2009) Diterpene content of the alga *Dictyota ciliolata* from a Moroccan lagoon. Phytochem Lett 2:211–215

Mao J-L, Miao Z-Q, Wang Z et al (2016) Arabidopsis ERF1 mediates cross-talk between ethylene and auxin biosynthesis during primary root elongation by regulating ASA1 expression. PLoS Genet 12(1):e1005760

Matanjun P, Mohamed S, Mustapha NM et al (2009) Nutrient content of tropical edible seaweeds, *Eucheuma cottonii, Caulerpa lentillifera* and *Sargassum polycystum*. J Appl Phycol 21:75–80

Mercier L, Lafitte C, Borderies G et al (2001) The algal polysaccharide carrageenans can act as an elicitor of plant defence. New Phytol 149:43–51

Michalak I, Chojnacka K (2015) Production of seaweed extracts by biological and chemical methods. In: Kim S-K, Chojnacka K (eds) Marine algae extracts: processes, products, applications. Wiley, Weinheim

Mihranyan A (2011) Cellulose from cladophorales green algae: from environmental problem to high-tech composite materials. J Appl Polym Sci 119(4):2449–2460

Montgomery JL, Allen VG, Pond KR et al (2001) Tasco-Forage: IV. Influence of a seaweed extract applied to tall fescue pastures on sensory characteristics, shelf-life, and vitamin E status in feedlot-finished steers. J Anim Sci 79:884–894

Mooney PA, van Staden J (1986) Algae and cytokinins. J Plant Physiol 123(1):1–21

Moore BG, Tischer RG (1964) Extracellular polysaccharides of algae: effects on life-support systems. Science 145(3632):586

Moubayed NMS, Al Houri HJ, AlKhulaifi MM et al (2017) Antimicrobial, antioxidant properties and chemical composition of seaweeds collected from Saudi Arabia (Red Sea and Arabian Gulf). Saudi J Biol Sci 24:162–169

Mowat JA (1965) A survey of results on the occurrence of auxins and gibberellins in algae. Bot Mar 8(1):149–155

Mulders KJ, Weesepoel Y, Bodenes P et al (2015) Nitrogen-depleted *Chlorella zofingiensis* produces astaxanthin, ketolutein and their fatty acid esters: a carotenoid metabolism study. J Appl Phycol 27(1):125–140

Munira S, Farenhorst A, Sapkota K et al (2018) Auxin herbicides and pesticide mixtures in groundwater of a Canadian Prairie Province. J Environ Qual 47(6):1462–1467

Muscatine L, Falkowski P, Porter J et al (1984) Fate of photosynthetically-fixed carbon in light and shade-adapted colonies of the symbiotic coral *Stylophora pistillata*. Proc R Soc Lond B 222:181–202

Napier R (2017) Adding to the agricultural toolkit: next generation auxins and anti-auxins. Impact 2017(7):69–71

Ngala BM, Valdes Y, dos Santos G et al (2016) Seaweed-based products from *Ecklonia maxima* and *Ascophyllum nodosum* as control agents for the root-knot nematodes *Meloidogyne chitwoodi* and *Meloidogyne hapla* on tomato plants. J Appl Phycol 28:2073–2082

Ngo TC, Dao DQ, Nguyen MT et al (2017) A DFT analysis on the radical scavenging activity of oxygenated terpenoids present in the extract of the buds of *Cleistocalyx operculatus*. RSC Adv 7 (63):39686–39698

Nimura K, Mizuta H (2002) Inducible effects of abscisic acid on sporophyte discs from *Laminaria japonica* Areschoug (Laminariales, Phaeophyceae). J Appl Phycol 14(3):159–163

Ohtaka K, Hori K, Kanno Y et al (2017) Primitive auxin response without TIR1 and aux/IAA in the charophyte alga *Klebsormidium nitens*. Plant Physiol 174(3):1621–1632

Ördög V, Stirk WA, Van Staden J et al (2004) Endogenous cytokinins in three genera of microalgae from the Chlorophyta 1. J Phycol 40(1):88–95

Ozioko FU, Chiejina NV, Ogbonna JC (2015) Effect of some phytohormones on growth characteristics of *Chlorella sorokiniana* IAM-C212 under photoautotrophic conditions. Afr J Biotechnol 14(30):2367–2376

Pardee KI, Ellis P, Bouthillier M et al (2004) Plant virus inhibitors from marine algae. Can J Bot 82:304–309

Patel HR, Patel M (2018) Role of auxins on rooting of different types of cuttings in Fig. Int J Curr Microbiol App Sci 7(3):1317–1322

Pathak J, Rajneesh, Maurya PK et al (2018) Cyanobacterial farming for environment friendly sustainable agriculture practices: innovations and perspectives. Front Environ Sci 6:7

Paula JCD, Vallim MA, Teixeira VL (2011) What are and where are the bioactive terpenoids metabolites from Dictyotaceae (Phaeophyceae). Rev Bras Farmacogn 21(2):216–228

Paulert R, Ebbinghaus D, Urlass C et al (2010) Priming of the oxidative burst in rice and wheat cell cultures by ulvan, a polysaccharide from green macroalgae, and enhanced resistance against powdery mildew in wheat and barley plants. Plant Pathol 59:634–642

Pavela R, Benelli G (2016) Essential oils as ecofriendly biopesticides? Challenges and constraints. Trends Plant Sci 21(12):1000–1007

Peng K, Li J, Jiao K et al (2018) The bioeconomy of microalgal biofuels. In: Jacob-lopes E, Queiroz Zepka L, Queiroz M (eds) Energy from microalgae. Green energy and technology. Springer, Cham, pp 157–169

Pilar GJ, Olegario BR, Rafael RR (2016) Occurrence of jasmonates during cystocarp development in the red alga *Grateloupia imbricata*. J Phycol 52(6):1085–1093

Piotrowska-Niczyporuk A, Bajguz A, Kotowska U et al (2018) Growth, metabolite profile, oxidative status, and Phytohormone levels in the green alga *Acutodesmus obliquus* exposed to exogenous auxins and cytokinins. J Plant Growth Regul 37:1–16

Prabhavathi V, Rajam MV (2007) Mannitol-accumulating transgenic eggplants exhibit enhanced resistance to fungal wilts. Plant Sci 173:50–54

Quareshy M, Prusinska J, Li J et al (2017) A cheminformatics review of auxins as herbicides. J Exp Bot 69(2):265–275

Ramya SS, Vijayanand N, Rathinavel S (2015) Foliar application of liquid biofertilizer of brown alga *Stoechospermum marginatum* on growth, biochemical and yield of *Solanum melongena*. Int J Recycl Org Waste Agric 4(3):167–173

Rao DLN, Burns RG (1990) The effect of surface growth on blue-green algae and bryophytes on some microbiological, biochemical and physical soil properties. Biol Fert Soils 9:239–244

Rasoulpour R, Afsharifar A, Izadpanah K (2018) Antiviral activity of prickly pear (*Opuntia ficus-indica* (L.) Miller) extract: Opuntin B, a second antiviral protein. Crop Prot 112:1–9

Richardson JW, Johnson MD, Zhang X et al (2014) A financial assessment of two alternative cultivation systems and their contributions to algae biofuel economic viability. Algal Res 4:96–104

Richmond A (1988) Spirulina. In: Borowitzka A, Borowitzka L (eds) Microalgal biotechnology. Cambridge University Press, Cambridge, pp 83–121

Rodrigues DB, Flores ÉM, Barin JS et al (2014) Production of carotenoids from microalgae cultivated using agroindustrial wastes. Food Res Int 65:144–148

Rodriguez-Garcia A, Hosseini S, Martinez-Chapa SO et al (2017) Multi-target activities of selected alkaloids and terpenoids. Mini Rev Org Chem 14(4):272–279

Rogers SL, Burns RG (1994) Changes in aggregate stability, nutrient status, indigenous microbial populations, and seedling emergence, following inoculation of soil with *Nostoc muscorum*. Biol Fert Soils 18:209–215

Roychoudhury P, Kaushik BD, Krishnamurthy GSR et al (1979) Effect of blue-green algae and *Azolla* application on the aggregation status of the soil. Curr Sci 48:454–455

Ruggiero B, Koiwa H, Manabe Y et al (2004) Uncoupling the effects of abscisic acid on plant growth and water relations. Analysis of sto1/nced3, an abscisic acid-deficient but salt stress-tolerant mutant in Arabidopsis. Plant Physiol 136(2):3134–3147

Ryu B-G, Kim W, Heo S-W et al (2015) Advanced treatment of residual nitrogen from biologically treated coke effluent by a microalga-mediated process using volatile fatty acids (VFAs) under stepwise mixotrophic conditions. Bioresour Technol 191:488–495

Saha KC, Mandal LN (1979) Effect of algal growth on the availability of phosphorus, iron and manganese in rice soil. Plant Soil 52:139–146

Sahayaraj K, Kalidas S (2011) Evaluation of nymphicidal and ovicidal effect of seaweed, *Padinapavonica* (Linn.) (Pheophyceae) on cotton pest, *Dysdercuscingulatus* (Fab.). Indian J Geo Mar Sci 40:125–129

Saifullah M, Stephen MM, Khattak B (2007) Isolation of *Trichoderma harzianum* and *in vitro* screening for its effectiveness against root-knot nematodes (*Meloidogyne* sp.) from Swat, Pakistan. Pak J Nematol 25:313–322

Saito M, Endo K, Kobayashi K et al (2018) High myristic acid content in the cyanobacterium Cyanothece sp. PCC 8801 results from substrate specificity of lysophosphatidic acid acyltransferase. Biochim Biophys Acta Mol Cell Biol Lipids 1863(9):939–947

Sakakibara H (2006) Cytokinins: activity, biosynthesis, and translocation. Annu Rev Plant Biol 57:431–449

Sakamoto Y, Mori K, Matsuo Y et al (2017) Breeding of a new potato variety 'Nagasaki Kogane' with high eating quality, high carotenoid content, and resistance to diseases and pests. Breed Sci 67(3):320–326

Sangeetha P, Babu S, Rangasamy R (2011) Potential of green alga *Chaetomorpha litorea* (Harvey) for biogas production. Int J Curr Sci 1:24–29

Sarwar M (2015) The killer chemicals for control of agriculture insect pests: the botanical insecticides. Int J Chem Biomol Sci 1(3):123–128

Sathasivam R, Radhakrishnan R, Hashem A et al (2017) Microalgae metabolites: a rich source for food and medicine. Saudi J Biol Sci 26(4):709–722

Schoenwaelder ME (2002) The occurrence and cellular significance of physodes in brown algae. Phycologia 41(2):125–139

Seethapathy P, Jayaraman R, Palani N et al (2016) Botanicals in eco-friendly post harvest disease management. Innov Farm 1(3):67–71

Sharma R, Khokhar MK, Jat RL et al (2012) Role of algae and cyanobacteria in sustainable agriculture system. Wudpecker J Agric Res 1(9):381–388

Sharma HS, Fleming C, Selby C et al (2014) Plant biostimulants: a review on the processing of macroalgae and use of extracts for crop management to reduce abiotic and biotic stresses. J Appl Phycol 26:465–490

Shields RJ, Lupatsch I (2012) Algae for aquaculture and animal feeds. Technikfol Theorie Praxis 21:2337

Shimonaga T, Fujiwara S, Kaneko M et al (2007) Variation in storage α-polyglucans of red algae: amylose and semi-amylopectin types in Porphyridium and glycogen type in Cyanidium. Mar Biotechnol 9(2):192

Silverstone AL, Sun T-P (2000) Gibberellins and the green revolution. Trends Plant Sci 5(1):1–2

Singh DT, Nirmala K, Modi DR et al (1987) Genetic transfer of herbicide resistance gene(s) from Gloeocapsa spp. to Nostoc muscorum. Mol Gen Genet 208:436–438. https://doi.org/10.1007/BF00328135

Singh R, Parihar P, Singh M et al (2017) Uncovering potential applications of cyanobacteria and algal metabolites in biology, agriculture and medicine: current status and future prospects. Front Microbiol 8:515

Skjanes K, Rebours C, Lindblad P (2013) Potential for green microalgae to produce hydrogen, pharmaceuticals and other high value products in a combined process. Crit Rev Biotechnol 33 (2):172–215

Skriptsova A (2015) Fucoidans of brown algae: biosynthesis, localization, and physiological role in thallus. Russ J Mar Biol 41(3):145–156

Solangi SK, Qureshi ST, Solangi MK et al (2018) Estimation of auxins and cytokinins requirements in sugarcane soma clones for effective in vitro regeneration procedure. Int J Plant Physiol Biochem 10(2):10–18

Solovchenko A, Lukyanov A, Solovchenko O et al (2014) Interactive effects of salinity, high light, and nitrogen starvation on fatty acid and carotenoid profiles in *Nannochloropsis oceanica* CCALA 804. Eur J Lipid Sci Technol 116(5):635–644

Souza JM, Yokoya NS (2016) Effects of cytokinins on physiological and biochemical responses of the agar-producing red alga *Gracilaria caudata* (Gracilariales, Rhodophyta). J Appl Phycol 28 (6):3491–3499

Spolaore P, Joannis-Cassan C, Duran E et al (2006) Commercial applications of microalgae. J Biosci Bioeng 101(2):87–96

Stadnik MJ, De Freitas MB (2014) Algal polysaccharides as source of plant resistance inducers. Trop Plant Pathol 39:111–118

Stengel DB, Connan S (2015) Natural products from marine algae. Springer, New York

Stick RV (2001) Carbohydrates: the sweet molecules of life. Elsevier, Amsterdam

Stirk W, Novák O, Strnad M et al (2003) Cytokinins in macroalgae. Plant Growth Regul 41 (1):13–24

Stirk W, Bálint P, Tarkowská D et al (2013) Hormone profiles in microalgae: gibberellins and brassinosteroids. Plant Physiol Biochem 70:348–353

Subhashini D, Kaushik BD (1981) Amelioration of sodic soils with blue-green algae. Aust J Soil Res 19:361–366

Sultana V, Baloch GN, Ara J et al (2011) Seaweeds as an alternative to chemical pesticides for the management of root diseases of sunflower and tomato. J Appl Bot Food Qual 84:162–168

Sushchik NN, Kalacheva GS, Gladyshev MI (2001) Secretion of free fatty acids by prokaryotic and eukaryotic algae at optimal, supraoptimal, and suboptimal growth temperatures. Microbiology 70(5):542–547

Svircev Z (2005) Mikroalge i cijanobakterije u biotehnologiji. Prirodnomatematic˘ki fakultet, Novi Sad

Takahashi N, Phinney BO, MacMillan J (2012) Gibberellins. Springer Science & Business Media, New York

Takaichi S (2011) Carotenoids in algae: distributions, biosyntheses and functions. Mar Drugs 9 (6):1101–1118

Tarakhovskaya E, Maslov YI, Shishova M (2007) Phytohormones in algae. Russ J Plant Physiol 54 (2):163–170

Tena N, Lobo-Prieto A, Aparicio R et al (2018) Storage and preservation of fats and oils. Reference module in food science. Elsevier, Amsterdam

Trentacoste EM, Martinez AM, Zenk T (2015) The place of algae in agriculture: policies for algal biomass production. Photosynth Res 123(3):305–315

Tsai C-H, Chen Y-C, Chen L-D et al (2008) A traditional Chinese herbal antilithic formula, Wulingsan, effectively prevents the renal deposition of calcium oxalate crystal in ethylene glycol-fed rats. Urol Res 36(1):17

Tuchy L, Chowańska J, Chojnacka K (2013) Seaweed extracts as biostimulants of plant growth: review. Chemik 67(7):636–641

Tvrzicka E, Kremmyda L-S, Stankova B et al (2011) Fatty acids as biocompounds: their role in human metabolism, health and disease-a review. Part 1: classification, dietary sources and biological functions. Biomed Pap 155(2):117–30

Uji T, Matsuda R, Takechi K et al (2016) Ethylene regulation of sexual reproduction in the marine red alga *Pyropia yezoensis* (Rhodophyta). J Appl Phycol 28(6):3501–3509

Valerio M, Lovelli S, Sofo A et al (2017) Root and leaf abscisic acid concentration impact on gas exchange in tomato (*Lycopersicon esculentum* Mill) plants subjected to partial root-zone drying. Ital J Agron 12(1):25–32

Vallim MA, De Paula JC, Pereira RC et al (2005) The diterpenes from Dictyotacean marine brown algae in the tropical Atlantic American region. Biochem Syst Ecol 33(1):1–16

Vaskovsky V, Smirnova G, Shashkov A et al (2015) Polysaccharides of algae 67. Carrageenan from Pacific red alga *Turnerella mertensiana* (Gigartinales, Rhodophyta). Russ Chem Bull 64 (5):1163–1167

VegaAlga—Result In Brief (2018) Algae-based fertiliser turns vegetable farming green. https://cordis.europa.eu/result/rcn/227612_en.html. Accessed 11 Aug 2018

Veillette M, Giroir-Fendler A, Faucheux N et al (2017) Esterification of free fatty acids with methanol to biodiesel using heterogeneous catalysts: from model acid oil to microalgae lipids. Chem Eng J 308:101–109
Venkataraman GS (1978) Indian experience with algal ponds. In: Conference on the State of the Art of Bioconversion of Organic Residues for Rural Communities. Guatemala City (Guatemala), 13 November 1978
Vonshak A (1986) Laboratory techniques for the culturing of microalgae. In: Richmond A (ed) Handbook for microalgal mass culture. CRC, Boca Raton, pp 117–145
Wang S, Zhong FD, Zhang YJ et al (2004) Molecular characterization of a new lectin from the marine algae *Ulva pertusa*. Acta Biochim Biophys Sin 36:111–117
Wang L, Yao Y, Sang W et al (2015) Structural features and immunostimulating effects of three acidic polysaccharides isolated from *Panax quinquefolius*. Int J Biol Macromol 80:77–86
Whitehead LF, Douglas AE (2003) Metabolite comparisons and the identity of nutrients translocated from symbiotic algae to an animal host. J Exp Biol 206(18):3149–3157
Whitton BA (2000) Soils and rice-fields. In: Whitton BA, Potts M (eds) The ecology of cyanobacteria: their diversity in time and space. Springer, Cham, pp 233–255
Win TT, Barone GD, Secundo F et al (2018) Algal biofertilizers and plant growth stimulants for sustainable agriculture. Ind Biotechnol 14(4):203–211
Woodson WR (2018) Pollination signals and flower senescence. Annu Plant Rev 6:294–311
Wu Y, Jenkins T, Blunden G et al (1998) Suppression of fecundity of the root-knot nematode, *Meloidogyne javanica*, in monoxenic cultures of *Arabidopsis thaliana* treated with an alkaline extract of *Ascophyllum nodosum*. J Appl Phycol 10:91–94
Yang Z, Liu J, Tischer SV et al (2016) Leveraging abscisic acid receptors for efficient water use in Arabidopsis. Proc Natl Acad Sci 113(24):6791–6796
Yeong TK, Jiao K, Zeng X et al (2018) Microalgae for biobutanol production – technology evaluation and value proposition. Algal Res 31:367–376
Yokoya NS, Stirk WA, Van Staden J et al (2010) Endogenous cytokinins, auxins, and abscisic acid in red algae from Brazil. J Phycol 46(6):1198–1205
Yoshida K, Igarashi E, Mukai M et al (2003) Induction of tolerance to oxidative stress in the green alga, *Chlamydomonas reinhardtii*, by abscisic acid. Plant Cell Environ 26(3):451–457
Zbakh H, Chiheb H, Bouziane H et al (2012) Antibacterial activity of benthic marine algae extracts from the Mediterranean coast of Morocco. J Microbiol Biotechnol Food Sci 2(1):219
Zhang H, Huang D, Cramer WA (1999) Stoichiometrically bound β-carotene in the Cytochromeb6f complex of oxygenic photosynthesis protects against oxygen damage. J Biol Chem 274 (3):1581–1587
Zhang C, Tanabe K, Tamura F et al (2007) Roles of gibberellins in increasing sink demand in Japanese pear fruit during rapid fruit growth. Plant Growth Regul 52(2):161
Zhao L, Feng C, Wu K et al (2017) Advances and prospects in biogenic substances against plant virus: a review. Pest Biochem Physiol 135:15–26
Zhekisheva M, Zarka A, Khozin-Goldberg I et al (2005) Inhibition of astaxanthin synthesis under high irradiance does not abolish triacylglycerol accumulation in the green alga Haematococcus pluvialis (Chlorophyceae). J Phycol 41(4):819–826

Chapter 15
Deriving Economic Value from Metabolites in Cyanobacteria

Carole A. Llewellyn, Rahul Vijay Kapoore, Robert W. Lovitt, Carolyn Greig, Claudio Fuentes-Grünewald, and Bethan Kultschar

Abstract This chapter focuses on the challenges associated with achieving economic value from metabolites derived from cyanobacteria. Significant advances have been made in cyanobacterial biotechnology in the last few years. However, the field is still immature, and many challenges remain. We start with a critical overview of the main technologies associated with cultivation, cell disruption and metabolite extraction. Then, we provide an overview of current significant metabolite groups from cyanobacteria relevant to industry covering phycobilins, carotenoids, polysaccharides, peptides, lipids, mycosporine-like amino acids, polyhydroxyalkanoates, cyanotoxins and platform chemicals, and the potential for stable isotopes production. We cover metabolites that are already in the market and those with future potential with a focus on spirulina (*Arthrospira*) the most commercially developed species of cyanobacteria. As large-scale cultivation and down-stream processing techniques continue to develop further, combining this with a systems biology and biorefinery approach will ensure that the best economic and environmental sustainability value can be achieved.

15.1 Introduction

Cyanobacteria, as biological cell factories, are becoming increasingly attractive to industry. This is driven by society's need to move away from petrochemical carbon-based chemicals towards a low-carbon bioeconomy. As large chemical companies increasingly move to biotechnology-based production processes cyanobacteria are proving to be attractive bio-based candidates adding to the repertoire of current workhorse bacteria and yeast microorganisms. Considering that ancient deposits of

---

C. A. Llewellyn (✉) · R. V. Kapoore · C. Greig · C. Fuentes-Grünewald · B. Kultschar
Department of Biosciences, College of Science, Swansea University, Swansea, UK
e-mail: c.a.llewellyn@swansea.ac.uk

R. W. Lovitt
College of Engineering, Swansea University, Swansea, UK

A. Hallmann, P. H. Rampelotto (eds.), *Grand Challenges in Algae Biotechnology*, Grand Challenges in Biology and Biotechnology,
https://doi.org/10.1007/978-3-030-25233-5_15

cyanobacteria likely formed the earth's oil deposits used in petrochemical-based oil refinery it is perhaps surprising that the development of cyanobacteria in biotechnology has not occurred earlier. A diverse range of markets for algal and cyanobacterial metabolites are rapidly emerging. Ultimately, the replacement of the petrochemical-based refinery with an alternative renewable and sustainable biorefinery approach including with the use of cyanobacteria could contribute towards an urgently needed low-carbon economy.

As photosynthetic microorganisms, cyanobacteria bring advantage compared to current biotechnological microorganisms by having the ability to convert CO_2, using light energy, into organic carbon-based metabolites. Like prokaryotic cells, they bring advantage compared to eukaryotic algae having simple growth requirements and in the ease with which they can be genetically engineered. Significant advances have been made in cyanobacterial biotechnology research in the last few years. There are already reviews covering industrially useful metabolites from cyanobacteria; covering metabolic engineering of cyanobacteria (Johnson et al. 2016; Kanno et al. 2017; Liu et al. 2016; Luan and Lu 2018; Xue and He 2015), and the potential as a chassis in industrial biotechnology (Al-Haj et al. 2016). However, the cyanobacteria biotechnology field is still immature and challenges remain.

Although difficult to ascertain, it is estimated that there are up to 8000 species of cyanobacteria, making up approximately 25% of our planet's primary production (Guiry 2012). Inhabiting a wide diversity of environments from marine, brackish, freshwater and soil environments they range from the tiny unicellular picocyanobacteria (<2 μm) *Synechococcus* and *Prochlorococcus* that dominate in oligotrophic regions of the global ocean to the long filamentous cyanobacteria, e.g. *Oscillatoria*, *Anabaena* and *Arthrospira platensis* (spirulina) found in lakes and in arid regions of desert.

Cyanobacteria have evolved over approximately 2.5 billion years shaping evolution and ecological change (Shih 2015). The evolutionary success of cyanobacteria is linked to their ability to survive under a wide range of conditions including high UV, salinity, pH and extreme temperatures (Nabout et al. 2013). They can also survive under low nitrogen conditions and play an important role in their ability to fix nitrogen. The survival of cyanobacteria and their ability to counteract adverse conditions and damaging stress is enabled by cyanobacteria being able to adjust their metabolism and metabolite composition. This has resulted in cyanobacteria containing a diverse range of metabolites, many of which are unique.

Of all cyanobacteria, the aquatic filamentous cyanobacteria species known as spirulina (*A. platensis* and *A. maxima*) has been highly commercially successful and is currently one of the most widely marketed superfoods. Spirulina is cultivated on a large scale in the USA, China and Europe. Currently, spirulina is primarily marketed in tablet, capsules, flakes, powder and a liquid form for use as a dietary supplement and as a whole food. Spirulina is also used as a feed supplement in aquarium, aquaculture and poultry industries. The global spirulina market is expected to register a CAGR (Compound Annual Growth Rate) of 10% during the forecast period and is expected to grow from an estimated US$700 million in 2016 to more than US$2000 million by 2026. Moreover, the EU government regulations have become stringent and have banned the usage of synthetic food colours in the food

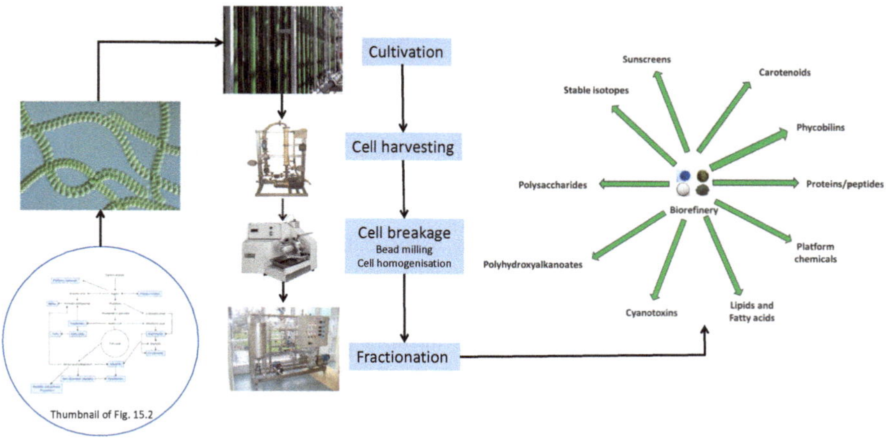

Fig. 15.1 Schematic summarising of key processes to producing metabolites from cyanobacteria. Cell metabolism and metabolic network pathways (inset from Fig. 15.2) need optimisation/engineering to improve metabolite yield. Large-scale cultivation needs optimising to maximise production and avoid contamination. Downstream processing involving cell harvesting and cell breakage needs optimising for economically viable and sustainable large-scale production. Extracts need fractionation enabling a range of products (biorefinery) thus maximising the economic value of the whole process

industry, which is expected to boost the sales of spirulina-based natural food colours (Research 2017).

This chapter addresses key processes and economic challenges associated with producing metabolites from cyanobacteria. A summary of the processes leading to fractionation of metabolites forming the basis of this chapter is summarised in Fig. 15.1. We begin the chapter with a brief overview on cultivation of cyanobacteria and costings associated with raceway production of spirulina. We follow this with a critical review of downstream processing covering large-scale applicable technologies associated with cell disruption to release metabolites and metabolite separation. We move on to discuss key groups of metabolites with commercial value with a focus on spirulina. We discuss the importance of taking a holistic systems biology approach and conclude with a discussion on the value of biorefinery to maximise the range of extracted metabolites.

15.2 Up and Downstream Processing of Cyanobacteria to Obtain Metabolites

15.2.1 Upstream Challenges and Techno-economics

Cultivation systems from basic raceways to high-tech efficient glass photobioreactors (PBRs) have improved in the last decade (Ajayan et al. 2012; Anyanwu et al. 2018). The amount of biomass that can be produced varies from a few grams per litre in raceways to several grams in PBRs. The cultivation system chosen depends on the quality of the biomass required and on the final product/ metabolites to be produced. Detailed up-to-date information regarding the technical specification and production costs of a spirulina raceway plant located in Europe is shown in Tables 15.1 and 15.2, respectively. Using this data for 4 hectares (ha) located in Southern Europe (e.g. Spain), 60 tons/ha/year of spirulina (Table 15.1) can be produced with a production cost of 4 €/kg of dry biomass (Table 15.2). However, these costings assume that the main consumables used during the cultivation process such as fertilisers, CO_2 and water are derived from waste from other industries and therefore have no costs associated with them. These costings compare favourably to several authors and private companies who have highlighted that to be profitable in terms of economic cost, a cultivation facility must produce dry biomass with a cost below US$5/kg at the current exchange rate. The main expenditure in terms of initial investment is equipment purchased (for upstream and downstream process) and the construction expenses, accounting both nearly a 50% of the total CAPEX (Table 15.2).

In assigning appropriate large-scale cultivation system, there are many considerations to be made to maximise production of biomass. An adequate supply of carbon dioxide and control of pH are critical as are the correct balance of nutrients (El-Sayed and El-Sheekh 2018). These need to be addressed alongside consideration of adequate light supply and appropriate temperature. A key challenge associated with maintaining large-scale production and ensuring economic viability is the minimisation of risks associated with contamination from bacteria, viruses, fungi and protozoan predators (Wang et al. 2013). Spirulina has developed as a major industrial strain largely because of its resistance to contamination. Research into the development of the most robust and contamination-resistant strains is required to promote the commercial development of cyanobacteria.

15.2.2 Downstream Processing Challenges

15.2.2.1 Cell Concentration

The first challenge to overcome in deriving economic value from cyanobacteria once they have been cultivated is concentrating up the typically low concentration of cells. Separating cells from water is a major issue as it is expensive and poses a major challenge in obtaining high-value materials (Fasaei et al. 2018). Typically, from autotrophically grown cultures, cell concentrations above 2 g/L are rare as cells become light limited at higher concentrations. These relatively dilute cell suspensions must first be concentrated to a pumpable viscosity (100–200 g/L) to reduce the volume being handled and increase the value of the material (Table 15.3). Once the cells are in a concentrated form there are many developed techniques in other biotechnological processes such as flocculation, centrifugation and microfiltration that can be applied (‘t Lam et al. 2018). Up to 80% of total production costs can fall on the downstream processing (DSP) operations (Davis et al. 2011; Walsh 2010).

Table 15.1 Parameter values associated with a 4-ha raceway cultivating spirulina

Technical parameters	Value
Biomass productivity	16 g/m^2/day
CO_2 usage	9.15 Kg/kg of algae biomass
Water evaporation	10 L/m^2/day
Mixing power consumption	1 W/m^3
Labour	4 People/ha
Production days	300 days
Land area	4 ha
Volume/Surface area ratio	0.25 m^3/m^2
CO_2 fixation efficiency	0.2
Dilution rate (due to evaporation)	L/day
Total culture volume	1,000,000 m^3
Total biomass production	60 ton/ha/year
Total CO_2 consumption	549 ton/ha/year
Total water evaporation	30,000 ton/ha/year
Water cost	0 €/kg
CO_2 cost	0 €/kg
Nutrients cost	0 €/kg
Fertilisers usage	0.30 Kg/kg algae biomass
Power cost	0.10 €/kWh
Power for harvesting and others	0.11 kWh/m^3 harvest
Photobioreactor cost	10 €/m^3
Equipment capacity	
Medium preparation unit	20 m^3/h
Sterilisation process	0 m^3/h
Air blower	0 m^3/h
Photobioreactors	10 m^3
Sedimenter	20 m^3/h
Harvest storage tank	2 m^3
Decanter	2 m^3/h
Harvest pump	2 m^3/h
CO_2 supply unit	586 Kg/h

Important factors governing separation are physical properties of the cells (mechanical properties, density, size, morphology) and the suspending growth medium (i.e. ionic strength, pH and viscosity) (Vandamme et al. 2013). These main factors are described by the Stokes equation:

$$v = d^2 \frac{(\rho_p - \rho_f)}{18\mu} g$$

where v is the particle velocity (m/s), g is the gravitational acceleration (m/s^2), d is the diameter of the spherical particle, ρ_p is the mass density of the particles (kg/m^3),

Table 15.2 Production costs associated with a 4-ha raceway cultivating spirulina

Equipment Costs	€
Medium preparation unit	10,815
Sterilisation process	–
Air blower	–
Photobioreactors	72,757
Sedimenter	20,000
Harvest storage tank	600
Decanter	40,000
Harvest pump	1000
CO_2 supply unit	13,477
Fix Capital Costs (CAPEX)	
Major purchased equipment	158,649
Installation costs	31,729
Instrumentation and control	31,729
Piping	47,594
Electrical	15,864
Buildings	15,864
Land improvements	7932
Service facilities	31,729
Land	–
Engineering and supervision	31,729
Construction expenses	102,328
Contractors fee	10,232
Contingency	33,977
Total fix capital	519,363
Fix Capital Costs per year (Assuming lifetime of 10 years)	
Depreciation	51,143
Property tax (@ 0.01 depreciation)	511
Insurance (@ 0.006 depreciation)	306
Purchase tax (@ 0.16 of items 1–12/10)	8309
Direct Production Costs (OPEX)	
Fertilisers (kg)	–
Water (m^3)	–
Carbon dioxide (kg)	–
Utilities	–
Water (m^3)	–
Power mixing (kWh)	7200
Power harvesting and others (kWh)	65
Labour and others	
Labour	480,000
Supervision (@ 0.2 labour)	19,200
Payroll charges (@ 0.25 (labour + supervision))	31,200
Maintenance (@ 0.04 MEC)	253

(continued)

Table 15.2 (continued)

Equipment Costs	€
Operating supplies (@ 0.004 items 1–5)	4
General plant overheads (@ 0.55 (labour + supervision + maintenance)	151,084
Tax (@ 0.16 items 1–7, 11 and 12)	6
Contingency (@ 0.05 items 1–7)	71,392
Marketing (@ 0.05 items 1–13)	1298
Total utilities	7265
Total labour and others	683,762
Total fix capital per year	60,271
Total direct production costs	691,027
Total production costs	751,298
The unit cost (per kg dry weight) of producing biomass	4

ρ_f is the mass density of the fluid (kg/m^3) and μ is the dynamic viscosity (kg/m*s) of the fluid. The particle velocity (v) will be vertically, downwards if $\rho_p > \rho_f$, upwards if $\rho_p < \rho_f$.

Several operations, as shown in Table 15.3, can be employed to carry out the concentration/dewatering process; the choice is governed by properties of the culture (cells and fluid) and by the expense of the processing in both capital and running costs. The most common is the use of flocculation/flotation, centrifugation or filtration. Flocculation uses chemical (or electrochemical) additives, which causes the particles to agglomerate, increasing their size and settling rate (Vandamme et al. 2013). Some species are autofloculating and avoid the need for flocculants addition and this often occurs in nutrient-deficient conditions or at extremes of pH. In some situations, the application of combined operations in a hybrid process is also advantageous, e.g. floatation and flocculation, flocculation and centrifugation (Fasaei et al. 2018).

Floatation can also be considered, as with flocculation additives are required to improve efficiency. This technique is used effectively in water treatment systems, however, in cell harvesting the main disadvantage is the contamination of the concentrated biomass, which results in potential devaluation. Typical materials used are inorganic salts of iron and aluminium and charged polymers include chitosan and polyacrylamide. Their potential is also limited in saline environments where high salt interferes with the colloidal properties such as the interaction between the cells and floc chemicals (Grima et al. 2003).

Centrifugation is another alternative. Within the centrifuge, high gravitational forces are driven by the density and size differences of the particle and the suspending media and so works well with dense (<1.05 g/mL) large cells <10 μm. This can, therefore, be applied to most algae. Small algae containing large quantities of oil (therefore relatively low density) give slurries and pastes up to 200 g/L (Milledge and Heaven 2011). The high shear forces associated with centrifugation can also damage the cells especially delicate structures such as flagella. They may also be used in conjunction with flocculation to enhance the

Table 15.3 Cell concentration operations

Operation	Separation mechanism	Desired dry conc. (g/L)	Energy requirement	Advantages	Disadvantages	Comments	Refs.
Sedimentation	Gravity, density, particle size and suspending medium	120–180	Virtually zero kWh/m^3	Simple and low capital investment	Slow, dependent on species and physiological status	Widely used in wastewater industry, slow and does not work well with small organisms <5 μ	Fasaei et al. (2018)
Flocculation	Gravity, density, particle size and suspending medium	100–150	Mixing power	Simple, can be rapid and low capital investment	Contamination of biomass, does not work well in salty or brackish environments, dependent on species and physiological status	Widely used in water industry, product contamination, cost of flocculants is significant at a low cell concentration	Slade and Bauen (2013) and Vandamme et al. (2013)
Electro-flocculation	Gravity, density, charge size and suspending medium	100–150	Power consumption and mixing	Use of electrode metal as a flocculant	Contamination of biomass with metals	Applied on a large scale for clean water production, potential product contamination, solution property dependent	Fasaei et al. (2018)
Flotation	Buoyancy, size charge and density	100–150	Mixing energy, 5–20 kWh/m^3	Relatively fast	Contamination with enhancement chemicals (flocs) and compressed air generation	Applied on a large scale in clean water processing, potential product contamination, solution property dependent	Fasaei et al. (2018) and Kim et al. (2005)

Centrifugation	Gravity field, density and particle size	150–250	High power for driving centrifuge, 25–50 kWh/m^3	Well-tried technology, work well with large cells (<10 μ) and widely applied	Safety and maintenance, not feasible for small cells and the density of medium a factor often requires preconcentration	Commonly used in a wide range of industries, especially oil–water separations, high capital investment and maintenance costs, high cost for the small organism at low concentrations	Dassey and Theegala (2013) and Milledge and Heaven (2011)
Cross-flow filtration	Pressure difference, particle size, charge and rheology	100–150	Pump energy, 5–20 kWh/m^3	Potential for sterile operation allowing water reformulation and reuse can be relatively low energy and rapid	Requires careful management, cleaning procedures and recycling of filters	Cheap, safe operation but can be problematic in high-viscosity systems	Gerardo et al. (2014) and Slade and Bauen (2013)
Spray drying	Evaporation and suspending medium	900–950	Heating and air compression, require preconcentration to 100 g/L, 600–800 kWh/m^3	High concentrations of materials required to minimise heat losses	Relatively large-scale operation required and loss of materials through heat treatment	Large-scale operation, high capital costs, potential explosive hazards	Slade and Bauen (2013)
Freeze drying	Sublimation and suspending medium	950–980	Cooling, vacuum, 200–300 kWh/m^3	High concentrations of materials required	Relatively low productivity, large-scale operation and scale-up is problematic	High capital and running costs	Bastiaens et al. (2017) and 't Lam et al. (2018)

efficiency of the separation process. The most common systems are based on disc stack centrifuges that can be emptied in situ.

Filtration systems potentially offer the best solution as it not only allows concentration of cells to about 120–150 g/L dry weight but also allows the possibility of sterilisation of water for reuse, a major consideration for reducing costs (Gerardo et al. 2014). Processes based on cross-flow filtration can be used on a large scale and is simple and safe with low capital and energy costs compared to centrifugation. The main problem with membranes is that membrane fouling needs to be managed correctly to ensure high membrane productivity and longevity to reduce operational and capital costs substantially. Indeed, economic cleaning and maintenance of membranes are critical (Sioutopoulos and Karabelas 2012).

In general, all of these processes are capable of concentrating cells to high viscous fluids or pastes up to about 25% solids. Above this level, the material becomes very difficult to handle as they cannot be pumped easily because of their non-Newtonian fluid behaviour, which requires specialist pumping and mixing (Gerardo et al. 2014). In drying processes, powders are typically obtained after cell washing, and in some recovery processes, it is a prerequisite for subsequent extraction (i.e. Supercritical Fluid Extraction) (Bastiaens et al. 2017).

15.2.2.2 Cell Disruption

Disrupting cyanobacterial cells to release metabolites requires careful consideration in the process to deriving economic value. Perhaps the largest advance in algal cell biorefining would be an effective low-energy cell disruption method. The algal cell wall is a very strong structure that can be affected by environmental and physiological conditions (Safi et al. 2014a, b). Cell wall composition and strength varies considerably between different groups and species of microalgae and cyanobacteria; some species are so resilient that this acts as a barrier to their commercial exploitation (Burczyk and Dworzanski 1988). Disruption techniques appropriate to the species (Alhattab et al. 2019) and mild enough to ensure effective product recovery (Phong et al. 2018) are a key objective for the economic development of microalgal metabolites. Traditional scalable methods, reviewed in Table 15.4, cover freeze–thaw and liquid shear devices such as high-pressure homogenisers and bead milling (Gunerken et al. 2015). Other potential methods, also reviewed in Table 15.4, are still in their infancy such as microwave, colloid grinding of dry powders, sonication or the use of enzymes to dissolve or weaken the cell wall (Bleakley and Hayes 2017; Gunerken et al. 2015; Kuddus et al. 2013; ‘t Lam et al. 2018). A plethora of methods have been used to disrupt wet spirulina cells for the extraction of phycocyanin, including freeze–thaw, mechanical cell disruption, sonication and enzymatic methods (Eriksen 2008; Sekar and Chandramohan 2008; Vernès et al. 2015).

Most physical disruption methods use large quantities of energy to achieve disruption (Table 15.4). In this situation, mechanical force is applied to disrupt the cell walls. This can take the form of phase change and crystal formation (freeze–thaw), liquid shear stress imparted liquid high-pressure homogenisers (Naghdi et al.

Table 15.4 Cell disruption mechanisms

Technique	Disruption mechanism	Energy requirement	Advantages	Disadvantages	Comments	Refs.
Freeze–thaw	Ice crystals cause cell disruption	90–100 (kWh)/t	Simple	Phase change, low capital cost, high energy cost	Simple technology, easily achieved	Gunerken et al. (2015)
Homogenisation	Liquid shear and pressure shock	50–150 (kWh)/t	Scalable	Heating, high capital cost	Highly dependent on morphology and structure of a cell	Naghdi et al. (2016) and Safi et al. (2014a)
Bead milling	Liquid shear	40–80 (kWh)/t	High yield, scalable	Heating, use of organic solvents	Very dependent on morphology and structure of a cell	Gunerken et al. (2015)
Grinding of dry powders	Ridged dry structures become fragile or disrupted	10 (kWh)/t	Relative rapid using colloid mills	Have to dry material first	Very dependent on morphology and structure of a cell	D'Hondt et al. (2017)
Sonication	Differential pressures causing cavitation	High	No additions	Hard to scale-up, heating and possible oxidation	Very good laboratory or monitoring process, can be used as complementary other disruption processes	Safi et al. (2014b) and Sarada et al. (1999)
Microwave	Heating of water molecules	Low	Effective disruption, short reaction times, reduced solvent usage	Heating	Widely used in cooking processes	Kapoore et al. (2018)
Enzymatic/chemical treatments	Digestion of walls causing loss of strength	Negligible mixing required	Relative low energy	Cost of enzyme, disposal of chemical waste, requires specific enzymes that can cause contamination of the final product	Very dependent on morphology and structure of a cell	Gunerken et al. (2015)
Osmotic shock	Walls disruption caused by a rapid change of pressure	Low	Pressure difference caused by high salts	Removal of contaminating materials	Very dependent on morphology and structure of a cell	Gunerken et al. (2015)

2016; Safi et al. 2014a, b), bead milling, sonication and mechanical grinding (bead milling of cell powders). The downside to these mechanical methods is the amount of energy needed to disrupt the cell wall and is typically associated with heating of the biomass, which potentially increases denaturation and degradation of products (Phong et al. 2018).

The addition of chemicals such as solvents or chaotropic agents can achieve disruption but the disrupting substances would then need to be removed (Bastiaens et al. 2017; Grima et al. 2003; Gunerken et al. 2015). Osmotic shock may also be used for some materials such as wall-associated polysaccharides (Gunerken et al. 2015). More recently, microwave and pulsed electrical fields (PEF) have been employed in the laboratory (Kapoore 2014; Kapoore et al. 2018), however, these methods are still being developed for large-scale operation so as to avoid heating and associated degradation. Microwaves have been used for obtaining a variety of bioactive compounds from a range of microalgal classes, where microwave generates high-frequency waves [ranging from 300 MHz (100 cm) to 300 GHz (0.1 cm)] with wavelengths of 0.001–1 m. This results in the vibration of water and other polar molecules within wet biomass, thereby resulting in temperature increases in the intracellular liquids, which subsequently causes the water to evaporate and exert pressure on the cell walls leading to cell disruption (Kapoore et al. 2018). In contrast, a PEF is a non-thermal process hence more energy efficient (and, therefore, have lower running costs) than the more traditional thermal processes. In PEF, a number of high electric field pulses are applied across biomass, where the electric field strength (typically 10–60 kV/cm) and the number of pulses required are dependent on the biomass and product of interest characteristics.

15.2.2.3 Recovery and Purification of Metabolites

Once the cells have been disrupted, the next stage is the recovery of the components; a holistic approach needs to be taken to allow full exploitation of the metabolites. The properties of disrupted cell materials are similar to other microbial disruptates apart from the fact that the proportions of membrane and wall-associated materials, proteins, pigments, etc. are higher than that encountered with yeast, fungi or bacteria. This complicates the separations as complex stable emulsions can be formed. One key consideration, prior to disruption, is the suspending medium, as it is often beneficial to wash the cells to remove potential contaminants that interfere with the separations (pH, ionic strength, salts, metals, etc.), which are dominated by the surface/colloidal properties of the materials such as hydrophobicity and charge and medium composition such as ionic strength and pH. Thus, the major challenge in recovery systems is to simplify these processes to few high-resolution processes so reducing the overall costs ('t Lam et al. 2018). Much of the work in this area has concentrated on spirulina and purification of phycobiliproteins and other antioxidant materials (Catchpole et al. 2018; Goto et al. 2015; Machmudah et al. 2018; Mendes et al. 2003; Sahena et al. 2009). Key operations are summarised in Table 15.5.

Table 15.5 Major unit operations used for large-scale purification of metabolites

Technique	Separation mechanism	Energy requirement	Advantages	Disadvantages	Comments	Refs.
Membrane Pressure-driven systems	Size and charge	5–20 kWh/t	Scalable, potential for sterile operation	Membrane fouling and cleaning	Well-developed technology	Chaiklahan et al. (2011) and Chakdar and Pabbi (2016)
Electro-dialysis		20 kWh/t		Membrane longevity		
Forward osmosis		5 kWh/t		Use of cleaning chemicals		
Absorption	Surface charge and hydrophobicity		Scalable	Expensive	Well-developed technology, but absorbent can be expensive to produce and maintain	Bermejo et al. (2006)
Salt precipitation	Protein surface charge	Energy required to make concentrated salt	Scalable	Large quantities of pure salt that need to be recycled	General protein purification techniques, potential environmental effects if not recycled	Bastiaens et al. (2017)
Supercritical CO_2	Solubility	Energy for compressing gas	Scalable, able to extract non-water-soluble materials, aided by mechanical cell disruption	Expensive, safety considerations concerning gas release, complex high-pressure equipment, Preparations have to be low moisture, i.e. dried materials	A good technology compatible with food, careful consideration of materials in conjunction with its value and preprocessing requirements	Catchpole et al. (2018), Goto et al. (2015), Machmudah et al. (2018), Mendes et al. (2003) and Sahena et al. (2009)
Solvent extraction	Solubility	Energy for solvent recovery (distillation)	Scalable, high yields, relatively low cost	Toxic, emulsion formation, limited by the type of solvent used, increased cost due to process flammability regulations	Limited use in modern food and cosmetic applications due to safety and regulatory aspects of products	Borowitzka (2013), Lee et al. (2010) and Mubarak et al. (2015)

(continued)

Table 15.5 (continued)

Technique	Separation mechanism	Energy requirement	Advantages	Disadvantages	Comments	Refs.
Chromatography	Surface charge and hydrophobicity		Scalable with difficulty	Possible use of solvents	Slow, limited to high-end, high-purity systems	Bermejo et al. (2006)
Electrophoresis	Charge and size	Electrical current	Not easily scalable	Not easily scalable, fluid system difficult to replicate in large scale	Powerful general laboratory techniques for analysis, a good analytical method which has not succeeded in a large operation	Kamble et al. (2018)

Membranes One of the most attractive techniques for the recovery and purification of metabolites is the use of pressure-driven membrane separation. These are classified on the basis of the size of the particles to be filtered. Microfiltration has been widely applied for cell harvesting, exopolysaccharide separation from whole cells, water reuse as well as clarification of disrupted cells to obtain protein, pigment solutions and metabolites while enriching the membrane fraction/wall fractions. A range of ultrafiltration membranes (Molecular weight cut off 2000–1,000,000 Da) have been used to concentrate and purify proteins, peptides and polysaccharides using size and charge selectivity in conjunction with dialysis, while nanofiltration can separate metabolic intermediates (MW $>$ 150 Da) from the water. Finally, reverse osmosis ($<$150 Da) can be used to purify water and to concentrate salt solutions allowing the possible reconstitution and formulation of growth media (Gerardo et al. 2013, 2014; Silkina et al. 2017). There are many advantages to using these methods as they are typically based on physicality requiring few additions (i.e. pH adjustment) and can be carried out safely with relatively low energy requirements. The critical area for successful application of membranes is in their management through careful cleaning and avoidance of fouling with appropriate pretreatments (such as prefiltration). These operations are ideally suited to relatively low concentrations of material where rheological problems are avoided (Gerardo et al. 2013).

Salt Precipitation This is a common low-cost technique for the purification of proteins. Typically, highly soluble salts (e.g. NH_4SO_4) are used in concentrated solutions. The advantage of this process is that it is relatively simple and if the precipitation is good, no further steps are required (i.e. centrifugation to enhance sedimentation). Once purified, the salts are removed by dialysis (Bastiaens et al. 2017).

Chromatography and Adsorption There are numerous methods and types of chromatography such as ion exchange, surface adsorption, partition and size exclusion. The mechanism used for separation mainly depends on the molecular characteristics and the interaction type. The most useful are based on ion exchange using salt or pH elution. Reverse-phase chromatography can also be used and is especially useful with low-molecular-weight metabolites. Here, the sorbent materials interact with the metabolite through hydrophobic interaction and then are altered by changing the polarity of the solvent by additions of a simple solvent such as methanol or acetonitrile. The major problem with this approach is the reuse or disposal of solvents or salts (Chakdar and Pabbi 2016). Surface adsorption chromatography is based upon adsorption of materials into a solid phase containing an active surface. These are based on both natural and synthetic materials, which are engineering to bind specific chemicals. The resolving power of chromatographic and adsorption methods in many situations outweighs the costs and productivity of such processes. The costs are largely associated with the packing materials, which can be extremely expensive and designed for pharmaceutical operations in many cases. Perhaps the most useful is expanded bed chromatography and similar variants, which are applicable to large-scale operation (Bermejo et al. 2006). The other major costs are the extensive use of solvents and high purity water when used on a production scale. The productivity of such systems is typically low compared to membrane technology. The whey protein industry is a good example of large-scale protein purification.

Supercritical CO_2 Extraction This technique requires special equipment and is currently quite expensive to manufacture, especially for lab scale where techniques are initially developed. Carbon dioxide specifically behaves as a supercritical fluid above its critical temperature (31.1 °C) and critical pressure (7.39 MPa), expanding to fill its container like gas but with a density like that of a liquid.

Supercritical CO_2 is becoming an important commercial and industrial solvent due to its role in chemical extraction in addition to its low toxicity and environmental impact (Goto et al. 2015). The relatively low temperature of the process and the stability of CO_2 allows most compounds to be extracted with little damage or denaturing. In addition, the solubility of many extracted compounds in CO_2 varies with pressure, permitting selective extractions. The extractability, selectivity and the solubility can be controlled by manipulating algal cells (level of disruption) and the conditions (pressure and temperature). These are often materials and substance specific (Goto et al. 2015). The optimum will depend on the purpose of the extraction. For an analytical extraction to determine, say, the antioxidant content of a polymer, then the essential factors are complete extraction in the shortest time. However, for the production of an essential oil extract from a plant, then a quantity of CO_2 used will be a significant cost, and 'complete' extraction not required, a yield of 70–80% perhaps being sufficient to provide economic returns. In another case, the selectivity may be more important, and a reduced rate of extraction will be preferable if it provides greater discrimination and selectivity (Catchpole et al. 2018). This technique is widely employed so far for the extraction of functional compounds from cyanobacterial biomass (Mallikarjun Gouda et al. 2015; Montero et al. 2005; Yang 2016).

15.3 Metabolites

Recovered metabolites need to be in sufficiently high concentration and in pure enough quantity. Furthermore, any cyanobacterial-derived product has either to compete on price or to differentiate itself from the alternative source in the market place. Fundamentally, there are two overarching factors that determine the economic viability of metabolites from cyanobacteria; firstly, there needs to be a market for the metabolite, i.e. a market pull, and secondly, the metabolite needs to be produced at a competitive price.

There is a wide range of useful and potentially useful organic carbon-based primary and secondary metabolites produced from cyanobacteria. Many of these are synthesised through complex metabolic and sometimes unique networks of biosynthetic pathways (Lau et al. 2015; Liu et al. 2016; Rastogi and Sinha 2009; Singh et al. 2017). These metabolites have applications in a wide range of industrial sectors including for bulk or commodity chemicals, speciality chemicals, consumer care products and life science products. Examples are for the biofuel, pharmaceutical, nutraceuticals, healthcare, feed and food industries. Primary metabolites such as bulk proteins, carbohydrates and lipids are most suited for use as commodity chemicals, biofuels and the food and feed industry. Secondary metabolites are reported to have industrial and medical uses including as colourings, antioxidants, anti-inflammatories, anticancer agents, antimicrobials, antivirals and antimalarials. Secondary metabolites produced in cyanobacteria can be classified according to their biosynthetic origin covering non-ribosomal peptides, polyketides, ribosomal peptides, alkaloids, isoprenoids and fatty acids or according to their function such as toxicity, photoprotection and antioxidant activity (Kultschar and Llewellyn 2018).

The chemical composition of cyanobacteria is not an intrinsic constant factor but varies over a wide range, both depending on species and on cultivation conditions. It is possible to accumulate the desired high-value products in microalgae to a large extent by changing environmental factors like temperature, illumination, pH, CO_2 supply, salt and nutrients. With an understanding of cell physiology and metabolic pathways, cells can be engineered to increase the production of secondary metabolites, thus improving yields and lowering costs for use in industry.

A schematic summarising the main biosynthetic pathways from the uptake of CO_2 via the tricarboxylic acid cycle to the key groups of metabolites is shown in Fig. 15.2. Whilst a wide range of metabolites within cyanobacteria (including cyanotoxins, phycobiliproteins, carotenoids, and lipids) have been well studied our knowledge on metabolic pathways and networks to maximise yields using a systems approaches to increase yields of metabolites and to produce them consistently remains a challenge. Furthermore, a large reservoir of unknown novel compounds with biological activity is still to be explored (Christaki et al. 2015). Notably to date metabolic in silico modelling has focussed on high volume low-value products rather than low volume high-value products. This is primarily because our knowledge of biosynthetic pathways and, in particular, their regulation in cyanobacteria is still in its infancy. Empirical analysis combined with the use of increasingly available genome-scale models for understanding physiology and metabolic engineering application will hasten speed to generate further economic value from cyanobacteria.

There is much unexplored potential for expanding markets for metabolites found in spirulina. A list and retail value of currently marketed metabolites identified in *Arthrospira platensis* are compiled in Table 15.6. Below we discuss key groups of metabolites in cyanobacteria, focused on spirulina, with emphasis on economics and the potential for new markets.

15.3.1 Phycobilins

Phycobilins are currently economically one of the most lucrative groups of metabolites from cyanobacteria. Phycobilins, encompassing phycocyanin, phycoerythrin and allophycocyanin are light-harvesting tetrapyrrole pigments that have a diversity of industrial uses. They are uniquely found in cyanobacteria (including spirulina),

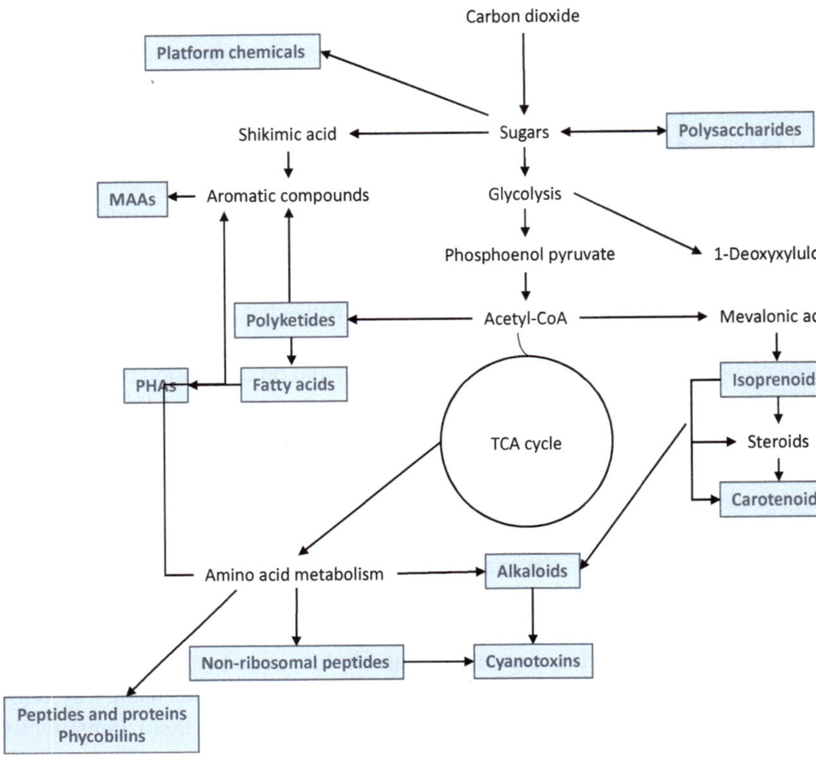

Fig. 15.2 Simple schematic linking key biosynthetic pathways from the uptake of CO_2 via the TCA cycle (tricarboxylic acid cycle/Krebs cycle) to key groups of metabolites discussed herein

rhodophytes (such as *Porphyridium*, *Rhodella* and *Galdieria*), the cryptophytes and glaucophytes. As water-soluble coloured pigments, they are endowed with properties, including unique absorbance and fluorescence, that make them commercially attractive. The existing and potential use of phycobilins has been widely reviewed (Borowitzka 2013; Chakdar and Pabbi 2016; Eriksen 2008; Sekar and Chandramohan 2008). In summary, phycobilins have application as natural colourants and for health-promoting properties in food and drink (e.g. phycocyanin 'Lina Blue' sold by DIC Corp., Japan), as possible antioxidants in the cosmetic industry, as fluorescent tags for use in flow cytometry and immunology (Glazer 1994), as photosensitisers in photodynamic therapy for treatment of cancers (Hu et al. 2008) and for other therapeutic benefits such as anti-inflammatory and hypolipidemic activities (Deng and Chow 2010). The current total market value for phycobiliprotein products (including fluorescent agents) is estimated to be greater than US$60 million and is expanding. Based upon their purity they command a value of between US$0.13–25/mg (Borowitzka 2013; Cuellar-Bermudez et al. 2015). Purity can be determined by measuring the ratio of absorbance at wavelengths of 620 and 280 nm (A620/A280 absorbance ratio) with food grade having a ratio of $\geq$0.7, reactive grade ~3.9 and analytical grade $\geq$4.0 (Patil et al. 2006). Purity and associated higher activity and value will be achieved as downstream processing techniques as described in Sect. 15.2.2 are improved.

Table 15.6 Retail value of currently marketed metabolites identified in *Arthrospira platensis*

Metabolites		Av.Q	RV	Cat. No.	References
Phycobili proteins	Phycocyanin	145[a]	10.10	P2172-10MG	Ajayan et al. (2012), Chen et al. (2010) and Ho et al. (2018)
	Allophycocyanin	46[a]	12.45	A7472-.5ML	Ajayan et al. (2012), Patel et al. (2005) and Tavanandi et al. (2018)
	Phycoerythrin	39	109	52412-1MG-F	Ajayan et al. (2012) and Patel et al. (2005)
Chlorophyll	Chlorophyll *a*	12.29	115	C5753-1MG	Leema et al. (2010)
	Chlorophyll *b*	0.32	127	C5878-1MG	Hynstova et al. (2018)
Carotenoids	β,β-Carotene	3.76	2.75	1065480-7X100MG	Leema et al. (2010)
	Zeaxanthin	2.50	386	14681-1MG-F	Miki et al. (1986)
	β-Cryptoxanthin	0.70	417	C6368-1MG	"
	Diatoxanthin	0.50	531	79449-1MG	"
	Diadinoxanthin	0.06	531	08379-1MG	Sommella et al. (2018)
	Antheraxanthin	0.03	431	([b]No. 0231)	"
	Canthaxanthin	0.54	75.10	11775-1MG	Hynstova et al. (2018)
	Echinenone	1.20	400	73341-1MG	Miki et al. (1986)
	Astaxanthin	0.41	1.74	SML0982-50MG	Hynstova et al. (2018)
	Lutein	1.75	299	07168-1MG	Leema et al. (2010)
	3-Hydroxyechinenone	0.90	431	([b]No. 0296)	Miki et al. (1986)
	Myxoxanthophyll	1.50	431	Estimated price	"
	Oscillaxanthin	0.40	431	Estimated price	"
Enzymes	RuBisCO	47[c]	5.85	R8000-1UN	Raven (2013)
	Carbonic anhydrase	0.0073[d]	1.90	C2624-100MG	Ores et al. (2016)

(continued)

Table 15.6 (continued)

Metabolites		Av.Q	RV	Cat. No.	References
Lipids	Palmitic (C16:0)	54.60	0.003	P0500-10G	Prates et al. (2018)
	Gamma-linolenic (C18:3n-6)	29.63	4.5	L2378-10MG	"
	Linoleic (C18:2)	17.75	1.74	L1376-10MG	"
	Capric (C10:0)	6.67	0.0002	C1875-100G	"
	Palmitoleic (C16:1)	6.03	0.72	P9417-100MG	"
	Oleic (C18:1)	4.82	0.0234	O1008-1G	"
	Stearic (C18:0)	2.08	0.015	S4751-1G	"
	Myristoleic (C14:1)	0.51	1.05	M3525-100MG	"
	Myristic (C14:0)	0.36	0.002	M3128-10G	"
	Eicosadienoic (C20:2)	0.23	2.2	E3127-25MG	"
Vitamins	Stearic	1.95	0.02	S4751-1G	Finamore et al. (2017)
	Vitamin B12 (Cobalamin)	0.00	0.171	V2876-100MG	"
	Vitamin B2 (Riboflavin)	0.0380	0.003	R4500-5G	"
	Vitamin B6 (Pyridoxine)	0.0065	0.007	P5669-5G	"
	Vitamin B3 (Niacin)	0.1400	1.405	1461003-200MG	"
	Vitamin B9 (Folic acid)	0.0499	0.01	F7876-1G	"
	Vitamin E (Tocopherol)	51.00	0.002	T3634-10G	"

Av.Q Average published quantity (mg/g of biomass dry weight)
RV Retail value of purified product (£/mg, Sigma, 2018)
[a]Typical values (NB: can vary >5 fold with strain/conditions (Tarko et al. 2012))
[b]Price from CaroteNature GmbH
[c]Estimate calculated from Rubisco/total protein average for cyanobacteria
[d]Estimate calculated from units per mg for retail product
Note: Table provides a comparative price index based on values for purchase from a biochemical supplier at a specific time point. It should be used as a comparative rather than a realistic price index, which does not reflect bulk market prices on which profit projections for planned biorefineries will depend. This may be particularly relevant when 'natural' compounds are considerably different in price to synthetic compounds, e.g. currently astaxanthin is considerably more expensive and canthaxanthin and lutein less so when naturally derived. It can, however, indicate promising areas for further analysis

15.3.2 Carotenoids

Carotenoids, encompassing carotenes and xanthophylls, are lipid-soluble tetraterpenes. They are essential components of all photosynthetic organisms and show a wide range of diversity within cyanobacteria. Commercially, carotenoids are used in cosmetics and in the feed and food industry particularly as antioxidants and colourants and are increasingly incorporated as nutritional dietary supplements. They have also been shown to prevent chronic disorders including cardiovascular diseases by preventing inflammatory response (Gammone et al. 2017). The global market for carotenoids reached US$1.5 billion in 2017 and should reach US$2.0 billion by 2022, at a CAGR (Compound Annual Growth Rate) of 5.7% for the period of 2017–2022 (McWilliams 2018). Individual carotenoids have different values according to their availability and market demand. A list of comparative prices for individual carotenoids found in *Arthrospira* is summarised in Table 15.6.

Global markets for carotenoids are currently focused around astaxanthin, beta-carotene, canthaxanthin, lutein, zeaxanthin and lycopene. *Arthrospira, Dunaliella* and *Haematococcus* are the three main commercial species producing these carotenoids. *Arthrospira* in common with *Dunaliella and Haematococcus* is rich in β-carotene. *Arthrospira,* as well as other cyanobacteria, is also a rich source of zeaxanthin. Zeaxanthin is important commercially as it can be used to prevent age-related macular degeneration (Abdel-Aal et al. 2013; Basu et al. 2001). In addition, cyanobacteria including *Arthrospira* produce some unusual and unique carotenoids that have yet to be fully explored in terms of their function and potential in biotechnology. Unusual carotenoids include the hydroxylated caloxanthin, nostoxanthin and the keto-carotenoids (e.g. echinenone and canthaxanthin). Both hydroxylated and ketolated carotenoids are known for their strong antioxidant properties (Sathasivam and Ki 2018). In addition, freshwater cyanobacteria produce unique monocyclic glycosidic derivatives such as myxoxanthophyll (Graham and Bryant 2009; Takaichi and Mochimaru 2007). Glycosylated carotenoids play a role in the regulation of membrane fluidity although their full role is still unknown, and they are yet to be fully explored in terms of their economic potential. Generally, carotenoids that are common across plants and algae command a lower market value than carotenoids that are more unique to cyanobacteria (e.g. β-carotene as shown in Table 15.6). Ascertaining functionality of more novel carotenoids remains open to research. The challenge, however, remains with bringing a new carotenoid to market in establishing the safety of the product and obtaining approval by regulatory authorities.

Carotenoid cleavage products (apocarotenoids) such as retinal are utilised in a range of food, pharmaceutical, cosmetics and nutraceutical industries (Ahrazem et al. 2016). Understanding the synthesis and breakdown of carotenoids is of importance to enable increased production of individual carotenoids and related cleavage products.

15.3.3 Polysaccharides

A. platensis biomass is mainly composed of proteins and polysaccharides with cell wall polysaccharides representing 10% of the biomass, which contains heteropolysaccharides composed of at least five different monosaccharides. These monosaccharides are linked together with glycosidic bonds and are characterised by the presence of pentoses, as well as their anionic nature (due to acidic sugars such as glucuronic and/or galacturonic acids) and anionic organic (acetyl, pyruvil) and inorganic (sulphate and phosphate) substituents. These anionic characteristics of polysaccharides display a range of interesting bioactive functionalities (Bernaerts et al. 2018; Challouf et al. 2011). Two types of polysaccharides obtained from *A. platensis* (storage and cell wall related) exhibit different functions. Storage polysaccharides are mainly responsible for energy storage, providing a substrate for metabolic processes and survival, whereas cell wall-related polysaccharides (composed of cell wall polysaccharides and extracellular polymeric substances) are responsible for providing structural rigidity. The total polysaccharides (storage and cell wall related) obtained from *A. platensis* contains sulphate, protein and uronic acids (5–20%, 6% and 7%, respectively) (Bernaerts et al. 2018; Raposo et al. 2013), and monosaccharides, uronic acids and sulphate composition in cell wall polysaccharides and extracellular polymeric substances of *A. platensis* have been described recently (Bernaerts et al. 2018).

A plethora of papers and reports have been published where cyanobacterial polysaccharides including exopolysaccharides are claimed to have biological and therapeutic benefits. However, there is still a lack of rigorous scientific, medical and cytotoxicity studies. In particular, the relationship between structural diversity and therapeutic/biological potential is lacking and much is still to be done (Bajpai et al. 2014; Challouf et al. 2011). There are several potential bioactives still to be explored, whereas with the known novel bioactives, efforts should be directed towards their commercial reality, as none of them has yet been found to have applications as pharmaceuticals (Borowitzka 2013). Considering the therapeutic potential of sulphated polysaccharides, it would be interesting to explore the production of oral vaccines and to consider the development of nanoformulations/encapsulations (using nanoparticles) in order to improve bioavailability.

15.3.4 Proteins and Peptides

The value of microalgal proteins can be enhanced by releasing the activity of specific sequences encrypted in the proteins by enzyme hydrolysis or microbial fermentation producing bioactive peptides (Daliri et al. 2017). Research on bioactive peptides continues to increase, and the therapeutic peptide market value was projected to reach US$25 billion by 2018 (Daliri et al. 2018). Microalgae can provide a useful natural source of these metabolites, and many have already been identified (Fan et al.

2014). Bioactive peptides can act as active ingredients for dietary supplements, cosmetic/dermatological products and pharmaceuticals, with a wide range of potential activities including anti-inflammatory, antidiabetic, antihypertensive, antimicrobial and antioxidative actions (Barberis et al. 2018).

Using *A. platensis* for illustration: Protein hydrolysates were found to contain an α-helical hydrophobic 18aa peptide, which interacts with negatively charged bacterial membranes giving it antibacterial properties like those of innate immune peptides (Sun et al. 2016).

Predictive analysis (Minkiewicz et al. 2008) for the enzymatic release of peptides from the abundant protein RuBisCO from *A. platensis* has identified potential nutraceutical activities including DPP-4 inhibitors used in the management of hyperglycaemia in type 2 diabetes, and ACE (angiotensin converting enzyme) inhibitors used in the management of hypertension (Selvaraj et al. 2017).

Effective utilisation of this potential will require cost-effective strategies for industrial-scale production. Valorization can be further improved by separating fractions before the hydrolysis phase to enable controlled production from lower value components, and development of innovative separation techniques tuned for specific metabolites (Carrasco-Castilla et al. 2012).

Recent success in the transformation of *A. platensis* has enabled the additional potential for value from microalgal proteins via the production of recombinant proteins and edible vaccines (Dehghani et al. 2018; Specht and Mayfield 2014).

15.3.5 Nonribosomal Peptides

A large proportion of biomolecules produced by cyanobacteria are peptides with the majority synthesised by specialised nonribosomal peptide synthetases (NRPS) or by a mixed enzymatic polyketide–NRPS system. NRPs are diverse in structure due to the combination of domains, which incorporate and modify proteinogenic and nonproteinogenic amino acids (Kehr et al. 2011).

Many species have been screened for the presence of genes encoding NRPS and polyketide synthase (PKS) (Welker and Von Döhren 2006) and it was found that certain species such as spirulina lack these specific genes. As toxin production is usually linked to NRPS and PKS, spirulina does not produce harmful cyanotoxins and therefore can be considered safe for use (Kwei et al. 2011).

15.3.6 Lipids and Fatty Acids

Cyanobacteria are low in lipids, compared to eukaryotic microalgae, however, they are relatively rich in polyunsaturated fatty acids (PUFAs). Total lipid in spirulina is regularly reported at about the 6% level (Chaiklahan et al. 2011; Habib 2008; Pohndorf et al. 2016; Sajilata et al. 2008) although this can be influenced by growth conditions (Chen et al. 2010). The major lipid classes in cyanobacterial membranes are monogalactosyl diacylglycerol (MGDG), monoglucosyl diacylglycerol (GlcDG), digalactosyl diacylglycerol (DGDG), sulfoquinovosyl diacylglycerol (SQDG) and phosphatidylglycerol (PG) (Sato and Murata 1988). The fatty acids are mainly unbranched chains containing 14, 16 or 18 carbon atoms and 0, 1, 2 or 3 double bonds (Sato and Murata 1988). Among them palmitic (0 double bonds, i.e. saturated), palmitoleic acid (1 double bond, i.e. monounsaturated) and gamma-linoleic acid (GLA; 3 double bonds) are the most commonly found. GLA has been reported at 0.42% in spirulina biomass (Sajilata et al. 2008). Overall cyanobacteria are quite variable in their fatty acid composition and marked differences can occur within the same genus.

Fatty acids are essential for health and demand by consumers for healthy eating is driving market growth for these important metabolites. The fatty acids market size is likely to exceed US$20 billion in 2023, growing at a CAGR of over 7% from 2016 to 2023 (Global Market Insights 2015).

15.3.7 Cyanotoxins

Many bloom-forming cyanobacteria produce highly toxic secondary metabolites known as cyanotoxins. They can be classified as hepatotoxins, cytotoxins, dermatoxins, neurotoxins or endotoxins (Rastogi and Sinha 2009) and are an important group of chemicals, which impact on aquatic ecosystems, human and animal health (Zanchett and Oliveira-Filho 2013).

Whilst most research has been conducted into their toxicological properties, there is also much interest in their industrial biotechnological application such as for biocides (algaecides, fungicides, herbicides and insecticides) and as pharmaceuticals (Rastogi and Sinha 2009; Vijayakumar and Menakha 2015).

The toxic traits of these cyanotoxins can also be exploited in drug discovery as anticancer agents such as apratoxin-a produced by *Lyngbya majuscula.* This toxic secondary metabolite possesses cytotoxicity in vitro against different human cell lines including epidermal KB and colon LoVo carcinoma cell lines (Luesch et al. 2001).

Another important application of cyanotoxins is as analytical standards for use in water quality testing. With the rise in global temperatures and increased eutrophication an increase in cyanotoxin producing species such as *Anabaena, Microcystis, Oscillatoria, Aphanizomenon* and *Lyngbya* will be seen which will, in turn, increase the demand for analytical standards (Manning and Nobles 2017). The cost of cyanotoxins as analytical standards are summarised in Table 15.7.

The most abundant cyanotoxins include the microcystins, which are hepatotoxic cyclic peptides produced by *Microcystis, Anabaenopsis, Nostoc* and *Anabaena* species. These toxins inhibit eukaryotic serine/threonine protein phosphatases, PP1 and PP2A (Yadav et al. 2011). There are over 50 variants of microcystin, and they are recognised based on the protein amino acids at the variable positions. The most potent being microcystin LR, which can be extracted from *Microcystis aeruginosa* using

Table 15.7 Cost of the cyanotoxins available as analytical standards from Cyano Biotech GmbH accessed in 2018

Cyanotoxin	Cost in Euro for 10 μg (Cyano Biotech GmbH, 2018)
Microcystin LR	109
Microcystin RR	128
Microcystin LR-RR-YR	339
Nodularin	140
(±) Anatoxin-a	140
Cyclindrospermopsin	144

sonic disruption in water to yield 200 μg/L (Silva-Stenico et al. 2011). Other forms include the less abundant microcystin RR and microcystin LR-RR-YR. Microcystin LR contains leucine and arginine at the protein amino acid positions (Merwe 2015), whereas microcystin RR is the arginine–arginine analogue of microcystin LR.

The neurotoxic anatoxins, such as anatoxin-a, are biocidal alkaloids that bind to nicotinic acetylcholine receptors irreversibly and are produced by *Anabaena flos-aquae, Aphanizomenon flos-aquae* and *Cylindrospermum.* Anatoxin-a can be extracted from *Anabaena flos-aquae* using ultrasonication and 75% methanol containing 0.002 M HCl to yield 4.27 μg/mg dry weight (Pawlik-Skowrońska et al. 2012).

Cylindrospermopsin also possesses biocidal activity and is known to be produced by *Cylindrospermopsis raciborskii, Aphanizomenon ovalisporum*, and *Anabaena bergii* (Gupta et al. 2013). Extraction can be achieved from *Aphanizomenon ovalisporum* by ultrasonication in Milli-Q water followed by centrifugation and the addition of 0.1% TFA to the supernatant produced a yield of 3.68 μg/mg dry weight (Guzmán-Guillén et al. 2012; Haque et al. 2017).

Whilst there are safety challenges associated with working with such potent toxins the rewards in terms of pharmaceutical and in drug discovery have the potential to be high.

15.3.8 Sunscreens

Mycosporine-like amino acids (MAAs) in cyanobacteria (although not in spirulina) are reported as being a group of metabolites with the most potential in industrial biotechnology. This is primarily because of the large market demand in the sun-screen industry and the drive by industry to replace existing less sustainable sources of organic synthetic and inorganic sunscreens. Covering early pioneering work (Dunlap et al. 1998) to recent comprehensive reviews on the topic (Derikvand et al. 2017; Kageyama and Waditee-Sirisattha 2018), there have been a plethora of patents on MAAs for use as sunscreens in the sun care industry.

Shinorine, porphyra-334 and mycosporine glycine are the MAAs most commonly associated with cyanobacteria with shinorine reported to be the most dominant (Kageyama and Waditee-Sirisattha 2018). The filamentous and heterocystous

N2-fixing cyanobacterial species *Anabaena* sp., *Nostoc commune* and *Scytonema* sp. are widely reported to contain shinorine, and a water soluble brownish compound with an absorption maximum at 315 nm was found to be induced by UV-B and released into the medium only in *Scytonema* sp. (Sinha et al. 2001). Glycosylated MAAs have been found in the terrestrial cyanobacterium *Nostoc commune* (Nazifi et al. 2015).

MAAs appear to be multifunctional, and mycosporine–glycine is known for its antioxidant properties (Rastogi and Sinha 2009). Mycosporine–glycine is able to quench singlet oxygen and hydroxyl radicals before oxidative stress response genes and antioxidant enzymes are induced (Suh et al. 2003). Photo-ageing of the skin is often caused by free radicals that are produced upon exposure to UV radiation. Treatment of skin with MAAs counteracts the process of free radical production: application of a cream containing 0.005% Porphyra-334 encapsulated in liposomes significantly improved both firmness and smoothness of the skin after 4 weeks by 10% and 12%, respectively, comparable to creams containing synthetic UV filters (Daniel et al. 2004).

Whilst the strength of using MAAs as natural sunscreens compounds is clear in terms of high extinction coefficient ($\varepsilon = 28{,}100 - 50{,}000$ L/mol cm) (Wada et al. 2015) and photochemical stability, there remains the same overall challenge as for other metabolites in scale-up and yield to bring them to market. The main challenge is the low cellular concentrations of MAAs. Despite the vast literature on MAAs, few studies address the issue of the quantities of MAAs that can be reasonably produced for commercial markets. For the red macroalgae *Porphyra* typically 20 mg/g dry weight can be produced (Hartmann et al. 2017). For microalgae, levels of MAAs are approximately 7 mg/g dry weight (Llewellyn and Airs 2010).

Therefore, for example 4000 L cultivation facility production would be in the order of a few kilograms per year. Incorporated as indicated (Daniel et al. 2004) at 0.005%, this would produce in the order of one tonne formulated product per year. Compare this to the billion tonnes of formulated sunscreen products sold each year and it is clear that deriving economic value from MAAs on a large scale is challenging and is, therefore, more likely for the foreseeable future to be viable for niche high-end markets.

15.3.9 Polyhydroxyalkanoates (PHAs)

The worldwide growing concern of plastic pollution demands research focused on the production of biodegradable plastics, as the global plastics production reached around 335 million tonnes in 2016 (European Bioplastics 2017), and is expected to double in the next 10–15 years. PHAs, fundamental building blocks of bioplastic, are well known for their biodegradability, biocompatibility, thermoplasticity, hydrophobicity, piezoelectricity and stereospecificity (Carpine 2017; Singh and Mallick 2017).

PHAs can be obtained via different chemical and/or biotechnological processes (mostly fermentative bacteria and higher plants). The current market share of bioplastics is only approximately 2% of the plastics market and is growing by 20–100% per year (Carpine 2017; Markl et al. 2018). The well-known natural PHA producers exploited for commercial production of polyhydroxy butanoate (PHB) include *Bacillus, Nocardia, Pseudomonas, Rhizobium, Ralstonia eutropha, Alcaligenes* and *Azotobacter* and numerous plant-based expression platforms. According to European bioplastics (European Bioplastics 2017), the production capacity of PHB is growing steadily (34 K tonnes in 2014 and it is forecasted to grow at 100 K tonnes in 2019). More details on bioplastic production plants are well reviewed in the recent literature (Carpine 2017; Singh and Mallick 2017). However, the market for PHB derived from bacteria and higher plants is too scarce due to the associated high production cost and other ethical considerations (Carpine 2017). In this context, cyanobacteria are emerging as an alternative host system due to their minimal nutrient requirements and photoautotrophic nature (Kamravamanesh et al. 2017).

PHAs derived from cyanobacteria are further classified into three groups according to chain length with varying thermal and mechanical properties. PHB is the most common, and the only PHA produced so far under photoautotrophic conditions. PHB accumulation has been reported in many cyanobacterial species under photoautotrophic and/or mixotrophic growth conditions with organic substrates such as acetate (Monshupanee et al. 2016; Panda et al. 2006; Stal et al. 1990; Takahashi et al. 1998), glucose, fructose (Budde et al. 2011), propionate (Lama et al. 1996; Samantaray and Mallick 2014), and valerate (Troschl et al. 2017). Several cyanobacterial strains have been screened for the accumulation of PHBs under normal cultivation conditions (Ansari and Fatma 2016; Bhati et al. 2010; Gopi et al. 2014), and attempts made to enhance production using different growth media (Carpine et al. 2015), photoperiods (Arias et al. 2018), nutrient limitation (nitrogen and phosphorous) (Arias et al. 2018; Carpine 2017; Kamravamanesh et al. 2017), and by using a two-stage cultivation approach (Monshupanee et al. 2016). Recently, a kinetic/dynamic model of PHB production by *Synechocystis* PCC6803 from CO_2 has been proposed and developed (Carpine et al. 2018).

To date, most of the cyanobacterial strains have been reported to accumulate PHB at $<10\%$ of dry cell weight (DCW) under photoautotrophic conditions (Bhati et al. 2010), whereas higher PHB content has only been detected in the case of heterotrophy or genetic modification of the strain (Kamravamanesh et al. 2017). The thermophilic cyanobacterium, *Synechococcus* sp. MA19 was reported to have achieved PHB at 27% of DCW whereas *Synechocystis* sp. PCC 6714 produced PHB at 37% DCW with CO_2 as the only carbon source (Kamravamanesh et al. 2018). These are encouraging developments and are beginning to bridge the gap in comparison with the highest level obtained in heterotrophic microorganisms in fermentation reactors (PHB at 78% DCW) (Markl et al. 2018).

There are few studies to date detailing genes contributing to PHB biosynthetic pathways and their regulation in cyanobacterial strains (Carpine 2017; Hauf et al. 2015; Singh and Mallick 2017), and recombinant DNA technology approaches provide potential to further enhance PHB accumulation. In addition, further refinements and challenges to establish an economically sustainable production platform include strain selection, cultivation systems, contamination issues, harvesting approaches, extraction technologies and recovery and purity of PHB from cyanobacterial strains.

15.3.10 Isoprenoids (Terpenes)

Isoprenoids have diverse and important biotechnological applications. Many isoprenoids have been used for pharmaceuticals, nutraceuticals, flavours, cosmetics, food additives and biofuels. Isoprenoids are usually classified into groups according to the number of carbons: hemiterpenes (C5), monoterpenes (C10), sesquiterpenes (C15), diterpenes (C20), triterpenes (C30) and tetraterpenes (carotenoids, C40). A wide variety of isoprenoids are synthesised in different cyanobacterial species via the MEP (methylerythritol phosphate) pathway (Pattanaik and Lindberg 2015), and the potential of metabolic engineering to increase yields is currently being explored (Lin and Pakrasi 2018; Vavitsas et al. 2018).

Some cyanobacteria also produce odorous monoterpenes (Li et al. 2012), sesquiterpenes (Agger et al. 2008) and hopanoids (Saenz et al. 2012). Thus, the cyanobacteria may provide a potential platform for the large-scale photosynthetic production of various isoprenoids. Understanding isoprenoid metabolic pathway regulation enables optimisation of the photochemical performance of cyanobacteria to improve yields of target carotenoids and other valuable biotechnology metabolites such as retinal (Sect. 15.3.2). More specifically more detailed knowledge on the routing and regulation of polyprenyl precursors at critical branch points of the biosynthetic network of isoprenoid formation is required (Sasso et al. 2012).

Limonene, a monocyclic terpene, is receiving increasing attention as a platform chemical (Sect. 15.3.12). Lin and coworkers (2017) reported a feasible strategy to engineer *Synechocystis* 6803 for improving or achieving the desired photosynthetic production of isoprenoids with minimal genetic interventions. These authors reported a 2.3-fold increase in limonene yield (6.7 mg/L of limonene) using this approach. The current market for limonene is ~US$160 million/year with a market selling price ~US$7/kg (Wu and Maravelias 2018).

15.3.11 Platform Chemicals

There are several reports and reviews covering the potential of genetic engineering of cyanobacterial strains for improving the yield of various commodity chemicals (Angermayr et al. 2015; Lin et al. 2017; Niederholtmeyer et al. 2010; Zhang et al. 2017). Briefly, accumulation of large titres of ethanol, butanediol, lactic acid, D-Lactic acid and isobutyraldehyde can be achieved in cyanobacterial strains by tapping pyruvate metabolism, whereas large titres of sucrose, glycerol and mannitol can be recovered via rechannelling of carbon from sugar metabolism. Other important pathways of intermediary metabolism explored so far include tapping of acetyl-CoA (for 3-hydroxybutyrate) or the isoprenoid biosynthesis pathway or the mevalonate pathway (for limonene and other isoprenoids, Sect. 15.3.10). Such tapping of the cyanobacterial metabolism at various branching points allows accumulation of large titres of the desired platform and commodity chemicals and is well reviewed (Angermayr et al. 2015).

15.3.12 Stable Isotopes

Stable isotope-labelled compounds are widely used for scientific research and clinical purposes, including metabolomics, proteomics and structural studies. These compounds can be prepared by chemical synthesis, or in vivo by growing simple organisms on inorganic sources of a stable isotope. Labelling using natural substrates is slow and inefficient, but high enrichment of labelled biochemical compounds can be achieved with enriched stable isotope substrates. However, the large-scale sustainable application of this concept is lacking and requires the careful selection of organisms and carefully controlled procedures (Zachleder et al. 2018).

As growth of microalgae can be achieved under strictly controlled conditions, stable isotopes from relatively cheap inorganic C, H and N sources (^{13}C, ^{15}N and ^{2}H from $^{13}CO_2$, $^{15}NO_3$ and $^{2}H_2O$) can be incorporated successfully into microalgae during photosynthesis to produce stable isotope-labelled biochemicals (amino acids, carbohydrates, lipids and nucleic acids) that are widely used for scientific research and clinical purposes (Kapoore 2014; Pulz and Gross 2004). Recent developments in sensitive analytical techniques such as GC-MS, LC-MS and NMR allow stable isotope-labelled sugars and amino acids to be used as tracers in pharmacology and clinical applications (e.g. gastrointestinal or breath diagnosis tests) (Butler et al. 2017; Schellekens et al. 2011).

The stable isotope of carbon ^{13}C is widely used in metabolic flux analysis and microalgae can produce many different ^{13}C labelled compounds (Griffiths et al. 2016), with many examples of commercial use including ^{13}C- triglycerides, ^{13}C-galactose and ^{13}C-xylose (Chakdar et al. 2012; Rizwan et al. 2018). The stable isotope of nitrogen ^{15}N is widely used in quantitative proteomics (Lehmann 2017). *Spirulina maxima* has been advocated as a versatile and sustainable platform for the production of stable isotope-labelled sugars and amino acids (Levert and Xia 2001).

Compared to that of ^{13}C-enriched compounds, deuterated compounds can be produced in vivo with fewer constraints due to their ease of incorporation. Metabolic labelling using D_2O is more economical than D_2 gas-based chemical synthesis, as D_2 gas is fivefold higher in price than D_2O. The only constraint to this metabolic labelling is the tolerance capacity of selected organisms. Many microalgal species are highly tolerant, especially where gradual adaptation protocols are employed, unlike the more sensitive higher plants and mammalian cells. Cyanobacteria have high potential as a commercial source of deuterated compounds compared to less efficient chemical production (which then necessitates further purification steps), as they can be fully deuterated by growth in D_2O (Zachleder et al. 2018). There are many applications—as a tool in quantitative proteomics and metabolomic flux analysis, and in medicine, pharmacology and clinical fields, where deuterated drugs are found to be more resistant to metabolic changes thereby permitting lower doses of the drug with fewer side effects (Schmidt 2017; Yang 2016). There are recent examples of cyanobacterial production of deuterated chlorophylls, lutein, carotene and phycocyanin (Garg et al. 2017; Zachleder et al. 2018).

The reported market so far for stable isotope-labelled biochemicals manufactured from microalgae is in excess of US$13 million, with prices in the range of US $260–5900/g for labelled amino acids and US$28/mg for labelled nucleic acids (Griffiths et al. 2016; Milledge 2011; Pulz and Gross 2004; Radmer 1996; Spolaore et al. 2006). Spectra Stable Isotopes (Columbia, MD, USA) produces >400 g/year labelled PUFAs from microalgae by sparging $^{13}CO_2$ and sells at US$38,000/g (Spolaore et al. 2006).

Although stable isotope biochemicals have been produced from cyanobacteria for around three decades, to harness its full potential, an in-depth understanding of cell behaviour and response to a high proportion of stable isotopes is required. The analytical methods such as GC-MS, LC-MS and NMR that are currently employed give a relatively superficial view of the dynamics, and a wider use of modern analytical techniques such as micro Raman spectroscopy and Nano-SIMS will be required (Zachleder et al. 2018).

15.4 Systems Biology

From successful engineering projects involving *Escherichia coli* and yeast, it is clear that improving the productivity of metabolites requires deep physiological, genetic and metabolic characterisation. Systems biology and metabolic engineering efforts undertaken in cyanobacteria until now have been hampered by significant knowledge gaps. This is a main contributory reason for the relatively small number of target chemicals explored so far and the low productivities. For cyanobacteria, we have gaps in our knowledge of metabolic regulation, metabolic pathway networks, metabolic engineering and synthetic biology. Systems biology is still in its infancy and requires further development and evaluation. Optimisation of metabolism in cyanobacteria has barely begun (Gudmundsson and Nogales 2015). A systems-level metabolic optimisation is required to achieve the true potential of cyanobacteria in biotechnology (Gudmundsson and Nogales 2015). In recent years, CRISPR genome editing technology has revolutionised the field of biotechnology. Recently CRISPR technology has been applied to three model cyanobacteria; *Synechococcus, Synechocystis* and *Anabaena* (Ungerer and Pakrasi 2016). The manipulation of whole metabolic networks by CRISPR- and sRNA-based technologies with multiple parallel targets will further stimulate the use of cyanobacteria for diverse applications in basic research and biotechnology (Hagemann and Hess 2018).

15.5 Biorefinery Approaches

Key to the success of developing algal production systems is the recovery of high-value products alongside recovery of the maximum number of products. This enables the maximum benefit and value from the produced biomass. To date, most commercial processes have focused on a single-product approach but there have been recent reports on utilising cyanobacteria as a biorefinery and producing multiple products (Kapoore et al. 2018). A biorefinery concept is modelled on the petroleum refinery, in which crude oil is converted into fuels and chemicals that provide multiple products and revenue streams.

A biorefinery aims to produce a spectrum of marketable products utilising different biochemical fractions including lipid, protein, vitamins, carbohydrate and pigments through a sustainable process. The genetic engineering of cyanobacteria, where CO_2 can be directly converted into a range of high-value marketable products holds tremendous potential and is the most important aspect of creating a sustainable platform. Such sustainable platforms can yield a range of high-value products from potential biofuels, bioplastics precursors to several commodity chemicals such as alcohols (ethanol, isopropanol, 1-butanol, 2-methly-1-butanol), alkanes, alkenes, aldehydes (isobutyraldehyde), isoprenoids (limonene), sugars (glucose, fructose, sucrose, mannose), organic acids (lactic acid, 3-hydroxypropionic acid) and others (2,3-butanediol) (Johnson et al. 2016).

The options for biorefining must be to exploit and enhance the value of the biomass and meet the specifications required for a range of products and must result in optimising the value of the products in a practical manner. This, to summarise, depends on the cost of production and the market price of the products and is dependent on many factors including, social and environmental, economic and geopolitical concerns. The aim must be to make robust generic systems of purification and isolation of materials that are possible and appropriate for the scale of operation and markets for the products.

Key features to consider within a biorefinery approach includes selection of ideal cyanobacterial strain with the most commercial potential (Kapoore et al. 2018); cost-effective cultivation conditions for improved growth rates and product yield (Zhang et al. 2017); selecting commodity chemicals with multiple applications; techno-economic evaluation of the whole biorefinery process chain (Ruiz et al. 2016); mitigating concerns about the limitations between maintaining cellular health and production of desired products (Oliver and Atsumi 2015) and lastly mitigating toxicity issues derived from the increased co-production of undesired products (Zhang et al. 2017). Overall, synthetic biology and metabolic engineering have tremendous potential to overcome these bottlenecks and creating a sustainable-engineered cyanobacterial biorefinery for a range of commodity chemicals.

A challenge in the economic production of metabolites is the requirement at the cultivation stage for cheap sources of nutrients, these can often be derived from waste materials (mineral nutrients such as P and N and CO_2 from combustion). Once cultivation has been achieved within the biorefinery, the distinct stages of cell concentration, cell disruption, separation of major components, purification and conversion into a wide spectrum of products all need careful consideration. Considerable care is required to optimise the options of such systems to arrive at the 'best' process. From Fig. 15.1, it can be seen that this allows a flexible approach to production being capable of switching the product spectrum. This alleviates some of the constraints caused by capital investment in the process equipment. Furthermore, regional economics, demands and resources all play a role in the best strategy and spectrum of products to be produced. Overall, a biorefinery offers an advantage in being an adaptable and flexible system, which is able to react to market conditions and demands.

15.6 Conclusion

We have highlighted the key processes and challenges associated with deriving economic value from metabolites derived from cyanobacteria. We have covered costs associated with cultivation, downstream cell processing and metabolites extraction options and key groups of metabolites with economic value. Whilst it is recognised that cyanobacteria have enormous potential in industrial biotechnology their full potential has not yet been reached. To reach this full potential requires a holistic strategy addressing challenges associated with cultivation, processing and systems biology approaches. We conclude that multidisciplinary expertise across all aspects of the biotechnological and engineering processes are required to derive maximum economic value from cyanobacteria. Whilst our focus has been on spirulina, it should be recognised that there is a wide diversity of cyanobacterial species presenting an enormous untapped resource. As scale-up of cultivation continues to progress, we will obtain a better understanding on the wider economic feasibility of using cyanobacteria in industry. It is clear, a biorefinery approach to deriving a multi-stream of metabolites makes best economic and environmental sense.

Acknowledgements CAL, RL, CGF, CG acknowledge Innovate-UK PhycoPigment funding, CAL RVK, CFG acknowledge Interreg NWE, ALG-AD funding. BK acknowledges BBSRC iCASE funding. CGF thanks Ms. C. Gabriel Castro, for kindly sharing information on spirulina production. Finally, we thank the three reviewers for their helpful comments.

References

Abdel-Aal E-SM, Akhtar H, Zaheer K et al (2013) Dietary sources of lutein and zeaxanthin carotenoids and their role in eye health. Nutrients 5(4):1169–1185. https://doi.org/10.3390/nu5041169

Agger SA, Lopez-Gallego F, Hoye TR et al (2008) Identification of sesquiterpene synthases from *Nostoc punctiforme PCC 73102* and *Nostoc* sp. strain PCC 7120. J Bacteriol 190 (18):6084–6096. https://doi.org/10.1128/jb.00759-08

Ahrazem O, Gómez-Gómez L, Rodrigo M et al (2016) Carotenoid cleavage oxygenases from microbes and photosynthetic organisms: features and functions. Int J Mol Sci 17(11):1781. https://doi.org/10.3390/ijms17111781

Ajayan KV, Selvaraju M, Thirugnanamoorthy K (2012) Enrichment of chlorophyll and phycobiliproteins in *Spirulina platensis* by the use of reflector light and nitrogen sources: an in-vitro study. Biomass Bioenergy 47:436–441. https://doi.org/10.1016/j.biombioe.2012.09.012

Al-Haj L, Lui YT, Abed RM et al (2016) Cyanobacteria as chassis for industrial biotechnology: progress and prospects. Life (Basel) 6(4):E42. https://doi.org/10.3390/life6040042

Alhattab M, Kermanshahi-Pour A, Brooks MS-L (2019) Microalgae disruption techniques for product recovery: influence of cell wall composition. J Appl Phycol 31(1):61–88. https://doi.org/10.1007/s10811-018-1560-9

Angermayr SA, Gorchs Rovira A, Hellingwerf KJ (2015) Metabolic engineering of cyanobacteria for the synthesis of commodity products. Trends Biotechnol 33(6):352–361. https://doi.org/10.1016/j.tibtech.2015.03.009

Ansari S, Fatma T (2016) Cyanobacterial polyhydroxybutyrate (PHB): screening, optimization and characterization. PLoS One 11(6):e0158168. https://doi.org/10.1371/journal.pone.0158168

Anyanwu R, Rodriguez C, Durrant A et al (2018) Optimisation of tray drier microalgae dewatering techniques using response surface methodology. Energies 11(9):2327. https://doi.org/10.3390/en11092327

Arias DM, Uggetti E, García-Galán MJ et al (2018) Production of polyhydroxybutyrates and carbohydrates in a mixed cyanobacterial culture: effect of nutrients limitation and photoperiods. New Biotechnol 42:1–11. https://doi.org/10.1016/j.nbt.2018.01.001

Bajpai VK, Rather IA, Lim J et al (2014) Diversity of bioactive polysaccharide originated from marine sources: a review. Indian J Geo-Mar Sci 43(10):1857–1869

Barberis SE, Origone AL, Adaro MO et al (2018) Chapter 5—Bioactive peptides as functional food ingredients. In: Grumezescu AM, Holban AM (eds) Role of materials science in food bioengineering. Academic, London, pp 147–186. https://doi.org/10.1016/B978-0-12-811448-3.00005-X

Bastiaens L, Van Roy S, Thomassen G et al (2017) 14—Biorefinery of algae: technical and economic considerations. In: Gonzalez-Fernandez C, Muñoz R (eds) Microalgae-based biofuels and bioproducts. Woodhead, Duxford, pp 327–345

Basu HN, Vecchio AJD, Flider F et al (2001) Nutritional and potential disease prevention properties of carotenoids. J Am Oil Chem Soc 78(7):665–675. https://doi.org/10.1007/s11746-001-0324-x

Bermejo R, Felipe MA, Talavera EM et al (2006) Expanded bed adsorption chromatography for recovery of phycocyanins from the microalga *Spirulina platensis*. Chromatographia 63 (1):59–66. https://doi.org/10.1365/s10337-005-0702-9

Bernaerts TMM, Gheysen L, Kyomugasho C et al (2018) Comparison of microalgal biomasses as functional food ingredients: focus on the composition of cell wall related polysaccharides. Algal Res 32:150–161. https://doi.org/10.1016/J.ALGAL.2018.03.017

Bhati R, Samantaray S, Sharma L et al (2010) Poly-β-hydroxybutyrate accumulation in cyanobacteria under photoautotrophy. Biotechnol J 5(11):1181–1185. https://doi.org/10.1002/biot.201000252

Bleakley S, Hayes M (2017) Algal proteins: extraction, application, and challenges concerning production. Foods 6(5):33. https://doi.org/10.3390/foods6050033

Borowitzka MA (2013) High-value products from microalgae—their development and commercialisation. J Appl Phycol 25(3):743–756. https://doi.org/10.1007/s10811-013-9983-9

Budde CF, Riedel SL, Hübner F et al (2011) Growth and polyhydroxybutyrate production by *Ralstonia eutropha* in emulsified plant oil medium. Appl Microbiol Biotechnol 89 (5):1611–1619. https://doi.org/10.1007/s00253-011-3102-0

Burczyk J, Dworzanski J (1988) Comparison of sporopollenin-like algal resistant polymer from cell wall of *Botryococcus*, *Scenedesmus* and *Lycopodium clavatum* by GC-pyrolysis. Phytochemistry 27(7):2151–2153. https://doi.org/10.1016/0031-9422(88)80115-8

Butler RN, Kosek M, Krebs NF et al (2017) Stable isotope techniques for the assessment of host and microbiota response during gastrointestinal dysfunction. J Pediatr Gastroenterol Nutr 64(1):8. https://doi.org/10.1097/MPG.0000000000001373

Carpine R (2017) Cyanobacteria for PHB production. PhD thesis, Università degli Studi di Napoli Federico II

Carpine R, Olivieri G, Hellingwerf K et al (2015) The cyanobacterial route to produce poly-β-hydroxybutyrate. Chem Eng Trans 43:289–294. https://doi.org/10.3303/CET1543049

Carpine R, Raganati F, Olivieri G et al (2018) Poly-β-hydroxybutyrate (PHB) production by *Synechocystis* PCC6803 from CO_2: model development. Algal Res 29:49–60. https://doi.org/10.1016/J.ALGAL.2017.11.011

Carrasco-Castilla J, Hernández-Álvarez AJ, Jiménez-Martínez C et al (2012) Use of proteomics and peptidomics methods in food bioactive peptide science and engineering. Food Eng Rev 4 (4):224–243. https://doi.org/10.1007/s12393-012-9058-8

Catchpole O, Moreno T, Montañes F et al (2018) Perspectives on processing of high value lipids using supercritical fluids. J Supercrit Fluids 134:260–268. https://doi.org/10.1016/j.supflu.2017.12.001

Chaiklahan R, Chirasuwan N, Loha V et al (2011) Separation and purification of phycocyanin from *Spirulina* sp. using a membrane process. Bioresour Technol 102(14):7159–7164. https://doi.org/10.1016/j.biortech.2011.04.067

Chakdar H, Pabbi S (2016) Cyanobacterial phycobilins: production, purification, and regulation. In: Shukla P (ed) Frontier discoveries and innovations in interdisciplinary microbiology. Springer India, New Delhi, pp 45–69

Chakdar H, Jadhav SD, Dhar DW et al (2012) Potential applications of blue green algae. J Sci Ind Res 71:13–20

Challouf R, Trabelsi L, Ben Dhieb R et al (2011) Evaluation of cytotoxicity and biological activities in extracellular polysaccharides released by cyanobacterium *Arthrospira platensis*. Braz Arch Biol Technol 54(4):831–838. https://doi.org/10.1590/S1516-89132011000400024

Chen H-B, Wu J-Y, Wang C-F et al (2010) Modeling on chlorophyll a and phycocyanin production by *Spirulina platensis* under various light-emitting diodes. Biochem Eng J 53(1):52–56. https://doi.org/10.1016/j.bej.2010.09.004

Christaki E, Bonos E, Florou-Paneri P (2015) Chapter 14—Innovative microalgae pigments as functional ingredients in nutrition. In: Kim S-K (ed) Handbook of marine microalgae. Academic, Boston, pp 233–243

Cuellar-Bermudez SP, Aguilar-Hernandez I, Cardenas-Chavez DL et al (2015) Extraction and purification of high-value metabolites from microalgae: essential lipids, astaxanthin and phycobiliproteins. Microb Biotechnol 8(2):190–209. https://doi.org/10.1111/1751-7915.12167

D'hondt E, Martín-Juárez J, Bolado S et al (2017) Cell disruption technologies. In: Gonzalez-Fernandez C, Muñoz R (eds) Microalgae-based biofuels and bioproducts, 1st edn. Elsevier, New York, pp 133–154

Daliri EB-M, Oh DH, Lee BH (2017) Bioactive peptides. Foods 6(5):32. https://doi.org/10.3390/foods6050032

Daliri EB-M, Lee BH, Oh DH (2018) Current trends and perspectives of bioactive peptides. Crit Rev Food Sci Nutr 58(13):2273–2284. https://doi.org/10.1080/10408398.2017.1319795

Daniel S, Cornelia S, Fred Z (2004) UV-A sunscreen from red algae for protection against premature skin aging. Cosmet Toiletries Manuf Worldw 129:139–143

Dassey AJ, Theegala CS (2013) Harvesting economics and strategies using centrifugation for cost effective separation of microalgae cells for biodiesel applications. Bioresour Technol 128:241–245. https://doi.org/10.1016/j.biortech.2012.10.061

Davis R, Aden A, Pienkos PT (2011) Techno-economic analysis of autotrophic microalgae for fuel production. Appl Energy 88(10):3524–3531. https://doi.org/10.1016/j.apenergy.2011.04.018

Dehghani J, Adibkia K, Movafeghi A et al (2018) Stable transformation of *Spirulina (Arthrospira) platensis*: a promising microalga for production of edible vaccines. Appl Microbiol Biotechnol 102(21):9267–9278. https://doi.org/10.1007/s00253-018-9296-7

Deng R, Chow T-J (2010) Hypolipidemic, antioxidant, and antiinflammatory activities of microalgae spirulina. Cardiovasc Ther 28(4):e33–e45. https://doi.org/10.1111/j.1755-5922.2010.00200.x
Derikvand P, Llewellyn CA, Purton S (2017) Cyanobacterial metabolites as a source of sunscreens and moisturizers: a comparison with current synthetic compounds. Eur J Phycol 52(1):43–56. https://doi.org/10.1080/09670262.2016.1214882
Dunlap WC, Chalker BE, Bandaranayake WM et al (1998) Nature's sunscreen from the Great Barrier Reef, Australia. Int J Cosmet Sci 20(1):41–51. https://doi.org/10.1046/j.1467-2494.1998.171734.x
El-Sayed AE-K, El-Sheekh M (2018) Outdoor cultivation of *Spirulina platensis* for mass production. Not Sci Biol 10(1):38–44. https://doi.org/10.15835/nsb10110177
Eriksen NT (2008) Production of phycocyanin-a pigment with applications in biology, biotechnology, foods and medicine. Appl Microbiol Biotechnol 80(1):1–14. https://doi.org/10.1007/s00253-008-1542-y
European Bioplastics, nova-institute (2017) Bioplastics market data. https://www.european-bioplastics.org/market/
Fan X, Bai L, Zhu L et al (2014) Marine algae-derived bioactive peptides for human nutrition and health. J Agric Food Chem 62(38):9211–9222. https://doi.org/10.1021/jf502420h
Fasaei F, Bitter JH, Slegers PM et al (2018) Techno-economic evaluation of microalgae harvesting and dewatering systems. Algal Res 31:347–362. https://doi.org/10.1016/j.algal.2017.11.038
Finamore A, Palmery M, Bensehaila S et al (2017) Antioxidant, immunomodulating, and microbial-modulating activities of the sustainable and ecofriendly spirulina. Oxidative Med Cell Longev 2017:3247528. https://doi.org/10.1155/2017/3247528
Gammone MA, Pluchinotta FR, Bergante S et al (2017) Prevention of cardiovascular diseases with carotenoids. Front Biosci 9:165–171
Garg H, Loughlin PC, Willows RD et al (2017) The $C2^1$-formyl group in chlorophyll *f* originates from molecular oxygen. J Biol Chem 292(47):19279–19289. https://doi.org/10.1074/jbc.M117.814756
Gerardo ML, Zacharof MP, Lovitt RW (2013) Strategies for the recovery of nutrients and metals from anaerobically digested dairy farm sludge using cross-flow microfiltration. Water Res 47 (14):4833–4842. https://doi.org/10.1016/j.watres.2013.04.019
Gerardo ML, Oatley-Radcliffe DL, Lovitt RW (2014) Integration of membrane technology in microalgae biorefineries. J Membr Sci 464:86–99. https://doi.org/10.1016/j.memsci.2014.04.010
Glazer AN (1994) Phycobiliproteins—a family of valuable, widely used fluorophores. J Appl Phycol 6(2):105–112. https://doi.org/10.1007/BF02186064
Global Market Insights (2015) Fatty acids market size
Gopi K, Balaji S, Muthuvelan B (2014) Isolation purification and screening of biodegradable polymer PHB producing cyanobacteria from marine and fresh water resources. Iran J Energy Environ 5(1):94–100. https://doi.org/10.5829/idosi.ijee.2014.05.01.14
Goto M, Kanda H, Wahyudiono et al (2015) Extraction of carotenoids and lipids from algae by supercritical CO_2 and subcritical dimethyl ether. J Supercrit Fluids 96:245–251. https://doi.org/10.1016/j.supflu.2014.10.003
Graham JE, Bryant DA (2009) The biosynthetic pathway for myxol-2′ fucoside (myxoxanthophyll) in the cyanobacterium *Synechococcus sp.* strain PCC 7002. J Bacteriol 191(10):3292–3300. https://doi.org/10.1128/jb.00050-09
Griffiths M, Harrison STL, Smit M et al (2016) Major commercial products from micro- and macroalgae. In: Bux F, Chisti Y (eds) Algae biotechnology, Green energy and technology. Springer, Cham, pp 269–300
Grima EM, Belarbi E-H, Fernández FGA et al (2003) Recovery of microalgal biomass and metabolites: process options and economics. Bitechnol Adv 20(7–8):491–515. https://doi.org/10.1016/S0734-9750(02)00050-2
Gudmundsson S, Nogales J (2015) Cyanobacteria as photosynthetic biocatalysts: a systems biology perspective. Mol BioSyst 11(1):60–70. https://doi.org/10.1039/c4mb00335g

Guiry MD (2012) How many species of algae are there? J Phycol 48(5):1057–1063. https://doi.org/10.1111/j.1529-8817.2012.01222.x

Gunerken E, D'hondt E, Eppink MH et al (2015) Cell disruption for microalgae biorefineries. Biotechnol Adv 33(2):243–260. https://doi.org/10.1016/j.biotechadv.2015.01.008

Gupta V, Ratha SK, Sood A et al (2013) New insights into the biodiversity and applications of cyanobacteria (blue-green algae)—prospects and challenges. Algal Res 2:79–97. https://doi.org/10.1016/j.algal.2013.01.006

Guzmán-Guillén R, Prieto Ortega AI, Moreno I et al (2012) Development and optimization of a method for the determination of *Cylindrospermopsin* from strains of *Aphanizomenon* cultures: intra-laboratory assessment of its accuracy by using validation standards. Talanta 100:356–363. https://doi.org/10.1016/j.talanta.2012.07.087

Habib MAB (2008) A review on culture, production and use of spirulina as food for humans and feeds for domestic animals and fish (FAO fisheries and aquaculture circular, no. 1034). Food and Agriculture Organization of the United Nations, Rome

Hagemann M, Hess WR (2018) Systems and synthetic biology for the biotechnological application of cyanobacteria. Curr Opin Biotechnol 49:94–99. https://doi.org/10.1016/j.copbio.2017.07.008

Haque F, Banayan S, Yee J et al (2017) Extraction and applications of cyanotoxins and other cyanobacterial secondary metabolites. Chemosphere 183:164–175. https://doi.org/10.1016/j.chemosphere.2017.05.106

Hartmann A, Murauer A, Ganzera M (2017) Quantitative analysis of mycosporine-like amino acids in marine algae by capillary electrophoresis with diode-array detection. J Pharm Biomed Anal 138:153–157. https://doi.org/10.1016/j.jpba.2017.01.053

Hauf W, Watzer B, Roos N et al (2015) Photoautotrophic polyhydroxybutyrate granule formation is regulated by cyanobacterial phasin PhaP in *Synechocystis sp.* strain PCC 6803. Appl Environ Microbiol 81(13):4411–4422. https://doi.org/10.1128/AEM.00604-15

Ho S-H, Liao J-F, Chen C-Y et al (2018) Combining light strategies with recycled medium to enhance the economic feasibility of phycocyanin production with *Spirulina platensis*. Bioresour Technol 247:669–675. https://doi.org/10.1016/j.biortech.2017.09.165

Hu L, Huang B, Zuo M-M et al (2008) Preparation of the phycoerythrin subunit liposome in a photodynamic experiment on liver cancer cells. Acta Pharmacol Sin 29:1539. https://doi.org/10.1111/j.1745-7254.2008.00886.x

Hynstova V, Sterbova D, Klejdus B et al (2018) Separation, identification and quantification of carotenoids and chlorophylls in dietary supplements containing *Chlorella vulgaris* and *Spirulina platensis* using high performance thin layer chromatography. J Pharm Biomed Anal 148:108–118. https://doi.org/10.1016/j.jpba.2017.09.018

Johnson TJ, Gibbons JL, Gu L et al (2016) Molecular genetic improvements of cyanobacteria to enhance the industrial potential of the microbe: a review. Biotechnol Prog 32(6):1357–1371. https://doi.org/10.1002/btpr.2358

Kageyama H, Waditee-Sirisattha R (2018) Chapter 5—Mycosporine-like amino acids as multifunctional secondary metabolites in cyanobacteria: from biochemical to application aspects. In: Atta ur R (ed) Studies in natural products chemistry, vol 59. Elsevier, New York, pp 153–194

Kamble SP, Vikhe GP, Chamle DR (2018) Extraction and purification of phycoerythrin—a natural colouring agent from *Spirulina platensis*. J Pharm Chem Biol Sci 6(2):78–84

Kamravamanesh D, Pflügl S, Nischkauer W et al (2017) Photosynthetic poly-β-hydroxybutyrate accumulation in unicellular cyanobacterium *Synechocystis sp.* PCC 6714. AMB Express 7 (1):143. https://doi.org/10.1186/s13568-017-0443-9

Kamravamanesh D, Kovacs T, Pflügl S et al (2018) Increased poly-β-hydroxybutyrate production from carbon dioxide in randomly mutated cells of cyanobacterial strain *Synechocystis* sp. PCC 6714: mutant generation and characterization. Bioresour Technol 266:34–44. https://doi.org/10.1016/j.biortech.2018.06.057

Kanno M, Carroll AL, Atsumi S (2017) Global metabolic rewiring for improved CO_2 fixation and chemical production in cyanobacteria. Nat Commun 8:14724. https://doi.org/10.1038/ncomms14724

Kapoore RV (2014) Mass spectrometry based hyphenated techniques for microalgal and mammalian metabolomics. PhD Thesis, University of Sheffield

Kapoore RV, Butler T, Pandhal J et al (2018) Microwave-assisted extraction for microalgae: from biofuels to biorefinery. Biology 7(1):18. https://doi.org/10.3390/biology7010018

Kehr JC, Picchi DG, Dittmann E (2011) Natural product biosyntheses in cyanobacteria: a treasure trove of unique enzymes. Beilstein J Org Chem 7:1622–1635. https://doi.org/10.3762/bjoc.7.191

Kim SG, Choi A, Ahn CY et al (2005) Harvesting of *Spirulina platensis* by cellular flotation and growth stage determination. Lett Appl Microbiol 40(3):190–194. https://doi.org/10.1111/j.1472-765X.2005.01654.x

Kuddus M, Singh P, Thomas G et al (2013) Recent developments in production and biotechnological applications of C-phycocyanin. Biomed Res Int 2013:9. https://doi.org/10.1155/2013/742859

Kultschar B, Llewellyn C (2018) Secondary metabolites in cyanobacteria. In: Vijayakumar R, Raja SSS (eds) Secondary metabolites: sources and applications. InTech, London, pp 23–36

Kwei CK, Lewis D, King K et al (2011) Molecular classification of commercial spirulina strains and identification of their sulfolipid biosynthesis genes. J Microbiol Biotechnol 21(4):359–365. https://doi.org/10.4014/jmb.1008.08016

Lama L, Nicolaus B, Calandrelli V et al (1996) Effect of growth conditions on endo- and exopolymer biosynthesis in *Anabaena cylindrica* 10 C. Phytochemistry 42(3):655–659. https://doi.org/10.1016/0031-9422(95)00985-X

Lau N-S, Matsui M, Abdullah A-A (2015) Cyanobacteria: photoautotrophic microbial factories for the sustainable synthesis of industrial products. Biomed Res Int 2015:9. https://doi.org/10.1155/2015/754934

Lee JY, Yoo C, Jun SY et al (2010) Comparison of several methods for effective lipid extraction from microalgae. Bioresour Technol 101(Suppl 1):S75–S77. https://doi.org/10.1016/j.biortech.2009.03.058

Leema JTM, Kirubagaran R, Vinithkumar NV et al (2010) High value pigment production from *Arthrospira (Spirulina) platensis* cultured in seawater. Bioresour Technol 101(23):9221–9227. https://doi.org/10.1016/j.biortech.2010.06.120

Lehmann WD (2017) A timeline of stable isotopes and mass spectrometry in the life sciences. Mass Spectrom Rev 36(1):58–85. https://doi.org/10.1002/mas.21497

Levert JM, Xia J (2001) Modeling the growth curve for *Spirulina (Arthrospira) maxima*, a versatile microalga for producing uniformly labelled compounds with stable isotopes. J Appl Phycol 13 (4):359–367. https://doi.org/10.1023/A:1017924422164

Li Z, Hobson P, An W et al (2012) Earthy odor compounds production and loss in three cyanobacterial cultures. Water Res 46(16):5165–5173. https://doi.org/10.1016/j.watres.2012.06.008

Lin PC, Pakrasi HB (2018) Engineering cyanobacteria for production of terpenoids. Planta 249:145–154. https://doi.org/10.1007/s00425-018-3047-y

Lin P-C, Saha R, Zhang F et al (2017) Metabolic engineering of the pentose phosphate pathway for enhanced limonene production in the cyanobacterium *Synechocystis* sp. PCC 6803. Sci Rep 7 (1):17503. https://doi.org/10.1038/s41598-017-17831-y

Liu L, Pohnert G, Wei D (2016) Extracellular metabolites from industrial microalgae and their biotechnological potential. Mar Drugs 14(10):191. https://doi.org/10.3390/md14100191

Llewellyn CA, Airs RL (2010) Distribution and abundance of MAAs in 33 species of microalgae across 13 classes. Mar Drugs 8(4):1273–1291. https://doi.org/10.3390/md8041273

Luan G, Lu X (2018) Tailoring cyanobacterial cell factory for improved industrial properties. Biotechnol Adv 36(2):430–442. https://doi.org/10.1016/j.biotechadv.2018.01.005

Luesch H, Moore RE, Paul VJ et al (2001) Isolation of dolastatin 10 from the marine cyanobacterium Symploca species VP642 and total stereochemistry and biological evaluation of its analogue symplostatin 1. J Nat Prod 64(7):907–910. https://doi.org/10.1021/np010049y

Machmudah S, Diono W, Kanda H et al (2018) Supercritical fluids extraction of valuable compounds from algae: future perspectives and challenges. Eng J 22(5):13–30. https://doi.org/10.4186/ej.2018.22.5.13

Mallikarjun Gouda K, Udaya Sankar K, Sarada R et al (2015) Supercritical CO_2 extraction of functional compounds from spirulina and their biological activity. J Food Sci Technol 52 (6):3627–3633. https://doi.org/10.1007/s13197-014-1426-3

Manning SR, Nobles DR (2017) Impact of global warming on water toxicity: cyanotoxins. Curr Opin Food Sci 18:14–20. https://doi.org/10.1016/j.cofs.2017.09.013

Markl E, Grünbichler H, Lackner M (2018) Cyanobacteria for PHB bioplastics production: a review. IntechOpen, Algae

Mcwilliams A (2018). The global market for carotenoids (FOD025F). In McWilliams A (ed) (FOD025F ed.): BCC Research

Mendes RL, Nobre BP, Cardoso MT et al (2003) Supercritical carbon dioxide extraction of compounds with pharmaceutical importance from microalgae. Inorg Chim Acta 356:328–334. https://doi.org/10.1016/s0020-1693(03)00363-3

Merwe DVD (2015) Cyanobacterial (blue-green algae) toxins. In: Handbook of toxicology of chemical warfare agents. Elsevier, New York, pp 421–429

Miki W, Yamaguchi K, Konosu S (1986) Carotenoid composition of *Spirulina maxima*. Bull Jpn Soc Sci Fish 52(7):1225–1227. https://doi.org/10.2331/suisan.52.1225

Milledge JJ (2011) Commercial application of microalgae other than as biofuels: a brief review. Rev Environ Sci Biotechnol 10(1):31–41. https://doi.org/10.1007/s11157-010-9214-7

Milledge JJ, Heaven S (2011) Disc stack centrifugation separation and cell disruption of microalgae: a technical note. Environ Nat Resour Res 1(1):17–24. https://doi.org/10.5539/enrr.v1n1p17

Minkiewicz P, Dziuba J, Iwaniak A et al (2008) BIOPEP database and other programs for processing bioactive peptide sequences. J AOAC Int 91(4):965–980

Monshupanee T, Nimdach P, Incharoensakdi A (2016) Two-stage (photoautotrophy and heterotrophy) cultivation enables efficient production of bioplastic poly-3-hydroxybutyrate in auto-sedimenting cyanobacterium. Sci Rep 6(1):37121. https://doi.org/10.1038/srep37121

Montero O, Macías-Sánchez MD, Lama CM et al (2005) Supercritical CO_2 extraction of β-carotene from a marine strain of the cyanobacterium *Synechococcus* species. J Agric Food Chem 53 (25):9701–9707. https://doi.org/10.1021/jf051283n

Mubarak M, Shaija A, Suchithra TV (2015) A review on the extraction of lipid from microalgae for biodiesel production. Algal Res 7:117–123. https://doi.org/10.1016/j.algal.2014.10.008

Nabout JC, Da Silva Rocha B, Carneiro FM et al (2013) How many species of Cyanobacteria are there? Using a discovery curve to predict the species number. Biodivers Conserv 22 (12):2907–2918. https://doi.org/10.1007/s10531-013-0561-x

Naghdi FG, González LMG, Chan W et al (2016) Progress on lipid extraction from wet algal biomass for biodiesel production. J Microb Biotechnol 9(6):718. https://doi.org/10.1111/1751-7915.12360

Nazifi E, Wada N, Asano T et al (2015) Characterization of the chemical diversity of glycosylated mycosporine-like amino acids in the terrestrial cyanobacterium *Nostoc commune*. J Photochem Photobiol B Biol 142:154–168. https://doi.org/10.1016/j.jphotobiol.2014.12.008

Niederholtmeyer H, Wolfstadter BT, Savage DF et al (2010) Engineering cyanobacteria to synthesize and export hydrophilic products. Appl Environ Microbiol 76(11):3462–3466. https://doi.org/10.1128/AEM.00202-10

Oliver JWK, Atsumi S (2015) A carbon sink pathway increases carbon productivity in cyanobacteria. Metab Eng 29:106–112. https://doi.org/10.1016/j.ymben.2015.03.006

Ores JDC, Amarante MCA, Kalil SJ (2016) Co-production of carbonic anhydrase and phycobiliproteins by spirulina sp. and *Synechococcus nidulans*. Bioresour Technol 219:219–227. https://doi.org/10.1016/j.biortech.2016.07.133

Panda B, Jain P, Sharma L et al (2006) Optimization of cultural and nutritional conditions for accumulation of poly-β-hydroxybutyrate in *Synechocystis* sp. PCC 6803. Bioresour Technol 97 (11):1296–1301. https://doi.org/10.1016/j.biortech.2005.05.013

Patel A, Mishra S, Pawar R et al (2005) Purification and characterization of C-Phycocyanin from cyanobacterial species of marine and freshwater habitat. Protein Expr Purif 40(2):248–255. https://doi.org/10.1016/j.pep.2004.10.028

Patil G, Chethana S, Sridevi AS et al (2006) Method to obtain C-phycocyanin of high purity. J Chromatogr A 1127(1):76–81. https://doi.org/10.1016/j.chroma.2006.05.073

Pattanaik B, Lindberg P (2015) Terpenoids and their biosynthesis in cyanobacteria. Life 5 (1):269–293. https://doi.org/10.3390/life5010269

Pawlik-Skowrońska B, Toporowska M, Rechulicz J (2012) Simultaneous accumulation of anatoxin-a and microcystins in three fish species indigenous to lakes affected by cyanobacterial blooms. Oceanol Hydrobiol Stud 41(4):53–65. https://doi.org/10.2478/s13545-012-0039-6

Phong WN, Show PL, Ling TC et al (2018) Mild cell disruption methods for bio-functional proteins recovery from microalgae-Recent developments and future perspectives. Algal Res 31:506–516. https://doi.org/10.1016/j.algal.2017.04.005

Pohndorf RS, Camara ÁS, Larrosa AP et al (2016) Production of lipids from microalgae *Spirulina* sp.: influence of drying, cell disruption and extraction methods. Biomass Bioenergy 93:25–32. https://doi.org/10.1016/j.biombioe.2016.06.020

Prates DDF, Radmann EM, Duarte JH et al (2018) Spirulina cultivated under different light emitting diodes: enhanced cell growth and phycocyanin production. Bioresour Technol 256:38–43. https://doi.org/10.1016/j.biortech.2018.01.122

Pulz O, Gross W (2004) Valuable products from biotechnology of microalgae. Appl Microbiol Biotechnol 65(6):635–648. https://doi.org/10.1007/s00253-004-1647-x

Radmer RJ (1996) Algal diversity and commercial algal products. Bioscience 46(4):263–270. https://doi.org/10.2307/1312833

Raposo MFDJ, De Morais RMSC, Bernardo De Morais AMM (2013) Bioactivity and applications of sulphated polysaccharides from marine microalgae. Mar Drugs 11(1):233–252. https://doi.org/10.3390/md11010233

Rastogi RP, Sinha RP (2009) Biotechnological and industrial significance of cyanobacterial secondary metabolites. Biotechnol Adv 27(4):521–539. https://doi.org/10.1016/j.biotechadv.2009.04.009

Raven JA (2013) Rubisco: still the most abundant protein of Earth? New Phytol 198(1):1–3. https://doi.org/10.1111/nph.12197

Research PM (2017) Global market study on spirulina: powder product form segment anticipated to dominate the global market in terms of both value and volume during 2016–2026 (F. Innovation, Trans). pp 229

Rizwan M, Mujtaba G, Memon SA et al (2018) Exploring the potential of microalgae for new biotechnology applications and beyond: a review. Renew Sust Energ Rev 92:394–404. https://doi.org/10.1016/J.RSER.2018.04.034

Ruiz J, Olivieri G, De Vree J et al (2016) Towards industrial products from microalgae. Energy Environ Sci 9(10):3036–3043. https://doi.org/10.1039/C6EE01493C

Saenz JP, Waterbury JB, Eglinton TI et al (2012) Hopanoids in marine cyanobacteria: probing their phylogenetic distribution and biological role. Geobiology 10(4):311–319. https://doi.org/10.1111/j.1472-4669.2012.00318.x

Safi C, Charton M, Ursu AV et al (2014a) Release of hydro-soluble microalgal proteins using mechanical and chemical treatments. Algal Res 3:55–60. https://doi.org/10.1016/j.algal.2013.11.017

Safi C, Ursu AV, Laroche C et al (2014b) Aqueous extraction of proteins from microalgae: effect of different cell disruption methods. Algal Res 3:61–65. https://doi.org/10.1016/j.algal.2013.12.004

Sahena F, Zaidul ISM, Jinap S et al (2009) Application of supercritical CO_2 in lipid extraction—a review. J Food Eng 95(2):240–253. https://doi.org/10.1016/j.jfoodeng.2009.06.026

Sajilata MG, Singhal RS, Kamat MY (2008) Fractionation of lipids and purification of γ-linolenic acid (GLA) from *Spirulina platensis*. Food Chem 109(3):580–586. https://doi.org/10.1016/j.foodchem.2008.01.005

Samantaray S, Mallick N (2014) Production of poly(3-hydroxybutyrate-co-3-hydroxyvalerate) co-polymer by the diazotrophic cyanobacterium *Aulosira fertilissima* CCC 444. J Appl Phycol 26(1):237–245. https://doi.org/10.1007/s10811-013-0073-9

Sarada R, Pillai MG, Ravishankar GA (1999) Phycocyanin from *Spirulina sp.*: influence of processing of biomass on phycocyanin yield, analysis of efficacy of extraction methods and stability studies on phycocyanin. Process Biochem 34(8):795–801. https://doi.org/10.1016/s0032-9592(98)00153-8

Sasso S, Pohnert G, Lohr M et al (2012) Microalgae in the postgenomic era: a blooming reservoir for new natural products. FEMS Microbiol Rev 36(4):761–785. https://doi.org/10.1111/j.1574-6976.2011.00304.x

Sathasivam R, Ki JS (2018) A review of the biological activities of microalgal carotenoids and their potential use in healthcare and cosmetic industries. Mar Drugs 16(1):E26. https://doi.org/10.3390/md16010026

Sato N, Murata N (1988) [24] Membrane lipids. In: Methods enzymology, vol 167. Elsevier, Amsterdam, pp 251–259

Schellekens RC, Stellaard F, Woerdenbag HJ et al (2011) Applications of stable isotopes in clinical pharmacology. Br J Clin Pharmacol 72(6):879–897. https://doi.org/10.1111/j.1365-2125.2011.04071.x

Schmidt C (2017) First deuterated drug approved. Nat Biotechnol 35(6):493–494. https://doi.org/10.1038/nbt0617-493

Sekar S, Chandramohan M (2008) Phycobiliproteins as a commodity: trends in applied research, patents and commercialization. J Appl Phycol 20(2):113–136. https://doi.org/10.1007/s10811-007-9188-1

Selvaraj G, Kaliamurthi S, Cakmak ZE et al (2017) RuBisCO of microalgae as potential targets for nutraceutical peptides: a computational study. Sci Alert 16(4–6):130–144. https://doi.org/10.3923/biotech.2017.130.144

Shih PM (2015) Cyanobacterial evolution: fresh insight into ancient questions. Curr Biol 25(5): R192–R193. https://doi.org/10.1016/j.cub.2014.12.046

Silkina A, Zacharof M-P, Hery G et al (2017) Formulation and utilisation of spent anaerobic digestate fluids for the growth and product formation of single cell algal cultures in heterotrophic and autotrophic conditions. Bioresour Technol 244:1445–1455. https://doi.org/10.1016/j.biortech.2017.05.133

Silva-Stenico ME, Silva CSP, Lorenzi AS et al (2011) Non-ribosomal peptides produced by Brazilian cyanobacterial isolates with antimicrobial activity. Microbiol Res 166(3):161–175. https://doi.org/10.1016/j.micres.2010.04.002

Singh AK, Mallick N (2017) Advances in cyanobacterial polyhydroxyalkanoates production. FEMS Microbiol Lett 364(20). https://doi.org/10.1093/femsle/fnx189

Singh R, Parihar P, Singh M et al (2017) Uncovering potential applications of cyanobacteria and algal metabolites in biology, agriculture and medicine: current status and future prospects. Front Microbiol 8:515. https://doi.org/10.3389/fmicb.2017.00515

Sinha RP, Klisch M, Helbling EW et al (2001) Induction of mycosporine-like amino acids (MAAs) in cyanobacteria by solar ultraviolet-B radiation. J Photochem Photobiol B 60(2–3):129–135. https://doi.org/10.1016/S1011-1344(01)00137-3

Sioutopoulos DC, Karabelas AJ (2012) Correlation of organic fouling resistances in RO and UF membrane filtration under constant flux and constant pressure. J Membr Sci 407:34–46. https://doi.org/10.1016/j.memsci.2012.03.036

Slade R, Bauen A (2013) Micro-algae cultivation for biofuels: cost, energy balance, environmental impacts and future prospects. Biomass Bioenergy 53:29–38. https://doi.org/10.1016/j.biombioe.2012.12.019

Sommella E, Conte GM, Salviati E et al (2018) Fast profiling of natural pigments in different spirulina (*Arthrospira platensis*) dietary supplements by DI-FT-ICR and evaluation of their antioxidant potential by pre-column DPPH-UHPLC assay. Molecules 23(5):1152. https://doi.org/10.3390/molecules23051152

Specht E, Mayfield S (2014) Algae-based oral recombinant vaccines. Front Microbiol 5(60). https://doi.org/10.3389/fmicb.2014.00060

Spolaore P, Joannis-Cassan C, Duran E et al (2006) Commercial applications of microalgae. J Biosci Bioeng 101(2):87–96. https://doi.org/10.1263/JBB.101.87

Stal LJ, Heyer H, Jacobs G (1990) Occurrence and role of poly-hydroxy-alkanoate in the cyanobacterium *Oscillatoria limosa*. In: Novel biodegradable microbial polymers. Springer, Dordrecht, pp 435–438

Suh HJ, Lee HW, Jung J (2003) Mycosporine glycine protects biological systems against photodynamic damage by quenching singlet oxygen with a high efficiency. Photochem Photobiol 78 (2):109–113. https://doi.org/10.1562/0031-8655(2003)078<0109:MGPBSA>2.0.CO;2
Sun Y, Chang R, Li Q et al (2016) Isolation and characterization of an antibacterial peptide from protein hydrolysates of *Spirulina platensis*. Eur Food Res Technol 242(5):685–692. https://doi.org/10.1007/s00217-015-2576-x
Takahashi H, Miyake M, Tokiwa Y et al (1998) Improved accumulation of poly-3-hydroxybutyrate by a recombinant cyanobacterium. Biotechnol Lett 20(2):183–186. https://doi.org/10.1023/A:1005392827791
't Lam GP, Vermue MH, Eppink MHM et al (2018) Multi-product microalgae biorefineries: from concept towards reality. Trends Biotechnol 36(2):216–227. https://doi.org/10.1016/j.tibtech.2017.10.011
Takaichi S, Mochimaru M (2007) Carotenoids and carotenogenesis in cyanobacteria: unique ketocarotenoids and carotenoid glycosides. Cell Mol Life Sci 64(19–20):2607–2619. https://doi.org/10.1007/s00018-007-7190-z
Tarko T, Duda-Chodak A, Kobus M, (2012) Influence of growth medium composition on synthesis of bioactive compounds and antioxidant properties of selected strains of Arthrospira cyanobacteria. Czech J Food Sci 30(3):258–267
Tavanandi HA, Chandralekha Devi A, Raghavarao K (2018) A newer approach for the primary extraction of allophycocyanin with high purity and yield from dry biomass of *Arthrospira platensis*. Sep Purif Technol 204:162–174. https://doi.org/10.1016/j.seppur.2018.04.057
Troschl C, Meixner K, Drosg B (2017) Cyanobacterial PHA production—review of recent advances and a summary of three years' working experience running a pilot plant. Bioengineering 4(2):26. https://doi.org/10.3390/bioengineering4020026
Ungerer J, Pakrasi HB (2016) Cpf1 is a versatile tool for CRISPR genome editing across diverse species of cyanobacteria. Sci Rep 6:39681. https://doi.org/10.1038/srep39681
Vandamme D, Foubert I, Muylaert K (2013) Flocculation as a low-cost method for harvesting microalgae for bulk biomass production. Trends Biotechnol 31(4):233–239. https://doi.org/10.1016/j.tibtech.2012.12.005
Vavitsas K, Fabris M, Vickers CE (2018) Terpenoid metabolic engineering in photosynthetic microorganisms. Genes (Basel) 9(11):520. https://doi.org/10.3390/genes9110520
Vernès L, Granvillain P, Chemat F et al (2015) Phycocyanin from *Arthrospira platensis*. Production, extraction and analysis. Curr Biotechnol 4(4):481–491
Vijayakumar S, Menakha M (2015) Pharmaceutical applications of cyanobacteria—a review. J Acute Med 5(1):15–23. https://doi.org/10.1016/j.jacme.2015.02.004
Wada N, Sakamoto T, Matsugo S (2015) Mycosporine-like amino acids and their derivatives as natural antioxidants. Antioxidants 4(3):603–646. https://doi.org/10.3390/antiox4030603
Walsh G (2010) Biopharmaceutical benchmarks 2010. Nat Biotechnol 28:917. https://doi.org/10.1038/nbt0910-917
Wang H, Zhang W, Chen L et al (2013) The contamination and control of biological pollutants in mass cultivation of microalgae. Bioresour Technol 128:745–750. https://doi.org/10.1016/j.biortech.2012.10.158
Welker M, Von Döhren H (2006) Cyanobacterial peptides—nature's own combinatorial biosynthesis. FEMS Microbiol Rev 30(4):530–563. https://doi.org/10.1111/j.1574-6976.2006.00022.x
Wu W, Maravelias CT (2018) Synthesis and techno-economic assessment of microbial-based processes for terpenes production. Biotechnol Biofuels 11(1):294. https://doi.org/10.1186/s13068-018-1285-7
Xue Y, He Q (2015) Cyanobacteria as cell factories to produce plant secondary metabolites. Front Bioeng Biotechnol 3:57. https://doi.org/10.3389/fbioe.2015.00057
Yadav S, Sinha R, Tyagi M et al (2011) Cyanobacterial secondary metabolites. Int J Pharm Bio Sci 2(2):B144
Yang J (2016) Deuterium: discovery and applications in organic chemistry. Elsevier, New York, p 130

Zachleder V, Vítová M, Hlavová M et al (2018) Stable isotope compounds—production, detection, and application. Biotechnol Adv 36(3):784–797. https://doi.org/10.1016/j.biotechadv.2018.01.010
Zanchett G, Oliveira-Filho EC (2013) Cyanobacteria and cyanotoxins: from impacts on aquatic ecosystems and human health to anticarcinogenic effects. Toxins 5(10):1896–1917. https://doi.org/10.3390/toxins5101896
Zhang A, Carroll AL, Atsumi S (2017) Carbon recycling by cyanobacteria: improving CO_2 fixation through chemical production. FEMS Microbiol Lett 364(16). https://doi.org/10.1093/femsle/fnx165

Chapter 16
European Union Legislation and Policies Relevant for Algae

Felix Leinemann and Valentina Mabilia

Abstract As a relative newcomer on the European market, the algae industry needs to face several challenges, including a regulatory environment that is often fragmented and unequally implemented. In this chapter, we look at the European policy and regulatory framework applicable to algae from their production or collection to their processing into food, feed, pharmaceuticals, chemicals or biofuels. Some European Union policies are relevant for all algae, namely those that deal with aquatic resources, renewable biological resources and anthropogenic pressures on the environment. Other policies and regulations are only applicable to one phase of the value chain, such as the Common Fisheries Policy or rules on food packaging. Other rules only apply to specific products, such as the European Union Regulation on the Registration, Evaluation, Authorisation and Restriction of Chemicals, or the rules on additives in feed. Needless to say, not every single rule will apply to every single algae producer or processor, depending on their production methods, final products and size of the operation. Given the industry's potential, the European Commission is currently investigating the main regulatory, financial, technological and structural challenges it is facing.

16.1 Introduction

This chapter looks at the EU legislation and policies applicable to algae throughout their collection, cultivation, processing and marketing. These rules belong to different policy areas and are implemented through different legal instruments.

F. Leinemann · V. Mabilia (✉)
Blue Economy Sectors, Aquaculture and Maritime Spatial Planning, DG Maritime Affairs and Fisheries (MARE), European Commission, Brussels, Belgium
e-mail: valentina.mabilia@ec.europa.eu

A. Hallmann, P. H. Rampelotto (eds.), *Grand Challenges in Algae Biotechnology*, Grand Challenges in Biology and Biotechnology,
https://doi.org/10.1007/978-3-030-25233-5_16

The algae industry[1] is still relatively small in the European Union. Only Ireland and France appear amongst the top 10 countries in terms of quantity of seaweed harvested from the wild and there are no European countries in the top 10 for the production of farmed seaweed (FAO 2018). The EU-wide production of both macro- and microalgae was 0.23 million tonnes wet mass in 2015 (roughly corresponding to 0.027 million tonnes dry weight) (Camia et al. 2018). The production of microalgae is less than 500 tonnes of dry weight per year according to industry sources (Verdelho Vieira 2017), with a theoretical estimated potential of 50 million tonnes of dry weight per year (Skarka 2012). Data provided by the industry claims that the EU algae biomass sector currently employs 14,000 people and is worth 1.69 billion euros (European Commission 2018a).

16.2 Policies

The European Union does not have a single policy dedicated to algae, but rather a mosaic of policies that touch upon the topic from different angles. Some EU policies are relevant for all algae: seagrasses, seaweeds and microalgae provide a host of ecosystem services including carbon sequestration, coastal defence and removal of dissolved nutrients (Chopin 2014). They are at the bottom of the food web and as a consequence have a deep influence on what kind of marine life develops in certain areas (Guo et al. 2016). Like the rest of our marine life, they are currently subject to cumulative pressures from climate change and other anthropogenic pressures (Unsworth et al. 2014). In this sense, the most important pieces of EU legislation that apply to algae are the Habitats Directive (Council Directive 92/43/EEC), the Marine Strategy Framework Directive (Directive 2008/56/EC) and the Water Framework Directive (Directive 2000/60/EC).

As aquatic resources, they fall under the European Commission's Maritime and Fisheries policy. The Blue Growth Strategy (European Commission 2012) had initially focused on five promising sectors, including aquaculture and marine biotechnology. This has evolved into the concept of "Blue Bioeconomy", a sector that includes the algae industry. In order to better understand the status of the Blue Bioeconomy, the European Commission has set up the Blue Bioeconomy Forum, a group of stakeholders, which will work on a roadmap for the sector over the course of the next year, outlining the main challenges for its development and the possible European solutions.

As renewable biological resources, algae are also a small but growing part of the European Commission's Bioeconomy Strategy (European Commission 2018d). The strategy is a high-level document setting out the commission's priorities for economic activities based on renewable biological resources from land and sea. It is the

[1]We use "algae industry" as an umbrella term to define economic activities using algae to provide products or services. That includes both macro- and microalgae.

strategic direction for future EU policy in this area, underlining sustainability, circularity and multidisciplinarity.

The strategy is related and fits within the objectives of the EU decarbonisation strategy, which seeks to improve the productivity of aquatic and marine resources in order to capture the full range of opportunities of the bio-economy for tackling climate change (Communication 2018/773).

Given their properties and their potential, algae have also been the subject of many research and innovation projects funded under the EC's Research policy.

The legal and policy framework becomes more complex when we consider the different commercial uses algae can be put to. These uses include products for human consumption, animal feed, pharmaceuticals, cosmetics, fertilisers, raw materials for the chemicals industry, bioremediation and algae-based fuels (Sharma and Sharma 2017).

16.3 Harvest and Production

Macroalgae in the EU are either harvested from the wild or farmed in aquaculture facilities. Microalgae are almost exclusively farmed (EUMOFA 2018). Figure 16.1 summarises the relations between pieces of EU legislation that apply to the production and harvest of algae in the EU.

The Commission's Common Fisheries Policy includes policies on both capture fisheries and aquaculture (including freshwater aquaculture), thereby covering all algae production. The commission issued strategic guidelines for the sustainable development of EU aquaculture in 2013, and the Member States have been asked to set up Multiannual National Strategic Plans for the promotion of sustainable aquaculture. Some Member States (France, Spain, Ireland, the Netherlands, Germany,

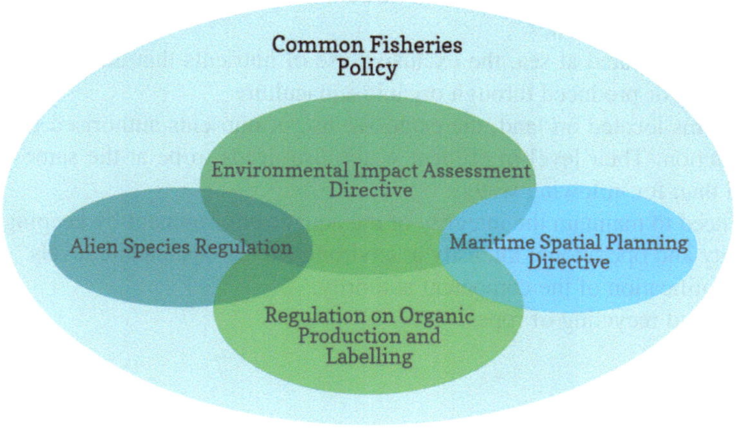

Fig. 16.1 EU legislation covering the harvest and production of algae

Italy and Portugal) have included algae in their strategies to a greater extent, whilst others just mentioned them in passing (European Commission 2018c).

Alien species legislation (Council Regulation (EC) No 708/2007) also applies to algae. Aquaculture operators who wish to farm species outside of their natural range have to apply for a permit. Each country's competent authorities can manage the process directly or through an advisory committee. Provisions are less stringent for algae farmed in closed facilities or whose movement poses no risk to the environment.

In order to set up new algae farms, an Environmental Impact Assessment (Directive 2011/92/EU) may also be required, depending on the threshold or case-by-case assessment of each Member State. This assessment will be compulsory as of 1st January 2021, for all farms producing more than 20 tonnes per year and wishing to obtain an organic certification (Regulation (EU) 2018/848).

Producers aiming to obtain the "organic" label also have to comply with the EU rules on the production and labelling of organic products. These rules have been recently modified and a new regulation will apply as of 1st January 2021 (ibid). The main requirements for algae farms include:

- The obligation to place the farms in locations that are not subject to contamination and that would be considered suitable for the collection of molluscs and bivalves for human consumption.[2]
- The separation of organic and non-organic production units.
- The need to conduct an environmental assessment for new algae farms producing more than 20 tonnes per year.
- A sustainable management plan detailing measures to minimise any negative impact on the environment, measures taken against predators and a waste reduction schedule.
- The use of sustainable practices throughout the production cycle.
- The regular introduction of juvenile wild algae in order to maintain a wide gene pool.
- The limitation of fertiliser use to those authorised for organic production and then only in indoor facilities.
- For farms located at sea, the exclusive use of nutrients that are either naturally occurring or produced through organic aquaculture.
- For farms located on land, the exclusive use of nutrients authorised for organic production. Their level in effluent water also needs to be at the same or lower level than for inflowing water.
- The need to maintain the integrity of the aquatic environment by keeping culture density and operational intensity at environmentally sustainable levels.
- The notification of the competent authority.
- Reuse and recycling of ropes.

[2]As defined in Regulation (EC) No 854/2004 or Regulation (EU) 2017/625

Collected wild algae can be labelled as organic if:

- The existing biomass is estimated before algae collection.
- The areas where they grow have high ecological status[3] or have been classified as suitable for the collection of molluscs and bivalves for human consumption by the Member States' competent authority.[4]
- The collection does not affect significantly the stability of the natural ecosystem, the maintenance of the species in the collection area, or have any other significant impact on the aquatic environment.
- The operators maintain records of all of the above.

Existing rules on organic production and labelling explicitly apply to microalgae as well as seaweeds (Regulation 2016/673).

The Maritime Spatial Planning Directive (Directive 2014/89/EU) is only relevant when algae farming or harvest happens at sea. The Directive requires Member States to produce spatial plans for their marine waters by 2021, in order to "contribute to the sustainable development of energy sectors at sea, of maritime transport, and of the fisheries and aquaculture sectors, and to the preservation, protection and improvement of the environment, including resilience to climate change impact" (Directive 2014/89/EU art. 5). In addition, the directive promotes the coexistence of relevant activities and uses, including by encouraging multiple purpose uses, where it is sustainable.

The European Committee for Standardisation (CEN) is currently working on the creation of a standard for algae and algae products (CEN/TC 454), which will provide definitions, classification and requirements for algae processing in order to ensure the quality of algae products in the EU.

16.4 Food, Feed and Pharmaceuticals

Algae being farmed or harvested for food, feed and pharmaceuticals have to comply with very strict standards in order to be considered fit for purpose. The EU regulatory framework in these areas is complex, as can be seen in Figs. 16.2 and 16.3, showing the relations between the main pieces of legislation in these areas.

Algae being farmed or harvested for use as food have to comply with EU food safety legislation (Regulation (EC) No 178/2002). That includes:

- Rules on the traceability of pre-packaged food (Directive 2011/91/EU), which makes it compulsory to identify the lot to which a foodstuff belongs.
- Rules on food hygiene (Regulation (EC) No 852/2004), which set standards to be met for the premises, transport conditions, waste, water supply, personal hygiene of workers and heat treatment processes.

[3]As defined in Directive 2000/60/EC.

[4]In accordance with Regulation (EC) No 854/2004 or Regulation (EU) 2017/625.

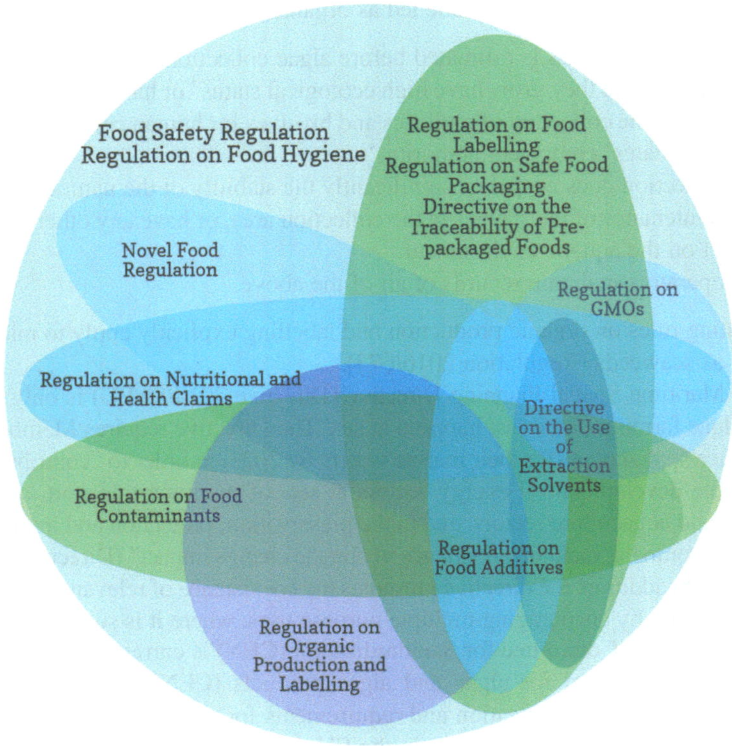

Fig. 16.2 EU legislation covering algae as food

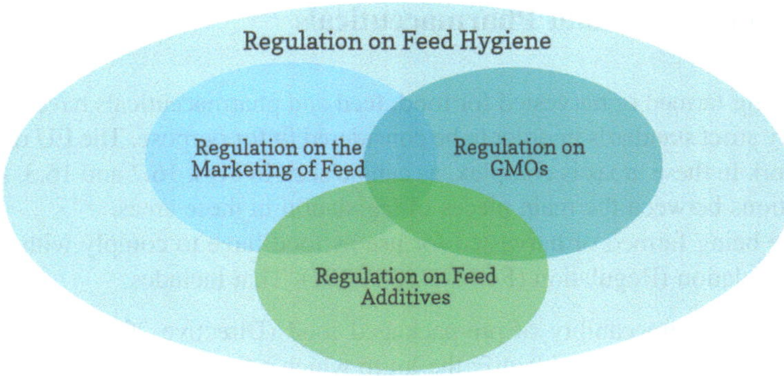

Fig. 16.3 EU legislation covering algae as feed

- Rules on safe food packaging (Regulation (EC) No 1935/2004 and Commission Regulation (EC) No 2023/2006), including those on plastic (Commission Regulation (EC) No 1895/2005), cellulose (Commission Directive 2007/42/EC) and recycled plastic packaging (Commission Regulation (EC) No 282/2008) where relevant.
- Rules on contaminants in foods (Commission Regulation (EC) No 1881/2006), which set the maximum levels allowed in food.
- Rules on the use of extraction solvents (Directive 2009/32/EC) and food additives (Regulation (EC) No 1333/2008).
- Rules on food labelling (Regulation (EU) No 1169/2011), unless the algae are being sold unprocessed.
- Rules on nutritional and health claims (Regulation (EC) No 1924/2006).
- Rules on GMOs (Regulation (EC) No 1829/2003).

Given that in the EU, the consumption of algae as food was not significant before 1997 (Le bras et al. 2014), only a few species listed in Table 16.1 are considered traditional foods. Most species are considered novel foods and have to be authorised in accordance with the relevant Regulation (Regulation (EU) 2015/2283), regardless of whether they are being used as foods or food additives. This is often considered one of the main regulatory barriers for new producers, since only a small number of species is currently authorised (Probst et al. 2015). The authorisation process, managed by the European Food Safety Agency (EFSA) can last several years and the average cost is between 50,000 and 200,000 euros (van Loveren and

Table 16.1 Algae species considered as traditional foods and authorised as novel foods in the EU

Traditional foods	Authorised novel foods
Chlorella pyrenoidosa	Algal oil from *Ulkenia* sp.
Spirulina spp.	Astaxanthin-rich *Haematococcus pluvialis*
Ecklonia bicyclis	*Odontella aurita*
Fucus vesiculosus	Dried *Tetraselmis chuii*
Palmaria palmata	Fucoidan extract from *Fucus vesiculosus*
Hizikia fusiforme	Fucoidan extract from *Undaria pinnatifida*
Chondrus crispus	*Ecklonia cava* phlorotannins
Laminaria digitata	
Laminaria japonica, Saccharina japonica	
Porphyra tenera	
Ascophyllum nodosum	
Ulva lactuca	
Himanthalia elongata	
Fucus serratus	
Saccharina latissima, Laminaria saccharina	
Undaria pinnatifida	
Gracilaria verrucosa	
Enteromorpha sp.	
Alaria esculenta	

Source: European Commission (2018b)

Unamunzaga 2018). The procedure was reformed in 2015 and the new authorisation process is designed to be faster, but given that the new rules only entered into force in 2018 it is too soon to judge their effects.

The EU issued a recommendation (EU recommendation 2018/464) to address the potential presence of heavy metal particles or other contaminants in edible seaweed. This is considered an issue for brown seaweeds in particular (see, e.g. Bouga and Combet 2015). The recommendation sets up a list of species in which levels of arsenic, cadmium, iodine, lead and mercury have to be monitored.

The European Union also introduced new rules on the production and labelling of organic products (Regulation (EU) 2018/848), which require the operator processing organic algae to:

- Systematically identify the critical processing steps.
- Take precautionary measures.
- Implement suitable cleaning measures, monitor their effectiveness and keep records of those operations.
- Guarantee that non-organic products are not placed on the market with an indication referring to organic production.
- Process organic and non-organic products separately.
- Use specifically authorised cleaning and disinfection products.
- Respect a conversion period of 6 months or a full production cycle (whichever is longer) when converting a non-organic production unit into an organic one.
- Notify the competent authorities.

For many years, algae have been used as feed in aquaculture hatcheries thanks to their nutritionally complete profile of amino acids, vitamins, minerals and essential fatty acids. They can be used as feed additives and have been included in the catalogue of feed materials established by Commission Regulation (EU) 68/2013.

The European Commission has established the European Union Register of Feed Additives, a list of additives and their relevant authorisation regulations. The register has only informative purposes and does not replace Community legal acts.

Algae being farmed or harvested for use as animal feed need to comply with similarly extensive rules, including:

- Rules on feed hygiene (Regulation (EC) No 183/2005), to keep the risk of biological, chemical and physical contamination of feed, animals and animal products as low as reasonably achievable.
- Rules on the marketing of feed (Regulation (EC) No 767/2009) and the related implementing acts.
- Rules on feed additives (Regulation (EC) No 1831/2003), which are currently being reviewed.

Regulation (EC) No. 1831/2003 in particular states in Article 24(6): "the person who, for the first time, places on the market a feed material that is not listed in the catalogue shall immediately notify its use to the representatives of the European feed business sectors referred to in Article 26(1). The representatives of the European feed business sectors shall publish a register of such notifications on the Internet and update the Register on a regular basis". Authorisations are granted for specific animal species, specific conditions of use and for 10 years periods.

For potential uses of genetically modified algae, the regulations on food safety and labelling of genetically modified food and genetically modified feed EC 1829/2003 and 1830/2003 apply.

Algae have increasingly been the object of research aimed at developing new pharmaceuticals, given they are a source of a wide range of bioactive compounds (Dmytryk et al. 2017). Medicinal products in the EU may be authorised either by a centralised procedure, run by the EU Medicines Agency or a mutual recognition procedure, which involves having the medicinal product authorised in a single Member State at first and then starting a process for the recognition of that authorisation in other Member States (Directive 2001/83/EC and Regulation (EC) No 726/2004). Clinical trials are also regulated by EU legislation through the Clinical trials Directive (Directive 2001/20/EC), which will be substituted by a Regulation in 2020 (Regulation (EU) No 536/2014). In order to overcome some of the existing bottlenecks in the biodiscovery process and shorten the time to market (the process to develop, test and authorise new medicinal products is notoriously lengthy) the EU has funded projects like PharmaSea (funded through Horizon 2020), focussing specifically on compounds of marine origin.

16.5 Chemicals (Including Fertilisers and Cosmetics)

Algae are also utilised as sources of fine chemicals, botanical extracts and active substances in several applications, for example beta-carotene, astaxanthin, phycocyanin, fucoxanthin, EPA, as well as numerous specialty high-value applications in pharmaceutics, medicine and cosmetics.

The main relevant piece of legislation for chemicals in the EU is the Regulation on the Registration, Evaluation, Authorisation and Restriction of Chemicals, also known as REACH (Regulation (EC) No. 1907/2006). This is a comprehensive set of rules that apply to chemical substances and compounds, placing the responsibility of managing the associated risks onto the industry. Anyone who manufactures or imports 1 tonne or more of a substance needs to register it with the European Chemicals Agency (ibid. Article 6). The regulation grants some exemptions: for instance, substances manufactured or imported for the purposes of research and development do not need to be registered for 5 years. On the contrary, substances that pose high risks cannot be used until an authorisation is issued and the applicants are asked to analyse the feasibility of substitution (ibid. Article 55).

Separate pieces of legislation cover some specific uses of chemicals, as shown in Fig. 16.4.

Another important piece of legislation that concerns companies using algae to produce chemicals is the Regulation on classification, labelling and packaging of substances and mixtures, also known as CLP (Regulation (EC) No 1272/2008). The Regulation incorporates the classification criteria and labelling rules agreed at UN

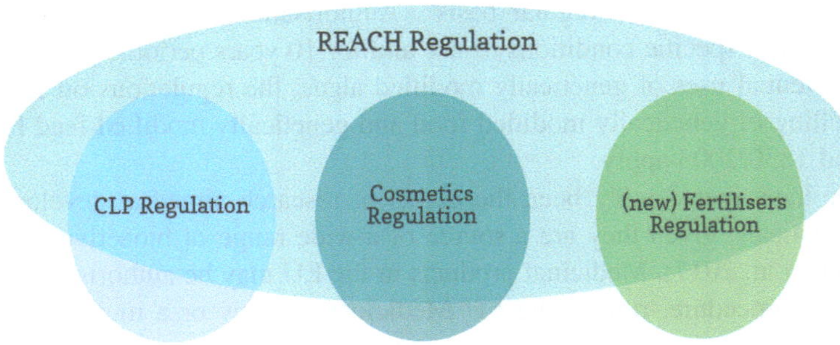

Fig. 16.4 EU legislation covering chemicals

level, the Globally Harmonised System of Classification and Labelling of Chemicals (GHS). It introduces new classification criteria (ibid. Article 18), hazard symbols (ibid. Article 19) (pictograms) and labelling phrases.

The Regulation requires companies to classify, label and package appropriately their hazardous chemicals. They also need to notify these actions to the European Chemicals Agency (ibid. Article 40), which maintains a Classification and Labelling Inventory, before they are allowed to place the products on the market.

Cosmetic products, including those containing algae and their derivatives, are subject to specific rules (Regulation (EC) No. 1223/2009). The main requirements for cosmetics are:

- None of the products or their ingredients can be tested on animals.
- The product should be safe for human health when used under normal or reasonably foreseeable conditions.
- A "responsible person" needs to be identified for every cosmetic product placed on the market. This must be a legal or natural person within the EU who is required to keep certain product information available. This information must be easily accessible to the competent authorities and can be requested in the case of medical emergency.
- Labels must include the name and address of the "responsible person", nominal content given by weight or by volume, the cosmetic product's shelf life, any precautions to be observed, the batch number of manufacture, the product's function and its ingredients in descending order of weight.

Fertilisers are another category of chemical products subject to an additional subset of rules (Regulation (EC) No. 2003/2003), although the current regulation only applies to mineral fertilisers. The regulation has recently been reviewed and a new one, widening its application to other fertilisers (potentially including those sourced from algae), is expected to be adopted soon.

The European Commission's industrial policy applies to all of these sectors and is designed to support European companies through broader instruments like the Investment Plan for Europe, the Single Market Strategy, the Digital Single Market strategy, the Circular Economy package, Energy Union, Capital Markets Union, the New Skills Agenda for Europe and the Trade for All Strategy.

16.6 Energy and Trade

Algae raised substantial interest as potential sources of advanced biofuels because of the potential advantages they offer: location at sea or on non-fertile lands, high yield and growth, photosynthetic efficiency, existence of marketable co-products and services (Rocca et al. 2015). Significant research efforts have been made in this field, including through EU funding.[5] The results are encouraging, but the technology is not yet mature enough: there are still technological, environmental and energetic challenges to overcome, on top of the question of economic viability (ibid.). Once these issues are solved, algal biofuels will have to comply with:

- The EU standard for biodiesel, the EU standard for bioethanol or the EU standard for bioethanol.
- The Renewable Energy Directive, which sets out precise sustainability criteria for biofuels to be considered renewable energy sources (Directive (EU) 2018/2001).

Finally, for the purposes of external trade algae have been assigned a customs code number (Regulation (EEC) No. 2658/87) under Chap. 12 "oil seeds and oleaginous fruits; miscellaneous grains, seeds and fruit; industrial or medicinal plants; straw and fodder". In other words, for these purposes algae are considered "plants", even though such a definition is scientifically inaccurate.

16.7 Conclusion: Challenges and Opportunities

As a relative newcomer, the algae industry faces several challenges in Europe, including regulatory, environmental, financial and market-related ones.

Regulatory barriers include the lack of harmonised legislation for certain aspects, differences in the implementation of EU law across the continent, the existence of rules that do not take into account the sector's specificities and the complexity and high burden of compliance. The applicable legislation was often not designed with algae in mind, and both producers and manufacturers need to be aware of many different rules.

Environmental challenges relate to the industry's own environmental sustainability, its potential to provide remediation services but also the way it is negatively influenced by pollution and climate change.

[5]CYANOFACTORY, MIRACLES, BIOMIC Fuels, BIOFAT, BIOMARA, to quote but a few projects.

Financial challenges include the difficulty for companies to find investors in a sector, which is still relatively unknown and considered high risk, but also the ones inherent in the nature of many algae start-ups, which may lack the skills to secure the funding they need.

Market challenges come from the lack of consumer awareness and understanding of algae products, lack of reliable data on supply and demand in Europe.

Nevertheless, positive developments are ongoing: regulations are changing and adapting, awareness of environmental concerns is rising, ecosystem services are being mapped by the EU's biodiversity strategy and consumer preferences move towards more ethical products.

This is an industry that shows great potential because of its sustainability credentials, its role in ensuring decarbonisation and its potential as a reliable source of food, feed and biomaterials. That is why the European Commission is trying to find out how to facilitate its development. The EC's climate and energy targets, its policies on biodiversity and the circular economy should all contribute to securing a sound future for this industry. Funding gaps, regulatory and market challenges are being explored further through the work of the Blue Bioeconomy Forum.

References

Blue Bioeconomy Forum (2018). https://webgate.ec.europa.eu/maritimeforum/en/frontpage/1349. Accessed 25 Mar 2019

Bouga M, Combet E (2015) Emergence of seaweed and seaweed-containing foods in the UK: focus on labeling, iodine content, toxicity and nutrition. Foods 4:240–253

Camia A, Robert N, Jonsson R, Pilli R, García-Condado S, López-Lozano R, van der Velde M, Ronzon T, Gurría P, M'Barek R, Tamosiunas S, Fiore G, Araujo R, Hoepffner N, Marelli L, Giuntoli J (2018) Biomass production, supply, uses and flows in the European Union. First results from an integrated assessment. Publications Office of the European Union, Luxembourg, p 72

Chopin T (2014) Seaweeds: top mariculture crop, ecosystem service provider. Global Aquaculture Advocate. https://www2.unb.ca/chopinlab/articles/files/Chopin%202014%20GAA%20Seaweeds%20Top%20mariculture%20crop,%20ecosystem%20service%20provider%202.pdf. Accessed 25 Mar 2019

Commission Directive 2007/42/EC of 29 June 2007 relating to materials and articles made of regenerated cellulose film intended to come into contact with foodstuffs

Commission Implementing Regulation (EU) 2016/673 of 29 April 2016 amending Regulation (EC) No 889/2008 laying down detailed rules for the implementation of Council Regulation (EC) No 834/2007 on organic production and labelling of organic products with regard to organic production, labelling and control

Commission Regulation (EC) No 1881/2006 of 19 December 2006 setting maximum levels for certain contaminants in foodstuffs

Commission Regulation (EC) No 1895/2005 of 18 November 2005 on the restriction of use of certain epoxy derivatives in materials and articles intended to come into contact with food

Commission Regulation (EC) No 2023/2006 of 22 December 2006 on good manufacturing practice for materials and articles intended to come into contact with food

Commission Regulation (EC) No 282/2008 of 27 March 2008 on recycled plastic materials and articles intended to come into contact with foods

Communication COM (2018) 773 from the Commission to the European Parliament, the European Council, the Council, the European Economic and Social Committee, the Committee of the Regions and the European Investment Bank. A Clean Planet for all. A European strategic long-term vision for a prosperous, modern, competitive and climate neutral economy
Communication from the Commission to the European Parliament, the Council, the European Economic and Social Committee and the Committee of the Regions (2012) Blue Growth opportunities for marine and maritime sustainable growth
Communication from the Commission to the European Parliament, the Council, the European Economic and Social Committee and the Committee of the Regions (2013) Strategic Guidelines for the sustainable development of EU aquaculture
Council Directive 92/43/EEC of 21 May 1992 on the conservation of natural habitats and of wild fauna and flora
Council Regulation (EEC) No 2658/87 of 23 July 1987 on the tariff and statistical nomenclature and on the Common Customs Tariff
Council Regulation (EC) No 708/2007 of 11 June 2007 concerning use of alien and locally absent species in aquaculture
Directive (EU) 2018/2001 of the European Parliament and of the Council of 11 December 2018 on the promotion of the use of energy from renewable sources
Directive 2000/60/EC of the European Parliament and of the Council of 23 October 2000 establishing a framework for Community action in the field of water policy
Directive 2001/20/EC of the European Parliament and of the Council of 4 April 2001 on the approximation of the laws, regulations and administrative provisions of the Member States relating to the implementation of good clinical practice in the conduct of clinical trials on medicinal products for human use
Directive 2001/83/EC of the European Parliament and of the Council of 6 November 2001 on the Community code relating to medicinal products for human use
Directive 2008/56/EC of the European Parliament and of the Council of 17 June 2008 establishing a framework for community action in the field of marine environmental policy
Directive 2009/32/EC of the European Parliament and of the Council of 23 April 2009 on the approximation of the laws of the Member States on extraction solvents used in the production of foodstuffs and food ingredients
Directive 2011/91/EU of the European Parliament and of the Council of 13 December 2011 on indications or marks identifying the lot to which a foodstuff belongs
Directive 2011/92/EU of the European Parliament and of the Council of 13 December 2011 on the assessment of the effects of certain public and private projects on the environment
Directive 2014/89/EU of the European Parliament and of the Council of 23 July 2014 establishing a framework for maritime spatial planning
Dmytryk A, Łukasz T, Katarzyna C (2017) Algae as source of pharmaceuticals. In: Tripathi B, Kumar D (eds) Prospects and challenges in algal biotechnology. Springer, Singapore
EU Register for feed additives. https://ec.europa.eu/food/safety/animal-feed/feed-additives/eu-register_en. Accessed 25 Mar 2019
European Commission (2018a) The 2018 Annual economic report on EU Blue Economy. Publications Office of the European Union, Luxembourg, p 75
European Commission (2018b) https://ec.europa.eu/food/safety/novel_food/authorisations_en. Accessed 25 Mar 2019
European Commission (2018c) https://ec.europa.eu/fisheries/cfp/aquaculture/multiannual-national-plans. Accessed 25 Mar 2019
European Commission (2018d) A sustainable bioeconomy for Europe: strengthening the connection between economy, society and the environment: Updated Bioeconomy Strategy. Publications Office of the European Union, Luxembourg
European Federation of Pharmaceutical Industries and Associations (2017) The pharmaceutical industry in figures. EFPIA, Brussels, p 6

European Market Observatory for Fisheries and Aquaculture Products (2018) Blue bioeconomy: situation report and perspectives. Publications Office of the European Union, Luxembourg, p 13
European Parliament (2018). http://www.europarl.europa.eu/legislative-train/theme-new-boost-for-jobs-growth-and-investment/file-review-of-the-fertilising-products-regulation. Accessed 25 Mar 2019
FAO (2018) The state of world fisheries and aquaculture 2018—meeting the sustainable development goals. FAO, Rome, p 25
Guo F, Kainz M, Sheldon F, Bunn S (2016) The importance of high quality algal food sources in stream food webs—current status and future perspectives. Freshw Biol 61:815–831
Le Bras Q, Ritter L, Fasquel D, Lesueur M, Lucas S, Gouin S (2014 Etude de la consommation des algues alimentaires en France. Programme IDEALG Phase 1. Etude nationale. Les publications du Pôle halieutique AGROCAMPUS OUEST n°35, p 13
Pharmasea project. http://www.pharma-sea.eu/. Accessed 25 Mar 2019
Probst L, Frideres L, Pedersen B, Amato F (2015) Sustainable, safe and nutritious food, new nutrient sources. Publications Office of the European Union, Luxembourg, p 11
Regulation (EC) No 1223/2009 of the European Parliament and of the Council of 30 November 2009 on cosmetic products
Regulation (EC) No 1272/2008 of the European Parliament and of the Council of 16 December 2008 on classification, labelling and packaging of substances and mixtures
Regulation (EC) No 1333/2008 of the European Parliament and of the Council of 16 December 2008 on food additives
Regulation (EC) No 178/2002 of the European Parliament and of the Council of 28 January 2002 laying down the general principles and requirements of food law, establishing the European Food Safety Authority and laying down procedures in matters of food safety
Regulation (EC) No 1829/2003 of the European Parliament and of the Council of 22 September 2003 on genetically modified food and feed
Regulation (EC) No 183/2005 of the European Parliament and of the Council of 12 January 2005 laying down requirements for feed hygiene
Regulation (EC) No 1831/2003 of the European Parliament and of the Council of 22 September 2003 on additives for use in animal nutrition
Regulation (EC) No 1907/2006 of the European Parliament and of the Council of 18 December 2006 concerning the Registration, Evaluation, Authorisation and Restriction of Chemicals (REACH), establishing a European Chemicals Agency
Regulation (EC) No 1924/2006 of the European Parliament and of the Council of 20 December 2006 on nutrition and health claims made on foods
Regulation (EC) No 1935/2004 of the European Parliament and of the Council of 27 October 2004 on materials and articles intended to come into contact with food
Regulation (EC) No 2003/2003 of the European Parliament and of the Council of 13 October 2003 relating to fertilisers
Regulation (EC) No 726/2004 of the European Parliament and of the Council of 31 March 2004 laying down Community procedures for the authorisation and supervision of medicinal products for human and veterinary use and establishing a European Medicines Agency
Regulation (EC) No 767/2009 of the European Parliament and of the Council of 13 July 2009 on the placing on the market and use of feed
Regulation (EC) No 852/2004 of the European Parliament and of the Council of 29 April 2004 on the hygiene of foodstuffs
Regulation (EU) 2015/2283 of the European Parliament and of the Council of 25 November 2015 on novel foods
Regulation (EU) 2018/848 of the European Parliament and of the Council of 30 May 2018 on organic production and labelling of organic products
Regulation (EU) No 68/2013 on the Catalogue of feed materials
Regulation (EU) No 1169/2011 of the European Parliament and of the Council of 25 October 2011 on the provision of food information to consumers

Regulation (EU) No 536/2014 of the European Parliament and of the Council of 16 April 2014 on clinical trials on medicinal products for human use
Rocca S, Agostini A, Giuntoli J, Marelli L (2015) Biofuels from algae: technology options, energy balance and GHG emissions; Insights from a literature review. Publications Office of the European Union, Luxembourg, p 4, 61
Sharma P, Sharma N (2017) Industrial and biotechnological applications of algae: a review. J Adv Plant Biol (1):1, 01–25
Skarka J (2012) Microalgae biomass potential in Europe land availability as a key issue. Technikfolgenabschätzung Theor Prax 21:72–79
Unsworth RKF, van Keulen M, Coles RG (2014) Seagrass meadows in a globally changing environment. Mar Pollut Bull 83(2):383–386
Van Loveren H, Unamunzaga C (2018) Algae as sources for novel foods. European Commission, Directorate General for Energy
Verdelho Vieira V (2017) Value and size algae biomass sector in Europe in 2016: current relevance and future potential for Biorefinery. Brussels. https://algaebiorefineryconference.eu/wp-content/uploads/2017/11/S03_VITOR-VERDELHO_Value-of-the-Algae-Biomass-sector-in-Europe-and-the-current-share-of-Biorefinery-and-future-potential.pdf. Accessed 25 Mar 2019

MIX
Papier aus verantwortungsvollen Quellen
Paper from responsible sources
FSC® C105338

If you have any concerns about our products,
you can contact us on
ProductSafety@springernature.com

In case Publisher is established outside the EU,
the EU authorized representative is:
Springer Nature Customer Service Center GmbH
Europaplatz 3, 69115 Heidelberg, Germany

Printed by Libri Plureos GmbH
in Hamburg, Germany